Fundamentals of Dynamics and Analysis of Motion

MARCELO R. M. CRESPO DA SILVA

Professor Emeritus
Department of Mechanical, Aerospace,
and Nuclear Engineering
Rensselaer Polytechnic Institute

Dover Publications, Inc.
Mineola, New York

Marcelo R. M. Crespo da Silva Professor Crespo da Silva received his undergraduate engineering degree in 1963 from the *Universidade do Brasil* in Rio de Janeiro, Brazil (now, the *Federal University of Rio de Janeiro*). After working in the electronics industry in Brazil from 1964 to 1965, he went to Stanford University, where he received his M.Sc. and Ph.D. degrees. At Stanford, he changed his interests to Dynamics, and his doctoral thesis dealt with satellite attitude dynamics. From 1969 to 1970 he was a National Research Council postdoctoral research associate at NASA Ames Research Center in California working on satellite dynamics. He joined the Rensselaer faculty in 1986 after rising through the ranks from assistant professor to professor at the University of Cincinnati, which he joined in 1971 as a faculty member in the aerospace engineering and engineering mechanics department. On a sabbatical leave during the first half of 1983 he continued research work on helicopter rotor blade dynamics at the Aeromechanics Laboratory at Ames Research Center. Professor Crespo da Silva's main interests are dynamics (including structural dynamics, satellite dynamics, and helicopter dynamics) and nonlinear oscillations. He has a number of publications in these areas, and has received research grants from the National Aeronautics and Space Administration (NASA), the U.S. Air Force, the U.S. Army Research Office, and the National Science Foundation. He is a member of the American Academy of Mechanics, senior member of the American Institute of Aeronautics and Astronautics, and a fellow of the American Society of Mechanical Engineers.

Bibliographical Note

Fundamentals of Dynamics and Analysis of Motion, first published by Dover Publications, Inc., in 2016, is an expanded and updated version of *Intermediate Dynamics, Complemented with Simulations and Animations,* originally published in 2003 by McGraw-Hill, Inc., New York. It includes additional material and, for the convenience of the reader, also includes several optional programs written in widely used software (one of which is free) for solving the problems that require numerical computations. The new title more accurately reflects all the material in this book.

International Standard Book Number

ISBN-13: 978-0-486-79737-3
ISBN-10: 0-486-79737-6

Manufactured in the United States by RR Donnelley
79737601   2016
www.doverpublications.com

Contents

Preface

Dynamics is a discipline that consists of the prediction and study of the motion of bodies under the influence of forces. It began to evolve as a rigorous scientific discipline based on solid mathematical principles during the seventeenth century when the famous book of Sir Isaac Newton (1642-1727), entitled *Philosophiæ Naturalis Principia Mathematica*[1], was published in 1687. In that classical work he laid the foundations for dynamics by postulating his three famous laws of motion, and his universal law of gravitation. The original work of Newton was directed to the motion of objects that were modeled as "point masses." A number of great scientists, including Leonhard Euler (1707-1783), Joseph Louis Lagrange (1736-1813), and William Rowan Hamilton (1805-1865), carried dynamics much further with their time-lasting contributions.

This book is aimed at an intermediate-level or first-year graduate course in dynamics for engineering and physics students. It is also suitable for self-study and as a reference book, and allows instructors to design their own course contents using selected parts of it. It covers Newtonian dynamics, and the basics of *Analytical Dynamics* in Chapter 7, which includes the necessary concepts of *calculus of variations* needed for that branch of dynamics. As in the other chapters, many examples of application of the theory are also included in Chapter 7.

It is assumed that the reader has had a course in statics, calculus, and basic ordinary differential equations. A working knowledge of these subjects is a prerequisite for a solid understanding of dynamics, and the approach used here is based on what an analyst actually needs and does to investigate and analyze dynamics problems.

This book stresses the use of *fundamentals* for setting up and solving dynamics problems, rather than the indiscriminate use of relatively elaborate "formulas" (which most humans will forget), the mere use of which does not contribute to the understanding of the subject matter.

A variety of problems is included in each chapter and, for each problem, a systematic approach to the formulation and analysis of the differential equations of motion of dynamical systems is presented. The methodology used here may be applied to a wide range of problems, from very simple to complex ones, encountered in engineering and in basic research.

The differential equations that govern the motion of dynamical systems are gen-

---

[1] *The Mathematical Principles of Natural Philosophy.* "Natural Philosophy" is what is called "Physical Science" today.

erally nonlinear and cannot be solved analytically. However, under some circumstances, much information can be extracted from them even without finding their solutions. Such cases are presented in this book. In actual engineering practice, computers are generally used for integrating them numerically, and this is also presented in this book, especially for problems for which either the motion cannot be investigated analytically or analytical methods that are beyond the scope of this book are needed for such investigations. These include "perturbation methods" for dealing with nonlinear oscillation problems, which are taught at the masters and doctoral levels. The implementation of the numerical integration of differential equations of motion is illustrated in this book with three modern software that are used at many universities and research institutions. They are relatively easy to use and require very little programming. This book is not tied to any of them, though. **It is emphasized that using any of them is optional**, and the readers may use any one of the three or any other appropriate numerical software of their choosing for performing the numerical tasks that necessarily go hand in hand with the analysis of most dynamical systems.

The three software programs whose use are illustrated in this book are Scilab, Matlab, and Simulink, and tutorials on them are presented in Appendix A. Matlab is a commercial software, Simulink is a "toolbox" of Matlab[2], and Scilab is a free and open source software[3] developed and published by Scilab Enterprises. It is essentially similar to Matlab for many applications, including dynamics. Any of them will do numerical computations and easily generate plots. As described in *http://www.scilab.org*, "Scilab is free and open source software for numerical computation providing a powerful computing environment for engineering and scientific applications." It has essentially the same utilities and nearly the same syntax as Matlab, and it is a free software, which is a big advantage. It has a *help* facility that describes all its commands, with examples. Matlab is also a software for numerical computation, but it has the disadvantage of being costly to individuals, except for the "Student Edition" that is offered to students at a substantially reduced price. In addition to the extensive amount of information provided in its manuals, Matlab also has a *help* facility. As described in *http://ewh.ieee.org/r1/ct/sps/PDF/MATLAB/chapter8.pdf*, "Simulink provides a graphical user interface (GUI) for building models as block diagrams."

Both Scilab and Matlab require some programming, but such programming generally involves only a few simple and self-explanatory lines. All Scilab and Matlab programs prepared by the author for the problems whose solution require numerical

---

[2]Simulink and Matlab are registered trademarks of The MathWorks, Inc.

[3]Scilab is available for download free of charge at www.scilab.org. It is a registered trademark.

computation are listed in Appendix E.

> If you are interested in using either Scilab or Matlab, see
> Appendix E for information on how to obtain the programs
> listed in that Appendix.

There are several comment lines in each program listed in Appendix E, which
are used for communicating information to those readers who might be interested
in such.[4] The **optional** Simulink blocks for the same problems are shown throughout the book. They may be of interest only to those readers who might prefer to
use Simulink. Scilab saves files using the extension ".sci", Matlab saves them with
the extension ".m", and Simulink models with the extension ".mdl". The programs
animate_nbars.sci, *animate3d_nbars.sci*, *animate_nbars.m*, and *animate3d_nbars.m*
(see Appendix E) for Scilab and Matlab, respectively, are for animating two- and
three-dimensional motion, as their name suggest. They will display, on the computer screen, drawings that resemble certain systems represented by connected lines,
and animate their motion with data generated by the integration of the differential
equations that govern that motion. Both are used to animate the solution of a
variety of problems solved in this book, including point mass problems.

Chapter 1 presents essential material that is of fundamental importance for the
study of subsequent chapters. It also introduces the notation used throughout the
book. The book then progresses from kinematics and dynamics of a point mass
(Chapter 2) to a system of point masses (Chapter 4), and then to continuum rigid
bodies, which are bodies with an infinite number of point masses that are constrained
so that the distance between any two of their points is constant.

Unlike in other books, kinematics and dynamics are not separated into different
chapters, except in Chapter 3, which deals only with kinematic analysis of mechanisms whose motion takes place either in a plane or in parallel planes. In that
chapter, the motion of one of the parts of the mechanism is specified without any
concern for determining the external forces that cause it. The material in Chapter 3
is a natural extension of the application of the analysis involving velocity and acceleration of a point, and also of the use of rotating unit vectors that first appears in
Chapter 2.

The first two sections of Chapter 4 deal with important properties of the motion
of a system of point masses. Such properties involve the motion of the center of
mass of the system, and the relation between the moment of the forces applied

---

[4]Comments in Scilab are anything that appears after a double slash, //, in a line, and comments
in Matlab are anything that appears after a percentage symbol in a line.

to the system and the time rate of change of angular momentum of the system. The concept of angular momentum for a system of particles is introduced in that chapter. Chapter 4 constitutes a transition to the study of the motion of rigid bodies. In addition, two classical important problems in celestial mechanics and spacecraft trajectory dynamics are treated in reasonable detail in Chapter 4. Thus, several important problems in mechanical engineering and in aerospace engineering, and the fundamentals of vibration analysis, have been presented up to that part of the book.

Chapter 5 deals with dynamics of rigid bodies in "simpler" planar motion so that the analysis of rotational motion involves only one moment of inertia of the body. Such motions are defined on p. 315 in that chapter. Dynamics of rigid bodies that are able to move in any manner in three-dimensional space is presented in Chapter 6. Examples 6.8.3 and 6.8.4 in Section 6.8 deal with important problems in spacecraft dynamics, further introducing the reader to that field of study.

Chapter 7 covers the basics of Analytical Dynamics and most of that work was developed by Lagrange. In contrast with Newtonian Dynamics, the Analytical Dynamics methodology does not involve the formulation of acceleration. Its fundamental quantities are the kinetic energy of the motion and the *virtual work* (which is defined in that chapter) done by the forces applied to the system, and both are scalar quantities. Such a methodology is actually simpler to apply to problems than the Newtonian methodology, and *classical integrals of the motion* (or *constants of the motion*) are easily obtained (when they exist, of course!) using that methodology. Free-body diagrams are rarely needed with that methodology. As in other chapters, many examples of application of the Analytical Dynamics methodology are also included in Chapter 7.

The subject of vibrations/oscillations is introduced throughout the book, starting in Chapter 1, as a natural application of analysis of motion. Equilibrium solutions, linearization for small motions about any equilibrium, the concept of stability, and methods for analyzing the stability of equilibrium solutions (Sections 1.17 and 1.18) are also presented in that chapter. Sections 1.17 (the Routh-Hurwitz stability criterion) and 1.18 (an introduction to the Lyapunov stability method) are new to this book and are not in the McGraw-Hill edition that was published in 2003. Several examples throughout the book deal with the motion of dynamical systems, both linear and nonlinear, frequently accompanied by numerical integration of the differential equations that govern the motion.

Chapter 8 is short and complements the rest of the book with additional material on vibrations/oscillations, including the analysis of the response of a linear system to a sinusoidal forcing function. Two classical problems are presented in Chapter 8, namely the analysis of the *Foucault pendulum*, and of a *vibration absorber*. The Foucault pendulum was devised to demonstrate the rotation of the earth. It also brings into evidence when one can either reasonably neglect such an effect and when one must take it into consideration when motion relative to a rotating planet as viewed by an observer fixed to that planet is observed for a longer time. Vibration absorbers are practical devices used in engineering, and the one considered in Chapter 8 is a particular type of such a device.

The book has several appendices. The first one, Appendix A, consists of tutorials for those users who want to use either Scilab, Matlab, or Simulink. Appendices B and C consist of important material dealing with sequential rotations, and properties of the inertia matrix of a body. Some of the "Lab Assignment Problems" that the author used in exams are included in Appendix D. Appendix E consists of a listing of the Scilab and Matlab programs written by the author for solving the problems in this book that require numerical computation. Several references for more advanced studies are listed in Appendix G.

The MKS (meter-kilogram-second) International System of units is used in this book. As per standard convention, these units are abbreviated as m, kg, and s, respectively. The unit of force in this system, which is the newton, is abbreviated, in accordance with international convention, as N; $1 \text{ N} = 1 \text{ kg·m/s}^2$. In some of the problems, lengths are also quoted either in centimeters (cm) or in millimeters (mm), and mass in grams. In some examples, especially those involving space applications in Chapter 4, the equivalent distance in miles is also given after the distance is quoted in km (for kilometer). In such cases, velocity may be quoted in both km/h (for kilometers per hour) and in miles/hour.

I am grateful to several anonymous reviewers for their useful suggestions, and to all those at Dover Publications who were involved in the publication of this book. I dedicate this book to my wife Linda Clotilde d'Escoffier Crespo da Silva.

October 2015   Marcelo R. M. Crespo da Silva

CHAPTER 1

Essential Material for Dynamics

The fundamental quantities that appear in Newtonian Dynamics for analyzing the motion of a point mass (or mass particle[1]) are the *absolute acceleration* of the particle and the *resultant force acting on the particle*, and both are vector quantities. The absolute acceleration is, by definition, a vector that is equal to the time derivative of the *absolute velocity* of the particle (which is, in turn, the time derivative of the *absolute position vector* for the particle). The resultant force is determined directly from the free-body diagram for the particle, which is a diagram showing all the individual forces acting on the particle. These quantities, and the meaning of the adjective *absolute*, are introduced in Chapter 2.

This chapter presents the basic material that is essential in the study of dynamics. It includes a basic review of vectors and differential equations. Reviews of free-body diagrams, Newton's laws of motion, and Newton's law of universal gravitation are also included in this chapter. Details of the calculation of the gravitational force acting on a body are presented in Section 1.8, together with a discussion of the meaning of weightlessness.

1.1   Vector and Scalar Quantities: A Brief Review

A *scalar* quantity is one that can be represented by a number, a variable, or a function of one or more variables. Examples of scalar quantities are mass, temperature, energy of a body, and the speed of a car. Such quantities may be functions of a scalar variable (or variables), such as time. In such a case one has a *scalar function*.

A *vector* is a quantity that has a magnitude (i.e., a "length") as well as a direction. In addition, to qualify as vectors, such quantities must obey the *parallelogram*

---

[1]A *mass particle*, which is simply referred to as a *particle*, is an idealized point having a constant mass.

rule of addition and be *commutative* (i.e., the result of the sum of two vectors must be independent of the order of summation). For example, to locate a point B relative to a point A in space one needs to specify not only the distance from A to B but also the direction of the line segment from A to B. The vector (which specifies distance and direction) that indicates the location of point B relative to point A is called the *position vector* of B relative to A. Additional examples of vectors are the velocity and acceleration of a point and the force acting on a body. Notice that, contrary to everyday speech, there is a distinction in dynamics between the *velocity* and *speed* of a point. *Speed* is a scalar quantity, always positive, that is equal to the magnitude of the velocity vector. The *velocity* of a point is a vector and, as such, contains information about the direction of the motion of the point.

In books and technical journals, vectors are commonly represented by boldface letters (e.g., $\mathbf{A}, \mathbf{v}$), but in this book they are represented by a letter (or by two letters written side by side) with an arrow on top (e.g., $\vec{r}, \vec{v}, \overrightarrow{AB}$). The representation adopted here is the one everyone uses either on their own or in classrooms because of the obvious inconvenience of using boldface symbols in those situations.

A *unit vector* is a vector of unit magnitude and no dimensions (i.e., no units). A unit vector is represented in this book by a symbol with a caret on top (e.g., $\hat{r}$, $\hat{e}_1$). Clearly, for any vector $\vec{r}$, whose magnitude is $|\vec{r}|$, the vector $\vec{r}/|\vec{r}|$ is the same as the unit vector $\hat{r}$ in the direction of $\vec{r}$.

Any vector $\vec{r}$ in three-dimensional space may be expressed in terms of any three independent basis vectors. However, it is most convenient to choose the basis vectors to be three orthogonal unit vectors, each one of which is chosen simply to indicate a reference direction in space. Such sets of unit vectors are called *unit vector triads*, or simply *triads*.

The unit vector triad $\{\hat{x}, \hat{y}, \hat{z}\}$ shown in Fig. 1.1 may be rotating in space, as would be the case if it were fixed to a maneuvering aircraft or to a rotating part in a machine, for example. The quantities x, y, and z are called the *scalar components* of the vector $\vec{r}$ along the directions defined by $\hat{x}$, $\hat{y}$, and $\hat{z}$, respectively, while the quantities $x\hat{x}$, $y\hat{y}$, and $z\hat{z}$ are called the *vector components* of $\vec{r}$. Notice that the magnitude of any vector $\vec{r}$ written in terms of three orthogonal unit vectors $\hat{x}$, $\hat{y}$, and $\hat{z}$ as $\vec{r} = x\hat{x} + y\hat{y} + z\hat{z}$ is, by the Pythagorean theorem, equal to the square root of the sum of the squares of its scalar components, i.e., $|\vec{r}| = \sqrt{x^2 + y^2 + z^2}$.

The scalar components of a vector may be either positive or negative and are often also referred to as the *measure numbers* of $\vec{r}$ in the triad that is used for expressing $\vec{r}$. The same vector $\vec{r}$ may also always be expressed in terms of a single component as $\vec{r} = r\hat{r}$, where $\hat{r}$ is a unit vector in the direction of the vector $\vec{r}$ itself. If the scalar component r is positive, it is the same as the magnitude $|\vec{r}|$ of the vector $\vec{r}$. If the unit vector $\hat{r}$ were chosen in the direction opposite to the vector $\vec{r} = r\hat{r}$, then the quantity r would be negative (and, of course, would be the negative of the

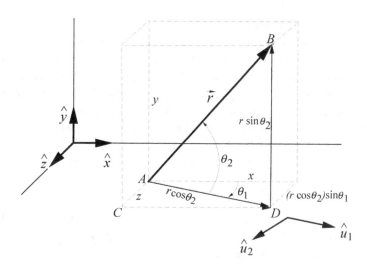

Figure 1.1: An arbitrary vector $\vec{r}$.

magnitude of $\vec{r}$).

You should keep in mind that the choice of unit vectors used for writing an expression for a vector is simply that – an individual *choice*. Different people will, in general, choose different unit vector triads to use, and the directions of the three unit vectors of the chosen triad may be either fixed in space or rotating. Different choices of unit vectors simply lead to different expressions for the same vector $\vec{r}$. Some judicious choices of unit vectors lead to simpler and more convenient expressions for a desired vector $\vec{r}$, and, as will be seen later, to simpler expressions for the derivatives of $\vec{r}$ when one needs to work with them.

In summary, to write an expression for a vector, you need to choose a set of unit vectors to work with. But that is not enough. You also need to choose a set of scalar variables to represent the components of the vector. In the previous example, where the expression written to represent a vector $\vec{r}$ was $\vec{r} = x\hat{x} + y\hat{y} + z\hat{z}$, the variables chosen for such representation were the three scalar quantities x, y, and z, which were distances from the chosen reference point A shown in Fig. 1.1, measured along three perpendicular axes. Such variables are called *rectangular coordinates* (or *Cartesian coordinates*).

In dynamics, the variables that are chosen to describe a motion are frequently called the *motion coordinates*. In the expression $\vec{r} = x\hat{x} + y\hat{y} + z\hat{z}$, for the position vector of point B relative to point A, the motion coordinates are the distances x, y, and z shown in Fig. 1.1.

There are a number of coordinates one can use for expressing a vector. They include Cartesian, polar, cylindrical, and spherical coordinates, or any combination of them. *Spherical coordinates* consist of the magnitude $|\vec{r}|$ of the vector $\vec{r}$ (which we will denote here as r) and the angles θ_1 and θ_2 shown in Fig. 1.1. In terms of such variables, the coordinates x, y, and z shown in that figure are simply obtained by projecting the vector $\vec{r}$ along the $\hat{x}$, $\hat{y}$, and $\hat{z}$ directions, respectively. By looking at Fig. 1.1, the following expressions can then be written by inspection of the figure: $x = (r\cos\theta_2)\cos\theta_1$, $y = r\sin\theta_2$, and $z = (r\cos\theta_2)\sin\theta_1$. If we still use the unit vector triad $\{\hat{x}, \hat{y}, \hat{z}\}$, the expression for the vector $\vec{r}$ shown in Fig. 1.1 is then

$$\vec{r} = r(\cos\theta_2\cos\theta_1)\hat{x} + r(\cos\theta_2\sin\theta_1)\hat{z} + r(\sin\theta_2)\hat{y} \tag{1.1.1}$$

Cylindrical coordinates consist of the two distances $|\overrightarrow{AD}|$ and $y = |\overrightarrow{DB}|$ and the angle θ_1 shown in Fig. 1.1. Using the unit vector triad $\{\hat{x}, \hat{y}, \hat{z}\}$, the expression for $\vec{r}$ in cylindrical coordinates is (see Fig. 1.1)

$$\vec{r} = |\overrightarrow{AD}|(\cos\theta_1)\hat{x} + |\overrightarrow{AD}|(\sin\theta_1)\hat{z} + y\hat{y} \tag{1.1.2}$$

> The choice of the coordinates (such as rectangular, cylindrical, or spherical) to formulate any dynamics problem and of the unit vectors used for expressing the vectors that appear in the formulation is based on convenience for each problem under consideration.

Polar coordinates involve a distance and an angle and, therefore, appear in two-dimensional problems. In the illustration shown in Fig. 1.1, one would have a two-dimensional problem when either θ_1 or θ_2 is constant for all times. When θ_1 is constant, the polar coordinates that specify the position of point B relative to point A (which is simply the vector $\vec{r} = \overrightarrow{AB}$) in Fig. 1.1 are the distance from A to B and the angle θ_2. Conversely, when θ_2 is constant, the polar coordinates that specify the position of point D relative to point A (which is the vector $\overrightarrow{AD}$) in the same figure are the distance from A to D and the angle θ_1.

As mentioned earlier, different choices of unit vectors may certainly be made for writing an expression for a vector. Another (among many others) perfectly valid expression for the vector $\vec{r}$ shown in Fig. 1.1, in which the unit vector triad $\{\hat{u}_1, \hat{u}_2, \hat{y} = \hat{u}_2 \times \hat{u}_1\}$ is used, is (see Fig. 1.1)

$$\vec{r} = \underbrace{r(\cos\theta_2)\hat{u}_1}_{\text{This is } \overrightarrow{AD}} + \underbrace{r(\sin\theta_2)\hat{y}}_{\text{This is } \overrightarrow{DB}} \tag{1.1.3}$$

In Fig. 1.1, the unit vector $\hat{u}_1$ was chosen to be parallel to the projection $\overrightarrow{AD}$ of $\vec{r}$ on the x-z plane, and directed from A to D. The angle θ_1 in that figure specifies the direction of $\hat{u}_1$ relative to the $\hat{x}$ direction. As will be seen in later chapters where more advanced material and a number of solved example problems are presented, the considerations just discussed will play an important role in problems involving the formulation of the equations that govern the motion of any dynamical system.

1.2 Transformation of Triads

As seen in Section 1.1, the representation of any vector $\vec{r}$ in terms of a set of orthogonal unit vectors is not unique since any set of three orthogonal unit vectors (or a *triad*) may be chosen to represent $\vec{r}$. The complexity (or simplicity) of the expression for a vector $\vec{r}$ in terms of any chosen triad (whose unit vectors you name, say, $\hat{x}$, $\hat{y}$, $\hat{z}$) will depend on the particular spatial orientation of the chosen triad. The simplest representation of a vector $\vec{r}$ is obtained by writing $\vec{r} = r\hat{r}$. In such a case, the unit vector $\hat{r}$ may be thought of as being one of the three unit vectors of some orthogonal triad. When necessary, to avoid confusion and to convey to the reader which triad is used to represent some vector $\vec{r}$, each scalar component of $\vec{r}$ will be identified with a subscript indicating the unit vector that is associated with that component. In such cases, the same symbol that is used for identifying the vector in question will be used in this book for identifying all its components. For example, consider a vector $\vec{r}$ and a triad $\{\hat{x}_1, \hat{x}_2, \hat{x}_3\}$ that was chosen (at this point, for no special reason at all) to represent $\vec{r}$. The expression for $\vec{r}$ in this triad might then be written as $\vec{r} = r_{x_1}\hat{x}_1 + r_{x_2}\hat{x}_2 + r_{x_3}\hat{x}_3$ (or even, $\vec{r} = r_1\hat{x}_1 + r_2\hat{x}_2 + r_3\hat{x}_3$ if there were no other triads for which the subscripts 1, 2, and 3 were used). In terms of another triad $\{\hat{n}, \hat{t}, \hat{z}\}$, for example, one might write $\vec{r} = r_n\hat{n} + r_t\hat{t} + r_z\hat{z}$ for the same vector $\vec{r}$.

The elements of a triad can be expressed in terms of the elements of any other triad by means of a transformation of triads. Such a transformation is nothing more than the expression for the projection of a unit vector in the direction of another unit vector, and such projections were illustrated in Section 1.1. To obtain such relationships between the unit vectors of two triads, start by drawing a figure showing the orientation of each triad of interest. This is illustrated in Fig. 1.2 where two unit vector triads $\{x\} = \{\hat{x}_1, \hat{x}_2, \hat{x}_3\}$ and $\{y\} = \{\hat{y}_1, \hat{y}_2, \hat{y}_3 = \hat{x}_3\}$ are shown.

Figure 1.2 shows how to obtain the expression for the unit vector[2] $\hat{y}_1$ in terms of the unit vectors $\hat{x}_1$ and $\hat{x}_2$. For this, one simply draws the projections of the

---

[2]Notice that there is no specific location or "origin" for unit vectors; such locations are irrelevant since unit vectors are only needed to define specific *directions* in space. Thus, any of the unit vectors shown in Fig. 1.2 could have been drawn anywhere on the page. However, for clarity, they are generally drawn intersecting at a common point. Such a practice also makes it easier to visualize the projections of a unit vector of one of the triads along the direction of a unit vector of the other triad.

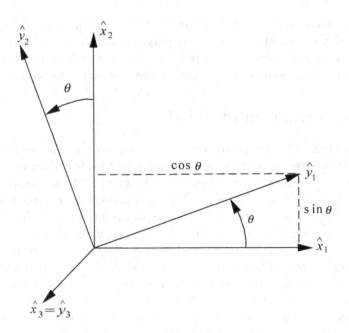

Figure 1.2: A simple transformation of triads.

unit vector $\hat{y}_1$ along the directions parallel to the unit vectors $\hat{x}_1$ and $\hat{x}_2$. These projections are the dashed lines shown in Fig. 1.2. In terms of $\hat{x}_1$ and $\hat{x}_2$, the vector components of the unit vector $\hat{y}_1$ in the directions of $\hat{x}_1$ and $\hat{x}_2$ are then $|\hat{y}_1|(\cos\theta)\hat{x}_1 = (\cos\theta)\hat{x}_1$ and $|\hat{y}_1|(\sin\theta)\hat{x}_2 = (\sin\theta)\hat{x}_2$, respectively. The unit vector $\hat{y}_1$ is then equal to the vector sum of those components. A similar construction may also be made to obtain the vector components of $\hat{y}_2$ in terms of $\hat{x}_1$ and $\hat{x}_2$. The following expressions relating the unit vectors $\hat{y}_1$ and $\hat{y}_2$ to the unit vectors $\hat{x}_1$ and $\hat{x}_2$ may then be written by inspection of Fig. 1.2 (such relations are called *transformation of triads*).

$$\hat{y}_1 = (|\hat{y}_1|\cos\theta)\hat{x}_1 + (|\hat{y}_1|\sin\theta)\hat{x}_2 = \quad \hat{x}_1\cos\theta + \hat{x}_2\sin\theta \qquad (1.2.1)$$

$$\hat{y}_2 = -(|\hat{y}_2|\sin\theta)\hat{x}_1 + (|\hat{y}_2|\cos\theta)\hat{x}_2 = -\hat{x}_1\sin\theta + \hat{x}_2\cos\theta \qquad (1.2.2)$$

Notice that if Eqs. (1.2.1) and (1.2.2) are written in matrix form as[3]

$$\begin{bmatrix} \hat{y}_1 \\ \hat{y}_2 \end{bmatrix} = \begin{bmatrix} \cos\theta & \sin\theta \\ -\sin\theta & \cos\theta \end{bmatrix} \begin{bmatrix} \hat{x}_1 \\ \hat{x}_2 \end{bmatrix} \triangleq [A] \begin{bmatrix} \hat{x}_1 \\ \hat{x}_2 \end{bmatrix}$$

---

[3]The symbol $\triangleq$, which appears throughout the book, denotes "equal to by definition." Here, it is defining the matrix $[A]$.

the inverse transformation from $\{\hat{y}_1, \hat{y}_2\}$ to $\{\hat{x}_1, \hat{x}_2\}$ (i.e., the equations expressing $\{\hat{x}_1, \hat{x}_2\}$ in terms of $\{\hat{y}_1, \hat{y}_2\}$) is readily obtained by performing the matrix multiplication involving the inverse $[A]^{-1}$ of the matrix $[A]$, as shown in the following equation. The result is the same as that given by Eqs. (1.2.3) and (1.2.4), which were written by inspecting Fig. 1.2 a second time.

$$\begin{bmatrix} \hat{x}_1 \\ \hat{x}_2 \end{bmatrix} = [A]^{-1} \begin{bmatrix} \hat{y}_1 \\ \hat{y}_2 \end{bmatrix} = \begin{bmatrix} \cos\theta & -\sin\theta \\ \sin\theta & \cos\theta \end{bmatrix} \begin{bmatrix} \hat{y}_1 \\ \hat{y}_2 \end{bmatrix} = [A]^T \begin{bmatrix} \hat{y}_1 \\ \hat{y}_2 \end{bmatrix}$$

As seen in this equation, the inverse $[A]^{-1}$ of a transformation matrix $[A]$ involving two unit vector triads is always equal to the transpose $[A]^T$ of $[A]$. Such a result is valid for any transformation matrix involving the unit vectors of two triads.

$$\hat{x}_1 = \hat{y}_1 \cos\theta - \hat{y}_2 \sin\theta \tag{1.2.3}$$
$$\hat{x}_2 = \hat{y}_1 \sin\theta + \hat{y}_2 \cos\theta \tag{1.2.4}$$

For a more elaborate example of transformation of triads, consider the triads $\{\hat{x}_1, \hat{x}_2, \hat{x}_3\}$ and $\{\hat{z}_1, \hat{z}_2, \hat{z}_3\}$ shown in Fig. 1.3. As shown in Fig. 1.3a, the triad $\{\hat{z}_1, \hat{z}_2, \hat{z}_3\}$ is obtained by first rotating $\{\hat{x}_1, \hat{x}_2, \hat{x}_3\}$ counterclockwise about the $\hat{x}_3$ direction by an angle θ_1 to obtain the triad $\{\hat{y}_1, \hat{y}_2, \hat{y}_3 = \hat{x}_3\}$, and then by rotating the triad $\{\hat{y}_1, \hat{y}_2, \hat{y}_3\}$ counterclockwise about the $\hat{y}_1$ direction by an angle θ_2 to obtain the triad $\{\hat{z}_1 = \hat{y}_1, \hat{z}_2, \hat{z}_3\}$. The two individual rotations are shown in Fig. 1.3b.

From Fig. 1.3b, it follows that:

$$\hat{y}_1 = \hat{x}_1 \cos\theta_1 + \hat{x}_2 \sin\theta_1 \tag{1.2.5}$$
$$\hat{y}_2 = -\hat{x}_1 \sin\theta_1 + \hat{x}_2 \cos\theta_1 \tag{1.2.6}$$
$$\hat{y}_3 = \hat{x}_3 \tag{1.2.7}$$

$$\hat{z}_1 = \hat{y}_1 \tag{1.2.8}$$
$$\hat{z}_2 = \hat{y}_2 \cos\theta_2 + \hat{y}_3 \sin\theta_2 \tag{1.2.9}$$
$$\hat{z}_3 = -\hat{y}_2 \sin\theta_2 + \hat{y}_3 \cos\theta_2 \tag{1.2.10}$$

Equations (1.2.5) to (1.2.10) may then be combined to yield the following representation of the transformation between the two unit vector triads $\{\hat{x}_1, \hat{x}_2, \hat{x}_3\}$ and

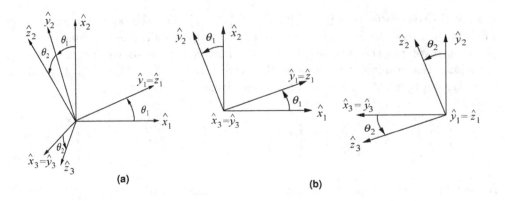

Figure 1.3: A more elaborate transformation of triads.

$\{\hat{z}_1, \hat{z}_2, \hat{z}_3\}$:

$$\hat{z}_1 = \hat{x}_1 \cos\theta_1 + \hat{x}_2 \sin\theta_1 \tag{1.2.11}$$

$$\hat{z}_2 = \hat{y}_2 \cos\theta_2 + \hat{y}_3 \sin\theta_2 = -\hat{x}_1 \sin\theta_1 \cos\theta_2 + \hat{x}_2 \cos\theta_1 \cos\theta_2 + \hat{x}_3 \sin\theta_2 \tag{1.2.12}$$

$$\hat{z}_3 = \hat{y}_3 \cos\theta_2 - \hat{y}_2 \sin\theta_2 = \hat{x}_1 \sin\theta_1 \sin\theta_2 - \hat{x}_2 \cos\theta_1 \sin\theta_2 + \hat{x}_3 \cos\theta_2 \tag{1.2.13}$$

These expressions for $\hat{z}_1$, $\hat{z}_2$, and $\hat{z}_3$ are the transformations that relate the two unit vector triads $\{\hat{x}_1, \hat{x}_2, \hat{x}_3\}$ and $\{\hat{z}_1, \hat{z}_2, \hat{z}_3\}$. The need for such transformations appears frequently in dynamics when one has to convert the expression of a vector written in terms of a particular unit vector triad into another unit vector triad. These operations will show up in a number of solved examples, and proposed problems throughout the remainder of the book.

1.3   Basic Operations with Vectors

In addition to the sum of two vectors (where the resulting vector is called the *resultant* of the two vectors) and the multiplication of a vector by a scalar, other basic operations that appear in dynamics are the dot product, cross product, and the time derivative of vectors.

Dot Product:

The *dot product* of two vectors $\vec{a}$ and $\vec{b}$ (written as $\vec{a} \bullet \vec{b}$ and read as "a dot b") is, by definition, a scalar quantity given as

$$\vec{a} \bullet \vec{b} = |\vec{a}||\vec{b}| \cos\theta \tag{1.3.1}$$

where $|\vec{a}|$ and $|\vec{b}|$ are the magnitudes of $\vec{a}$ and $\vec{b}$, respectively, and θ (with $0 \leq \theta \leq 2\pi$) is the angle between the two vectors, as shown in Fig. 1.4. The figure also shows the cross product of two vectors, which is presented after Example 1.3.1.

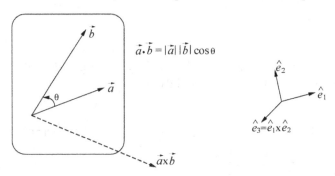

Figure 1.4: Two vectors $\vec{a}$ and $\vec{b}$ and a triad for illustrating the dot and the cross products.

The dot product of two vectors $\vec{a}$ and $\vec{b}$ is a scalar that is proportional to the projection of either one of the vectors in the direction of the other. This is seen by observing Fig. 1.4 and noticing that the projection of vector $\vec{b}$, for example, in the direction of $\vec{a}$ is equal to $(|\vec{b}| \cos \theta)\hat{a}$, where $\hat{a} = \vec{a}/|\vec{a}|$. The quantity $|\vec{b}| \cos \theta$ is the scalar component of $\vec{b}$ along $\vec{a}$. Using the definition of the dot product, such a projection may be simply obtained as $(\vec{b} \bullet \hat{a})\hat{a}$. Similarly, the projection of $\vec{a}$ in the direction of $\vec{b}$ may be obtained as $(\vec{a} \bullet \hat{b})\hat{b}$.

If the vectors $\vec{a}$ and $\vec{b}$ shown in Fig. 1.4 are expressed in terms of an orthogonal triad $\{e\} = \{\hat{e}_1, \hat{e}_2, \hat{e}_3\}$ as $\vec{a} = a_{e_1}\hat{e}_1 + a_{e_2}\hat{e}_2 + a_{e_3}\hat{e}_3$ and $\vec{b} = b_{e_1}\hat{e}_1 + b_{e_2}\hat{e}_2 + b_{e_3}\hat{e}_3$, one obtains

$$\begin{aligned}
\vec{a} \bullet \vec{b} = \vec{b} \bullet \vec{a} &= a_{e_1}\hat{e}_1 \bullet (b_{e_1}\hat{e}_1) + a_{e_1}\hat{e}_1 \bullet (b_{e_2}\hat{e}_2) + \ldots + a_{e_3}\hat{e}_3 \bullet (b_{e_3}\hat{e}_3) \\
&= a_{e_1}b_{e_1} + a_{e_2}b_{e_2} + a_{e_3}b_{e_3}
\end{aligned} \tag{1.3.2}$$

since $\hat{e}_1 \bullet \hat{e}_1 = \hat{e}_2 \bullet \hat{e}_2 = \hat{e}_3 \bullet \hat{e}_3 = 1$ and $\hat{e}_1 \bullet \hat{e}_2 = \hat{e}_1 \bullet \hat{e}_3 = \hat{e}_2 \bullet \hat{e}_3 = 0$. By using the dot product of $\vec{a}$ with $\vec{b}$, the cosine of the angle θ between two known vectors $\vec{a}$ and $\vec{b}$ can be readily obtained.

Example 1.3.1 Determine the projection of the vector $\vec{a} = \hat{x}_1 - 5\hat{x}_2 + 2\hat{x}_3$ (where $\hat{x}_1, \hat{x}_2$, and $\hat{x}_3$ are orthogonal unit vectors) along the direction $\overrightarrow{AB}$ of the line that passes through the points A and B whose rectangular coordinates are $A(x_1 = 1, x_2 = 2, x_3 = 3)$ and $B(x_1 = 6, x_2 = 10, x_3 = 7)$. The axes x_1, x_2, and x_3 are parallel to the unit vectors $\hat{x}_1$, $\hat{x}_2$, and $\hat{x}_3$, respectively.

a. What is the scalar component of the projection of $\vec{a}$ along $\overrightarrow{AB}$?

b. Does the projection of $\vec{a}$ along $\overrightarrow{AB}$ point in the direction from A to B or from B to A?

■ **Solution**

a. Letting $\vec{c}$ denote the projection of $\vec{a}$ in the direction of $\vec{b} = \overrightarrow{AB}$, the vector $\vec{c}$ is obtained with the following calculations.[4]

$$
\begin{aligned}
\vec{b} &= (6-1)\hat{x}_1 + (10-2)\hat{x}_2 + (7-3)\vec{x}_3 = 5\hat{x}_1 + 8\hat{x}_2 + 4\hat{x}_3 \\
\hat{b} &= \frac{\vec{b}}{|\vec{b}|} = \frac{5\hat{x}_1 + 8\hat{x}_2 + 4\hat{x}_3}{\sqrt{5^2 + 8^2 + 4^2}} \\
\vec{c} &= \left[\vec{a} \bullet \hat{b}\right]\hat{b} = \left[(\hat{x}_1 - 5\hat{x}_2 + 2\hat{x}_3) \bullet \left(\frac{5\hat{x}_1 + 8\hat{x}_2 + 4\hat{x}_3}{\sqrt{105}}\right)\right]\frac{5\hat{x}_1 + 8\hat{x}_2 + 4\hat{x}_3}{\sqrt{105}} \\
&= \frac{5 - 40 + 8}{\sqrt{105}}\frac{5\hat{x}_1 + 8\hat{x}_2 + 4\hat{x}_3}{\sqrt{105}} \approx -1.29\hat{x}_1 - 2.06\hat{x}_2 - 1.03\hat{x}_3
\end{aligned}
$$

The scalar component of the projection of $\vec{a}$ along $\overrightarrow{AB}$ is the quantity $\vec{a} \bullet \hat{b} = -27/\sqrt{105}$.

b. Since the result obtained in this example is negative, the projection vector $\vec{c}$ points in the $\overrightarrow{BA}$ direction (i.e., in the direction from B to A, instead of A to B).

Cross Product:

The *cross product* of two vectors $\vec{a}$ and $\vec{b}$ (written as $\vec{a} \times \vec{b}$ and read as "a cross b") making an angle θ with each other (see Fig. 1.4) is, by definition, a vector whose magnitude is equal to $|\vec{a}||\vec{b}|| \sin \theta|$ and whose direction is perpendicular to the two vectors and is given by the *right-hand rule*.

---

[4]The symbol $\approx$, which appears throughout the book, denotes "approximately equal to."

The Right-Hand Rule

Open your right hand and keep your right thumb extended. Now, keeping the right thumb extended, curl your fingers in the direction indicated by the curved arrow (shown in Fig. 1.4) that defines the angle θ measured from $\vec{a}$ (i.e., the first vector in the expression $\vec{a} \times \vec{b}$) to $\vec{b}$ (i.e., the second vector in the expression $\vec{a} \times \vec{b}$). The rule defines the direction of the cross product $\vec{a} \times \vec{b}$ as being the direction indicated by your right thumb. For this, always use $\theta \leq \pi$; when doing so, $\sin\theta \geq 0$, and the magnitude of the cross product is equal to $|\vec{a}||\vec{b}|\sin\theta$.

The right-hand rule is also known as the *screwdriver rule* since the direction of $\vec{a} \times \vec{b}$ is that which a regular screw, turned by a screwdriver, would travel if the screwdriver were turned by an angle θ (always using $\theta \leq \pi$) to align $\vec{a}$ with $\vec{b}$.

Simply by using the right-hand rule, you can immediately conclude that $\vec{b} \times \vec{a} = -\vec{a} \times \vec{b}$.

If the vectors $\vec{a}$ and $\vec{b}$ are expressed in terms of their components in a triad $\{\hat{e}_1, \hat{e}_2, \hat{e}_3\}$ with, say, $\hat{e}_3 \overset{\Delta}{=} \hat{e}_1 \times \hat{e}_2$ (as shown in Fig. 1.4), the cross product $\vec{a} \times \vec{b}$ is then obtained as

$$
\begin{aligned}
\vec{a} \times \vec{b} = {} & a_{e_1}\hat{e}_1 \times (b_{e_1}\hat{e}_1 + b_{e_2}\hat{e}_2 + b_{e_3}\hat{e}_3) \\
& + a_{e_2}\hat{e}_2 \times (b_{e_1}\hat{e}_1 + b_{e_2}\hat{e}_2 + b_{e_3}\hat{e}_3) \\
& + a_{e_3}\hat{e}_3 \times (b_{e_1}\hat{e}_1 + b_{e_2}\hat{e}_2 + b_{e_3}\hat{e}_3) \\
= {} & (a_{e_2}b_{e_3} - a_{e_3}b_{e_2})\hat{e}_1 + (a_{e_3}b_{e_1} - a_{e_1}b_{e_3})\hat{e}_2 + (a_{e_1}b_{e_2} - a_{e_2}b_{e_1})\hat{e}_3
\end{aligned}
$$

$$(1.3.3)$$

To determine the cross products $\hat{e}_1 \times \hat{e}_2$, $\hat{e}_2 \times \hat{e}_3$, etc., that show up in Eq. (1.3.3), see the unit vector triad shown in Fig. 1.4 and use the right-hand rule; as disclosed by that figure, $\hat{e}_1 \times \hat{e}_2 = \hat{e}_3$, $\hat{e}_1 \times \hat{e}_3 = -\hat{e}_2$, etc.

The cross product may also be determined in terms of the following determinant, if one prefers to use determinants to do so:

$$
\vec{a} \times \vec{b} = \begin{vmatrix} \hat{e}_1 & \hat{e}_2 & \hat{e}_1 \times \hat{e}_2 = \hat{e}_3 \\ a_{e_1} & a_{e_2} & a_{e_3} \\ b_{e_1} & b_{e_2} & b_{e_3} \end{vmatrix}
$$

$$(1.3.4)$$

For the calculation of a cross product using determinants to give the correct result given by Eq. (1.3.3), the following rules must be followed:

■ The first two unit vectors written in positions 1-1 and 1-2 in the first row of the determinant are any two of the three unit vectors of triad $\{e\}$, but the third unit vector in the first row of that determinant must be equal to the cross product of the unit vector written in position 1-1 with the unit vector written in position 1-2.

■ To form the cross product $\vec{a} \times \vec{b}$, the second and third rows of the determinant should contain the components of $\vec{a}$ and $\vec{b}$, respectively, along the unit vectors that appear in the first row. Thus, if the unit vectors $\hat{e}_3$ and $\hat{e}_2$ shown in Fig. 1.4 are listed first, the cross product $\vec{a} \times \vec{b}$ may also be correctly obtained as

$$\vec{a} \times \vec{b} = \begin{vmatrix} \hat{e}_3 & \hat{e}_2 & \hat{e}_3 \times \hat{e}_2 = -\hat{e}_1 \\ a_{e_3} & a_{e_2} & -a_{e_1} \\ b_{e_3} & b_{e_2} & -b_{e_1} \end{vmatrix} \qquad (1.3.5)$$

Since the cross product is a vector, the three components of the cross product of two vectors $\vec{a}$ and $\vec{b}$ depend on the specific triad $\{e\}$ that one chooses to work with (the vector obtained as the result of the cross product *does not* depend on any triad, of course; to see this, look back at the definitions of the magnitude and direction of the cross product of two vectors).

Example 1.3.2 Determine the unit vector $\hat{n}$ that is perpendicular to the plane defined by the vectors $\vec{a} = \hat{x}_1 - 5\hat{x}_2 + 2\hat{x}_3$ and $\vec{b} = \vec{AB} = 5\hat{x}_1 + 8\hat{x}_2 + 4\hat{x}_3$ of Example 1.3.1. The unit vector $\hat{x}_3$ is defined as $\hat{x}_3 = \hat{x}_1 \times \hat{x}_2$.

■ **Solution**

The perpendicular to the vectors $\vec{a}$ and $\vec{b}$ is parallel to the cross product $\vec{a} \times \vec{b}$ and also to the cross product $\vec{b} \times \vec{a} = -\vec{a} \times \vec{b}$. Therefore, either one of the unit vectors in the direction of $\vec{a} \times \vec{b}$ or $-\vec{a} \times \vec{b}$ is a solution. We then have

$$\begin{aligned}
\vec{a} \times \vec{b} &= (\hat{x}_1 - 5\hat{x}_2 + 2\hat{x}_3) \times (5\hat{x}_1 + 8\hat{x}_2 + 4\hat{x}_3) \\
&= \underbrace{\hat{x}_1 \times (5\hat{x}_1)}_{=0} + \underbrace{\hat{x}_1 \times (8\hat{x}_2)}_{=8\hat{x}_3} + \underbrace{\hat{x}_1 \times (4\hat{x}_3)}_{=-4\hat{x}_2} \underbrace{-5\,\hat{x}_2 \times (5\hat{x}_1)}_{=-5\hat{x}_3} \underbrace{-5\,\hat{x}_2 \times (8\hat{x}_2)}_{0} \\
&\quad \underbrace{-5\,\hat{x}_2 \times (4\hat{x}_3)}_{=4\hat{x}_1} + \underbrace{2\,\hat{x}_3 \times (5\hat{x}_1)}_{=5\hat{x}_2} + \underbrace{2\,\hat{x}_3 \times (8\hat{x}_2)}_{=-8\hat{x}_1} + \underbrace{2\,\hat{x}_3 \times (4\hat{x}_3)}_{=0} \\
&= -36\hat{x}_1 + 6\hat{x}_2 + 33\hat{x}_3
\end{aligned}$$

$$\frac{\vec{a} \times \vec{b}}{|\vec{a} \times \vec{b}|} = \frac{-36\hat{x}_1 + 6\hat{x}_2 + 33\hat{x}_3}{\sqrt{36^2 + 6^2 + 33^2}} \approx -0.732\hat{x}_1 + 0.122\hat{x}_2 + 0.671\hat{x}_3$$

The unit vector $\hat{n}$ that is perpendicular to the plane defined by the vectors $\vec{a}$ and $\vec{b}$ is then either $\hat{n} = -0.732\hat{x}_1 + 0.122\hat{x}_2 + 0.671\hat{x}_3$ or $\hat{n} = 0.732\hat{x}_1 - 0.122\hat{x}_2 - 0.671\hat{x}_3$. Either answer is valid.

Time Derivatives of Vectors

The derivatives of vectors obey the *product rule*, so, for any two vectors $\vec{A}$ and $\vec{B}$ and any scalar function of time $f(t)$,

$$\frac{d}{dt}\left[f(t)\vec{A}\right] = \left[\frac{d}{dt}f(t)\right]\vec{A} + f(t)\frac{d}{dt}\vec{A}$$

$$\frac{d}{dt}\left(\vec{A}\bullet\vec{B}\right) = \left(\frac{d}{dt}\vec{A}\right)\bullet\vec{B} + \vec{A}\bullet\left(\frac{d}{dt}\vec{B}\right)$$

$$\frac{d}{dt}\left(\vec{A}\times\vec{B}\right) = \left(\frac{d}{dt}\vec{A}\right)\times\vec{B} + \vec{A}\times\left(\frac{d}{dt}\vec{B}\right)$$

Presentation of the time derivative $d\vec{A}/dt$ of any vector $\vec{A}$ is postponed to Chapter 2.

1.4 Moment of a Vector about a Point

The moment of a vector $\vec{v}$ about a point O in space, denoted by $\vec{M}_O$ (with the subscript O denoting the point about which the moment is taken), is, by definition, a vector that is equal to the cross product of the vector from O to *any* point P on the *line of action* of $\vec{v}$ (line AA in Fig. 1.5), and the vector $\vec{v}$, i.e.,

$$\vec{M}_O \overset{\Delta}{=} \overrightarrow{OP} \times \vec{v} \qquad (1.4.1)$$

Since (see Fig. 1.5)

$$\left|\vec{M}_O\right| = \left|\overrightarrow{OP}\times\vec{v}\right| = |\overrightarrow{OP}|\,|\vec{v}|\,|\sin\phi| \qquad (1.4.2)$$

and, as disclosed by Fig. 1.5, the quantity $|\overrightarrow{OP}|\,|\sin\phi|$ is the perpendicular distance from point O to the line of action of $\vec{v}$ (line AA), *the moment $\vec{M}_O$ is independent of the choice of point P on line AA.*

Example 1.4.1 The rectangular coordinates (in meters) of three points A, B, and C are: $A(x_1 = 1, x_2 = 2, x_3 = 3)$, $B(x_1 = 6, x_2 = 10, x_3 = 7)$, and $C(x_1 = 6, x_2 = 13, x_3 = 20)$. Determine the perpendicular distance between point C and line AB.

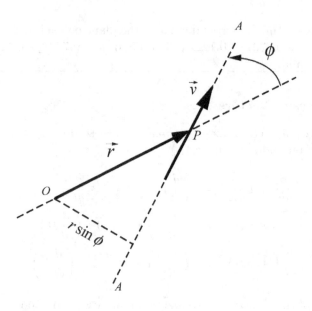

Figure 1.5: Moment of a vector about a point.

■ **Solution**

By letting $\hat{x}_1$, $\hat{x}_2$, and $\hat{x}_3$ denote three orthogonal unit vectors parallel to the x_1, x_2, and x_3 axes, respectively, the perpendicular distance d from point A to line AB is then determined, as presented in this section, as follows.

$$\begin{aligned}
\overrightarrow{AB} &= (6-1)\hat{x}_1 + (10-2)\hat{x}_2 + (7-3)\hat{x}_3 = 5\hat{x}_1 + 8\hat{x}_2 + 4\hat{x}_3 \\
\overrightarrow{CB} &= (6-6)\hat{x}_1 + (10-13)\hat{x}_2 + (7-20)\hat{x}_3 = -3\hat{x}_2 - 13\hat{x}_3 \\
d &= \left| \overrightarrow{CB} \times \frac{\overrightarrow{AB}}{|\overrightarrow{AB}|} \right| = \left| \frac{-15\hat{x}_2 \times \hat{x}_1 + 92\hat{x}_2 \times \hat{x}_3 - 65\hat{x}_3 \times \hat{x}_1}{\sqrt{5^2 + 8^2 + 4^2}} \right| \\
&= \frac{\sqrt{15^2 + 92^2 + 65^2}}{\sqrt{5^2 + 8^2 + 4^2}} \approx 11.09 \text{ meters (m)}
\end{aligned}$$

Notice that the vector $\overrightarrow{AB}/|\overrightarrow{AB}|$ that appears in the expression for the distance d is the unit vector in the direction of $\overrightarrow{AB}$. The distance d may also be determined as

$$d = \left| \overrightarrow{CA} \times \frac{\overrightarrow{AB}}{|\overrightarrow{AB}|} \right|$$

and, as discussed in this section, both calculations give the same result.

1.5   Newton's Laws of Motion

Newton's three laws of motion, which were formulated for a single particle of constant mass, are the fundamental tools for the mathematical formulation of dynamics problems. They are presented here.

> **Newton's First Law**
> If the resultant $\vec{F}$ of all the forces acting on a particle is zero, the particle will move in space with constant velocity $\vec{v}$.

A vector is constant if its magnitude is constant *and* its direction does not change with time. Therefore, for a particle to move in space with constant velocity $\vec{v}$, its *speed* (i.e., the magnitude of its velocity) must be constant and its trajectory must be a straight line.

> **Newton's Second Law**
> A particle of mass m subjected to a system of forces whose resultant is $\vec{F}$ moves in such a way that the time rate of change of its absolute linear momentum vector, which is defined as the product $m\vec{v}$ (where $\vec{v}$ is the absolute velocity of the particle), is equal to the resultant force $\vec{F}$.

Mathematically, Newton's second law for a particle is expressed as follows:

$$\boxed{\vec{F} = \tfrac{d}{dt}\left(m\vec{v}\right)} = m\tfrac{d}{dt}\vec{v} \overset{\Delta}{=} m\vec{a} \tag{1.5.1}$$

where $\vec{a}$ is the *absolute acceleration* of the particle.

Newton's laws postulate the existence of a reference point that is not accelerating (such a point is called an *inertial point*), and of an *inertial reference frame*. This requires the following explanations.

> A *reference frame* is a collection of points forming a three-dimensional space where the distance between all pairs of such points is constant. A reference frame is said to be *inertial* if all its points are inertial (which also implies that no line connecting two of its points changes direction in space).

The velocity of a point P, relative to another point O, is always associated with a reference frame; otherwise the concept of velocity is meaningless. The absolute velocity $\vec{v}$ of a particle P is, by definition, the time rate of change of the *absolute position vector*[5] $\vec{r} = \overrightarrow{OP}$ of P, as measured by an observer fixed to an inertial reference frame. Calculations involving the time derivative of vectors to determine velocity and acceleration (which is the time derivative of velocity) for analyzing dynamics problems are dealt with in detail in Chapter 2.

In practice, approximations are used for selecting a reference point and a reference frame that play the role of "inertial" in the analysis of the motion of a dynamical system. For example, in problems dealing with the motion of a planet around the sun, treating the sun as inertial and tracking the direction of the rotating line that joins the centers of the sun and the planet with an angle measured from a line in the plane of the motion that is nonrotating is an "approximation" that has proven to yield accurate results for the resulting motion. Similarly, the same is done for the analysis of the motion of a spacecraft relative to a planet, such as the earth. On the other hand, consider the motion of a projectile and of a simple pendulum relative to the earth. For a short-range projectile, the analysis of the motion may be done by disregarding the rotation of the earth about its own axis. However, the results obtained with such an approximation may not be accurate, and, thus, unacceptable, for long-range projectiles. For the case of a simple pendulum, its motion during a relatively "small" time interval may be reasonably approximated by neglecting the rotation of the earth, but the result of such an analysis may not be realistic in other cases. An important example of such a case, which is the Foucault pendulum, used for experimentally demonstrating in 1851 that the earth rotates about its own axis, is analyzed in Chapter 8.

Newton's second law is only valid for objects whose speed is small when compared to the speed of light, which is approximately $300,000$ km/s ($\approx 186,000$ miles/s); that is 1.08 billion kilometers per hour. Problems where speeds approach the speed of light belong to the realm of *relativistic mechanics* and are not treated in this book.

The vector equation (1.5.1) yields a set of scalar equations that involve the variables that were chosen for describing the motion of the particle and may also involve unknown forces. After such equations are combined to eliminate the unknown forces that appear in them, such as internal reaction forces, a set of *scalar differential equations of motion* for the particle are then obtained, and the problem one now has

---

[5]The *position vector* of a point P relative to a reference point O is simply the vector $\overrightarrow{OP}$ from O to P. A position vector $\overrightarrow{OP}$ is said to be an *absolute position vector* when the reference point O is inertial. Thus, the absolute position vector of a point P is a vector from an inertial point O to point P.

consists of extracting, from such equations, as much information as possible about the unknown motion of the particle. This may include numerical integration of such equations to simulate the motion of the system for desired initial conditions and/or analytical investigations to generate information that allows one to understand as much as possible the dynamic behavior of the system.

In practice, the analysis of the motion governed by the differential equations obtained as just discussed includes the investigation of small motions about constant solutions (which are called *equilibrium solutions*) by linearizing the differential equations about those solutions. The linearization process, which involves a Taylor series expansion of all the nonlinear terms that appear in the differential equations, is presented in Section 1.14.

One should be aware that dynamical systems may exhibit a variety of phenomena that are not disclosed by linearized differential equations, and that such phenomena may cause a physical system to exhibit undesirable behavior, or even physical damage in the system. An experiment describing one of such phenomena is presented in Section 1.19.

When analyzing the motion of a system of particles interacting with each other, Newton's second law is still applicable to each individual particle provided that we isolate each particle from all others and include the force of interaction due to the other particles in the free-body diagram for the particle under observation. In such cases, Newton's third law plays a significant role in the analysis.

Newton's Third Law
The mutual forces of action and reaction between two mass particles i and j are collinear with the particles, equal in magnitude, and opposite in direction.

Expressed mathematically, Newton's third law states that

$$\vec{F}_{ij} = -\vec{F}_{ji} \tag{1.5.2}$$

where $\vec{F}_{ij}$ is the force on particle i due to particle j.

1.6 Kepler's Laws and Newton's Law of Universal Gravitation

Johannes Kepler (1571-1630) was an assistant to the Dutch astronomer Tycho Brahe (1546-1601) at the time of Brahe's death. He was entrusted in 1601 with the vast and precise amount of observation data on the motion of the planets that Brahe had gathered and collected. Using mathematical tools of the time, he analyzed the data for many years. He published his first two laws of planetary motion in 1609, and the third law in 1619.

Kepler's First Law
Each planet moves in an ellipse with the sun at one of its foci.

Kepler's Second Law
The motion of each planet is such that the radius vector from the sun to the planet sweeps out equal areas in equal times.

Kepler's Third Law
The square of the period of a planet in its motion around the sun is proportional to the cube of the semi-major axis of its elliptical orbit around the sun.

An important question whose answer was found by Newton is the following: Given two bodies that are subjected only to the forces of attraction between them, how does the force of attraction depend on the distance between the bodies so that the motion of one body relative to the other is an ellipse? The answer to such a question is now known as Newton's law of universal gravitation.

Newton's Law of Universal Gravitation
Any two particles of masses m_1 and m_2, respectively, attract each other with a force whose magnitude is proportional to the product of their masses and inversely proportional to the square of the distance r between them.

Mathematically, Newton's law of universal gravitation is expressed as

$$|\vec{F}| = \frac{Gm_1m_2}{r^2} \tag{1.6.1}$$

where G is a universal gravitational constant, whose value is $G \approx 6.67 \times 10^{-11}$ cubic meters per kilogram per second squared [$m^3/(kg{\cdot}s^2)$].

The magnitude of the gravitational force acting on a particle of mass m located on the surface of a spherical homogeneous planet of radius R and mass M is equal to $|\vec{F}| = (GM/R^2)m \overset{\Delta}{=} mg$ (see Sections 1.7 and 1.8). The constant g is called the *acceleration of gravity* at the surface of the attracting planet. For the earth, $g \approx 9.81\,\mathrm{m/s}^2 \approx 32.2\,\mathrm{ft/s}^2$.

1.7   Force of Attraction due to a Spherical Body

Newton's laws were formulated for particles, which are defined as material points that have mass but no dimension. Mass particles are, of course, idealizations. Several other researchers extended the work of Newton to both discrete and continuum system of particles.

Using Newton's law of universal gravitation, which was presented in Section 1.6, let us determine the mutual force of attraction between a spherical body of radius R and mass density ρ kg/m^3, and a particle P of mass m located at a distance $D \geq R$ from the center O of the spherical body. For simplicity, it will also be assumed that the mass density ρ of the spherical body is constant, thus, equal to $M/(4\pi R^3/3)$, where M is the mass of the body. Such a body could be a model for a planet. Of course, planets are neither perfectly spherical nor homogeneous, but such idealizations turn out to be good first approximations for calculating the forces of attraction due to them. More refined approximations are, of course, used in more precise calculations involving the motion of the planets and the motion of artificial satellites over a long time.

A point Q on a spherical portion of the body with radius $r \leq R$ and the mass particle P are shown in Fig. 1.6. To calculate the mutual force of attraction between the body and P, the body is divided into an infinite number of mass elements, each having the infinitesimal volume $dV = (r \sin\theta\, d\phi)(r\, d\theta)\, dr$ [since as seen from Fig. 1.6, the lengths of the sides of the infinitesimal volume element are dr, $rd\theta$, and $(r \sin\theta)\, d\phi$]. The mass dM of the element located at point Q in Fig. 1.6 is then $dM = \rho dV = [M/(4\pi R^3/3)]dV$. When ϕ is varied from 0 to 2π, the mass element dM covers a strip on a cone of angle θ. After that, varying θ from 0 to π makes that strip generate a thin shell of radius r. Then, varying r from 0 to R causes that shell to generate the solid sphere. This is what the triple integral in Eq. (1.7.3) will do.

Why did we divide the sphere into infinitesimal volume elements? Simply because we know how to calculate the force of attraction between two point masses, which is given by Newton's law of universal gravitation. Letting $s = |\overrightarrow{QP}|$ denote the distance between Q and P and $\hat{s} = \overrightarrow{QP}/s$ be the unit vector from Q to P (see Fig. 1.6), the infinitesimal force acting on P due to the attraction of the mass

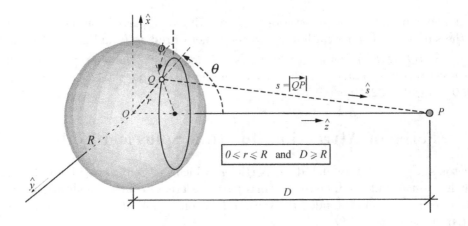

Figure 1.6: A portion of a spherical body of mass M, and a material particle P of mass m.

element at Q is then

$$d\vec{F} = -\frac{Gm\,dM}{s^2}\hat{s} = -\frac{Gm\,dM}{s^3}\overrightarrow{QP} = -\frac{GmM}{4\pi R^3/3}\frac{r^2(\sin\theta)\,dr\,d\theta\,d\phi}{s^3}\overrightarrow{QP} \quad (1.7.1)$$

The resultant force $\vec{F}$ acting on P due to attraction of the entire spherical body is simply equal to the integral, over the spherical body, of $d\vec{F}$.

By expressing the vector $\overrightarrow{QP}$ in terms of the quantities r, θ, ϕ, and D, and of the unit vector triad $\{\hat{x}, \hat{y}, \hat{z}\}$ (see Fig. 1.6) as

$$\overrightarrow{QP} = \overrightarrow{QO} + \overrightarrow{OP} = -[(r\sin\theta)(\hat{x}\cos\phi + \hat{y}\sin\phi) + (r\cos\theta)\hat{z}] + D\hat{z} \quad (1.7.2)$$

and by noticing that

$$s = |\overrightarrow{QP}| = \sqrt{(r\sin\theta\cos\phi)^2 + (r\sin\theta\sin\phi)^2 + (D - r\cos\theta)^2}\,,$$

which gives $s = \sqrt{D^2 + r^2 - 2Dr\cos\theta}$, the expression for the force $\vec{F}$ then becomes

$$\vec{F} = -\frac{GMm}{4\pi R^3/3}\int\limits_{r=0}^{R}\int\limits_{\theta=0}^{\pi}\int\limits_{\phi=0}^{2\pi}\frac{(D - r\cos\theta)\hat{z} - (\hat{x}\cos\phi + \hat{y}\sin\phi)r\sin\theta}{(D^2 + r^2 - 2Dr\cos\theta)^{3/2}}r^2(\sin\theta)\,d\phi\,d\theta\,dr$$

$$(1.7.3)$$

The integrals involving the $\hat{x}$ and $\hat{y}$ components in Eq. (1.7.3) are zero because $\int_0^{2\pi}\cos\phi\,d\phi = \int_0^{2\pi}\sin\phi\,d\phi = 0$. Therefore, the resultant force is directed along the line joining the particle P and the center O of the spherical body. Physically, this occurs because the sum of the components of the force $d\vec{F}$ along the perpendicular to line OP is zero due to the spherical symmetry of the body centered at O. In other words, since the attracting sphere has constant density, for each point Q where a mass element $dM = \rho\,dV$ is located, there is another point Q^* on the body (because the body is spherical), located at $\phi = 180°$ from the position shown in Fig. 1.6, along the plane perpendicular to line OP, and, thus the resultant of these two forces is along line OP.

The integrations associated with the $\hat{z}$ component in Eq. (1.7.3) may be readily performed by recognizing that $\sin\theta\,d\theta = -d(\cos\theta)$, and then letting $u = \cos\theta$. The resulting integrals can be found in a table of integrals. They are evaluated below using integration by parts. Since

$$v \overset{\Delta}{=} \int \frac{du}{(D^2 + r^2 - 2Dru)^{3/2}} = \frac{1}{Dr\sqrt{D^2 + r^2 - 2Dru}}$$

and

$$\int \frac{u\,du}{(D^2 + R^2 - 2Dru)^{3/2}} \overset{\Delta}{=} \int u\,dv = uv - \int v\,du$$

$$= \frac{u}{Dr\sqrt{D^2 + r^2 - 2Dru}} + \frac{\sqrt{D^2 + r^2 - 2Dru}}{D^2 r^2}$$

Equation (1.7.3) then yields, after the integration with respect to $u = \cos\theta$ and ϕ,

$$\vec{F} = \hat{z}\frac{6\pi GMm}{4\pi R^3} \int_{r=0}^{R} \left\{ \frac{1}{\sqrt{D^2 + r^2 - 2Dr\cos\theta}}\left[\frac{1}{r} - \frac{\cos\theta}{D}\right] \right.$$

$$\left. - \frac{\sqrt{D^2 + r^2 - 2Dr\cos\theta}}{D^2 r} \right\}_{\theta=0}^{\pi} r^2\,dr$$

This yields

$$\vec{F} = -\frac{GMm}{D^2}\hat{z} \tag{1.7.4}$$

By comparing Eq. (1.7.4) with Eq. (1.6.1), it is seen that an attracting spherical body of constant mass density (or one with mass distribution with ϕ symmetry, with ϕ as shown in Fig. 1.6) behaves as if it were a point with mass M concentrated at the center of the sphere. This conclusion is then applicable for two of such spheres attracting each other; i.e., their mutual force of attraction is directed along the line that joins their centers.

Let us now use these results to make some important observations about the force of gravity, weight, and what is commonly referred to in practice as *weightlessness*. These are important physical concepts that can be (and often are) easily misunderstood. They are discussed in Section 1.8.

1.8　Gravitational Force, Weight, and the Meaning of Weightlessness

A gravitational force is a force due to the mutual attraction between two masses. For most applications, the main *gravitational force* acting on a body B is the force acting on that body due to the gravitational attraction of the planet the body is near to. A body of mass m located on the surface of the earth, for example, is always subjected to the mutual force of attraction between the earth and the body. If the earth is approximated as a homogeneous sphere of mass M_E and radius R_E, the magnitude of that mutual force of attraction is equal to $GM_E m/R_E^2 = mg$, where $g \overset{\Delta}{=} GM_E/R_E^2 \approx 9.81 \, \text{m/s}^2$ is called the *acceleration of gravity* at the earth's surface.[6] Such mutual force of attraction is *always* acting on a body subjected to the earth's attraction, even if the body is in free fall or in orbit around the earth. If a body B is in contact with the earth, the body is also subjected to a reaction force $\vec{N}$, which is normal to the earth's surface at the point of contact. If the effects of the spin of the earth about its own axis and of the motion of the earth in space are neglected, the acceleration of B is zero and the magnitude $|\vec{N}|$ of the normal force is equal to the gravitational force mg. The magnitude $|\vec{N}|$ is the force that would be measured by a scale if body B were on the scale, and the scale were fastened to the earth.

If you imagine yourself being the body B described, the normal reaction $\vec{N}$ is the force you feel on your body when you are in contact with the earth, whether you stand, sit, or lie down. Now, imagine yourself inside an elevator, standing on a scale nailed to the floor of the elevator. The scale would read the value of the magnitude $|\vec{N}|$ of the normal force $\vec{N}$, which would be $|\vec{N}| = mg$ if the elevator were not accelerating. If the elevator were in free fall toward the earth's surface, with an acceleration equal to $g \, \text{m/s}^2$, you would also be in free fall with the same acceleration. In such a case, the scale underneath you would read $|\vec{N}| = 0$ (because both you and the elevator are falling toward the ground with the same acceleration and the same speed) and, therefore, you would feel as if you were weightless even though the gravitational force of attraction of the earth, with magnitude equal to

---

[6]The inhomogeneity of the earth and its departure from a perfect sphere cause the acceleration of gravity to vary slightly from the value given, depending on where the mass is located on the earth's surface.

$GM_Em/(R_E + h)^2 \approx GM_Em/R_E^2 \overset{\Delta}{=} mg$, where h is the height from the earth's surface, would still be acting on your body (but you would not feel that force "pressing" your body).

From this discussion, it is seen that, although a body resting on the earth's surface and one in free fall are subjected to the same gravitational force mg, the sensation of "weightlessness" (which is characterized by the lack of the pressing sensation felt on the body) can only be explained if *weight* is associated with the normal reaction force applied by the earth to the body (instead of the gravitational force mg, which is always nonzero).

1.9   Equivalent Force Systems

The concept of equivalent force systems is introduced in statics courses. Two sets of forces are said to be equivalent to each other if they have the same resultant $\vec{F}$, and also the same moment $\vec{M}_P$ about the same point P, which is any point chosen as desired. A given set of forces can be replaced by its equivalent system, which consists of the force $\vec{F}$ whose line of action passes through P, and the moment $\vec{M}_P$. The resultant $\vec{F}$ of a system of forces $\vec{F}_1, \vec{F}_2, \ldots, \vec{F}_n$ is, by definition, the vector sum $\vec{F} = \vec{F}_1 + \vec{F}_2 + \ldots, +\vec{F}_n$.

Two equivalent force systems have the same effect on a body they are acting on, i.e., they produce the same translational and rotational motions on that body. Figures 1.7b and 1.7c illustrate two force systems applied to a body that are each equivalent to the force system in Fig. 1.7a, where $F_1 = 8$ newtons (N), $F_2 = 20$ N, $F_3 = 10\sqrt{3}$ N, $L = 2$ m, and $h = 1$ m .

Inspecting Fig. 1.7a, one can see that the resultant force $\vec{F}$ of the system is

$$\vec{F} = \vec{F}_1 + \vec{F}_2 + \vec{F}_3 = \left[\left(20\cos\frac{\pi}{6} - 10\sqrt{3} \right) \hat{x} + \left(20\sin\frac{\pi}{6} - 8 \right) \hat{y} \right] = 2\hat{y} \text{ N}$$

and this is the force that will appear in any other system that is equivalent to the one in Fig. 1.7a.

The sum of the moments about point A of each force of the given system is equal to zero (in fact, the moment about A of each force is zero). Therefore, the given system is equivalent to a single force $\vec{F} = 2\hat{y}$ whose line of action passes through point A. This is the single-force system shown in Fig. 1.7b. You will also obtain a single-force equivalent system if you choose to take moments about any point that is in the line of action of the resultant $\vec{F}$ shown in Fig. 1.7b.

If we choose point C to take moments about, we obtain

$$\vec{M}_C = \overrightarrow{CB} \times \vec{F}_1 + \overrightarrow{CA} \times \vec{F}_2 + \overrightarrow{CD} \times \vec{F}_3$$

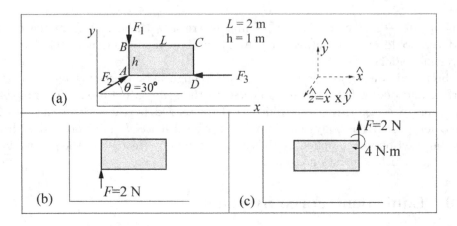

Figure 1.7: An example of equivalent force systems.

$$= \underbrace{(-2\hat{x}) \times (-8\hat{y})}_{=16\hat{z}} + \underbrace{(-2\hat{x} - \hat{y}) \times \left[20\left(\hat{x}\cos\frac{\pi}{6} + \hat{y}\sin\frac{\pi}{6}\right)\right]}_{=-20\hat{z}+10\sqrt{3}\,\hat{z}}$$

$$+ \underbrace{(-\hat{y}) \times \left(-10\sqrt{3}\,\hat{x}\right)}_{=-10\sqrt{3}\,\hat{z}} = -4\hat{z} \text{ N·m}$$

Thus, the force-moment system in Fig. 1.7c is also equivalent to the ones in Figs. 1.7a and 1.7b.

1.10   Center of Gravity and Center of Mass

The concept of *center of gravity of a body* of mass m is associated with the concept of equivalent force systems that was reviewed in Section 1.9. In contrast, the center of mass of a body has nothing to do with any forces acting on the body. These concepts will be briefly discussed here using Fig. 1.8 where two bodies are subjected to their mutual forces of attraction. The body centered at O is spherical, for simplicity, with its mass M being much greater than that of any other body in its vicinity. This spherical body might approximate the earth, for example. Its force of attraction exerted on other bodies, such as on body B shown in Fig. 1.8, is what is called a *gravitational force*. The figure is not drawn to scale, of course, but that is not relevant to this presentation.

The *center of mass* of a body is a point C whose position is defined using a weighted average of the position of each mass element of the body. Using Fig. 1.8,

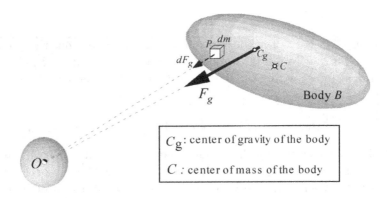

Figure 1.8: Distinction between the *center of gravity* and the *center of mass* of a body.

the position vector $\overrightarrow{OC}$ (i.e., the vector from O to C) of the center of mass C of body B is defined as

$$\overrightarrow{OC} = \frac{\int\limits_{body} \overrightarrow{OP}\, dm}{m} \tag{1.10.1}$$

The center of mass of any body is a property of the body. It does not depend on any forces acting on the body, and it does not depend on how the body is oriented in space. The center of mass of a body plays a significant role in the formulation of the differential equations that govern the motion of a body. This is first addressed in detail in Chapter 4.

In contrast, the location of the *center of gravity* of a body depends on the gravitational force acting on each part of the body and, thus, on the orientation of the body. Points on body B in Fig. 1.8 that are closer to the attracting center O are subjected to a greater gravitational force than points that are farther away. In that figure, each infinitesimal element of body B, of mass dm, that is at a distance $|\overrightarrow{OP}|$ from the center O of the attracting body at O is subjected to the gravitational force $d\vec{F}_g$ shown in the figure. The resultant $\vec{F}_g$ is calculated according to Newton's law of universal gravitation and is also shown in Fig. 1.8.

$$\vec{F}_g = \int\limits_{body} d\vec{F}_g = -\int\limits_{body} \frac{GM\, dm}{|\overrightarrow{OP}|^2} \frac{\overrightarrow{OP}}{|\overrightarrow{OP}|} \tag{1.10.2}$$

> The center of gravity of a body of mass m is, by definition, the point C_g in Fig. 1.8 that would be subjected to the same resultant $\vec{F}_g$ of the gravitational forces acting on the body if the entire mass of the body were concentrated at C_g.

The determination of the center of gravity of a body involves two steps:

Step 1. Replace the resultant of the gravitational forces acting on each part of the body with a *single equivalent force*. Such a force is the vector $\vec{F}_g$ given by Eq. (1.10.2). When this step is completed, the line of action of that force is then known. Such a calculation is illustrated in Example 1.10.1 at the end of this section.

Step 2. Choose C_g to be the point on the line of action determined in step 1 that would be subjected to the same resultant $\vec{F}_g$ of the gravitational forces acting on the body if the entire mass of the body were concentrated at that point. That is, C_g is the point located by the distance $|\overrightarrow{OC_g}|$, along the line of action of $\vec{F}_g$, defined as

$$|\vec{F}_g| = \frac{GMm}{|\overrightarrow{OC_g}|^2} \qquad \text{thus} \qquad |\overrightarrow{OC_g}| = \sqrt{\frac{GMm}{|\vec{F}_g|}} \qquad (1.10.3)$$

In many applications, including most of those considered in this book, the gravitational field of the earth is approximated as that of an infinite plane (exceptions to this are presented in Chapter 4). In such an ideal case, the force on a body due to gravity does not depend on the body's position, and its magnitude near the surface of the earth is simply approximated as mg where $g = GM_E/R_E^2$, as discussed in Section 1.8 (p. 22). In practice, such an approximation is justified when a body does not move too far from the line that is perpendicular to the local vertical. In such cases, the center of gravity coincides with the center of mass. It is important to keep in mind, however, the conceptual difference between the center of gravity and the center of mass because one must not confuse the two when a distinction must be made, such as in more advanced problems dealing with the motion of satellites and space structures. The distinction between the center of mass and the center of gravity becomes crucial when one investigates the rotational motion of orbiting satellites, for example. However, the determination of the center of gravity is neither necessary nor relevant in these or in many other problems. As shown in Chapters 5 and 6, what is important in rotational dynamics is the moment, about the center of mass of the body, of the forces that are applied to the body, and the *angular momentum* of the system (defined in Chapter 4).

Example 1.10.1 The body shown in Fig. 1.9 is a dumbbell that consists of two small spheres, each of mass $m/2$, connected at their centers A and B by a rigid massless bar of length L. Point C is the center of mass of the dumbbell, and the orthogonal $\hat{x}$ and $\hat{y}$ unit vectors, with $\hat{x}$ parallel to line OC, are included in the figure for use in later calculations. The dumbbell is able to rotate in the plane of the figure, and its orientation relative to line OC is tracked by the angle $\theta(t)$, where t is time.

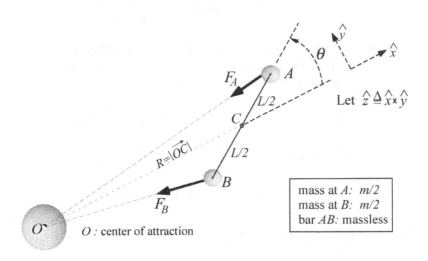

Figure 1.9: Figure for Example 1.10.1.

A gravitational *center of attraction*, represented by the sphere centered at a point O, exerts the forces $\vec{F}_A$ and $\vec{F}_B$ on the spheres at A and B, respectively (see Fig. 1.9), with magnitudes $F_A = (\mu m/2)/|\overrightarrow{OA}|^2$ and $F_B = (\mu m/2)/|\overrightarrow{OB}|^2$, respectively, where μ is a constant. For $R = 4$ m, $L = 2$ m, $m = 1$ kg, and $\mu = 1.8$ m^3/s^2, determine:

a. The resultant gravitational force $\vec{F}_g = \vec{F}_A + \vec{F}_B$ acting on the dumbbell, expressed in terms of the angle θ and the unit vectors $\hat{x}$ and $\hat{y}$

b. The resultant moment, about C, due to the forces $\vec{F}_A$ and $\vec{F}_B$ acting on the dumbbell

c. The distance from B to the point of intersection of the resultant $\vec{F}_g$ with line AB, and the distance from O to that same point

d. The distance from O to the center of gravity C_g of the dumbbell

■ **Solution**

a. The resultant force $\vec{F}_g$ is simply

$$\vec{F}_g = \vec{F}_A + \vec{F}_B \;=\; -\frac{\mu m/2}{|\overrightarrow{OA}|^2}\,\underbrace{\frac{\overrightarrow{OA}}{|\overrightarrow{OA}|}}_{\substack{\text{Unit vector}\\\text{from } O \text{ to } A}} \;-\; \frac{\mu m/2}{|\overrightarrow{OB}|^2}\,\underbrace{\frac{\overrightarrow{OB}}{|\overrightarrow{OB}|}}_{\substack{\text{Unit vector}\\\text{from } O \text{ to } B}}$$

$$= \;-\frac{\mu m}{2|\overrightarrow{OA}|^3}\,\overrightarrow{OA} - \frac{\mu m}{2|\overrightarrow{OB}|^3}\,\overrightarrow{OB}$$

This expression for $\vec{F}_g$ shows that we now need expressions for the vectors $\overrightarrow{OA}$ and $\overrightarrow{OB}$ and for their magnitudes. This is done by examining Fig. 1.9, which leads us to write the following equations

$$\overrightarrow{OA} \;=\; \overrightarrow{OC} + \overrightarrow{CA} = \left(R + \frac{L}{2}\cos\theta\right)\hat{x} + \frac{L}{2}\left(\sin\theta\right)\hat{y}$$

$$\overrightarrow{OB} \;=\; \overrightarrow{OC} + \overrightarrow{CB} = \left(R - \frac{L}{2}\cos\theta\right)\hat{x} - \frac{L}{2}\left(\sin\theta\right)\hat{y}$$

and from these, in turn, we can determine the magnitudes

$$|\overrightarrow{OA}| \;=\; \left[\left(R + \frac{L}{2}\cos\theta\right)^2 + \left(\frac{L}{2}\sin\theta\right)^2\right]^{1/2} = \left(R^2 + \frac{L^2}{4} + RL\cos\theta\right)^{1/2}$$

$$|\overrightarrow{OB}| \;=\; \left[\left(R - \frac{L}{2}\cos\theta\right)^2 + \left(\frac{L}{2}\sin\theta\right)^2\right]^{1/2} = \left(R^2 + \frac{L^2}{4} - RL\cos\theta\right)^{1/2}$$

Finally, this then yields the following expression for the resultant force $\vec{F}_g$:

$$\vec{F}_g = -\mu\frac{m}{2}\left[\frac{\left(R + \frac{L}{2}\cos\theta\right)\hat{x} + \left(\frac{L}{2}\sin\theta\right)\hat{y}}{\left(R^2 + \frac{L^2}{4} + RL\cos\theta\right)^{3/2}}\right.$$

$$\left.+\frac{\left(R - \frac{L}{2}\cos\theta\right)\hat{x} - \left(\frac{L}{2}\sin\theta\right)\hat{y}}{\left(R^2 + \frac{L^2}{4} - RL\cos\theta\right)^{3/2}}\right] \qquad (1.10.1a)$$

For $R = 4$ m, $L = 2$ m, $m = 1$ kg, $\theta = 30°$, and $\mu = 1.8$ m^3/s^2, this yields

$$\vec{F}_g \approx -0.126\hat{x} + 0.01\hat{y}\;\text{ N}$$

b. The resultant moment, about the center of mass C, due to the forces $\vec{F}_A$ and $\vec{F}_B$ acting on the dumbbell is (see Fig. 1.9)

$$\vec{M}_C = \overrightarrow{CA} \times \vec{F}_A + \overrightarrow{CB} \times \vec{F}_B = \frac{L}{2} \left(\hat{x} \cos\theta + \hat{y} \sin\theta \right) \times \vec{F}_A$$
$$- \frac{L}{2} \left(\hat{x} \cos\theta + \hat{y} \sin\theta \right) \times \vec{F}_B$$

and this gives, with $\hat{z} = \hat{x} \times \hat{y}$ as defined in Fig. 1.9,

$$\vec{M}_C = -\frac{\mu m L R}{4} \left[\frac{1}{\left(R^2 + \frac{L^2}{4} - RL\cos\theta \right)^{3/2}} \right.$$

$$\left. - \frac{1}{\left(R^2 + \frac{L^2}{4} + RL\cos\theta \right)^{3/2}} \right] (\sin\theta)\,\hat{z} \qquad (1.10.1b)$$

For $R = 4$ m, $L = 2$ m, $m = 1$ kg, $\theta = 30°$, and $\mu = 1.8$ m^3/s^2, this yields

$$\vec{M}_C \approx -0.041\hat{z} \text{ N·m}$$

which, for $\theta = 30°$, is a clockwise moment (see the unit vectors $\hat{x}$ and $\hat{y}$ in Fig. 1.9, and then observe the direction of $\hat{z} = \hat{x} \times \hat{y}$ using the right-hand rule). Notice that such a moment causes the angle θ in Fig. 1.9 to decrease. This behavior is expected since, as seen from Fig. 1.9, the force acting at B is larger than that acting at A, and this creates the clockwise moment $\vec{M}_C$. The moment $\vec{M}_C$ will cause AB to rotate.

c. The two-force system $\vec{F}_A$ and $\vec{F}_B$ acting on the dumbbell is equivalent to the force $\vec{F}_g$ applied at the center of mass C of the dumbbell and the moment $\vec{M}_C$. That system is also equivalent to a single force $\vec{F}_g$ in some direction in space that passes through O and intersects line AB at some point C_1 for which the moment $\vec{M}_{C_1}$, about such a point C_1, is equal to zero. Such a situation is shown in Fig. 1.10, where C_1 is the point of intersection of the resultant $\vec{F}_g$ with line AB. Another point C_g on the *line of action* of such a resultant force is also shown in that figure and is used for answering the question in part d. The moment about any point on that line of action of $\vec{F}_g$ is equal to zero.

Since point C_1 is on the line of action of $\vec{F}_g$, we then have

$$\vec{M}_{C_1} = \overrightarrow{C_1 A} \times \vec{F}_A + \overrightarrow{C_1 B} \times \vec{F}_B$$
$$= (L - D)\left(\hat{x} \cos\theta + \hat{y} \sin\theta \right) \times \vec{F}_A - D\left(\hat{x} \cos\theta + \hat{y} \sin\theta \right) \times \vec{F}_B = 0$$

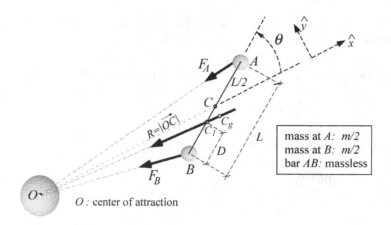

Figure 1.10: The single-force equivalent to the two forces $\vec{F}_A$ and $\vec{F}_B$ in Fig. 1.9.

and this yields

$$
\vec{M}_{C_1} = -\frac{\mu m R}{2} \left[\frac{D}{(R^2 + L^2/4 - RL\cos\theta)^{3/2}} \right.
$$

$$
\left. -\frac{L - D}{(R^2 + L^2/4 + RL\cos\theta)^{3/2}} \right] (\sin\theta)\,\hat{z} = 0 \quad (1.10.1c)
$$

By solving for the distance D, we then obtain

$$
D = \frac{L}{1 + \left(\dfrac{R^2 + L^2/4 + RL\cos\theta}{R^2 + L^2/4 - RL\cos\theta} \right)^{\frac{3}{2}}}
$$

The location of point C_1 along line AB defines the orientation of the resultant force $\vec{F}_g$ shown in Fig. 1.10. With this information, any other desired point on that line can now be located.

For $R = 4$ m, $L = 2$ m, and $\theta = 30°$, we obtain

$$
D \approx 0.43 \text{ m}
$$

i.e., for $\theta = 30°$, C_1 is a point on line AB, between B and C (see Fig. 1.10), located 0.43 m from B and 0.57 m from the center of mass C. For the given data, its distance $|\overrightarrow{OC_1}|$ from point O, obtained by geometry from Fig. 1.10, is

$$|\overrightarrow{OC_1}| = \sqrt{\left[R - \left(\frac{L}{2} - D\right)\cos\theta\right]^2 + \left[\left(\frac{L}{2} - D\right)\sin\theta\right]^2} \approx 3.52 \text{ m}$$

Thus, the answers to this part are the distances D and $|\overrightarrow{OC_1}|$.

d. Having determined the location of point C_1, one can now locate any other point on the line of action of the resultant gravitational force $\vec{F}_g$. For point C_g (see Fig. 1.10), which is defined to be the center of gravity of the dumbbell as

$$|\vec{F}_g| \stackrel{\Delta}{=} \frac{\mu m}{|\overrightarrow{OC_g}|^2} \qquad (1.10.1\text{d})$$

we obtain for the distance $|\overrightarrow{OC_g}|$, with the magnitude of $\vec{F}_g$ obtained from Eq. (1.10.1a),

$$|\overrightarrow{OC_g}| = \sqrt{\frac{\mu m}{|\vec{F}_g|}} \approx 3.78 \text{ m} \qquad (1.10.1\text{e})$$

1.11   Free-Body Diagrams and Forces

When formulating a dynamics problem using the Newtonian approach (be it for a single mass particle, an element of a machine, or an airplane in flight, for example) it is invaluable to draw a *free-body diagram* for the object under consideration. The free-body diagram for an object P is obtained by isolating it from its surroundings and representing, by vectors, all forces acting on P due to its interaction with the surrounding medium. The forces that are encountered in actual situations can be classified into two basic types: *field forces*, such as the force acting on an object due to the gravitational attraction of the earth or other massive objects, and *contact forces*, such as the force due to a spring connected to the object or the force due to friction. The free-body diagram for a body with finite dimensions should include these two types of forces and all the moments (or *torques*)[7] that might be applied to the body.

---

[7]The word *torque* is generally used synonymously with *moment*. It is also used by some authors for the moment of a pair of forces that are equal in magnitude and opposite in direction. Such a pair of forces is known as a *couple*.

The moment, about a point P that coincides with a particle, of any force applied to the particle is zero. Therefore, since particles have no dimension, they can only be subjected to forces, not moments. The moment, about other points in space, of a force applied to a particle may, of course, be different from zero.

Let us look at a couple of examples to illustrate how to draw a proper free-body diagram.

Example 1.11.1 Consider the pendulum shown in Fig. 1.11. The pendulum consists of a bar AB of constant length L and mass m_1, supported by a pin at point A, and of a particle of mass m_2 located at point B. A moment M_{bar} is applied to the bar at A. That moment is due either to a motor (which is not shown in the figure) or to friction at the pin (if there is friction), or both. Moments are represented by a curved arrow (which represents the curved fingers of a hand, as in the right-hand rule), or by a straight arrow (since moments are vectors). However, since forces are always represented by straight arrows in free-body diagrams, it is wise to avoid using the same type of straight arrows to represent moments (so that they are not misunderstood as a force). The angle θ measured from the vertical to the bar shown in Fig. 1.11 is used for tracking the orientation of the bar. The system is close enough to the surface of the earth (i.e., L is much smaller than the radius of the earth) so that every point of the pendulum is subjected to the same constant gravitational force (which is a very reasonable approximation). It is assumed that there are no other forces applied to the system. The free-body diagrams for the bar AB and for the particle at B are obtained by isolating the bar, and the particle at B, with the cuts indicated by the dashed lines shown in Fig. 1.11.

To draw the free-body diagram for bar AB, start by drawing the bar. A vertical line and the angle θ that was chosen to track the direction of the bar at all times are also included in the diagram shown in Fig. 1.11. Their inclusion in the diagram is optional, but many authors include such information for completeness. Such a practice will prove to be convenient later, when formulating the differential equation that governs the motion of the system. The *differential equation of motion* (or equations, for the more general case when there is more than one) is the equation that allows the analyst to investigate and predict the motion of the system.

Next, put the forces in the free-body diagram of the bar. At point A on the bar there is a reaction force of unknown magnitude and direction. Forces in free-body diagrams are generally split into components along orthogonal directions. In Fig. 1.11 those directions were chosen to be parallel and perpendicular to the bar. The orthogonal arrows labeled A_1 and A_2 represent such components for the reaction force at A. The directions of the orthogonal arrows representing these components are totally arbitrary. At this stage of the formulation of the solution to a problem, it

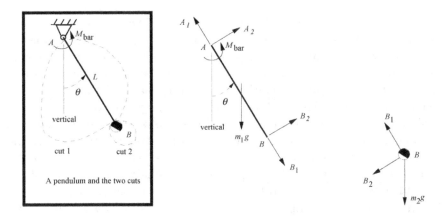

Figure 1.11: A pendulum and the free-body diagrams for each of its bodies.

does not matter if the vector $\vec{A}_1$ is drawn pointing away from the bar (as shown in the middle part of Fig. 1.11) or toward the bar. When values for the scalar components A_1 and A_2 of the reaction force at A are eventually determined, they will be either positive or negative. A positive value for A_1, for example, simply means that the direction of the arrow labeled A_1 in Fig. 1.11 is indeed in the direction drawn in the figure, while a negative value means it is in the opposite direction.

At point B of bar AB there is also a force of unknown magnitude and direction applied to the bar. That force is also split into two orthogonal components (along the same directions chosen previously) that are represented by the arrows labeled B_1 and B_2, respectively. As before, the directions for the orthogonal arrows representing these components are totally arbitrary.

The only remaining force acting on bar AB is the gravitational force due to the attraction of the earth, which always acts along the *local vertical*. Here, and elsewhere in the book (except when noted), the gravitational field will be approximated as constant so that a gravitational force acting on a body of mass m is simply a vector of magnitude mg. The gravitational force acting on the bar[8] is represented

---

[8]The gravitational force acting on the bar is, of course, distributed along the bar. Each infinitesimal bar element of mass dm_1 is subjected to the gravitational force whose magnitude is $g\,dm_1$. These forces are equivalent to (i.e., they can be replaced by) a resultant force, and the moment of the distributed forces, applied at any point P chosen on the bar. The resultant force is equal to the integral, over the length of the bar, of $g\,dm_1$, or simply $m_1 g$ when g is approximated as a constant. The moment of the distributed forces depends on the choice of point P, and such a moment is zero if P is chosen to be the center of gravity of the bar. Under the assumption of a constant gravitational field, the center of gravity of the bar coincides with its center of mass. For a homogeneous bar, for example, the center of mass is at the middle of the bar.

by the arrow labeled $m_1 g$ shown in the free-body diagram in Fig. 1.11.

The final item to add to the free-body diagram for the bar is the moment M_{bar} discussed earlier. That moment is represented by the curved arrow labeled M_{bar} that is added to the diagram (the counterclockwise sense was chosen arbitrarily; it could also have been drawn clockwise). If the quantity M_{bar} is positive, the moment is counterclockwise, as drawn. If M_{bar} is negative, the negative value of M_{bar} will automatically reverse the physical interpretation of the counterclockwise arrow, making it a clockwise moment. Thus, in either case, representing the moment by that counterclockwise arrow labeled M_{bar} will account for the actual physical situation (and we do not have to draw another free-body diagram with that arrow reversed). The free-body diagram for the bar is now complete.

The free-body diagram for the point mass at B is also shown in Fig. 1.11. The forces applied to the point mass are the gravitational force due to the attraction of the earth (which is the vertical force with magnitude equal to $m_2 g$) and the reaction force applied by bar AB. The latter is represented by the negative [according to Newton's third law (Section 1.5)] of the forces B_1 and B_2 that were drawn in the free-body for the bar. Therefore, the corresponding arrows for these forces should be drawn in the free-body diagram for the point mass in opposite directions to those that appear in the free-body diagram for the bar. The labels for these forces must be the same ones that were used in the free-body diagram for the bar.[9] This completes the free-body diagram for the point mass.

Since the point mass is just that, a point (i.e., it has no dimensions), no moment can ever be applied to it. That is why there is no reaction moment at point B, only forces. If the point mass were replaced by a finite body, the reactions at the connection between that body and bar AB would consist of the forces labeled B_1 and B_2 and a moment (which we might label M_B) if the connection were a weld or a pin with friction.

The two free-body diagrams for the pendulum shown in Fig. 1.11 involve five unknowns (assuming that M_{bar} is either a known function of time or a function of these five unknowns), namely the angle θ and the four reaction forces A_1, A_2, B_1, and B_2. To determine them, we are then going to need to generate five equations involving these five unknowns. The generation of such equations is one of the objectives of *dynamics*, the other being the determination of the desired unknowns using those same equations. This will be done in examples starting in Chapter 2. The equations involving such unknowns are generated by using Newton's laws of motion, and their extension to finite bodies (which was developed by Euler).

It should be noted that for electrically charged particles moving in an electro-

---

[9]We could also keep the same directions for the arrows and attach a minus sign to the corresponding labels. Either method accomplishes the same purpose.

magnetic field with flux density $\vec{B}$, an electromagnetic force is also induced on the particle. It is known that the electromagnetic force is in the direction perpendicular to the relative velocity $\vec{v}$ between the two charged particles and the magnetic flux density $\vec{B}$. If q is the charge of each particle, the electromagnetic force is determined as $q\vec{v} \times \vec{B}$. Such forces are not considered in this book.

Example 1.11.2 Consider a bar of length L, of constant mass density m/L (where m is the mass of the bar), in contact with a vertical wall and a horizontal floor as shown in Fig. 1.12a. The bar is moving [i.e., the angle $\theta = \theta(t)$ is changing with time t] on the vertical plane, and there is friction at the contacts of the bar with the wall and the floor. The initial conditions for the motion are $\theta(0) = \theta_0$ and $(d\theta/dt)_{t=0} = w_0$, where w_0 may be either positive or negative, so that point B of the bar may be moving either to the left or right. When θ increases, point B moves to the left with velocity v_B (relative to the fixed reference represented by the floor and the wall), and point A moves up the wall with velocity v_A as indicated in the figure. The free-body diagram for the bar is shown in Fig. 1.12b.

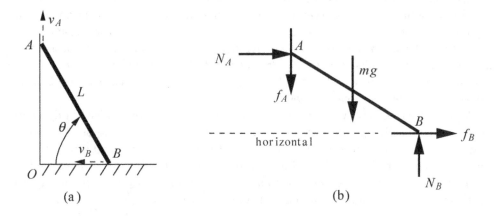

(a)   (b)

Figure 1.12: (a) A bar in contact with a wall and the floor, and (b) the free-body diagram for the bar.

The forces that appear in the free-body diagram are:

■ The gravitational force mg, which passes through the center of mass of the bar (located at the geometric center of the uniform bar).

- The normal reaction forces N_A (applied by the vertical wall to the bar) and N_B (applied by the floor to the bar); these forces are perpendicular to the surfaces the bar is in contact with.
- The friction forces f_A and f_B acting at the contact points A and B, respectively.

The friction force that acts on a body that translates along a surface is *always* directed opposite to the velocity of the point of contact of the body relative to the contact surface.[10] Therefore, the friction forces f_A and f_B shown in Fig. 1.12 are directed opposite to the velocities of point A relative to the wall and of point B relative to the floor, respectively. By inspecting the figure, it is seen that the directions of f_A and f_B depend on whether θ increases or decreases. The directions indicated in Fig. 1.12b are for the case when the angle θ, measured from the floor to bar AB, is increasing, i.e., when $d\theta/dt > 0$. The directions of f_A and f_B shown in Fig. 1.12 are reversed when $d\theta/dt < 0$.

There are two types of friction, *dry friction* and *viscous friction*, and several mathematical models for them. The most common model for dry friction is *Coulomb friction*, which is named for Charles Augustin de Coulomb (1736-1806), who was a French engineer who became well known for his experimental work with friction.

For Coulomb friction between two surfaces in contact with each other, and moving relative to each other, the magnitude of the friction force is proportional to the magnitude of the normal force of contact between the surfaces. The proportionality coefficient is called the *Coulomb friction coefficient* (or the *coefficient of dry friction*). This is a reasonably good model for the friction force between two dry surfaces in contact with each other.

Viscous friction is provided by a *viscous medium*, such as oil, when a body moves in that medium. The magnitude of the viscous friction force acting on the body is a function of the magnitude of the velocity of the body relative to that medium. If the dependency on the velocity is linear, the friction is called *linear viscous friction*.

The following are the expressions for the friction forces f_A and f_B that appear in the free-body diagram for the bar considered in this example.

For Coulomb friction:

$$f_A = \mu_A |N_A| sgn\left(\frac{d\theta}{dt}\right) \qquad (1.11.1)$$

$$f_B = \mu_B |N_B| sgn\left(\frac{d\theta}{dt}\right) \qquad (1.11.2)$$

---

[10]Rolling of a rigid body (with or without slipping) is a special case that is considered in Section 5.4 of Chapter 5.

where $\mu_A > 0$ and $\mu_B > 0$ are the kinetic coefficients of Coulomb friction at the contact points A and B, respectively, in Fig. 1.12. Notice that the use of the *sign function* (denoted either as *sign* or *sgn* in mathematical expressions) in Eqs. (1.11.1) and (1.11.2) automatically handles the change in the direction of the motion, thus making those expressions valid for all times. Also notice that the expressions depend on $d\theta/dt$ in a nonlinear manner since the sign function does not vary linearly with its argument. The sign function is, by definition, equal to $+1$ when its argument is positive, equal to -1 when its argument is negative, and undefined when its argument is zero. One may arbitrarily define $sgn(0)$ to be equal to zero, and this is commonly done.

For linear viscous friction:

$$f_A = c_A \frac{d}{dt}(L \sin\theta) = c_A L \frac{d\theta}{dt}\cos\theta \tag{1.11.3}$$

$$f_B = c_B \frac{d}{dt}(-L\cos\theta) = c_B L \frac{d\theta}{dt}\sin\theta \tag{1.11.4}$$

where $c_A > 0$ and $c_B > 0$ are the viscous friction coefficients at A and B, respectively.

For *linear viscous friction* acting on a point P of a body that is translating in a viscous medium, the friction force is linearly proportional to the velocity of P relative to the medium (all the necessary details about the proper formulation of velocity are presented in Chapter 2), and the direction of the friction force is opposite to that velocity. For point A in Fig. 1.12, for example, the velocity of A is the time derivative of the vector from O to A (see Example 2.5.3, p. 109), and this gives the expression for f_A given by Eq. (1.11.3) when friction is of the viscous type. The viscous friction force f_A would be provided by a *viscous damper* device (see the note on p. 50) that would be connected to points A and O.

Notice that, according to Eq. (1.11.3), f_A is positive (i.e., the viscous friction force at A points downward, as shown in Fig. 1.12) when $d\theta/dt > 0$ (because the term $c_A L \cos\theta$ in that expression is not negative for $0 \le \theta \le \pi/2$), and is negative when $d\theta/dt < 0$. Also, according to the expression for f_B given by Eq. (1.11.4), the viscous friction force at B is positive (i.e., that friction points to the right, as shown in Fig. 1.12) when $d\theta/dt > 0$ and negative when $d\theta/dt < 0$. Therefore, the expressions for the viscous friction forces given by Eqs. (1.11.3) and (1.11.4) are valid for all times. The use of the sign function is not needed for the case of viscous friction.

1.12   Chain Rule of Differentiation and Exact Differentials

> The material presented in this section is indispensable when energy methods are used. Energy methods are extremely useful and versatile. Potential energy is determined using the mathematics presented in this section. You may skip this section for now, but you will need to study it later to fully understand how to determine potential energy (from the simplest to the most complicated cases) and the basics that constitute the foundations of energy methods.

The chain rule of differentiation is frequently used for manipulating a differential equation of motion (basic material on the latter is presented in Section 1.13). To review the chain rule, let $x_1, x_2, \ldots, x_n$ be variables that depend on time t in an unknown manner (i.e., an expression for x_1, x_2, ..., as a function of time, is not known). Also, let $f(x_1, x_2, \ldots, x_n)$ be some known function of these variables.

For example, we might have $f = x_1^2 \sin x_2$, or $f = x_1^2 + K_1 x_2^2 + K_2 x_1 \sin x_2$, or any other function. In dynamics, x_1, x_2, ... might stand for variables that were chosen to represent the position and velocity of a particle or the angular orientation and angular velocity of a bar, while f might stand for the sum of the kinetic and potential energies of the motion. Suppose we now want to obtain the total time derivative (which we will refer to from now on simply as the time derivative) df/dt of the function f. According to the chain rule of differentiation, we then have

$$\boxed{\frac{df}{dt} = \frac{\partial f}{\partial x_1}\frac{dx_1}{dt} + \frac{\partial f}{\partial x_2}\frac{dx_2}{dt} + \ldots + \frac{\partial f}{\partial x_n}\frac{dx_n}{dt}} \qquad (1.12.1)$$

The coefficients of $dx_1/dt, dx_2/dt, \ldots$ are called the *partial derivatives* of f with respect to $x_1, x_2, \ldots$. As suggested by Eq. (1.12.1), the coefficient of dx_1/dt is obtained by holding $x_2, \ldots, x_n$ constants (so that $dx_2/dt = \ldots = dx_n/dt = 0$) and then taking the derivative of f with respect to x_1. A similar procedure is followed for obtaining the coefficients of $dx_2/dt, \ldots$. For example, for $f = x_1^2 \sin x_2 + K x_2^2$, we have

$$\frac{\partial f}{\partial x_1} = 2x_1 \sin x_2$$

$$\frac{\partial f}{\partial x_2} = x_1^2 \cos x_2 + 2K x_2$$

which gives

$$\frac{df}{dt} = (2x_1 \sin x_2)\frac{dx_1}{dt} + (x_1^2 \cos x_2 + 2K x_2)\frac{dx_2}{dt}$$

Another operation that will appear in dynamics is the *differential* of a function f. That is simply df and, with $f = f(x_1, x_2, \ldots, x_n)$, it is simply obtained as

$$df = \frac{\partial f}{\partial x_1}dx_1 + \frac{\partial f}{\partial x_2}dx_2 + \ldots + \frac{\partial f}{\partial x_n}dx_n \qquad (1.12.2)$$

The quantity df/dt is a *finite quantity*, while df is an *infinitesimal quantity*. Since df, as given by Eq. (1.12.2), was obtained by starting with a function f, it is clearly the "d of a function f," and, for this reason, it is said to be an *exact differential*. When necessary, an exact differential is written as $d(f)$, which should be read as "d of f." The integral of an exact differential $d(f)$ is, except for a constant of integration, the function f.

Suppose, now, that we have the opposite problem (which is a problem that will appear in dynamics, as you will see later): Given an expression for some infinitesimal quantity dg, as

$$dg = a_1 dx_1 + a_2 dx_2 + \ldots + a_n dx_n \qquad (1.12.3)$$

where $a_1, a_2, \ldots, a_n$ are functions of $x_1, x_2, \ldots, x_n$, we need to determine if there is a function g such that "dg is the d of a g." The answer is yes only if dg can be integrated, i.e., if dg is an exact differential. As indicated by Eq. (1.12.2), if dg is an exact differential, the coefficients of $dx_1, dx_2, \ldots$ in the expression for dg have to be of the following form:

$$
\begin{aligned}
a_1 &= \frac{\partial g}{\partial x_1} \\
a_2 &= \frac{\partial g}{\partial x_2} \\
&\vdots \\
a_n &= \frac{\partial g}{\partial x_n}
\end{aligned}
$$

$$(1.12.4)$$

Since $\partial^2 g/\partial x_i \partial x_j = \partial^2 g/\partial x_j \partial x_i$ for continuous functions (and we will only deal with continuous functions), the test to determine if dg is an exact differential consists of checking to see if the relationship $\partial a_i/\partial x_j = \partial a_j/\partial x_i$ is satisfied for all the coefficients that appear in the expression for the given differential dg.

As an example, consider the differential $dg = 2x_1(\sin x_2)dx_1 + (x_1^2 \cos x_2 + 2Kx_2)dx_2$, where K is a constant. We already know that $dg = d(x_1^2 \sin x_2 + Kx_2^2)$ and, thus, that $g = x_1^2 \sin x_2 + Kx_2^2$ plus a constant, since this $g(x_1, x_2)$ was one of the functions used in a previous example. But if we did not have that information, the test to see whether the given dg can be integrated to yield a function $g(x_1, x_2)$ consists of checking to see if $\partial(2x_1 \sin x_2)/\partial x_2$ is equal to $\partial(x_1^2 \cos x_2 + 2Kx_2)/\partial x_1$. For this example we have

$$\frac{\partial}{\partial x_2}(2x_1 \sin x_2) = 2x_1 \cos x_2$$

$$\frac{\partial}{\partial x_1}(x_1^2 \cos x_2 + 2Kx_2) = 2x_1 \cos x_2$$

Since the given expression for dg passes the test, a function g exists. We then must have for this example,

$$dg = 2x_1 \sin x_2 dx_1 + (x_1^2 \cos x_2 + 2Kx_2)dx_2 = \frac{\partial g}{\partial x_1}dx_1 + \frac{\partial g}{\partial x_2}dx_2$$

which implies that

$$\frac{\partial g}{\partial x_1} = 2x_1 \sin x_2$$

$$\frac{\partial g}{\partial x_2} = x_1^2 \cos x_2 + 2Kx_2$$

The function $g(x_1, x_2)$ is determined by integrating these two equations. By integrating the first equation with respect to x_1 (you could also start by integrating the second equation with respect to x_2; no particular order is relevant), we obtain $g = x_1^2 \sin x_2 + h(x_2)$, where $h(x_2)$ is a function of x_2 only. This then gives

$$\frac{\partial g}{\partial x_2} = x_1^2 \cos x_2 + 2Kx_2 = x_1^2 \cos x_2 + \frac{dh}{dx_2}$$

from which we obtain

$$\frac{dh}{dx_2} = 2Kx_2$$

and, therefore,

$$h(x_2) = \int 2Kx_2 \, dx_2 = Kx_2^2 + \text{ an arbitrary constant of integration}$$

1.13   Some Basics on Differential Equations

> The material and techniques presented in this section are indispensable
> for analyzing the motion of dynamical systems, as done in many of the
> examples presented in this book. This section may also be skipped
> for now, but you may need to study it later to be able to analyze the
> motion governed by the differential equations that you will obtain when
> you investigate a dynamical system.

As mentioned earlier, application of Newton's second law to a moving body yields
several scalar equations that involve unknown reaction forces and the variables that
were chosen to investigate the motion of the body. By appropriately combining
such equations to eliminate the unknown reaction forces, one then obtains a set of
scalar equations that only involve the unknown variables that describe the motion.
The variables and their time derivatives (certainly the second time derivative, and
maybe the first time derivative also) will appear in such equations. If only one
variable is needed to describe the motion, only one of such equations is obtained;
two equations are obtained if there are two variables, and so on. Such equations
are *differential equations*, and the task we now have is to integrate them to obtain
information about the motion so that we can predict how the body or bodies are
going to move in space.

In general, it is impossible to integrate the differential equations of motion analytically because they are nonlinear (i.e., they may involve products of the variables
and of their time derivatives, trigonometric functions of the variables, and other
nonlinear relations). Finding exact analytical solutions to nonlinear equations, in
general, has so far been beyond the capacity of the human brain. However, there
are several cases when it is possible to extract, analytically, important information
about the motion from certain types of nonlinear differential equations. There are
also several methods for finding approximate solutions to such equations. Approximate methods to deal with nonlinear differential equations can be found in some of
the references listed in Appendix G (see Reference 18, for example). If you want to
deepen your knowledge of the subject matter, look for courses and/or books that
deal with matters such as nonlinear oscillations and applications such as dynamics
of gyroscopic instruments, vehicle (space and otherwise) dynamics, and dynamics
of flexible bodies (such as structural dynamics).

Simple analytical methods that allow one to extract important information from
several types of differential equations of motion that appear in dynamics (and seen in
a number of examples presented throughout this book) are presented in this section.
More advanced (and, thus, more powerful) methods are covered in Chapter 7.

A single dot over a symbol is used throughout the book to denote the first derivative with respect to time of the variable represented by that symbol, and the second derivative with respect to time is represented by two dots over the symbol. This is a very common notation used in books and in technical journals for the derivative of a function with respect to time.

CASE 1

For some problems the acceleration of a body may be a known function of time $f(t)$, and the differential equation of motion may be of the following form:

$$\ddot{x} + f(t) = 0 \qquad (1.13.1)$$

Examples of such simple cases are a body translating along a nonrotating straight line and subjected to a known force that is proportional to $f(t)$, or a bar that is rotating on a plane and is subjected to a known moment that is also proportional to a function $f(t)$. Of course, we would not change the independent variable from t to x in Eq. (1.13.1) by using the chain rule in the $\ddot{x}$ term, followed by an integration attempt with respect to x as done in Case 4, because the resulting term $\int f(t)dx$ cannot be integrated [except for the simple case when $f(t)$ is a constant]. In this case, we proceed by integrating Eq. (1.13.1) with respect to time to obtain

$$\int \ddot{x}dt + \int f(t)dt = \dot{x} + \int f(t)dt = \text{constant} \overset{\Delta}{=} C_0 \qquad (1.13.2)$$

Assuming that the integration that appears in this equation can be done analytically, another integration then yields the following solution for $x(t)$ (with C_1 denoting another constant of integration).

$$\int \dot{x}dt + \int \left[\int f(t)dt \right] dt = x + \int \left[\int f(t)dt \right] dt = C_0 t + C_1 \qquad (1.13.3)$$

The case when $f(t)$ is a constant leads to the well-known expressions for the velocity and for the position of a particle moving with constant acceleration along a straight line, which are presented in introductory physics courses.

CASE 2

Another classical and relatively simple case that is sometimes encountered in dynamics involves the type of system exemplified in Case 1 but for which the force applied to the body is a function of velocity only. In such a case, the differential equation of motion is of the following form:

$$\ddot{x} + f(\dot{x}) = 0 \tag{1.13.4}$$

This equation may be rearranged in two ways:

1. Rearranging it as $d\dot{x}/f(\dot{x}) + dt = 0$ to yield (with C_0 being a constant of integration)

$$t = C_0 - \int \frac{d\dot{x}}{f(\dot{x})} \tag{1.13.5}$$

2. Making use of the chain rule (with $\ddot{x} = \dot{x} d\dot{x}/dx$) to rearrange it as $(\dot{x} d\dot{x})/f(\dot{x}) = -dx$, and then integrating the result to obtain (with C_1 being a constant of integration):

$$\int \frac{\dot{x} d\dot{x}}{f(\dot{x})} + x = C_1 \tag{1.13.6}$$

CASE 3 LINEAR DIFFERENTIAL EQUATIONS WITH CONSTANT COEFFICIENTS (AN INTRODUCTION TO THE LINEAR VIBRATION PROBLEM)

Now, let us review how to find the solution to linear differential equations with constant coefficients. Such equations appear in engineering practice in the investigation of small motions of many dynamical systems about an equilibrium. A linear second-order differential equation with constant coefficients is of the form

$$m\ddot{x} + c\dot{x} + kx = f(t) \tag{1.13.7}$$

where m, c, and k are constants and $f(t)$ is a known function of time. An example of a system whose motion is governed by such an equation is a mass-spring-damper system suspended vertically and subjected to the force of gravity and to a known external force $F(t)$ applied to the mass. For a mass-spring-damper system, m, c, and k are, respectively, the mass, the viscous damping coefficient, and the spring stiffness, and $f(t) = mg + F(t)$, with g being the acceleration of gravity.

Both sides of Eq. (1.13.7) can, of course, be divided by m to make the coefficient of $\ddot{x}$ equal to one, as in Eqs. (1.13.14) and (1.13.17) that appear later. A far better procedure is to divide Eq. (1.13.7) by k and, for $k/m > 0$, define a nondimensional time τ as $kt^2/m = \tau^2$ (thus, $\tau = t\sqrt{k/m}$) to re-write the differential equation as $d^2x/d\tau^2 + c^* dx/d\tau + x = f(\tau\sqrt{m/k})/k$, where $c^* = (c/k)\sqrt{k/m} = c/\sqrt{mk}$. The advantage of nondimensionalization lies in the fact that the three coefficients m, k, and c appear lumped together in a single nondimensional coefficient c^* (thus eliminating the need for specifying individual values for m, k, and c to analyze the motion). For $c/m > 0$, the quantity c^* is called the *nondimensional damping coefficient* of the system.

The solution to Eq. (1.13.7) can be decomposed into two parts as $x(t) = x_p(t) + x_h(t)$, where $x_p(t)$ is called the *particular solution* and $x_h(t)$ is called the *homogeneous solution*.[11] The homogeneous solution is, by definition, the solution for $f(t) = 0$ in Eq. (1.13.7).

In this brief review, only the case for which $f(t) = \text{constant} \stackrel{\Delta}{=} A$ is considered. For $f(t) = \text{constant} \stackrel{\Delta}{=} A$, the particular solution is a constant. With $x = \text{constant} \stackrel{\Delta}{=} x_e$, Eq. (1.13.7) yields $x_e = A/k$. As mentioned earlier, a particular solution that is constant is called the *equilibrium solution* (for the sake of clarity, the subscript e is used for indicating an equilibrium solution). Notice that in the case when there is no spring, i.e., when $k = 0$, the particular solution is (with $c \neq 0$) $\dot{x}_p = \text{constant} = A/c$, which gives $x_p = x_p(t) = At/c + x_p(0)$.

With $k \neq 0$ and $f(t) = \text{constant} \stackrel{\Delta}{=} A$, substitution of $x(t) = x_e + x_h(t) = A/k + x_h(t)$ in Eq. (1.13.7) yields the following linear *homogeneous* differential equation for the homogeneous solution $x_h(t)$ [which is the deviation of $x(t)$ from the equilibrium solution $x_e = A/k$].

$$m\ddot{x}_h + c\dot{x}_h + kx_h = 0 \qquad (1.13.8)$$

The solution to a homogeneous linear differential equation with constant coefficients, i.e., Eq. (1.13.8), is an exponential of the form $x_h = Be^{st}$. This is so because the exponential function is the only function for which a sum involving its derivatives is proportional to the function itself. By substituting such a solution into Eq. (1.13.8), the following algebraic equation (which is called the *characteristic equation*) is obtained for the exponent s:

$$ms^2 + cs + k = 0 \qquad (1.13.9)$$

The solutions to the characteristic equation are

$$s = -\frac{c}{2m} + \sqrt{\left(\frac{c}{2m}\right)^2 - \frac{k}{m}} \stackrel{\Delta}{=} s_1 \qquad s = -\frac{c}{2m} - \sqrt{\left(\frac{c}{2m}\right)^2 - \frac{k}{m}} \stackrel{\Delta}{=} s_2 \quad (1.13.10)$$

---

[11]This decomposition, which is only valid for linear differential equations, is called the *superposition principle*.

and this gives[12], when $s_1 \neq s_2$ [i.e., when $[c/(2m)]^2 \neq k/m$]

$$x_h(t) = A_1 e^{s_1 t} + A_2 e^{s_2 t} \qquad (1.13.11)$$

where A_1 and A_2 are constants of integration that are determined by making use of the initial conditions $x(0)$ and $\dot{x}(0)$.

The solutions $s = s_1$ and $s = s_2$ for the exponents s are real when $4mk < c^2$ and complex conjugate numbers when $4mk > c^2$. When either the real part of a complex conjugate solution for s_1 and s_2, or one of the real solutions, is positive, the solution given by Eq. (1.13.11) grows (because of the positive exponent) exponentially as time increases. Obviously, such a motion cannot be kept in a small neighborhood of the equilibrium solution $x_e = A/k$ [i.e., the deviation $x_h(t)$ from the equilibrium $x_e = A/k$ cannot be kept arbitrarily small] by simply choosing arbitrarily small nonzero initial conditions $x_h(0)$ and $\dot{x}_h(0)$. The equilibrium $x_e = A/k$ is said to be *unstable* in these cases.

In a linear system (i.e., a system whose motion is governed by linear differential equations) instability implies that the motion goes to infinity as time increases. However, this is not necessarily the case for nonlinear systems. [See Example 1.13.3 and several of the problems (e.g., Problems 1.29 and 1.30) in Section 1.20. Other examples also appear elsewhere in the book.] That is why the definition of instability, as stated earlier, says that an equilibrium is unstable if the motion $x(t)$ that is started in a small vicinity of that equilibrium cannot be kept arbitrarily close to that equilibrium, as time passes, simply by changing the initial conditions of the motion. This definition covers linear and nonlinear systems.

When $k/m > [c/(2m)]^2$, the exponents s_1 and s_2 given by Eqs. (1.13.10), and the constants A_1 and A_2 in Eq. (1.13.11), are complex conjugate numbers. In such a case, the solution written in the form of Eq. (1.13.11) is a complex function. When this happens, it is more convenient to write the exponents s_1 and s_2 as $s_1 = \sigma + i\omega$ and $s_2 = \sigma - i\omega$, where $i = \sqrt{-1}$, $\sigma = -c/(2m)$, and $\omega = \sqrt{k/m - [c/(2m)]^2}$. The quantity ω is real if $c^2 < 4mk$ and imaginary if $c^2 > 4mk$. Since, as discovered by Euler,

$$e^{i\omega t} = \cos(\omega t) + i\sin(\omega t) \qquad\qquad e^{-i\omega t} = \cos(\omega t) - i\sin(\omega t) \qquad (1.13.12)$$

---

[12]For the special case when $[c/(2m)]^2$ is exactly equal to k/m, which may rarely occur, the homogeneous solution is of the form $x_h = (A_1 + A_2 t)e^{-ct/(2m)}$.

the homogeneous solution takes the form shown in Eq. (1.13.13), where $B_1 = A_1 + A_2$ and $B_2 = i(A_1 - A_2)$ are now real quantities.

$$x_h = e^{\sigma t}[B_1 \cos(\omega t) + B_2 \sin(\omega t)] \qquad (1.13.13)$$

This solution shows that, when $k/m > [c/(2m)]^2$ (i.e., when the exponents s_1 and s_2 are complex conjugates), the homogeneous solution is oscillatory, with frequency equal to ω, and with an "amplitude" that decreases with time when $c/m > 0$ (which is the case of an actual damper). The equilibrium $x_e = A/k$ is said to be *asymptotically stable* in this case, and such motion decays toward the equilibrium with the passage of time. Asymptotic stability is a desirable feature in practice.

For the case when $c = 0$ (i.e., no damping), the solution $x(t) = x_e + x_h(t) = x_e + B_1 \cos(\omega t) + B_2 \sin(\omega t)$ is periodic [i.e., it repeats itself every $t = 2\pi/\omega$ seconds, when ω is expressed in radians per second (rad/s)], with frequency equal to $\omega = \sqrt{k/m}$. The equilibrium solution $x_e = k/A$ is said to be *stable* in this case, and $x(t)$ simply consists of a perennial oscillation about the equilibrium $x = x_e$. For a stable motion, the amplitude of the oscillations $x_h(t) = x(t) - x_e$ can be made arbitrarily small by simply changing the initial conditions of the motion. The frequency $\omega = \sqrt{k/m}$ is called the *undamped natural frequency* of the oscillation about that equilibrium.

Oscillations that are governed by the linear differential equation (1.13.7) belong to a class of problems that is called *linear vibrations*. The field of vibrations is classified into two categories: *linear vibrations* and *nonlinear vibrations* (or *nonlinear oscillations*).

Although the subject of nonlinear oscillations is beyond the scope of this book, one should be aware that there are circumstances for which nonlinearities in the differential equations of motion play a dominant role in the motion. Nonlinear terms in the differential equation (in general, equations) of motion are actually present in every system (even for a simple system like a pendulum). Except for rare simple cases, linear differential equations start to appear only after nonlinear differential equations are *linearized* for small motions about a particular solution (such as an equilibrium solution) exhibited by the system. The linearization process is presented in Section 1.14, and you will encounter it in several problems solved in this book.

In several examples in this book, numerical integration is used for solving the differential equations of motion of a system, which are generally nonlinear.

Summary of Main Ideas in Case 3

1. The solution of a linear differential equation with constant coefficients, $m\ddot{x} + c\dot{x} + kx = A$, where A is a constant and $k \neq 0$, is of the form $x(t) = x_e + Be^{st}$, where $x_e = \text{constant} = A/k$ is the equilibrium solution.

2. The exponents s are the roots $s = s_1$ and $s = s_2$ of the equation $ms^2 + cs + k = 0$, which is called the *characteristic equation*.

3. The equilibrium $x = x_e$ is:

- *Unstable* if the real part of *any* of the two exponents s_1 or s_2 is positive. This happens if either c/m or k/m is negative. It also happens when $c = 0$ and k/m is negative.
- *Asymptotically stable* if the real part of the exponents s_1 and s_2 is negative. For this to happen, c/m and k/m have to be positive. The motion is a damped oscillation in this case.
- *Stable* if $c = 0$ and k/m is positive. In this case the exponents s_1 and s_2 are purely imaginary. The motion is an undamped oscillation with frequency $\omega = \sqrt{k/m}$.

Therefore, the check for stability or instability simply consists of looking at the signs of the coefficients k/m and c/m.

Example 1.13.1 The motion of a dynamical system is governed by the following differential equation:

$$\ddot{x} + (c - 4)\dot{x} + kx = 16$$

a. Determine the value of k if the system exhibits the equilibrium solution $x_e = 8$. For what range of values of c is the equilibrium stable, and for what range is it unstable? For what range of c is the equilibrium asymptotically stable?

b. Repeat part a for $x_e = -8$.

c. Choose your own value of c for the equilibrium to be asymptotically stable, and then determine the value of k for the system to oscillate with a frequency equal to $\omega = 1.5$. What is the value of the equilibrium x_e in this case?

■ **Solution**

a. Since $x_e = 16/k = 8$, it follows that $k = 2$. The equilibrium is asymptotically stable [i.e., the motion $x(t)$ approaches the equilibrium $x = \text{constant} = x_e$ as time goes to infinity] if the coefficients of $\ddot{x}$, $\dot{x}$, and x of the linear differential equation have the same sign, and unstable otherwise. Therefore, since $k = 2$ is positive, the equilibrium $x_e = 8$ is asymptotically stable if $c > 4$, and unstable if $c < 4$. The case $c = 4$ is a special case. In this case, the equilibrium is classified as *stable* (instead of asymptotically stable) because, with $k > 0$, the motion is an undamped oscillation about the equilibrium.

b. For $x_e = -8 = 16/k$ we obtain $k = -2$. This does not change the characteristic equation for the given differential equation, which still is $s^2 + (c - 4)s + k = 0$. However, since k is negative, the equilibrium $x_e = 16/k = -8$ is unstable. The roots of the characteristic equation are $2 - c/2 \pm \sqrt{(2 - c/2)^2 + 2}$.

c. Since we want the equilibrium to be asymptotically stable, we must have $c > 4$ and $k > 0$. Let $c = 5$, for example. The differential equation is then $\ddot{x} + \dot{x} + kx = 16$. Since the solution to the differential equation is of the form $x = 16/k + Be^{st}$, the characteristic equation is $s^2 + s + k = 0$, and its roots are $s = -0.5 \pm \sqrt{0.25 - k}$. The solution $s = -0.5 \pm i\omega$, with $\omega = 1.5$, then implies $0.25 - k = -1.5^2$, which gives $k = 2.5$. The equilibrium solution in this case is $x_e = 16/k = 6.4$. The motion in this case tends to $x = x_e = 6.4$ as time goes to infinity. More details on the linear vibration problem are presented in Section 1.16.

CASE 4

For a number of problems in dynamics (such as problems involving a frictionless pendulum and a frictionless spring-mass system), the differential equation that governs the motion of a body is of the following form:

$$\ddot{x} + f(x) = 0 \qquad\qquad (1.13.14)$$

Equation (1.13.14) can be integrated analytically if the function $f(x)$ is of the form $f(x) = Ax + B$, where A and B are constants. An example of a mechanical system whose motion is governed by such an equation with $f(x) = Ax + B$ is a mass moving along a straight line and connected to a massless linear spring, with the other end of the spring connected to a fixed point. The same equation, with $A = 0$, also appears in the investigation of the short-range motion of a projectile subjected only to the gravitational force of attraction of the earth. For the case when $A = 0$, the solution for $x(t)$ is immediately obtained as $x(t) = -Bt^2/2 + C_1t + C_2$, where

C_1 and C_2 are constants of integration that are determined by the initial conditions $x(0)$ and $\dot{x}(0)$.

For any function $f(x)$ (except for a constant), it is useless to try integrating Eq. (1.13.14) with respect to time by first multiplying both sides of the equation by dt to obtain

$$\int \ddot{x}\, dt + \int f(x)dt = \text{constant} \quad (\textit{what not to do unless } f(x) \textit{ is constant})$$

The first term in this equation integrates to $\dot{x}$, but the second term cannot be integrated (unless f is a constant). The operation that leads to this equation (i.e., first, a multiplication by dt, and then the attempt to integrate) simply put us "against a corner," so to speak, from which there is no way out.

In an attempt to integrate Eq. (1.13.14) once, what if we try the chain rule to change the independent variable from t to x (which is the argument of the function f)? By recognizing that $\ddot{x}$ is simply a notation for $d(dx/dt)/dt$, we can use the chain rule to write it as

$$\frac{d}{dt}\left(\frac{dx}{dt}\right) = \left[\frac{d}{dx}\left(\frac{dx}{dt}\right)\right]\frac{dx}{dt}$$

By proceeding in this manner, Eq. (1.13.14) is then rewritten as

$$\dot{x}\frac{d\dot{x}}{dx} + f(x) = 0 \tag{1.13.15}$$

which now yields

$$\int \dot{x}\frac{d\dot{x}}{dx}dx + \int f(x)dx = \text{constant}$$

or

$$\frac{\dot{x}^2}{2} + \int f(x)dx = \text{constant} \overset{\Delta}{=} C_0 \tag{1.13.16}$$

The integration attempt to obtain $\dot{x}$ as a function of x was now successful (at least in principle, because the expression for $f(x)$ may be so complicated that no known analytical form for the integral exists). For some dynamics problems the form of $f(x)$ is such that it can be integrated analytically. In such cases, Eq. (1.13.16) is very useful since, without even knowing the solution for $x(t)$, it allows us to analytically find the candidates x_m for the maximum and minimum values of $x(t)$. Such values are simply obtained by setting $\dot{x} = 0$ in Eq. (1.13.16) and then solving the resulting algebraic equation for x_m.

Notice that Eq. (1.13.16), which relates $\dot{x}$ to x, is a first-order differential equation for $x(t)$. The constant of integration C_0 is determined by using the initial conditions $x(0) = x_0$ and $\dot{x}(0) = \dot{x}_0$, as $C_0 = \dot{x}_0^2/2 + \left[\int f(x)\,dx\right]_{x=x_0}$.

Note: When there is *viscous friction* (or viscous damping) in the system (and this is very common in practice), Eq. (1.13.14) becomes (when the damping is linear)

$$\ddot{x} + c\dot{x} + f(x) = 0 \tag{1.13.17}$$

where c is a constant. A *linear viscous damper* is a device that *reacts* to motion by applying, to the body it is connected to, a force proportional to velocity $\dot{x}$ (or a moment that is proportional to angular velocity, for the case of a torsional viscous damper). An example of such a device is the shock absorber that is present in every car (and mounted in parallel with the big springs that are located underneath the car). The case when $f(x)$ is of the form $Ax + B$, which is a simpler but important case that appears in practice, was considered separately. It is important to notice that the analytical technique based on the chain rule, which was presented in this section, will not lead us anywhere in the attempt to integrate Eq. (1.13.17) when $c \neq 0$. The reason for this is that application of the chain rule now yields, after the integration attempt,

$$\frac{\dot{x}^2}{2} + c\int \dot{x}\,dx + \int f(x)dx = \text{constant}$$

The problem with this integration attempt is that the second term in this equation cannot be integrated because it is not known how the integrand $\dot{x}$ depends on the unknown function $x(t)$. In such cases, when $f(x)$ is a nonlinear function of x, one would rely mostly on numerical integration of Eq. (1.13.17).

Consider now Examples 1.13.2 to 1.13.4. For each example, k, L, m, and g are given positive constants.

Example 1.13.2 The motion of a pendulum of length L m (such as that shown in Fig. 1.11, on p. 33, with $M_{\text{bar}} = 0$) is governed by the following differential equation:

$$\ddot{\theta} + k\sin(\theta) = 0$$

where $k = g/L$ and $g = 9.81$ m/s^2 is the acceleration of gravity. The variable $\theta(t)$ is the angle the pendulum makes with the vertical.

   a. Determine all the equilibrium values of $\theta(t)$.

   b. Determine the maximum and minimum values of $\theta(t)$ if the motion of the pendulum is started with the initial conditions $\theta(0) = 56.6°$ and $\dot{\theta}(0) = 1.53\sqrt{k}$ rad/s.

■ Solution

a. At an equilibrium we must have $\theta = \text{constant} \overset{\Delta}{=} \theta_e$, with θ_e being the solutions to the following algebraic equation that is obtained from the given differential equation of motion.

$$\sin \theta_e = 0$$

The solutions for θ_e are then $\theta_e = 0, \pm\pi, 2\pi, \ldots$. (Of all these equilibrium solutions, only two are physically distinct from each other, i.e., $\theta_e = 0$ and $\theta_e = \pi$.)

b. By applying the chain rule procedure presented in this section to the given differential equation, we obtain the following result:

$$\frac{\dot{\theta}^2}{2} + k \int \sin \theta \, d\theta = \frac{\dot{\theta}^2}{2} - k \cos \theta = \text{constant} \overset{\Delta}{=} C_1 = \frac{\dot{\theta}^2(0)}{2} - k \cos \theta(0) \approx 0.62k$$

The maximum and minimum values of θ, which we will denote as θ_m, occur when $\dot{\theta} = 0$. With $\dot{\theta} = 0$ we then obtain

$$\cos \theta_m = -0.62$$

and this yields $\theta_m \approx 128.3°$ and $\theta_m \approx -128.3°$ (or $\theta_m = 360 - 128.3 = 231.7°$). The values of $\ddot{\theta} = -k \sin \theta$ at these values of θ are approximately

- $-0.78k$ (which is negative because k is positive) when $\theta = 128.3°$. This indicates that $\theta = 128.3°$ is a maximum value.
- $0.78k$ (which is positive) when $\theta = -128.3°$ (or $231.7°$). This points to a minimum value of θ.

Therefore, we conclude the following:

$$\theta_{\max} = 128.3° \qquad \theta_{\min} = -128.3°$$

Thus, the pendulum oscillates between these two values of the angle θ. Next, Example 1.13.3 sheds more light on the motion of the pendulum (especially after observing Fig. 1.13).

Example 1.13.3 What are the maximum and minimum values of $\theta(t)$ for Example 1.13.2 if the motion of the pendulum is started with the initial conditions $\theta(0) = 0$ and $\dot{\theta}(0) = 2.1\sqrt{k}$ rad/s? Plot $\dot{\theta}$ versus θ, and use the plot to explain the resulting motion of the pendulum.

■ **Solution**

For these new initial conditions we obtain:

$$\frac{\dot{\theta}^2}{2} + k \int \sin\theta \, d\theta = \frac{\dot{\theta}^2}{2} - k\cos\theta = \text{constant} \stackrel{\triangle}{=} C_2 = \frac{\dot{\theta}^2(0)}{2} - k\cos\theta(0) = 1.205k$$

From this equation, the value of $\theta = \theta_m$ for which $\dot{\theta} = 0$ is obtained as $\cos\theta_m = -1.205$. Since the cosine of an angle is bounded between -1 and 1 we then conclude that there are no minimum and maximum values of θ for the given initial conditions.

If the angle θ does not exhibit maximum and minimum values, what is the pendulum doing then? The answer to such a question may be obtained by plotting $\dot{\theta}$ versus θ (such a plot is known in dynamics and in vibrations as a *phase plane plot*), with $\dot{\theta}$ given by the following relationship:

$$\frac{\dot{\theta}^2}{2k} - \cos(\theta) = \text{constant} \stackrel{\triangle}{=} C$$

Such a plot, when drawn for several values of the constant C, actually provides an understanding of the types of motion the system may exhibit.

Let us plot the nondimensional angular velocity $\dot{\theta}/\sqrt{k}$ (or, simply, $d\theta/d\tau$, where $\tau = t\sqrt{k}$ is nondimensional time) versus θ so that we do not have to concern ourselves with a specific value for k. The phase plane plot for the pendulum, for θ in the range $-2\pi \leq \theta \leq 2\pi$, is shown in Fig. 1.13 (the plot repeats itself outside of that range). The three points marked with a • correspond to the equilibrium values $\theta_e = -2\pi$, 0, and 2π (which were determined in Example 1.13.2 and are physically the same as $\theta_e = 0$).

As indicated in Fig. 1.13, the curves $C = $ constant were generated for the following values of C: 0.62 (obtained in Example 1.13.2), 1.205 (obtained in this example), and also 1 and -0.8. Notice that, as indicated by the equation $\dot{\theta}^2/(2k) - \cos(\theta) = \text{constant} \stackrel{\triangle}{=} C$, the curves that exhibit maximum and minimum values of θ (i.e., *the closed curves that exhibit a value of $\dot{\theta} = 0$*) are those for which C is between -1 and 1, which is the range of $\cos\theta$ when θ is real. It is also worth observing that since a function $\theta(t)$ increases when $\dot{\theta}(t)$ is positive, and decreases when $\dot{\theta}(t)$ is negative, any curve $\dot{\theta}$ versus θ in a phase plane plot is *always* traced (with the passage of time)

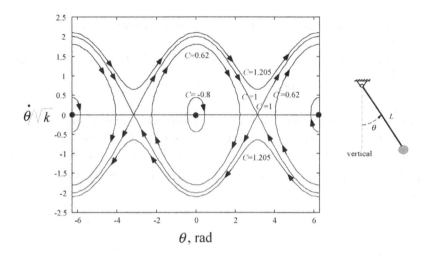

Figure 1.13: Plot of $\dot{\theta}(t)/\sqrt{k}$ versus θ for a pendulum.

in the direction of the arrows shown in Fig. 1.13 (this observation is true for any other dynamical system). This can be easily observed if the differential equation of motion for the system is integrated numerically, and the result displayed as the integration proceeds with time.

Figure 1.13 discloses that the motion $\theta(t)$ is oscillatory (thus, it exhibits a maximum and a minimum value) only if the initial conditions are such that $C < 1$. This happens when the initial condition pair of values $(\dot{\theta}(0)/\sqrt{k}, \theta(0))$ fall inside the region bounded by the curves that pass through the points $(\dot{\theta}(0)/\sqrt{k} = 0, \theta(0) = -\pi)$ and $(\dot{\theta}(0)/\sqrt{k} = 0, \theta(0) = \pi)$. As determined in Example 1.13.2, $\theta = \pm\pi$ are also equilibrium points for the pendulum. Such motions are oscillations about the equilibrium $\theta_e = 0$, which is, thus, a stable equilibrium.

If the initial conditions are such that $C > 1$ [which happens when the initial condition pair of values $(\dot{\theta}(0)/\sqrt{k}, \theta(0))$ fall outside the region just mentioned], then the angular velocity $\dot{\theta}$ of the pendulum never reaches zero and, as a consequence, the angle θ keeps either increasing or decreasing (i.e., the pendulum shown in the rightmost part of Fig. 1.13 keeps rotating either counterclockwise or clockwise). A motion started with initial conditions that fall on the curve $C = 1$ ends up at the equilibrium point $\theta = \pi$, but it takes an infinite time for it to reach that equilibrium. As disclosed by Fig. 1.13, the equilibrium $\theta_e = \pi$ is unstable (because the motion cannot be kept in a small neighborhood of that equilibrium when that equilibrium is perturbed). The determination of the stability of small motions about these equilibrium points for this system, using linearization, is presented in Example 1.18.3 in Section 1.15.

These conclusions about the motion can be verified by an actual experiment with a pendulum.

A more detailed and enlightening visualization concerning the motion of a system governed by the general differential equation $\ddot{x} + c\dot{x} + f(x) = 0$, where $f(x)$ is any function (linear or nonlinear) of the motion variable x, is obtained if we inspect the three-dimensional plot involving x, $\dot{x}$, and the function $C(x, \dot{x}) \triangleq \dot{x}^2/2 + \int f(x)\,dx$. Notice that,

$$\frac{d}{dt}C(x, \dot{x}) = \dot{x}\ddot{x} + \dot{x}f(x) = -c\dot{x}^2$$

and, thus, C = constant if there is no damping in the system (in which case, $c = 0$). Let us look at such a three-dimensional plot for the pendulum of Examples 1.13.2 and 1.13.3, for which $C = C(\theta, \dot{\theta}) = \dot{\theta}^2/(2k) - \cos\theta$, where $k = g/L$ (g being the acceleration of gravity, and L the length of the pendulum). A three-dimensional plot of C versus θ and $d\theta/d\tau$, where $\tau = t\sqrt{k}$ is a nondimensional time, is shown in Fig. 1.14.[13]

The surface seen in Fig. 1.14 is like a mountain terrain that exhibits peaks and valleys, except that the plot in that figure is not a physical plot like that at all because its θ and $\dot{\theta}$ axes are angular orientation and angular velocity, respectively, instead of Cartesian coordinates that specify position in space. The peaks of that surface (which appear at $\theta_e = \pm\pi, \pm 3\pi, \ldots$) correspond to unstable equilibrium points of the pendulum, while the lowest point on the valleys (which appear at $\theta_e = 0, \pm 2\pi, , \ldots$) correspond to a stable equilibrium point.

The curves shown in Fig. 1.13 are cuts on the three-dimensional surface of Fig. 1.14 for different values of C. They are the level curves C = constant for that surface.

---

[13]A plot similar to the one shown in Fig. 1.14 may be readily generated with either Scilab or Matlab using the following commands typed in their respective command windows.
In Scilab:

```
X=-2*%pi:0.5:2*%pi; Xdot=X; [x, xdot]=meshgrid(X, Xdot); C=xdot.∧2-cos(x);
mesh(x, xdot, C); xgrid; xlabel('$\theta$', 'fontsize', 4);
ylabel('$d\theta/d\tau$','fontsize', 4); zlabel('C','fontsize',4)
```

In Matlab:

```
X=-2*pi:0.5:2*pi; Xdot=X; [x,xdot]=meshgrid(X,Xdot); C=xdot.∧2-cos(x); mesh(x,xdot,C);
set(gca, 'box', 'on', 'xlim', [-8 ,8], 'ylim', [-8,8]); gcac=get(gca, 'children');
set(gcac,'EdgeColor','black');
xlabel('\theta', 'fontsize', 12); ylabel('d\theta/d\tau', 'fontsize', 12); zlabel('C', 'fontsize', 12)
```

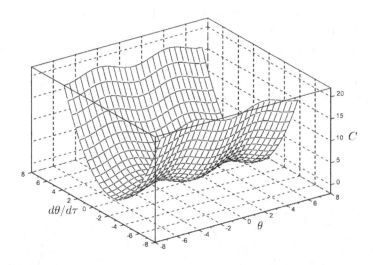

Figure 1.14: The surface $C(\theta, \dot{\theta}) = \dot{\theta}^2/(2k) - \cos\theta$. (Notice that $\tau = t\sqrt{k}$, and thus, $d\theta/d\tau = (1/\sqrt{k})\dot{\theta}$.)

If the plot shown in Fig. 1.14 were an actual physical plot of a mountain terrain, a small ball placed either at the peaks or at the bottom of a valley of such a terrain would stay there forever. If the ball were displaced from a peak, it would move away and it would never be able to stay close to that peak for all times. Such peaks are the physical analogy to the unstable equilibrium points of Fig. 1.13, while the bottom of the valleys of the physical terrain are the physical analogy to the stable equilibrium points of that figure; the ball in this analogy would oscillate back and forth near the bottom of a curved valley of the physical terrain. If there were friction in the physical terrain (which is equivalent to having friction in the dynamical system under investigation), the ball would roll along the physical surface on the terrain and eventually would end up at the bottom of one of the curved valleys of the terrain. The stable equilibrium points would then be *asymptotically stable* (instead of just stable) in such a case. Dynamicists love to come up with such analogies. They actually help to interpret difficult motions and to explain the concept of stability.

Returning to Fig. 1.14, where $C(\theta, \dot{\theta}) = \dot{\theta}^2/(2k) - \cos\theta$ is plotted versus θ and $\dot{\theta}$, keep in mind that an undamped motion takes place along a level curve $C = $ constant in that plot. If there were viscous damping in the pendulum, so that the damping moment is $-c\dot{\theta}$ where c [in N·m/(rad/s)] is the damping coefficient, we would have $d(C(\theta, \dot{\theta}))/dt = -c\dot{\theta}^2$. In such a case, the phase plane plot $\dot{\theta}$ versus θ would start at a given initial point $\dot{\theta}(0)$-$\theta(0)$ and end up at the asymptotically stable equilibrium point in a spiraling manner.

Figure 1.15 shows the shape of the three-dimensional surface $C(\theta, \dot{\theta}) = \dot{\theta}^2/(2k) - \cos\theta$ near a stable equilibrium point (Fig. 1.15a), and near an unstable equilibrium point of the pendulum (Fig. 1.15b); some level curves on the plane $(d\theta/d\tau)$-θ, which correspond to $C = $ constant, are also shown in the figure.

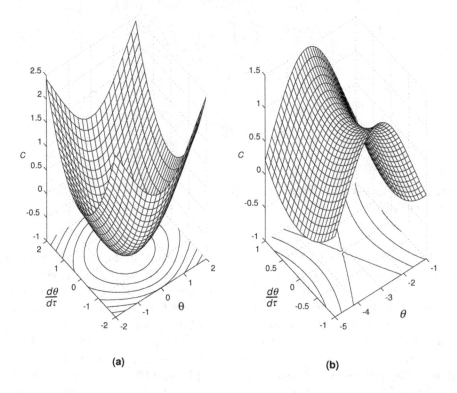

(a)   (b)

Figure 1.15: Closeup view of the surface $C(\theta, \dot{\theta})$ (a) near a center and (b) near a saddle point.

As seen in Fig. 1.15, the three-dimensional surface $C(\theta, \dot{\theta}) = \dot{\theta}^2/(2k) - \cos\theta$ resembles a cup in the neighborhood of a stable equilibrium, and the cuts $C = $ constant (i.e., the level curves) are closed curves in such a case. For this reason, a stable equilibrium of an undamped system is called a *center*. It is called a stable *focus* if it is an asymptotically stable equilibrium; if c were negative, it would be called an unstable focus. The situation is quite different in the neighborhood of the unstable equilibrium point $\theta_e = \pi$ for the pendulum. The three-dimensional surface

$C(\theta, \dot{\theta}) = \dot{\theta}^2/(2k) - \cos\theta$ now resembles a saddle in the vicinity of that unstable point for the pendulum. Because of such resemblance, the unstable equilibrium in this example is called a *saddle point*.

Figure 1.15 discloses that the stability of an equilibrium, i.e., stable, asymptotically stable, or unstable, can be determined by examining the behavior of the function $C(\theta, \dot{\theta})$) near that equilibrium. The function $C(\theta, \dot{\theta}) - C(\theta_e, 0)$ is positive in the neighborhood of the equilibrium $\theta_e = 0$, and sign variable in the neighborhood of the equilibrium $\theta_e = \pi$. In addition, for $c > 0$, the time derivative of C, $dC/dt = -c\dot{\theta}^2$, is never positive. This implies that, if the equilibrium $\theta_e = 0$ is perturbed, the value of $C(\theta, \dot{\theta}) - C(\theta_e, 0)$, whose sign is always opposite to the sign of dC/dt in that neighborhood, always decreases. Therefore, such a motion approaches the equilibrium $\theta_e = 0$ as time t increases.

In contrast, the function $C(\theta, \dot{\theta}) - C(\theta_e, 0)$ is sign variable in the neighborhood of the equilibrium $\theta_e = \pi$. Therefore, in such a case, there is always a neighborhood near the equilibrium $\theta_e = \pi$ where both $C(\theta, \dot{\theta}) - C(\theta_e, 0)$ and $dC/dt = -c\dot{\theta}^2$ have the same sign. This implies that the equilibrium $\theta_e = \pi$ is unstable since the perturbed motion cannot be kept in the neighborhood of such an equilibrium.

These observations constitute the basis of the stability method devised by the Russian mathematician Aleksandr Mikhailovich Lyapunov (1857-1918), which is briefly presented in Section 1.18. The Lyapunov method is applicable to linear and to nonlinear dynamical systems.

The constant term $C(\theta_e, 0)$ is actually irrelevant. The only important factor that determines the character of the equilibrium is whether the function $C(\theta, \dot{\theta})$ is increasing, decreasing, or sign variable near the equilibrium. Functions that exhibit such a behavior are called positive definite, negative definite, or sign variable, respectively. As illustrated in Fig. 1.15, a function that is positive definite near an equilibrium resembles a cup near that neighborhood; it would resemble an inverted cup if it were negative definite.

Example 1.13.4 The motion of a mechanical system is governed by the following differential equation:

$$m\ddot{x} + k(x - L) = mg$$

The system consists of a block of mass m kg connected to a spring of unstretched length L m and stiffness (or *spring constant*) k N/m, and is constrained to move along the vertical. The other end of the spring is connected to a fixed point. The system is shown in Fig. 1.16.

The variable $x(t)$ in Fig. 1.16 is the actual length of the stretched spring. For $mg/(kL) = 3$, determine the maximum and minimum values of $x(t)$ when the motion

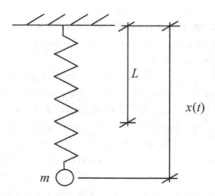

Figure 1.16: A spring-mass system constrained to move along the vertical.

of the system is started with the initial conditions $x(0) = L$ m and $\dot{x}(0) = 2L\sqrt{k/m}$ m/s.

■ Solution

By making use of the chain rule to write $\ddot{x}$ as $\ddot{x} = (d\dot{x}/dx)\dot{x}$, the given differential equation can then be integrated once, as shown earlier in this section, to yield

$$\frac{m}{2k}\dot{x}^2 + \int \left(x - L - \frac{mg}{k}\right) dx = \frac{m}{2k}\dot{x}^2 + \frac{1}{2}\left(x - L - \frac{mg}{k}\right)^2 = \text{constant} \overset{\Delta}{=} C_3$$

$$= \frac{m}{2k}\dot{x}^2(0) + \frac{1}{2}\left[x(0) - L - \frac{mg}{k}\right]^2 = 2L^2 + \frac{1}{2}\left(\frac{mg}{k}\right)^2$$

The maximum and minimum values of x occur when $\dot{x} = 0$. By letting x_m denote such values, this equation then yields, with $\dot{x}_m = 0$,

$$\frac{1}{2}\left(x_m - L - \frac{mg}{k}\right)^2 = 2L^2 + \frac{1}{2}\left(\frac{mg}{k}\right)^2$$

This equation can be readily solved for x_m, and the solutions are

$$x_m = L + \frac{mg}{k} \pm L\sqrt{\left(\frac{mg}{kL}\right)^2 + 4}$$

For $mg/(kL) = 3$, we then obtain $x_m \approx 7.61L$ and $0.39L$. These are, respectively, the maximum and minimum values of $x(t)$ (the corresponding values of $\ddot{x}$ are $\ddot{x} = -3.61k L/m$ when $x = 7.61L$, and $\ddot{x} = 3.61k L/m$ when $x = 0.39L$).

Since $x(t)$ has maximum and minimum values, the motion consists of an oscillation between $x_{\min} \approx 0.39L$ and $x_{\max} \approx 7.61L$. Since the differential equation

$m\ddot{x}+k(x-L) = mg$ is linear, the frequency of the oscillation (which is an undamped oscillation) is equal to $\sqrt{k/m}$ rad/s (see Case 3, p. 43).

1.14   Taylor Series Expansion and Linearization

Consider a given continuous function $f(x)$ of a variable x. The idea of a Taylor series expansion [named after the English mathematician Brook Taylor (1685-1731)] of the function $f(x)$ is to approximate the function in the vicinity of a desired value of x, say $x = x^*$, by a polynomial series in $x_s \stackrel{\Delta}{=} x - x^*$ as

$$f(x) = f(x^* + x_s) \approx A_0 + A_1 x_s + A_2 x_s^2 + A_3 x_s^3 + \ldots \qquad (1.14.1)$$

and then truncate the series at some specified degree of x_s. The new variable $x_s = x - x^*$, which is the x-deviation from the value $x = x^*$, is called the *perturbation* variable. The Taylor series approximation of the function $f(x)$ provides a reasonable approximation for the function only in a small vicinity of $x_s = 0$.

The value of the coefficient A_0 in Eq. (1.14.1) is obtained by simply setting $x_s = 0$ (i.e., $x = x^*$) in that equation, and this yields [which requires that the given function $f(x)$ be continuous at $x = x^*$]

$$A_0 = [f(x^* + x_s)]_{x_s=0} = f(x^*)$$

The value of the coefficient A_1 is determined by simply taking the derivative of both sides of Eq. (1.14.1) with respect to x_s and then evaluating the result at $x_s = 0$ to obtain [which requires that the first derivative of the given function $f(x)$ also be continuous at $x = x^*$]

$$A_1 = \left[\frac{d}{dx_s} f(x^* + x_s) \right]_{x_s=0} = \left[\frac{d}{dx} f(x) \right]_{x=x^*}$$

By continuing to take higher derivatives of Eq. (1.14.1), and evaluating them at $x_s = 0$, the values of the remaining coefficients $A_2, A_3, \ldots$, are then obtained as

$$A_k = \frac{1}{k!} \left[\frac{d^k}{dx_s^k} f(x^* + x_s) \right]_{x_s=0} = \frac{1}{k!} \left[\frac{d^k}{dx^k} f(x) \right]_{x=x^*} \qquad (k=1,2,\ldots) \;\; (1.14.2)$$

Thus, the approximation of $f(x)$ by an nth-degree polynomial requires that the function $f(x = x^* + x_s)$ and its first n derivatives be continuous.

Following the same procedure, the Taylor series expansion of a function of several variables may be obtained. Consider, for example, a continuous function $f(x_1, x_2)$

of two variables x_1 and x_2, with continuous partial derivatives with respect to each variable. Its expansion about a pair of values $x_1 = x_1^*$ and $x_2 = x_2^*$ is written as

$$f(x_1, x_2) = f(x_1^*, x_2^*) + A_1(x_1 - x_1^*) + A_2(x_2 - x_2^*)$$
$$+ B_{11}(x_1 - x_1^*)^2 + B_{12}(x_1 - x_1^*)(x_2 - x_2^*) + B_{22}(x_2 - x_2^*)^2 + \ldots$$
$$(1.14.3)$$

where the "$\ldots$" indicates the cubic and higher-order terms $(x_1 - x_1^*)^3$, $(x_1 - x_1^*)^2(x_2 - x_2^*)$, $(x_1 - x_1^*)^4$, etc.

The coefficients $A_1, A_2, \ldots$ that appear in Eq. (1.14.3) are determined as

$$A_1 = \left[\frac{\partial f}{\partial x_1}\right]_{x_1 = x_1^*, x_2 = x_2^*} \qquad A_2 = \left[\frac{\partial f}{\partial x_2}\right]_{x_1 = x_1^*, x_2 = x_2^*}$$

$$B_{11} = \frac{1}{2}\left[\frac{\partial^2 f}{\partial x_1^2}\right]_{x_1 = x_1^*, x_2 = x_2^*} \qquad B_{12} = \left[\frac{\partial^2 f}{\partial x_1 \partial x_2}\right]_{x_1 = x_1^*, x_2 = x_2^*}$$

$$B_{22} = \frac{1}{2}\left[\frac{\partial^2 f}{\partial x_2^2}\right]_{x_1 = x_1^*, x_2 = x_2^*}$$
$$(1.14.4)$$

A function $f(x)$ is said to have been linearized about $x = x^*$, when only the constant term $A_0 = f(x^*)$ and the linear term $A_1 x_s$, where $x_s = x - x^*$ is small, are kept in its Taylor series approximation given by Eq. (1.14.1). Linearization is not valid if $A_1 = 0$; the behavior of $f(x)$ in such a case is nonlinear even in a small neighborhood of $x = x^*$ (see Problem 1.31 in Section 1.20).

Similarly, linearization (about $x_1 = x_1^*, x_2 = x_2^*, \text{etc.}$) of a function $f(x_1, x_2, \ldots)$ of several variables $x_1, x_2, \ldots$ implies keeping only the constant term $f(x_1^*, x_2^*, \ldots)$ and the linear terms in $x_{s_1} \stackrel{\Delta}{=} x_1 - x_1^*, x_{s_2} \stackrel{\Delta}{=} x_2 - x_2^*, \ldots, x_{s_n} \stackrel{\Delta}{=} x_n - x_n^*$ in its Taylor series expansion about the point under consideration.

Example 1.14.1 Linearize $f_1 = \sin\theta$ and $f_2 = \cos\theta$ about $\theta = \theta^*$. In particular, what are the linearized approximations for $\sin\theta$ and $\cos\theta$ about $\theta = 0$ and $60°$?

■ **Solution**
With $\theta \stackrel{\Delta}{=} \theta^* + \theta(s)$, we write

$$\sin\theta = \sin(\theta^* + \theta_s) = A_0 + A_1\theta_s + \ldots$$

$$A_0 = \sin\theta^* \qquad A_1 = \left[\frac{d}{d\theta}\sin(\theta)\right]_{\theta = \theta^*} = \cos\theta^*$$

Thus, the linearized approximation for $\sin\theta$ about any arbitrary value $\theta = \theta^*$ is $\sin\theta = \sin(\theta^* + \theta_s) = \sin\theta^* + \theta_s\cos\theta^*$.

Similarly, for $\cos\theta$ we have:

$$\cos\theta = \cos(\theta^* + \theta_s) = B_0 + B_1\theta_s + \ldots$$

$$B_0 = \cos\theta^* \qquad B_1 = \left[\frac{d}{d\theta}\cos(\theta)\right]_{\theta=\theta^*} = -\sin\theta^*$$

For $\theta^* = 0$ and $60°$ ($\pi/3$ rad), the linearized approximations for $\sin\theta$ and $\cos\theta$ are:

For $\theta^* = 0$:

$$\sin\theta = \theta + \ldots \qquad\qquad \cos\theta = 1 + \ldots$$

For $\theta^* = \pi/3$ rad:

$$\sin\theta = \sin\frac{\pi}{3} + \left(\cos\frac{\pi}{3}\right)\left(\theta - \frac{\pi}{3}\right) = \frac{\sqrt{3}}{2} + \frac{1}{2}\left(\theta - \frac{\pi}{3}\right) + \ldots$$

$$\cos\theta = \cos\frac{\pi}{3} - \left(\sin\frac{\pi}{3}\right)\left(\theta - \frac{\pi}{3}\right) = \frac{1}{2} - \frac{\sqrt{3}}{2}\left(\theta - \frac{\pi}{3}\right) + \ldots$$

where "..." denote the nonlinear terms in the perturbation $\theta - \theta^*$ [i.e., the terms involving $(\theta - \theta^*)^2$, $(\theta - \theta^*)^3$, etc.] that are disregarded.

The function $\sin\theta$, and its approximations $\sin\theta^* + (\cos\theta^*)(\theta - \theta^*)$ and $\sin\theta^* + (\cos\theta^*)(\theta - \theta^*) - (\sin\theta^*)(\theta - \theta^*)^2/2$, for $\theta^* = \pi/3$, are shown in Fig. 1.17. Since $-1 \leq \sin\theta \leq 1$, the Taylor series approximations are only valid in the region where their values are between -1 and 1. It can be seen in Fig. 1.17 that the linear approximation is valid only in a very small neighborhood of $\theta^* = \pi/3$, while the quadratic approximation is valid in a larger neighborhood of $\theta^* = \pi/3$.

Example 1.14.2 Expand $f(x) = \sqrt{1-x}$ in Taylor series about $x = 0$, and keep terms up to the quadratic order in the expansion.

■ Solution

The expansion of $f(x) = \sqrt{1-x} = (1-x)^{1/2}$ about $x = 0$ is

$$\sqrt{1-x} = 1 + A_1 x + A_2 x^2 + \ldots$$

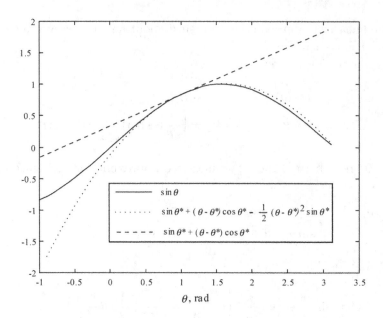

Figure 1.17: The function $\sin\theta$ and two of its Taylor series approximations about $\theta^* = \pi/3$ rad.

where

$$
A_1 = \left[\frac{df}{dx}\right]_{x=0} = \left[\frac{1}{2}(1-x)^{1/2-1}\frac{d}{dx}(1-x)\right]_{x=0} = -\left[\frac{1}{2}(1-x)^{-1/2}\right]_{x=0} = -\frac{1}{2}
$$

$$
A_2 = \frac{1}{2}\left[\frac{d^2f}{dx^2}\right]_{x=0} = \frac{1}{2}\left\{\frac{d}{dx}\left[\left(-\frac{1}{2}(1-x)^{-1/2}\right)\right]\right\}_{x=0}
$$

$$
= \frac{1}{2}\left[-\frac{1}{4}(1-x)^{-3/2}\right]_{x=0} = -\frac{1}{8}
$$

Therefore, the Taylor series of $\sqrt{1-x}$ about $x = 0$ is

$$
\sqrt{1-x} = 1 - \frac{x}{2} - \frac{x^2}{8} + \dots
$$

Example 1.14.3 Show that the Taylor series expansion of a product $F(x) = f_1(x)f_2(x)$ of two functions, about $x = x^*$, is equal to the product of the Taylor series expansions of $f_1(x)$ and $f_2(x)$ [the functions $f_1(x)$ and $f_2(x)$ are assumed to be continuous and have continuous derivatives at $x = x^*$ so that their Taylor series exist]. Then use this result to obtain the Taylor series expansion of $F(x) = (\cos x)/(1+x)$, about $x = 0$, keeping up to quadratic terms in the final expansion.

■ Solution

According to Eqs. (1.14.1) and (1.14.2), the series expansion of $f(x)$ about $x = x^*$ is, with $x_s \overset{\Delta}{=} x - x^*$,

$$f(x) = f(x^*) + \left[\frac{df}{dx}\right]_{x=x^*} x_s + \frac{1}{2}\left[\frac{d^2 f}{dx^2}\right]_{x=x^*} x_s^2 + \dots$$

Or, with $f(x)$ being the product of two functions, i.e., $f(x) = f_1(x)f_2(x)$,

$$f(x) = f_1(x^*)f_2(x^*) + \left[\frac{df_1}{dx}f_2 + f_1\frac{df_2}{dx}\right]_{x=x^*} x_s$$
$$+ \frac{1}{2}\left[\frac{d^2 f_1}{dx^2}f_2 + 2\frac{df_1}{dx}\frac{df_2}{dx} + f_1\frac{d^2 f_2}{dx^2}\right]_{x=x^*} x_s^2 + \dots \qquad (1.14.3a)$$

On the other hand, the product of the expansions of $f_1(x)$ and $f_2(x)$ yields the following result:

$$f_1(x)f_2(x) = \left\{ f_1(x^*) + \left[\frac{df_1}{dx}\right]_{x=x^*} x_s + \frac{1}{2}\left[\frac{d^2 f_1}{dx^2}\right]_{x=x^*} x_s^2 + \dots \right\}$$
$$\times \left\{ f_2(x^*) + \left[\frac{df_2}{dx}\right]_{x=x^*} x_s + \frac{1}{2}\left[\frac{d^2 f_2}{dx^2}\right]_{x=x^*} x_s^2 + \dots \right\}$$

Since each of the expansions of $f_1(x)$ and $f_2(x)$ were truncated to a certain order (i.e., the quadratic terms), we must not keep terms higher than that order when multiplying the two expansions that appear in the product. Thus, if only the quadratic terms are kept in the final result, we then obtain

$$f(x) = f_1(x^*)f_2(x^*) + \left[\frac{df_1}{dx}f_2 + f_1\frac{df_2}{dx}\right]_{x=x^*} x_s$$
$$+ \frac{1}{2}\left[\frac{d^2 f_1}{dx^2}f_2 + 2\frac{df_1}{dx}\frac{df_2}{dx} + f_1\frac{d^2 f_2}{dx^2}\right]_{x=x^*} x_s^2 + \dots \qquad (1.14.3b)$$

By comparing Eqs. (1.14.3b) and (1.14.3a) it is seen that they are the same. If one desires to obtain the expansion of $f_1(x)f_2(x)$ with terms up to x_s^3, for example, then both functions $f_1(x)$ and $f_2(x)$ must be expanded to that order (i.e., to contain terms up to x_s^3) and the product truncated so that terms containing powers higher than x_s^3 are disregarded. This assures that the expansions are mathematically consistent. The main objective of this example was to present the following result:

> The Taylor series expansion of a product $F(x) = f_1(x)f_2(x)$ of two functions, about $x = x^*$, is equal to the product of the Taylor series expansions of $f_1(x)$ and $f_2(x)$.

This result is very useful when determining the Taylor series expansion of a product of several functions. That is done simply by multiplying the expansions of the individual functions, *and appropriately truncating the resulting expression to the order that is consistent with the individual expansions.*

To expand the function $f(x) = (\cos x)/(1 + x)$ about $x = x^*$, we then use the result just presented. For $x^* = 0$, the expansions for $\cos(x)$ and $(1 + x)^{-1}$ are

$$\cos x = 1 - \frac{x^2}{2} + \ldots$$
$$(1 + x)^{-1} = 1 - x + x^2 + \ldots$$

and this yields the following expansion for $(\cos x)/(1 + x)$:

$$\frac{\cos x}{1 + x} = \left[1 - \frac{x^2}{2} + \ldots\right]\left[1 - x + x^2 + \ldots\right] = 1 - x + x^2 - \frac{x^2}{2} + \ldots$$
$$= 1 - x + \frac{x^2}{2} + \ldots$$

1.15   Stability Analysis and Linearization

The differential equations of motion of dynamical systems are, in general, nonlinear. In contrast with linear differential equations, nonlinear differential equations may exhibit several equilibrium solutions, and the stability of each equilibrium needs to be addressed.

An equilibrium is considered to be *stable* if it is possible to choose pairs of initial conditions $x_s(0)$ and $\dot{x}_s(0)$, so that the perturbed motion $x_s(t) \overset{\Delta}{=} x(t) - x_e$ can be kept in an arbitrarily small neighborhood of that equilibrium. Otherwise, the equilibrium is *unstable*. An equilibrium is *asymptotically stable* if, in addition to being stable, the perturbation $x_s(t)$ goes to zero as time goes to infinity.

The solution to nonlinear differential equations of motion cannot, in general, be determined analytically. That makes it difficult to obtain general answers that can provide a physical understanding about the behavior of the system so that appropriate changes could be made to the system to modify that behavior to conform to specifications. Much (but not all, as alluded to in Section 1.19) can be learned about the behavior of a dynamical system from linearized equations about an equilibrium. The knowledge so acquired is always used in engineering design situations. Such situations include the design of control systems to affect the behavior of a given physical system.

In contrast, the theory of linear differential equations is well developed, especially linear differential equations with constant coefficients, whose exact solutions are easily obtained in terms of known functions. For this reason, the linearization process (which involves truncation of Taylor series up to the linear terms only) is widely used in practice for the investigation of small oscillations of dynamical systems about a particular solution of interest that is exhibited by the system.

The linearization of a differential equation $\ddot{x} = f(x, \dot{x})$ about an equilibrium solution $x = \text{constant} \stackrel{\Delta}{=} x_e$ involves two steps.

Step 1. The determination of the differential equation's constant solutions $x = x_e$. This involves solving the algebraic nonlinear equation $f(x_e, \dot{x}_e = 0) = 0$ for x_e (which may have to be done numerically). Notice that since the resulting algebraic equation $f(x_e, \dot{x}_e = 0) = 0$ is nonlinear, the system may exhibit several equilibrium solutions $x_{e_1}, x_{e_2}, \ldots$ (in contrast with a linear equation, which exhibits only one solution).

Step 2. The expansion of every nonlinear term in the differential equation in Taylor series about each constant solution determined in step 1.

When a nonlinear differential equation $\ddot{x} = f(x, \dot{x})$ is linearized about an equilibrium solution $x \stackrel{\Delta}{=} x_e = \text{constant}$, the linearized differential equation that is obtained is always of the form given by Eq. (1.13.7), i.e., $m\ddot{x}_s + c\dot{x}_s + kx_s = f(t)$, where $x_s(t) = x(t) - x_e$ is called the *perturbation* about the equilibrium x_e. The coefficients of the linearized differential equation depend on the value of x_e. Linearization is not valid when $k = 0$; the nonlinear terms dominate the motion in such cases.

The information obtained from the linearized differential equation is used in the design of dynamical systems (such as the suspension system for a car or an airplane landing gear). The analysis presented in Case 3 of Section 1.13 can be applied to the linearized differential equation to determine the stability of each equilibrium. However, as mentioned earlier, one should be aware that nonlinearities in the original differential equations may give rise to phenomena that are not predicted by the linearized differential equations (see Section 1.19, and Example 2.7.8, p. 162).

It should also be noted that stability analysis based only on the linearized differential equations may not be conclusive when any of the roots of the characteristic equation are purely imaginary. Fortunately, these cases are rare in actual practice, but if such a case does arise, one should either do a few numerical tests that involve integrating the original differential equations with different initial conditions to observe the response or resort to a method that accounts for nonlinearities such as *Lyapunov's method*. If no further tests are done to verify the conclusions about stability in an inconclusive situation obtained from linearization, the equilibrium should be labeled *infinitesimally stable*, instead of stable. This serves as a warning that stability determination was done from a linearized analysis and no further stability test was performed with the original nonlinear differential equation(s) of motion. *The nonlinearities do not change the conclusion on stability based on linearized differential equations when the equilibrium is either asymptotically stable or unstable.*

1.16   Multiple Degree-of-Freedom Linear Systems

The general form of a set of n second-order linear homogeneous differential equations with constant coefficients, involving the variables $x_1(t)$, $x_2(t)$,, $\ldots$, $x_n(t)$, is

$$M_{11}\ddot{x}_1 + \ldots + M_{1n}\ddot{x}_n + C_{11}\dot{x}_1 + \ldots + C_{1n}\ddot{x}_n + K_{11}x_1 + \ldots + K_{1n}x_n = 0$$
$$M_{21}\ddot{x}_1 + \ldots + M_{2n}\ddot{x}_n + C_{21}\dot{x}_1 + \ldots + C_{2n}\ddot{x}_n + K_{21}x_1 + \ldots + K_{2n}x_n = 0$$
$$\vdots$$
$$M_{n1}\ddot{x}_1 + \ldots + M_{nn}\ddot{x}_n + C_{n1}\dot{x}_1 + \ldots + C_{nn}\ddot{x}_n + K_{n1}x_1 + \ldots + K_{nn}x_n = 0$$

$$(1.16.1)$$

Generally, such a set of linear equations are obtained only after linearizing, about an equilibrium solution, the nonlinear differential equations that govern the motion of a dynamical system.

For compactness, the set of equations of (1.16.1) is written in matrix form as

$$[M]\,\underline{\ddot{x}} + [C]\,\underline{\dot{x}} + [K]\,\underline{x} = \underline{0} \qquad (1.16.2)$$

where $\underline{x}$ and $\underline{0}$ are $n \times 1$ column matrices defined as $\underline{x} = [x_1\ x_2\ \ldots x_n]^T$, $\underline{0} = [0\ 0\ \ldots 0]^T$ (where the superscript T denotes *transpose*), while $[M]$, $[C]$, and $[K]$ are $n \times n$ matrices of the coefficients in the n equations of (1.16.1).

Since the exponential function is the only one for which its derivatives are proportional to the function itself, the general solution to Eq. (1.16.2) is $\underline{x} = \underline{b}_0 e^{st}$. In such a solution, s is constant and $\underline{b}_0 = [b_{01}\ b_{02}\ \ldots\ b_{0n}]^T$ is an $n \times 1$ column matrix whose elements are determined by the initial conditions of the motion. Substitution of the general solution into Eq. (1.16.2) yields

$$\left[Ms^2 + Cs + K\right]\underline{b}_0 e^{st} = \underline{0} \qquad (1.16.3)$$

If the matrix $[M] s^2 + [C] s + [K]$ has an inverse, the only solution to Eq. (1.16.3) is $\underline{b}_0 = \underline{0}$. Therefore, the only way to obtain a nonzero solution is for the determinant of the matrix $[M] s^2 + [C] s + [K]$ to be equal to zero, i.e.,

$$\left| [M] s^2 + [C] s + [K] \right| = 0 \qquad (1.16.4)$$

This equation is a polynomial equation of degree $2n$ in s. It is the *characteristic equation* associated with the equations of (1.16.1), and it can be put in the form (where $m = 2n$)

$$s^m + a_{m-1} s^{m-1} + \ldots + a_1 s + a_0 = 0 \qquad (1.16.5)$$

The solutions to this polynomial equation (which are called the *eigenvalues* of the matrix $[M] s^2 + [C] s + [K]$) may be real or complex. Complex roots occur in complex conjugate pairs, i.e.,

$$s = \sigma + i\omega \qquad s = \sigma - i\omega \qquad (1.16.6)$$

where $i = \sqrt{-1}$ and σ and ω are real.

If none of the eigenvalues are repeated, the m exponential terms of the form $e^{s_k t}$ are independent of each other, and the general solution to the equations of (1.16.1) is a sum of the form $\sum_{k=1}^{m} b_k e^{(\sigma_k + i\omega_k)t} = \sum_{k=1}^{m} e^{\sigma_k t} [\cos(\omega_k t) + i \sin(\omega_k t)]$. However, if any eigenvalue is repeated, it can be readily shown that the solution for $\underline{x}(t)$ corresponding to that eigenvalue is of the form $t^p e^{st}$, where p is an integer whose value depends on the multiplicity of that repeated eigenvalue. The solution corresponding to an eigenvalue s with multiplicity 2 (i.e., repeated twice) turns out to be of the form $t e^{st}$; $t^2 e^{st}$ for an eigenvalue of multiplicity 3; and so on.

From the point of view of stability, the important point to notice, since the trigonometric functions are bounded, is that any equilibrium solution of a system for which its nonlinear differential equations were linearized to yield the equations of (1.16.1) is unstable if the real part of *any* of the eigenvalues $s_1, s_2, \ldots, s_m$ is positive. The equilibrium is asymptotically stable if the real parts of *all* of them are negative. If there is no damping in the system, the equilibrium is stable [*infinitesimally stable* if the equations of (1.16.1) were obtained by linearization] if all the eigenvalues are purely imaginary. Nowadays, the eigenvalues of a matrix and the roots of a characteristic equation may be readily obtained numerically with no difficulty. In Matlab, for example, the commands *eig* and *roots*, respectively, will do so. In Scilab, the commands are *spec* and *roots*, respectively. For example, $spec([1, 2; 3, 4])$ (in Scilab) and $eig([1, 2; 3, 4])$ (in Matlab) give the eigenvalues of the 2 by 2 matrix $[1, 2; 3, 4]$, and $roots([3, 2, 1, 4]$ in either software yields the roots of the polynomial $3x^3 + 2x^2 + x + 4$.

Although it is a simple task to numerically obtain the roots of a polynomial with an appropriate software package, there are many cases for which at least some of the coefficients of a characteristic equation depend on one or more unknown parameters. This presents a major difficulty for polynomials of degree greater than 3, for which no analytical solution for their roots can be found in general. Fortunately, a method has been devised by Edward J. Routh (1831-1907) and published in 1877, and independently devised in a different form by Adolf Hurwitz (1859-1919) and published in 1895, that determines whether the solutions for any of the roots of a polynomial equation of the form of Eq. (1.16.5) have positive real parts without having to find the roots. Such a method is known as the *Routh-Hurwitz stability criterion*.

1.17   The Routh-Hurwitz Stability Criterion

According to this criterion, first make sure that the coefficient of the highest power of s in the characteristic equation is positive. This is represented by Eq. (1.16.5) after dividing Eq. (1.16.4) by that coefficient so that the coefficient of the highest power in s is equal to one, for convenience. Then arrange the coefficients of the resulting equation, Eq. (1.16.5), in an $m \times m$ matrix as

$$[S] = \begin{bmatrix} a_{m-1} & 1 & 0 & 0 & 0 & \cdots & 0 \\ a_{m-3} & a_{m-2} & a_{m-1} & 1 & 0 & \cdots & 0 \\ a_{m-5} & a_{m-4} & a_{m-3} & a_{m-2} & a_{m-1} & \cdots & \vdots \\ \vdots & \vdots & a_{m-5} & a_{m-4} & a_{m-3} & \cdots & \vdots \\ \vdots & \vdots & \vdots & \vdots & a_{m-5} & \cdots & \vdots \\ \vdots & \vdots & \vdots & \vdots & \cdots & \ddots & a_2 \\ \vdots & \vdots & \vdots & \vdots & \vdots & \cdots & a_0 \end{bmatrix} \qquad (1.17.1)$$

Stop constructing the $[S]$ matrix when you see all the coefficients $a_{m-1}, a_{m-2}, \ldots, a_0$ appear along its diagonal.

According to the Routh-Hurwitz criterion, the equilibrium corresponding to the equations of Eq. (1.16.1) is asymptotically stable if all the principal minor determinants Δ_1, Δ_2, ..., Δ_m of $[S]$ are positive, and unstable if any of them is negative. These determinants are the following:

$$\Delta_1 = a_{m-1} \qquad \Delta_2 = \begin{vmatrix} a_{m-1} & 1 \\ a_{m-3} & a_{m-2} \end{vmatrix} \qquad \Delta_3 = \begin{vmatrix} a_{m-1} & 1 & 0 \\ a_{m-3} & a_{m-2} & a_{m-1} \\ a_{m-5} & a_{m-4} & a_{m-3} \end{vmatrix}$$

$$\Delta_4 = \begin{vmatrix} a_{m-1} & 1 & 0 & 0 \\ a_{m-3} & a_{m-2} & a_{m-1} & 1 \\ a_{m-5} & a_{m-4} & a_{m-3} & a_{m-2} \\ a_{m-7} & a_{m-6} & a_{m-5} & a_{m-4} \end{vmatrix} \qquad \cdots \qquad \Delta_m = \left| [S] \right| \qquad (1.17.2)$$

The Routh-Hurwitz criterion is a test for asymptotic stability, and for instability. It fails if any of the determinants Δ_i $(i = 1, 2, \ldots, m)$ is equal to zero. It turns out that a necessary, but not sufficient, condition for asymptotic stability is that all the coefficients of the characteristic equation have the same sign. If such a condition is not satisfied the equilibrium is unstable; but if it is satisfied, the equilibrium may or may not be asymptotically stable and one needs to apply the Routh-Hurwitz test.

Example 1.17.1 The characteristic equation associated with the linearization of the differential equations of motion of a dynamical system about its equilibrium is $s^2 + cs + k = 0$, where $k \neq 0$. Under what conditions is the equilibrium asymptotically stable, and under what conditions is it unstable?

■ Solution
You already saw such a system on p. 43. The $[S]$ matrix for the given characteristic equation is

$$[S] = \begin{bmatrix} c & 1 \\ 0 & k \end{bmatrix}$$

and this yields $\Delta_1 = c$ and $\Delta_2 = ck$. Therefore, the equilibrium is asymptotically stable if $c > 0$ and $ck > 0$, i.e., if $c > 0$ and $k > 0$. The equilibrium is unstable if either c or k is negative. The criterion fails if $c = 0$. But the roots of the characteristic polynomial in such a case are immediately found to be $s = \pm\sqrt{-k}$ and this allows us to conclude that the equilibrium is unstable if $k < 0$ and infinitesimally stable when $k > 0$ (infinitesimally stable because it was stated that the given characteristic equation is associated with a linearized differential equation).

Example 1.17.2 The motion of a linear system about its equilibrium is governed by the differential equation

$$\frac{d^4x}{dt^4} + 5\frac{d^3x}{dt^3} + 9\frac{d^2x}{dt^2} + 9\frac{dx}{dt} + kx = 0$$

where $k > 0$ is a constant. Determine the range of the parameter k for the equilibrium $x = 0$ to be asymptotically stable. For what range of k is the equilibrium unstable?

■ **Solution**

The characteristic equation in this case is

$$s^4 + 5s^3 + 9s^2 + 9s + k = 0$$

and this yields the following matrix for testing the stability:

$$[S] = \begin{bmatrix} 5 & 1 & 0 & 0 \\ 9 & 9 & 5 & 1 \\ 0 & k & 9 & 9 \\ 0 & 0 & 0 & k \end{bmatrix}$$

The principal minor determinants of this matrix are

$$\Delta_1 = 5 \qquad \Delta_2 = \begin{vmatrix} 5 & 1 \\ 9 & 9 \end{vmatrix} = 5 \times 9 - 1 \times 9 = 36$$

$$\Delta_3 = \begin{vmatrix} 5 & 1 & 0 \\ 9 & 9 & 5 \\ 0 & k & 9 \end{vmatrix} = 5 \times (9 \times 9 - 5k) - 1 \times (9 \times 9 - 5 \times 0) = 324 - 25k$$

$$\Delta_4 = \begin{vmatrix} 5 & 1 & 0 & 0 \\ 9 & 9 & 5 & 1 \\ 0 & k & 9 & 9 \\ 0 & 0 & 0 & k \end{vmatrix} = k\Delta_3$$

Since $k > 0$, these expressions immediately disclose that the only condition for the equilibrium to be asymptotically stable is $324 - 25k > 0$, which yields $0 < k < 12.96$. The equilibrium is unstable if $k > 12.96$. The Routh-Hurwitz criterion fails if $k = 12.96$, and this case has to be dealt with separately.

For $k = 12.96$ the roots of the characteristic equation are[14] $s \approx \pm 1.342i$ and $s \approx -2.5 \pm 0.975i$. Since the solution for $x(t)$ in this case is of the form $x =$

---

[14] As an exercise in algebra, notice that when $k = 12.96$ two of the roots of the characteristic equation become purely imaginary and, thus, the characteristic equation in this case may be put in the factored form $(s^2 + \omega_1^2)(s - \sigma - i\omega_2)(s - \sigma + i\omega_2) = 0$. This expands to $s^4 - 2\sigma s^3 + (\sigma^2 + \omega_1^2 + \omega_2^2)s^2 - 2\sigma\omega_1^2 s + \omega_1^2(\sigma^2 + \omega_2^2) = 0$. Comparing this expression with the characteristic equation $s^4 + 5s^3 + 9s^2 + 9s + k = 0$ yields $-2\sigma = 5$, $(\sigma^2 + \omega_1^2 + \omega_2^2) = 9$, $-2\sigma\omega_1^2 = 9$, and $\omega_1^2(\sigma^2 + \omega_2^2) = k$, which, in turn, yields $\sigma = -5/2$, $\omega_1^2 = 9/5$, $\omega_2^2 = 19/20$.

$B_1 \cos(1.342t) + B_2 \sin(1.342t) + e^{-2.5t}[B_3 \cos(0.975t) + B_4 \sin(0.975t)]$ (where B_1, B_2, B_3, and B_4 are constants determined by the initial conditions of the motion), the equilibrium is stable in this case (instead of asymptotically stable) since the exponential term dies out as $t \to \infty$ but the motion tends to a harmonic oscillation.

Note: If we had $k \geq 0$, the equilibrium for the case $k = 0$ would be the particular solution $\dot{x} = \text{constant} = 0$. The differential equation for $\dot{x}$ in such a case is $d^3\dot{x}/dt^3 + 5d^2\dot{x}/dt^2 + 9d\dot{x}/dt + 9\dot{x} = 0$, for which the characteristic equation is $s^3 + 5s^2 + 9s + 9 = 0$. It can be readily verified that one of the roots of this characteristic equation is $s = -3$, and the characteristic equation then becomes $(s + 3)(s^2 + 2s + 3) = 0$. The other two roots are then $s = -1 \pm \sqrt{2}i$, which indicates that the solution for $\dot{x}$ is of the form $\dot{x} = A_1 e^{-3t} + e^{-t}\left[A_2 \cos(\sqrt{2}t) + A_3 \sin(\sqrt{2}t)\right]$, where A_1, A_2, and A_3 are constants determined by the initial conditions of the motion. As $t \to \infty$, the exponential terms die out and $\dot{x}$ tends to zero, which discloses that the particular solution $\dot{x} = 0$ is asymptotically stable (but $x(t)$ tends to a constant).

Example 1.17.3 The motion of a linear system about its equilibrium is governed by the following set of differential equations, where $c_1 = c_2 \overset{\Delta}{=} c$, with $c > 0$, $k_1 \neq 0$, and $k_2 \neq 0$.

$$\begin{bmatrix} 1 & 0 \\ 0 & 1 \end{bmatrix}\begin{bmatrix} \ddot{x} \\ \ddot{y} \end{bmatrix} + \begin{bmatrix} c_1 & -G \\ G & c_2 \end{bmatrix}\begin{bmatrix} \dot{x} \\ \dot{y} \end{bmatrix} + \begin{bmatrix} k_1 & 0 \\ 0 & k_2 \end{bmatrix}\begin{bmatrix} x \\ y \end{bmatrix} = \begin{bmatrix} 0 \\ 0 \end{bmatrix} \qquad (1.17.3a)$$

Determine the effect of the parameters k_1, k_2, c, and G on the stability of the equilibrium of the system, and show the results on a plot of k_2 versus k_1 for $c = 0$ and $c > 0$.

■ Solution
The equilibrium of the system is $x = y = 0$, and the characteristic equation associated with the given differential equations is

$$\begin{vmatrix} s^2 + cs + k_1 & -Gs \\ Gs & s^2 + cs + k_2 \end{vmatrix} = s^4 + 2cs^3 + (k_1 + k_2 + G^2 + c^2)s^2 + c(k_1 + k_2)s + k_1 k_2 = 0$$

which yields the following 4×4 matrix for testing the stability of the equilibrium:

$$[S] = \begin{bmatrix} 2c & 1 & 0 & 0 \\ c(k_1 + k_2) & k_1 + k_2 + G^2 + c^2 & 2c & 1 \\ 0 & k_1 k_2 & c(k_1 + k_2) & k_1 + k_2 + G^2 + c^2 \\ 0 & 0 & 0 & k_1 k_2 \end{bmatrix}$$

The principal minor determinants of this matrix are

$$\Delta_1 = 2c$$
$$\Delta_2 = 2c(k_1 + k_2 + G^2 + c^2) - c(k_1 + k_2) = c[k_1 + k_2 + 2(G^2 + c^2)]$$
$$\Delta_3 = c^2[2(k_1 + k_2)(k_1 + k_2 + G^2 + c^2) - 4k_1k_2 - (k_1 + k_2)^2]$$
$$= c^2\left[(k_1 - k_2)^2 + 2(k_1 + k_2)(G^2 + c^2)\right]$$
$$\Delta_4 = k_1k_2\Delta_3$$

The equilibrium is asymptotically stable if $\Delta_1 > 0$, $\Delta_2 > 0$, $\Delta_3 > 0$, and $\Delta_4 > 0$, and unstable if any of these conditions is not satisfied. If $c = 0$, which implies that there is no damping in the system, the Routh-Hurwitz criterion fails. In this case it is necessary to look at the roots of the characteristic equation to determine the conditions for stability or instability of the equilibrium. It will prove to be instructive if we look at such a case first.

The case $c = 0$:
The characteristic equation for this case is of the form $s^4 + a_2s^2 + a_0 = 0$, and its roots are

$$s^2 = -\frac{a_2}{2} \pm \sqrt{\frac{a_2^2}{4} - a_0} \tag{1.17.3b}$$

If the roots of Eq. (1.17.3b) are complex, the square root of such a root will also be complex, but with a positive real part.[15] The equilibrium is unstable in such a case. The equilibrium is stable only if Eq. (1.17.3b) yields two negative real roots, and this happens only if $a_2 > 0$, $a_0 > 0$, and $a_2^2 - 4a_0 > 0$. The equilibrium is unstable if any of these conditions is not satisfied. For the problem at hand we have

$$a_2 = k_1 + k_2 + G^2 \tag{1.17.3c}$$
$$a_0 = k_1k_2$$
$$a_2^2 - 4a_0 = (k_1 - k_2)^2 + 2G^2(k_1 + k_2) + G^4$$

The condition $a_0 > 0$ implies that both k_1 and k_2 must be either negative or positive. In the k_1-k_2 space shown in Fig. 1.18a, that means that a necessary condition for the equilibrium to be stable is that (k_1, k_2) points be either in the first or third quadrant of that space.

The stable region is further reduced by the straight line $a_2 = k_1 + k_2 + G^2 = 0$ and by the curve $a_2^2 - 4a_0 = (k_1 - k_2)^2 + 2G^2(k_1 + k_2) + G^4 = 0$. A sketch of

---

[15]This is so because any complex number is of the form $A(\cos\alpha \pm i\sin\alpha) = Ae^{\pm i\alpha}$, where A and α are real numbers, with $A > 0$ and $0 \le \alpha \le \pi$, and $i = \sqrt{-1}$. The square root of $Ae^{\pm i\alpha}$ is then $\sqrt{A}e^{\pm i\alpha/2} = \sqrt{A}[\cos(\alpha/2) \pm i\sin(\alpha/2)]$, whose real part $\sqrt{A}\cos(\alpha/2)$ is never negative.

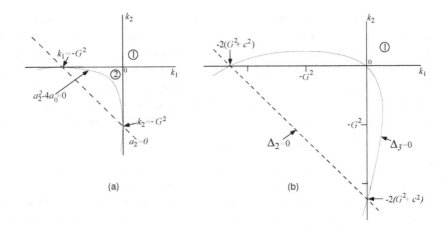

Figure 1.18: Stability chart for Example 1.17.3; (a) for $c = 0$, (b) for $c > 0$.

such a curve and of the straight line $a_2 = 0$ is shown in Fig. 1.18a. It is a simple matter to check which of the sides separated by these boundaries correspond to the conditions $a_2 > 0$ and $a_2^2 - 4a_0 > 0$; for this, just use the point $(k_1 = 0, k_2 = 0)$, for example. We then conclude that only points in the regions marked ① and ② in Fig. 1.18a correspond to a stable equilibrium. Also, notice that the terms $G\dot{y}$ and $G\dot{x}$ in the given differential equations are not damping terms, and that as $G \to 0$ the stable region marked ② shrinks and finally disappears when $G = 0$. Thus, the parameter G in the undamped case is entirely responsible for the stabilization of the equilibrium for points in that region of the parameter space k_1-k_2. Also notice that the sign of G is irrelevant, and that for $G = 0$ the given differential equations reduce to the simple case of two uncoupled oscillators.

The case $c > 0$:
The stability chart for this case is obtained in the same manner described earlier and is shown in Fig. 1.18b. Both parts of Fig. 1.18 are drawn to the same scale. The Routh-Hurwitz test now is applicable to this case, and asymptotic stability requires that $\Delta_1 > 0$ (which implies $c > 0$), $\Delta_2 > 0$, $\Delta_3 > 0$, and $\Delta_4 > 0$. Since $\Delta_4 = k_1 k_2 \Delta_3$, the latter condition restricts the asymptotic stability region to the first and third quadrants of the k_1-k_2 parameter space. The remaining conditions will restrict the asymptotic stability region even further.

The line $\Delta_2 = 0$ is located to the left of the line $a_2 = 0$ shown in 1.18a, but the curve $\Delta_3 = 0$, which passes through the points $(k_1 = -2G^2 - 2c^2, k_2 = 0)$, $(k_1 = 0, k_2 = -2G^2 - 2c^2)$, and the origin $(k_1 = 0, k_2 = 0)$, eliminates region ② from the stability chart when there is damping in the system, leaving only region ①, which now yields an asymptotically stable (instead of just stable) equilibrium.

From this analysis, we see that damping in the system (even infinitesimal damping) destabilized region ② shown in Fig. 1.18a, which may sound as a startling conclusion since, after all, one would expect damping to make a stable equilibrium asymptotically stable. Surprisingly as it may seem, such systems do exist and the linear first-derivative G-terms that appear along the anti-diagonal of the matrix $\begin{bmatrix} c_1 & -G \\ G & c_2 \end{bmatrix}$ in Eq. (1.17.3a) are called *gyroscopic coupling*. Notice the sign change in the G-coupled terms. In contrast with the elements c_1 and c_2 along the diagonal of that matrix, they are not damping terms. For a classical celestial mechanics problem where gyroscopic coupling appears in the differential equations of motion of the system, see Problem 4.13 in Chapter 4 (p. 310). Also, two important examples of gyroscopic coupling involving satellites in orbit around a planet are presented in Chapter 6. One involves a spin-stabilized satellite (see Example 6.8.4, p. 482) and the other is a gravity-gradient-stabilized satellite (see Problem 6.24, p. 497). Another example involving gyroscopic coupling is the Foucault pendulum, which is analyzed in Chapter 8.

1.18   Introduction to the Lyapunov Stability Method

Referring to Fig. 1.15 again (p. 56), it is seen that it is possible to find pairs of initial conditions $\theta(0)$-$\dot{\theta}(0)$ inside a circular neighborhood of the center (or focus, for the case $c > 0$), with radius δ, for which the perturbed motion will never leave any chosen circular region of radius $\epsilon > \delta$ around the equilibrium $\theta_e = 0$ for any arbitrarily small value that one wishes to choose for ϵ. Such an equilibrium is stable for $c = 0$ and asymptotically stable for $c > 0$. Such is not the case for the equilibrium $\theta_e = \pi$, and that equilibrium is, then, unstable. Based on these observations, the following are the definitions of stability for a dynamical system whose differential equations of motion are of the general form $\ddot{\underline{x}} = \underline{f}(\underline{x}, \dot{\underline{x}}; t)$, where $\underline{x} = [x_1, x_2, \dots, x_n]^T$ is the $n \times 1$ column matrix whose elements are the n coordinates $x_1, x_2, \dots, x_n$ that were chosen to analyze the motion.

Stability   An equilibrium $\underline{x} = \text{constant} \overset{\Delta}{=} \underline{x}_e$ is stable in the sense of Lyapunov if, for any chosen positive number ϵ, it is possible to find a positive number δ so that a motion started with the initial conditions $\underline{x}(t = 0) \overset{\Delta}{=} \underline{x}_0$ and $\dot{\underline{x}}(t = 0) \overset{\Delta}{=} \dot{\underline{x}}_0$ in the region $||\underline{q}_0 - \underline{q}_e|| < \delta$ where $\underline{q} = [x_1, x_2, \dots, x_n, \dot{x}_1, \dot{x}_2, \dots, \dot{x}_n]^T$, the inequality $||\underline{q} - \underline{q}_e|| < \epsilon$ (where $\underline{q}_e$ stands for $\underline{q}$ evaluated at $\underline{x} = \underline{x}_e$) is always satisfied for $t \geq 0$ no matter how small one chooses ϵ to be.[16] In other words, an equilibrium is Lyapunov stable if the perturbed motion is able to stay in any small neighborhood of that equilibrium.

Asymptotic stability   The equilibrium $\underline{x}_e$ is asymptotically stable if, in addition to being stable, the perturbed motion $\underline{x}(t) - \underline{x}_e$ approaches the equilibrium $\underline{x}_e$ as $t \to \infty$.

Instability   An equilibrium $\underline{x}_e$ is unstable if it is not stable.

Since in general it is not known how small the radius δ is, the concept of stability, as defined, is a *local property* of any equilibrium $\underline{x}_e$. If the radius δ can be extended to the entire region of the space $\underline{x}$, the concept of stability is a global one and the equilibrium $\underline{x}_e$ is said to be *stable in the large* (or *globally stable*). For example, the nonlinear system whose motion is governed by the differential equation $\ddot{x} + x + x^3 = 0$ exhibits only one equilibrium $x_e = 0$ (which is the only real solution to $x_e + x_e^3 = 0$), and that equilibrium is globally stable since any perturbation of its equilibrium results in a motion that yields a closed curve $\dot{x}$ versus x about that equilibrium for any set of initial conditions. Generally, the motion of nonlinear systems is governed by differential equations that exhibit more than one equilibrium and, in such cases, stability would be local.

The *region of attraction* of an equilibrium $\underline{q}_e$ is the region of the state space $\underline{q} = [x_1, x_2, \ldots, x_n, \dot{x}_1, \dot{x}_2, \ldots, \dot{x}_n]^T$ where a motion started inside it yields a trajectory that surrounds the equilibrium that is inside that region. For example, the region of attraction of the equilibrium $\theta_e = 0$ shown in Fig. 1.13 (p. 53) is the region enclosed by the curves that pass through the equilibrium points $(\theta = -\pi, \dot{\theta} = 0)$ and $(\theta = \pi, \dot{\theta} = 0)$ if there is no damping in the system (i.e., if $c = 0$). With damping, the region of attraction of that equilibrium is even larger since there are initial conditions outside that region that yield a state trajectory that ends up at the equilibrium $\theta_e = 0$.

The Lyapunov method is based on examining the sign definiteness of a chosen continuous function $\mathcal{V}(\underline{x}, \dot{\underline{x}})$, and of its total time derivative $d\mathcal{V}/dt$, in the neighborhood of each equilibrium of the system. Such a function is called a Lyapunov function if it is positive definite, and $d\mathcal{V}/dt \leq 0$. Figure 1.15a (p. 56) shows the geometric representation of a function that behaves as positive definite in the neighborhood of an equilibrium, in which case it resembles a cup in that neighborhood. Basically, the Lyapunov method is summarized in the following theorems, which we

---

[16]The notation $||\underline{q}||$, where $\underline{q} = [q_1, q_2, \ldots, q_{2n}]T$ means the "norm" of $\underline{q}$. A norm is a function that assigns a positive value, which is some measure of length, to an $m \times 1$ column matrix. For example, $||\underline{y}|| = max|y_i|$ $(i = 1, 2, \ldots, m)$, and $||\underline{y}|| = \sqrt{y_1^2 + y_2^2 + \ldots + y_m^2}$ are two acceptable definitions for the norm $||\underline{y}||$. The latter, known as the Euclidean norm, is generally the one used with the notion of stability.

apply here to the neighborhood of an equilibrium $\underline{x} = \text{constant} = \underline{x}_e$ of a system whose motion is governed by a set of differential equations $\ddot{\underline{x}} = \underline{f}(\underline{x}, \dot{\underline{x}})$.

Stability theorem   If there exists a Lyapunov function $\mathcal{V}(\underline{x}, \dot{\underline{x}})$ in the neighborhood of the equilibrium $\underline{x}_e$, then that equilibrium is stable. Generally when such a stability test is performed, the equilibrium is labeled Lyapunov stable.

Asymptotic stability theorem   If the time derivative $d\mathcal{V}/dt = [\partial \mathcal{V}/\partial \underline{x}]\dot{\underline{x}} + [\partial \mathcal{V}/\partial \dot{\underline{x}}]\underline{f}$ of the function $\mathcal{V}$ mentioned in the previous item is negative, that equilibrium is asymptotically stable.

Instability theorem   The equilibrium $\underline{x}_e$ is unstable if $\mathcal{V}$ and $d\mathcal{V}/dt$ can have the same sign in any region in the neighborhood of that equilibrium.

Although we restricted ourselves here to the stability of an equilibrium of the system $\ddot{\underline{x}} = \underline{f}(\underline{x}, \dot{\underline{x}})$, the Lyapunov method is actually more general since it is applicable to the stability of any particular solution $\underline{x}(t) = \underline{x}_p(t)$ a system $\ddot{\underline{x}} = \underline{f}(\underline{x}, \dot{\underline{x}}; t)$ may exhibit. For a more in-depth study of Lyapunov's method, see References 9 and 12 listed in Appendix G. The difficulty with Lyapunov's method is finding an acceptable testing function for some problems. For many dynamical systems, though, the Hamiltonian (presented in Chapter 7, p. 528, and in Section 7.5.3) is a suitable testing function to use.

The English mathematician James Joseph Sylvester (1814-1897), who made many contributions to mathematics, developed the necessary and sufficient conditions for a quadratic function V of m variables $x_1, x_2, \ldots, x_m$ to be positive definite. Such functions can be written as $V = \frac{1}{2}\underline{x}^T S \underline{x}$, where the symmetric matrix $S = \frac{\partial^2 V}{\partial \underline{x}^2}$ is called the *Hessian* of V. The conditions for S to be positive definite are the same as the conditions for the eigenvalues of S to be positive. Such conditions are known as Sylvester's criterion, and it requires that all the principal minor determinants of S be positive. The specific form of the Hessian of the function V is

$$\frac{\partial^2 V}{\partial \underline{x}^2} = \begin{bmatrix} \dfrac{\partial^2 V}{\partial x_1^2} & \dfrac{\partial^2 V}{\partial x_1 \partial x_2} & \cdots & \dfrac{\partial^2 V}{\partial x_1 \partial x_m} \\[2ex] \dfrac{\partial^2 V}{\partial x_2 \partial x_1} & \dfrac{\partial^2 V}{\partial x_2^2} & \cdots & \dfrac{\partial^2 V}{\partial x_2 \partial x_m} \\[2ex] \vdots & \vdots & & \\[2ex] \dfrac{\partial^2 V}{\partial x_m \partial x_1} & \dfrac{\partial^2 V}{\partial x_m \partial x_2} & \cdots & \dfrac{\partial^2 V}{\partial x_m^2} \end{bmatrix}$$

and its principal minor determinants are

$$\Delta_1 = \frac{\partial^2 V}{\partial x_1^2} \qquad \Delta_2 = \begin{vmatrix} \dfrac{\partial^2 V}{\partial x_1^2} & \dfrac{\partial^2 V}{\partial x_1 \partial x_2} \\ \dfrac{\partial^2 V}{\partial x_2 \partial x_1} & \dfrac{\partial^2 V}{\partial x_2^2} \end{vmatrix}$$

$$\Delta_3 = \begin{vmatrix} \dfrac{\partial^2 V}{\partial x_1^2} & \dfrac{\partial^2 V}{\partial x_1 \partial x_2} & \dfrac{\partial^2 V}{\partial x_1 \partial x_3} \\ \dfrac{\partial^2 V}{\partial x_2 x_1} & \dfrac{\partial^2 V}{\partial x_2^2} & \dfrac{\partial^2 V}{\partial x_2 \partial x_3} \\ \dfrac{\partial^2 V}{\partial x_3 x_1} & \dfrac{\partial^2 V}{\partial x_3 \partial x_2} & \dfrac{\partial^2 V}{\partial x_3^2} \end{vmatrix}$$

$$\vdots$$

$$\Delta_n = \begin{vmatrix} \dfrac{\partial^2 V}{\partial \underline{x}^2} \end{vmatrix}$$

Example 1.18.1 The motion of a mechanical system is governed by the differential equation

$$\ddot{x} + c\dot{x} + x + 2x^2 - 5x^3 + kx\dot{x} = 0.2 \qquad (1.18.1\text{a})$$

where c and k are constants, with $c \geq 0$.

a. Determine all the equilibrium values of $x(t)$, and linearize the differential equation about each equilibrium.

b. Determine which equilibrium solutions are stable, and which ones are unstable. What are the conditions, if any, to be imposed on c and k for the system equilibrium solution (or solutions, as the case may be) to be asymptotically stable?

c. What is the frequency of the small oscillations about each stable equilibrium exhibited by the system? Does the value of k affect the damping in the system?

d. Analyze the stability of each equilibrium of the system using the Lyapunov method.

■ Solution

a. At equilibrium, $x = \text{constant} \overset{\triangle}{=} x_e$. The values of x_e are obtained by solving the following algebraic equation, which is obtained from the given differential equation with all the time derivative terms equated to zero.

$$x_e + 2x_e^2 - 5x_e^3 = 0.2 \qquad (1.18.1\text{b})$$

The roots of this cubic equation may be found by trial and error with a hand-held calculator. They may also be found using the *roots* command in either Scilab or Matlab, or by plotting the function $x_e + 2x_e^2 - 5x_e^3 - 0.2$ versus x_e, and the roots read from the plot. One may also find one of the roots by the iterative trial-and-error process, and then factoring out the polynomial to reduce the rest of the calculations to finding, analytically, the roots of the resulting quadratic equation. One of the roots is found to be $x_e \approx 0.167$. By dividing the polynomial $5x_e^3 - 2x_e^2 - x_e + 0.2$ by $x_e - 0.167$, we find that $5x_e^3 - 2x_e^2 - x_e + 0.2 \approx (x_e - 0.167)(5x_e^2 - 1.165x_e - 1.195)$, and the roots of the resulting quadratic polynomial are $x_e \approx -0.386$ and 0.619. The equilibrium solutions are then

$$x_e \approx 0.167, \ -0.386, \ \text{and} \ 0.619$$

b. To determine the stability of each equilibrium point exhibited by the system, start by linearizing the given differential equation about each of the equilibrium solutions. For this, let

$$x(t) \overset{\triangle}{=} x_e + x_s(t) \qquad (1.18.1\text{c})$$

In terms of the perturbation variable $x_s(t)$, the differential equation becomes

$$\ddot{x}_s + c\dot{x}_s + x_e + x_s + 2(x_e + x_s)^2 - 5(x_e + x_s)^3 + k\dot{x}_s(x_e + x_s) - 0.2 = 0 \qquad (1.18.1\text{d})$$

Linearization now involves disregarding the nonlinear term $\dot{x}_s x_s$ and the x_s^2 and x_s^3 terms that are obtained after the quadratic and cubic powers are expanded (which, for a polynomial, is the same as obtaining the Taylor series expansion of those terms). Noticing that the constant term $x_e + 2x_e^2 - 5x_e^3 - 0.2$ is zero because that is how x_e is determined from the equilibrium equation, the following linearized equation is then obtained[17]:

$$\ddot{x}_s + (c + kx_e)\,\dot{x}_s + \left(1 + 4x_e - 15x_e^2\right) x_s = 0 \qquad (1.18.1e)$$

Specifically, we then have for each equilibrium:

$$\text{for } x_e = 0.167 : \qquad \ddot{x}_s + (c + 0.167k)\,\dot{x}_s + 1.25x_s = 0 \qquad (1.18.1f)$$

$$\text{for } x_e = -0.386 : \qquad \ddot{x}_s + (c - 0.386k)\,\dot{x}_s - 2.779x_s = 0 \qquad (1.18.1g)$$

$$\text{for } x_e = 0.619 : \qquad \ddot{x}_s + (c + 0.619k)\,\dot{x}_s - 2.271x_s = 0 \qquad (1.18.1h)$$

The equilibrium solutions $x_e = -0.386$ and $x_e = 0.619$ are unstable because the coefficients of $\ddot{x}_s$ and x_s in the linearized differential equations for motion about them, namely, Eqs. (1.18.1g) and (1.18.1h), have opposite signs (see the Summary of Main Ideas in Case 3, Section 1.13, p. 47). Such unstable motions start growing exponentially with time. Although the exponential approximation is not valid for large times because the original differential equation is nonlinear (and the exponential solution is only an approximation that was obtained by linearization), such motions cannot be kept in a small neighborhood of the equilibrium solutions $x_e = -0.386$ and 0.619.

Equation (1.18.1f) is the differential equation that governs the linearized motion about the equilibrium $x_e = 0.167$. Therefore, that equilibrium is infinitesimally stable if $c + 0.167k = 0$ (the equilibrium point $x_e = 0.167$ is a *center*, as discussed in Example 1.13.3), asymptotically stable if $c + 0.167k > 0$ (the equilibrium point that was a center is called a *stable focus* in such a case), and unstable if $c + 0.167k < 0$ (the equilibrium point $x_e = 0.167$ is called an *unstable focus* in this case). In the latter case (i.e., if $c + 0.167k < 0$), the system does not exhibit any stable equilibrium solution (and this feature is not a desirable situation in practice). The stability of the equilibrium $x_e = 0.167$ can be verified by numerically integrating Eq. (1.18.1a) with initial conditions near that equilibrium, such as $x(0) = 0.15$ and 0.015 and $\dot{x}(0) = 0$.

c. Small motions about the equilibrium $x_e = 0.167$ are undamped oscillations with frequency $\omega = \sqrt{1.25} \approx 1.118$ when $c + 0.167k = 0$. Such oscillations die out with the passage of time when $c + 0.167k > 0$, and the value of k contributes to the damping of such motions because that value affects the coefficient of $\dot{x}_s$ in the linearized differential equation of motion.

---

[17]Notice that, when more than one equilibrium solution is found, it is wise to obtain the linearized differential equation without using any specific value for x_e, and then use specific values of x_e afterward. Otherwise, one would end up repeating all the same mathematical steps for each equilibrium solution.

When $c + 0.167k > 0$, the frequency ω of the small damped oscillations about the asymptotically stable equilibrium $x_e = 0.167$ is obtained directly from the characteristic equation for Eq. (1.18.1f), which is $s^2 + (c+0.167k)s + 1.25 = 0$. The roots of this polynomial equation are

$$s = -\frac{1}{2}(c + 0.167k) \pm \sqrt{\frac{1}{4}(c + 0.167k)^2 - 1.25}$$

The roots are complex conjugates only when $(c + 0.167k)^2/4 < 1.25$ and, in such a case, the frequency of the damped oscillations about $x_e = 0.167$ is $\omega = [1.25 - (c + 0.167k)^2/4]^{1/2}$.

All these conclusions may be verified by numerical integration of the nonlinear differential equation of motion, Eq. (1.18.1a). Try it yourself to see the motion of the system when you use different initial conditions to start the motion. Try initial conditions near the three equilibrium points and observe the motion $x(t)$, and also see what happens as you watch the motion in the $\dot{x}$-x plane.

d. The stability of each equilibrium of the system may be readily determined by the Lyapunov method using the function $\mathcal{H} = \dot{x}^2/2 + x^2/2 + 2x^3/3 - 5x^4/4 - 0.2x$, which is obtained from the terms in Eq. (1.18.1a) that can be integrated with respect to x. Since the term $\dot{x}^2/2$ is never negative, the Lyapunov method consists of simply testing the function $\mathcal{U}(x) = x^2/2 + 2x^3/3 - 5x^4/4 - 0.2x$ for positive definiteness near each equilibrium of the system. At equilibrium, $d\mathcal{U}/dx = x + 2x^2 - 5x^3 - 0.2 = 0$, whose solutions were determined earlier. The Hessian of $\mathcal{U}$ is $d^2\mathcal{U}/dx^2 = 1 + 4x - 15x^2$, which gives $[d^2\mathcal{U}/dx^2]_{x_e=0.167} \approx 1.25$, $[d^2\mathcal{U}/dx^2]_{x_e=-0.386} \approx -2.79$, and $[d^2\mathcal{U}/dx^2]_{x_e=0.619} \approx -2.271$. Since $d\mathcal{H}/dt = -(c + kx)\dot{x}^2$, as can be easily verified, and $\mathcal{H} = \dot{x}^2/2 + \mathcal{U}$ is sign variable near the equilibrium solutions $x_e = -0.386$ and 0.619, these equilibria are unstable since $\mathcal{H}$ and $d\mathcal{H}/dt$ can have the same sign. The equilibrium $x_e = 0.167$ is asymptotically stable if $c + 0.167k > 0$, and unstable if $c + 0.167k < 0$. These are the same conclusions obtained earlier, but with much less effort using Lyapunov's method. The equilibrium $x_e = 0.167$ is Lyapunov stable if $c + 0.167k = 0$ (instead of just infinitesimally stable as concluded earlier).

Example 1.18.2 The motion of a mechanical system is governed by the differential equation

$$\ddot{x} + \dot{x} + x(2 - \cos x) = k + \sin t \qquad (1.18.2a)$$

where k is a constant. The term $k + \sin(t)$ is due to an external force that is applied to the system. For $k = 0$ and -8, determine all the equilibrium values of $x(t)$, linearize the differential equation about each equilibrium, and determine whether those equilibrium solutions are stable, asymptotically stable, or unstable.

■ Solution
At equilibrium, $x = \text{constant} \triangleq x_e$, and x_e can be constant only in the absence of the $\sin(t)$ term that appears in the differential equation (i.e., only when the excitation is constant). Therefore, the values of x_e are determined by solving the following algebraic equation, which is obtained from the given differential equation with all the time derivative terms equated to zero, and leaving out the $\sin(t)$ excitation.

$$x_e(2 - \cos x_e) = k \qquad (1.18.2b)$$

■ The case $k = 0$
For this case we have $x_e = 0$, and $\cos x_e = 2$. Since $x(t)$ (and, therefore, x_e of course) must be real, there is no solution to the equation $\cos x_e = 2$. Therefore, the only equilibrium solution when $k = 0$ is $x_e = 0$.

Normally we let $x(t) = x_e + x_s(t)$ to expand the differential equation of motion for small x_s. However, since $x_e = 0$, there is no need to introduce a new symbol x_s in this case. The expansion of $\cos x$ about $x = 0$ is $\cos x = 1 + \dots$, and the linearized differential equation in this case is then

$$\ddot{x} + \dot{x} + x = \sin t$$

Since the coefficients of the $\ddot{x}$, $\dot{x}$, and x terms have the same sign, the equilibrium $x_e = 0$ is asymptotically stable.

■ The case $k = -8$
For this case we have to solve the equation $x_e(2 - \cos x_e) = -8$. Since an analytical solution to this nonlinear equation cannot be obtained, we must then resort to a numerical solution. This may be done by a few iterations with a hand-held calculator or by plotting the function $x_e(2 - \cos x_e) + 8$ versus x_e and then extracting the values of x_e for which the function is zero. Such a plot is shown in Fig. 1.19. The following roots of the equation $x_e(2 - \cos x_e) + 8 = 0$ are obtained from that plot: $x_e \approx -2.74$, -5.19, and -6.87.

To linearize the given differential equation about each equilibrium solution, let

$$x(t) \triangleq x_e + x_s(t) \qquad (1.18.2c)$$

In terms of the perturbation variable $x_s(t)$, Eq. (1.18.2a) becomes

$$\ddot{x}_s + \dot{x}_s + (x_e + x_s)\left[2 - \cos(x_e + x_s)\right] = k + \sin t \qquad (1.18.2d)$$

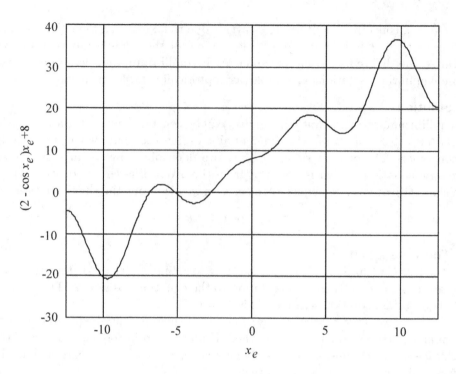

Figure 1.19: Plot for determining the roots of $x_e(2 - \cos x_e) + 8$.

By expanding $\cos(x_e + x_s)$ in Taylor series for small x_s as

$$\cos(x_e + x_s) = \cos x_e + x_s \left[\frac{d}{dx_s} \cos(x_e + x_s) \right]_{x_s = 0} + \ldots = \cos x_e - x_s \sin x_e + \ldots$$

linearization of Eq. (1.18.2d) for small deviations θ_s from the equilibrium then yields [noticing that the constant term $x_e(2 - \cos x_e) + 8$ is zero] the following differential equation:

$$\ddot{x}_s + \dot{x}_s + (2 - \cos x_e + x_e \sin x_e)\, x_s = 0 \qquad (1.18.2e)$$

We then have for each equilibrium:

$$\begin{aligned}
&\text{for } x_e = -2.74: && \ddot{x}_s + \dot{x}_s + 3.99 x_s = \sin t && (1.18.2f)\\
&\text{for } x_e = -5.19: && \ddot{x}_s + \dot{x}_s - 3.07 x_s = \sin t && (1.18.2g)\\
&\text{for } x_e = -6.87: && \ddot{x}_s + \dot{x}_s + 4.97 x_s = \sin t && (1.18.2h)
\end{aligned}$$

These linearized differential equations disclose that the equilibrium solutions $x_e = -2.74$ and -6.87 are asymptotically stable. In these cases the free (i.e., without the $\sin t$ excitation) motion $x(t)$ is a decaying exponential that tends to the corresponding equilibrium $x = x_e$ when the motion is started with small values of $\dot{x}(0)$ and $x(0) - x_e$. When the $\sin t$ excitation is present, the linearized motion about each of these asymptotically stable equilibrium solutions consists of the sum of the homogeneous solution and the particular solution due to the excitation $\sin t$. The time-dependent part of the homogeneous solution dies out with the passage of time, and eventually the response $x(t)$ is the sum of the corresponding equilibrium solution and the sinusoidal function due to the $\sin t$ excitation (see Section 8.2 for a presentation on the response to a sinusoidal excitation).

The equilibrium $x_e = -5.19$ is unstable, and a motion that is started near that equilibrium does not stay near it.

Example 1.18.3 Let us investigate further the motion of the pendulum of length L that was addressed in Examples 1.13.2 and 1.13.3. The differential equation of motion for the pendulum in terms of the angle θ measured from the vertical is repeated here, where $k = g/L$ (with $g = 9.81$ m/s^2 being the acceleration of gravity).

$$\ddot{\theta} + k\sin(\theta) = 0 \tag{1.18.3a}$$

As determined in Example 1.13.2, this system exhibits the equilibrium solutions $\theta_e = 0, \pi, 2\pi, \ldots$, of which only two are physically distinct from each other, namely, $\theta_e = 0$ and π rad. Linearize the differential equation about each of these two equilibrium solutions, and determine whether they are stable, asymptotically stable, or unstable.

■ Solution
To linearize the differential equation about each equilibrium solution, let

$$\theta(t) \stackrel{\Delta}{=} \theta_e + \theta_s(t) \tag{1.18.3b}$$

and expand, for small θ_s, the nonlinear term $k\sin\theta$ that appears in the differential equation.

The expansion for $\sin\theta$ (see Example 1.14.1, with $\theta^* = \theta_e$ in that example) is

$$\begin{aligned} \sin\theta &= \sin(\theta_e + \theta_s) = \sin\theta_e + \theta_s\cos\theta_e + \ldots \\ &= \theta_s\cos\theta_e + \ldots \qquad \text{(because } \sin\theta_e = 0\text{)} \end{aligned} \tag{1.18.3c}$$

In terms of the perturbation variable $\theta_s(t)$, the linearized differential equation is then

$$\ddot{\theta}_s + (k\cos\theta_e)\,\theta_s = 0 \tag{1.18.3d}$$

Specifically, we then have for each equilibrium:

$$\text{for } \theta_e = 0: \qquad \ddot{\theta}_s + k\theta_s = 0 \qquad\qquad (1.18.3e)$$

$$\text{for } \theta_e = \pi: \qquad \ddot{\theta}_s - k\theta_s = 0 \qquad\qquad (1.18.3f)$$

Since $k > 0$, the equilibrium $\theta_e = 0$ is infinitesimally stable, and the equilibrium $\theta_e = \pi$ is unstable. Further investigation should be done to ascertain stability of the equilibrium $\theta_e = 0$, with θ governed by the nonlinear Eq. (1.18.3a). This was actually done in Examples 1.13.2 and 1.13.3. The equilibrium $\theta_e = 0$ would be asymptotically stable if there were damping in the system.

It is worth noticing that the investigation of the same nonlinear system presented in this example, and in Examples 1.13.2 and 1.13.3, by different methods provided results that complement each other. That is what one does in practice when investigating the motion of dynamical systems. Generally, a combination of methods (including numerical integration of the nonlinear differential equations that govern the motion) is used to obtain as much information as possible about the dynamic behavior of any particular system. Such work helps one to understand the behavior of the system, to predict its motion, and to take appropriate measures to control it when it is necessary to do so.

The plot of $\dot{\theta}/\sqrt{k}$ versus θ shown in Fig. 1.13 (p. 53) for Example 1.13.3 provides a graphical visualization of the dynamical behavior of the nonlinear system investigated here. That figure shows that a motion started with initial conditions in the vicinity of the point $(\dot{\theta}(0) = 0, \theta = 0)$, which is the equilibrium point that is stable, stays in that vicinity as time increases. Such a motion is an oscillation whose amplitude can be made as small as desired. In contrast, this is not the case for a motion that is started in the vicinity of the point $(\dot{\theta}(0) = 0, \theta = \pi)$, which, as determined earlier and also in this example, is the unstable equilibrium point.

Notice that the solution to the linearized differential equation about $\theta_e = \pi$ [namely, $\ddot{\theta}_s - k\theta_s = 0$, where $\theta_s = \theta - \pi$ is the deviation of $\theta(t)$ from $\theta_e = \pi$] tends to infinity as time increases. A linearized differential equation is generated under the assumption that the perturbed motion $\theta_s(t) = \theta(t) - \theta_e$ is small. Therefore, the *only* correct conclusion that the linearized differential equation $\ddot{\theta}_s - k\theta_s = 0$ provides is that the equilibrium $\theta_e = \pi$ is unstable. Important detail about the motion that is started with initial conditions near the unstable equilibrium point $\theta_e = \pi$ [i.e., the motion started with both $\dot{\theta}(0)$ and $\theta(0) - \pi$ small] is disclosed by observing Fig. 1.13. When looking at that figure, notice that no matter how small those initial conditions are in this case, $\theta(t)$ never stays in the neighborhood of the equilibrium $\theta_e = \pi$ (and that is why that equilibrium is classified as unstable).

It is important to notice that, in contrast to a linear system (i.e., a system whose motion is governed by a linear, and not just linearized, differential equation), a motion started close to an unstable equilibrium of a nonlinear system may not grow indefinitely because the nonlinearities in the system may limit the motion to finite values.

Example 1.18.4 The motion of a mechanical system is governed by the differential equation

$$\ddot{x} + c\dot{x} + x - x^3 + kx^2\dot{x} = 0 \qquad (1.18.4a)$$

where c and $k \neq 0$ are constants, with $c \geq 0$. Determine all the equilibrium solutions $x = \text{constant} \stackrel{\Delta}{=} x_e$, and specify whether they are stable, asymptotically stable, or unstable. If both c and k were zero, the quantity $H \stackrel{\Delta}{=} \dot{x}^2/2 + x^2/2 - x^4/4$ would be constant. Show that $\dot{H} = -(c + kx^2)\dot{x}^2$ for all cases.

■ Solution

The equilibrium solutions are $x_e = 0$, $x_e = 1$, and $x_e = -1$. With $x(t) = x_e + x_s(t)$, linearization of Eq. (1.18.4a) about $x_e = 1$ and $x_e = -1$ yields $\ddot{x}_s + (c+k)\dot{x}_s - 2x_s = 0$. Therefore, since the coefficient of x_s is negative, those equilibrium solutions are unstable.

Linearization of Eq. (1.18.4a) about $x_e = 0$ yields $\ddot{x}_s + c\dot{x}_s + x_s = 0$. The equilibrium is asymptotically stable if $c > 0$. When $c = 0$, linearization indicates that the equilibrium is infinitesimally stable. However, a simple numerical integration of Eq. (1.18.4a) with a small value of $x(0)$ and with $\dot{x}(0) = 0$, say, shows that the stability of the equilibrium $x_e = 0$ depends on the sign of k when $c = 0$. In such a case, that equilibrium is actually asymptotically stable if $k > 0$, unstable if $k < 0$, and stable if $k = 0$. The explanation for this actually lies in the fact that $\dot{H} = -(c + kx^2)\dot{x}^2$, which shows that for $k > 0$ the term kx^2 adds damping to the system since x^2 is never negative. With $c > 0$, $c + kx^2 \approx c$ since x^2 can be made as small as desired, in which case $\dot{H} \approx -c\dot{x}^2$, and the equilibrium is asymptotically stable. However, when $c = 0$, the term kx^2 in the expression for $\dot{H}$ prevails. Since $\dot{H} < 0$ when a motion is started with $k > 0$, H decreases with time. Since H is not negative when x is small (because the $x^2/2$ term is greater than the $x^4/4$ term in such a case), H must go to zero as $t \to \infty$. This then shows that the equilibrium $x_e = 0$ is also asymptotically stable when $c = 0$ and $k > 0$. The same argument discloses that the equilibrium $x_e = 0$ is unstable if $c = 0$ and $k < 0$, and Lyapunov stable if $c = k = 0$.

1.19 An Interesting Experiment

You can build a very simple system to see the effects of nonlinearities. Just take any spring and measure its unstretched length, which we will call L_u. Now, attach a mass m at one end of the spring and attach the other end to a fixed point. Keep a hand on the mass, and let the spring stretch along the vertical. Choose a proper mass until the equilibrium length of the spring (i.e., the length of the spring when there is no motion) is about $4L_u/3$. Do not worry if it is not exactly $4L_u/3$; the value $4L_u/3$ will produce the most dramatic effect in this demonstration, but any value that is close to $4L_u/3$ is good enough (how close is "close" is actually answered only by a nonlinear analysis).

To do the experiment, trying to keep the spring vertical, grab the mass and stretch the spring a little bit from its equilibrium. Then release it, and watch the motion. The mass should start moving up and down, as expected. But after a little while the angle the spring makes with the vertical starts to increase (let's call it θ), while the up-and-down amplitude of the motion decreases. It decreases until the spring swings laterally only, essentially without the up-and down motion. Then, the angle starts to decrease, and the up-and-down motion starts to appear again; and this sequence of events repeats itself.

You can repeat the experiment by changing the value of the mass from that "critical value" that you found for your spring. Change it by increasing, and also by decreasing, the value of the mass. What you will find is that there is a range of values of m for which the described phenomenon occurs. For values of mass outside that range, the motion takes place essentially along the vertical, where it was started.

The motion just described happens to be a nonlinear phenomenon. It cannot be predicted by linearizing the differential equations of motion for small values of θ and for the up-and-down swing of the spring.

This simple experiment demonstrates a nonlinear phenomenon that can occur in many dynamical systems. Given the appropriate circumstances, it occurs with systems like satellites, ships, airplanes, buildings, and machines. In ships, for example, it can cause undesirable (and dangerous) large rolling motions due to nonlinear interactions with the pitch motion of the ship. In three-axis stabilized satellites, it can cause undesirably large roll-yaw motions also due to nonlinear interactions with pitch motions. The phenomenon that is demonstrated with this simple spring-mass experiment is called *nonlinear resonance*. For the analysis of such motions, see References 16 and 18, for example, listed in Appendix G.

1.20   Problems

Unless indicated otherwise, for Problems 1.1 to 1.9, $\{\hat{x}_1, \hat{x}_2, \hat{x}_3\}$ is a unit vector triad with $\hat{x}_2 = \hat{x}_1 \times \hat{x}_3$.

1.1   It is claimed that the vectors $\vec{a} = \hat{x}_1 + 2\hat{x}_2 + 3\hat{x}_3$ and $\vec{b} = -5\hat{x}_1 + 4\hat{x}_2 - \hat{x}_3$ are perpendicular to each other. Determine whether the claim is true or false.

1.2 Determine the value of the constant k so that the vectors $\vec{a} = 2\hat{x}_1 + 5\hat{x}_2 + 3\hat{x}_3$ and $\vec{b} = 5\hat{x}_1 + k\hat{x}_2 + 3\hat{x}_3$ are perpendicular to each other.

1.3 Determine the angle between the vectors $\vec{a} = 2\hat{x}_1 + 5\hat{x}_2 + 3\hat{x}_3$ and $\vec{b} = 5\hat{x}_1 + 8\hat{x}_2 + 3\hat{x}_3$.

1.4 Determine the value of the constant k so that the angle between the vectors $\vec{a} = 4\hat{x}_1 + k\hat{x}_2 + \hat{x}_3$ and $\vec{b} = 0.5\hat{x}_1 + \hat{x}_2 + \hat{x}_3$ is $60°$.

1.5 Determine the cross product $\vec{p}$ of the vectors $\vec{a} = \hat{x}_1 + 2\hat{x}_2 + 5\hat{x}_3$ and $\vec{b} = 3\hat{x}_1 + 2.5\hat{x}_2 + 4.2\hat{x}_3$. What is the magnitude of $\vec{p}$? What would the answers be if $\hat{x}_2$ were chosen as $\hat{x}_2 = \hat{x}_3 \times \hat{x}_1$?

1.6 Determine the angle between the vectors $\vec{a}$ and $\vec{b}$ given in Problem 1.5.

1.7 Determine the vector $\vec{c}$ that is equal to the projection of the vector $\vec{a} = \hat{x}_1 + 3\hat{x}_2 + 4\hat{x}_3$ along the direction of the vector $\vec{b} = 3\hat{x}_1 + 2.5\hat{x}_2 + 4.2\hat{x}_3$.

1.8 Determine the perpendicular distance between point P whose rectangular coordinates are $x_1 = 4, x_2 = 3, x_3 = 2$, and the vector $\vec{a}$ given in Problem 1.7. The axes x_1, x_2, and x_3 are parallel to the unit vectors $\hat{x}_1$, $\hat{x}_2$, and $\hat{x}_3$, respectively, and the point $M(x_1 = 5, x_2 = 7, x_3 = 9)$ is in the line of action of $\vec{a}$.

1.9 Determine the moment of the vectors $\vec{a}$ and $\vec{b}$, given in Problem 1.7, about point P whose coordinates are given in Problem 1.8. Point $N(x_1 = 2, x_2 = 8, x_3 = 12)$ is in the line of action of $\vec{b}$.

1.10 Consider the vector $\overrightarrow{AB}$ from point $A(x = 1, y = 2)$ to point $B(x = 5, y = 0)$. Define, in a sketch, a set of convenient unit vectors that you choose to work with (clearly naming them with any letter of your choice, and using hats on the top of those letters) and then determine the moment $\vec{M}_O$ of the vector $\overrightarrow{AB}$ about the origin of the coordinate system.

1.11 Consider the vector $\overrightarrow{AB}$ from point $A(x = 1, y = 2)$ to point $B(x = 5, y = 0)$, and the vector from point $P(x = 1, y = 0)$ to point $Q(x = 5, y = 1)$. Define, in a sketch, a set of convenient unit vectors that you choose to work with (clearly naming them with any letter of your choice, and using hats on the top of those letters) and then determine the expression for the vector that is equal to the projection of $\overrightarrow{AB}$ along the unit vector from P to Q.

1.12 Consider the vector $\overrightarrow{AB}$ from point $A(x = 1, y = 2)$ to point $B(x = 5, y = 0)$. Define, in a sketch, a set of convenient unit vectors that you choose to work with (clearly naming them with any letter of your choice, and using hats on the top of those letters) and then determine the coordinates of all points Q on the y-axis for which the magnitude of the moment $\vec{M}_Q$ of the vector $\overrightarrow{AB}$ about Q is equal to $|\vec{M}_Q| = 12$.

1.13 A *couple* is a pair of forces $\vec{F}$ and $-\vec{F}$ separated by a distance D. Show that the moment of a couple is independent of the point about which the moment is taken.

1.14 Determine the time derivative of the following functions $f_1 = \tan x + (1 + y^2)^{-3/2}$ and $f_2 = (\sin x)/[1 + \cos^2(xy)]$, where x and y are functions of time t.

1.15 In Problem 1.14, $x(t) = \cos(k_1 t)$ and $y(t) = k_2 t^2$, where $k_1 = 1$ rad/s and $k_2 = 1$ s^{-2}. Determine the values of $\dot{f}_1$ and $\dot{f}_2$ when $t = 12.5$ minutes.

1.16 Which of the following differentials is an exact differential? Determine the function g for those cases where dg is an exact differential.

    a. $dg = (2xy + 4)y^2 \, dx + (3xy + 8)xy \, dy$
    b. $dg = x^2 \dot{x} dt$
    c. $dg = 6y^2 \, dx + 3xy \, dy$
    d. $dg = 3y^2 \, dx + 6xy \, dy$

1.17 Determine the function $P(x, \theta)$ for $dg = P \, dx - k(\sin \theta)d\theta$ to be an exact differential. What is the function $g(x, \theta)$ in such a case? The parameter k is a constant.

1.18 The motion $\theta(t)$ of a mechanical system is governed by the nonlinear differential equation $\ddot{\theta} - \Omega^2 (\sin \theta) \cos \theta + k \sin \theta = 0$, where $k > 0$ is a constant. An example of such a system is a pendulum that makes an angle θ with the vertical and is driven by a motor forcing it to spin at a constant angular velocity Ω about a vertical axis; for such a pendulum, $k = g/L$, where g is the acceleration of gravity, and L is the length of the pendulum. For $k/\Omega^2 = 2$ and $1/2$, determine, analytically, the maximum and minimum values of $\theta(t)$ when the motion is started with the initial conditions $\theta(0) = 0.1$ rad and $\dot{\theta}(0) = 0.5\sqrt{k}$ rad/s.

1.19 Expand each of the following functions in Taylor series about $x = 0$, keeping two terms in the expansion. The parameter k is a constant.

    a. $f(x) = k \tan x$

b. $f(x) = \dfrac{k}{\sqrt{1+x}}$

c. $f(x) = \dfrac{k \tan x}{\sqrt{1+x}}$

1.20 Determine the equilibrium solution $x = \text{constant} \overset{\Delta}{=} x_e$ of each one of the given nonlinear differential equations, and then expand each equation in Taylor series about each of their equilibrium solutions by "perturbing the equilibrium" as $x(t) \overset{\Delta}{=} x_e + \epsilon x_s(t)$, where ϵ is just a bookkeeping constant parameter that is introduced to keep track of orders of magnitude in the Taylor series expansion. Keep terms up to order ϵ^2 in the expansions, and then set $\epsilon = 1$ when writing your answers. The coefficient k that appears in the equations is an arbitrary constant. What are the linearized differential equations for $x_s(t)$?

a. $\ddot{x} + \sin x + \dfrac{k\dot{x}}{1-x} = 0$

b. $\ddot{x} + x \cos x + \dfrac{k\dot{x}}{(1-x)^2} = 0$

1.21 Determine, analytically, all the equilibrium solutions for the rotating pendulum described in Problem 1.18. How many physically different equilibrium solutions do such a pendulum exhibit when $\Omega^2 > k$ and when $\Omega^2 < k$? For each of these cases, determine the stability of each equilibrium of the system by linearizing the differential equation that governs the motion, and also by using Lyapunov's method.

1.22 The motion $x(t)$ of a dynamical system is governed by the differential equation $\ddot{x} + x - x^3 = 0$, with $x(0) = 0, \dot{x}(0) = 0.4$. Determine, analytically, the maximum and minimum values of $x(t)$. Verify your answer by sketching the phase plane plot $\dot{x}$ versus x and showing the phase plane trajectory for the given initial conditions.

1.23 The motion of a dynamical system is governed by the differential equation $\ddot{x} + x + x \sin x = 0$, with $x(0) = 0, \dot{x}(0) = 0.4$. Integrate the differential equation once to obtain an equation of the form $\dot{x}^2 + g(x) = \text{constant} \overset{\Delta}{=} C$, and then determine the maximum and minimum values of x from a plot of $g(x) - C$ versus x.

1.24 Solve, analytically, the following differential equation for $\dot{x}(t)$: $\ddot{x} + t\dot{x} = 0$. The initial conditions are $x(0) = x_0, \dot{x}(0) = \dot{x}_0$.

1.25 Analyze the stability of the equilibrium (i.e., indicate whether the equilibrium is stable, asymptotically stable, or unstable) of each of the six systems whose motion is governed by the given differential equation.

    a. $\ddot{x} + 2x = 0$
    b. $\ddot{x} - 2x = 0$
    c. $\ddot{x} + 0.4\dot{x} + 2x = 0$
    d. $\ddot{x} + 0.4\dot{x} - 2x = 0$
    e. $\ddot{x} + \dot{x} + 5x = 6$
    f. $\ddot{x} - \dot{x} + 5x = 6$

1.26 Determine, analytically, the damped and undamped frequencies of oscillation for each system in Problem 1.25. Indicate which systems, if any, do not oscillate.

1.27 Solve, analytically, each of the differential equations given in Problem 1.25 for $x(t)$ when the motion is started with the initial conditions $x(0) = 0.2$ and $\dot{x}(0) = 0.1$.

1.28 The motion $x(t)$ of a mechanical system is governed by the differential equation $\ddot{x} + 2\dot{x} + 17x = 0$. The motion is started with the initial conditions $x(0) = 0$ and $\dot{x}(0) = 1$. Determine, analytically, the maximum value of $x(t)$.

1.29 The motion $x(t)$ of a dynamical system is governed by the differential equation $\ddot{x} - x + x^3 = 0$.

    a. Analyze the stability of each equilibrium of the system, indicating whether the equilibrium is asymptotically stable, stable, or unstable.
    b. Determine the maximum and minimum values of $x(t)$ when the motion is started with $x(0) = 0.2$ and $\dot{x}(0) = 0$. Verify your answer by sketching the phase plane plot $\dot{x}$ versus x and showing the phase plane trajectory for the given initial conditions.

1.30 The motion $x(t)$ of a mechanical system is governed by the differential equation $\ddot{x} + x + 2x^2 - 5x^3 = 0.2$. This is the same system considered in Example 1.18.1 (p. 77), with $c = 0$ and $k = 0$ in that example, and for this case $\dot{x}^2/2 + x^2/2 + 2x^3/3 - 5x^4/4 - 0.2x = \text{constant} \overset{\Delta}{=} H$. Draw the phase plane plot $\dot{x}$ versus x for several values of H appropriately chosen by you, and then indicate in which region of that plot the pair of initial conditions $\dot{x}(0)$ and $x(0)$ yields an oscillatory motion about the stable equilibrium solution of the system (which is $x_e \approx 0.167$, as already determined in Example 1.18.1). For $\dot{x}(0) = 0$, for what range of $x(0)$ is the resulting motion oscillatory?

1.31 The motion $\theta(t)$ of a mechanical system is governed by the differential equation $\ddot{\theta} + \theta - \sin\theta = 0$. Show from a plot of $\theta - \sin\theta$ versus θ that $\theta = 0$ is the only equilibrium of the system. Plot $\dot{\theta}$ versus θ and show, from the plot, that the motion is oscillatory. Show that linearization of the differential equation about the equilibrium yields $\ddot{\theta} + k\theta = 0$, where the "stiffness" k is $k = 0$, and that such an equation fails, in this case, to exhibit the actual behavior disclosed by the plot of $\dot{\theta}$ versus θ for the given system. Analyze the stability of the equilibrium in this case.

1.32 Analyze the stability of the equilibrium of the system whose motion $x(t)$ is governed by the differential equation $\ddot{x} + x + k\dot{x}^3 = 0$. Is the stability obtained from linearization conclusive? What can you conclude by looking at the behavior of the function $H = \dot{x}^2/2 + x^2/2$ when $k \neq 0$?

1.33 Determine the stability of the equilibrium solution $x_e = 0$ for the system whose motion is governed by the differential equation $d^2x/dt^2 + kx^3 = 0$.

1.34 Analyze the stability of each equilibrium for the system given in Example 1.18.2, when the system is not driven by the $\sin t$ term, using the Lyapunov method. The equilibrium solutions for $k = 0$ and $k = -8$ are given in that example.

The following problems involve numerical integration of differential equations of motion. Appendix A presents tutorials on three software you may want to use, but you may use any other appropriate software of your choice.

1.35 Integrate numerically the differential equation in Problem 1.18, and then check the analytical answer you have obtained for that problem.
Hint: Work with the nondimensional time $\tau = t\sqrt{k}$, so that the differential equation becomes $d^2\theta/d\tau^2 - (\Omega^2/k)(\sin\theta)\cos\theta + \sin\theta = 0$ (and thus, there is no need to know values for Ω and k, but only for the ratio k/Ω^2).

1.36 Integrate numerically the nonlinear differential equation for Example 1.18.1, i.e., Eq. (1.18.1a), (p. 77), and plot $\dot{x}$ versus x for $c = k = 0$, with $\dot{x}(0) = 0$ and with several values (of your choice) of the initial condition $x(0)$ near the three equilibrium points of the system (see Example 1.18.1 for the values that were determined for the equilibrium solutions $x = \text{constant} \stackrel{\Delta}{=} x_e$).

1.37 With the numerical integration scheme you used for Problem 1.36, verify the effect the value of k has on the motion. For this, take $c = 0$ and $\dot{x}(0) = 0$ and do the numerical integration for two different values (of your choice) of k, one positive and one negative, with $x(0) = 0.3$ or any other value of your choice that is near the equilibrium $x_e = 0.167$. Also, maintaining the same values of c and k, do the integration again with a value of $x(0)$ near the other two equilibrium points (see Example 1.18.1). Do some numerical experiments of your own using other values of $c + 0.167k$ to verify the conclusion that the system has no stable equilibrium points when $c + 0.167k < 0$.

1.38 In case you have not done so already, check the answer to Problem 1.28 using any numerical software of your choice.

1.39 The following is a block diagram for numerically integrating a certain second-order differential equation.

    a. What is the differential equation that is integrated?
    b. What is the value of the output when $t = 0.2$ time units? Choose your own values for any data you need. To answer this part, you can either solve the equation using any numerical software of your choice, or solve the equation analytically, which happens to be possible to do.

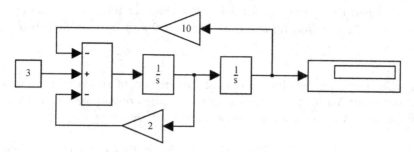

Figure 1.20: Figure for Problem 1.39.

1.40 The following code is for numerically solving a differential equation defined by the function *rhs_of_diff_eqs*. The solution is generated by the call to a routine named *ode* as *ode(initial_conditions, t0, t, rhs_of_diff_eqs)*, which keeps calling the function *rhs_of_diff_eqs* until the final time $t = 5$ is reached. It generates the solution, starting at $t = t0 = 0$, and stores it in the variable *y_and_ydot* at the points $t = 0, 0.1, 0.2, \ldots, 5$.

$$
\begin{aligned}
y\_and\_ydot &= ode([0.8; 0], 0, [0 : 0.1 : 5], rhs\_of\_diff\_eqs); \\
function\ dydt &= rhs\_of\_diff\_eqs(t, y); \\
dydt(1) &= y(2); \quad dydt(2) = -0.1 * y(2) - 4 * y(1) - 2 * y(1) \wedge 3; \\
endfunction
\end{aligned}
$$

Without consulting Appendix A, what do you think is the differential equation that is solved by *ode*?

CHAPTER 2

Kinematics and Dynamics of Point Masses

Kinematics is the study of the velocity and acceleration of particles, and of points in bodies of finite dimension, without regard to the forces that cause their motion. It is a study of the dependence of those quantities on the coordinates (and of their time derivatives) necessary to locate a point or body in space at any instant of time. In kinematics, consideration is only given to the dependence of velocity and acceleration on the variables (and their time derivatives) needed to describe the motion. In dynamics, differential equations of motion of a system must be formulated taking into account the forces acting on the system, and *integrated* so that the motion can be predicted. When the motion is known (which is the case of a "pure" kinematics problem), such as the motion of a car moving with known speed along a given track or the prescribed motion of an element of a machine, a kinematic analysis includes finding the position, the velocity, and/or the acceleration of one point relative to another.

> Implementation of numerical integration of differential equations of motion is illustrated in this book with Scilab, Matlab, and Simulink. Tutorials are presented in Appendix A. Any other appropriate numerical software may be used, of course.

2.1 How to Formulate a Dynamics Problem

When a particle P is free to move in three-dimensional space, three time-dependent variables are needed to locate its position at any time t. Those three variables may be, for example, the distances from a reference point O in space, taken along three orthogonal lines originating at O; or they might be the distance from O to P and two angles that specify the orientation in space of the line that passes through O

and P. There are many possible choices one can make and, as far as locating the particle in space is concerned, any set of such variables is satisfactory. The number of such variables is reduced by one if the motion is constrained to a surface in space and by two if the motion is constrained to a line. For example, if a particle is constrained to move on the surface of a sphere, only two variables are needed to locate it on that surface. Possible choices for those variables are two angles.

The particle is said to have as many *degrees of freedom* as the number of time-dependent variables needed to completely specify its position in space. In dynamics, such variables are called the *motion coordinates*. Thus, a particle that is free to move in three-dimensional space, with no constraints, has three degrees of freedom.

Two particles moving independently of each other have six degrees of freedom, while the number of degrees of freedom is reduced to five if the particles are connected by a rigid bar because the six variables that are needed for two points moving independently of each other would then be related by one *equation of constraint*. The constraint equation in this case is the equation stating that the length of the bar is a constant. For a rigid bar, the five independent variables that are used to describe the motion might be three rectangular coordinates for a chosen point on the bar and two angles for specifying the orientation of the bar. If the bar were constrained to move on a plane, the motion of the bar relative to that plane would have three degrees of freedom. In such a case, the variables that one might choose to investigate the motion might be two rectangular coordinates for a point on the bar and one orientation angle.

A particle moving on a straight line has one degree of freedom. This is the simplest problem in dynamics. For two-dimensional motion the simplest problem in dynamics consists of the motion of a particle subjected to a constant force (i.e., constant in magnitude and direction). The classical projectile problem, which is widely studied in basic physics courses, is one of these problems.

For each dynamics problem under consideration, the application of Newton's second law to a material particle of mass m, i.e.,

$$\vec{F} = \frac{d}{dt}\,(m\vec{v}) = m\frac{d\vec{v}}{dt} \tag{2.1.1}$$

requires that

1. We isolate the particle from its surroundings and draw a free-body diagram for it, choose a set of unit vectors and variables (the *motion coordinates*) to work with, and then write an expression for the resultant $\vec{F}$ of the forces (i.e., field forces and contact forces) acting on the particle.

2. We develop an expression (in terms of the time-dependent variables that are

chosen to describe the motion) for the *absolute velocity* $\vec{v}$ of the particle and for its *absolute acceleration* $\vec{a} = d\vec{v}/dt$.

The quantity $m\vec{v}$, with $\vec{v}$ being the absolute velocity of the particle, is called the *absolute linear momentum* of the particle.

When use is made of Newton's second law by following the two steps just presented, a set of scalar equations involving unknown reaction forces, and the motion coordinates (and their time derivatives), is then obtained. Such scalar equations are obtained when the corresponding components along the directions of the unit vectors on both sides of the vector equation $\vec{F} = md\vec{v}/dt$ are equated to each other. By a process of elimination of the unknown reaction forces in those scalar equations, a set of equations that only involve the motion coordinates and their time derivatives, and applied forces, can be obtained. The latter equations are the *differential equations of motion* for a specific problem under consideration; they are the equations (or equation, if there is only one degree of freedom) that need to be used for investigating the motion.

The difficulty encountered in integrating the differential equations of motion for any problem being investigated will be directly related to the original choice made for the motion coordinates, i.e., for the time-dependent variables chosen to study the motion. Thus, some care should be exercised in each problem concerning the choice of such variables. For example, when analyzing the motion of a pendulum, or the motion of an orbiting satellite relative to an attracting planet, one might wish to use polar coordinates for the motion variables, while a set of rectangular coordinates might be chosen to analyze the motion of a short-range projectile. A variety of problems that illustrate the degree of difficulty inherent with different choices of motion variables is presented in Section 2.5 (see, for example, Example 2.5.10, p. 129).

Intuitively, we know that the velocity of a point P depends on one's frame of reference. Take, for example, the case of an observer who is stationary inside a moving car and looking at a point P fixed to the windshield of the car when the windshield is also stationary relative to the car. For that observer, the velocity of that point is zero. However, for other observers outside the car, the velocity of that same point is not zero.

As mentioned in Chapter 1, Newton's laws call for the existence of an inertial reference frame, and both the velocity and the acceleration used in his laws must be referred to that frame. The difficulty here is that neither an inertial point (i.e., a point with zero acceleration) nor a nonrotating line seem to exist in nature! In practice, approximations are then made. For most engineering dynamics problems the earth is treated as inertial, which means that the effect of its motion around the sun and of the spin about its own axis are neglected. Whether such approximations yield acceptable results in practice depends on how long the motion being studied is tracked. Very accurate prediction of the motion of a long-range projectile relative to the earth, for example, may require accounting for the rotational motion of the earth about its own axis when formulating the differential equations of motion of the projectile.

A classical experiment with an apparatus known as the *Foucault pendulum* is widely used for demonstrating the effect of the rotation of the earth on motion relative to the earth. The experiment was devised by Jean-Bernard-Léon Foucault (1819-1868), who was a very skilled French experimental physicist. He was the first to provide experimental proof of the rotation of the earth, and the pendulum used for this bears his name. He performed his experiment in Paris, France, in 1851, and repeated it several times after that. A Foucault pendulum can be seen in a number of science museums. Basically, it consists of a ball attached to a long, thin metal string that hangs from the ceiling of a tall room in a building. The experiment demonstrates that the oscillating pendulum rotates with respect to the earth due to the rotation of the earth about its own axis. Analysis of the motion discloses that the period of the rotation is equal to $24/|\sin \lambda|$ hours, where λ is the latitude of the pendulum's location on the earth. The rotation is clockwise in the northern hemisphere and counterclockwise in the southern hemisphere. At the poles, where $\lambda = \pm 90°$, the pendulum rotates $360/4 = 90°$ every $24/4 = 6$ hours (h). In Washington, D.C. (United States), Paris (France), and Rio de Janeiro (Brazil), a rotation of $90°$ takes approximately 9 h and 30 minutes (min), 8 h, and 15 h and 25 min, respectively. The analysis of the motion of the Foucault pendulum is presented in a separate section in Chapter 8.

A Note on Simple Cases Involving Eq. (2.1.1)   If the resultant force $\vec{F}$ acting on a mass particle is equal to zero, Newton's second law yields $m\vec{v} = $ constant (or simply, $\vec{v} = $ constant), which is Newton's first law. This is also known as the *principle of conservation of linear momentum*, but this integrated form of Newton's second law is an obvious one and is of limited use when only a single particle is involved. A similar principle involving a system of particles is presented in Chapter 4.

If the resultant force $\vec{F}$ acting on a mass particle is either a constant or a known function that only involves the time t, i.e., when $\vec{F} = \vec{F}(t)$, Eq. (2.1.1) may be readily integrated to yield

$$\int\limits_{t=t_0}^{t} \vec{F}\, dt = \int\limits_{t=t_0}^{t} \frac{d\,(m\vec{v})}{dt}\, dt = [m\vec{v}]_{t=t} - [m\vec{v}]_{t=t_0} \tag{2.1.2}$$

Equation (2.1.2) states that the change in the linear momentum $m\vec{v}$ of the particle from any time t_0 to any other time t is equal to the integral, over time, of the resultant force $\vec{F}(t)$ acting on the particle. Specific examples are found in elementary dynamics books.

An interesting integrated form of Eq. (2.1.2) is obtained when the system is subjected to a force that is applied only for a very short time Δt, which is short enough so that the particle essentially does not change its position during that time. Such forces are called *impulsive forces*, and the manner in which they change with time during the short time Δt is generally unknown. If the force starts to be applied at some time t_0, Eq. (2.1.2) still yields, of course,

$$\int\limits_{t=t_0}^{t_0+\Delta t} \vec{F}\, dt = [m\vec{v}]_{t=t_0+\Delta t} - [m\vec{v}]_{t=t_0} \overset{\Delta}{=} m\,\Delta\vec{v} \tag{2.1.3}$$

The integral on the left side of Eq. (2.1.3), in the limit as $\Delta t \to 0$, is called an *impulse*. In practice, an impulse can only be approximated, of course, by a short-duration large force. Such forces appear in impact problems that are addressed in first courses in dynamics. They are also approximations for short-duration propulsive forces that are imparted to satellites, for example, to change the velocity of their center of mass by a desired amount $\Delta\vec{v}$ in a very short time to correct their space trajectory or change their orbit around a celestial body.

2.2   A Note on the Reaction Force due to Springs

A spring is a particular type of flexible body and, as such, it deforms when subjected to an applied force. Many physical systems incorporate a spring attached to a mass. During the deformation process, a spring reacts to deformation with an internal force, and that force appears in the free-body diagram for the mass it is attached to. The study of deformation of flexible objects belongs to the discipline called *theory of elasticity*, which is beyond the scope of this book. Early discoveries in elasticity were made by Robert Hooke (1635-1703) who was a contemporary of Newton. In most dynamics problems, the mass of the spring is neglected, and such a practice will be followed here.

There are two types of springs: *extensional springs* (which are simply referred to as *springs*) and *torsional springs*. Both are sketched in Fig. 2.1, with a block P and a bar, respectively, connected to them. In Fig. 2.1a, L_u is the length of the spring when it is unstretched, while r is its actual length when it is stretched from

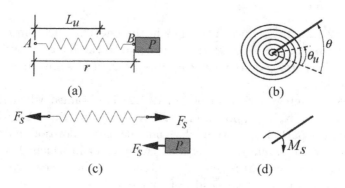

(a)   (b)

(c)   (d)

Figure 2.1: (a) Extensional and (b) torsional springs. (c) Free-body diagrams showing a massless spring and the spring reaction force for the extensional spring. (d) Free-body diagram showing the spring reaction moment for the torsional spring.

that position. In Fig. 2.1b, the torsional spring is unstretched when $\theta = \theta_u$, where θ is measured from some arbitrary line taken as reference. Figures 2.1c and 2.1d show how the spring reaction force F_s and the torsional spring reaction moment M_s appear in free-body diagrams for block P and the bar, respectively. The complete free-body diagram for a massless spring is also shown in Fig. 2.1c. For such a spring, the right-hand side of the equation $\vec{F} = d(m\vec{v})/dt$ is equal to zero when applied to the spring. That is why the free-body diagram for a massless spring simply consists of the forces $\vec{F}_s$ and $-\vec{F}_s$ acting at both ends of the spring. Thus, that diagram does not offer any additional useful information beyond what is already in the free-body diagram for the block that is attached to the spring.

The reaction spring force F_s is a function of the stretch $r - L_u$, and the reaction moment M_s for the torsional spring is a function of $\theta - \theta_u$. For *linear springs*, these functions are simply:

For a linear extensional spring: $\qquad F_s \;=\; k_1(r - L_u)$

$$(2.2.1)$$

For a linear torsional spring: $\qquad M_s \;=\; M_1(\theta - \theta_u)$

where k_1 and M_1 are constants that are determined experimentally.

The constants k_1 (in newtons per meter) and M_1 (in newton-meters per radian) are each called the *spring stiffness* (or the *spring constant*), and they are both positive quantities. Notice that the direction of the spring force F_s in Eq. (2.2.1) is automatically reversed, as expected, when the spring is compressed (i.e., when $r - L_u < 0$). The same occurs for the reaction moment due to the torsional spring.

It should be kept in mind that for real springs, the linear relationships given in Eq. (2.2.1) are only valid for some range of $|r - L_u|$ and $|\theta - \theta_u|$, respectively. For large values of $|r - L_u|$ for the extensional spring, it becomes harder to stretch (or to compress) the spring. This behavior is modeled with the better approximation $F_s = k_1(r - L_u) + k_3(r - L_u)^3$, where $k_3 > 0$. Such a spring is called a *hardening spring*.

2.3 A Moment Equation Obtained from Newton's Second Law

As indicated in Section 2.1, a number of mathematical manipulations may be required involving the scalar equations that are obtained from $\vec{F} = d(m\vec{v})/dt$ to eliminate unknown reaction forces to finally obtain the differential equations of motion for the particle. For some problems, such an elimination process can be shortened (and, sometimes, considerably shortened) by manipulating Newton's second law as shown in this section. For a single mass particle, the methodology presented here does not provide any new information beyond what is already obtained from $\vec{F} = d(m\vec{v})/dt$. But it may provide significant shortcuts as previously mentioned. This is illustrated in detail in a number of solved examples in the book.

If we calculate the moment $\vec{M}_O = \vec{r} \times \vec{F}$, about an inertial point O, of the resultant $\vec{F}$ of the forces acting on a particle P of mass m, and make use of Newton's second law $\vec{F} = d(m\vec{v})/dt$, where $\vec{v} = d\vec{r}/dt$, the following equality is obtained:

$$\vec{M}_O = \vec{r} \times \vec{F} = \vec{r} \times \frac{d}{dt}(m\vec{v}) = \frac{d}{dt}[\vec{r} \times (m\vec{v})] \overset{\Delta}{=} \frac{d\vec{H}_O}{dt} \qquad (2.3.1)$$

The equality $d[\vec{r} \times (m\vec{v})]dt = \vec{r} \times [d(m\vec{v})/dt]$ is true because the extra term $(d\vec{r}/dt) \times (m\vec{v})$ in $d[\vec{r} \times (m\vec{v})]/dt = (d\vec{r}/dt) \times (m\vec{v}) + \vec{r} \times [d(m\vec{v})/dt]$ is equal to zero [since $d\vec{r}/dt = \vec{v}$, and $\vec{v} \times (m\vec{v}) = 0$].

The vector $\vec{H}_O \overset{\Delta}{=} \vec{r} \times (m\vec{v})$, which is the moment about an inertial point O of the absolute linear momentum $m\vec{v}$ of particle P, is defined as the *absolute angular momentum* of P about the inertial point O (the prefix *absolute* refers to the fact that $\vec{v}$ is the absolute velocity of P); the quantity $m\vec{v}$ is called the *absolute linear momentum* of P.

Equation (2.3.1) shows that for those problems where the moment $\vec{M}_O$ is zero, the absolute angular momentum $\vec{H}_O$ is constant, i.e., its magnitude is constant, and its direction is fixed in inertial space (see Example 2.5.11, p. 137). In some problems, it may also happen that only one or two of the components of $\vec{M}_O$ (instead of all three), in terms of a set of inertial directions, is zero. For such cases, although the angular momentum $\vec{H}_O$ is not conserved (i.e., not constant for all times), those inertial components of $\vec{H}_O$ for which the inertial components of $\vec{M}_O$ are zero remain constant in time.

It turns out that Eq. (2.3.1), which was obtained here for a point mass by simply doing the cross product of $\vec{r}$ with both sides of the basic equation $\vec{F} = d(m\vec{v})/dt$ for that point mass, is of fundamental importance in dynamics, especially for a system of point masses, including rigid and flexible bodies. Such a mathematical operation is done again in Chapter 4 (for a system of point masses), and in Chapters 5 and 6 (for rigid bodies). The resulting equation, which involves the moment of the forces acting on the system and the absolute angular momentum of the system (defined in those chapters), is called *Euler's vector equation*.

Let us now look at the problem of how to find the expressions for the absolute velocity and acceleration of a point so that we can start analyzing specific problems in dynamics.

2.4   Absolute Position, Velocity, and Acceleration of a Point

By definition, the absolute velocity $\vec{v}$ of a point P is the time derivative, in an inertial reference frame, of its absolute position vector. As shown in Fig. 2.2, the *position vector $\vec{r}$* of a point P, relative to a reference point O in space, is simply the vector drawn from O to P. The position vector $\vec{r}$ is said to be *absolute* when the reference point O has zero acceleration. Such a reference point is called an *inertial point*, and point O is labeled as such in the figure.

To introduce the definition of the time derivative of a vector,[1] two inertial position vectors for point P at different times are shown in Fig. 2.2, namely $\vec{r}(t)$ (i.e.,

---

[1] We are concerned here with the time derivative of $\vec{r}(t)$ in an inertial frame because that is what is needed for correctly using Newton's second law.

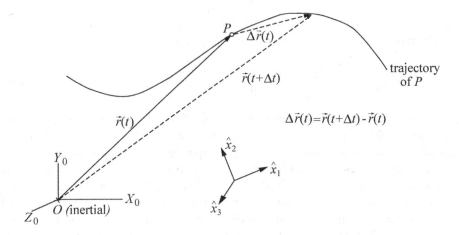

Figure 2.2: A position vector $\vec{r}$ at two instants of time, and the change $\Delta\vec{r}$ of the position vector.

"vector r at time t"), and $\vec{r}(t+\Delta t)$ (i.e., "vector r at time t plus delta t"). They correspond to the absolute position vector of P at two instants of time t and $t+\Delta t$, after a time increment Δt. In a dynamics problem, the trajectory of P shown in Fig. 2.2 is unknown, and it is determined by integrating a set of differential equations that are obtained from Newton's second law.

An inertial reference frame is shown in Fig. 2.2 as being represented by the orthogonal lines labeled X_0, Y_0, and Z_0. A set of orthogonal unit vectors $\{\hat{x}_1, \hat{x}_2, \hat{x}_3\}$ is also shown in the figure; the need for them will start to be explained in the paragraph immediately following Eq. (2.4.2).

The *absolute velocity* $\vec{v}(t)$ of point P at any instant of time t is, by definition, equal to the time derivative, in an inertial frame, of the absolute position vector of P. The definition of the derivative of a scalar function (which is introduced in calculus) is extended to vectors simply as

$$\boxed{\vec{v}(t) = \frac{d\vec{r}(t)}{dt}} \overset{\Delta}{=} \lim_{\Delta t \to 0} \frac{\vec{r}(t+\Delta t) - \vec{r}(t)}{\Delta t} \tag{2.4.1}$$

As can be seen in Fig. 2.2, the velocity $\vec{v}(t)$ of point P is always tangent to the trajectory described by the point during its motion in space. This is so because, as Δt goes to zero, the direction of the change $\Delta\vec{r} = \vec{r}(t+\Delta t) - \vec{r}(t)$ of the position vector becomes, in the limit as $\Delta t \to 0$, tangent to the trajectory at the location of P at time t.

The *absolute acceleration* $\vec{a}(t)$ of point P at any instant of time t is defined as the time derivative of the absolute velocity $\vec{v}(t)$ of P, i.e.,

$$\boxed{\vec{a}(t) = \frac{d\vec{v}(t)}{dt}} \triangleq \lim_{\Delta t \to 0} \frac{\vec{v}(t + \Delta t) - \vec{v}(t)}{\Delta t} = \frac{d^2\vec{r}(t)}{dt^2} \qquad (2.4.2)$$

Equations (2.4.1) and (2.4.2) are just the definitions of velocity $\vec{v}$ and acceleration $\vec{a}$. The more difficult task consists in carrying out the derivatives to obtain, for each problem, explicit expressions for the components of $\vec{v}$ and $\vec{a}$ along the three orthogonal unit vectors of a triad that is chosen to continue the work. Let us say that a triad $\{x\} = \{\hat{x}_1, \hat{x}_2, \hat{x}_3\}$, illustrated in Fig. 2.2, has been chosen. The unit vectors $\hat{x}_1, \hat{x}_2$, and $\hat{x}_3$ may be either nonrotating or drawn on a rotating body. Different choices are generally made for different problems, as will be illustrated in Section 2.5. Since Newton's second law requires using absolute velocity and absolute acceleration, a set of nonrotating axes is also shown in Fig. 2.2; the axes are labeled X_0, Y_0, and Z_0.

The absolute position vector $\vec{r}$ is of the general form

$$\vec{r} = r_{x_1}\hat{x}_1 + r_{x_2}\hat{x}_2 + r_{x_3}\hat{x}_3 \qquad (2.4.3)$$

where the scalar components r_{x1}, r_{x2}, and r_{x3} are functions of the variables that are chosen to describe the motion.

The use of different coordinates to express a vector was illustrated in the review presented in Section 1.1 of Chapter 1. As indicated in that section, such variables may be rectangular, cylindrical, or spherical coordinates (or even a combination of these classical coordinates), or any other convenient set of variables. The specific choice depends on the particular problem to be analyzed.

As mentioned earlier in this section, the form of the differential equations of motion that will be obtained (and, therefore, the difficulty encountered in integrating them) will be directly related to the choice made for such variables.

The time derivative of $\vec{r}$ is simply calculated as

$$\begin{aligned} \frac{d\vec{r}}{dt} &= \dot{r}_{x_1}\hat{x}_1 + \dot{r}_{x_2}\hat{x}_2 + \dot{r}_{x_3}\hat{x}_3 + r_{x_1}\frac{d\hat{x}_1}{dt} + r_{x_2}\frac{d\hat{x}_2}{dt} + r_{x_3}\frac{d\hat{x}_3}{dt} \\ &\triangleq \left[\frac{d\vec{r}}{dt}\right]_{\{x\}} + r_{x_1}\frac{d\hat{x}_1}{dt} + r_{x_2}\frac{d\hat{x}_2}{dt} + r_{x_3}\frac{d\hat{x}_3}{dt} \end{aligned} \qquad (2.4.4)$$

where a dot over a scalar function denotes the time derivative of the scalar function represented by that symbol. The time derivative of any unit vector (or of any vector of constant magnitude) is zero only if the direction of the unit vector is not changing with time (i.e., if the unit vector is not rotating). The translation of a unit vector in space is of no relevance at all because unit vectors are introduced only to allow us to keep track of directions.

The combination of the first three terms in the expression for $d\vec{r}/dt$ [Eq. (2.4.4)], which is renamed $[d\vec{r}/dt]_{\{x\}}$, is the time derivative of $\vec{r}$ *as seen by an observer fixed at O and rotating the same way as the triad* $\{x\}$. The remaining three terms in Eq. (2.4.4), i.e., the ones involving the time derivative of unit vectors, are due to the rotation of the triad $\{x\}$.

Before we make use of rotating unit vectors, let us look at some example problems that are formulated using nonrotating unit vectors. Rotating unit vectors are dealt with in Section 2.6, and a number of examples, using such unit vectors, are presented in Section 2.7.

2.5 Solved Example Problems

This section starts with the simplest type of problem in dynamics, which involves motion of a material point along a nonrotating straight line. The difficulty of the problems is then increased. As the level of difficulty is increased, the amount of mathematical manipulations involved in obtaining the differential equations that govern the motion is gradually increased. The motion is analyzed using the methods presented in Chapter 1, but often the differential equations have to be integrated numerically because no analytical solution is possible. Such situations are very typical in actual practice.

Example 2.5.1 A body P of mass $m = 10$ kg is thrown from a height $h = 200$ m above the earth's surface with an initial velocity $[v]_{t=0} = v_0 = 200$ km/h directed upward. The body is modeled as a point mass and is subjected only to the gravitational force of the earth; its motion takes place along the vertical. Determine:

a. The maximum height reached by the body P, measured from the ground
b. The speed (i.e., the magnitude of the velocity vector) of P when it hits the ground
c. The speed of P as a function of its height measured from the ground
d. The time it takes P to hit the ground

■ Solution

The free-body diagram for P is shown in Fig. 2.3. The position of the body is tracked by the vector from an inertial point O on the ground, using the distance $y(t)$ from O to P as the coordinate of the motion. A unit vector $\hat{r}$, for expressing the vector $\overrightarrow{OP}$, is also shown in the figure. Notice that such choices are up to the individual, and you could have made different ones. You could have chosen to work with a unit vector in the direction opposite to $\hat{r}$ and to track the body from another inertial point, such as the launching point, for example. But the choices that were made are as good as any other. Since the motion takes place along the vertical, only one unit vector is needed; it could be chosen to be pointing either upward or downward. The force $\vec{F}$ acting on P is indicated by the heavy arrowed line shown in Fig. 2.3, while the position vector $\vec{r}$ is indicated with a broken arrowed line to clearly distinguish it from a force. It is a good practice to adopt such a convention so that others do not misinterpret your free-body diagram.

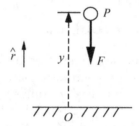

Figure 2.3: Free-body diagram for Example 2.5.1.

In terms of the unit vector $\hat{r}$ and of the motion coordinate y shown in Fig. 2.3, the expression for the absolute position vector $\vec{r} = \overrightarrow{OP}$ (with point O being inertial) is $\vec{r} = y\hat{r}$. Therefore, the absolute velocity and acceleration of P are

$$\vec{v} \;=\; \frac{d\vec{r}}{dt} = \dot{y}\hat{r} \qquad\qquad (2.5.1a)$$

$$\vec{a} \;=\; \frac{d\vec{v}}{dt} = \ddot{y}\hat{r} \qquad\qquad (2.5.1b)$$

We now need an expression for the gravitational force F acting on the particle. For this, a decision has to be made; namely, how should the force F be modeled? Engineers and scientists are always facing such decisions about modeling. In this case, we have two choices. The simplest choice is to approximate F as mg, with g being the constant acceleration of gravity. Such a choice is good if the particle stays "close enough" to the earth's surface during its motion along the vertical. Given that the initial speed is only (!) 200 km/h, that is probably going to be the case (but, at this stage of the formulation, we do not know that for sure). The other choice, which is more accurate, is to use $F = GM_Em/(R_E + y)^2$, as given by Newton's law of universal gravitation, where M_E and R_E are the mass and the radius of the earth, respectively. Let's do the problem both ways, starting with the simplest choice for F as $F = mg$. When both analyses are finished, a plot is presented showing how the answers to the questions that were asked depend on how F was modeled.

With $\vec{F} = -mg\hat{r}$ (notice that the minus sign goes together with the direction of the unit vector that was chosen; if we had chosen to work with a unit vector in the opposite direction, a plus sign would show up here and a minus sign would show up in the expressions for the position vector $\vec{r}$, for the velocity $\vec{v}$, and the acceleration $\vec{a}$), $\vec{F} = m\vec{a}$ then gives the following differential equation for $y(t)$:

$$\ddot{y} = -g \tag{2.5.1c}$$

This differential equation is of the type that was reviewed in Section 1.13 of Chapter 1, and its solution is

$$y = -\frac{gt^2}{2} + t\dot{y}(0) + y(0) \tag{2.5.1d}$$

where, for this problem, $\dot{y}(0) = v_0 = 200$ km/h and $y(0) = h = 200$ m.

Since $\ddot{y} = d\dot{y}/dt = \dot{y}(d\dot{y}/dy)$ (see the presentation in Section 1.13 of Chapter 1, about changing variables by using the chain rule), we also obtain from Eq. (2.5.1c)

$$\frac{\dot{y}^2}{2} + gy = \text{constant} = \left[\frac{\dot{y}^2}{2} + gy\right]_{t=0} = \frac{v_0^2}{2} + gh \tag{2.5.1e}$$

All the questions that were asked are then answered from the expressions obtained for y and $\dot{y}^2$ in Eqs. (2.5.1d) and (2.5.1e).

a. We can immediately determine the maximum height reached by P by setting $\dot{y} = 0$ in Eq. (2.5.1e) that involves $\dot{y}^2$ because, when there are no bounds imposed on the independent variable t, the maximum (or minimum) of a function $y(t)$ occurs at the point when the first derivative of the function is zero. For the present problem, since the second derivative $\ddot{y} = -g$ is negative, the point where $\dot{y} = 0$ corresponds to the maximum value y_{max} of y. For the given initial conditions, Eq. (2.5.1e) yields

$$y_{\text{max}} = h + \frac{v_0^2}{2g} \approx 357 \text{ m}$$

Since the maximum height is only 157 m above the initial point (i.e., 357 m above the earth's surface), and the radius of the earth is $R_E \approx 6378$ km, the approximation of a constant acceleration of gravity should be perfectly acceptable. But what if the answer were much higher? Would the answer be acceptable then? This question will be addressed when we use $F = GM_E m / (R_E + y)^2$.

Another way to determine the maximum height y_{max} is to use the expression $\dot{y} = -gt + \dot{y}(0)$, which is obtained by integrating $\ddot{y} = -g$ once, set $\dot{y} = 0$ to obtain the time t when $\dot{y} = 0$, which is $t = \dot{y}(0)/g \approx 5.66$ s, and then find y_{max} from the solution $y = -gt^2/2 + t\dot{y}(0) + y(0)$ as $y_{max} = [y]_{t=5.66} \approx 357$ m. As you see, there is more than one way to solve this problem. But no matter which way you prefer, the solution is always obtained by integrating the differential equation of motion (which is $\ddot{y} = -g$ for this particular problem).

b. We can also immediately determine the speed of P when it hits the ground by making use of the expression that involves y and $\dot{y}^2$ [Eq. (2.5.1e)]. By setting $y = 0$ in that equation we now obtain

$$\left[\dot{y}^2\right]_{\text{when } y=0} = v_0^2 + 2gh$$

or

$$\left[\dot{y}\right]_{\text{when } y=0} = \pm\sqrt{v_0^2 + 2gh} \approx \pm 83.7 \text{ m/s} \approx \pm 301 \text{ km/h}$$

Looking at the directions of the position vector $\vec{r}$ and of the unit vector $\hat{r}$ in Fig. 2.3, we conclude that the velocity of P when it strikes the ground is $[\vec{v}]_{\text{when } y=0} = -301\hat{r}$ km/h (the magnitude of the velocity is 301 km/h).

c. The speed of P as a function of its height from the ground, obtained from Eq. (2.5.1e), is $v = |\pm \sqrt{v_0^2 - 2g(y - h)}|$, i.e., $v = \sqrt{v_0^2 - 2g(y - h)}$.

d. The time it takes P to hit the ground is more easily obtained when the differential equation $\ddot{y} = -g$ is integrated directly with respect to time. This was possible to do for this differential equation, and the solution is given by Eq. (2.5.1d). By setting $y = 0$ in that solution, and letting t_f denote the time when $y = 0$, we then have to solve the following quadratic equation for t_f:

$$-\frac{gt_f^2}{2} + t_f\dot{y}(0) + h = 0$$

Two solutions for the time t_f are obtained:

$$t_f = \frac{\dot{y}(0)}{g} \pm \sqrt{\left(\frac{\dot{y}(0)}{g}\right)^2 + \frac{2h}{g}}$$

Obviously, only the positive solution makes physical sense. For the given initial conditions, we then obtain $t_f \approx 14.2$ s.

Notes:

■ If you look at the solution obtained for the maximum value of y, $y_{max} = h + v_0^2/(2g)$, you see that the maximum height reached by the particle increases as we increase the initial upward speed $v_0 = [\dot{y}]_{y=h}$. This is certainly reasonable, and it is what intuition tells us. However, it is not reasonable to expect that the approximation $F = $ constant $= mg$ remains valid as y increases.

■ Another thing that the expression $y_{max} = h + v_0^2/(2g)$ indicates is that for y_{max} to be infinite the initial speed $v_0 = [\dot{y}]_{t=0}$ will have to be infinite. Physically, $y_{max} = \infty$ should be interpreted as being a very big distance, which is many times the radius of the earth. Since it certainly would take an infinite amount of time for the distance y to be infinite, that means that the particle would never return to the vicinity of the earth after it was launched. The physical interpretation of this is that the particle would have "escaped" the gravitational attraction of the earth. The minimum launch speed for this to happen is called the *escape speed*. Notice, though, that it would be physically impossible to launch a space probe to make it leave forever the neighborhood of the launching point if an infinite launching speed would be required! Satellites could never have been launched into deep space if that were the case. Let us then look at this problem again in Example 2.5.2 with the objective of determining the maximum distance y_{max} and the minimum value of the initial speed $v_0 = [\dot{y}]_{t=0}$ that is necessary to impart to the particle for it to escape the gravitational attraction of the earth.

Example 2.5.2 Determine the maximum distance y_{max} for the problem in Example 2.5.1 when $v_0 = \dot{y}(0) = 10,000$ km/h and when $v_0 = 20,000$ km/h. What is the value of the escape speed for a body launched from the earth's surface, and from a height $y(0) = h = 400$ km above the earth's surface? The earth may be approximated as a homogeneous sphere with radius $R_E \approx 6378$ km.

■ **Solution**

With $h = 0$, the solution $y_{max} = h + v_0^2/2g$ found in Example 2.5.1 yields $y_{max} \approx 393$ km when $v_0 = 10,000$ km/h, and $y_{max} \approx 1573$ km when $v_0 = 20,000$ km/h. Let us check these answers with $F = GM_E m/(R_E + y)^2$, which is the expression that has to be used for calculating the escape speed.

With $\vec{F} = -GM_E m/(R_E + y)^2)\hat{r} = -mgR_E^2/(R_E + y)^2\hat{r}$, $\vec{F} = m\vec{a}$ now gives the following second-order differential equation for $y(t)$:

$$\ddot{y} = -\frac{gR_E^2}{(R_E + y)^2} \tag{2.5.2a}$$

Although not impossible, it is now more difficult to integrate this differential equation with respect to time. Instead of trying to do that analytically or integrating it numerically, let us use again the method presented in Section 1.13 of Chapter 1. Again, with $\ddot{y} = d\dot{y}/dt = \dot{y}(d\dot{y}/dy)$, we now obtain from Eq. (2.5.2a)

$$\int \dot{y}\frac{d\dot{y}}{dy}\,dy + \int \frac{gR_E^2 dy}{(R_E + y)^2} = \text{constant}$$

or

$$\frac{\dot{y}^2}{2} - \frac{gR_E^2}{R_E + y} = \text{constant} = \frac{\dot{y}^2(0)}{2} - \frac{gR_E^2}{R_E + y(0)} \tag{2.5.2b}$$

Notice that this equation discloses that to have $\dot{y} = 0$ when y is infinite (i.e., to escape the gravitational field of the earth by launching the particle with the smallest required initial speed), we must impart to the body, at the launching height $y = h$, an initial speed equal to $\dot{y}(0) = \sqrt{2gR_E/(1 + h/R_E)}$ since the constant of integration in this situation is equal to zero. Notice that the positive solution for $\dot{y}(0)$ was used since we want the body to move away from the earth, as indicated by the convention for the directions shown in Fig. 2.3. This critical speed, which is called the *escape speed* (and is denoted as v_{esc}) at the launching height h, is the minimum initial speed that should be imparted to the body for it to escape the gravitational field of the attracting planet it is launched from. Any initial value of $\dot{y}$ that is bigger than v_{esc} will cause the body to escape the gravitational field of the attracting planet and have a nonzero "speed at infinity" (which is certainly what is desired in practice). For a body launched from the surface of the earth (i.e., $h = 0$), the escape speed is $v_{\text{esc}} \approx 40,270$ km/h $\approx 25,000$ miles/hour (mi/h). For $h = 400$ km, the escape speed is $v_{\text{esc}} \approx 39,000$ km/h $\approx 24,280$ mi/h.

The maximum value of y, $y_{\max}$, is obtained by setting $\dot{y} = 0$ in Eq. (2.5.2b). For $y(0) = 0$, Eq. (2.5.2b) yields $y_{\max} \approx 419$ km when $\dot{y}(0) = v_0 = 10,000$ km/h, and $y_{\max} \approx 2090$ km when $\dot{y}(0) = v_0 = 20,000$ km/h.

Figure 2.4 shows a plot of $y_{\max}/R_E$ versus the initial speed $\dot{y}(0) = v_0$ obtained with $F = mgR_E^2/(R_E + y)^2$ and with the approximation $F = mg$, for $y(0) = h = 0$. Observe how the approximation $F = mg$ becomes worse and worse as v_0 is increased.

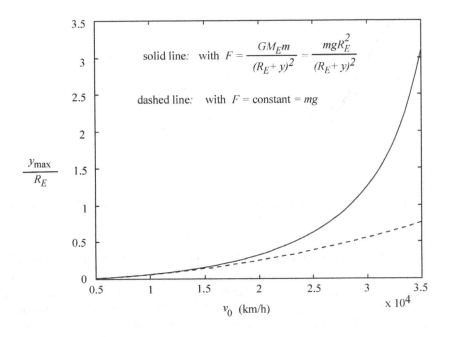

Figure 2.4: Maximum height $y_{\max}$ reached by a particle thrown from the earth's surface for Example 2.5.2.

Example 2.5.3 As shown in Fig. 2.5a, a bar AB of length $L = 1.5$ m is forced to slide against a fixed (nonrotating) vertical wall so that the angle θ it makes with the horizontal changes with time as $\theta = \pi/2 - k_1 t \sin(k_2 t)$, where $k_1 = 1$ rad/s and $k_2 = 0.08$ rad/s. Point O in Fig. 2.5 is inertial, i.e., its absolute acceleration is equal to zero. Determine the absolute velocity and acceleration of points A and B when the bar strikes the ground (i.e., when $\theta = 0$), and the velocity of point B relative to point A at that instant.

■ Solution

Since the angle θ is known for all times, there are no differential equations of motion to be formulated and, therefore, this type of problem is a kinematics problem. The absolute velocities of points A and B are equal to the time derivatives of the absolute position vectors from the inertial point O to points A and B, respectively. Also, the absolute accelerations of those points are equal to the time derivatives of their respective absolute velocities.

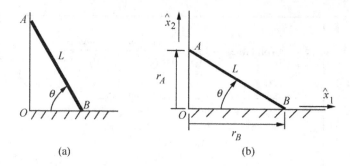

(a)   (b)

Figure 2.5: The system and the absolute position vectors of A and B for Example 2.5.3.

Since points A and B move along straight lines, the natural way to determine their absolute velocity and acceleration is to express their absolute position vectors $\vec{r}_A = \overrightarrow{OA}$ and $\vec{r}_B = \overrightarrow{OB}$ in terms of the known angle θ and of the unit vectors $\hat{x}_1$ and $\hat{x}_2$ shown in Fig. 2.5b, as given in Eq. (2.5.3a). These expressions are simply geometrical relationships that are obtained by inspection of Fig. 2.5.

$$\vec{r}_A = (L\sin\theta)\hat{x}_2 \qquad\qquad \vec{r}_B = (L\cos\theta)\hat{x}_1 \qquad (2.5.3a)$$

Once the expressions for the absolute position vectors $\vec{r}_A$ and $\vec{r}_B$ are written down, the calculation of the absolute velocity and acceleration of points A and B simply involve taking the first and second time derivatives of $\vec{r}_A$ and $\vec{r}_B$. By proceeding in this manner, we then obtain

$$\vec{v}_A = \frac{d\vec{r}_A}{dt} = L\dot{\theta}\left(\cos\theta\right)\hat{x}_2 \qquad \vec{a}_A = \frac{d\vec{v}_A}{dt} = L\left(\ddot{\theta}\cos\theta - \dot{\theta}^2\sin\theta\right)\hat{x}_2 \qquad (2.5.3b)$$

$$\vec{v}_B = \frac{d\vec{r}_B}{dt} = -L\dot{\theta}\left(\sin\theta\right)\hat{x}_1 \qquad \vec{a}_B = \frac{d\vec{v}_B}{dt} = -L\left(\ddot{\theta}\sin\theta + \dot{\theta}^2\cos\theta\right)\hat{x}_1 \qquad (2.5.3c)$$

For $\theta = \pi/2 - t\sin(0.08t)$, the first zero-crossing, $\theta = 0$, is found to occur when $t \approx 4.48$ s. This may be obtained by a few iterations with a calculator, or from the plot of $\theta(t)$ versus t shown in Fig. 2.6. Therefore, the bar hits the floor when $t \approx 4.48$ s.

With $\theta = \pi/2 - t\sin(0.08t)$, we obtain $\dot{\theta} = -\sin(0.08t) - 0.08t\cos(0.08t)$. Therefore, the value of $\dot{\theta}(t)$ when $\theta = 0$ is

$$\left[\dot{\theta}\right]_{t=4.48\text{ s}} = -\sin(0.08 \times 4.48) - 0.08 \times 4.48\cos(0.08 \times 4.48) \approx -0.686 \text{ rad/s}$$

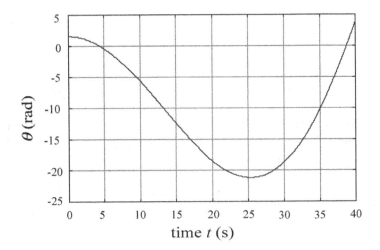

Figure 2.6: Plot of $\theta(t) = \pi/2 - t\sin(0.08t)$ for Example 2.5.3.

With $L = 1.5$ m, this gives the following velocities at the instant when $\theta = 0$:

$$[\vec{v}_A]_{\theta=0} = -[1.5 \times 0.686 \times \cos(0)]\,\hat{x}_2 \approx -1.03\hat{x}_2 \text{ m/s} \qquad [\vec{v}_B]_{\theta=0} = 0$$

The minus sign in the answer for $\vec{v}_A$ means that the velocity of A at that instant is opposite to the $\hat{x}_2$ unit vector; i.e., point A is moving down.

The velocity of B relative to A, denoted as $\vec{v}_{B/A}$, is, by definition, equal to $\vec{v}_B - \vec{v}_A$. Therefore,

$$\left[\vec{v}_{B/A}\right]_{\theta=0} = 1.03\hat{x}_2 \text{ m/s}$$

The respective accelerations when $\theta = 0$ are

$$[\vec{a}_A]_{\theta=0} \approx -0.21\hat{x}_2 \text{ m/s}^2 \qquad [\vec{a}_B]_{\theta=0} = -0.707\hat{x}_1 \text{ m/s}^2$$

Example 2.5.4 A rectangular block is forced by a bar AB to move along a fixed straight groove that is inclined at a constant angle $\alpha = 30°$ with the horizontal as shown in Fig. 2.7a. Point P is the point of contact of bar AB with the block, and the distance h shown in the figure is the constant perpendicular distance from the fixed point A to the groove. The bar is driven by a motor so that the angle θ shown in the figure is changing with time t as $\theta = 2\pi/3 - k_1\cos(k_2t)$ rad, where $k_1 = 5\pi/12$

rad and $k_2 = \pi/2$ rad/s. Determine the absolute velocity and absolute acceleration (magnitude and direction) of the block when $t = 1$ s, 2 s, and 4 s.

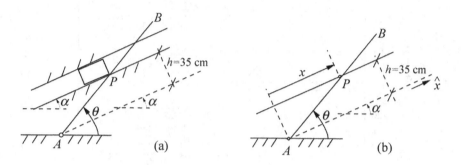

Figure 2.7: The system and the absolute position vector of P for Example 2.5.4.

■ Solution

Since the block is only translating, each point of the body has the same velocity. The same holds for the acceleration of each of its points. We then choose to work with point P to determine the velocity and acceleration of the block. Notice that this problem is also a kinematics problem, as the one in Example 2.5.3, since the motion of the block is already known. Also, we are not concerned here with determining the forces that are necessary to impart that specified motion to the block.

Since point P moves along a straight line, this problem is also readily solved by introducing the variable x and the unit vector $\hat{x}$ shown in Fig. 2.7b to express the absolute position vector $\vec{r}$ of point P in terms of the known angle $\theta(t)$ and the known distance $h = 35$ cm as

$$\vec{r} = x\hat{x} = \frac{h}{\tan(\theta - \alpha)}\hat{x} \qquad (2.5.4a)$$

This expression is simply a geometrical relationship that is obtained by inspection of Fig. 2.7. Once the expression for the absolute position vector $\vec{r}$ is written down, the problem is readily solved simply by taking the first and second time derivatives of that expression. For this problem, we then obtain

$$\vec{v} = \frac{d\vec{r}}{dt} = \dot{x}\hat{x} = -\frac{h}{\tan^2(\theta - \alpha)}\frac{d}{dt}\tan(\theta - \alpha)\hat{x} = -\frac{h\dot{\theta}}{\sin^2(\theta - \alpha)}\hat{x} \qquad (2.5.4b)$$

$$\vec{a} = \frac{d\vec{v}}{dt} = \ddot{x}\hat{x} = \left\{ -\frac{h\ddot{\theta}}{\sin^2(\theta - \alpha)} + \frac{2h\dot{\theta}}{\sin^3(\theta - \alpha)}\frac{d}{dt}\sin(\theta - \alpha)\right\}\hat{x}$$

$$= \left\{ -\frac{h\ddot{\theta}}{\sin^2(\theta - \alpha)} + \frac{2h\dot{\theta}^2\cos(\theta - \alpha)}{\sin^3(\theta - \alpha)}\right\}\hat{x} \qquad (2.5.4c)$$

The answers to the questions that were asked for the various values of t are then calculated using these expressions. They are given in the following table, where $\dot{\theta} = [5\pi^2 \sin(\pi t/2)]/24$ and $\ddot{\theta} = [5\pi^3 \cos(\pi t/2)]/48$. The absolute velocity and absolute acceleration of P are $\dot{x}\hat{x}$ and $\ddot{x}\hat{x}$, respectively.

t (s)	θ (degrees)	$\dot{\theta}$ (rad/sec)	$\ddot{\theta}$ (rad/s^2)	$\dot{x}$ (m/s)	$\ddot{x}$ (m/s^2)
1	120	2.06	0	-0.72	0
2	195	0	-3.23	0	16.88
4	45	0	3.23	0	-16.88

Example 2.5.5 A block of mass m kg slides on a face of a plane that is inclined at a constant angle $\alpha = 30°$ with the horizontal as shown in Fig. 2.8. The motion of

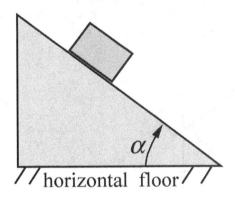

Figure 2.8: The system for Example 2.5.5.

the block takes place on a vertical plane, and it consists of translation only.[2] The block is initially at rest at a distance of 4 m from the bottom of the plane. Obtain the differential equation of motion for the block, and determine the time it takes the block to arrive at the bottom of the inclined plane, for the following two cases:

a. Friction in the system is viscous, with friction coefficient c N/(m/s), and $c/m = 0.8$ s^{-1}.

b. Friction in the system is of the Coulomb type; the static and kinetic friction coefficients are $\mu_s = 0.3$ and $\mu_k = 0.25$, respectively.

---

[2]The motion of a body is only a translation if lines that are "painted" on the body do not change orientation during the motion.

■ **Solution**

Since there is no rotation involved, the block moves as if it were a point mass, i.e., all points of the body have the same velocity and the same acceleration. The free-body diagram and the absolute position vector $\vec{r} = \overrightarrow{OP}$ for a point P of the block are shown in Fig. 2.9. The position vector $\vec{r} = \overrightarrow{OP}$ is shown in a dashed arrowhead line to distinguish it from forces, which are shown as solid arrowhead lines. The inertial reference point O was arbitrarily chosen at the bottom of the inclined plane; it could have been chosen at any other fixed location on the inclined plane or elsewhere. The two orthogonal unit vectors $\hat{x}_1$ and $\hat{x}_2$ shown in that figure were also chosen for expressing all the vectors that appear in the formulation of the equations for this problem. Any other set of orthogonal unit vectors could have been chosen. The rationale behind the particular choice of $\hat{x}_1$ shown in the figure is that the expression for the absolute acceleration of P will only have a component along $\hat{x}_1$ because the motion of P takes place along a straight line on the inclined plane. The direction for $\hat{x}_1$ could have been chosen to be opposite to the one shown in the figure, and the expression obtained for the acceleration of P would still be a simple one.

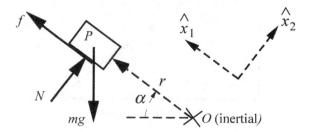

Figure 2.9: Free-body diagram and the position vector $\vec{r}$ for Example 2.5.5.

In terms of $\hat{x}_1$, the absolute position vector $\vec{r}$, absolute velocity $\vec{v}$, and absolute acceleration $\vec{a}$ of P are, respectively,

$$\vec{r} = r\hat{x}_1 \qquad \vec{v} = \frac{d\vec{r}}{dt} = \dot{r}\hat{x}_1 \qquad \vec{a} = \frac{d\vec{v}}{dt} = \ddot{r}\hat{x}_1 \qquad (2.5.5a)$$

The resultant force $\vec{F}$ acting on the block is the sum of the gravitational force, the normal force $N\hat{x}_2$ applied by the plane to the block, and the friction force $\vec{f}$, which was arbitrarily drawn with an arrow pointing to the left in Fig. 2.9. Therefore, the expression for $\vec{F}$ is

$$\vec{F} = mg(-\hat{x}_1 \sin\alpha - \hat{x}_2 \cos\alpha) + N\hat{x}_2 + f\hat{x}_1 \qquad (2.5.5b)$$

From Newton's second law, $\vec{F} = m\vec{a}$, the following two equations are then obtained:

$$N - mg\cos\alpha \quad = \quad 0 \tag{2.5.5c}$$

$$m\ddot{r} \quad = \quad f - mg\sin\alpha \tag{2.5.5d}$$

These two equations involve three unknowns, namely, $r(t)$, N, and the friction force f. A third equation is obviously necessary for completing the formulation of the problem. The missing equation involves the friction force $f\hat{x}_1$, and to generate such an equation it is necessary to know which type of friction force is involved. Let us consider both viscous friction and dry (Coulomb) friction. The friction force for both cases is directed opposite to the velocity $\vec{v}$ when the block translates relative to the inclined plane.

Viscous friction. $\vec{f} = -c\vec{v} = -c\dot{r}\hat{x}_1$, which gives $f = -c\dot{r}$.

Coulomb friction. Either $|f| \leq \mu_s|N|$ if there is no motion, or $f = -\mu_k|N|\,\mathrm{sgn}\,\dot{r}$ if the block is moving.

Case a: Viscous Friction
For this case the differential equation of motion, obtained from Eq. (2.5.5d), is

$$\ddot{r} + \frac{c}{m}\dot{r} = -g\sin\alpha \tag{2.5.5e}$$

This is a linear non-homogeneous differential equation with constant coefficients, and for $c \neq 0$ its solution is (see Case 3 in Section 1.13 of Chapter 1, p. 43):

$$r = r(0) - \frac{mg\sin\alpha}{c}t + \frac{m}{c}\left[\dot{r}(0) + \frac{mg\sin\alpha}{c}\right]\left[1 - e^{-ct/m}\right] \tag{2.5.5f}$$

For $r(0) = 4$ m, $\dot{r}(0) = 0$, $\alpha = 30°$, and $c/m = 0.8$ s^{-1} this solution yields, after a few iterations with a calculator with $r = 0$, $t \approx 1.54$ s.

Case b: Coulomb Friction
For this case, we first need to determine if the block will move. This is done by calculating the friction force under the assumption that the block does not move (i.e., that it is in static equilibrium) and then checking whether the assumption is true or false by comparing the magnitude of the friction force with the magnitude $\mu_s|N|$ of the maximum possible friction force. If it is found that $|f|/|N| > \mu_s$, the block is actually moving.

If the block is in static equilibrium (i.e., if $r = $ constant), Eq. (2.5.5d) yields

$$f - mg\sin\alpha = 0 \qquad \text{or} \qquad f = mg\sin\alpha \tag{2.5.5g}$$

which gives

$$\frac{|f|}{|N|} = \frac{mg\sin\alpha}{mg\cos\alpha} = \tan\alpha \approx 0.577$$

Since this ratio is greater than μ_s, the block is moving. In addition, since the value found for f is positive, the block starts to move down the inclined plane (because the friction force was drawn upward in the free-body diagram shown in Fig. 2.8, and an upward friction force implies $f > 0$). With the direction convention used in Fig. 2.8, the friction force is then

$$f = -\mu_k|N|\operatorname{sgn}\dot{r} = -\mu_k mg(\cos\alpha)\operatorname{sgn}\dot{r} \qquad (2.5.5h)$$

and the differential equation of motion for this case, obtained from Eq. (2.5.5d), is

$$\ddot{r} + \mu_k g(\cos\alpha)\operatorname{sgn}\dot{r} = -g\sin\alpha \qquad (2.5.5i)$$

This differential equation is nonlinear because of the $\operatorname{sgn}\dot{r}$ term, which is not a linear function of its argument $\dot{r}$. However, the initial conditions of the motion are such that the block starts moving down the plane. Since the direction of the motion does not change, $\operatorname{sgn}\dot{r} < 0$ while the block keeps moving down. Therefore, the motion of the body is governed, at all times, by the following equation, which is a linear differential equation.

$$\ddot{r} = g(\mu_k\cos\alpha - \sin\alpha)$$

The solution to this linear non-homogeneous differential equation is simply

$$r = r(0) + t\dot{r}(0) + \frac{g(\mu_k\cos\alpha - \sin\alpha)t^2}{2}$$

For $r(0) = 4$ m, $\dot{r}(0) = 0$, $\alpha = 30°$, and $\mu_k = 0.25$ this solution yields, with $r = 0$, $t \approx 1.7$ s.

Example 2.5.6 Solve Example 2.5.5 when the initial conditions are $r(0) = 4$ m and $\dot{r}(0) = 1.5$ m/s.

■ **Solution**

For the case of viscous friction, the solution (which is valid for all times) was determined in Example 2.5.5 as

$$r = r(0) - \frac{mg\sin\alpha}{c}t + \frac{m}{c}\left[\dot{r}(0) + \frac{mg\sin\alpha}{c}\right]\left[1 - e^{-ct/m}\right] \qquad (2.5.6a)$$

For $r(0) = 4$ m, $\dot{r}(0) = 1.5$ m/s, $\alpha = 30°$, and $c/m = 0.8$ s^{-1}, the solution yields (after a few iterations with a calculator with $r = 0$) $t \approx 1.85$ s.

For the case of Coulomb friction, the differential equation of motion was found to be

$$\ddot{r} + \mu_k g(\cos \alpha)\text{sgn}\,\dot{r} = -g \sin \alpha \qquad (2.5.6b)$$

Since the function $\text{sgn}\,\dot{r}$ depends on $\dot{r}$ in a nonlinear manner, this differential equation is nonlinear. But, since $\text{sgn}\,\dot{r} = \text{constant} = 1$ when $\dot{r} > 0$ and $\text{sgn}\,\dot{r} = \text{constant} = -1$ when $\dot{r} < 0$, this differential equation can be solved analytically in a "piecewise manner", i.e., by solving it separately when $\dot{r} > 0$ and when $\dot{r} < 0$. For these two possibilities, the solutions are then

$$\text{for } \dot{r} > 0: \qquad r = r(0) + t\dot{r}(0) - \frac{g(\sin \alpha + \mu_k \cos \alpha)t^2}{2} \qquad (2.5.6c)$$

$$\text{for } \dot{r} < 0: \qquad r = r(0) + t\dot{r}(0) - \frac{g(\sin \alpha - \mu_k \cos \alpha)t^2}{2} \qquad (2.5.6d)$$

Since the block is forced to start moving up the plane with $\dot{r}(0) = 1.5$ m/s, the solution for $\dot{r} > 0$ discloses that for $r(0) = 4$ m, $\alpha = 30°$, and $\mu_k = 0.25$, the block keeps moving up the plane until $\dot{r}$ becomes zero. That happens when

$$t = \frac{\dot{r}(0)}{g(\sin \alpha + \mu_k \cos \alpha)} \approx 0.21 \text{ s}$$

The corresponding value of r is found to be

$$r(0.21) = r(0) + \frac{\dot{r}^2(0)}{2g(\sin \alpha + \mu_k \cos \alpha)} \approx 4.16 \text{ m}$$

The question to be answered now is whether the dry friction coefficient is large enough to prevent the block from moving down the plane after $t = 0.21$ s. If the block permanently stopped at that location, the friction force f shown in Fig. 2.9 would be $f = mg \sin \alpha$, as determined in Example 2.5.5. Since the normal force is $N = mg \cos \alpha$, the ratio $|f|/|N| = \tan \alpha \approx 0.577$ would then be greater than the value of the static coefficient of friction μ_s, which is physically impossible. Therefore, the block only stops momentarily (i.e., when $\dot{r}$ becomes zero at $t = 0.21$ s as previously determined), and then proceeds to travel down the plane. Notice that the block would permanently stop at the top of its trajectory up the plane if $\alpha \leq \arctan \mu_s$ (which, for $\mu_s = 0.3$, is $\alpha \leq 16.7°$).

We now have to determine how long it takes the block to travel 4.16 m down the inclined plane, starting from rest. By using the new initial conditions $r(0) = 4.16$ m, and $\dot{r}(0) = 0$ in the solution previously obtained when $\dot{r}(t) < 0$, namely,

$$r = r(0) + t\dot{r}(0) - \frac{g(\sin\alpha - \mu_k\cos\alpha)t^2}{2} \qquad (2.5.6e)$$

we then find that $r = 0$ occurs when

$$t = \sqrt{\frac{2r(0)}{g(\sin\alpha - \mu_k\cos\alpha)}} \approx 1.73 \text{ s}$$

The total time for the block to travel from $r(0) = 4$ m to $r = 4.16$ m up the plane and then pass the position when $r = 0$ is, therefore, $t_{\text{total}} \approx 0.21 + 1.73 = 1.94$ s.

Example 2.5.7 The same block of Example 2.5.6 is now connected to the bottom of the inclined plane by a linear spring of unstressed length $L_u = 1$ m and stiffness k N/m, where $k/m = 10$ s^{-2}. The system and its free-body diagram are shown in Fig. 2.10; the reaction forces f and f_s are due to friction and the spring, respectively. The block is given an initial displacement $r(0) = 0.86$ m, with zero initial velocity.

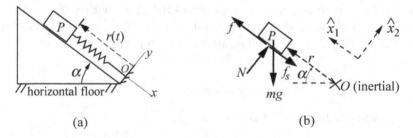

(a)   (b)

Figure 2.10: The system, and its free-body diagram, for Example 2.5.7.

Consider, separately, both cases of viscous friction and Coulomb friction, with the friction coefficients given in parts a and b of Example 2.5.5. That is, when friction is viscous, the friction coefficient is c N/(m/s), with $c/m = 0.8$ s^{-1}, and when friction is of the Coulomb type, the static and kinetic coefficients of friction are $\mu_s = 0.3$ and $\mu_k = 0.25$, respectively.

a. What is the differential equation that governs the motion of the block for each case?

b. Determine the value of the equilibrium distance $r(t) = \text{constant} \overset{\Delta}{=} r_e$. Notice that for the case of viscous friction there is a single value of r_e, while for the case of Coulomb friction there is a range of values of r_e when the block does not move.

c. Plot $r(t)$ versus t for each case, when $r(0) = 0.86$ m.

■ Solution

• **Viscous friction:**

a. Referring to Fig. 2.10b, the forces acting on the block are the spring force f_s, the gravitational force, and the friction force f. Therefore, the resultant force acting on the block is

$$\vec{F} = mg(-\hat{x}_1 \sin \alpha - \hat{x}_2 \cos \alpha) + N\hat{x}_2 + f\hat{x}_1 - k(r - L_u)\hat{x}_1 \qquad (2.5.7a)$$

and Newton's second law yields the following equations:

$$N - mg \cos \alpha = 0 \qquad (2.5.7b)$$
$$m\ddot{r} = f - k(r - L_u) - mg \sin \alpha \qquad (2.5.7c)$$

For viscous friction, the expression for the friction force f shown in Fig. 2.10 is $f = -c\dot{r}$, and the differential equation of motion is then

$$\ddot{r} + \frac{c}{m}\dot{r} + \frac{k}{m}(r - L_u) + g \sin \alpha = 0 \qquad (2.5.7d)$$

b. Equation (2.5.7d) is a linear non-homogeneous differential equation with constant coefficients and it can be solved analytically. As shown in Section 1.13 of Chapter 1 (see Case 3, p. 43), its solution is

$$r(t) = r_e + e^{\sigma t}[B_1 \cos(\omega t) + B_2 \sin(\omega t)] \qquad (2.5.7e)$$

where $r_e = L_u - mg(\sin \alpha)/k$ is the equilibrium solution, $\sigma = -c/(2m)$, and $\omega = \sqrt{k/m - [c/(2m)]^2}$. For $\alpha = 30°$, $L_u = 1$ m, $k/m = 10$ s^{-2} and $c/m = 0.8$ s^{-1}, we obtain $r_e \approx 0.51$ m, $\sigma = -0.4$ s^{-1}, and $\omega \approx 3.14$ rad/s. Such a motion is a damped oscillation with frequency $\omega \approx 3.14$ rad/s. For the initial conditions $r(0) = 0.86$ m and $\dot{r}(0) = 0$, as specified in the problem statement, the constants B_1 and B_2 are $B_1 = r(0) - r_e \approx 0.35$ m and $B_2 = [\dot{r}(0) - \sigma B_1]/\omega \approx 0.045$ m.

• **Coulomb friction:**

a. If the block does not move, the magnitude of the friction force for this case cannot be greater than its maximum value $\mu_s N = \mu_s mg \cos \alpha$. When the block moves, the expression for the friction force f shown in Fig. 2.10 is $f =$

$-\mu_k|N|\text{sgn}\,\dot{r} = -(\mu_k mg\cos\alpha)\text{sgn}\,\dot{r}$. The differential equation of motion for this case is then

$$\ddot{r} + \frac{k}{m}(r - L_u) + g\sin\alpha + \mu_k g(\cos\alpha)\text{sgn}\,\dot{r} = 0 \qquad (2.5.7\text{f})$$

Equation (2.5.7f) is nonlinear because $\text{sgn}\,\dot{r}$ is a nonlinear function of $\dot{r}$. Notice, however, that for each interval of the motion for which $\dot{r}$ does not change sign (i.e., $\text{sgn}\,\dot{r}$ is either $+1$ or -1) such a differential equation is linear and can be easily solved. Figure 2.11 shows how the resultant of the spring force and of the component of the gravitational force along the inclined plane, normalized by mg (i.e., $k(r - L_u)/(mg) + \sin\alpha$), change with the distance r(t).

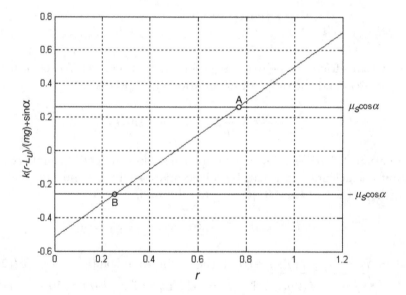

Figure 2.11: The force $k(r - L_u)/(mg) + \sin\alpha$ versus r for Example 2.5.7.

The block will not move in the region where $|k(r - L_u)/(mg) + \sin\alpha| \leq \mu_s\cos\alpha$ (i.e., region between points A and B in the figure) if one tries to start a motion in that region with $\dot{r} = 0$. The motion will be impending when $|k(r - L_u)/(mg) + \sin\alpha| = \mu_s\cos\alpha \approx 0.26$ for $\mu_s = 0.3$ and $\alpha = 30°$. For point A, $r \approx 0.764$ m, and for point B, $r \approx 0.255$ m. When the block starts to move, the magnitude of the friction force drops to $\mu_k mg\cos\alpha$ and remains essentially constant during the motion. For $r(0) = 0.86$ m, $k(r - L_u)/(mg) + \sin\alpha \approx 0.357$, which is greater than $\mu_s \cos\alpha \approx 0.26$. The block then starts to move down the inclined plane. The friction force is equal to $\mu_k mg \cos\alpha$, and the differential equation of motion for this case is then

$$\ddot{r} + \frac{k}{m}(r - L_u) + g\sin\alpha - \mu_k g(\cos\alpha) = 0 \qquad (2.5.7g)$$

With $r(0) = 0.86$ and $\dot{r}(0) = 0$, the solution to this equation is

$$r = L_u - (\sin\alpha - \mu_k \cos\alpha)mg/k + A\cos(\sqrt{k/m}\,t)$$

where $A = r(0) - [L_u - (\sin\alpha - \mu_k \cos\alpha)mg/k] \approx 0.138$ m.

The block "enters" the region AB shown in Fig. 2.11 and, since it does so with a nonzero velocity, it continues to move down the inclined plane. The friction force continues to be equal to $\mu_k mg \cos\alpha$, opposing the motion, and $\dot{r} = -A\sqrt{k/m}\sin(\sqrt{k/m}\,t) = 0$ when $\sqrt{k/m}\,t = \sqrt{10}\,t = \pi$ (i.e., when $t \approx 0.993$ s). The value of r at that time is $r = \text{constant} \approx 0.584$ m, and that is the equilibrium for this case of Coulomb friction.

c. Figure 2.12 shows plots of the solutions for $r(t)$ versus t for both cases of viscous friction and Coulomb friction, as determined earlier.

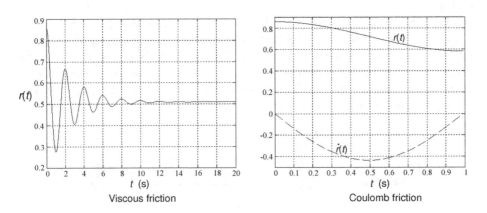

Figure 2.12: Solutions for $r(t)$ versus t for viscous friction and for Coulomb friction, for Example 2.5.7.

Example 2.5.8 involves a simple two-degree of freedom system (i.e., a system for which two variables that are independent of each other are needed to describe its motion) for which Newton's second law immediately yields the two differential equations of motion for the system.

Example 2.5.8 A small ball P, modeled as a mass particle, is thrown from a point A with an initial velocity as shown in Fig. 2.13. Neglecting air resistance, and assuming that the only force acting on the ball is the gravitational force of attraction of the earth, determine the value of v_0 for the ball to hit a target at point B shown in the figure. What is the maximum height reached by the ball?

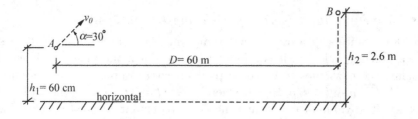

Figure 2.13: Figure for Example 2.5.8.

■ **Solution**

Because of the small distances involved, the gravitational force acting on particle P is approximated as a constant (i.e., constant direction and constant magnitude). The free-body diagram for P is shown in Fig. 2.14. Two variables are needed to describe the motion, and rectangular coordinates for point P are the most convenient ones for this problem. By arbitrarily choosing point A to be the inertial reference point, those variables are the distances $x(t)$ and $y(t)$ shown in Fig. 2.14.

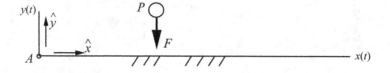

Figure 2.14: Free-body diagram for Example 2.5.8.

In terms of the unit vectors $\hat{x}$ and $\hat{y}$ shown in Fig. 2.14, we then have

Resultant force acting on P : $\qquad \vec{F} = -mg\hat{y}$

Absolute position vector for the mass particle P : $\qquad \vec{r} = x\hat{x} + y\hat{y}$

Absolute velocity of P :
$$\vec{v} = \frac{d\vec{r}}{dt} = \dot{x}\hat{x} + \dot{y}\hat{y}$$

Absolute acceleration of P :
$$\vec{a} = \frac{d\vec{v}}{dt} = \ddot{x}\hat{x} + \ddot{y}\hat{y}$$

Newton's second law $\vec{F} = m\vec{a}$ then yields the following differential equations that govern the motion of the ball:

$$\ddot{x} = 0 \qquad \text{and} \qquad \ddot{y} = -g$$

The solutions to these simple differential equations are

$$\dot{x} = \text{constant} = \dot{x}(0) = v_0 \cos\alpha \qquad x = (v_0 \cos\alpha)t + x(0)$$
$$\dot{y} = \dot{y}(0) - gt = v_0 \sin\alpha - gt \qquad y = (v_0 \sin\alpha)t - \frac{gt^2}{2} + y(0)$$

With $x(0) = y(0) = 0$, these solutions yield, with $x = 60$ m, $y = 2.6 - 0.6 = 2$ m, and $\alpha = 30°$,

$$v_0 \approx 26.9 \text{ m/s}$$

The maximum value of y (i.e., the maximum height) occurs when $\dot{y} = 0$; the corresponding value of $y(t)$ is maximum because $\ddot{y} = -g$ is negative. Since $\dot{y} = v_0 \sin\alpha - gt$, the maximum height is reached when $t = (v_0 \sin\alpha)/g \approx 1.37$ s, and this gives

$$y_{\max} \approx 9.2 \text{ m}$$

$\implies$ The Scilab program *ballistics.sci* and the Matlab program *ballistics.m* listed in Appendix E are functions (i.e., equivalent to a subroutine in Fortran) that numerically integrate the differential equations of motion using Scilab and Matlab, respectively. Both animate the motion. You can change them to include the effect of aerodynamic drag on the body (see Problem 2.19 in Section 2.9). All the programs listed in Appendix E are functions, and an example of their use is given in the comment lines at the beginning of each one of them. For the function *ballistics*, one simply types either in the Scilab console or in the Matlab command window (with Δt=0.01 in Scilab, and 0.002 in Matlab, say)
`my_output = ballistics([0:`Δ`t:2.5], 0, -9.81, 0, 10, 0, 10);`
This integrates the differential equations with the initial conditions $x(0) = 0$, $\dot{x}(0) = 10$, $y(0) = 0$, and $\dot{y}(0) = 10$. The speed of the animation depends on the step size of the time array `t=[0:`Δ`t:2.5]` and on the CPU "speed" of your computer. Different step sizes for Scilab and Matlab are suggested because of the different execution speeds of the two software.

> **NOTE**: The value of Δt to choose for an animation depends on the speed of your computer. For slowing down an animation, decrease the suggested values given in this book. Conversely, increase them if you want to speed up the animation.

Those who have Simulink and prefer to use it may want to construct the block diagram model[3] shown in Fig. 2.15, setting the *Simulation* → *Parameters* menu

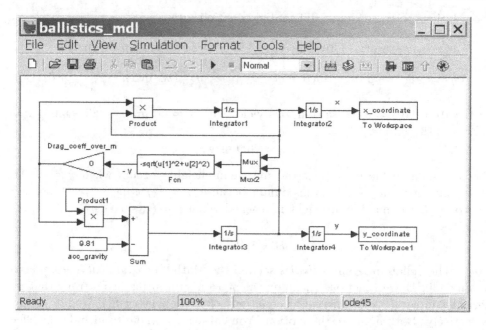

Figure 2.15: A Simulink block diagram for the system of Example 2.5.8.

for this model with time t of the form $[0:\Delta t:2.5]$ and using the same initial conditions mentioned in the previous paragraph. The blocks labeled *To Workspace* and *To Workspace1* send the arrays *x_coordinate* and *y_coordinate*, which store the x- and y-coordinates of point P shown in Fig. 2.14, to the Matlab workspace. After the simulation stops, an animation of the motion is generated by typing

---

[3]The Simulink block diagrams shown in the book may also be obtained from the Dover website mentioned on pp. 596 and 645. **Those who do not like to deal with block diagrams may ignore the references to Simulink in this book.**

animate_nbars(x_coordinate, y_coordinate) in the Matlab workspace. For additional details, see the tutorials in Appendix A.

Example 2.5.9 involves a more elaborate two-degree of freedom system that consists of a block sliding on a sliding block. For this system, unknown reaction forces have to be eliminated from the equations that are obtained from Newton's second law to obtain the two differential equations of motion for the system.

Example 2.5.9 A block P_2 of mass m_2 slides on a face of a block P_1, which is on an inclined plane ABC, of mass m_1. Block P_1 is able to slide on a horizontal surface as shown in Fig. 2.16. Both blocks are constrained to move on the same vertical plane, which is the plane of the paper in the figure. Friction in the system is viscous, with friction coefficients c_2 between P_2 and P_1, and c_1 between P_1 and the horizontal surface. The spring DE connecting P_2 to P_1 is linear, with unstressed length L_u m and stiffness k N/m. Obtain the differential equations of motion for the system in terms of the variables $x(t)$ and $r(t)$ shown in the figure and integrate them numerically, animating the motion of the system. Notice that the blocks translate only, as if they were point masses.

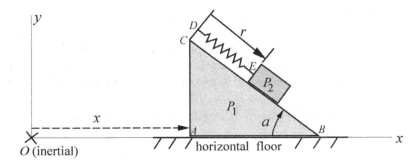

Figure 2.16: A block sliding on a sliding block for Example 2.5.9.

■ Solution

The free-body diagrams for P_1 and P_2 are shown in Fig. 2.17. The position vectors chosen to track the bodies, and two sets of orthogonal unit vectors $\{\hat{x}_1, \hat{x}_2\}$ and $\{\hat{y}_1, \hat{y}_2\}$, are also shown in that figure. The analysis begins by writing expressions for the absolute position, the absolute acceleration, and the resultant force acting on each block. By inspecting Fig. 2.17, the following expressions are written

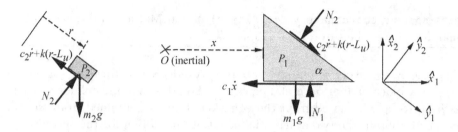

Figure 2.17: Free-body diagrams and unit vectors for Example 2.5.9.

For block P_1:

$$\vec{r}_1 \;=\; x\hat{x}_1 \qquad \vec{a}_1 = \frac{d^2\vec{r}_1}{dt^2} = \ddot{x}\hat{x}_1 \tag{2.5.9a}$$

$$\vec{F}_1 \;=\; (N_1 - m_1 g)\hat{x}_2 - c_1\dot{x}\hat{x}_1 + [c_2\dot{r} + k(r - L_u)]\,\hat{y}_1 - N_2\hat{y}_2 \tag{2.5.9b}$$

For block P_2:

$$\vec{r}_2 \;=\; x\hat{x}_1 + |\overrightarrow{AC}|\hat{x}_2 + r\hat{y}_1 \quad ; \quad \vec{a}_2 = \frac{d^2\vec{r}_2}{dt^2} = \ddot{x}\hat{x}_1 + \ddot{r}\hat{y}_1 \tag{2.5.9c}$$

$$\vec{F}_2 \;=\; N_2\hat{y}_2 - m_2 g\hat{x}_2 - [c_2\dot{r} + k(r - L_u)]\,\hat{y}_1 \tag{2.5.9d}$$

Notice that no matter which set of unit vectors is used in this problem, i.e., either $\{\hat{x}_1, \hat{x}_2\}$ or $\{\hat{y}_1, \hat{y}_2\}$, a transformation between these two sets will always be required to extract the differential equations of motion from Newton's second law for each block.

From $\vec{F}_1 = m_1\vec{a}_1$ and $\vec{F}_2 = m_2\vec{a}_2$, the following scalar equations are obtained if we choose to work with the unit vector set $\{\hat{x}_1, \hat{x}_2\}$ for block P_1 and $\{\hat{y}_1, \hat{y}_2\}$ for block P_2.

For block P_1:

$$m_1\ddot{x} = -c_1\dot{x} + [c_2\dot{r} + k(r - L_u)]\cos\alpha - N_2\sin\alpha \tag{2.5.9e}$$

$$N_1 - m_1 g - [c_2\dot{r} + k(r - L_u)]\sin\alpha - N_2\cos\alpha = 0 \tag{2.5.9f}$$

For block P_2:

$$m_2\left(\ddot{r} + \ddot{x}\cos\alpha\right) = m_2 g\sin\alpha - c_2\dot{r} - k(r - L_u) \tag{2.5.9g}$$

$$m_2\ddot{x}\sin\alpha = N_2 - m_2 g\cos\alpha \tag{2.5.9h}$$

Notice that $x(t)$ does not appear by itself in Eqs. (2.5.9e) to (2.5.9h) and it is irrelevant where the inclined plane ABC is located when one starts the motion (see Fig. 2.16).

Equation (2.5.9g) does not involve any of the unknown reactions N_1 or N_2 and, therefore, it is one of the differential equations of motion for the system. The second differential equation that governs the motion of the system is obtained by combining Eqs. (2.5.9e) and (2.5.9h) to eliminate the unknown reaction N_2. This yields the following differential equation:

$$(m_1 + m_2 \sin^2 \alpha)\ddot{x} + c_1\dot{x} - [c_2\dot{r} + k(r - L_u)]\cos \alpha + m_2 g(\sin \alpha)\cos \alpha = 0 \quad (2.5.9i)$$

Equations (2.5.9g) and (2.5.9i) are two coupled linear differential equations with constant coefficients, and they can be solved analytically. The solution to a set of coupled linear differential equations is shown in Section 1.16 of Chapter 1 and in an example in Chapter 8; it is a natural extension of the material that was presented under Case 3 in Section 1.13 of Chapter 1.

Equations (2.5.9g) and (2.5.9i) are solved numerically here. For this to be done, numerical integration routines require that the expressions for $\ddot{x}$ and $\ddot{r}$ first be written in terms of $x, r, \dot{x}$, and $\dot{r}$. Inspection of Eq. (2.5.9i) discloses that $\ddot{x}$ is already in the required form. The required form for $\ddot{r}$ is then obtained by simply substituting into Eq. (2.5.9g) the expression for $\ddot{x}$ given by Eq. (2.5.9i). Take the time to verify that the resulting expression that is obtained for $\ddot{r}$ is

$$\left(m_1 + m_2 \sin^2 \alpha\right)\ddot{r} = c_1\dot{x}\cos \alpha + (m_1 + m_2)\left\{-\left[\frac{c_2\dot{r} + k(r - L_u)}{m_2}\right] + g\sin \alpha\right\}$$

$$(2.5.9j)$$

$\Longrightarrow$ For Scilab users, the program *inclined_plane.sci* listed in Appendix E integrates Eqs. (2.5.9i) and (2.5.9j), and animates the motion. For Matlab users, the program that does the same thing is *inclined_plane.m*, which is also listed in Appendix E. In the animated figure, the length AB is set at 3 length units (3 m when g is in m/s^2). An example of their use as given in the comment lines at the beginning of each program, using the initial conditions $x(0) = 1$, $\dot{x}(0) = 0$, $r(0) = 2$, and $\dot{r}(0) = 0$ is[4]

```
my_output = inclined_plane([0 : 0.01 : 20], 0.1, 5, 0, 0, %pi/4, 9.81, 1, 1, 0, 2, 0);
```

The variable my_output is a *len* by 5 matrix whose columns are $t, x, \dot{x}, r$, and $\dot{r}$, where *len* is the number of time points in the array $[0 : 0.01 : 20]$. It is of interest

---

[4]This illustration is for Scilab, where %pi denotes π. For Matlab, replace %pi with pi in the same command, and a step size $\Delta t = 0.005$ is suggested for the time array.

only if one wishes to generate plots, such as plot(my_output(:, 1), my_output(:, 2)) to plot x versus t, for example. Change the step size of the array $[0 : 0.01 : 20]$ if the animation runs too fast or too slow on your computer.

Those who have Simulink and prefer to use it may want to construct the block diagram model shown in Fig. 2.18, setting the *Simulation* → *Parameters* menu for

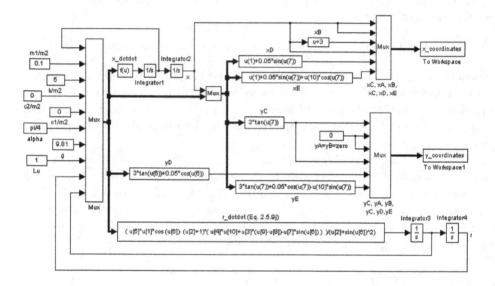

Figure 2.18: A Simulink model for Example 2.5.9. The expression for $\ddot{x}$ in the $f(u)$ block labeled x_dotdot is (-u[5]*u[1]+(u[4]*u[10]+u[3]*(u[9]-u[8])-u[7]*sin(u[6]))*cos(u[6]))/(u[2]+sin(u[6])^2).

this model with time t of the form $[0:\Delta t:20]$, and using the same initial conditions mentioned in the previous paragraph. The blocks labeled *To Workspace* and *To Workspace1* send the arrays *x_coordinates* and *y_coordinates*, which contain the x- and y-coordinates of points A, B, C, D, and E shown in Fig. 2.16, to the Matlab workspace. If you want, click on the tab that says *Workspace I/O* and put a check mark on the little square that says *Time*, if it is not already checked by default, so that the model also sends the time to the workspace. After the simulation stops, an animation of the motion is generated by typing

animate_nbars(x_coordinates, y_coordinates)

in the Matlab command window.

It is suggested that you use numerical integration (with any software you prefer to work with) to investigate the effect of changes in the input parameters $m_1/m_2, k/m_2$, etc., on the motion of the system. Choose your own values for those parameters and initial conditions for the motion. For your chosen values, what is the equilibrium displacement for $r(t)$ and the frequencies of the oscillations? Obtain them from plots of $r(t)$ and $x(t)$ versus time t.

Suggested Problem
For this example, verify, analytically, that the quantity

$$E \overset{\Delta}{=} (m_1 + m_2)\frac{\dot{x}^2}{2} + m_2\left[\frac{\dot{r}^2}{2} + \dot{x}\dot{r}\cos\alpha\right] - m_2 g r \sin\alpha + \frac{k}{2}(r - L_u)^2$$

remains constant during the motion if $c_1 = c_2 = 0$. *Hint*: Take the time derivative of E and use the expressions for $\ddot{x}$ and $\ddot{r}$ from the differential equations of motion obtained in this example to show that $\dot{E} = 0$. The quantity E happens to be equal to the sum of the *kinetic energy* and the *potential energy* of the motion. These quantities are introduced in Section 2.8.

Example 2.5.10 illustrates the degree of difficulty associated with different choices of variables for analyzing a dynamics problem. It also shows that the number of mathematical manipulations in solving a problem depends on the choice of unit vector triad one decides to work with.

Example 2.5.10 Consider the pendulum discussed in Section 1.11 of Chapter 1 and shown in Fig. 1.11. For convenience, that figure is repeated here as Fig. 2.19. Free-body diagrams for different parts of the system are shown in separate parts of the figure (explanations about the diagrams are given in the solution).

The pendulum consists of a small body (modeled as a point mass) of mass $m = 0.2$ kg attached to one end of a bar AB of length $L = 60$ cm, which is approximated as being massless.[5] The other end of the bar is attached to an inertial frictionless pin. The motion of the system takes place on a vertical plane, and the only forces acting on the point mass are due to the gravitational attraction of the earth and the force due to contact with the bar. The bar is subjected to a known moment M_{bar}, which is applied by a motor that is not shown in the figure.

a. Obtain the differential equation of motion for the pendulum in terms of the angle θ. What is the advantage of working with the angle θ, instead of either x or y shown in Fig. 2.19?

---

[5]Dynamics of rigid bodies, taking their distributed mass into account, is treated in Chapters 5 and 6.

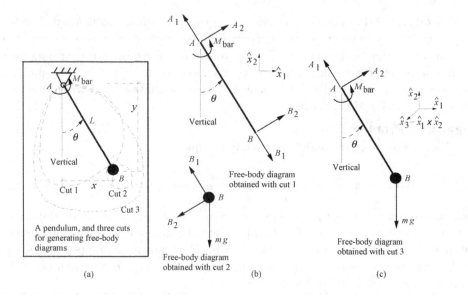

Figure 2.19: Free-body diagrams for a pendulum with a massless bar for Example 2.5.10.

b. Obtain the equations for calculating the reaction forces A_1, A_2, B_1, and B_2 shown in the free-body diagrams, and explain how to calculate them using such equations.

c. For $M_{\text{bar}} = 0$, determine the maximum value $\theta_{\max}$ of the angle θ if the motion is started with the initial conditions $\theta(0) = 0$ and $\dot{\theta}(0) = 2$ rad/s.

d. A sensor in the system measures the angle $\theta(t)$, and the signal that is generated is input to the motor at A so that $M_{\text{bar}} = -K\theta(t)$, where $K = 4$ N·m/rad. For the same initial conditions $\theta(0) = 0$ and $\dot{\theta}(0) = 2$ rad/s that were used in part c, determine the new value of $\theta_{\max}$.

e. With $M_{\text{bar}} = 0$ and $\theta(0) = 0$, plot the nondimensional angular velocity $\dot{\theta}(t)\sqrt{L/g}$ versus $\theta(t\sqrt{g/L})$ (using $-360 \le \theta \le 360°$) for $\dot{\theta}(0)\sqrt{L/g} = 1.5$, 2, and 2.5. Superimpose all three plots in the same graph, and describe the motion in a few words based on the observation of the plots.

■ Solution

a. For this problem, point B moves along a circle centered at the inertial point A (see Fig. 2.19) and the absolute velocity of point B is simply $L\dot{\theta}$ directed along the perpendicular to bar AB. If, instead of working with the angle θ, we work with the variable x and also with the unit vectors $\hat{x}_1$ and $\hat{x}_2$ shown in

Fig. 2.19, inspection of that figure discloses that the expression for the absolute position vector of B is $\vec{r} = \overrightarrow{AB} = x\hat{x}_1 - \sqrt{L^2 - x^2}\,\hat{x}_2$. With such choices, the expression for the absolute velocity of B becomes $\vec{v} = \dot{x}\hat{x}_1 + (x\dot{x}/\sqrt{L^2 - x^2})\hat{x}_2$. Such an expression is much more complicated than the one written in terms of the angle θ. This will yield an even more complicated expression for the acceleration $\vec{a} = d\vec{v}/dt$ if we work with the variable x, and, in such a case, the resulting differential equation of motion, obtained from $\vec{F} = m\vec{a}$, will be much more difficult to deal with than the differential equation written in terms of the angle θ. Also notice that the expression for the velocity in terms of the variable x will have a division-by-zero problem if x approaches L (i.e., if bar AB approaches the horizontal). The first problem alone, namely, the additional complexity of the differential equation, gives us enough justification to work with the variable θ. The same difficulties previously mentioned will occur if we choose to work with the variable y.

Three cuts are shown in Fig. 2.19a. The cut identified as Cut 1 allows us to see the internal reaction forces acting on the bar, so that we can answer the questions asked in part b about those forces. Cut 2 allows us to draw the free-body diagram for the point mass located at the end B of the bar. The free-body diagrams for the two bodies that compose the pendulum are shown in Fig. 2.19b. The two diagrams involve five unknowns: θ, A_1, A_2, B_1, and B_2. To determine all these unknowns, five equations are then needed. For a two-dimensional problem, $\vec{F} = d(m\vec{v})/dt$ for a point mass yields two scalar equations. The other three equations for a massless body are obtained as indicated in the following box (the dynamics of translating and rotating rigid bodies, taking into account their masses, is dealt with in Chapters 5 and 6).

For a *massless body* subjected to an arbitrary number n of forces $\vec{F}_1, \vec{F}_2, \ldots, \vec{F}_n$, the same equations of statics apply (even though the acceleration of points of the body is not zero), i.e.,

Summation of forces $= 0$ (in two dimensions this yields two equations)

and

Summation of moments (about any point) $= 0$ (in two dimensions this yields one equation)

Now, why is the free-body diagram for Cut 3 shown in Fig. 2.19c? Strictly speaking, neither that cut nor the free-body diagram for that cut are actually needed because all the unknowns for this problem already appear in the free-body diagrams for Cuts 1 and 2. However, Cut 3 happens to be the most convenient cut for obtaining the differential equation of motion of the pendulum. The reason for this is that the moment about point A of all the unknown reaction forces that appear in the free-body diagram for that cut is zero. Since point A is inertial, we can apply the moment equation, Eq. (2.3.1), to the free-body diagram for Cut 3 to obtain the differential equation of motion for the pendulum after performing only a line or so of mathematical manipulations. For this problem, Eq. (2.3.1) becomes $\vec{M}_A = md(\overrightarrow{AB} \times \vec{v})/dt$. We could work only with the free-body diagram for Cut 3 if the reaction forces A_1, A_2, B_1, and B_2 were not asked for.

Before we proceed to obtain the equations that will allow us to calculate the four reaction forces A_1, A_2, B_1, and B_2, let us see how the free-body diagram for Cut 3 yields the differential equation of motion with just a few mathematical manipulations.

The moment $\vec{M}_A$ is simply (see Fig. 2.19, and use the right-hand rule)

$$\vec{M}_A = M_{\text{bar}}\hat{x}_3 - (mgL\sin\theta)\hat{x}_3$$

Since (also, see Fig. 2.19, and use the right-hand rule)

$$\overrightarrow{AB} \times \vec{v} = L(L\dot{\theta})\hat{x}_3 = L^2\dot{\theta}\hat{x}_3$$

the equation $\vec{M}_A = md(\overrightarrow{AB} \times \vec{v})/dt$ yields the following differential equation of motion:

$$\ddot{\theta} + \frac{g}{L}\sin\theta = \frac{M_{\text{bar}}}{mL^2} \qquad (2.5.10a)$$

This is the equation that was asked for.

b. Let us now generate the equations for calculating the reaction forces A_1, A_2, B_1, and B_2. Referring to Fig. 2.19 we have, by working with the nonrotating unit vectors $\hat{x}_1$ and $\hat{x}_2$:

■ Resultant force acting on the mass particle at B (see the free-body diagram for Cut 2):

$$\vec{F} = (B_1\cos\theta - B_2\sin\theta - mg)\hat{x}_2 - (B_1\sin\theta + B_2\cos\theta)\hat{x}_1$$

■ Absolute position vector of point B:

$$\vec{r} = \overrightarrow{AB} = L(\hat{x}_1\sin\theta - \hat{x}_2\cos\theta)$$

■ Absolute velocity of B:

$$\vec{v} = \frac{d\vec{r}}{dt} = L\dot{\theta}(\hat{x}_1 \cos\theta + \hat{x}_2 \sin\theta)$$

■ Absolute acceleration of B:

$$\vec{a} = \frac{d\vec{v}}{dt} = L(\ddot{\theta}\cos\theta - \dot{\theta}^2 \sin\theta)\hat{x}_1 + L(\ddot{\theta}\sin\theta + \dot{\theta}^2 \cos\theta)\hat{x}_2$$

Newton's second law $\vec{F} = m\vec{a}$ then yields the following scalar equations for the point mass at B:

$$mL(\ddot{\theta}\cos\theta - \dot{\theta}^2 \sin\theta) = -B_1 \sin\theta - B_2 \cos\theta \qquad (2.5.10b)$$
$$mL(\ddot{\theta}\sin\theta + \dot{\theta}^2 \cos\theta) = B_1 \cos\theta - B_2 \sin\theta - mg \qquad (2.5.10c)$$

For the massless bar AB, let us take moments about point A since this results in one simple equation that involves only the unknown force B_2. By equating to zero the summation of the forces acting on bar AB, and the summation of the moments about point A, the following equations are obtained for the free-body diagram for Cut 1.

$$A_1 - B_1 = 0 \qquad (2.5.10d)$$
$$A_2 + B_2 = 0 \qquad (2.5.10e)$$
$$M_{\text{bar}} + B_2 L = 0 \qquad (2.5.10f)$$

Notice that the moment equation, Eq. (2.5.10f), only involves one unknown because we chose to work with force components in the directions parallel and perpendicular to bar AB and took moments about point A. Because of that choice, three of the unknown reaction forces pass through point A and, thus, their moment about A is equal to zero. We could also have taken moments about point B of the massless bar to obtain a simple equation that would only involve the unknown A_2, which is $A_2 = M_{\text{bar}}/L$.

The five equations that will answer all the questions that were asked for in parts b to d are Eqs. (2.5.10b) through (2.5.10f) or, better still, Eqs. (2.5.10a), (2.5.10b) [or (2.5.10c)], (2.5.10d), (2.5.10e), and (2.5.10f).

Having the differential equation of motion, we can calculate the reaction forces and also answer the question in parts b to d. The reaction forces B_2 and A_2 are simply determined from Eqs. (2.5.10e) and (2.5.10f) as $B_2 = -A_2 = -M_{\text{bar}}/L$. To calculate the reaction forces A_1 and B_1, the nonlinear differential equation of motion, Eq. (2.5.10a), has to be integrated first to obtain $\theta(t)$. This can be done numerically. Once this is done, the reaction force B_1 can be calculated by using either Eq. (2.5.10b) or Eq. (2.5.10c). The reaction force A_1 is then determined from Eq. (2.5.10d) as $A_1 = B_1$. The numerical integration of the differential equation and the calculation of the reaction forces is a relatively simple process, as illustrated earlier.

c, d. Let us now determine the maximum value of the angle $\theta(t)$, as asked for in parts c and d. Since part c is the same as part d with $K = 0$, let us find the value for part d first.

With $M_{\text{bar}} = -K\theta$, integration of Eq. (2.5.10a) once with respect to θ yields, according to the analytical steps presented in Section 1.13 of Chapter 1 (see Case 4, p. 48):

$$\frac{\dot{\theta}^2}{2} - \frac{g}{L}\cos\theta + \frac{K\theta^2}{2mL^2} = \text{constant} = \frac{\dot{\theta}^2(0)}{2} - \frac{g}{L}\cos\theta(0) + \frac{K\theta^2(0)}{2mL^2}$$

$$(2.5.10\text{g})$$

The maximum and minimum values of θ are obtained by setting $\dot{\theta} = 0$ in Eq. (2.5.10g). With $\theta(0) = 0$, $\dot{\theta}(0) = 2$ rad/s, $m = 0.2$ kg, $L = 0.6$ m, and $g = 9.81$ m/s^2, we obtain

$$\frac{\dot{\theta}^2}{2} - \frac{9.81}{0.6}\cos\theta + \frac{K\theta^2}{2 \times 0.2 \times 0.6^2} = \frac{2^2}{2} - \frac{9.81}{0.6}\cos(0) = -14.35 \quad (2.5.10\text{h})$$

■ $K = 0$. For this case, Eq. (2.5.10h) yields $\theta \approx \pm 0.5$ rad $\approx \pm 28.6°$ when $\dot{\theta} = 0$. From Eq. (2.5.10a), it follows that $\ddot{\theta} = -g(\sin\theta)/L - K\theta/(mL^2)$ is negative when $\theta = 28.6°$ (thus, a maximum value for θ) and positive when $\theta = -28.6°$ (thus, a minimum value for θ). Therefore, $\theta_{\max} = 28.6°$, $\theta_{\min} = -\theta_{\max} = -28.6°$, and θ oscillates, in this case, between $\theta_{\min}$ and $\theta_{\max}$.

■ $K = 4$ N·m/rad. For this case, the values of θ for which $\dot{\theta} = 0$ are obtained from Eq. (2.5.10h), after a few iterations with a calculator, as $\theta \approx \pm 0.236$ rad $\approx \pm 13.5°$. Again, from $\ddot{\theta} = -g(\sin\theta)/L - K\theta/(mL^2)$, it follows that $\ddot{\theta}$ is negative when $\theta = 13.5°$, and positive when $\theta = -13.5°$. Therefore, $\theta_{\max} = 13.5°$, and $\theta_{\min} = -\theta_{\max} = -13.5°$. In this case, θ oscillates between these two values.

e. The plot of the nondimensional angular velocity $d\theta/d\tau = \dot{\theta}(t)\sqrt{L/g}$, where $\tau = t\sqrt{g/L}$ is nondimensional time, versus $\theta(\tau)$ for any value of the nondimensional quantity $K/(mgL)$, is generated by using Eq. (2.5.10g). For this, that equation is first rearranged as

$$\frac{L}{g}\frac{\dot{\theta}^2}{2} - \cos\theta + \frac{K\theta^2}{2mgL} = \text{constant} \triangleq E = \frac{L}{g}\frac{\dot{\theta}^2(0)}{2} - \cos\theta(0) + \frac{K\theta^2(0)}{2mgL}$$

The plots for $K = 0$ (i.e., $M_{\text{bar}} = 0$), for three values of E, namely, $E = 0.125$, $E = 1$, and $E = 2.125$, are shown in Fig. 2.20. Although the plot can easily be done by hand, Fig. 2.20 was generated with Scilab. It can also be easily generated with Matlab.[6]

Inspection of Fig. 2.20, discloses that the motion is not oscillatory when $E > 1$, with the value of the constant E being determined by the initial conditions $\theta(0)$ and $\sqrt{L/g}\dot{\theta}(0)$. For $K = 0$, the pendulum continuously rotates around its pivot point A (see Fig. 2.19, p. 130) when $E > 1$.

Note

Look at Eqs. (2.5.10b) and (2.5.10c) for the point mass at B. Both of those equations involve $\ddot{\theta}$, the reaction $B_2 = -M_{\text{bar}}/L$, and the unknown reaction B_1. More mathematical manipulations are now required to eliminate B_1 to obtain the differential equation of motion for the system. This can be done by multiplying the first equation by $\cos\theta$, the second by $\sin\theta$, and then adding the results to finally obtain the differential equation of motion for this system, which is Eq. (2.5.10a) that was obtained earlier by a simpler procedure.

These extra manipulations were necessary to "extract" the differential equation of motion from Eqs. (2.5.10b) and (2.5.10c) because $\ddot{\theta}$ appeared in both components

---

[6]The following Scilab commands were used. For explanations, see the tutorials in Appendix A.

```
th=-400:1:400; thprime1 = sqrt(2 * (cos(%pi * th/180) + 0.125));
thprime2 = sqrt(2 * (cos(%pi * th/180) + 1));thprime3 = sqrt(2 * (cos(%pi * th/180) + 2.125));
plot(th,thprime1,th,-thprime1,th,thprime2,th,-thprime2,th,thprime3,th,-thprime3);
xgrid; xlabel(['$\theta$', ' (degrees)'], 'fontsize', 4);
ylabel(['$d\theta/d\tau=(d\theta/dt)*\sqrt(L/g)$'], 'fontsize', 4);
xstring(-42,2.51,'$E=2.125$'); xstring(-18,2,'1'); xstring(-38,1.45,'0.125');
xstring(-18,-2.3,'1'); xstring(-48,-2.75,'2.125');
```

The equivalent Matlab commands are:
```
th=-400:1:400; thprime1 = sqrt(2 * (cos(pi * th/180) + 0.125));
thprime2 = sqrt(2 * (cos(pi * th/180) + 1));thprime3 = sqrt(2 * (cos(pi * th/180) + 2.125));
plot(th,thprime1,th,-thprime1,th,thprime2,th,-thprime2,th,thprime3,th,-thprime3);
axis([-400 400 -3 3]); grid; a=gca;
set(a, 'ytick', [-3, -2.5, -2, -1.5, -1, -0.5, 0, 0.5, 1, 1.5, 2, 2.5, 3]);
xlabel('\theta (degrees)', 'fontsize', 12);
ylabel('d\theta/d\tau=(d\theta/dt)*\surd(L/g)', 'fontsize', 12);
text(-42,2.6, '{\it E}=2.125'); text(-18,2.1, '1'); text(-38,1.6, '0.125');
text(-18,-2.15, '1'); text(-48,-2.6, '2.125');
```

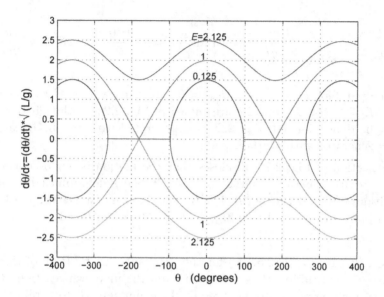

Figure 2.20: $d\theta/d\tau$ versus θ for a simple pendulum $[d^2\theta/d\tau^2 + \sin\theta = 0]$.

of the expression that was obtained for the acceleration $\vec{a}$. There would be no need for those extra steps if we had chosen to work with two orthogonal unit vectors where $\ddot{\theta}$ appears only in one of the components of $\vec{a}$. Such unit vectors are actually the ones parallel and perpendicular to bar AB for this problem. Since such unit vectors are rotating, it is then necessary to learn how to obtain the time derivative of rotating unit vectors when one decides to work with them. For a variety of problems, working with rotating unit vectors is advantageous because the expressions for velocity and acceleration turn out to be simpler and the number of manipulations to obtain the differential equations of motion can be reduced, even substantially in many cases.[7] The problem of determining the time derivative of rotating unit vectors is addressed in Section 2.6.

---

[7]This should be expected because, as mentioned earlier, point B in Fig. 2.19 simply moves along a circle centered at point A. This can also be seen by looking at the expression $\vec{v} = L\dot{\theta}(\hat{x}_1 \cos\theta + \hat{x}_2 \sin\theta)$ that was obtained for the absolute velocity of point B and noticing that $\hat{x}_1 \cos\theta + \hat{x}_2 \sin\theta$ is the unit vector normal to line AB (and pointing in the same direction of the vector labeled A_2 shown in Fig. 2.19). If we denote such a unit vector by $\hat{n}$, say, the absolute velocity of point B is simply $\vec{v} = L\dot{\theta}\hat{n}$, which gives $\vec{a} = d\vec{v}/dt = L\ddot{\theta}\hat{n} + L\dot{\theta}(d\hat{n}/dt)$. But, now, one needs to know how to determine the time derivative of a rotating unit vector.

Example 2.5.11 To investigate the translational motion of a satellite subjected to the gravitational attraction of a planet, the satellite is modeled as a particle P of mass m, while the planet, whose mass M is much bigger than that of the satellite, is treated as a spherical inertial body centered at a point O and with radius R. If the only forces acting on this two-body system are due to their mutual attraction, show that the motion of P in the three-dimensional space x-y-z shown in Fig. 2.21 takes place in a single plane. The motion of P is started with the initial conditions $\vec{r}(0) = \vec{r}_0$ and $\vec{v}(0) = \vec{v}_0$, where $\vec{r}(t) = \overrightarrow{OP}$ and $\vec{v} = d\vec{r}/dt$ (see Fig. 2.21).

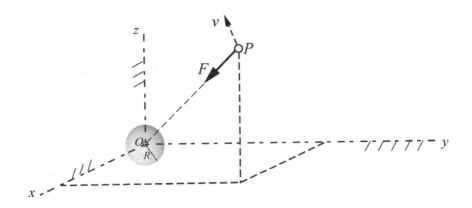

Figure 2.21: Free-body diagram for a body P subjected to a central gravitational force for Example 2.5.11.

■ **Solution**

The solution to this problem is strikingly simple if you notice that the moment about the inertial point O of the force $\vec{F}$ acting on P is equal to zero, i.e., $\vec{M}_O = \vec{r} \times \vec{F} = 0$. Therefore, as shown in Section 2.3,

$$\vec{H}_O = m\vec{r} \times \vec{v} = \text{constant} = m\vec{r}(0) \times \vec{v}(0) = m\vec{r}_0 \times \vec{v}_0$$

Since the cross product $\vec{r} \times \vec{v}$ is a vector that is normal to the plane defined by $\vec{r}$ and $\vec{v}$, this equation discloses that the normal to that plane does not change direction. Therefore, this proves that the motion of P takes place in a single plane in space.

This conclusion is a classical result in dynamics, and the system considered here is a model for the motion of any planet in our solar system around the sun (which is a good model since the mass of the sun is more than 700 times greater than the mass of all the planets together) or for the motion of an artificial satellite around a spherical celestial body. A detailed study of the two-body system is presented in Chapter 4.

2.6   Velocity and Acceleration Again and Angular Velocity

Let us now look at the problem of obtaining expressions for velocity and acceleration using rotating unit vectors. As indicated by Eq. (2.4.4) in Section 2.4, which is repeated below as Eq. (2.6.1), this requires knowing how to calculate the time derivative of a unit vector.

$$
\begin{aligned}
\frac{d\vec{r}}{dt} &= \dot{r}_{x_1}\hat{x}_1 + \dot{r}_{x_2}\hat{x}_2 + \dot{r}_{x_3}\hat{x}_3 + r_{x_1}\frac{d\hat{x}_1}{dt} + r_{x_2}\frac{d\hat{x}_2}{dt} + r_{x_3}\frac{d\hat{x}_3}{dt} \\
&\triangleq \left[\frac{d\vec{r}}{dt}\right]_{\{x\}} + r_{x_1}\frac{d\hat{x}_1}{dt} + r_{x_2}\frac{d\hat{x}_2}{dt} + r_{x_3}\frac{d\hat{x}_3}{dt}
\end{aligned}
\tag{2.6.1}
$$

The derivative of a rotating unit vector $\hat{x}$ (or of any vector whose magnitude is constant) is a vector that is always perpendicular to the unit vector $\hat{x}$. This can be shown by making use of the fact that the dot product $\vec{x} \bullet \vec{x}$ is constant; specifically,

$$
\hat{x} \bullet \hat{x} = 1
$$

By taking the time derivative of this equation and by noticing that $(d\hat{x}/dt) \bullet \hat{x} = \hat{x} \bullet (d\dot{x}/dt)$, we obtain

$$
\frac{d\,(\hat{x} \bullet \hat{x})}{dt} = 2\frac{d\hat{x}}{dt} \bullet \hat{x} = 0
$$

which proves that $d\hat{x}/dt$ is normal to $\hat{x}$, since two vectors are normal to each other when their dot product is equal to zero.

To develop the general expression for the time derivative of any unit vector, consider a triad $\{\hat{x}, \hat{y}, \hat{z} \triangleq \hat{x} \times \hat{y}\}$ rotating about a line AA as shown in Fig. 2.22. To clearly specify directions, we need to bring a reference frame into the picture. For this, let the orientation of line AA be fixed in some reference frame, which, for now, is arbitrarily chosen. Such a reference frame is represented by the orthogonal lines labeled X_0, Y_0, and Z_0 in the figure. Figure 2.22a illustrates the two-dimensional case, where the axis about which the triad is rotating is perpendicular to the direction of one of the unit vectors ($\hat{z}$ in this illustration). Figure 2.22b illustrates the general case of three-dimensional motion.

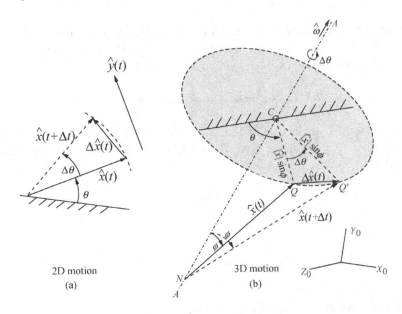

Figure 2.22: Time derivative of a unit vector and the angular velocity vector $\vec{\omega} = \dot{\theta}\hat{\omega}$.

To determine the time derivative of any of the unit vectors of the triad $\{\hat{x}, \hat{y}, \hat{z} \overset{\Delta}{=} \hat{x} \times \hat{y}\}$, in the chosen reference frame, let us arbitrarily consider one of the unit vectors of the triad, say $\hat{x}$. Let ϕ be the angle between the unit vector $\hat{\omega}$ (which is, by definition, a unit vector parallel to the axis of rotation AA) and the unit vector $\hat{x}$. The two-dimensional case shown Fig. 2.22a corresponds to $\phi = \pi/2$.

As shown in Fig. 2.22, the tip of the unit vector $\hat{x}$ sweeps out an arc of a circle of radius equal to $|\hat{x}| \sin \phi$, centered at a point C on the axis AA, while the unit vector $\hat{x} = \overset{\rightarrow}{NQ}$ sweeps out a portion of a right cone in space. At time t, the orientation of $\hat{x}$ is $\hat{x}(t)$ as shown in Fig. 2.22. After an infinitesimal instant of time Δt is elapsed, the radius CQ rotates by an angle $\Delta\theta$ to CQ', and the new orientation of $\hat{x}$ is along the vector $\hat{x}(t + \Delta t)$, which is also shown in Fig. 2.22. The vector $\Delta\hat{x}(t) = \hat{x}(t + \Delta t) - \hat{x}(t)$ is perpendicular to the unit vector $\hat{\omega}$ along line AA and is nearly perpendicular to $\hat{x}$ (it tends to the perpendicular to $\hat{x}$ as $\Delta t \to 0$). As seen from Fig. 2.22, its magnitude $|\Delta\hat{x}|$ is

$$|\Delta\hat{x}| = |\hat{x}(t + \Delta t) - \hat{x}(t)|$$

As $\Delta t \to 0$, the angle of rotation $\Delta\theta$ also approaches zero, and $\Delta\hat{x}(t)$ becomes tangent to the arc swept by the tip of $\hat{x}(t)$; the length of that arc is equal to

$(|\hat{x}| \sin \phi) \Delta \theta$, which is equal to the magnitude of the cross product $(\Delta \theta) \hat{\omega} \times \hat{x}$ [or the magnitude of the cross product $(\Delta \theta) \hat{x} \times \hat{\omega} = -(\Delta \theta) \hat{\omega} \times \hat{x}$]. Thus, as $\Delta t \to 0$, the magnitude of $d\hat{x}$ becomes

$$|d\hat{x}(t)| = |(d\theta) \hat{\omega} \times \hat{x}| = |(d\theta) \hat{x} \times \hat{\omega}| \qquad (2.6.2)$$

Since, as disclosed by inspection of Fig. 2.22, the vector $d\hat{x}(t)$ is in the direction of the cross product $\hat{\omega} \times \hat{x}$ (rather than the cross product $\hat{x} \times \hat{\omega}$), we can then write from Eq. (2.6.2):

$$d\hat{x}(t) = (d\theta) \hat{\omega} \times \hat{x} \qquad (2.6.3)$$

or

$$\frac{d\hat{x}(t)}{dt} = \dot{\theta} \, \hat{\omega} \times \hat{x} \overset{\Delta}{=} \vec{\omega} \times \hat{x} \qquad (2.6.4)$$

where $\dot{\theta}$ denotes $d\theta/dt$. The vector $\vec{\omega}$, defined as

$$\vec{\omega} = \lim_{\Delta t \to 0} \frac{\Delta \theta}{\Delta t} \, \hat{\omega} = \dot{\theta} \, \hat{\omega} \qquad (2.6.5)$$

is called the *angular velocity* (in the chosen reference frame, as illustrated in Fig. 2.22) of the triad $\{\hat{x}, \hat{y}, \hat{z}\}$, associated with the rotation of that triad about axis AA. The direction of the vector $\vec{\omega}$ is, by definition, the direction of the infinitesimal rotation vector $(d\theta) \hat{\omega}$, given by the right-hand rule,[8] which was presented in Chapter 1.

The magnitude $|\vec{\omega}|$ of the angular velocity $\vec{\omega}$ is called *angular speed*. The scalar component $\dot{\theta}$, in $\vec{\omega} = \dot{\theta} \hat{\omega}$, which may be positive or negative, is often also called angular velocity. It is also referred as such throughout the book and, when it is known to be a positive quantity, it is referred to as angular speed.

Notice that for the case of planar motion, which is illustrated in Fig. 2.22a with a rotation about a fixed axis parallel to $\hat{z} \overset{\Delta}{=} \hat{x} \times \hat{y}$ (with $\hat{y}$ being perpendicular to $\hat{x}$), one does not even have to introduce the angular velocity vector $\vec{\omega} = \dot{\theta} \hat{z}$ to obtain the result $d\hat{x}/dt = \dot{\theta} \hat{y}$. For such cases, this result can be inferred simply by inspection

---

[8]In dynamics, it is advisable to use a curved line *with only one arrow* to indicate an angle. Using two arrows to indicate an angle is only fine either for angles that are constants or for statics work, where variables do not change with time. But in dynamics lines can rotate and, therefore, the rotation sense (i.e., clockwise or counterclockwise) needs to be taken into account. Two arrows would certainly make it impossible to apply the right-hand rule to determine the direction of angular velocity. *Just as one would not use two arrows for drawing a vector, two arrows should not be used for indicating an angle that can change with time.*

of a simple figure similar to Fig. 2.22.[9] Try to keep such a figure in mind, since by doing so you should always be able to write, for planar problems, the expression for the derivative of a unit vector simply by inspection of a similar figure.

Since the unit vector $\hat{x}$ in Fig. 2.22 is only one of the three unit vector of the $\{\hat{x}, \hat{y}, \hat{z}\}$ triad that is rotating in space with angular velocity $\vec{\omega}$ in the reference frame under consideration, we can then write the following expressions for the time derivative, in that reference frame, of the remaining unit vectors $\hat{y}$ and $\hat{z}$.

$$\frac{d\hat{y}}{dt} = \vec{\omega} \times \hat{y} \qquad\qquad \frac{d\hat{z}}{dt} = \vec{\omega} \times \hat{z} \qquad\qquad (2.6.6)$$

Therefore, Eq. (2.6.1), in which $\vec{r} = r_{x_1}\hat{x}_1 + r_{x_2}\hat{x}_2 + r_{x_3}\hat{x}_3$, can be written as follows:

$$\frac{d\vec{r}}{dt} = \left[\frac{d\vec{r}}{dt}\right]_{\{x\}} + \vec{\omega} \times (r_{x_1}\hat{x}_1 + r_{x_2}\hat{x}_2 + r_{x_3}\hat{x}_3)$$

or simply,

$$\boxed{\frac{d\vec{r}}{dt} = \left[\frac{d\vec{r}}{dt}\right]_{\{x\}} + \vec{\omega} \times \vec{r}} \qquad\qquad (2.6.7)$$

It is important to notice that $\vec{\omega}$ in Eq. (2.6.7) is the angular velocity, relative to the particular reference frame one is using, of the $\{x\}$ triad that was chosen to express the vector $\vec{r}$, and not the angular velocity of anything else. If the reference frame being used is inertial, the time derivative $d\vec{r}/dt$ given by Eq. (2.6.7) is called the *absolute time derivative* of the vector $\vec{r}$.

The relationship expressed by Eq. (2.6.7) is of fundamental importance in dynamics for vectors that are expressed in terms of rotating triads. That relationship will be used throughout the book, as it is used by all dynamicists.

The first term in the right-hand side of Eq. (2.6.7), $[d\vec{r}/dt]_{\{x\}}$, is the time derivative of the vector $\vec{r}$ as seen by an observer fixed to the rotating triad $\{x\} = \{\hat{x}_1, \hat{x}_2, \hat{x}_3\}$. As indicated by Eq. (2.6.1), that term only involves the time derivative of the scalar components of $\vec{r}$. The second term, i.e., the one involving the omega cross operation $\vec{\omega} \times \vec{r}$, is present because of the use of a rotating unit vector triad

---

[9]To see this, rotate the unit vector $\hat{x}$ by an infinitesimal angle $d\theta$. As you do so, the direction of the vector $d\hat{x}$ is the direction followed by the "tip of the unit vector" $\hat{x}$. Give that direction a name; the name used in Fig. 2.22 is $\hat{y}$. The length of the arc that is swept is $(d\theta)|\hat{x}| = d\theta$. This, then, is how one can immediately conclude that $d\hat{x}/dt = \dot{\theta}\hat{y}$.

for expressing the vector $\vec{r}$. It is the additional contribution to $d\vec{r}/dt$ due to the rotation of the $\{\hat{x}, \hat{y}, \hat{z}\}$ triad, which is seen by an observer fixed to the X_0-Y_0-Z_0 frame of reference.

When does one choose to use rotating triads, and when does one choose nonrotating triads? The choice is only a matter of convenience for each particular problem. The resulting expressions for the position vector, the velocity, and the acceleration will have different forms for different triads. Some forms for such quantities will have fewer and simpler terms than others. More often than not, simpler forms for such expressions are obtained when rotating triads are used for expressing them. Also, some forms will require additional mathematical manipulations in the scalar equations obtained from $\vec{F} = m\vec{a}$ to obtain the differential equation (or equations, when more than one motion coordinate is involved) of motion for the system. This was illustrated in Example 2.5.10. Such additional manipulations that are required near the end of a formulation can be avoided by a judicious choice of triads at the beginning of the formulation.

Several examples that are formulated using rotating triads are presented in Section 2.7 and throughout the book. The determination of the angular velocity vector in the general case that involves several angles that change with time, even in the case of three-dimensional motion, is actually not a difficult matter (see the last sentence in the boxed text in Section B.1 of Appendix B). Section B.2 presents the general "formula" for the acceleration of a point, which contains five different types of terms. As indicated in that appendix, the use of that formula for solving problems is actually unnecessary and, therefore, is avoided in this book. Instead, a repetitive but basic approach is used throughout, where absolute velocity and acceleration are formulated from first principles for each problem.

2.7   More Solved Example Problems

Example 2.7.1 The magnitude of the vector $\vec{r} = \overrightarrow{OP}$ shown in Fig. 2.23 is constant and equal to $r = 2$ m. The vector rotates in a plane, and its direction relative to a nonrotating line on that plane is tracked by the angle θ.

    a. Using the nonrotating unit vectors $\hat{x}_1$ and $\hat{x}_2$, determine the time derivative of $\vec{r}$ in the reference frame defined by the plane of the paper.

b. Convert the answer obtained in part a to the unit vectors $\hat{r}$ and $\hat{n}$, where $\hat{r}$ is parallel to $\vec{r}$ and $\hat{n}$ is perpendicular to $\hat{r}$ (see Fig. 2.23), to show that the expression for $d\vec{r}/dt$ is now a very simple one and equal to $d\vec{r}/dt = r\dot{\theta}\hat{n} = 2\dot{\theta}\hat{n}$ (see Section 1.2, in Chapter 1, if a review on how to convert from one set of unit vectors to another is needed).

c. Obtain the result $d\vec{r}/dt = r\dot{\theta}\hat{n} = 2\dot{\theta}\hat{n}$ by completely avoiding the nonrotating unit vectors $\hat{x}_1$ and $\hat{x}_2$ and working directly with the rotating unit vectors $\hat{r}$ and $\hat{n}$.

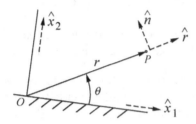

Figure 2.23: Figure for Example 2.7.1.

■ Solution

a. By inspecting Fig. 2.23, we can write for the vector $\vec{r}$,

$$\vec{r} = 2\left(\hat{x}_1 \cos\theta + \hat{x}_2 \sin\theta\right)$$

Therefore, the time derivative of $\vec{r}$ in the reference frame defined by the plane of the paper is

$$\vec{v} = \frac{d\vec{r}}{dt} = 2\dot{\theta}\left(\hat{x}_2 \cos\theta - \hat{x}_1 \sin\theta\right) \quad \text{m/s}$$

b. To convert the expression found in part a to the unit vectors $\hat{r}$ and $\hat{n}$, inspect Fig. 2.23 again. As shown in Section 1.2 of Chapter 1, the relations between the unit vectors $\{\hat{x}_1, \hat{x}_2\}$ and $\{\hat{r}, \hat{n}\}$ are simply obtained by projecting the unit vectors of one set along the directions of the unit vectors of the other set. As disclosed by Fig. 2.23,

$$\begin{aligned}
\hat{x}_1 &= \hat{r}\cos\theta - \hat{n}\sin\theta \\
\hat{x}_2 &= \hat{n}\cos\theta + \hat{r}\sin\theta
\end{aligned}$$

By substituting these relations into the expression that was obtained for $d\vec{r}/dt$, we then find

$$\vec{v} = \frac{d\vec{r}}{dt} = 2\dot{\theta}(\cos^2\theta + \sin^2\theta)\hat{n} = 2\dot{\theta}\hat{n}$$

This expression is much simpler than the one that was developed in terms of the nonrotating unit vectors $\{\hat{x}_1, \hat{x}_2\}$. It is also an expression that one could, from now on, write simply by inspecting a figure similar to Fig. 2.23. The reasons for this are:

1. The time derivative $d\vec{r}/dt$ of the constant magnitude vector $\vec{r}$ must be perpendicular to $\vec{r}$.
2. $d\vec{r}$ is perpendicular to $\vec{r}$ and points in the direction followed by the tip of $\vec{r}$ as $\vec{r}$ is turned slightly in the sense indicated by the curved arrow used for the angle θ.
3. For a vector $\vec{r}$ of constant magnitude, the magnitude of $d\vec{r}$ is equal to $|\vec{r}||d\theta|$, which is the length of an arc of a circle of radius $|\vec{r}|$.

By using the nonrotating unit vectors $\hat{x}_1$ and $\hat{x}_2$, we did not even have to mention and directly deal with angular velocity. But look at the number of manipulations that had to be made to get the simple one-component expression for $d\vec{r}/dt$.

c. We will now see a more convenient way of obtaining the simpler (and easy to visualize) expression for $d\vec{r}/dt$. In terms of the rotating unit vectors shown in Fig. 2.23, the vector $\vec{r}$ takes the following form, which is the simplest form of a vector because it only involves one component,

$$\vec{r} = 2\hat{r}$$

and the calculation of its time derivative in the reference frame defined by the plane of the paper now involves the angular velocity vector $\vec{\omega} = \dot{\theta}\hat{z}$, where $\hat{z} = \hat{r} \times \hat{n}$. We then have

$$\vec{v} = \frac{d\vec{r}}{dt} = (\dot{\theta}\hat{z}) \times (2\hat{r}) = 2\dot{\theta}\hat{n}$$

Of course there was a price, so to speak, to be paid for the simplicity inherent in this expression. The price is that one has to know how to obtain the expression for the absolute angular velocity $\vec{\omega}$ of a rotating line and also has to know about the omega cross operation that gives the time derivative of a unit vector. For planar motion, the time derivative of a unit vector may even be written by inspection of a figure, as already pointed out. These operations lead to fewer mathematical manipulations and are extremely convenient in dynamics.

Example 2.7.2 Obtain the second time derivative, in the reference frame defined by the plane of the paper, of the vector $\vec{r}$ of Example 2.7.1 using the unit vectors $\{\hat{r}, \hat{n}\}$ and the unit vectors $\{\hat{x}_1, \hat{x}_2\}$ shown in Fig. 2.23.

■ Solution
In terms of the triad $\{\hat{r}, \hat{n}, \hat{z} \overset{\Delta}{=} \hat{r} \times \hat{n}\}$, whose angular velocity is $\vec{\omega} = \dot{\theta}\hat{z}$, we then obtain (see Fig. 2.23, and use the right-hand rule to determine the direction of the cross product $\vec{\omega} \times \hat{n}$):

$$\vec{a} = \frac{d\vec{v}}{dt} = \frac{d}{dt}(2\dot{\theta}\hat{n}) = 2\ddot{\theta}\hat{n} + 2\dot{\theta}\frac{d\hat{n}}{dt} = 2\ddot{\theta}\hat{n} + 2\dot{\theta}(\dot{\theta}\hat{z} \times \hat{n}) = 2\ddot{\theta}\hat{n} - 2\dot{\theta}^2\hat{r} \quad \text{m/s}^2$$

Notice that the time derivative of the unit vector $\hat{n}$, which is equal to $-\dot{\theta}\hat{r}$, might also have been obtained by direct inspection of Fig. 2.23.

Again, this expression has two components, and each one of them has a reasonably simple form. The expression for $\vec{a}$ is more complicated and, thus, less convenient, if the triad $\{\hat{x}_1, \hat{x}_2, \hat{z} = \hat{x}_1 \times \hat{x}_2\}$ is used. The angular velocity of that triad, relative to the plane of the paper, is zero, and the following expression is obtained for the vector $\vec{a} = d\vec{v}/dt$:

$$\begin{aligned} \vec{a} &= \frac{d}{dt}\left[2\dot{\theta}\left(\hat{x}_2 \cos\theta - \hat{x}_1 \sin\theta\right)\right] \\ &= 2(\ddot{\theta}\cos\theta - \dot{\theta}^2 \sin\theta)\hat{x}_2 - 2(\ddot{\theta}\sin\theta + \dot{\theta}^2 \cos\theta)\hat{x}_1 \end{aligned}$$

Example 2.7.3 Usage of Polar Coordinates and of a Convenient Rotating Triad
For several problems in dynamics, it is convenient to use polar coordinates to locate a point in a plane. Using the polar coordinates r and θ and the unit vectors $\hat{r}$ and $\hat{n}$ shown in Fig. 2.23, obtain the general expressions for the absolute velocity and for the absolute acceleration of point P. Point O shown in the figure is an inertial point on the plane of the paper, and that plane is also inertial, i.e., it does not rotate.

■ Solution
Since the absolute position vector $\vec{r}$ for point P is simply (see Fig. 2.23)

$$\vec{r} = r\hat{r} \tag{2.7.3a}$$

and the absolute angular velocity $\vec{\omega}$ of the triad $\{\hat{r}, \hat{n}, \hat{z} \overset{\Delta}{=} \hat{r} \times \hat{n}\}$ is

$$\vec{\omega} = \dot{\theta}\hat{z} \tag{2.7.3b}$$

the absolute velocity of point P is then

$$\vec{v} = \frac{d\vec{r}}{dt} = \dot{r}\hat{r} + r\frac{d\hat{r}}{dt} = \dot{r}\hat{r} + \vec{\omega} \times (r\hat{r}) = \dot{r}\hat{r} + r\dot{\theta}\hat{n} \qquad (2.7.3c)$$

Figure 2.24 is a graphical visualization of the time derivative $d\vec{r}/dt$ of a rotating vector $\vec{r}$. It allows one to write the expression for the velocity $\vec{v} = d\vec{r}/dt$ without having to derive it again or to consult such a derivation.

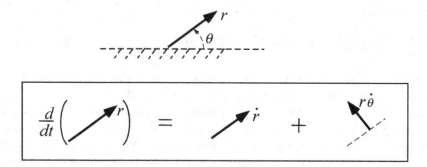

Figure 2.24: Graphical visualization of the time derivative $d\vec{r}/dt$ of a rotating vector $\vec{r}$.

The absolute acceleration of point P is determined by taking the time derivative of Eq. (2.7.3c) as shown here.

$$\vec{a} = \frac{d\vec{v}}{dt} = \ddot{r}\hat{r} + \dot{r}\frac{d\hat{r}}{dt} + (r\ddot{\theta} + \dot{r}\dot{\theta})\hat{n} + r\dot{\theta}\frac{d\hat{n}}{dt}$$

Since $d\hat{r}/dt = \omega \times \hat{r} = \dot{\theta}\hat{n}$, and $d\hat{n}/dt = \vec{\omega} \times \hat{n} = -\dot{\theta}\hat{r}$ (see Fig. 2.23 and use the right-hand rule), this expression for the acceleration then becomes

$$\vec{a} = \frac{d\vec{v}}{dt} = (\ddot{r} - r\dot{\theta}^2)\hat{r} + (r\ddot{\theta} + 2\dot{r}\dot{\theta})\hat{n} \qquad (2.7.3d)$$

As seen, these operations are really not difficult to perform when polar coordinates and the unit vectors chosen in this example are used. The advantage of using such rotating unit vectors lies in the simplicity of the expression that is obtained for the velocity, and the relatively compact expression that is obtained for the acceleration.

Notice that the expression for the velocity given by Eq. (2.7.3c) is the sum of the following two velocities:

1. The velocity $\dot{r}\hat{r}$ that would be obtained if θ were held constant (i.e., if P were moving along the radial line)
2. The velocity $r\dot{\theta}\hat{n}$ that would be obtained if r were held constant (i.e., if P were moving along a circle of radius r)

These observations make it relatively simple to obtain Eq. (2.7.3c) for the velocity $\vec{v} = d\vec{r}/dt$ simply by inspecting a figure such as Fig. 2.23 or Fig. 2.24.

As for the expression for the acceleration, the only terms that appear in Eq. (2.7.3d) that might be viewed as obvious are the terms $\ddot{r}\hat{r}$ (which would be the acceleration if P were moving along the radial line, with θ held constant), and $r\ddot{\theta}\hat{n}$ (which would be the acceleration if r were held constant so that P would be moving along a circle of radius r). The other two terms, $-r\dot{\theta}^2\hat{r}$ and $2\dot{r}\dot{\theta}\hat{n}$, are called the *centripetal acceleration* and the *Coriolis acceleration*, respectively. The negative of the centripetal acceleration is called *centrifugal acceleration*.

Figure 2.25 is a graphical visualization that summarizes the results obtained here. The larger box shows the time derivative of the radial $\hat{r}$ and of the transverse $\hat{\theta}$ components of the second time derivative of a vector $\vec{r}$. Each component of the vector $\vec{v} = d\vec{r}/dt$ yields two parts for each one of the components of the final result for $\vec{a} = d\vec{v}/dt$.

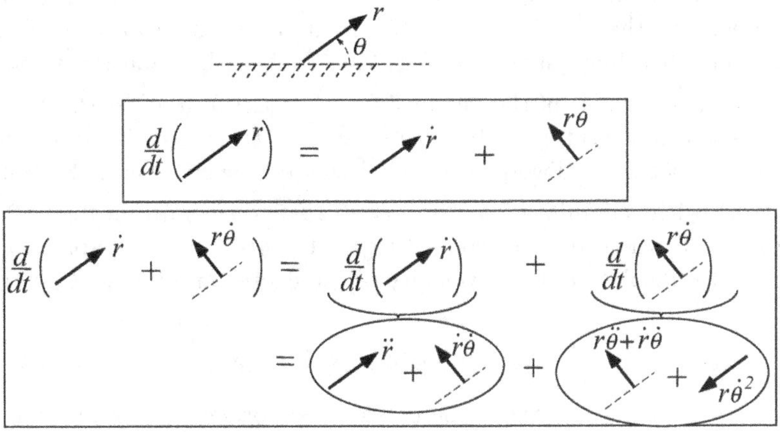

Figure 2.25: Graphical visualization of the first and second time derivatives of a rotating vector $\vec{r}$.

Example 2.7.4 A turntable rotates about a fixed axis that passes through its inertial center C as shown in Fig. 2.26. The angular velocity of the turntable, relative to the inertial plane of the paper, is $\Omega(t)$. Determine the absolute velocity $\vec{v}$ and the absolute acceleration $\vec{a}$ of a particle P that is able to move on the turntable. Express the answers in terms of the polar coordinates $r(t)$ and $\theta(t)$ that locate the particle relative to the turntable at any time t.

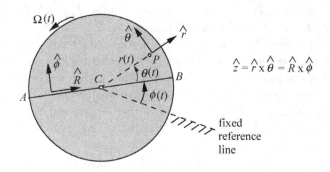

Figure 2.26: Figure for Example 2.7.4.

■ Solution

The absolute velocity of P is the velocity of P relative to C (since point C is inertial) taking into account the rotation of line CP relative to a nonrotating reference line (the "fixed reference line" shown in Fig. 2.26). It is the time derivative, in the inertial plane of the paper, of the vector $\vec{r} \overset{\Delta}{=} \overrightarrow{CP}$ drawn from C to P. We now need to choose a set of unit vectors to be able to express the vector $\vec{r}$ and carry out the time derivative operation. Two possible choices are shown in Fig. 2.26. Suppose we decide to work with the unit vector triad $(\hat{r}, \hat{\theta}, \hat{z} \overset{\Delta}{=} \hat{r} \times \hat{\theta})$ shown in Fig. 2.26, with $\hat{r}$ always pointing in the direction from C to P. In this case, the expression for the vector $\vec{r}$ and the absolute angular velocity of the chosen triad are, respectively,

$$\vec{r} = r\hat{r} \tag{2.7.4a}$$

$$\vec{\omega} = (\dot{\phi} + \dot{\theta})\hat{z} = (\Omega + \dot{\theta})\hat{z} \tag{2.7.4b}$$

Notice that $\dot{\phi} = \Omega$, since the angle ϕ tracks the orientation of any line fixed to the rotating table relative to the nonrotating reference line shown in Fig 2.26. Also, notice that the term $\dot{\theta}\hat{z}$ is the angular velocity of line CP relative to the rotating table (or, in other words, the angular velocity of line CP in the reference frame fixed to the rotating table).

With $\vec{r}$ and $\vec{\omega}$ given by Eqs. (2.7.4a) and (2.7.4b), respectively, the absolute velocity and absolute acceleration of P are then determined as follows.

$$\vec{v} = \frac{d\vec{r}}{dt} = \dot{r}\hat{r} + \vec{\omega} \times \vec{r} = \dot{r}\hat{r} + r(\Omega + \dot{\theta})\hat{\theta} \overset{\Delta}{=} v_r\hat{r} + v_\theta\hat{\theta} \tag{2.7.4c}$$

$$\vec{a} = \frac{d\vec{v}}{dt} = \ddot{r}\hat{r} + \left[\dot{r}(\Omega + \dot{\theta}) + r(\dot{\Omega} + \ddot{\theta})\right]\hat{\theta} + \vec{\omega} \times \vec{v}$$

$$= \left[\ddot{r} - r(\Omega + \dot{\theta})^2\right]\hat{r} + \left[(\dot{\Omega} + \ddot{\theta})r + 2\dot{r}(\Omega + \dot{\theta})\right]\hat{\theta} \overset{\Delta}{=} a_r\hat{r} + a_\theta\hat{\theta} \quad (2.7.4d)$$

These expressions for $\vec{v}$ and $\vec{a}$ are valid for any kind of motion of particle P on the rotating table. If the motion of P were to be constrained to a radial groove fixed to the rotating table, for example, we would simply have to set θ equal to a constant in Eqs. (2.7.4c) and (2.7.4d). If the motion of P were to be constrained to a circular groove of radius R and centered at C, we would, instead, set $r = \text{constant} = \overset{\Delta}{=} R$.

In formulating the expressions for the velocity and acceleration we could, of course, choose other unit vector triads to work with. If the triad $(\hat{R}, \hat{\phi}, \hat{z} = \hat{R} \times \hat{\phi})$ shown in Fig. 2.26 is chosen, the expression for the vector $\vec{r}$ becomes (see Fig. 2.26)

$$\vec{r} = r[\hat{R}\cos\theta + \hat{\phi}\sin\theta] \quad (2.7.4e)$$

while the absolute angular velocity of the new triad $\{\hat{R}, \hat{\phi}, \hat{z}\}$, which is denoted below as $\vec{\omega}_2$ so that it is not confused with the vector $\vec{\omega}$ used earlier, is now

$$\vec{\omega}_2 = \Omega\hat{z} \quad (2.7.4f)$$

In this case, the following expression for $\vec{v}$ is obtained:

$$\begin{aligned}
\vec{v} &= \frac{d\vec{r}}{dt} = \left[\frac{d\vec{r}}{dt}\right]_{\{\hat{R},\hat{\phi},\hat{z}\}} + \vec{\omega}_2 \times (\hat{R}r\cos\theta + \hat{\phi}r\sin\theta) \\
&= (\dot{r}\cos\theta - r\dot{\theta}\sin\theta)\hat{R} + (\dot{r}\sin\theta + r\dot{\theta}\cos\theta)\hat{\phi} + \vec{\omega}_2 \times (\hat{R}r\cos\theta + \hat{\phi}r\sin\theta) \\
&= [\dot{r}\cos\theta - r(\Omega + \dot{\theta})\sin\theta]\hat{R} + [\dot{r}\sin\theta + r(\Omega + \dot{\theta})\cos\theta]\hat{\phi} \\
&\overset{\Delta}{=} v_R\hat{R} + v_\phi\hat{\phi} = (v_r\cos\theta - v_\theta\sin\theta)\hat{R} + (v_r\sin\theta + v_\theta\cos\theta)\hat{\phi} \quad (2.7.4g)
\end{aligned}$$

Notice that, as indicated in Eq. (2.7.4g) and verified by inspection of Fig. 2.26, the expression $\vec{v} = v_R\hat{R} + v_\phi\hat{\phi}$ is the same as the projection of $\vec{v} = v_r\hat{r} + v_\theta\hat{\theta}$ along the orthogonal unit vector system $\{\hat{R}, \hat{\phi}\}$ shown in that figure.

The following expression is obtained for the acceleration $\vec{a}$:

$$\begin{aligned}
\vec{a} &= \frac{d}{dt}\vec{v} = \left[\frac{d}{dt}\vec{v}\right]_{\{\hat{R},\hat{\phi},\hat{z}\}} + \vec{\omega}_2 \times \vec{v} \\
&= \left\{[\ddot{r} - r\dot{\theta}(\Omega + \dot{\theta})]\cos\theta - [\dot{r}\dot{\theta} + \dot{r}(\Omega + \dot{\theta}) + r(\dot{\Omega} + \ddot{\theta})]\sin\theta\right\}\hat{R} \\
&\quad + \left\{[\ddot{r} - r\dot{\theta}(\Omega + \dot{\theta})]\sin\theta + [\dot{r}\dot{\theta} + \dot{r}(\Omega + \dot{\theta}) + r(\dot{\Omega} + \ddot{\theta})]\cos\theta\right\}\hat{\phi} + \vec{\omega}_2 \times \vec{v}
\end{aligned}$$

If we perform the cross product that appears in the expression for $\vec{a}$, collect items in $\cos\theta$ and $\sin\theta$, and group the terms $r\Omega^2 + r\dot{\theta}^2 + 2r\Omega\dot{\theta}$ that appear in the resulting expression for $\vec{a}$ as $r(\Omega+\dot{\theta})^2$, we obtain the following expression for the acceleration $\vec{a}$:

$$\vec{a} = \left\{[\ddot{r} - r\dot{\theta}(\Omega + \dot{\theta})^2]\cos\theta - [r(\dot{\Omega} + \ddot{\theta}) + 2\dot{r}(\Omega + \dot{\theta})]\sin\theta\right\}\hat{R}$$
$$+ \left\{[\ddot{r} - r\dot{\theta}(\Omega + \dot{\theta})^2]\sin\theta + [r(\dot{\Omega} + \ddot{\theta}) + 2\dot{r}(\Omega + \dot{\theta})]\cos\theta\right\}\hat{\phi}$$
$$\stackrel{\triangle}{=} a_R\hat{R} + a_\phi\hat{\phi} = (a_r\cos\theta - a_\theta\sin\theta)\hat{R} + (a_r\sin\theta + a_\theta\cos\theta)\hat{\phi} \qquad (2.7.4\text{h})$$

Also notice that, as indicated in Eq. (2.7.4h) and verified by inspection of Fig. 2.26, the expression $\vec{a} = a_R\hat{R} + a_\phi\hat{\phi}$ is the same as the projection of $\vec{a} = a_r\hat{r} + a_\theta\hat{\theta}$ along the orthogonal unit vector system $\{\hat{R}, \hat{\phi}\}$ shown in that figure.

This example showed in detail the symbolic manipulations involved in formulating expressions for velocity and for acceleration that are needed when using Newton's second law. Before we leave this example, it is worthwhile to look at the results for the velocity and the acceleration that were obtained and reflect on them because many of the steps done here appear in a large variety of situations in practice. Let us start by making some observations on the steps that led us to the two different forms for the velocity $\vec{v}$.

> *Observation* 1. The simplest expressions for $\vec{v}$ and $\vec{a}$ were obtained by using the triad for which one of the unit vectors, namely, $\hat{r}$, was along the position vector $\vec{r}$, i.e., the triad $\{\hat{r}, \hat{\theta}, \hat{z}\}$ shown in Fig. 2.26. In that triad, the two scalar components of $\vec{v}$, i.e., $v_r = \dot{r}$ and $v_\theta = r(\Omega + \dot{\theta})$, have the same simple physical interpretation mentioned in Example 2.7.3 (p. 145). Those observations are repeated here.

> *Observation* 2. The radial $\hat{r}$-component of the velocity, $v_r\hat{r} = \dot{r}\hat{r}$, would be the velocity $\vec{v}$ if the angles θ and ϕ shown in Fig. 2.26 were held constant; in such a case, particle P would simply travel along the straight line CP.

> *Observation* 3. The transverse $\hat{\theta}$-component of the velocity, $v_\theta\hat{\theta} = r(\Omega + \dot{\theta})\hat{\theta}$, would be the velocity $\vec{v}$ if the polar coordinate r shown in Fig. 2.26 (which, again, is the distance measured from C to P) were kept constant. In such a case, particle P would simply travel along a circle of radius r centered at point C. The velocity $\vec{v}$ is always the vector sum of the radial $\hat{r}$-component and of the transverse $\hat{\theta}$-component.

> *Observation* 4. As indicated in Eq. (2.7.4g), the components $v_R\hat{R}$ and $v_\phi\hat{\phi}$ of the velocity $\vec{v}$ are simply composed, as expected, of projections of the simpler components v_r and v_θ along the directions $\hat{R}$ and $\hat{\phi}$ as $v_R = v_r\cos\theta - v_\theta\sin\theta$ and $v_\phi = v_r\sin\theta + v_\theta\cos\theta$. Thus, a "better way" (meaning, one that involves

fewer mathematical manipulations and, therefore, fewer calculations) to obtain the expression given by Eq. (2.7.4g) would be to find the quantities v_r and v_θ first, and then convert the result into v_R and v_ϕ by simple projections. The same observation holds for the acceleration $\vec{a}$.

Example 2.7.5 A particle P is constrained to move on the surface of a nonrotating cylinder of radius R m as shown in Fig. 2.27. The x-, y-, and z-axes shown in that figure are fixed to the cylinder. Point C of the cylinder has a constant acceleration equal to $a_C \hat{y}$ relative to inertial space.

a. Obtain the expression for the absolute acceleration of P by making use of the cylindrical coordinates $\phi(t)$ and $z(t)$ and the conveniently chosen triad $\{\hat{R}, \hat{\phi}, \hat{z} = \hat{R} \times \hat{\phi}\}$, all of which are shown in Fig. 2.27. The x-, y-, and z-axes in that figure are inertial (i.e., they do not rotate relative to inertial space).

b. For $R = 1$ m and $a_C = 4$ m/s², plot the magnitude of the absolute acceleration $\vec{a}_P$ of point P versus the angle ϕ, if $\dot{\phi}$ = constant = 30 revolutions per minute (rpm) and the speed of P relative to the moving cylinder is constant and equal to 5 m/s.

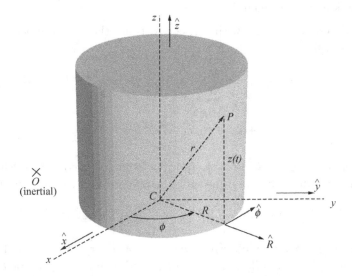

Figure 2.27: Particle constrained to move on a cylindrical surface.

■ **Solution**

a. With point O shown in Fig. 2.27 being inertial, the absolute position vector for P is

$$\overrightarrow{OP} = \overrightarrow{OC} + \overrightarrow{CP} = \overrightarrow{OC} + R\hat{R} + z\hat{z} \qquad (2.7.5a)$$

Since the x-axis shown in Fig. 2.27 is nonrotating, the absolute angular velocity of the triad $\{\hat{R}, \hat{\phi}, \hat{z} = \hat{R} \times \hat{\phi}\}$ is equal to[10] $\vec{\omega} = \dot{\phi}\hat{z}$. Therefore, the absolute velocity of point P is

$$\vec{v}_P = \frac{d}{dt}\overrightarrow{OP} = \underbrace{\frac{d}{dt}\overrightarrow{OC}}_{\text{call this } \vec{v}_C} + \underbrace{\dot{R}\hat{R}}_{\substack{=0 \\ \text{because } R=\text{constant}}} + \dot{z}\hat{z} + \vec{\omega} \times (R\hat{R} + z\hat{z})$$

$$= \vec{v}_C + \dot{z}\hat{z} + R\dot{\phi}\,\hat{\phi} \qquad (2.7.5b)$$

Since we already know that $d\vec{v}_C/dt$ is equal to the given acceleration $a_C\hat{y}$, the simplest way to deal with the term $\vec{v}_C$ in Eq. (2.7.5b) is to leave it as is in that equation [i.e., it is simply not wise to express $\vec{v}_C$ in components and then take the derivative of such components, because the final expression for $d\vec{v}_C/dt$ is already known beforehand].

By differentiating $\vec{v}_P$ with respect to time, the absolute acceleration of P is then obtained as

$$\vec{a}_P = \frac{d}{dt}\vec{v}_P = \underbrace{\frac{d}{dt}\vec{v}_C}_{\text{this is } \vec{a}_C} + \ddot{z}\hat{z} + R\ddot{\phi}\,\hat{\phi} + \vec{\omega} \times \left(\dot{z}\hat{z} + R\dot{\phi}\,\hat{\phi}\right)$$

$$= \vec{a}_C + \ddot{z}\hat{z} + R\ddot{\phi}\,\hat{\phi} - R\dot{\phi}^2\hat{R} \qquad (2.7.5c)$$

Since $\vec{a}_C$ is given as $\vec{a}_C = a_C\hat{y}$, and $\hat{y} = \hat{R}\sin\phi + \hat{\phi}\cos\phi$ (see Fig. 2.27), the following result is then obtained for the acceleration $\vec{a}_P$ of point P.

$$\vec{a}_P = (a_C\sin\phi - R\dot{\phi}^2)\hat{R} + (R\ddot{\phi} + a_C\cos\phi)\hat{\phi} + \ddot{z}\hat{z} \qquad (2.7.5d)$$

b. Now it is desired to plot $|\vec{a}_P|$ versus ϕ, for $R = 1$ m, $a_C = 4$ m/s^2, $\dot{\phi} =$ constant $= 30$ rpm $= 30 \times 2\pi/60$ rad/s, and when the speed of P relative to the moving cylinder is constant and equal to 5 m/s.

---

[10]If the cylinder were rotating with angular velocity $\vec{\Omega}$ relative to inertial space, the absolute angular velocity $\vec{\omega}$ of the $\{\hat{R}, \hat{\phi}, \hat{z}\}$ triad would be equal to $\vec{\omega} = \vec{\Omega} + \dot{\phi}\hat{z}$ since the x-axis, where the angle ϕ is measured from, was chosen to be fixed to the cylinder.

The magnitude of $\vec{a}_P$ is

$$|\vec{a}_P| = \sqrt{(a_C \sin \phi - R\dot{\phi}^2)^2 + (R\ddot{\phi} + a_C \cos \phi)^2 + \ddot{z}^2} \qquad (2.7.5e)$$

The values for $\ddot{\phi}$ and $\ddot{z}$ are needed in this expression. Since $\dot{\phi} = $ constant, it follows that $\ddot{\phi} = 0$. To calculate $\ddot{z}$, use the only information that has not been used so far, i.e., that the speed of P relative to the moving cylinder is constant and equal to 5 m/s. Since the velocity of any point of the cylinder is $\vec{v}_C = d\overrightarrow{OC}/dt$, the speed of P relative to the moving cylinder is the magnitude of the vector $\vec{v}_P - \vec{v}_C$. By making use of Eq. (2.7.5b), $|\vec{v}_P - \vec{v}_C|$ is then determined as

$$|\vec{v}_P - \vec{v}_C| = \sqrt{\dot{z}^2 + (R\dot{\phi})^2} = \text{constant} = 5 \text{ m/s}$$

This gives $\dot{z} = \text{constant} \approx \pm 3.89$ m/s. Therefore, $\ddot{z} = 0$, and $|\vec{a}_P|$ is

$$|\vec{a}_P| = \sqrt{4^2 + \pi^4 - 2 \times 4 \times \pi^2 \sin \phi} \approx \sqrt{113.4 - 79 \sin \phi} \text{ m/s}^2$$

The plot of $|\vec{a}_P|$ versus ϕ is shown in Fig. 2.28. Although particle P moves with a constant speed relative to the cylinder, its acceleration is not constant (and

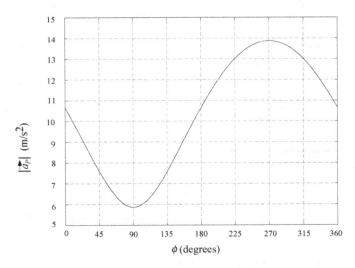

Figure 2.28: Plot of $|\vec{a}_P|$ versus ϕ for Example 2.7.5.

it would not be constant even if the cylinder were not accelerating) because the velocity of P is changing direction.

Example 2.7.6 The pendulum considered in Example 2.5.10 (p. 129) is connected by a bearing (represented by point A in Fig. 2.29) to a truck that is moving to the left along a straight horizontal track with a known acceleration (relative to the earth, which is approximated as inertial) equal to a_A m/s^2, also directed to the left. The motion of the pendulum is constrained to a vertical plane, and is started with the initial conditions $\theta(0) = 0$ and $\dot{\theta}(0) = 0$, and the only external force acting on the point mass at B is the gravitational force of attraction of the earth. An external moment M_{bar} is applied to the pendulum stem as shown in Fig. 2.29. The pendulum length is $L = 1$ m, and the mass of the particle at point B is $m = 1$ kg. Approximate the bar as massless.

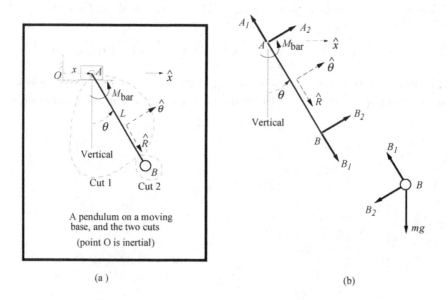

(a)   (b)

Figure 2.29: A pendulum mounted on a moving base for Example 2.7.6.

a. Obtain the differential equation of motion for the pendulum, and then determine the maximum and minimum values of $\theta(t)$ for the case when $M_{bar} = 0$ and $a_A = $ constant $= \sqrt{3}g \approx 17$ m/s^2.

b. Determine the maximum and minimum values for the axial reaction force acting at end B of bar AB for the case when $M_{bar} = 0$, and $a_A = $ constant $= \sqrt{3}g \approx 17 \text{ m/s}^2$.

Note: The questions in parts c and d cannot be answered analytically because the differential equation of motion is nonlinear and M_{bar} is neither zero nor a function that only involves the angle θ. For both cases, $M_{bar} = -c\dot{\theta}$, where $c = 1$ N·m/(rad/s) is due to *viscous friction* in the bearing at A. To answer the questions for parts c and d, use a computer to integrate the differential equation numerically.

c. Determine the maximum value of the angle θ, and the time it occurs, for the case when $a_A = $ constant $= 2g \text{ m/s}^2$.

d. During the first 5 s, point A is forced to accelerate to the left (see Fig. 2.29) as $a_A(t) = kt$ where $k = 5 \text{ m/s}^3$. Afterward, the acceleration of A is suddenly dropped to zero, but point A is forced to continue to travel to the left at a constant velocity. Plot $\theta(t)$ versus time t for $0 \leq t \leq 15$ s. What is the maximum value of θ, and when does it occur?

■ Solution

a. The free-body diagrams for the system consisting of the massless bar and the mass particle at B are shown in Fig. 2.29b. We will use the unit vector $\hat{x}$ to express the absolute acceleration of point A, which is $\vec{a}_A = \ddot{x}\hat{x} = -a_A\hat{x}$ where $a_A = $ constant (with a_A being positive, the minus sign, which is associated with the unit vector $\hat{x}$, accounts for the fact that point A is accelerating to the left). To deal with the particle at B, we will make use of the more convenient rotating unit vectors $\hat{R}$ and $\hat{\theta}$ shown in the figure. The absolute angular velocity of the triad $\{\hat{R}, \hat{\theta}, \hat{z} \overset{\Delta}{=} \hat{R} \times \hat{\theta}\}$ is $\vec{\omega} = \dot{\theta}\hat{z}$ (see Fig. 2.29, and use the right-hand rule to obtain $\vec{\omega}$). In terms of those unit vectors, we have for the particle at B:

■ Resultant force acting on the particle at B:

$$\vec{F} = mg(\hat{R}\cos\theta - \hat{\theta}\sin\theta) - B_1\hat{R} - B_2\hat{\theta}$$

■ Absolute position vector for the particle at B:

$$\vec{r}_B = \overrightarrow{OB} = \overrightarrow{OA} + \overrightarrow{AB} = x\hat{x} + L\hat{R}$$

■ Absolute velocity of the particle at B:

$$\vec{v}_B = \frac{d\vec{r}_B}{dt} = \dot{x}\hat{x} + \frac{d(L\hat{R})}{dt} = \dot{x}\hat{x} + \vec{\omega} \times (L\hat{R}) = \dot{x}\hat{x} + L\dot{\theta}\hat{\theta}$$

■ Absolute acceleration of the particle at B:

$$\vec{a}_B = \frac{d\vec{v}_B}{dt} = \ddot{x}\hat{x} + \left[L\ddot{\theta}\hat{\theta} + \vec{\omega} \times (L\dot{\theta}\hat{\theta})\right] = \ddot{x}\hat{x} + L\ddot{\theta}\hat{\theta} - L\dot{\theta}^2\hat{R}$$

Since $\ddot{x}\hat{x} = \ddot{x}(\hat{\theta}\cos\theta + \hat{R}\sin\theta)$, Newton's second law, $\vec{F} = m\vec{a}_B$, then yields the following scalar equations. Notice that, as indicated by Eq. (2.7.6a), only the unknown reaction force B_2 is needed for obtaining the differential equation of motion for the system.

$$m(L\ddot{\theta} + \ddot{x}\cos\theta) = -mg\sin\theta - B_2 \qquad (2.7.6a)$$
$$B_1 - mg\cos\theta = m(L\dot{\theta}^2 - \ddot{x}\sin\theta) \qquad (2.7.6b)$$

These equations involve the three unknowns θ, B_1, and B_2. Therefore, one more equation is needed. The additional equation is obtained by working with the free-body diagram for the bar. As indicated in Example 2.5.10 (p. 129), the additional equation that only involves the unknown B_2 can be readily obtained by equating to zero the summation of the moments, about any point, of the forces acting on the massless bar (the dynamics of rigid bodies, taking into account their distributed masses, is dealt with in later chapters, starting with Chapter 5). By taking moments about point A (because the unknowns A_1 and A_2 would not appear in the resulting equation), we obtain

$$B_2 = -\frac{M_{\text{bar}}}{L} \qquad (2.7.6c)$$

With B_2 given by Eq. (2.7.6c), Eq. (2.7.6a) becomes

$$L\ddot{\theta} + g\sin\theta + \ddot{x}\cos\theta = \frac{M_{\text{bar}}}{mL} \qquad (2.7.6d)$$

Equation (2.7.6d) does not involve any unknown reaction forces, and, therefore, it is the differential equation of motion for the system. Once a solution for $\theta(t)$ is obtained by integrating that equation, the axial force B_1 can be calculated by making use of Eq. (2.7.6b).

As indicated in Section 1.13 of Chapter 1, application of certain techniques to a differential equation of motion may yield a variety of information about the motion of the system without having to solve the equation analytically (which is not always possible to do, since differential equations of motion are generally nonlinear). For this problem, the maximum and minimum values of

θ can be determined analytically for the case when $M_{\text{bar}} = 0$ and $\vec{a}_A$ is either a constant of a function of θ only.

The maximum and minimum values for the angle θ occur when $\dot{\theta} = 0$. By making use of the chain rule of differentiation to express $\ddot{\theta}$ as

$$\ddot{\theta} = \frac{d\dot{\theta}}{dt} = \frac{d\dot{\theta}}{d\theta}\frac{d\theta}{dt} = \dot{\theta}\frac{d\dot{\theta}}{d\theta}$$

the differential equation of motion, Eq. (2.7.6d), may be rearranged and then integrated once, for $\ddot{x} = \text{constant} = -a_A$ and $M_{\text{bar}} = 0$, as

$$L \int \dot{\theta} \frac{d\dot{\theta}}{d\theta} d\theta + \int (g \sin\theta - a_A \cos\theta) \ d\theta = \text{ constant} \stackrel{\Delta}{=} C_1$$

or

$$\frac{L}{2}\dot{\theta}^2 - g\cos\theta - a_A \sin\theta = \text{ constant} = C_1 \qquad (2.7.6e)$$

By making use of the given initial conditions, the constant C_1 is then $C_1 = L\dot{\theta}^2(0)/2 - g\cos\theta(0) - a_A \sin\theta(0) = -g$, since $\dot{\theta}(0) = 0$ and $\theta(0) = 0$.

By setting $\dot{\theta} = 0$ in Eq. (2.7.6e), the maximum and minimum values of θ, $\theta \stackrel{\Delta}{=} \theta_m$, are, then, the solutions to the following algebraic equation:

$$g\cos\theta_m + a_A \sin\theta_m = g \qquad (2.7.6f)$$

Since $\sin\theta = \pm\sqrt{1 - \cos^2\theta}$, this equation yields the following quadratic equation for θ_m:

$$(g^2 + a_A^2)\cos^2\theta_m - 2g^2\cos\theta_m + g^2 - a_A^2 = 0 \qquad (2.7.6g)$$

whose solutions are

$$\cos\theta_m = 1 \quad \text{and} \quad \cos\theta_m = \frac{g^2 - a_A^2}{g^2 + a_A^2}$$

Therefore, $\theta_m = 0$ and, with $a_A = \sqrt{3}g$, $\theta_m = \arccos(-0.5) = \pm120°$. As can be verified from Eq. (2.7.6f), the positive solution $\theta_m = 120°$ is the one that corresponds to $a_A/g = \sqrt{3}$. The extra value $\theta_m = -120°$ was "created" when Eq. (2.7.6f) was squared to obtain Eq. (2.7.6g), and it does not satisfy Eq. (2.7.6f). The solution for θ_m could also have been obtained by using a calculator to do a few numerical iterations in Eq. (2.7.6f) to obtain $\theta_m = 120°$. For $a_A/g = \sqrt{3}$ we then have $\theta_{\min} = 0$ and $\theta_{\max} = 120°$.

b. For the axial force B_1, for the case when $M_{\text{bar}} = 0$ and $\ddot{x} = -a_A = \text{constant} = -\sqrt{3}g$ m/s^2, the following result is obtained by combining Eqs. (2.7.6b) and (2.7.6e).

$$B_1 = 3m(g\cos\theta + a_A\sin\theta) - 2mg \qquad (2.7.6h)$$

By setting $dB_1/d\theta = 0$, with $dB_1/d\theta$ obtained from Eq. (2.7.6h), the maximum value for B_1 is found to occur when $\tan\theta = a_A/g$. For $a_A/g = \sqrt{3}$, such a value of θ is $\theta = 60°$, which is equal to $\theta_{\max}/2$. The corresponding value for B_1 is $B_1 = 4mg \approx 39.2$ N for $m = 1$ kg. Since $d^2B_1/d\theta^2 = -3m(g\cos\theta + a_A\sin\theta)$ is negative for $\theta = 60°$, the value that was found for B_1 is indeed a maximum. Notice that $[B_1]_{\theta_{\min}} = [B_1]_{\theta=0} = mg$. This is the minimum value of B_1.

c and d. $\Longrightarrow$ For Scilab users, the program *pend_moving_base.sci* listed in Appendix E integrates the differential equation of motion with $M_{\text{bar}} = -c\dot{\theta}$, which is Eq. (2.7.6i). It generates a plot of θ versus t, and animates the motion. For Matlab users, the program *pend_moving_base.m* that is also listed in Appendix E will do the same.

$$\ddot{\theta} + \frac{g}{L}\sin\theta = \frac{a_A}{L}\cos\theta - \frac{c}{mL^2}\dot{\theta} \qquad (2.7.6i)$$

As indicated in the comment lines at the beginning of those programs, an example of their use (with $g/L = 9.81$, $c/(mL^2) = 1$, and the initial conditions $\theta(0) = 0$ and $\dot{\theta}(0) = 0$) is (with $\Delta t = 0.005$ in Scilab, and 0.002 in Matlab)

```
my_output = pend_moving_base([0 : Δt : 15], 9.81, 1, 0, 0);
```

The plots of θ versus time t for parts c and d are shown in Fig. 2.30. For part c, a reading taken from Fig. 2.30a yields $\theta_{\max} \approx 107°$, which occurs when $t \approx 0.7$ s. For part d, a reading from Fig. 2.30b yields $\theta_{\max} \approx 68°$, which occurs when $t = 5$ s.

The variable `my_output` is a *len* by 3 matrix whose columns are t, θ, and $\dot{\theta}$, where *len* is the number of time points in the array [0:Δt:15]. It is of interest if one wishes to generate other plots [such as a plot of $\dot{\theta}$ versus θ with the command `plot(my_output(:, 2), my_output(:, 3))` or even to obtain the maximum and minimum values of a variable, such as `max(my_output(:, 2))` or `min(my_output(:, 2))`]. The animated figure that is generated is based on an x-y coordinate system centered at point A in Fig. 2.29, with the $+x$ axis parallel to the unit vector $\hat{x}$ in that figure, and with the $+y$ axis at 90 degrees counterclockwise from the $+x$ axis. Therefore, the animation will show the pendulum as viewed by an observer located on the moving base at A.

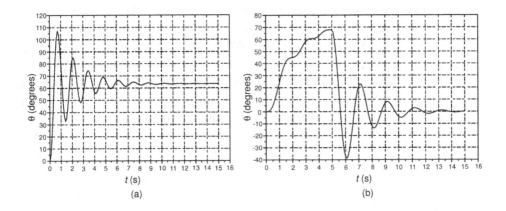

Figure 2.30: $\theta(t)$ versus t for (a) part c and (b) part d of Example 2.7.6.

Those who have Simulink and prefer to use it may want to construct the block diagram model shown in Fig. 2.31, setting the *Simulation* → *Parameters* menu for this model with time t of the form [0:Δt:15], and using the same

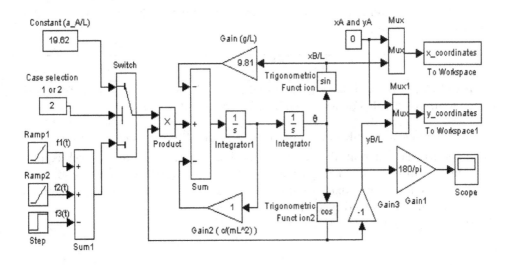

Figure 2.31: Simulation, for Simulink users, of a pendulum on a moving base for Example 2.7.6. For the "Case selection," choose 1 for part d or 2 for part c.

initial conditions mentioned in the previous paragraph. To see an animation of the motion after the simulation stops, type the following command in the Matlab command window: `animate_nbars(x_coordinates, y_coordinates)`.

The blocks named *Ramp1* and *Ramp2* in Fig. 2.31 generate a signal that varies linearly with time t as $f(t) = f(t_0) + k(t - t_0)$, with $f(t) = 0$ for $t < t_0$. The initial time t_0, the initial value $f(t_0)$, and the slope k are given as inputs to the blocks in a menu that appears when their icons are double-clicked with the computer mouse. Two ramp generators, and a step generator, are used for producing the desired input acceleration $a_A(t)$ for part d, as shown in Fig. 2.32.

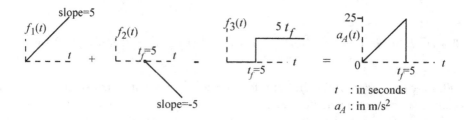

Figure 2.32: How to generate the acceleration input (for Simulink users) in Fig. 2.31 for part d of Example 2.7.6.

Example 2.7.7 Consider, again, the pendulum mounted on a moving base analyzed in Example 2.7.6. For $M_{\text{bar}} = 0$ and $\ddot{x} = \text{constant} = -a_A$:

  a. Determine the equilibrium values for the angle θ in terms of g and a_A.

  b. Linearize the differential equation of motion, Eq. (2.7.6d), about each equilibrium solution obtained in part a. For $a_A = 2g$, which equilibrium solutions are stable and which ones are unstable? Describe, in a few words, the motion $\theta(t)$ when it is started near the stable equilibrium.

  c. Repeat part b for $a_A = 0$.

■ Solution

a. For convenience, the differential equation of motion for the system is repeated here.

$$L\ddot{\theta} + g\sin\theta - a_A\cos\theta = \frac{M_{\text{bar}}}{mL} = 0 \qquad (2.7.7a)$$

The equilibrium values for the angle θ are, by definition, the constant values of $\theta \triangleq \theta_e$ that satisfy the differential equation of motion of the system. With $\theta = \text{constant} \triangleq \theta_e$, Eq. (2.7.7a) yields

$$g\sin\theta_e - a_A\cos\theta_e = 0 \qquad (2.7.7b)$$

or, simply,

$$\theta_e = \arctan\frac{a_A}{g}$$

For a given value of a_A/g, this equation yields two physically distinct values for θ_e, with the difference between those values being equal to $180°$. Physically, if the motion were started with $\theta(0) = \theta_e$ and $\dot{\theta}(0) = 0$, the angle $\theta(t)$ would remain constant and equal to θ_e for $t > 0$. However, if the equilibrium were perturbed by making the initial conditions different from these values (even slightly different), a motion would result.

b. To linearize the differential equation about each equilibrium, introduce a small perturbation $\theta_s(t)$ as $\theta(t) = \theta_e + \theta_s(s)$, expand each nonlinear term in the differential equation in Taylor series in $\theta_s(t)$, and disregard nonlinear terms in the expansion. Several examples of Taylor series expansion and linearization are given in Section 1.14 of Chapter 1.

The Taylor series expansions for $\sin\theta$ and $\cos\theta$ about $\theta = \theta_e$ are

$$\sin\theta = \sin\theta_e + \theta_s\cos\theta_e + \ldots \qquad (2.7.7c)$$
$$\cos\theta = \cos\theta_e - \theta_s\sin\theta_e + \ldots \qquad (2.7.7d)$$

By using these expansions, Eq. (2.7.7a) yields the following linearized differential equation for the perturbed motion $\theta_s(t)$:

$$\ddot{\theta}_s + \frac{g}{L}\left(\cos\theta_e + \frac{a_A}{g}\sin\theta_e\right)\theta_s = 0 \qquad (2.7.7e)$$

Or, since $\tan\theta_e = a_A/g$,

$$\ddot{\theta}_s + \frac{g}{L\cos\theta_e}\theta_s = 0 \qquad (2.7.7\text{f})$$

For $a_A = 2g$ we then have for each equilibrium:

$$\text{for } \theta_e \approx 63.4°: \qquad \ddot{\theta}_s + 2.24\frac{g}{L}\theta_s = 0 \qquad (2.7.7\text{g})$$

$$\text{for } \theta_e = 243.4°: \qquad \ddot{\theta}_s - 2.24\frac{g}{L}\theta_s = 0 \qquad (2.7.7\text{h})$$

Therefore, as discussed in Section 1.15 of Chapter 1, the equilibrium $\theta_e \approx$ 63.4° is infinitesimally stable and the equilibrium $\theta_e = 243.4°$ is unstable. It can be verified that the equilibrium $\theta_e \approx 63.4°$ is stable by using the Lyapunov method presented in Chapter 1 and/or numerically integrating Eq. (2.7.7a) with initial conditions near that equilibrium. Therefore, a motion $\theta(t)$ that is started near the equilibrium $\theta_e = 63.4°$ stays near that equilibrium as time passes. Such a motion consists of an undamped oscillation (as per the material that was first presented in Section 1.13) about that equilibrium, with frequency equal to $\omega = \sqrt{2.24g/L} \approx 1.5\sqrt{g/L}$ rad/s. In contrast, the angle $\theta(t)$ does not remain near the equilibrium $\theta_e = 243.4°$ even if it is started very close to that equilibrium.

Notice that the equilibrium $\theta_e = 63.4°$ becomes asymptotically stable (i.e., the motion dies out and returns to the equilibrium state as time increases) when viscous damping, for example, is added to the system. Equation (2.7.6i) (p. 158) is the differential equation of motion in such a case.

c. For $a_A = 0$, the equilibrium solutions are $\theta_e = 0$ and $\theta_e = 180°$. In this case, the linearized motion consists of small undamped oscillations only about the equilibrium $\theta_e = 0$, and the frequency of those small oscillations is $\omega = \sqrt{g/L}$ rad/s. The equilibrium $\theta_e = 180°$ is unstable.

Suggestion: Solve this problem by numerically integrating the nonlinear differential equation of motion with and without viscous damping, and compare the results with those obtained from the linearized differential equation.

Example 2.7.8 Analysis of the Motion of the Spring-pendulum in Two Dimensions

The motion of the spring-pendulum shown in Fig. 2.33a takes place on a vertical plane. The pendulum consists of a small body (modeled as a point mass P) of mass

m attached to the end of a massless linear spring of unstressed length equal to L_u m and stiffness k N/m. The other end of the spring is connected to a fixed frictionless pin, which is an inertial point O. The only forces acting on P are the gravitational attraction of the earth and the reaction force applied by the spring. Figure 2.33b shows the free-body diagram for P.

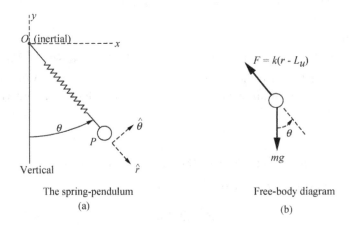

The spring-pendulum
(a)

Free-body diagram
(b)

Figure 2.33: The spring-pendulum for Example 2.7.8.

a. Obtain the differential equations of motion for the system in terms of the polar coordinates $r = |\overrightarrow{OP}|$ and θ shown in Fig. 2.33.

b. As indicated throughout the book, a dynamical system is said to be in equilibrium when the variables that are used for describing its motion are constants. Therefore, for the case of this pendulum, equilibrium means $r =$ constant (let us call that constant r_e) and $\theta =$ constant (let us call that constant θ_e). Show that there are two physically distinct equilibrium solutions that satisfy the differential equations of motion and that they are the following two pairs of values for θ_e and r_e: ($\theta_e = 0$, $r_e = L_u + mg/k$), and ($\theta_e = \pi$, $r_e = L_u - mg/k$).

c. Rewrite the differential equations obtained in part a in terms of the nondimensional variables $r_s \overset{\Delta}{=} (r - r_e)/r_e$, which is the deviation distance from the equilibrium distance r_e, nondimensionalized by r_e, and the angle θ.

d. Linearize the differential equations of motion about the equilibrium ($\theta_e = 0$, $r_e = L_u + mg/k$), and determine the frequencies of the small oscillations of the spring-pendulum about that equilibrium (if needed, review Sections 1.14 and 1.13 in Chapter 1 for the material to do this).

e. Since the differential equations for this system are nonlinear and no exact analytical solution can be obtained for them, numerically solve the differential equations in part c for $r_s(t)$ and $\theta(t)$. For this, use the nondimensional time τ defined as $\tau = t\sqrt{k/m}$. Integrate the differential equations with the initial conditions $r_s(0) = 0.05$, $\dot{r}_s(0) = 0$, $\theta(0) = r_s(0)/10$, $\dot{\theta}(0) = 0$, and plot $r_s(\tau)$ and $\theta(\tau)$ versus τ for $\sqrt{mg/(kr_e)} = 0.5$ and 0.3.

■ Solution

a. The free-body diagram for the mass particle P is shown in Fig. 2.33. For convenience, let us use the triad $\{\hat{r}, \hat{\theta}, \hat{z} \overset{\Delta}{=} \hat{r} \times \hat{\theta}\}$, where the unit vectors $\hat{r}$ and $\hat{\theta}$ are shown in the figure. The absolute angular velocity of that triad is

$$\vec{\omega} = \dot{\theta}\hat{z} \tag{2.7.8a}$$

We then have for point P where the mass particle is located:

■ Absolute position vector of P:

$$\vec{r} = \overrightarrow{OP} = r\hat{r} \tag{2.7.8b}$$

■ Absolute velocity of P:

$$\vec{v} = \frac{d\vec{r}}{dt} = \dot{r}\hat{r} + \vec{\omega} \times \vec{r} = \dot{r}\hat{r} + r\dot{\theta}\hat{\theta} \tag{2.7.8c}$$

■ Absolute acceleration of P:

$$\vec{a} = \frac{d\vec{v}}{dt} = \ddot{r}\hat{r} + (r\ddot{\theta} + \dot{r}\dot{\theta})\hat{\theta} + \vec{\omega} \times \vec{v} = (\ddot{r} - r\dot{\theta}^2)\hat{r} + (r\ddot{\theta} + 2\dot{r}\dot{\theta})\hat{\theta} \tag{2.7.8d}$$

■ Resultant force acting on P:

$$\vec{F} = mg(\hat{r}\cos\theta - \hat{\theta}\sin\theta) - k(r - L_u)\hat{r} \tag{2.7.8e}$$

Newton's second law $\vec{F} = m\vec{a}$ then yields the following scalar equations for the spring-pendulum. These equations involve only the variables r and θ and their time derivatives. Therefore, they are the differential equations of motion for the system.

$$m(\ddot{r} - r\dot{\theta}^2) = mg\cos\theta - k(r - L_u) \tag{2.7.8f}$$
$$m(r\ddot{\theta} + 2\dot{r}\dot{\theta}) = -mg\sin\theta \tag{2.7.8g}$$

Note: Equation (2.7.8g) could also have been obtained by using the moment equation $\vec{M}_O = d\,\vec{H}_O = d[\vec{r} \times (m\vec{v})]/dt$. This yields the following scalar equation, which is equivalent to Eq. (2.7.8g):

$$m\frac{d}{dt}(r^2\dot{\theta}) = -mgr\sin\theta$$

b. As indicated in the problem statement, the equilibrium solution corresponds to $r = \text{constant} \stackrel{\Delta}{=} r_e$ and $\theta = \text{constant} \stackrel{\Delta}{=} \theta_e$. The constant values r_e and θ_e have to satisfy the differential equations of motion. Therefore, by substituting this constant solution into Eqs. (2.7.8f) and (2.7.8g), the following equations are obtained for r_e and θ_e:

$$mg\cos\theta_e - k(r_e - L_u) = 0 \qquad (2.7.8h)$$
$$mg\sin\theta_e = 0 \qquad (2.7.8i)$$

The solutions to Eq. (2.7.8i) are $\theta_e = 0, \pi, 2\pi, 3\pi, \ldots$, but only two solutions $\theta_e = 0$ and $\theta_e = \pi$ are physically distinct.

For $\theta_e = 0$, the first equilibrium equation yields $r_e = L_u + mg/k$, and for $\theta_e = \pi$ we obtain $r_e = L_u - mg/k$. The second solution corresponds to an upside down pendulum and will not be considered further. The equilibrium $\theta_e = \pi$ happens to be unstable.

c. Let us now start to investigate what the spring-pendulum does when a motion is started with some initial conditions. The main interest in practice is to find out what kind of motion results when an equilibrium of the system is perturbed. For this, it is always convenient to redefine the motion variables as

$$variable\_name(t) = variable\_name_e + new\_variable\_name(t)$$

where *variable_name* and *variable_name_e* represent, respectively, the time-dependent variables being used for describing the motion (for this problem, they are r and θ) and their constant equilibrium values. The equilibrium under consideration here is $r_e = L_u + mg/k$ and $\theta_e = 0$. The subscript e in this book stands for *equilibrium*. The time-dependent *new_variable_name* stands for any convenient symbol associated with the perturbation of the variable *variable_name*. For the problem being analyzed here we might write $r(t) = r_e + r_s(t)$, or use any other symbol for the perturbation $r(t) - r_e$ of the variable $r(t)$. Since $\theta_e = 0$, there is no need to introduce a new symbol to replace θ.

When integrating differential equations in dynamics, it is convenient to nondimensionalize all variables for the problem, including the time t (time nondimensionalization was first discussed in Section 1.13 of Chapter 1). This will actually reduce the number of parameters in a problem to a minimum. In this problem, in addition to the four initial conditions for the two second-order differential equations, Eqs. (2.7.8f) and (2.7.8g), the quantities m, k, L_u, and g appear in those equations. Nondimensionalization of r and time t can actually reduce these four quantities to only one nondimensional parameter for this problem, which, as you will see in the rest of the solution, is $kr_e/(mg) = 1 + kL_u/(mg)$.

Let us then proceed by redefining the variable $r(t)$ in terms of a nondimensional new variable $r_s(t)$ as:

$$r(t) = r_e[1 + r_s(t)] \tag{2.7.8j}$$

The physical interpretation of the variable $r_s(t)$ is simply the elongation of the spring-pendulum, measured from its equilibrium length r_e and nondimensionalized by r_e.

Substitution of Eq. (2.7.8j) into Eqs. (2.7.8f) and (2.7.8g) yields

$$\ddot{r}_s - (1+r_s)\dot{\theta}^2 = \frac{g}{r_e}\cos\theta - \frac{k}{mr_e}(r_e + r_e r_s - L_u) = -\frac{g}{r_e}(1 - \cos\theta) - \frac{k}{m}r_s \tag{2.7.8k}$$

$$(1+r_s)\ddot{\theta} + 2\dot{r}_s\dot{\theta} = -\frac{g}{r_e}\sin\theta \tag{2.7.8l}$$

Notice that the four quantities m, k, g, and L_u now appear only as two parameters, k/m and $g/r_e = 1/(L_u/g + m/k)$, in these differential equations. The number of parameters can be further reduced to only one by nondimensionalizing the time t. This is done in part e.

d. For the linearization process, r_s and θ are now assumed to remain small enough so that $\sin\theta \approx \theta$ and $\cos\theta \approx 1$, and products of all the higher-order small terms ($r_s\ddot{\theta}, \dot{r}_s\dot{\theta}, \dot{\theta}^2, r_s\dot{\theta}^2$, and others) that will appear in the expansion of Eqs. (2.7.8k) and (2.7.8l) are neglected. The objective of doing this is to obtain the linearized differential equations, for which the solutions can be readily obtained (as shown in Case 3 in Section 1.13 of Chapter 1).

The differential equations that govern the linearized motions about the equilibrium that corresponds to $\theta_e = 0$ and $r_e = L_u + mg/k$ are then

$$\ddot{r}_s + \frac{k}{m}r_s = 0 \tag{2.7.8m}$$

$$\ddot{\theta} + \frac{g}{r_e}\theta = 0 \tag{2.7.8n}$$

These equations describe two independent undamped oscillations: a θ oscillation with frequency equal to $\sqrt{g/r_e}$, and an r_s oscillation with frequency equal to $\sqrt{k/m}$. In dynamics and vibrations, the frequencies of the linearized oscillations of a system are called the *natural frequencies* of oscillation. Linear oscillations are what engineers hope that systems will exhibit in practice because their character can be easily studied and predicted.

e. It is generally a good idea to nondimensionalize the time t to minimize the number of parameters one has to deal with when integrating the differential equations. With $\tau = t\sqrt{k/m}$, the time derivatives $\dot{r}_s, \ddot{r}_s, \dot{\theta}$, and $\ddot{\theta}$ are transformed as follows (using the chain rule)[11]:

$$\dot{r}_s = \frac{dr_s}{d\tau}\frac{d\tau}{dt} = \sqrt{\frac{k}{m}}\frac{dr_s}{d\tau} \qquad \ddot{r}_s = \frac{d\dot{r}_s}{dt} = \left[\frac{d}{d\tau}\left(\sqrt{\frac{k}{m}}\frac{dr_s}{d\tau}\right)\right]\frac{d\tau}{dt} = \frac{k}{m}\frac{d^2r_s}{d\tau^2}$$

$$\dot{\theta} = \sqrt{\frac{k}{m}}\frac{d\theta}{d\tau} \qquad \ddot{\theta} = \frac{k}{m}\frac{d^2\theta}{d\tau^2}$$

Therefore, Eqs. (2.7.8k) and (2.7.8l) become

$$\frac{d^2r_s}{d\tau^2} - (1+r_s)\left(\frac{d\theta}{d\tau}\right)^2 = -r_s - \frac{mg}{kr_e}(1-\cos\theta) \qquad (2.7.8\text{o})$$

$$(1+r_s)\frac{d^2\theta}{d\tau^2} + 2\frac{dr_s}{d\tau}\frac{d\theta}{d\tau} = -\frac{mg}{kr_e}\sin\theta \qquad (2.7.8\text{p})$$

Notice that now only one parameter, $mg/(kr_e)$, appears in the differential equations. This is a great advantage because the same response $r_s(\tau)$ and $\theta(\tau)$ obtained for any particular value of this parameter (which is the square of the ratio of the two natural frequencies of oscillation of the spring-pendulum) is exhibited for many spring-pendulum systems built with different springs and masses. One would have to know, though, the value of the time nondimensionalization factor $\sqrt{k/m}$ to relate $r_s(\tau)$ to $r_s(t)$ and $\theta(\tau)$ to $\theta(t)$ for each system.

$\implies$ The Scilab program *spring_pendulum.sci* numerically integrates Eqs. 2.7.8o) and (2.7.8p), generates plots of θ and r_s versus nondimensional time τ, and animates the motion. Its Matlab counterpart *spring_pendulum.m* does the same. Both are listed in Appendix E. As indicated in the comment lines at the beginning of those programs, an example of their use [with $mg/(kr_e) =$

---

[11]The frequency of a linearized oscillation is generally a convenient time nondimensionalization factor to use. Thus, for this problem, we could also have used the nondimensional time $\tau = t\sqrt{g/r_e}$.

0.5^2, and the initial conditions $r_s(0) = 0.1, \dot{r}_s(0) = 0, \theta(0) = 0.01$, and $\dot{\theta}(0) = 0$] is (try $\Delta t = 0.1$ for Scilab, and 0.02 for Matlab):

$$\text{my\_output} = \text{spring\_pendulum}([0 : \Delta t : 350], 0.5 \wedge 2, 0.1, 0, 0.01, 0);$$

The second parameter on the call to the function *spring_pendulum* is the value of $mg/(kr_e)$. If you are using either Scilab or Matlab, try changing the value of $mg/(kr_e)$ by small amounts (using 0.48^2 and 0.52^2, for example) to see how it affects the motion. It turns out that as the value of $\sqrt{mg/(kr_e)}$ moves away from the most critical 0.5 value, the nonlinear phenomenon (discussed in the sequel) starts to disappear, until the motion is eventually very close to that modeled by the linearized differential equations (2.7.8m) and (2.7.8n). Also, one may wish to change the step size of the time array (which is the middle value in the array $[0 : \Delta t : 350]$, just illustrated) to either slow down or speed up the animation. A smaller Δt will slow it down, and a larger Δt will speed it up.

Those who have Simulink and prefer to use it may want to construct the block diagram model shown in Fig. 2.34, setting the *Simulation $\rightarrow$ Parameters* menu for this model with time t of the form $[0:\Delta t:350]$, and using the same

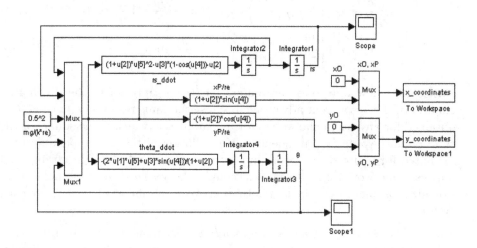

Figure 2.34: Simulation of the spring-pendulum in Example 2.7.8, for Simulink users.

initial conditions mentioned in the previous paragraph. To see an animation

of the motion after the simulation stops, type the following command in the
Matlab command window: `animate_nbars(x_coordinates, y_coordinates)`.
Figures 2.35 and 2.36 show the results of the numerical integration of Eqs.

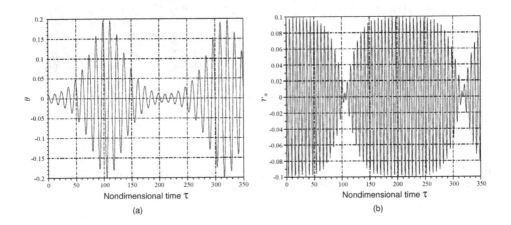

Figure 2.35: Results of the numerical integration of Eqs. (2.7.8o) and (2.7.8p) with
$mg/(kr_e) = (1/2)^2$, $r_s(0) = 0.1$, $\theta(0) = 0.01$, and $[dr_s/d\tau]_{\tau=0} = [d\theta/d\tau]_{\tau=0} = 0$
(Example 2.7.8).

(2.7.8o) and (2.7.8p) for $mg/(kr_e) = (0.5)^2$, and Fig. 2.37 for $mg/(kr_e) = (0.3)^2$. For Figs. 2.35 and 2.37 the integration was done with the initial conditions $r_s(0) = 0.1, \theta(0) = r_s(0)/10 = 0.01$, and $[dr_s/d\tau]_{\tau=0} = [d\theta/d\tau]_{\tau=0} = 0$. For Fig. 2.36 the initial conditions are $r_s(0) = 0.05, \theta(0) = r_s(0)/10 = 0.005$, and $[dr_s/d\tau]_{\tau=0} = [d\theta/d\tau]_{\tau=0} = 0$. Notice the beating phenomenon in Figs. 2.35 and 2.36, with the θ motion slowly growing to a value that is much larger than its initial value.

The beating phenomenon that is seen in Figs. 2.35 and 2.36 is a nonlinear phenomenon, and it is completely missed by the linearized differential equations $\ddot{r}_s + (k/m)r_s = 0$ and $\ddot{\theta} + (g/r_e)\theta = 0$. When such a nonlinear phenomenon occurs, the θ motion that is started with a very small amplitude slowly grows to a much larger amplitude that happens to be independent of the initial conditions in θ and depends only on the initial conditions in r_s. This is illustrated in Figs. 2.35 and 2.36. The smaller the initial conditions in θ, the longer it takes for the angle θ to grow to $\theta_{\max} \approx 2r_s(0)$ when $mg/(kr_e) = (1/2)^2$.

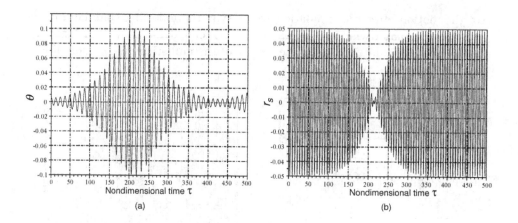

Figure 2.36: Results of the numerical integration of Eqs. (2.7.8o) and (2.7.8p) with $mg/(kr_e) = (1/2)^2$, $r_s(0) = 0.05$, $\theta(0) = 0.005$, and $[dr_s/d\tau]_{\tau=0} = [d\theta/d\tau]_{\tau=0} = 0$ (Example 2.7.8).

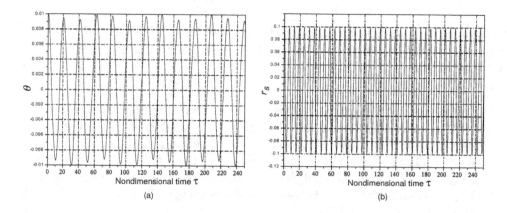

Figure 2.37: Results of the numerical integration of Eqs. (2.7.8o) and (2.7.8p) with $mg/(kr_e) = (0.3)^2$, $r_s(0) = 0.1$, $\theta(0) = 0.01$, and $[dr_s/d\tau]_{\tau=0} = [d\theta/d\tau]_{\tau=0} = 0$ (Example 2.7.8).

The nonlinear phenomenon just described can actually be observed with the experiment mentioned in Section 1.19 of Chapter 1. Such a phenomenon is known as a *nonlinear resonance phenomenon*. Since $r_e = L_u + mg/k$, the value of r_e when $mg/(kr_e) = (1/2)^2$ is $r_e = 4L_u/3$, which is the value that was mentioned in Section 1.19 of Chapter 1.

Analytical investigations of nonlinear oscillations using *perturbation methods*,[12] which are beyond the scope of this book, actually disclose that the
nonlinear phenomenon previously mentioned occurs for a certain range of the
parameter $\sqrt{mg/(kr_e)}$ (which is the ratio of the two natural frequencies of
the spring-pendulum) in the neighborhood of $\sqrt{mg/(kr_e)} = 1/2$ and disappears outside that range. The width of the nonlinear resonance range depends
on the value of the initial condition $r_s(0)$. The response shown in Fig. 2.37
for $\sqrt{mg/(kr_e)} = 0.3$ does not exhibit the nonlinear resonance phenomenon
that caused θ to slowly grow as seen in Figs. 2.35 and 2.36. The value of
$mg/(kr_e) = 0.3^2$ happens to be outside the region for which the nonlinearities
in the differential equations of motion dominate the dynamic behavior of the
system. The r_s and θ motions shown in Fig. 2.37 are very close to the response
predicted by the linearized differential equations. In practice, the nonlinear resonance phenomenon can be eliminated by either increasing or decreasing the
mass at the end of the spring.

Example 2.7.9 A particle P of mass $m = 0.5$ kg moves inside a 2 m horizontal thin
tube filled with oil. The tube is always in contact with a fixed horizontal table and
is driven by a motor with a constant counterclockwise angular velocity $\Omega = 5$ rad/s
relative to inertial space. End O of the tube is inertial, and the particle is always
in contact with a wall of the tube (see Fig. 2.38).

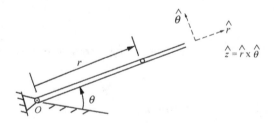

Figure 2.38: A constrained particle on a rotating tube for Example 2.7.9.

Because of the viscosity of the oil, P is subjected to a linear viscous friction
force with friction coefficient $c = 1$ N/(m/s). If P is released from rest relative to
the rotating tube (i.e., with zero velocity relative to the tube) at a distance $r_0 = 1$

---

[12]See Reference 18 listed in Appendix G.

m from point O, determine the time it takes for P to reach the edge of the tube at $r = 2$ m. Approximate the tube as massless.

■ Solution

By working with polar coordinates, and the tube-fixed unit vectors $\hat{r}$ and $\hat{\theta}$ shown in Fig. 2.38 (whose absolute angular velocity is $\vec{\omega} = \dot{\theta}\hat{z} = \Omega\hat{z}$, where $\hat{z} = \hat{r} \times \hat{\theta}$), the absolute position vector $\vec{r}$, absolute velocity $\vec{v}$, and absolute acceleration $\vec{a}$ of the particle are (with $\Omega = $ constant, so that $\dot{\Omega} = 0$):

$$\vec{r} = r\hat{r} \tag{2.7.9a}$$

$$\vec{v} = \frac{d\vec{r}}{dt} = \dot{r}\hat{r} + \omega \times \vec{r} = \dot{r}\hat{r} + \Omega r\hat{\theta} \tag{2.7.9b}$$

$$\vec{a} = \frac{d\vec{v}}{dt} = \ddot{r}\hat{r} + \frac{d(\Omega r)}{dt}\hat{\theta} + \omega \times \vec{v} = (\ddot{r} - \Omega^2 r)\hat{r} + 2\Omega\dot{r}\hat{\theta} \tag{2.7.9c}$$

The free-body diagrams for the particle and for the tube are shown in Fig. 2.39. The forces that appear in the free-body diagram for the particle are

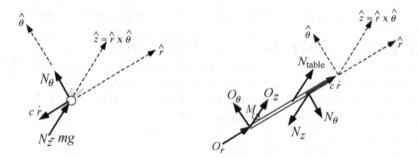

Figure 2.39: Free-body diagrams and the unit vectors $(\hat{r}, \hat{\theta}, \hat{z} = \hat{r} \times \hat{\theta})$ for Example 2.7.9.

1. The gravitational force, equal to $-mg\hat{z}$.
2. The linear viscous friction force, which is proportional to the velocity $\dot{r}\hat{r}$ of the particle relative to the rotating tube and acts in the direction opposite to $\dot{r}\hat{r}$. It is assumed that this is the total friction force that is acting on the particle, even if the particle touches the wall of the tube. With c being the coefficient of viscous friction, that force is then $\vec{f} = -c\dot{r}\hat{r}$.
3. A force with components in the $\hat{\theta}$ and $\hat{z}$ directions due to the contact between P and the rotating tube. These force components, $N_\theta\hat{\theta}$ and $N_z\hat{z}$, are shown in the free-body diagram.

The free-body diagram for the tube consists of the forces $-\vec{f} = c\dot{r}\hat{r}, -N_z\hat{z}$, $-N_\theta\hat{\theta}$, a force $N_{\text{table}}\hat{z}$ that is the resultant of the distributed force due to the contact between the tube and the table, and the support reaction forces $O_r\hat{r}, O_\theta\hat{\theta}$, and $O_z\hat{z}$ at point O. The moment $M\hat{z}$ applied by the motor to the tube is also shown in the diagram. It is assumed that there is no friction between the fixed table and the rotating tube (otherwise there would also be an additional force acting on the tube due to such a friction).

The vector sum of the forces acting on P is

$$\vec{F} = (N_z - mg)\hat{z} - c\dot{r}\hat{r} + N_\theta\hat{\theta} \tag{2.7.9d}$$

By applying Newton's second law, $\vec{F} = m\vec{a}$, to the particle, the following scalar equations are then obtained.

$$\ddot{r} + \frac{c}{m}\dot{r} - \Omega^2 r = 0 \tag{2.7.9e}$$

$$N_\theta = 2m\Omega\dot{r} \tag{2.7.9f}$$

$$N_z = mg \tag{2.7.9g}$$

Equation (2.7.9e) is the differential equation for the motion of P relative to the rotating tube since that equation does not involve any unknown reaction forces. That equation is a linear differential equation with constant coefficients, and its solution was first presented in Section 1.13 of Chapter 1. Equations (2.7.9f) and (2.7.9g) are simply algebraic equations that can be used for calculating, if desired, the components N_θ and N_z of the contact force between P and the tube. Notice that the calculation of N_θ requires the integration of the differential equation of motion first to determine $\dot{r}(t)$, which is needed in Eq. (2.7.9f) for finding that reaction force.

To determine the value of t for which $r = R = 2$ m, the differential equation of motion has to be solved for $r(t)$ first. The solution is given either by Eq. (1.13.11) or Eq. (1.13.13) in Chapter 1. With $m = 0.5$ kg, $c = 1$ N/(m/s), and $\Omega = 5$ rad/s, the two roots of the characteristic equation $s^2 + (c/m)s - \Omega^2 = 0$ are

$$s_1 = -\frac{c}{2m} + \sqrt{\Omega^2 + \left(\frac{c}{2m}\right)^2} \approx 4.1 \text{ s}^{-1}$$

$$s_2 = -\frac{c}{2m} - \sqrt{\Omega^2 + \left(\frac{c}{2m}\right)^2} \approx -6.1 \text{ s}^{-1}$$

Therefore, the general solution to Eq. (2.7.9e) is

$$r(t) = A_1 e^{s_1 t} + A_2 e^{s_2 t}$$

where A_1 and A_2 are constants determined from the initial conditions as

$$A_1 = \frac{\dot{r}_0 - s_2 r_0}{s_1 - s_2} \approx 0.6 \text{ m}$$

$$A_2 = \frac{s_1 r_0 - \dot{r}_0}{s_1 - s_2} \approx 0.4 \text{ m}$$

The value of the time t that corresponds to $r = R = 2$ m is obtained after a few numerical iterations with the solution found for $r(t)$ as $t \approx 0.285$ s (≈ 0.29 s). Notice that, when $t = 0.285$ s, the tube has turned by $\Omega t = 1.425$ rad $\approx 81.6°$. It is interesting to notice that if there were no friction between P and the tube (i.e., if $c = 0$), one would obtain $r(t) = \left[\dot{r}_0 \left(e^{\Omega t} - e^{-\Omega t}\right) + \Omega r_0 \left(e^{\Omega t} + e^{-\Omega t}\right)\right]/(2\Omega)$, and this would have yielded $\Omega t \approx 1.32$ rad $\approx 75.6°$ for $r = R = 2$ m. This shows, as one would have expected, that the friction force is responsible for keeping the particle on the tube for a longer time.

Example 2.7.10 Two particles P_1 and P_2 of masses m_1 and m_2, respectively, are connected by a massless inextensible string of length L that passes through a small hole on a fixed frictionless horizontal table as shown in Fig. 2.40.

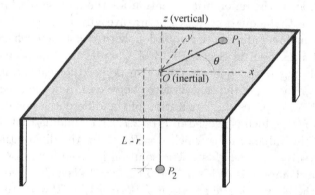

Figure 2.40: Dynamics of two constrained particles for Example 2.7.10.

The motion of the system is started with the initial conditions $r(0) = r_0$, $\dot{r}(0) = \dot{r}_0$, $\theta(0)$ arbitrary, and $\dot{\theta}(0) = \dot{\theta}_0$. Particle P_2 is constrained to move along the vertical.

    a. Obtain the differential equation of motion for particle P_1 in terms of the distance $r(t)$.

b. Determine the tension F_r in the string as a function of $r(t)$.

c. For $m_2 = 2m_1$, $L = 1$ m, $r(0) = 0.5$ m, $\dot{r}(0) = 1$ m/s, and $\dot{\theta}(0) = 2$ rad/s, what are the maximum and minimum values of $r(t)$ and of the tension in the string?

■ Solution

a. The free-body diagrams for both point masses are shown in Fig. 2.41. Orthogonal unit vectors $\hat{r}$, $\hat{\theta}$, and $\hat{z} = \hat{\theta} \times \hat{r}$ are also included in the figure. Line A-A is simply a nonrotating reference line that is used for measuring the angle $\theta(t)$ for the rotating line OP_1, while the force $-N\hat{z}$ is the normal reaction force applied by the table to particle P_1. The minus sign in $-N\hat{z}$ is associated with the unit vector $\hat{z}$. If the unit vector $\hat{z}$ were chosen to be pointing upward instead of downward, one would write $N\hat{z}$ for the normal reaction force and $-m_1g\hat{z}$ for the gravitational force acting on P_1.

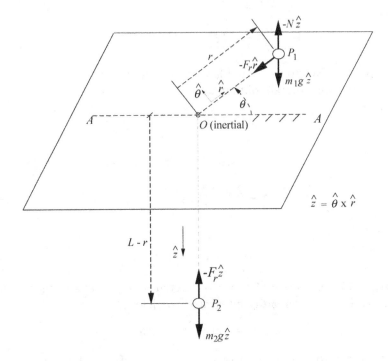

Figure 2.41: Free-body diagrams, motion variables, and unit vectors for Example 2.7.10.

The resultant forces acting on P_1 and P_2 are then

$$\text{Particle } P_1: \qquad \vec{F}_1 = -F_r \hat{r} + (m_1 g - N)\,\hat{z} \qquad (2.7.10\text{a})$$

$$\text{Particle } P_2: \qquad \vec{F}_2 = (m_2 g - F_r)\,\hat{z} \qquad (2.7.10\text{b})$$

To obtain the expression for the absolute acceleration for each particle, start with the respective absolute position vectors, which are $\vec{r}_1 = r\hat{r}$ (for particle P_1) and $\vec{r}_2 = (L - r)\hat{z}$ (for particle P_2). This gives the following expressions for the absolute velocity and the absolute acceleration of both particles.

$$\vec{v}_1 = \frac{d\vec{r}_1}{dt} = \dot{r}\hat{r} + \left(-\dot{\theta}\hat{z}\right) \times \vec{r}_1 = \dot{r}\hat{r} + r\dot{\theta}\hat{\theta} \qquad (2.7.10\text{c})$$

$$\begin{aligned} \vec{a}_1 &= \frac{d\vec{v}_1}{dt} = \ddot{r}\hat{r} + \left(r\ddot{\theta} + \dot{r}\dot{\theta}\right)\hat{\theta} + \left(-\dot{\theta}\hat{z}\right) \times \vec{v}_1 \\ &= \left(\ddot{r} - r\dot{\theta}^2\right)\hat{r} + \left(r\ddot{\theta} + 2\dot{r}\dot{\theta}\right)\hat{\theta} \end{aligned} \qquad (2.7.10\text{d})$$

$$\vec{v}_2 = \frac{d\vec{r}_2}{dt} = -\dot{r}\hat{z} \qquad (2.7.10\text{e})$$

$$\vec{a}_2 = \frac{d\vec{v}_2}{dt} = -\ddot{r}\hat{z} \qquad (2.7.10\text{f})$$

By applying Newton's second law to each individual particle, the following equations are immediately obtained.

Particle P_1:

$$N = m_1 g \qquad (2.7.10\text{g})$$

$$m_1\left(\ddot{r} - r\dot{\theta}^2\right) = -F_r \qquad (2.7.10\text{h})$$

$$r\ddot{\theta} + 2\dot{r}\dot{\theta} \equiv \frac{1}{r}\frac{d}{dt}\left(r^2\dot{\theta}\right) = 0 \qquad (2.7.10\text{i})$$

Particle P_2:

$$m_2\ddot{r} = F_r - m_2 g \qquad (2.7.10\text{j})$$

Equation (2.7.10i) is one of the two differential equations of motion for this system, and it can be readily integrated to yield

$$r^2\dot{\theta} = \text{constant} = r_0^2\dot{\theta}_0 \qquad (2.7.10\text{k})$$

Equations (2.7.10h) and (2.7.10j) may be combined to eliminate the tension F_r to yield the second differential equation of motion for the system, which is

$$(m_1 + m_2)\ddot{r} - m_1 r\dot{\theta}^2 + m_2 g = 0 \qquad (2.7.10\text{l})$$

By making use of Eq. (2.7.10k) to eliminate $\dot{\theta}$ from this equation, we immediately obtain the differential equation that only involves the variable $r(t)$, which is

$$(m_1 + m_2)\ddot{r} - \frac{m_1 r_0^4 \dot{\theta}_0^2}{r^3} + m_2 g = 0 \qquad (2.7.10\text{m})$$

Notice that Eqs. (2.7.10k) and (2.7.10m) disclose that if the initial condition $\dot{\theta}_0$ is zero, i.e., if the initial absolute velocity of P_1 has no component along the normal to line OP_1, the angle $\theta(t)$ and the acceleration of both particles remain constant.

b. To determine the string tension F_r as a function of $r(t)$, Eqs. (2.7.10j) and (2.7.10m) are combined to obtain the following result:

$$F_r = \frac{m_1 m_2}{m_1 + m_2}\left(g + \frac{r_0^4 \dot{\theta}_0^2}{r^3}\right) \qquad (2.7.10\text{n})$$

Notice that inspection of Eq. (2.7.10n) discloses that the maximum value of F_r occurs when $r(t)$ is minimum (and vice versa).

c. Let us now determine specific values for the maximum and minimum values of $r(t)$ and the tension F_r. The extremum (i.e., maximum and minimum) values for $r(t)$ may be obtained by first making use of the chain rule of differentiation to write $\ddot{r} = (d\dot{r}/dr)\dot{r}$, and then integrating Eq. (2.7.10m) once with respect to r, according to the procedure presented in Case 4 of Section 1.13 in Chapter 1, which is used throughout the book. This yields

$$\frac{1}{2}\left(m_1 + m_2\right)\dot{r}^2 + \frac{m_1 r_0^4 \dot{\theta}_0^2}{2r^2} + m_2 g r = \text{constant}$$

$$= \frac{1}{2}\left(m_1 + m_2\right)\dot{r}_0^2 + \frac{m_1 r_0^2 \dot{\theta}_0^2}{2} + m_2 g r_0 \qquad (2.7.10\text{o})$$

By letting r_m denote the extremum values of $r(t)$, the values of r_m are obtained by setting $\dot{r} = 0$ in Eq. (2.7.10o). For $m_2 = 2m_1$, $r(0) = 0.5$ m, $\dot{r}(0) = 1$ m/s, and $\dot{\theta}(0) = 2$ rad/s, this yields the following cubic equation for r_m:

$$g r_m^3 - \left(1 + \frac{g}{2}\right)r_m^2 + \frac{1}{16} = 0$$

The three roots of this cubic equation, determined with $g = 9.81$ m/s^2, are

$$r_m \approx 0.583 \qquad r_m \approx -0.096 \qquad r_m \approx 0.114$$

Noticing that with $r_m = 0$ the left-hand side of the cubic equation for r_m is equal to $1/16 = 0.0625$, it is reasonable to expect one root to be close to such a value. With $g = 9.81$ m/s^2, a root may be found with just a few iterations as $r_m \approx 0.114$. The other two roots may be found either in a similar manner or by factoring out the root $r_m \approx 0.114$ and then solving the resulting quadratic equation to obtain them. They may also be readily determined using the *root* command in either Scilab or Matlab; its syntax for determining the roots of the nth degree polynomial equation $c_n x^n + c_{n-1} x^{n-1} + \ldots\ldots + c_1 x + c_0 = 0$ is roots($[c_n, c_{n-1}, \ldots, c_1, c_0]$). For our equation, one would then use roots($[9.81, -(1 + 9.81/2), 0, 1/16]$), and the three roots will be displayed on the computer screen.

Since $r(t)$ cannot be negative, the maximum and minimum values of $r(t)$ are then

$$r_{\max} \approx 0.583 \text{ m} \qquad\qquad r_{\min} \approx 0.114 \text{ m}$$

Therefore, according to Eq. (2.7.10n), the maximum and minimum values of the tension F_r are (with m_1 in kilograms):

$$(F_r)_{\max} \approx 119 m_1 \text{ N} \qquad\qquad (F_r)_{\min} \approx 7.38 m_1 \text{ N}$$

$\implies$ For Scilab users, the program *two_particles_table.sci* listed in Appendix E numerically integrates Eqs. (2.7.10l) and (2.7.10k), using r/L instead of r. For Matlab users, the program *two_particles_table.m*, also listed in Appendix E, will do the same. Both generate a three-dimensional animation of the motion with the string $P_1 O P_2$ seen in Fig. 2.40 represented by two bars OP_1 and OP_2 whose coordinates at P_1 and P_2, using an x-y-z coordinate system centered at O, with x along line AA and pointing to the right, and z pointing upward, are $(r \cos\theta, r \sin\theta, 0)$ and $(0, 0, r - L)$, respectively. The comment lines at the beginning of both programs show an example of usage. Logic statements could be added to these programs to make the simulation stop if either P_1 or P_2 hits point O.

Those who have Simulink and prefer to use it may want to construct the block diagram model shown in Fig. 2.42, setting the *Simulation → Parameters* menu for this model with time t of the form [0:Δt:10], and using the initial conditions $r(0)/L = 0.5$, $\dot{r}(0) = 0$, $\theta(0) = 0$, and $\dot{\theta}(0) = 2$. In such a case, an animation of the motion is obtained by typing
animate3d_nbars(x_coordinates, y_coordinates, z_coordinates)
in the Matlab command window after the simulation stops.

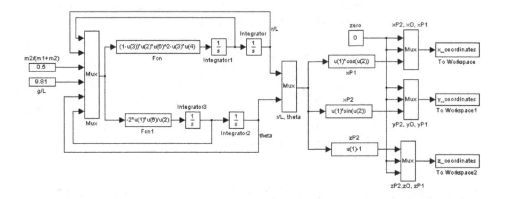

Figure 2.42: A Simulink model for the system of Example 2.7.10.

Example 2.7.11 The Central Force Problem for Modeling the Dynamics of Satellites

In Example 2.5.11 (p. 137), it was seen that the translational motion of a satellite P subjected only to the central gravitational attraction of a spherical planet takes place in a single plane. Let us now look at that problem again, also by neglecting the effect of any other force, including forces due to contact with the atmosphere. Compared to the dimensions of the earth, the satellite is modeled as a particle P of mass m and it is subjected to the central gravitational attraction of the earth, where the earth is modeled as an inertial homogeneous sphere of radius $R_E \approx 6378$ km and mass M_E. The free-body diagram for the satellite and the polar coordinates r and θ (where r is the distance from the planet's center at O to P and θ is measured from a fixed reference line in the plane of the motion) that are chosen to describe the motion are shown in Fig. 2.43. Two rotating unit vectors $\hat{r}$ and $\hat{\theta}$ are also shown in that figure. They will be used again, since it has been shown that they are very convenient for formulating the expression for the absolute acceleration of P, which is needed for obtaining the differential equations of motion for the system.

a. Obtain the differential equations of motion for P in terms of the variables r and θ.

b. Determine the maximum and minimum values of the distance r when the motion of P is started with the initial conditions $r(0) = r_0 = 6578$ km, $\dot{r}(0) =$

0, and $v(0) = v_0 = 28,000$ km/h, where $\vec{v} = d\vec{r}/dt$ is the absolute velocity of P. Will the satellite crash on the earth's surface?

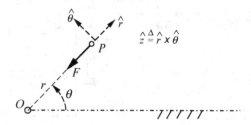

Figure 2.43: Motion coordinates and free-body diagram for Example 2.7.11.

■ Solution

a. Since $M_E >> m$, the effect of the mutual gravitation force of attraction $F = GM_E m/r^2$ (see Fig. 2.43) on the motion of the earth is extremely small and can certainly be neglected. If the motion of the earth in space is not considered either, point O in Fig. 2.43 is inertial, and we then have

■ Absolute position vector for particle P:

$$\vec{r} = \overrightarrow{OP} = r\hat{r}$$

■ Absolute velocity of P:

$$\vec{v} = \frac{d\vec{r}}{dt} = \dot{r}\hat{r} + r\frac{d\hat{r}}{dt} = \dot{r}\hat{r} + r\dot{\theta}\hat{\theta}$$

■ Absolute acceleration of P:

$$\vec{a} = \frac{d\vec{v}}{dt} = \ddot{r}\hat{r} + (r\ddot{\theta} + \dot{r}\dot{\theta})\hat{\theta} + \dot{r}\frac{d\hat{r}}{dt} + r\dot{\theta}\frac{d\hat{\theta}}{dt} = (\ddot{r} - r\dot{\theta}^2)\hat{r} + (r\ddot{\theta} + 2\dot{r}\dot{\theta})\hat{\theta}$$

■ Resultant force acting on P:

$$\vec{F} = -\frac{GM_E m}{r^2}\hat{r} = -\frac{mgR_E^2}{r^2}\hat{r}$$

Newton's second law $\vec{F} = m\vec{a}$ then yields the following scalar differential equations of motion for the satellite:

$$\ddot{r} - r\dot{\theta}^2 = -\frac{gR_E^2}{r^2} \qquad (2.7.11a)$$

$$r\ddot{\theta} + 2\dot{r}\dot{\theta} = \frac{1}{r}\frac{d}{dt}(r^2\dot{\theta}) = 0 \tag{2.7.11b}$$

b. Equation (2.7.11b) can be immediately integrated to yield

$$r^2\dot{\theta} = \text{constant} = \left[r^2\dot{\theta}\right]_{t=0} = r^2(0)\dot{\theta}(0)$$

Since $\dot{r}(0) = 0$, as stated in item b, and $\vec{v} = \dot{r}\hat{r} + r\dot{\theta}\hat{\theta}$ for all times, it follows that $r(0)\dot{\theta}(0) = v_0$. Therefore,

$$r^2\dot{\theta} = r_0 v_0$$

This equation gives us an expression for $\dot{\theta}$ in terms of the distance r. By substituting that expression in Eq. (2.7.11a), the following differential equation that only involves the variable $r(t)$ is obtained.

$$\ddot{r} - \frac{r_0^2 v_0^2}{r^3} + \frac{gR_E^2}{r^2} = 0 \tag{2.7.11c}$$

Since $\ddot{r} = d\dot{r}/dt = \dot{r}(d\dot{r}/dr)$, this differential equation can then be integrated once (as first shown in Case 4 in Section 1.13) as

$$\int_{\dot{r}=\dot{r}(0)}^{\dot{r}} \dot{r}\frac{d\dot{r}}{dr}dr - r_0^2 v_0^2 \int_{r=r_0}^{r} \frac{dr}{r^3} + gR_E^2 \int_{r=r_0}^{r} \frac{dr}{r^2} = 0$$

or

$$\frac{\dot{r}^2}{2} + \frac{r_0^2 v_0^2}{2r^2} - \frac{gR_E^2}{r} = \frac{v_0^2}{2} - \frac{gR_E^2}{r_0} \tag{2.7.11d}$$

The maximum and minimum values of the distance r are obtained by setting $\dot{r} = 0$ in this equation. Denoting such values by r_m, this results in a quadratic equation for r_m. Since $\dot{r}(0) = 0$, one of the solutions is $r_m = r_0 = 6578$ km. The other solution is found to be $r_m \approx 6541$ km. Therefore, $r_{\min} = 6541$ km and $r_{\max} = 6578$ km. Since $r_{\min} > R_E$, the satellite does not crash on the earth's surface.

Note: The central force problem of Example 2.7.11 is a particular case of the two-body problem of celestial mechanics, which is analyzed in detail in Chapter 4.

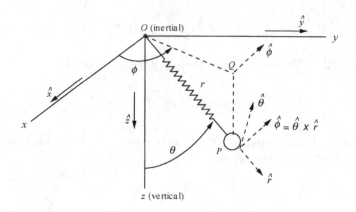

Figure 2.44: The spring-pendulum in three-dimensional space.

Example 2.7.12 The support point O of the spring-pendulum shown in Fig. 2.44 is inertial, and the orthogonal x-, y-, and z-axes shown in that figure are fixed to the earth which, again, is taken to be inertial.

a. Obtain the expression for the absolute velocity $\vec{v}_P$ and absolute acceleration $\vec{a}_P$ of the mass particle P in terms of the spherical coordinates $r(t) = |\overrightarrow{OP}|$, $\theta(t)$, and $\phi(t)$. Express the answers in the unit vector triad $\{\hat{r}, \hat{\phi}, \hat{\theta} \overset{\Delta}{=} \hat{r} \times \hat{\phi}\}$ shown in the figure, where $\hat{\phi}$ is perpendicular to the plane OPQ. Line OQ is the projection of OP on the xy plane (see Fig. 2.44).

b. What are the differential equations of motion for P if the spring is linear, with unstressed length L_u and stiffness k N/m, and the only external force acting on the system is due to gravity?

c. Show that the motion of P takes place on a vertical plane if it is started with $\dot{\phi} = 0$ (i.e., if it is started with a velocity that has no component along $\hat{\phi}$).

■ **Solution**

a. From Fig. 2.44, it is seen that the absolute position vector of point P is $\vec{r}_P = \overrightarrow{OP} = r\hat{r}$ while the absolute angular velocity $\vec{\omega}$ of the $\{\hat{r}, \hat{\phi}, \hat{\theta}\}$ triad is

$$\vec{\omega} = -\dot{\theta}\hat{\phi} - \dot{\phi}\hat{z} = -\dot{\theta}\hat{\phi} - \dot{\phi}(\hat{r}\cos\theta - \hat{\theta}\sin\theta) \qquad (2.7.12a)$$

The absolute velocity and acceleration of P are then obtained as

$$\vec{v}_P = \frac{d}{dt}\vec{r} = \dot{r}\hat{r} + \vec{\omega} \times \vec{r}_P = \dot{r}\hat{r} + r\dot{\theta}\hat{\theta} + (r\sin\theta)\dot{\phi}\,\hat{\phi}$$

$$\vec{a}_P = \frac{d}{dt}\vec{v}_P = \left[\ddot{r}\hat{r} + (r\ddot{\theta} + \dot{r}\dot{\theta})\hat{\theta} + (r\ddot{\phi}\sin\theta + \dot{r}\dot{\phi}\sin\theta + r\dot{\phi}\dot{\theta}\cos\theta)\hat{\phi}\right]$$

$$+\vec{\omega}\times\vec{v}_P = [\ddot{r} - r\dot{\theta}^2 - r\dot{\phi}^2\sin^2\theta]\hat{r} + [r\ddot{\theta} + 2\dot{r}\dot{\theta} - r\dot{\phi}^2(\sin\theta)\cos\theta]\hat{\theta}$$

$$+[r\ddot{\phi}\sin\theta + 2\dot{r}\dot{\phi}\sin\theta + 2r\dot{\phi}\dot{\theta}\cos\theta]\hat{\phi} \qquad (2.7.12b)$$

Notice the following properties about the expression for $\vec{v}_P$:

- The first term $\dot{r}\hat{r}$ is simply the velocity of P relative to O as if the angles θ and ϕ were constant.
- The second term, $r\dot{\theta}\hat{\theta}$, is the velocity of P relative to O as if r and ϕ were constant.
- The third and last term $(r\sin\theta)\dot{\phi}\hat{\phi}$ is the velocity of P relative to O as if r and θ were held constant. If only the second term were present, with the others being zero, the motion of P relative to O would be constrained to an arc of a circle centered at O and with radius r. The third term expresses the fact that P would simply move on a circle centered on the vertical z axis and of radius $r\sin\theta$, due to the angular velocity $\dot{\phi}$ about that vertical axis.

One could now make use of the knowledge gained by these observations to write the expression for the velocity $\vec{v}_P$ by inspection from a figure for similar future situations.

It is also useful to notice that the $\hat{\phi}$ component of the acceleration $\vec{a}_P$ can be expressed in a simpler form as

$$\frac{1}{r\sin\theta}\frac{d}{dt}\left(r^2\dot{\phi}\sin^2\theta\right)$$

Such an observation is very useful as it will allow us to immediately integrate one of the differential equations of motion for this system when the moment, about O, of the resultant force acting on the particle located at P has no component in the $\hat{\phi}$ direction.

b. The free-body diagram for the mass particle at P is the same as the one shown in Fig. 2.33 (p. 163). Therefore, with the resultant force $\vec{F}$ acting on P being equal to

$$\vec{F} = mg\hat{z} - k(r - L_u)\hat{r} = mg(\hat{r}\cos\theta - \hat{\theta}\sin\theta) - k(r - L_u)\hat{r}$$

the following differential equations of motion are obtained from Newton's second law, $\vec{F} = m\vec{a}_P$.

$$m(\ddot{r} - r\dot{\theta}^2 - r\dot{\phi}^2\sin^2\theta) = mg\cos\theta - k(r - L_u) \quad (2.7.12c)$$

$$r\ddot\theta + 2\dot r\dot\theta - r\dot\phi^2(\sin\theta)\cos\theta \;=\; -g\sin\theta \qquad (2.7.12\text{d})$$
$$r\ddot\phi\sin\theta + 2\dot r\dot\phi\sin\theta + 2r\dot\phi\dot\theta\cos\theta \;=\; 0 \qquad (2.7.12\text{e})$$

c. To show that the motion of P takes place on a vertical plane if it is started with $\dot\phi = 0$, one manipulates Eq. (2.7.12e) to write it as

$$\frac{1}{r\sin\theta}\frac{d}{dt}\left(r^2\dot\phi\sin^2\theta\right) = 0$$

or, simply,

$$r^2\dot\phi\sin^2\theta = \text{constant} = r^2(0)\dot\phi(0)\sin^2\theta(0)$$

If the motion is started with $\dot\phi = 0$ (i.e., with a velocity that has no component along $\hat\phi$), this equation yields $\phi = \text{constant}$, since r and θ cannot be zero for all times. This means that the motion of P takes place on the same vertical plane where it was started. Such a motion was investigated in Example 2.7.8 (p. 162).

Example 2.7.13 A point P of mass m slides on a circular frictionless groove of radius R. The groove is rotating with a known absolute angular velocity $\dot\phi \overset{\Delta}{=} \Omega(t)$ about an axis AA that passes through its center C as shown in Fig. 2.45. The center C of the groove is inertial, and axis AA is vertical. The unit vector triad $\{\hat r, \hat\theta, \hat n \overset{\Delta}{=} \hat r \times \hat\theta\}$ shown in the figure will be used in the solution to the problem.

a. Obtain the expressions for the absolute velocity and absolute acceleration of P by working with the variable θ, which is the angle that specifies the location of the particle on the rotating groove.

b. Obtain the expressions for the velocity and the acceleration of P relative to the groove.

c. What is the differential equation that governs the motion of P, and what are the equations for determining the reaction forces applied on P by the groove? Explain how to determine those reactions.

■ **Solution**

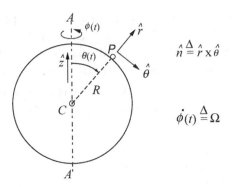

Figure 2.45: A particle moving on a rotating ring.

a. In terms of the rotating triad $\{\hat{r}, \hat{\theta}, \hat{n} \overset{\Delta}{=} \hat{r} \times \hat{\theta}\}$ shown in Fig. 2.45, the absolute position vector of P is simply $\vec{r}_P = \overrightarrow{CP} = R\hat{r}$. Since the direction of the axis AA from which the angle θ is measured is fixed in inertial space, the absolute angular velocity $\vec{\omega}$ of the $\{\hat{r}, \hat{\theta}, \hat{n}\}$ triad is equal to (see Fig. 2.45)

$$\vec{\omega} = \Omega \hat{z} + \dot{\theta}\hat{n} = \Omega(\hat{r}\cos\theta - \hat{\theta}\sin\theta) + \dot{\theta}\hat{n} \tag{2.7.13a}$$

Therefore, the absolute velocity and absolute acceleration of P are

$$\vec{v}_P = \underbrace{\frac{d}{dt}\vec{r}_P = \dot{R}\hat{r}}_{=0} + \vec{\omega} \times \vec{r}_P = R\dot{\theta}\hat{\theta} + (\Omega R \sin\theta)\,\hat{n} \tag{2.7.13b}$$

$$\begin{aligned}
\vec{a}_P &= \frac{d}{dt}\vec{v} = R\ddot{\theta}\hat{\theta} + (\dot{\Omega}R\sin\theta + \Omega R\dot{\theta}\cos\theta)\hat{n} + \vec{\omega} \times \vec{v}_P \\
&= R\left[\ddot{\theta} - \Omega^2(\sin\theta)\cos\theta\right]\hat{\theta} - R\left[\dot{\theta}^2 + \Omega^2\sin^2\theta\right]\hat{r} \\
&\quad + R\left[\dot{\Omega}\sin\theta + 2\Omega\dot{\theta}\cos\theta\right]\hat{n} \tag{2.7.13c}
\end{aligned}$$

b. The velocity and acceleration of P relative to the rotating groove are obtained by simply freezing the rotation of the groove. Therefore, the following results are obtained directly from Eqs. (2.7.13b) and (2.7.13c).

$$\vec{v}_{P/\text{groove}} = (\vec{v}_P)_{\Omega=0} = R\dot{\theta}\hat{\theta} \tag{2.7.13d}$$

$$\vec{a}_{P/\text{groove}} = (\vec{a}_P)_{\Omega=0} = R\ddot{\theta}\hat{\theta} - R\dot{\theta}^2\hat{r} \tag{2.7.13e}$$

Notice that the expression for the absolute velocity of P, given by Eq. (2.7.13b), is the sum of the velocity of P relative to the groove, which is the term $R\dot{\theta}\hat{\theta}$, and the absolute velocity of the point on the groove that coincides with P, which is the term $R\Omega(\sin\theta)\hat{n}$. The path of that point relative to the groove is simply a circle of radius $R\sin\theta$, perpendicular to axis AA.

c. The resultant $\vec{F}$ of all forces acting on particle P is the vector sum of the forces shown in Fig. 2.46. The force $\vec{f} \overset{\Delta}{=} f\hat{n}$ is an unknown lateral reaction force that keeps P on the groove and that maintains $\dot{\phi}$ equal to the prescribed function $\Omega(t)$.

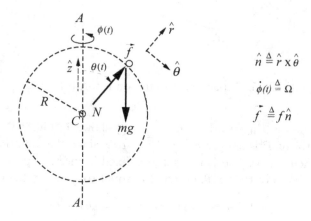

Figure 2.46: Free-body diagram for a particle moving on a rotating circular groove.

By inspecting Fig. 2.46, the following expression is obtained for $\vec{F}$.

$$\vec{F} = N\hat{r} + mg\left(\hat{\theta}\sin\theta - \hat{r}\cos\theta\right) + f\hat{n} \tag{2.7.13f}$$

Therefore, Newton's second law, $\vec{F} = m\vec{a}_P$, yields the following three equations that involve the three unknown variables θ, N, and f.

$$R\left[\ddot{\theta} - \Omega^2\left(\sin\theta\right)\cos\theta\right] = g\sin\theta \tag{2.7.13g}$$

$$N - mg\cos\theta = -mR\left(\dot{\theta}^2 + \Omega^2\sin^2\theta\right) \tag{2.7.13h}$$

$$f = mR\left(\dot{\Omega}\sin\theta + 2\Omega\dot{\theta}\cos\theta\right) \tag{2.7.13i}$$

Equation (2.7.13g), whose only unknown is $\theta(t)$, is the differential equation that governs the motion of P. It is a nonlinear differential equation because the terms $\sin\theta$ and $(\sin\theta)\cos\theta$ are nonlinear in θ.

If one needs to determine the reaction forces N and f, Eqs. (2.7.13h) and (2.7.13i) are the ones that allow that to be done. To do so, one integrates Eq. (2.7.13g) numerically (because an analytical solution cannot be obtained for that nonlinear differential equation) to obtain $\theta(t)$, and then uses the result of that integration to calculate N and f from Eqs. (2.7.13h) and (2.7.13i). This can be done with either Scilab, Matlab, or Simulink, or any other appropriate software. For more on this example, see Problem 2.41 in Section 2.9.

2.8 A Work-Energy Approach Obtained from Newton's Second Law

In some examples presented in this chapter, the fact that some of the second-order differential equations of motion could be integrated once (by making use of the chain rule of differentiation to prepare the differential equation to be integrated once) played a significant role in the analysis of the motion. As seen in Case 4 in Section 1.13 of Chapter 1, that integration was possible for differential equations of the form $\ddot{x} + f(x) = 0$, for which one obtains $\dot{x}^2/2 + \int f(x)dx = $ constant. The relatively trivial case for which the differential equation is of the form $\ddot{x} + f(t) = 0$, which can be readily solved to obtain $x(t)$, is excluded from this discussion. The existence of such *constants of the motion* (i.e., functions of the motion variable, and of its first time derivative, that remain constant during the motion while the variable, itself, is changing with time), or *integrals of motion* as they are more often referred to in dynamics, made it possible to reduce the problem of dealing with a second-order differential equation to one dealing with a first-order differential equation. The integrals of motion also made it possible to obtain important information about the motion of the system without even having to find a solution for the motion variable in terms of time t, which is a task that is often impossible to perform analytically.

Some examples presented in this chapter dealt with two-degree of freedom systems, i.e., systems that involve two independent variables that change with time in a manner that becomes known only after integrating two differential equations of motion. Under certain conditions, it is also possible to find integrals of motion for such systems, and for systems involving more than two degrees of freedom. This can be done by manipulating the second-order differential equations with the chain rule, but that would, in general, involve a substantial amount of hard work.

At the end of Example 2.5.9 (p. 129) you were asked to verify that a certain function was an integral of the motion for that problem when there was no damping in the system. You might have asked yourself, How was that function found? The answer to such a question is provided in this section in a simplistic manner using the material presented in Section 1.13 in Chapter 1. A powerful and general procedure is presented in Chapter 7.

In this section it is shown that, under certain conditions, an integrated form of the second-order differential equations of motion that results from direct application of Newton's second law, $\vec{F} = d(m\vec{v})/dt$, may readily be obtained without even having to generate those second-order differential equations. For the sake of generality, this is done here for the case of the motion of N point masses.

Consider N mass particles P_i $(i = 1, 2, \ldots, N)$, each of mass m_i, subjected to several forces whose resultant is $\vec{F}_i$. Let $\vec{r}_i(t)$ be the absolute position vector for particle P_i. Now define the quantity called the "infinitesimal work done by $\vec{F}$ during an infinitesimal displacement $d\vec{r}$ of P" (or, simply, the *infinitesimal work* done by $\vec{F}$) as being the following scalar quantity dW obtained from the following dot product:

$$dW \overset{\Delta}{=} \vec{F}_1 \bullet d\vec{r}_1 + \vec{F}_2 \bullet d\vec{r}_2 + \ldots + \vec{F}_N \bullet d\vec{r}_N \tag{2.8.1}$$

The determination of the expression for dW as indicated here is, in general, a simple task to perform because it only involves a summation and dot products of vectors. This is illustrated in the examples at the end of this section. To obtain the expression for dW we need to write an expression for the absolute position vector $\vec{r}_i$ for each particle P_i (where a force is applied) in terms of a set of variables chosen to locate each particle. For any particular problem, there will be some number n of such variables. For the motion of a single particle, for example, $n \leq 3$, with $n = 3$ being the case of unconstrained motion in three-dimensional space. By letting $q_1, q_2, \ldots, q_n$ denote those n variables, the result of the dot product operation given by Eq. (2.8.1) will always be an expression of the form shown in Eq. (2.8.2) (see the examples at the end of this section).

$$dW = \sum_{i=1}^{N} \vec{F}_i \bullet d\vec{r}_i = Q_{q_1}(q_1, q_2, \ldots, q_n, \dot{q}_1, \dot{q}_2, \ldots, \dot{q}_n; t)\, dq_1 + \ldots$$
$$+ Q_{q_n}(q_1, q_2, \ldots, q_n, \dot{q}_1, \dot{q}_2, \ldots, \dot{q}_n; t)\, dq_n$$
$$+ a_0(q_1, q_2, \ldots, q_n, \dot{q}_1, \dot{q}_2, \ldots, \dot{q}_n; t)\, dt \tag{2.8.2}$$

The coefficients $Q_{q_1}, Q_{q_2}, \ldots, Q_{q_n}$ of each *infinitesimal displacement* $dq_1, dq_2, \ldots, dq_n$ are, in general, functions of the motion variables $q_1, q_2, \ldots, q_n$ and of their time derivatives $\dot{q}_1, \dot{q}_2, \ldots, \dot{q}_n$ and may be even an "explicit" function of the time t (see the following explanation). Time derivatives $\dot{q}_1, \dot{q}_2, \ldots, \dot{q}_n$ will appear in the expression obtained for dW if there are forces that are velocity dependent, such as friction forces.

The appearance of the independent variable t in a function such as $f_1(q_1, q_2; t) = q_1 q_2^2 \sin(t)$, where q_1 and q_2 are unknown functions of t, is said to be *explicit*. The unknown functions $q_1(t)$ and $q_2(t)$ in this example are said to be *implicit* functions of t. The semicolon used here in $f_1(q_1, q_2; t) = q_1 q_2^2 \sin(t)$, and in Eq. (2.8.2), is a common notation that is used for distinguishing dependent variables (such as q_1 and q_2 in this example) from the independent variable t.

As illustrated in the examples at the end of this section, a term of the form $a_0 dt$ may appear in the expression obtained for dW when one or more of the variables $q_1, q_2, \ldots, q_n$ is a prescribed (i.e., known) function of time.

In general, the expression that is obtained for the infinitesimal work dW *is not* the exact differential of any function W, and, thus, it should be viewed simply as an infinitesimal quantity. When necessary for the sake of clarity, an infinitesimal quantity dW that is an exact differential is written in this book as $d(W)$ and should be read as "d of W"; otherwise it is written as dW and should be read as "d W." The test to determine whether an infinitesimal quantity dW is an exact differential $d(W)$ was presented in Section 1.12 of Chapter 1.

Some of the forces that are applied to any particle P_i may be perpendicular to the infinitesimal displacement $d\vec{r}_i$ for that particle. If that happens, there is no need to worry about those forces at all because the dot product of those forces with $d\vec{r}_i$ is simply equal to zero. Only the component of the forces that is parallel to $d\vec{r}_i$ contributes to the infinitesimal work dW.

In addition to the expression for dW obtained as indicated, an alternative expression for dW can also be obtained by making use of Newton's second law for each particle to replace the expression for each force $\vec{F}_i$ by $d(m\vec{v}_i)/dt$, where $\vec{v}_i = d\vec{r}_i/dt$. By doing so, the following expression for the infinitesimal work dW is also obtained:

$$dW \triangleq \sum_{i=1}^{N} \vec{F}_i \bullet d\vec{r}_i = \sum_{i=1}^{N} \left[\frac{d}{dt}(m_i \vec{v}_i) \right] \bullet \vec{v}_i \, dt = \sum_{i=1}^{N} [d(m_i \vec{v}_i)] \bullet \vec{v}_i$$

$$= d\left(\sum_{i=1}^{N} \frac{1}{2} m_i \vec{v}_i \bullet \vec{v}_i \right) \triangleq d(T) \tag{2.8.3}$$

In summary,

$$dW = d(T) \tag{2.8.4}$$

where the quantity

$$T \triangleq \sum_{i=1}^{N} \frac{1}{2} m_i \vec{v}_i \bullet \vec{v}_i \tag{2.8.5}$$

is called the *kinetic energy* of the motion of the entire system of point masses.

As indicated by the result expressed by Eq. (2.8.4), the infinitesimal work done by the forces that cause the displacements $d\vec{r}_i$ is equal to the change in the kinetic energy of the system during those displacements. This is known in dynamics as the *principle of work and energy*.

The right-hand side of Eq. (2.8.4) is an exact differential (i.e., it is the "d of T," as discussed in Section 1.12 of Chapter 1) and, thus, it can always be integrated to yield the function T plus a constant of integration. However, the left-hand side of that equation, with dW determined simply by performing dot product operations as indicated earlier, may not be integrated in general. It can only be integrated if the expression that is obtained for dW in that manner is an exact differential. That expression will never be an exact differential if any force $\vec{F}_i$ that contributes to dW is velocity dependent, for example. This is the case because only dq_i's, and not $d\dot{q}_i$'s, will appear in the expression that is obtained for dW as $dW = \vec{F}_1 \bullet d\vec{r}_1 + \vec{F}_2 \bullet d\vec{r}_2 + \ldots + \vec{F}_N \bullet d\vec{r}_N$ for any problem.

In general, dW is not an exact differential $d(W)$ of any function W. For the cases when dW is an exact differential, i.e., when dW is of the form $d(W)$, Eq. (2.8.4) can be readily integrated to yield $T - W = $ constant. When dW is not an exact differential, a function W that accounts for all the terms in dW does not exist.

When dW is an exact differential, the equation $T - W = $ constant does not involve any time derivative higher than the first derivative (i.e., velocities, but not accelerations) of any variable. By making use of $dW = d(T)$, when dW can be integrated, one is then able to obtain, in a relatively simple manner, an *integral of the motion* for the system, namely, $T - W = $ constant. It should be noted that the equation $T - W = $ constant is a first-order differential equation, and important information about the motion of the system may be readily obtained from it, even without having to integrate it again either analytically, when possible, or numerically.

For several problems presented in this chapter, an integral of the motion was obtained by manipulating the second-order differential equation of motion using the chain rule of differentiation to prepare that differential equation to be integrated once. For some of those problems dW is an exact differential and, thus, the integrated form of $dW = d(T)$ may be used for obtaining the same integrated equation, thus reducing the amount of mathematical manipulations involved in the process.

A note on the most general expression for dW:
The most general expression of dW, which is of the form given by Eq. (2.8.2), may always be split into the three parts given in the following equation, with one of the

parts being the term $a_0 dt$ that may or may not appear in dW.

$$dW = \underbrace{-\sum_{i=1}^{n} \frac{\partial}{\partial q_i} U(q_1, q_2, \ldots, q_n; t) + \text{what is left out of the expression of dW}}_{\text{This is generated from the terms } Q_{q_1} dq_1 + Q_{q_2} dq_2 + \ldots + Q_{q_n} dq_n \text{ in Eq. (2.8.2)}}$$

$$+ a_0 \, dt \tag{2.8.6}$$

The minus sign in front of the term $\frac{\partial U}{\partial q_1} dq_1 + \frac{\partial U}{\partial q_2} dq_2 + \ldots + \frac{\partial U}{\partial q_n} dq_n$ in Eq. (2.8.6) is a universal convention, and with such a convention, the function $U(q_1, q_2, \ldots, q_n; t)$ is called the *potential energy* of the motion. The development of expressions for potential energy is included in the examples at the end of this section. In dynamics, the quantity $T + U$ is called the *mechanical energy* of the motion.

The work-energy approach is actually the basis of a branch of dynamics that is called *Analytical Dynamics*, which is mostly based on work developed by Lagrange. In such an approach there is no need to formulate expressions for acceleration to obtain the differential equations of motion of dynamical systems. A number of other great masters, such as William Rowan Hamilton (1805-1865) and Karl Gustav Jacobi (1804-1851), devised very elegant and powerful theories that are part of analytical dynamics. *Hamilton's principle*, for example, is widely used by physicists and research engineers in investigations of the motion of dynamical systems, including flexible structures.

Chapter 7 covers the basics of *analytical dynamics*. At least some of that material is covered in intermediate dynamics courses. Using *Lagrange's equation*, and searching for *classical integrals of the motion*, as presented in that chapter, is actually simpler than working directly with $\vec{F} = m\vec{a}$.

Let us now look at several examples of application of the material presented in this section. Examples 2.8.1 and 2.8.2 involve a mass particle subjected to a constant gravitational force and to an inverse-square gravitational force, respectively. The differential equations of motion for both cases were formulated in Examples 2.5.8 (p. 122) and 2.7.11 (p. 179). For convenience, the two figures for those examples are repeated as Fig. 2.47.

Example 2.8.1 Infinitesimal Work and Potential Energy for a Constant Gravitational Force

For this example refer to Fig. 2.47a (which is the same as Fig. 2.14 of Example 2.5.8). With $\vec{F} = -mg\hat{y}$ and the absolute position vector for the particle equal to $\vec{r} = x\hat{x} + y\hat{y}$ [thus, $d\vec{r} = (dx)\hat{x} + (dy)\hat{y} + xd\hat{x} + yd\hat{y} = (dx)\hat{x} + (dy)\hat{y}$ because, for

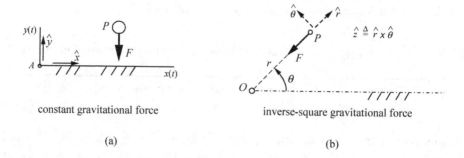

constant gravitational force   inverse-square gravitational force

(a)   (b)

Figure 2.47: Figure for Examples 2.8.1 and 2.8.2.

the chosen unit vectors, $d\hat{x} = d\hat{y} = 0$], the following expression is obtained for the infinitesimal work dW.

$$dW = -(mg\hat{y}) \bullet d\vec{r} = -mg\,dy$$

The same expression for dW is obtained for the simpler case of motion along a vertical straight line considered in Example 2.5.1 (p. 103), or for motion along any other path.

This expression for dW is clearly an exact differential $d(W)$ and, therefore, can be integrated to yield $W = -mgy + C$, where C is an arbitrary constant of integration (which can always be set to zero without any loss of generality). The quantity $U = mgy$, obtained from $dW = -d(U)$ in this case where dW is an exact differential that involves only one variable, is called the *potential energy* associated with the constant gravitational force $-mg\hat{y}$. Thus, the potential energy associated with a constant gravitational force whose magnitude is mg is simply equal to (except for an arbitrary constant of integration) the product of mg with the particle's upward vertical distance y from a fixed reference point.

For the motion of a projectile subjected only to the gravitational force, as considered in Example 2.5.8, we then obtain

$$T + U = \frac{m}{2}(\dot{x}^2 + \dot{y}^2) + mgy = \frac{1}{2}mv^2 + mgy = \text{constant} \stackrel{\Delta}{=} E$$

where $v = |\vec{v}| = |\dot{x}\hat{x} + \dot{y}\hat{y}| = \sqrt{\dot{x}^2 + \dot{y}^2}$ is the magnitude of the absolute velocity of the particle, and $T = mv^2/2$ is the kinetic energy of the motion. The value of the constant E is determined by the initial conditions of the motion, $\dot{x}(0), \dot{y}(0)$, and $y(0)$.

Example 2.8.2 Potential Energy for an Inverse-Square Gravitational Force

For an inverse-square gravitational force [see Fig. 2.47b, which is the same as Fig. 2.43 for Example 2.7.11 (p. 180)] we have

$$\vec{F} = -\frac{GM_E m}{r^2}\hat{r}$$

With the absolute position vector for the mass particle written as $\vec{r} = r\hat{r}$ (see Fig. 2.47b), the infinitesimal displacement vector $d\vec{r}$ is obtained as

$$d\vec{r} = (dr)\hat{r} + r d\hat{r} = (dr)\hat{r} + (rd\theta)\hat{\theta}$$

Therefore, the following expression is obtained for the infinitesimal work dW done by the force $\vec{F}$.

$$dW = \vec{F} \bullet d\vec{r} = -\frac{GM_E m}{r^2}dr$$

The expression for dW for this case is also an exact differential and, thus, it can be integrated to yield (with C being an arbitrary constant of integration which can be set to zero)

$$W = \frac{GM_E m}{r} + C$$

The quantity

$$U = -\frac{GM_E m}{r}$$

that is obtained from $dW = -d(U)$ in this case is the *potential energy* associated with the inverse-square gravitational force.

Notice that for Example 2.7.11 (p. 179), the work-energy method presented here immediately yields, without using Eqs. (2.7.11a) and (2.7.11b) obtained from $\vec{F} = m\vec{a}$ in that example,

$$T + U = \frac{1}{2}mv^2 - \frac{GM_E m}{r} = \frac{1}{2}m(\dot{r}^2 + r^2\dot{\theta}^2) - \frac{GM_E m}{r} = \text{constant} \triangleq E$$

where v is the speed of the particle. This equation is the same as Eq. (2.7.11d). The value of the constant E is determined by the initial conditions of the motion, $r(0), \dot{r}(0)$, and $\dot{\theta}(0)$.

Example 2.8.3 Infinitesimal Work and Potential Energy Associated with a Spring Force

Figure 2.48 shows a spring connecting two bodies and the force the spring exerts on those bodies when the attachment points with the spring are at a distance $r = |\vec{r}|$ apart, where $\vec{r}$ is the vector from one connection point to the other. The spring is approximated as massless.

The spring force acting on one of the bodies is the vector $\vec{F}_s$ and, therefore, the spring force acting on the other body is $-\vec{F}_s$ since the spring is massless. The vectors $\vec{r}_A = \overrightarrow{OA}$ and $\vec{r}_B = \overrightarrow{OB}$ are the absolute position vectors for the attachment points A and B, respectively. A unit vector $\hat{u}_1$, conveniently chosen to be parallel to line AB, is also shown in the figure. The direction of line AB may be changing with time as the bodies move in space.

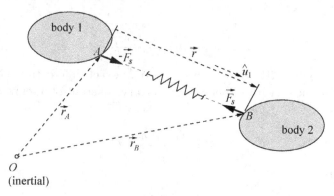

Figure 2.48: Reaction force of a spring.

Letting L_u be the unstretched length of the spring, the spring force $\vec{F}_s$ is then

$$\vec{F}_s = -F_s \hat{u}_1$$

where $F_s = k(r - L_u)$ if the spring is linear. If the spring is not linear, F_s is a nonlinear function of the spring deformation $r - L_u$.

Since the spring applies the forces $\vec{F}_s$ and $-\vec{F}_s$ to both bodies, the infinitesimal work due to the spring force is then, according to the definition of infinitesimal work given by Eq. (2.8.1),

$$
\begin{aligned}
dW &= \vec{F}_s \bullet d\vec{r}_B + (-\vec{F}_s) \bullet d\vec{r}_A = \vec{F}_s \bullet d(\vec{r}_B - \vec{r}_A) = \vec{F}_s \bullet d(r\hat{u}_1) \\
&= (-F_s \hat{u}_1) \bullet [(dr)\hat{u}_1 + r \, d\hat{u}_1] \\
&= -F_s dr \quad [\text{since } d\hat{u}_1 \text{ is perpendicular to } \hat{u}_1, \, \hat{u}_1 \bullet (d\hat{u}_1) = 0]
\end{aligned}
$$

This expression for dW is an exact differential because the spring force depends only on r. By writing $dW = -d(U)$ where, as indicated earlier, the negative sign is always used to abide by international convention, the function $U = U(r) = \int F_s(r)dr$ is the potential energy associated with the spring force.

For a linear spring with stiffness k N/m, the potential energy is obtained as

$$U = \int k(r - L_u) \, dr = \frac{k}{2}(r - L_u)^2 + C$$

where C is a constant of integration.

Since motion involving $r(t)$ always occur with changes in $r(t)$ [and, thus, with changes $d(U)$ in the potential energy U of the spring], the constant of integration in the expression for U is always irrelevant. Therefore, such a constant may always be set to zero without any loss of generality.

In this example, it has been shown, in detail, *how* the expressions for dW and the potential energy of a spring were obtained. By observing the final result $dW = -F_s dr$ for the infinitesimal work, which led to the expression for the potential energy U after an integration, it is seen that dW only depends on the quantity $r - L_u$, which is how much the spring has been either elongated or compressed. This observation makes it possible for one to write the expression for dW due to a linear spring in one of two simple ways:

■ By simply multiplying the spring force $k(r - L_u)$ by the infinitesimal displacement dr of the spring, with a minus sign in front of the result because the spring force is always opposed to the direction the spring deforms
■ By starting with the potential energy $U = \frac{1}{2}k(r - L_u)^2$, and then obtaining dW as $dW = -d(U) = -k(r - L_u)dr$

For a torsional spring shown in Fig. 2.1 (p. 98), the infinitesimal work done by the reaction moment M_s of the spring is $dW = -M_s d\theta$, which is also an exact differential because M_s only depends on the angle θ.

For a linear torsional spring with stiffness k N·m/rad, the infinitesimal work dW done by the spring moment and the potential energy U are then obtained as shown in the following equations, where θ_u is the value of θ when the torsional spring is unstretched.

$$dW = -k(\theta - \theta_u) \, d\theta = -d(U)$$
$$U = \int k(\theta - \theta_u) \, d\theta = \frac{k}{2}(\theta - \theta_u)^2 + C$$

As mentioned earlier, the quantity C in this expression is a constant of integration that is irrelevant, and may be set to zero with no loss of generality.

Example 2.8.4 Infinitesimal Work and Potential Energy for the
Spring-Pendulum

An undamped spring-pendulum is shown in Fig. 2.49. It is the same as the one
that was analyzed in Example 2.7.8 using Fig. 2.33 (p. 163). By using the results
obtained in Examples 2.8.1 (p. 191) and 2.8.3 (p. 194), one may write the following
expression for the potential energy for the motion of this system simply by inspecting
Fig. 2.49.

$$U = -mgr\cos\theta + \frac{k}{2}(r - L_u)^2$$

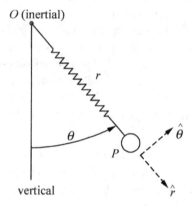

Figure 2.49: Figure for Example 2.8.4 (the unstressed length of the spring is L_u).

Since, in this example, the only forces that do work are the spring force and the
gravitational force (i.e., the reaction forces at the fixed support do not do work),
the expression for the infinitesimal work dW should then be

$$dW = -d(U) = [mg\cos\theta - k(r - L_u)]\,dr - (mgr\sin\theta)d\theta$$

An alternative way to verify this result is to start with the definition of dW, i.e.,
(see Fig. 2.49),

$$dW = \vec{F} \bullet d\vec{r}_P = \left[mg(\hat{r}\cos\theta - \hat{\theta}\sin\theta) - k(r - L_u)\hat{r}\right] \bullet \left[(dr)\hat{r} + (rd\theta)\hat{\theta}\right]$$

This gives $dW = [mg\cos\theta - k(r - L_u)]\,dr - (mgr\sin\theta)d\theta$, which is the same as the
expression that was previously written by inspecting Fig. 2.49.

Let us now verify whether the expression for dW is an exact differential. The test for this was presented in Section 1.12 of Chapter 1.

Since $\partial[mg\cos\theta - k(r - L_u)]/\partial\theta = \partial(-mgr\sin\theta)/\partial r = -mg\sin\theta$, dW is an exact differential. By denoting it as $-d(U)$, we then have

$$\frac{\partial U}{\partial r} = k(r - L_u) - mg\cos\theta$$

$$\frac{\partial U}{\partial\theta} = mgr\sin\theta$$

The first of these partial derivatives may be integrated with respect to r to yield $U = k(r - L_u)^2/2 - mgr\cos\theta + f_1(\theta)$, where $f_1(\theta)$ is a function of θ only. The second of these partial derivatives may be integrated with respect to θ to yield $U = -mgr\cos\theta + f_2(r)$, where $f_2(r)$ is a function of r only. By comparing the two expressions obtained for U, the potential energy U is then, except for a constant of integration which, again, is arbitrarily set to zero, $U = -mgr\cos\theta + k(r - L_u)^2/2$. This is the same expression as the one that was written by inspecting Fig. 2.49 and using the knowledge gained from Examples 2.8.1 and 2.8.3.

Notice that the kinetic energy for this system is $T = m\vec{v}_P \bullet \vec{v}_P/2 = m[\dot{r}^2 + (r\dot{\theta})^2]/2$. Since dW is an exact differential $-d(U)$ for this example, it follows that the total mechanical energy $T + U$ for the undamped spring-pendulum remains constant for all times, i.e., $mv_P^2/2 + k(r - L_u)^2/2 - mgr\cos\theta = \text{constant} \overset{\Delta}{=} E$, where $v_P = |\vec{v}_P| = \sqrt{\dot{r}^2 + r^2\dot{\theta}^2}$ and E is determined by the initial conditions of the motion. Notice, however, that the equation $T + U = \text{constant}$ for this system gives $\dot{r}^2 + r^2\dot{\theta}^2$ in terms of the two unknown variables r and θ and, therefore, neither the maximum nor minimum values of either r or θ can be determined from that equation alone because that is one equation in two unknowns.

To answer specific questions about r and θ, it is necessary to work with the two nonlinear differential equations of motion, Eqs. (2.7.8f) and (2.7.8g) (p. 164). Those nonlinear differential equations have to be integrated numerically so that the extremum values of r and θ can be determined,[13] and that was done in Example 2.7.8.

Example 2.8.5 Infinitesimal Work and Potential Energy for Example 2.5.9

Referring to Fig. 2.17 of Example 2.5.9 (p. 126), the only forces that do work on that system are the gravitational force $-m_2 g\hat{x}_2$, the viscous friction force $\vec{f}_1 = -c_1\dot{x}\hat{x}_1$

---

[13] As mentioned earlier, there are analytical methods called *perturbation methods* that go beyond linearization of the differential equations, for obtaining approximate solutions to some types of nonlinear differential equations (see, for example, Reference 18 listed in Appendix G).

acting on block P_1, as well as the spring force and viscous friction force acting between blocks P_1 and P_2 when the two blocks move relative to each other.

The gravitational force $-m_1 g \hat{x}_2$ and the normal forces $\vec{N}_1$ and $\vec{N}_2$ shown in Fig. 2.17 do not do work because they are always perpendicular to the displacement of the points where they are applied. The sum of the dot products of those forces with the associated displacements are equal to zero.

The infinitesimal work done by the gravitational force $-m_2 g \hat{x}_2$ and by the friction force acting on block P_1 are obtained as (see Fig. 2.17, p. 126).

$$
\begin{aligned}
[dW]_{\text{due to } -m_2 g \hat{x}_2} &= (-m_2 g \hat{x}_2) \bullet d\left[x \hat{x}_1 + r \hat{y}_1\right] = (-m_2 g \hat{x}_2) \bullet \left[(dx)\hat{x}_1 + (dr)\hat{y}_1\right] \\
&= (m_2 g \sin \alpha) dr \\
[dW]_{\text{due to } -c_1 \dot{x} \hat{x}_1} &= -c_1 \dot{x} \hat{x}_1 \bullet (dx)\hat{x}_1 = -c_1 \dot{x}\, dx
\end{aligned}
$$

Note: This expression for the infinitesimal work done by the gravitational force $-m_2 g \hat{x}_2$ could also have been obtained by making use of the knowledge acquired in Example 2.8.1 (p. 191) about the potential energy associated with a constant gravitational force. Since the vertical distance traveled by block P_2 is $r \sin \alpha$ (see Fig. 2.16, p. 125), and block P_2 is below the reference level that distance is measured from, which is a level that is independent of the motion coordinates, the potential energy associated with that force is $U = -m_2 g r \sin \alpha$. As seen in Example 2.8.1 (p. 191), the reason for the minus sign is that P_2 is below the reference level that is used. Thus, this gives the infinitesimal work as $dW = -d(U) = m_2 g (\sin \alpha) dr$.

The infinitesimal work done by the viscous friction force and by the spring force acting between blocks P_1 and P_2 (see Fig. 2.17, p. 126) is different from zero only when the two blocks move relative to each other. If we imagine ourselves sitting on block P_1, for example, and then look at block P_2, we then see that the infinitesimal displacement of P_2 relative to block P_1 is $(dr)\hat{y}_1$. The expression for the infinitesimal work done by the viscous friction force and the spring force acting between those two blocks must then be

$$
\begin{aligned}
[dW]_{\text{due to viscous friction between } P_1 \text{ and } P_2} &= -c_2 \dot{r}\, dr \\
[dW]_{\text{due to the spring connecting } P_1 \text{ and } P_2} &= -k\left(r - L_u\right) dr
\end{aligned}
$$

The infinitesimal work done by all forces acting on the system is equal to the sum of each dW obtained, i.e.,

$$
dW = \left[m_2 g \sin \alpha - k\left(r - L_u\right) - c_2 \dot{r}\right] dr - c_1 \dot{x}\, dx
$$

This expression for dW is of the general form given in Eq. (2.8.2) (p. 188), except that it does not contain a term in dt because neither $x(t)$ nor $r(t)$ are specified functions of time. Only the part $[m_2 g \sin \alpha - k\,(r - L_u)]\,dr$ in the expression for dW can be obtained from a potential energy function. The potential energy is then

$$U = \frac{1}{2}k\,(r - L_u)^2 - m_2 gr \sin \alpha$$

We still have, of course, $dW = d(T)$, where T is the kinetic energy of the system, whose expression follows, but dW cannot be integrated if $c_1 \neq 0$ and $c_2 \neq 0$ because neither $r(t)$ nor $x(t)$ are known functions of time.

$$
\begin{aligned}
T &= \frac{m_1}{2}(\dot{x}\hat{x}_1) \bullet (\dot{x}\hat{x}_1) + \frac{m_2}{2}(\dot{x}\hat{x}_1 + \dot{r}\hat{y}_1) \bullet (\dot{x}\hat{x}_1 + \dot{r}\hat{y}_1) \\
&= \frac{1}{2}m_1 \dot{x}^2 + \frac{1}{2}m_2(\dot{x}^2 + \dot{r}^2 + 2\dot{x}\dot{r}\cos\alpha)
\end{aligned}
$$

The total mechanical energy $T + U$ would be a constant during the motion only if there were no friction in the system, i.e., if $c_1 = c_2 = 0$.

The quantity $E \overset{\Delta}{=} T + U$ in this example is the one you were asked to verify, in the suggested problem (p. 129) at the end of Example 2.5.9 (p. 125), that remains constant when $c_1 = c_2 = 0$. In this example, it was shown how that quantity was obtained and why it remains constant during the motion of that system.

Example 2.8.6 The Pendulum on a Fixed Base of Example 2.5.10
Referring to Fig. 2.19 of Example 2.5.10 (p. 130), we have for the infinitesimal work dW and for the kinetic energy T:

$$
\begin{aligned}
dW &= M_{\text{bar}}d\theta - (mg \sin\theta)L\,d\theta \\
T &= \frac{1}{2}m(L\dot{\theta})^2
\end{aligned}
$$

In general, this expression for the infinitesimal work dW is not an exact differential because of the term $M_{\text{bar}}d\theta$. It would be an exact differential if the moment M_{bar} were either a constant or a known function that only involves the variable θ.

Note: For the case of Example 2.5.10 (p. 129), M_{bar} was given as a function of θ and, therefore, dW is an exact differential for that example. For that example, we then have

$$T - W = \frac{1}{2}m(L\dot{\theta})^2 - mgL \cos\theta - \int M_{\text{bar}}(\theta)\,d\theta = \text{constant}$$

The quantity $-mgL\cos\theta - \int M_{\text{bar}}(\theta)\,d\theta$ is the potential energy of the motion for Example 2.5.10. For $M_{\text{bar}} = -K\theta$, as specified in Example 2.5.10, the expression $T - W =$ constant yields Eq. (2.5.10g) (p. 134). That equation was obtained in Example 2.5.10 by using the methodology based on the chain rule presented as Case 4 in Section 1.13 of Chapter 1.

The preceding integral of the motion for Example 2.5.10 allowed us to analytically determine the maximum and minimum values of the angle θ without having to obtain an explicit solution for $\theta(t)$. Two methods were presented to obtain such an integral of the motion, namely, using the methodology presented in Section 1.13 and using the work-energy methodology presented in this section. The use of either methodology is entirely a matter of personal preference. Some readers may naturally favor one of the methodologies over the other.

Example 2.8.7 Potential Energy for the System in Example 2.7.6

The infinitesimal work done by the moment M_{bar} and by the gravitational force acting on the pendulum considered in Example 2.7.6 (see Fig. 2.29, p. 154) is $M_{\text{bar}}d\theta - (mgL\sin\theta)d\theta$. The only other forces for which the infinitesimal work is not zero in this example are an external force $\vec{F}_{\text{ext}}\hat{x}$ that must be applied to the block at A to make it move with a desired $x(t)$ (see Fig. 2.29), and the friction force that may exist while the block at A moves along the horizontal floor. The desired motion $x(t)$ was specified in Example 2.7.6 as a known constant acceleration $\ddot{x} \stackrel{\triangle}{=} -a_A$. Letting $F_{\text{block}}\hat{x}$ denote the vector sum of the external force $\vec{F}_{\text{ext}}\hat{x}$ and the friction force that may be acting on the block at A, the infinitesimal work done by all forces acting on the system is then determined as

$$
\begin{aligned}
dW &= (M_{\text{bar}} - mgL\sin\theta)\,d\theta + F_{\text{block}}dx \\
&= (M_{\text{bar}} - mgL\sin\theta)\,d\theta + F_{\text{block}}\dot{x}\,dt
\end{aligned}
$$

Since $\dot{x}$ is known, dx was expressed as $\dot{x}dt$ in the expression for dW. The resulting expression is not an exact differential because of the contributions from the unknown quantities M_{bar} and F_{block}. The only part of dW that can obtained for sure from a potential energy function is the exact differential $-mgL(\sin\theta)d\theta$, and the potential energy for this system is $U = -mgL\cos\theta$. Notice that, in this example, $T + U$ does not remain constant during the motion even if there is no friction in the system and $M_{\text{bar}} = 0$. The physical explanation for this is that energy has to be input into the system by an external force $F_{\text{ext}}\hat{x}$ to make the block at A move in a specified manner. If $M_{\text{bar}} = 0$ and if there is no friction in the system, the infinitesimal work dW is still $dW = -(mgL\sin\theta)\,d\theta + F_{\text{ext}}\dot{x}\,dt$.

Example 2.8.8 A small body P, modeled as a particle of mass $m = 0.02$ kg, is constrained to slide on a nonrotating frictionless parabolic vertical track $y = ax^2$ m (see Fig. 2.50), where $a = \frac{1}{3}$ m^{-1}. As shown in the figure, the particle is also connected to a spring, and the other end of the spring is attached to an inertial point O. The spring is linear, with stiffness $k = 9$ N/m and unstressed length $L_u = h = 1.2$ m. P is released at a point A, for which $x = x_A = -1.5$ m, with an initial velocity, down the track, of magnitude $v_A = 6$ m/s.

  a. Determine, analytically, the maximum $(y_{\max})$ and the minimum $(y_{\min})$ values of the height y reached by the particle. What are the corresponding values of the distance x?
  b. Plot the reaction force exerted on P by the track versus x. What is its maximum magnitude, and where does it occur?

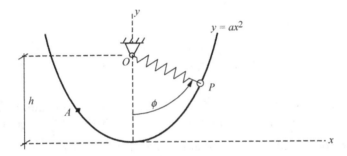

Figure 2.50: A mass particle constrained to move on a parabolic track.

■ Solution

  a. The analytical determination of the maximum and minimum values of y, $y_{\max}$ and $y_{\min}$, is possible here because there is no friction in the system. Otherwise, the nonlinear differential equation of motion for the system would have to be integrated numerically to obtain $y(t)$, from which the values of $y_{\max}$ and $y_{\min}$ would be determined.

  To determine $y_{\max}$ and $y_{\min}$, we need to obtain a first-order differential equation for y, and then set $\dot{y}$ to zero in such an equation. The most expeditious way to do so is to make use of the basic equation $dW = d(T)$ because the reaction force applied to particle P by the track is normal to the infinitesimal displacement of P at all times and, therefore, does not do any work. The only other forces acting on P are the spring force and the gravitational force, both of which contribute to dW in a manner that can be obtained from a potential energy function U. Therefore, we have for this problem, $T+U = \text{constant} \overset{\triangle}{=} E$, which is the total mechanical energy of the motion. The expressions for the kinetic energy T and for the potential energy U obtained by inspecting Fig 2.50 and by using the knowledge acquired in previous examples, are

$$T = \frac{1}{2}m\left(\dot{x}^2 + \dot{y}^2\right) = \frac{1}{2}m\dot{x}^2\left(1 + 4a^2x^2\right) \qquad (2.8.8a)$$

$$U = mgy + \frac{k}{2}\left[\sqrt{x^2 + (h-y)^2} - L_u\right]^2 \qquad (2.8.8b)$$

Since $T + U = \text{constant}$, we then have with $x^2 = y/a$,

$$\frac{1}{2}m\dot{x}^2\left(1 + 4a^2x^2\right) + mgy + \frac{k}{2}\left[\sqrt{\frac{y}{a} + (h-y)^2} - L_u\right]^2 = \text{constant} \overset{\triangle}{=} E \quad (2.8.8c)$$

where

$$E = \frac{1}{2}mv_A^2 + mgy_A + \frac{k}{2}\left[\sqrt{x_A^2 + (h - ax_A^2)^2} - L_u\right]^2 \approx 1.11$$

Notice that the expressions for T and U would be more complicated to deal with if the polar coordinates $|\overrightarrow{OP}|$ and ϕ shown in Fig. 2.50 were used.

The height y reaches a maximum value y_{max} when $\dot{y} = 0$. The value of y_{max} is then determined by setting $\dot{y} = 0$ in Eq. (2.8.8c), i.e., from

$$mgy_{max} + \frac{k}{2}\left[\sqrt{\frac{1}{a}y_{max} + (h - y_{max})^2} - L_u\right]^2 - 1.11 = 0$$

For the given values of $m, k, a, h,$ and L_u, the following value for y_{max} is obtained either with a few iterations with a calculator or from a plot of the left-hand side of the preceding equation versus y_{max}.

$$y_{max} \approx 0.88 \text{ m}$$

Since $y = ax^2$ and, thus, $x = \pm\sqrt{y/a}$, the maximum and minimum values of x are

$$x_{max} = \sqrt{y_{max}/a} \approx 1.62 \text{ m} \qquad\qquad x_{min} = -x_{max}$$

b. *Reaction force exerted on P by the track*. Since there is no friction, the reaction force applied by the parabolic track to the mass particle P consists only of a force N that is normal to the track at the point of contact with P. To determine N, we now need a free-body diagram for P and to use Newton's second law to obtain the expression for N. The free-body diagram for P is shown in Fig. 2.51. Two unit vectors, $\hat{x}$ and $\hat{y}$, which are parallel to the x- and y-axes, respectively, are also shown in that figure. They will be used in the following formulation.

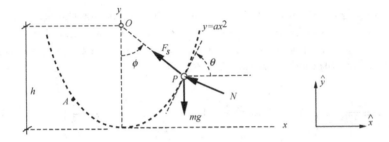

Figure 2.51: Free-body diagram for a mass particle constrained to move on a parabolic track.

In terms of the rectangular coordinates x and y, and of the unit vectors $\hat{x}$ and $\hat{y}$, the absolute acceleration of P and the resultant force acting on P are expressed as (see Fig. 2.51)

$$\vec{a}_P = \ddot{x}\hat{x} + \ddot{y}\hat{y} = \ddot{x}\hat{x} + 2a\left(x\ddot{x} + \dot{x}^2\right)\hat{y}$$

$$\vec{F} = k\left[\sqrt{x^2 + (h-y)^2} - L_u\right](\hat{y}\cos\phi - \hat{x}\sin\phi) - mg\hat{y} + N\left(\hat{y}\cos\theta - \hat{x}\sin\theta\right)$$

where (again, see Fig. 2.51)

$$\cos\phi = \frac{h-y}{\sqrt{x^2 + (h-y)^2}} \qquad \sin\phi = \frac{x}{\sqrt{x^2 + (h-y)^2}}$$

$$\cos\theta = \frac{dx}{\sqrt{(dx)^2 + (dy)^2}} = \frac{1}{\sqrt{1 + (dy/dx)^2}} = \frac{1}{\sqrt{1 + 4a^2x^2}}$$

$$\sin\theta = \frac{dy/dx}{\sqrt{1 + (dy/dx)^2}} = \frac{2ax}{\sqrt{1 + 4a^2x^2}}$$

From Newton's second law, $\vec{F} = m\vec{a}_P$, the following two equations are then obtained.

$$k\left[\sqrt{x^2 + (h-y)^2} - L_u\right]\cos\phi + N\cos\theta - mg = m\ddot{y} = 2ma\left(x\ddot{x} + \dot{x}^2\right)$$

$$\tag{2.8.8d}$$

$$-k\left[\sqrt{x^2 + (h-y)^2} - L_u\right]\sin\phi - N\sin\theta = m\ddot{x} \tag{2.8.8e}$$

Equations (2.8.8d) and (2.8.8e) can be readily solved for the normal force N. This may be done by multiplying Eq. (2.8.8d) by $\cos\theta$, and Eq. (2.8.8e) by $\sin\theta$, and, then, subtracting the two results. The resulting expression for N [where $\dot{x}^2$ is obtained from Eq. (2.8.8c)] is

$$N = \frac{1}{\sqrt{1 + 4a^2x^2}}\left\{mg + 2ma\dot{x}^2 - k\left[1 - \frac{L_u}{\sqrt{x^2 + (h - ax^2)^2}}\right](h + ax^2)\right\}$$

$$\tag{2.8.8f}$$

The plot of N versus x using this expression for N with the given values for a, m, k, L_u, and h, and with $\dot{x}^2$ given by Eq. (2.8.8c), is shown in Fig. 2.52. The value used for g is $g = 9.81$ m/s^2. The maximum magnitude of the reaction N, obtained from Fig. 2.52, is $N \approx 1.68$ N, and it occurs when $x = 0$, i.e., at the bottom of the track.

Example 2.8.9 For the problem in Example 2.8.8, determine the maximum and the minimum values of $x(t)$ if there is friction in the system. Friction is linear and viscous, with friction coefficient $c = 0.028$ N/(m/s).

■ **Solution**
The linear viscous friction force acting on particle P is equal to $-c\vec{v}_P$, where $\vec{v}_P = \dot{x}\hat{x} + \dot{y}\hat{y}$ is the velocity of the particle relative to the parabolic track (which is the same as the absolute velocity of P since the track is not accelerating). Therefore, the resultant force $\vec{F}$ acting on P is now

$$\vec{F} = k\left[\sqrt{x^2 + (h-y)^2} - L_u\right](\hat{y}\cos\phi - \hat{x}\sin\phi) - mg\hat{y} + N(\hat{y}\cos\theta - \hat{x}\sin\theta)$$
$$-c(\dot{x}\hat{x} + \dot{y}\hat{y})$$

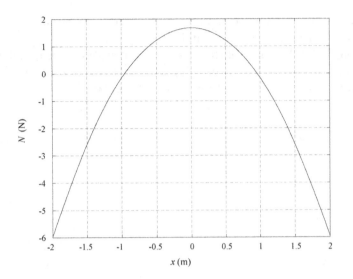

Figure 2.52: Plot of N versus x for a mass particle constrained to move on a parabolic track.

From Newton's second law, $\vec{F} = m\vec{a}_P = m\left(\ddot{x}\hat{x} + \ddot{y}\hat{y}\right)$, the following two equations are obtained for this case.

$$k\left[\sqrt{x^2 + (y - h)^2} - L_u\right]\cos\phi \quad + \quad N\cos\theta - c\dot{y} - mg = m\ddot{y} \qquad (2.8.9a)$$

$$-k\left[\sqrt{x^2 + (y - h)^2} - L_u\right]\sin\phi \quad - \quad N\sin\theta - c\dot{x} = m\ddot{x} \qquad (2.8.9b)$$

The differential equation of motion for P is obtained by eliminating the normal force N from these equations. This may be done by multiplying Eq. (2.8.9a) by $\sin\theta$ and Eq. (2.8.9b) by $\cos\theta$, and then adding the two results. With $y = ax^2$, this yields the following nonlinear differential equation of motion when the expressions for $\cos\phi$, $\sin\phi$, $\cos\theta$, and $\sin\theta$ given in Example 2.8.8 are used.

$$\left(1 + 4a^2x^2\right)\ddot{x} + 4a^2x\dot{x}^2 + \frac{k}{m}\left[1 - \frac{L_u}{\sqrt{x^2 + (ax^2 - h)^2}}\right]\left[2a\left(ax^2 - h\right) + 1\right]x$$

$$+\frac{c}{m}\dot{x}\left(1 + 4a^2x^2\right) + 2agx = 0 \qquad (2.8.9c)$$

For $c \neq 0$, $T + U$ is not constant, and the problem cannot be solved analytically any more. Since Eq. (2.8.9c) cannot be integrated analytically to yield $\dot{x}$ as a function of x, it will have to be integrated numerically.

Figure 2.53 shows the result of the numerical integration of Eq. (2.8.9c) with the initial conditions $x(0) = -1.5$ m, $\dot{x}(0) = 3\sqrt{2}$ m/s (which corresponds to $|\vec{v}_P(0)| = 6$ m/s, as given in Example 2.8.8), and $c = 0.028$ N/(m/s).

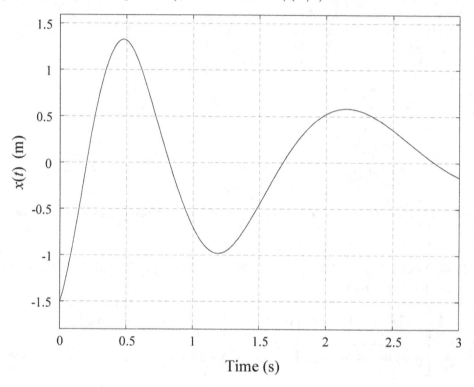

Figure 2.53: Result of the numerical integration of Eq. (2.8.9c).

$\Longrightarrow$ Those who have Scilab can do the numerical integration using the program *parabola_track.sci* listed in Appendix E. Matlab users can use the equivalent program *parabola_track.m*, which is also listed in Appendix E. Both programs generate a similar figure, and animate the motion. As indicated in the comment lines at the beginning of those programs, the syntax for using them is the same for both, and is simply (with $\Delta t = 0.005$ in Scilab and 0.001 in Matlab)

```
my_output=parabola_track([0 : Δt3],1/3,9/0.02,1.4, 1.2,1.2,9.81, ...
-1.5,3*sqrt(2));
```

The maximum and minimum values of $x(t)$ obtained from Fig. 2.53 are $x_{\max} \approx$

1.24 m and $x_{\min} = -1.5$ m, and they occur when $t \approx 0.5$ s and $t = 0$, respectively. Notice that, because of the damping in the system, $x \to 0$ as $t \to \infty$, and that $x = 0$ is the equilibrium value for $x(t)$ for this system.

Those who have Simulink and prefer to use it may want to construct the block diagram model shown in Fig. 2.54, setting the *Simulation → Parameters* menu for this model with time t of the form $[0:\Delta t:3]$, and using the initial conditions $x(0) = -1.5$ and $\dot{x}(0) = 3\sqrt{2}$. The expression for $\ddot{x}$, given by Eq. (2.8.9c), should be typed (as indicated in the tutorial presented in Appendix A) in the window that opens up on the computer screen by double-clicking the $f(u)$ block labeled *x_ddot*. An animation of the motion is obtained with a call to the function *animate_nbars* as described throughout the book.

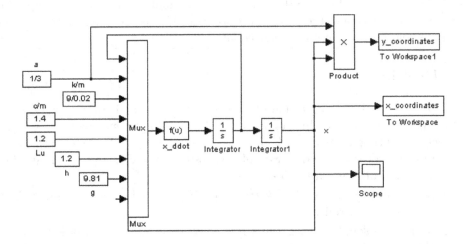

Figure 2.54: Simulink block diagram for integrating Eq. (2.8.9c).

Note: The differential equation of motion for this system may be obtained more expeditiously using $dW = d(T)$ as follows.[14] For this system, the only forces acting on P that contribute to dW are the spring force, the gravitational force, and the friction force. As mentioned in Example 2.8.8, the first two give a contribution to dW that can be obtained from a potential energy function U given by Eq. (2.8.8b). The contribution to dW from the friction force is

$$(dW)_{\text{friction}} = -c\left(\dot{x}\hat{x} + \dot{y}\hat{y}\right) \bullet \left[(dx)\,\hat{x} + (dy)\,\hat{y}\right] = -c\dot{x}\,dx - c\dot{y}\,dy = -c\dot{x}\left(1 + 4a^2x^2\right)dx$$

---

[14]See Chapter 7 for a methodology that can be applied to any dynamical system.

Therefore, $dW = d(T)$ yields, with $dW = -d(U) + (dW)_{\text{friction}}$, and T given by Eq. (2.8.8a) in Example 2.8.8:

$$-d\left(mgax^2 + \frac{k}{2}\left[\sqrt{x^2 + (h - ax^2)^2} - L_u\right]^2\right) - c\dot{x}\left(1 + 4a^2x^2\right)dx$$

$$= d\left(\frac{1}{2}m\dot{x}^2(1 + 4a^2x^2)\right) \tag{2.8.9d}$$

As already mentioned earlier, the expression for dW is not an exact differential because of the friction term, and, thus, it cannot be integrated. However, since the only unknown quantity that appears in Eq. (2.8.9d) is the variable $x(t)$, by manipulating it as $dW/dt = d(T)/dt$, Eq. (2.8.9c), which is the second-order differential equation that governs the motion of the system, is obtained. This is the most convenient and expeditious manner for obtaining that equation.

2.9   Problems

2.1 A point P moves along a nonrotating straight line, and the vector $\vec{r}$ from a fixed point O on that line to P is $\vec{r} = [k_1 t \sin(k_2 t)]\,\hat{x}$ m, where $\hat{x}$ is a unit vector pointing from O to P, $k_1 = 5$ m/s, and $k_2 = 1$ rad/s. Determine the magnitude and the direction of the absolute velocity and of the absolute acceleration of P when $t = 5$ s.

2.2 Repeat Problem 2.1 for $\vec{r} = \left[k_1 t^2 e^{-k_2 t} \cos(k_3 t)\right]\hat{x}$, where $k_1 = 1$ m/s^2, $k_2 = 1\,\text{s}^{-1}$, and $k_3 = 2$ rad/s.

2.3 Consider Problem 2.1 again, but with the straight line track rotating counter-clockwise on a fixed straight platform with a constant absolute angular velocity of magnitude equal to 0.2 rad/s. Determine the magnitude of the following quantities when $t = 5$ s:

    a. The absolute velocity and the absolute acceleration of P
    b. The velocity and the acceleration of P relative to an observer fixed to the rotating track

2.4 Determine the time of the first occurrence of a maximum of the scalar component of the vector $\vec{r} = r(t)\hat{x}$ for Problems 2.1 and 2.2. The answers may be found either with a few iterations with a calculator or by generating accurate plots of $r(t)$ versus t for both cases with a computer.

2.5 To model the effect of drag caused by air resistance, the acceleration $\ddot{r}(t)$ of a body moving along a straight line is approximated as $c_0 - c_1 \dot{r}^2$, where c_0 is a constant acceleration due to an engine on the body and c_1 is a constant.

Observers at a fixed point start to measure the distance r when the velocity $\dot{r}$ is clocked at $\dot{r} = \dot{r}_0$.

    a. Determine, analytically, the velocity $\dot{r}$ as a function of the distance traveled (r) for the cases when $c_1 \neq 0$ and $c_1 = 0$.

    b. After its engine is turned off (i.e., $c_0 = 0$), the body travels 0.8 km when its velocity is then determined to be $\dot{r} = \dot{r}_0/2$. Determine the value of c_1; be sure to specify its units.

2.6 Consider a body P of mass m thrown upward from the earth's surface, without rotation, and with an initial velocity $\dot{y}(0) = v_0$, where y is the distance from the earth's surface to P (as in Fig. 2.3 for Example 2.5.1, p. 104). The body is also subjected to a force $\vec{f}$ that opposes its velocity $\vec{v}$ as $\vec{f} = -c\vec{v}$, where $c > 0$ is constant. Determine, analytically, $y(t)$ in terms of the quantities m, c, g, and v_0, where g is the acceleration of gravity at the surface of the earth. Also, obtain the analytical expression for the maximum height reached by the body. Approximate the gravitational force on P as a constant. Plot $y_{\max}/[y_{\max}]_{c=0}$ (where $[y_{\max}]_{c=0} = v_0^2/(2g)$, as determined in Example 2.5.1) versus the nondimensional parameter $v_0 c/(mg)$ for $10^{-4} \leq v_0 c/(mg) \leq 1$.

2.7 Integrate numerically the differential equation of Problem 2.5 for $r(t)$, with $r(0) = 0$ and $\dot{r}(0) = 1$ m/s, and plot $r(t)$ versus t and $\dot{r}$ versus t (with $0 \leq t \leq 10$ s) for $c_1 = 0$, $c_1 = 0.01$, and $c_1 = 0.03$ m^{-1}. Use $c_0 = 5$ m/s^2 and superimpose all the $r(t)$ plots in the same graph and all the $\dot{r}(t)$ plots in another graph.

2.8 A body of mass m is launched vertically from $y = 0$ (see Fig. 2.3, p. 104) with a speed of 70 m/s. When it reaches its maximum height, a horizontal force of magnitude $f = 5mg$ is applied to it during Δt s. Some time later the body hits the horizontal plane $y = 0$ at a distance $D = 750$ m from where it was launched. Determine the time Δt, and the total time of flight counting from the instant it was launched.

2.9 Calculate the kinetic energy imparted to the body in Problem 2.8 due to the ignition of the rocket on board the body. What is the ratio between that energy and the kinetic energy that was imparted to the body when it was launched vertically? The mass of the body is $m = 40$ kg.

2.10 Determine the escape speed from the surface of the moon and from the surface of Mars. Treat both as uniform spheres, the moon with radius equal to 1740 km and Mars with radius equal to 3390 km, and disregard the gravitational attraction of any other celestial body. The acceleration of gravity at the surfaces of the moon and Mars are approximately equal to $g/6$ and $g/2.6$, respectively, where $g \approx 9.81$ m/s^2 is the acceleration of gravity at the surface of the earth.

2.11 An object of mass m kg is launched vertically from the surface of the earth with an initial velocity $v_0 = 6000$ km/h. In addition to the gravitational force of attraction of the earth, the object is also subjected to an upward constant vertical thrust T N during its entire trajectory. For $T/m = 1$ m/s^2, determine the maximum height reached by the object and the time it takes for it to reach that height. Clearly state all the approximations you decide to make when solving this problem, and justify them.

2.12 Repeat Problem 2.11 for $T/m = 4$ m/s^2 and $T/m = 6$ m/s^2. Use the constant gravitational force approximation, and also Newton's law of universal gravitation, and comment on the results.

2.13 The object in Prob. 2.11 is subjected to a variable thrust T that changes with time as $T = k_1 T_0 t e^{-k_2 t}$, where $T_0/m = 3$ m/s^2, $k_1 = 1$ s^{-1}, and $k_2 = 0.1$ s^{-1}. The initial launching speed at $y = 0$ is $v_0 = 210$ km/h. Determine the maximum value of $y(t)$ and the time it occurs. Clearly state all the approximations you make when solving this problem.

2.14 Write a brief explanation saying *how* you would determine the maximum distance $y(t)$ for Problem 2.11 (without determining it) if the thrust T were of the form $T = k_1 T_0 t y e^{-k_2 t}$, where T_0, k_1, and k_2 are constants.

2.15 A common theoretical model for the earth's atmosphere is one for which the air density changes exponentially with height y from the earth's surface as $\rho = \rho_0 e^{-ky}$, where $\rho_0 \approx 1.5$ kg/m^3 and $k \approx 0.14$ km^{-1}. Repeat Problem 2.11 if instead of the constant thrust T the object is subjected to an aerodynamic drag force $\vec{f}_{\text{drag}}$ that is always directed opposite to the velocity of the object and whose magnitude is modeled using the "exponential atmosphere" as $c\dot{y}^2 e^{-ky}$ where $k = 0.14$ km^{-1} and $c/m = 1.3$ km^{-1} for that particular object.

2.16 An object P of mass $m = 2$ kg is launched from the earth's surface with an initial velocity whose magnitude is $v_0 = 40$ m/s and whose direction makes an angle $\alpha = 30°$ with the horizontal as shown in Fig. 2.13, p. 122. The object is subjected to a constant thrust $T = 10$ N that makes a constant angle of $45°$ with the horizontal during the entire trajectory. Determine:

    a. The maximum height reached by the object
    b. The time it takes to reach the maximum height
    c. The distance along the horizontal plane the object strikes the ground, and the time it takes for that to happen
    d. The equation $y = y(x)$ for the trajectory (see Fig. 2.14, p. 122, for the definition of the variables x and y)

Clearly state all the approximations you make when solving this problem.

2.17 A cannon A located on a platform whose height is $h = 200$ m shoots projectiles with an initial velocity whose magnitude is $v_0 = 300$ m/s. Determine the angle of firing α for the projectile to hit a stationary object that is located at a distance of $R = 5000$ m from the cannon. The line of sight of the target (i.e., the line from the firing point to the target) is inclined at an angle $\beta = 20°$ with the horizontal, and the angle of firing is measured from such a line.

Suggestion: Using the constant gravitational force approximation, it is simpler to solve this problem if a rectangular x-y coordinate system is used, with x along the line of sight and y perpendicular to it. It is suggested that you also solve it with x and y along the horizontal and the vertical, respectively, so that you see that the first choice mentioned is simpler.

2.18 Show that the relation between R, v_0, α, and β for Problem 2.17 is

$$\frac{Rg}{v_0^2} = \frac{2 \left(\sin \alpha \right) \cos \left(\alpha + \beta \right)}{\cos^2 \beta}$$

which is an equation that could be programmed into an onboard computer for generating a display of the required firing angle for given input values of v_0, R, and β. See the suggestion in Problem 2.17.

2.19 A projectile P of mass m is launched from the earth's surface with a velocity $\vec{v}_0$ that makes an angle α degrees with the horizontal, and whose magnitude is $v_0 = |\vec{v}_0|$ m/s. In addition to the gravitational force, the projectile is subjected to a drag force $\vec{f}_{\text{drag}}$ whose magnitude is modeled as being proportional to the square of the magnitude v of the projectile's velocity $\vec{v}$ as $|\vec{f}_{\text{drag}}| = cv^2$, with c being a constant. The direction of the drag force is opposite to $\vec{v}$.

a. Use rectangular coordinates x and y, with x along the horizontal plane and y along the vertical, and obtain the differential equations that govern the motion of the projectile using the constant acceleration of gravity model. Clearly show the free-body diagram for P, properly labeling all quantities, and show all the steps of the formulation that lead to the differential equations. What are the initial conditions for your differential equations?

b. Since the differential equations cannot be solved analytically when $c \neq 0$, integrate them numerically with $v_0 = 100$ m/s and $\alpha = 40°$, and determine the projectile's range (i.e., the distance, measured along the horizontal, at the instant the projectile hits the ground) for $c/m = 10^{-4}$ and $c/m = 10^{-3}$ m^{-1}.

c. For $v_0 = 100$ m/s and $\alpha = 40°$, plot the projectile's range versus c/m.

Note: Those using Scilab or Matlab can do this problem by modifying either the program *ballistics.sci* or the program *ballistics.m* listed in Appendix E.

2.20   A block P of mass $m = 2$ kg is able to slide on an inclined plane that makes a constant angle α degrees with the horizontal. Friction between P and the plane is of the Coulomb type (i.e., dry friction). The static and kinetic coefficients of friction are $\mu_s = 0.25$ and $\mu_k = 0.2$, respectively. The angle α is equal to twice the value needed for P to start sliding down the plane when the block is started from rest. The block starts its motion at a distance of $D = 10$ m from the bottom of the inclined plane and with an initial velocity of 30 m/s up the plane. Determine:

    a. The total distance the block travels up the plane
    b. The time it takes the block to travel twice the distance determined in part a

Does the distance D affect the solution to this problem?

2.21   For the system described in Problem 2.20 determine the time it takes for P to reach the bottom of the inclined plane if its initial velocity is 30 m/s down the plane.

2.22   Determine the maximum distance the block of Problem 2.20 travels up the inclined plane if $\alpha = 15°$ and friction is viscous and linear, with friction coefficient $c = 0.2$ N/(m/s).

2.23   A block P of mass $m = 3$ kg is able to slide on an inclined plane that makes a constant angle $\alpha = 30°$ with the horizontal, as shown in Fig. 2.55. Friction

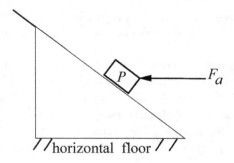

Figure 2.55: Figure for Problem 2.23.

between P and the plane is of the Coulomb type (i.e., dry friction). The block is initially at rest on the inclined plane at a distance of 10 m from the bottom of the plane. The static and kinetic coefficients of friction are $\mu_s = 0.3$ and

$\mu_k = 0.25$, respectively. A constant horizontal force $F_a = 10$ N is then applied to the block. Determine

    a. The total distance the block travels on the plane
    b. The time it takes for the block to reach the bottom of the inclined plane

2.24 Repeat Problem 2.23 for $F_a = 5$ N. Where is the block located 3 s after the application of the force F_a?

2.25 For those using either Scilab, Matlab, or Simulink: simulate and animate the motion of the block in Problems 2.20 and 2.21, automatically stopping the simulation if the block reaches either the top or the bottom of the inclined plane. The length of the inclined plane is 150 m.

2.26 Two blocks P_1 and P_2 of masses m_1 and m_2, respectively, are connected by an inextensible string of negligible mass, and the string passes through a small pulley at an inertial point O. Block P_1 moves on a fixed inclined vertical plane of length D, while P_2 moves along a vertical line as shown in Fig. 2.56a.

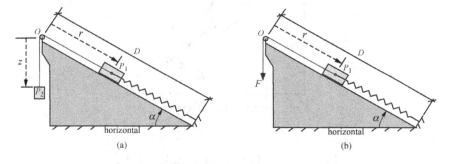

Figure 2.56: Figure for Problems 2.26 and 2.27.

Block P_1 is connected to the inclined plane by a linear spring of stiffness k N/m and unstressed length L_u m. Friction in the system can be neglected. Obtain the differential equation of motion of the system in terms of the distance $r(t)$, its equilibrium solution r_e, and the frequency ω of the oscillation of the blocks when they are displaced from their equilibrium. If the differential equation is nonlinear, linearize it about $r = r_e$ to find the frequency of the *linearized* oscillation, instead of the frequency of the oscillation. What are the values of r_e and ω if $\alpha = 30°$, $L_u = D$, $m_1 = 4m_2$, and $k = k_1 m_1 g$, where $k_1 = 1.25$ m^{-1}?

2.27 In the system described in Problem 2.26, block P_2 is replaced by a constant force F N, as shown in Fig. 2.56b. Obtain the differential equation of motion of the system in terms of the distance $r(t)$, its equilibrium solution r_e, and the frequency ω of the oscillation of the block when it is displaced from its equilibrium. If the differential equation is nonlinear, linearize it about $r = r_e$ to find the frequency of the *linearized* oscillation, instead of the frequency of the oscillation. What are the values of r_e and ω if $\alpha = 30°$, $L_u = D$, and $k = k_1 m_1 g$, where $k_1 = 1.25$ m^{-1} and $F = m_1 g/4$? Is this problem the same as Problem 2.26?

2.28 A block P of mass m kg is able to slide on a vertical inclined plane that makes a constant angle $\alpha = 30°$ with the horizontal.

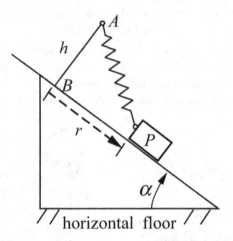

Figure 2.57: Figure for Problem 2.28.

The block is connected by a linear spring to a fixed point A above the inclined plane as shown in Fig. 2.57. The distance h in that figure is a known constant. There is no friction in the system. The spring stiffness is K N/m, and its unstressed length is $L_u = h$ m. The block is initially motionless at point B shown in Fig. 2.57. For $Kh/(mg) = 2.5$, determine:

a. If the block loses contact with the inclined plane during its motion (*Hint:* Look at the value of the reaction force applied by the plane to the block.)
b. The maximum value of the distance r measured from point B along the direction that points down the inclined plane

2.29 Determine the equilibrium solutions (in terms of h) for Problem 2.28, and the frequency of the small linearized oscillations about each stable equilibrium. How many equilibrium solutions does the system exhibit?

2.30 A block P of mass m kg is able to slide on a vertical inclined plane that makes
      a constant angle $\alpha = 30°$ with the horizontal as shown in Fig. 2.58. The
      plane is long enough for the block not to fall from it. Friction between the
      plane and the block is of the Coulomb type (i.e., dry friction), and the block is
      subjected to an applied force F_a that changes with time as $F_a(t) = kmt$, where
      $k = 0.8$ m/s^3, and is directed as shown in the figure, with $\theta = 20°$. The static
      and kinetic coefficients of friction are $\mu_s = 0.25$ and $\mu_k = 0.2$, respectively.
      The block is initially held at rest at a distance $r(0) = 40$ m from a point O on
      the inclined plane and then released when the force F_a is applied. Determine:

      a. The time when the block starts to move, and whether it starts moving
         up or down the plane
      b. The values of t when the block passes through point O
      c. The values of $r(t)$ at the end of 20 s, and the total distance traveled by
         the block during that time

      Describe, in detail, the motion during its first 20 s.

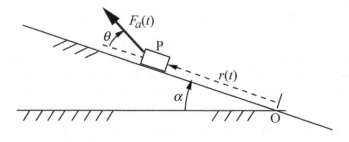

Figure 2.58: Figure for Problems 2.30 to 2.32.

2.31 Use numerical integration to determine $r(t)$ and $\dot{r}(t)$ if the block in Problem
      2.30 is subjected to a force $F_a = k_1 m \sin(k_2 t)$, where $k_1 = 8$ m/s^2 and $k_2 = 0.1$
      rad/s, for the first 40 s of the motion. Describe how the block moves (if it
      moves at all) during the first 5 s, and during the next 35 s. If it moves, how
      many times does it pass through point O during that time, and what are the
      corresponding values of t?

2.32 Solve Problem 2.30 when friction is viscous, instead of the Coulomb type. For
      this, consider linear viscous friction, with a friction coefficient equal to 0.2
      N/(m/s). Take the mass of the block to be $m = 1$ kg. Solve this problem
      analytically.

2.33 A small body P of mass m slides inside an open groove on a semi-circular
      inertial vertical ring of radius R as shown in Fig. 2.59. Point P is initially at

rest at the position when $\theta(0) = \theta_0$, where θ_0 is very small but non-zero (the particle would not move if θ_0 were equal to zero because that is a position of static equilibrium). Determine, analytically, the value of θ when the particle leaves the ring and how far away from point O it strikes the ground. The only external force acting on the particle is due to earth's gravity. Neglect friction between the ring and the particle.

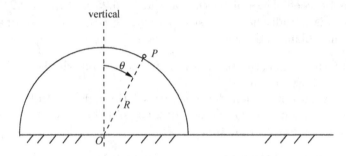

Figure 2.59: Figure for Problems 2.33 to 2.35.

2.34　The value of θ when the particle in Problem 2.33 leaves the ring cannot be determined analytically if there is friction between the ring and the particle. Use numerical integration to determine that value of θ (in degrees) for the case when friction is of the Coulomb type with a kinetic coefficient of friction $\mu_k = 0.2$.

2.35　Repeat Problem 2.34 for the case when friction is linear and viscous (instead of the Coulomb type) and proportional to the velocity of the particle relative to the ring, with a viscous friction coefficient equal to c N/(m/s). Determine the value of θ (in degrees) when the particle leaves the ring, for $(c/m)\sqrt{R/g} = 0.2$.

2.36　A block P of mass m is connected to a fixed pin O by a linear spring of stiffness k N/m and unstressed length L_u m. The block is constrained to move along a horizontal guide as shown in Fig. 2.60, and friction in the system can be neglected. The perpendicular distance from O to the guide is h m. The plane defined by the guide and the spring is vertical.

　　　Obtain the differential equation that governs the motion of the block in terms of the variable $x(t)$ shown in the figure. For this, clearly show the free-body diagram for P, properly labeling all quantities, and show all the steps of the formulation that lead to the differential equation. For $L_u = h$, $x(0) = 5h$, and $\dot{x}(0) = Ah\sqrt{k/m}$, with $A = 10$, determine (analytically) the maximum and the minimum values of x/h. Also, generate a plot of $x_{\max}/h$ and $x_{\min}/h$ versus $x'(0)/h$ for $L_u = h$ and $x(0) = 5h$ for $-10 \le A \le 10$, where the prime

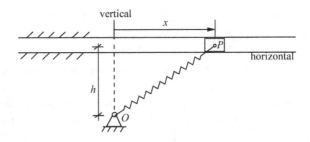

Figure 2.60: Figure for Problem 2.36.

denotes differentiation with respect to a nondimensional time τ defined as $\tau = t\sqrt{k/m}$.

2.37 What are the equilibrium solutions x_e for $x(t)$ for Problem 2.36 when $L_u \leq h$, and when $L_u > h$? How many equilibrium solutions does the system exhibit in each case? Is the linearized differential equation about each equilibrium you determined valid for each one of those cases? Explain in detail. If they are valid, what are such equations? Plot x_e/h versus L_u/h for $0 \leq L_u/h \leq 5$.

2.38 For $L_u = \sqrt{5}h$ and h, determine all the equilibrium solutions for the system described in Problem 2.36, and analyze their stability, indicating whether the equilibrium is infinitesimally stable, asymptotically stable, or unstable. If any of the linearized differential equations is not valid for describing the small motion about an equilibrium solution, plot $d(x/h)/d\tau$ versus x/h, where $\tau = t\sqrt{k/m}$, and indicate whether that particular equilibrium is stable, asymptotically stable, or unstable.

2.39 Numerically integrate the differential equation of motion for Problem 2.36, with $L_u = h$ and with linear viscous damping in the system, with damping coefficient c N/(m/s). Introduce a nondimensional time τ so that the only parameter that appears in the normalized differential equation to be integrated is $c/\sqrt{mk}$. What are the maximum and minimum values of x/h, for $c/\sqrt{mk} = 0$ and 0.2, when the motion is started with $x(0) = 5h$ and $\dot{x}(0) = 10h\sqrt{k/m}$? Compare the analytical answer obtained in Problem 2.36 (for $c = 0$) with the answer obtained from the numerical integration. Plot x/h versus normalized time, and $d(x/h)/d\tau$ for both cases. Choose an appropriate value for the final integration time that will allow you to clearly understand and describe the motion. Write a small paragraph describing the motion.

2.40 For the problem in Example 2.7.13 (p. 184), let $\dot{\phi} = \text{constant} \triangleq \Omega$ and the motion of P be started with the initial conditions $\theta(0) = \theta_0$ and $\dot{\theta}(0) = 0$, where

θ_0 is very small (so that $\cos\theta_0 \approx 1$) but not zero. Determine, analytically, the expression for the value of θ when the particle leaves the ring, and plot that value (in degrees) versus $\Omega^2 R/g$ for $0 \le \Omega^2 R/g \le 1$.

2.41   Numerically integrate the differential equation of motion for Example 2.7.13, which is Eq. (2.7.13g) (p. 186), to determine the nondimensional reaction forces $N/(mg)$ and $f/(mg)$ for the case when $\Omega = $ constant, where N and f are given by Eqs. (2.7.13h) and (2.7.13i), respectively. For this, work with the nondimensional time $\tau \overset{\Delta}{=} t\sqrt{g/R}$, and rewrite those equations using τ. For the numerical integration, use $\theta(0) = 0$ and a very small value for $\theta(0)$. Plot $f/(mg)$ and $N/(mg)$ versus τ for $\Omega^2 R/g = 0.5$ and 2 from $\tau = 0$ until the particle loses contact with the ring; superimpose the plots for each one of the two values of $\Omega^2 R/g$. Write a few lines about your observation of the results.

2.42   Determine all the equilibrium solutions for $\theta(t)$ for the system described in Example 2.7.13 (p. 184), and analyze their stability. For each equilibrium, indicate whether it is stable, asymptotically stable, or unstable. The groove is tubular so that P cannot leave it. How many equilibrium positions does the system have? Do they exist for any value of the angular velocity Ω?

Show a hand sketch of the phase plane plot $d\theta/d\tau$ (where $\tau = t\sqrt{g/R}$) versus θ for the system, for $R\Omega^2/g = 0.5$ and 2, and indicate whether the trajectories are traced clockwise or counterclockwise with the passage of time. Your sketch does not have to be an accurate plot.

2.43   The system shown in Fig. 2.61 is a double pendulum constructed of two homogeneous thin bars OA and AB of negligible mass and constant lengths L_1

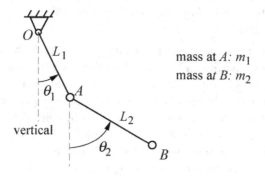

mass at A: m_1
mass at B: m_2

Figure 2.61: Figure for Problem 2.43.

and L_2, respectively, and two small bodies of masses m_1 and m_2, located at A and B as indicated in the figure. Bar OA is pivoted to an inertial point O, and the two bars are pinned at A and can rotate about that point relative to each other. The motion of the entire system takes place on a vertical plane, and friction in the system can be neglected.

    a. Obtain the differential equations of motion for the system in terms of the angles $\theta_1(t)$ and $\theta_2(t)$ shown in Fig. 2.61.

    b. Choose your own values for L_2/L_1 and m_2/m_1, and observe the motion by integrating the differential equations of motion numerically. Choose different combinations of initial conditions $\theta_1(0)$, $\theta_2(0)$, $\dot{\theta}_1(0)$, and $\dot{\theta}_2(0)$, so that you can observe different motions. Then write a few paragraphs describing your observations in detail.

2.44 Obtain the differential equation of motion, and solve part a of Example 2.8.8 (p. 201), if the parabolic track is spun about the vertical y-axis with a constant counterclockwise angular velocity $\Omega = 200$ rpm relative to inertial space. Do this without integrating the differential equation numerically. Write a paragraph describing the motion.

    Suggestion: The x- and y-axes shown in Figs. 2.50 (p. 201) and 2.51 (p. 203) are rotating with the parabola with absolute angular velocity $\Omega\hat{y}$. Use those axes and the rotating unit vector triad $\{\hat{x}, \hat{y}, \hat{z} \overset{\triangle}{=} \hat{x} \times \hat{y}\}$, with $\hat{x}$ and $\hat{y}$ parallel to those axes, to formulate the differential equation of motion.

    Hint: Verify that the following function H is a constant of the motion (i.e., that $dH/dt = 0$), and then use this observation to do the problem analytically. The methodology to obtain such a constant of the motion is presented in Chapter 7.

$$H = \frac{1}{2}m\dot{x}^2 \left(1 + 4a^2x^2\right) + mgy + \frac{k}{2}\left[\sqrt{x^2 + (h - ax^2)^2} - L_u\right]^2 - \frac{1}{2}m\Omega^2x^2$$

2.45 Solve the problem of Example 2.8.9 (p. 204) if friction in the parabolic track is of the Coulomb type, instead of viscous. The static and kinetic coefficients of friction are $\mu_s = 0.25$ and $\mu_k = 0.2$, respectively. Determine x_{max} and x_{min} from a plot of $x(t)$ versus t. Write a paragraph describing the motion.

    Suggestion: If either Scilab or Matlab is used, the respective programs named *parabola_ track* listed in Appendix E to integrate the differential equation of motion could be modified to solve this problem.

2.46 Solve the problem of Example 2.8.9 (p. 204) if the parabolic track is spun about the vertical y-axis with a counterclockwise constant angular velocity $\Omega = 200$ rpm relative to inertial space and friction between P and the rotating track is of the linear viscous type with coefficient c N/(m/s). Because of the friction term, the differential equation of motion, which is nonlinear, has to be integrated numerically to determine x_{max} and x_{min}, as done in Example 2.8.9.

Suggestion: If either Scilab or Matlab is used, the respective programs named *parabola_track* listed in Appendix E to integrate the differential equation of motion could be modified to solve this problem.

2.47 Determine all the equilibrium solutions for $x(t)$ for the body on the rotating parabolic track of Problem 2.46, and analyze their stability. Choose your own value for c, and use $m = 0.02$ kg, $a = \frac{1}{3}$ m^{-1}, $k = 9$ N/m, and $L_u = h = 1.2$ m. For each equilibrium, indicate whether it is stable, asymptotically stable, or unstable.

Show a hand sketch of the phase plane plot $\dot{x}$ versus x for the system, for $\Omega = 200$ rpm, and indicate whether the trajectories are traced clockwise or counterclockwise with the passage of time. Your sketch does not have to be an accurate plot. Show the equilibrium solutions in your sketch, indicating which ones are stable, asymptotically stable, and unstable.

2.48 Numerically integrate the differential equations of motion for the satellite problem of Example 2.7.11 (p. 179), and generate plots of $r(t)$ and $\theta(t)$ versus time t for $r(0) = 6578$ km, $\dot{r}(0) = 0$, and $r(0)\dot{\theta}(0) = v_0$. Generate the plots for $v_0 = 27,000$ km/h, $28,000$ km/h, and $30,000$ km/h. Determine, from the plots, the maximum and minimum values of the distance $r(t)$. Identify the cases for which the satellite crashes on the earth's surface and determine the time the crash occurs. For the cases where the satellite revolves around the earth, determine, from the plots, the time it takes the satellite to complete one revolution. The radius of the earth is $R_E \approx 6378$ km.

2.49 Consider the two-particle system described in Example 2.7.10 (p. 174) when particle P_2 shown in Fig. 2.40 is allowed to move in three-dimensional space.

a. Obtain the differential equations of motion for the system using the distance r and the angle θ shown in Fig. 2.40 and two appropriate angles to specify the position of P_2 in space.

b. Integrate numerically the equations obtained in part a and plot $r(t)$ versus t. For this, take $m_2 = 2m_1$, $L = 1$ m, $r(0) = 0.5$ m, $\dot{r}(0) = 1$ m/s, $\dot{\theta}(0) = 2$ rad/s, with OP_2 starting from rest at an angle of $20°$ with the vertical in any direction chosen by you (make sure you clearly specify that direction in your solution). What are the maximum and minimum values of r obtained from your plot?

CHAPTER 3

Kinematic Analysis of Planar Mechanisms

Since a particle has no dimension, it cannot rotate in space, but only translate. Rotation is a motion that involves change of orientation of a line. The material presented in this chapter consists of the study of position, velocity, and acceleration of mechanical systems that are composed of bars or bodies interconnected to each other and moving either along a plane or parallel planes. Figures in this chapter are drawn in a single plane, which is taken to be the reference frame for velocities and accelerations that are determined here. The translational motion of some of the points of the system and/or the rotational motion of some of its bars are assumed to be known, but calculation of the forces involved will not be of concern here. Knowing the motion of some points and/or bars of the system, the objective here is to determine velocity and acceleration of other points and/or angular velocity and angular acceleration of any of the bars of the system. This material is a direct application of the kinematic formulation that was presented in Chapter 2, and several classical mechanisms are considered in separate sections of this chapter.

3.1   Instantaneous Center of Rotation

When a rigid body[1] is in motion and the directions of the velocities of any two points of the body are known at a certain instant, one can find the direction of the velocity of any other point of the body at that instant from a graphical construction. At that instant the body appears to be rotating about a certain point in space. That point is called the *instantaneous center of rotation* or the *instantaneous center of zero velocity* for the body at that instant. If, in addition, the magnitude of the velocity of one of the points of the body is known at that same instant, the magnitude of the velocity of any other point of the body may also be determined for that instant

---

[1]A body is said to be rigid when the distance between any two points of the body is constant.

by making use of the instantaneous center of rotation. In general, the instantaneous center of rotation is moving in space.

The use of the instantaneous center of rotation may provide a quick way for determining the instantaneous (i.e., at a certain instant *only*) velocity of any other point of the body. The angular velocity of the body at the instant considered is also determined from such analysis. Depending on the specific situation, calculation of velocities using the instantaneous center of rotation may provide a quick and acceptable approximate answer to a problem.

The disadvantage of doing calculations using an instantaneous center of rotation is that such calculations are only valid for a specific instant. A specific figure that is used for calculating velocity at a certain instant using the instantaneous center of rotation at that instant is of no use for doing the same calculations at another instant. Also, the instantaneous center of rotation is not useful for determining accelerations.

The determination of the location of the instantaneous center of rotation for a body and the mathematical demonstration of its use in the calculation of velocities are now presented. Consider a rotating body for which the directions of the velocities $\vec{v}_A$ and $\vec{v}_B$ of two of its points A and B, respectively, relative to a reference point O are known as illustrated in Fig. 3.1. For the instant represented by the figure, the body is rotating clockwise and the magnitude of its angular velocity in the reference frame that is fixed to the plane of the paper is the quantity ω indicated in the figure. At that instant, the body appears to be rotating about an axis that passes through point C (thus, the name instantaneous center of rotation) shown in Fig. 3.1, which is located at the intersection of the perpendiculars to the velocity vectors $\vec{v}_A$ and $\vec{v}_B$.

To demonstrate the usefulness of point C, let $d_A = |\overrightarrow{AC}|$ and $d_B = |\overrightarrow{BC}|$ be the distances between points A and C and between points B and C, respectively. Now, look at the velocity $\vec{v}_{B/A}$ of point B relative to point A, which is defined as $\vec{v}_{B/A} \overset{\Delta}{=} \vec{v}_B - \vec{v}_A$. Since we are dealing with bodies for which the distance $L = |\overrightarrow{AB}|$ between points A and B is constant, the vector $\vec{v}_{B/A}$ is perpendicular to AB (because B has to be moving, relative to A, along an arc of a circle centered at A). Therefore, the component of $\vec{v}_{B/A}$ along line AB, which is $|\vec{v}_B|\cos\alpha_B - |\vec{v}_A|\cos\alpha_A$, has to be zero, i.e.,

$$|\vec{v}_B|\cos\alpha_B - |\vec{v}_A|\cos\alpha_A = 0 \qquad (3.1.1)$$

If we draw the perpendicular from C to line AB, we can write two expressions for the same distance, which are $d_A\cos\alpha_A$ and $d_B\cos\alpha_B$. We then have the following equation:

$$d_B\cos\alpha_B = d_A\cos\alpha_A \qquad (3.1.2)$$

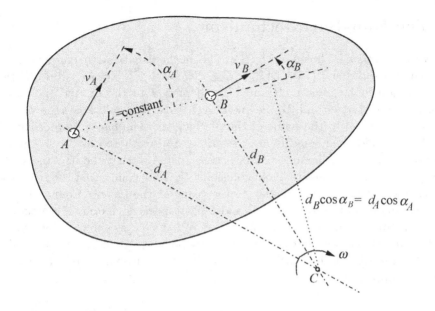

Figure 3.1: How to find the instantaneous center of rotation of a body.

Equations (3.1.1) and (3.1.2) immediately disclose that $|\vec{v}_A|/d_A = |\vec{v}_B|/d_B$ rad/s. Therefore, since points A and B are any two points on the body, we can then write

$$\frac{|\vec{v}_A|}{d_A} = \frac{|\vec{v}_B|}{d_B} = \ldots = \left(\frac{|\vec{v}_P|}{d_P}\right)_{\text{for any other point } P} = \text{constant} = \omega \qquad (3.1.3)$$

Equation (3.1.3) provides a convenient and expeditious way to determine the magnitude of the velocity of any point B on the body by the graphical construction illustrated in Fig. 3.1 when the directions of $\vec{v}_A$ and $\vec{v}_B$ and the magnitude of $\vec{v}_A$ are known. To do this, locate the instantaneous center of rotation by finding the intersection of the perpendiculars to $\vec{v}_A$ and $\vec{v}_B$ (when they are not parallel to each other), measure (or calculate by using the appropriate geometrical relations for each particular figure) the distances d_A and d_B, calculate $\omega = |\vec{v}_A|/d_A$, and then calculate $|\vec{v}_B|$ as $|\vec{v}_B| = \omega d_B$.

It is important to remember that the instantaneous center of rotation can only be used, when convenient, for calculating angular velocity and velocity at a particular instant. It cannot be used for finding accelerations.

3.2   The Four-Bar Mechanism

A *four-bar mechanism* is shown in Fig. 3.2. The mechanism actually consists of three moving bars (or *links*) AB (with $|\overrightarrow{AB}| = L_2$), BC (with $|\overrightarrow{BC}| = L_3$), and CD (with $|\overrightarrow{CD}| = L_4$) joined by pins at B and C, and pinned to the ground (i.e., to the plane of the paper) at A and D. Segment AD, with $\overrightarrow{AD}| = L_1$, is regarded as a fourth bar, giving rise to the name of the mechanism. The bars are approximated as rigid bodies, and their lengths are essentially approximated as L_2, L_3, and L_4, respectively. In practice, the mechanism is constructed so that the bars move along different, but parallel, planes to avoid collisions; the pins connecting the bars would be long like a nail but perpendicular to the plane of the paper. Points A and D are the centers of rotation of bars AB and CD, respectively. Point C_{BC} shown in Fig. 3.3 is the instantaneous center of rotation of bar BC; as shown in Section 3.1, it can be used for determining the velocity of any point on bar BC in terms of either $\dot{\theta}_2$ or $\dot{\theta}_4$. For the sake of illustration, the velocity $\vec{v}_{C/B}$ of point C relative to point B is also shown in Fig. 3.3.

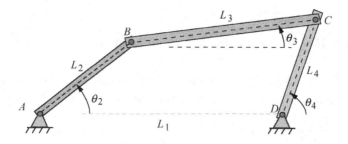

Figure 3.2: A four-bar mechanism (the angles θ_2, θ_3, and θ_4 are changing with time).

Either link AB or link CD is driven by a motor (which is not shown in the figure) to make either θ_2 or θ_4 a desired function of time t. In such a case, the motion of the mechanism is completely determined by the knowledge of either $\theta_2(t)$ or $\theta_4(t)$. Taking the angle $\theta_2(t)$ to be the known variable, the objective of the analysis is to determine how the other angles shown in Figs. 3.2 and 3.3 and their first and second time derivatives (i.e., the angular velocity and angular acceleration of the other links) depend on θ_2, $\dot{\theta}_2$, and $\ddot{\theta}_2$. There are no differential equations to be integrated here since the motion is specified. This type of analysis is called *kinematic analysis*, as opposed to dynamic analysis where the objective would have been to determine the motion of the system when known forces are applied to it and/or the reaction forces that any of the bars are subjected to during the motion.

The general kinematic analysis of four-bar mechanisms, which is based on equa-

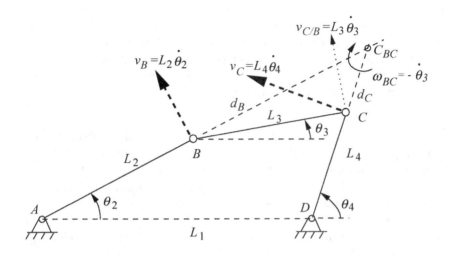

Figure 3.3: A four-bar mechanism and the instantaneous center of rotation of bar BC.

tions that are valid for all times, is presented in Section 3.2.1.

3.2.1 General Analysis of the Four-Bar Mechanism

The kinematic analysis of any mechanism involves first drawing a simple sketch of the mechanism in a *general* configuration, and then writing a set of geometric relations by inspection of the sketch. Such relations consist of equations for appropriate distances along perpendicular directions. The general sketch for the four-bar mechanism is shown in Fig. 3.2. The following two equations (which are called the *loop equations* for the four-bar mechanism) can be written by inspecting either that figure or Fig. 3.3. They are simply the scalar components along line AD, and along the perpendicular to line AD, of the vector identity $\overrightarrow{AD} = \overrightarrow{AB} + \overrightarrow{BC} + \overrightarrow{CD}$.

$$L_2 \cos\theta_2 + L_3 \cos\theta_3 - L_4 \cos\theta_4 = L_1 \tag{3.2.1}$$

$$L_4 \sin\theta_4 = L_2 \sin\theta_2 + L_3 \sin\theta_3 \tag{3.2.2}$$

The left- and right-hand sides of Eq. (3.2.1) represent two expressions for the same distance $|\overrightarrow{AD}|$, and Eq. (3.2.2) represents two expressions for the perpendicular distance from point C to line AD.[2] Notice that the mechanism was drawn in a *general configuration* in Figs. 3.2 and 3.3 for generating Eqs. (3.2.1) and (3.2.2). No specific values of the angles (especially 0 and 90°, which would eliminate some terms in the equations) should be specified in those equations because, if that is done, they would be valid only for a specific time (and, thus, no time derivatives could be taken for continuing the analysis).

> Once Eqs. (3.2.1) and (3.2.2) are obtained, the problem of determining the angular velocities and accelerations of any of the bars of the mechanism is essentially solved. That is done simply by taking the first and second time derivatives of those equations to obtain a set of equations that involve the angular velocities $\dot{\theta}_2$, $\dot{\theta}_3$, and $\dot{\theta}_4$ and the angular accelerations $\ddot{\theta}_2$, $\ddot{\theta}_3$, and $\ddot{\theta}_4$. This general procedure yields the equations that allow one to determine the angular velocities and angular accelerations of all the members of a four-bar mechanism in any configuration.

The following equations that involve angular velocities are obtained by taking the first time derivative of Eqs. (3.2.1) and (3.2.2).

$$L_2\dot{\theta}_2 \sin\theta_2 + L_3\dot{\theta}_3 \sin\theta_3 - L_4\dot{\theta}_4 \sin\theta_4 \;=\; 0 \qquad (3.2.3)$$
$$L_2\dot{\theta}_2 \cos\theta_2 + L_3\dot{\theta}_3 \cos\theta_3 - L_4\dot{\theta}_4 \cos\theta_4 \;=\; 0 \qquad (3.2.4)$$

Note: Equations (3.2.3) and (3.2.4) could also have been obtained by making use of the following vector identity involving the velocities of points C and B.

$$\vec{v}_C = \vec{v}_B + \vec{v}_{C/B}$$

The three velocity vectors that appear in this vector relationship are shown in Fig. 3.3. As mentioned earlier, the velocities $\vec{v}_C$ and $\vec{v}_B$ are the velocities of points C and B, respectively, in the reference frame fixed to the plane of the paper. In the rest of this chapter, such a velocity of a point, say $\vec{v}_C$ (with only one subscript) of point C, will be referred to as the *absolute velocity of point C*. Since point C is constrained to move along an arc of a circle centered at D and with radius $|\overrightarrow{DC}| = L_4$, and bar CD rotates with angular velocity $\dot{\theta}_4$, the absolute velocity $\vec{v}_C$ is perpendicular to line CD as drawn in Fig. 3.3. Similarly, the absolute velocity $\vec{v}_B$

---

[2]One could also have written, instead, equations for distances along different orthogonal lines. For example, if distances along line BC and along the perpendicular to line BC are chosen, one would obtain $L_2\cos(\theta_2-\theta_3)+L_3-L_4\cos(\theta_4-\theta_3) = L_1\cos\theta_3$ and $L_4\sin(\theta_4-\theta_3) = L_2\sin(\theta_2-\theta_3)+L_1\sin\theta_3$. This second choice, or any other convenient choice, yields equations that are equivalent to Eqs. (3.2.1) and (3.2.2) and accomplish the same purpose of those equations.

of point B is perpendicular to line AB as also shown in that figure. The velocity $\vec{v}_{C/B}$ is the velocity of point C relative to point B. Since, when viewed from point B, point C is constrained to move along an arc of a circle centered at B (because the length $|\overrightarrow{BC}| = L_3$ of bar BC is constant), and bar BC rotates in the plane of the paper with absolute angular velocity $\dot{\theta}_3$, the velocity $\vec{v}_{C/B}$ is the vector labeled $v_{C/B} = L_3\dot{\theta}_3$ in Fig. 3.3.

The vector identity $\vec{v}_C = \vec{v}_B + \vec{v}_{C/B}$ yields two scalar equations by projecting the vectors $\vec{v}_C$, $\vec{v}_B$, and $\vec{v}_{C/B}$ along two orthogonal directions. As can be verified by inspecting Fig. 3.3, the two equations (3.2.3) and (3.2.4) simply represent the components of this vector relationship along the directions perpendicular and parallel to line AD, respectively.[3]

Equations (3.2.3) and (3.2.4) are linear in the angular velocities and, thus, can be easily solved for $\dot{\theta}_3$ and $\dot{\theta}_4$ in terms of $\dot{\theta}_2$ and the angles θ_2, θ_3, and θ_4 when their values are obtained from Eqs. (3.2.1) and (3.2.2). Having the numerical values of the angles θ_2, θ_3, and θ_4, the solution for the angular velocities $\dot{\theta}_3$ and $\dot{\theta}_4$ can be easily obtained with a simple matrix inversion and a matrix multiplication as follows. The solution can also be readily obtained by hand or with an inexpensive calculator.

$$\begin{bmatrix} L_3\sin\theta_3 & -L_4\sin\theta_4 \\ L_3\cos\theta_3 & -L_4\cos\theta_4 \end{bmatrix} \begin{bmatrix} \dot{\theta}_3 \\ \dot{\theta}_4 \end{bmatrix} = \begin{bmatrix} -L_2\dot{\theta}_2\sin\theta_2 \\ -L_2\dot{\theta}_2\cos\theta_2 \end{bmatrix}$$

or

$$\begin{bmatrix} \dot{\theta}_3 \\ \dot{\theta}_4 \end{bmatrix} = \begin{bmatrix} L_3\sin\theta_3 & -L_4\sin\theta_4 \\ L_3\cos\theta_3 & -L_4\cos\theta_4 \end{bmatrix}^{-1} \begin{bmatrix} -L_2\dot{\theta}_2\sin\theta_2 \\ -L_2\dot{\theta}_2\cos\theta_2 \end{bmatrix} \tag{3.2.5}$$

The angular accelerations $\ddot{\theta}_3$ and $\ddot{\theta}_4$ of links BC and DC, respectively, are obtained by simply taking the time derivative of Eqs. (3.2.3) and (3.2.4). This yields

$$L_2(\ddot{\theta}_2\sin\theta_2 + \dot{\theta}_2^2\cos\theta_2) + L_3(\ddot{\theta}_3\sin\theta_3 + \dot{\theta}_3^2\cos\theta_3) = L_4(\ddot{\theta}_4\sin\theta_4 + \dot{\theta}_4^2\cos\theta_4) \tag{3.2.6}$$

$$L_2(\ddot{\theta}_2\cos\theta_2 - \dot{\theta}_2^2\sin\theta_2) + L_3(\ddot{\theta}_3\cos\theta_3 - \dot{\theta}_3^2\sin\theta_3) = L_4(\ddot{\theta}_4\cos\theta_4 - \dot{\theta}_4^2\sin\theta_4) \tag{3.2.7}$$

---

[3]Notice that a judicious choice of directions along which to project a vector equation may even yield an immediate solution to one of the variables in that vector equation. For example, if the vector equation $\vec{v}_C = \vec{v}_B + \vec{v}_{C/B}$ is projected along line DC (which is perpendicular to $\vec{v}_C$), the following equation is obtained (see Fig. 3.3).

$$0 = L_2\dot{\theta}_2\sin(\theta_4 - \theta_2) + L_3\dot{\theta}_3\sin(\theta_4 - \theta_3)$$

This equation immediately yields the solution for $\dot{\theta}_3$, which is the same solution that is obtained from Eqs. (3.2.3) and (3.2.4).

Note: Equations (3.2.6) and (3.2.7) could also have been obtained by making use of the following vector identity involving the accelerations of points C and B.

$$\frac{d}{dt}\vec{v}_C = \frac{d}{dt}\vec{v}_B + \frac{d}{dt}\vec{v}_{C/B}$$

This vector identity yields two scalar equations when each of the three acceleration vectors $\vec{a}_C = d\vec{v}_C/dt$, $\vec{a}_B = d\vec{v}_B/dt$, and $\vec{a}_{C/B} = d\vec{v}_{C/B}/dt$ are written in terms of components along line AD and the perpendicular to line AD. This is left for you, the reader, to do. If needed, look at Fig. 3.4, which is the same as the figure first presented in Example 2.7.3 in Chapter 2. The figure is a graphical visualization of the time derivative $\vec{v} = d\vec{r}/dt$ of a rotating vector $\vec{r}$ (shown in the smaller box) and also of the time derivative $\vec{a} = d\vec{v}/dt$ of $\vec{v}$ (shown in the larger box). It provides a quick reference for writing the expressions for velocity $\vec{v} = d\vec{r}/dt$ and acceleration $\vec{a} = d\vec{v}/dt$ of a point, using rotating reference directions. For a vector $\vec{r}$ of constant magnitude, $\dot{r} = \ddot{r} = 0$ in the figure. Such a figure will be helpful for writing expressions for $\vec{a}_C$, $\vec{a}_B$, $\vec{a}_{C/B}$, etc., in this section, in Example 3.2.2 (p. 233) and in other examples.

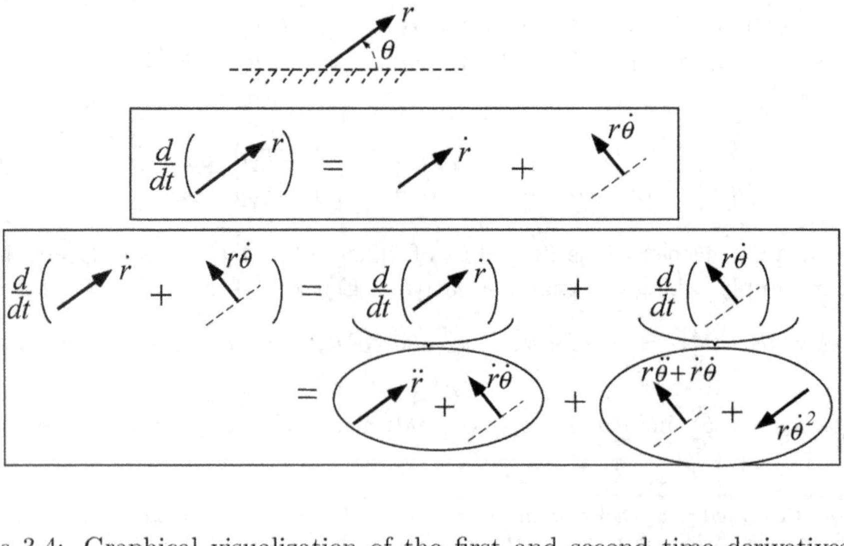

Figure 3.4: Graphical visualization of the first and second time derivatives of a rotating vector $\vec{r}$.

Equations (3.2.6) and (3.2.7) are linear in the angular accelerations and, therefore, can be easily solved for $\ddot{\theta}_3$ and $\ddot{\theta}_4$ in terms of the angular acceleration $\ddot{\theta}_2$, of the angular velocities $\dot{\theta}_2, \dot{\theta}_3$, and $\dot{\theta}_4$, and of the angles θ_2, θ_3, and θ_4 once their values are known. In matrix form, the solution for the angular accelerations $\ddot{\theta}_3$ and $\ddot{\theta}_4$ is obtained as follows. Again, the solution can readily be obtained by hand or with an inexpensive calculator.

$$
\begin{bmatrix} \ddot{\theta}_3 \\ \ddot{\theta}_4 \end{bmatrix} = \begin{bmatrix} L_3 \sin\theta_3 & -L_4 \sin\theta_4 \\ L_3 \cos\theta_3 & -L_4 \cos\theta_4 \end{bmatrix}^{-1}
$$
$$
\times \begin{bmatrix} L_4 \dot{\theta}_4^2 \cos\theta_4 - L_3 \dot{\theta}_3^2 \cos\theta_3 - L_2(\ddot{\theta}_2 \sin\theta_2 + \dot{\theta}_2^2 \cos\theta_2) \\ -L_4 \dot{\theta}_4^2 \sin\theta_4 + L_3 \dot{\theta}_3^2 \sin\theta_3 - L_2(\ddot{\theta}_2 \cos\theta_2 - \dot{\theta}_2^2 \sin\theta_2) \end{bmatrix}
$$

$$(3.2.8)$$

Of all the equations for this mechanism [Eqs. (3.2.1) and (3.2.2), (3.2.5), and (3.2.8)], Eqs. (3.2.1) and (3.2.2) are the most difficult to solve because they are non-linear in the unknown angles θ_3 and θ_4 (with the nonlinearities being the trigonometric functions). Although it is possible to find an analytical solution to those equations by making a number of algebraic manipulations and using trigonometric identities such as $\sin^2\alpha + \cos^2\alpha = 1$ for any angle α, a numerical solution is much more convenient and practical to obtain.

Numerical solutions for θ_3 and θ_4 can be obtained graphically by simply drawing the figure to scale and measuring the resulting angles. This approach is a simple one, but it is also one that may not give very accurate answers. Its main disadvantage, however, is that a new drawing has to be made when a new solution is desired for a different configuration of the mechanism at some other time. Another way to obtain a solution is to plot the expressions for $\sin\theta_3$ and $\cos\theta_3$ given by Eqs. (3.2.1) and (3.2.2) versus θ_4, with $0 \le \theta_4 \le 2\pi$, and then select, from the plot, only the values of $\sin\theta_3$ and $\cos\theta_3$ that satisfy the relationship $\sin^2\theta_3 + \cos^2\theta_3 = 1$. This strategy has the same disadvantage previously mentioned.

A more efficient and direct way for solving Eqs. (3.2.1) and (3.2.2) is given in Section 3.2.2, which is, then, illustrated by some examples. The kinematics of four-bar mechanisms is automated in Section 3.2.3 for those who want to use either Scilab, Matlab, or Simulink.[4]

3.2.2 Solution of Eqs. (3.2.1) and (3.2.2)

The solution methodology presented here makes use of the angles α, β, and γ shown in Fig. 3.5, and of simple trigonometric relations that are obtained by inspection of that figure.

First, determine the angle α by using the following expressions that are obtained

---

[4]Tutorials for all three are in Appendix A.

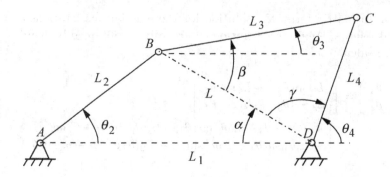

Figure 3.5: The angles α, β, and γ needed for calculating θ_3 and θ_4.

by inspection of Fig. 3.5:

$$\sin \alpha = \frac{L_2 \sin \theta_2}{L} \qquad \cos \alpha = \frac{L_1 - L_2 \cos \theta_2}{L} \qquad (3.2.9)$$

where

$$L = \sqrt{(L_2 \sin \theta_2)^2 + (L_1 - L_2 \cos \theta_2)^2} = \sqrt{L_1^2 + L_2^2 - 2L_1 L_2 \cos \theta_2} \quad (3.2.10)$$

Notice that point B in Fig. 3.5 will be below line AD, which is a physical possibility, if the value that is found for $\sin \alpha$ is negative.

For this mechanism, the angle α can be anywhere in the range $[-\pi, \pi]$. Therefore, to calculate the value of α, both $\sin \alpha$ and $\cos \alpha$ should be used [instead of either $\sin \alpha$ or $\cos \alpha$, or $\tan \alpha = (\sin \alpha)/(\cos \alpha)$] because we need to determine which quadrant α lies on if we want to draw the actual configuration of the mechanism based on these calculations. In many computer software, including Matlab, the function that does this is $atan2(argument\_1, argument\_2)$, where $argument\_1$ is the value for $\sin \alpha$ and $argument\_2$ is the value for $\cos \alpha$. But in Scilab, for example, such a function is $atan$, where $atan(argument)$ yields the two-quadrant arc-tangent and $atan(argument\_1, argument\_2)$ yields the four-quadrant arc-tangent.

Next, determine the angles β and γ by using the following expressions that are obtained by inspection of the triangle BCD in Fig. 3.5.

$$L_3 \sin \beta = L_4 \sin \gamma \qquad (3.2.11)$$

$$L_3 \cos \beta + L_4 \cos \gamma = L \qquad (3.2.12)$$

To determine β, start by combining Eqs. (3.2.11) and (3.2.12) as $(L_3 \cos \beta - L)^2 + (L_3 \sin \beta)^2 = L_4^2$ to obtain

$$\cos \beta = \frac{L_3^2 + L^2 - L_4^2}{2LL_3} \tag{3.2.13}$$

Notice that a real solution for β is possible only if $|L_3^2 + L^2 - L_4^2| \leq 2LL_3$. If this condition is violated for certain ranges of the angle $\theta_2(t)$, then the mechanism cannot operate in that range. The same is true when the calculation of the sine and cosine of any angle yields a value that is outside the range $[-1, 1]$.

If the value obtained for $\cos \beta$ is in the physical range $[-1, 1]$, then there are two possibilities for $\sin \beta$, each one corresponding to a possible configuration of the mechanism (i.e., with point C located either above or below line BD in Fig. 3.5). The solution to use is determined by the initial configuration that is chosen for the mechanism to start its motion.

Solution number 1: $\qquad \sin \beta = \sqrt{1 - \cos^2 \beta} \qquad$ (3.2.14)

Solution number 2: $\qquad \sin \beta = -\sqrt{1 - \cos^2 \beta} \qquad$ (3.2.15)

The angle β corresponding to these two solutions is then determined by using the values that are found for $\sin \beta$ and $\cos \beta$.

Having determined α and β, the angle θ_3 is simply determined as

$$\theta_3 = \beta - \alpha \tag{3.2.16}$$

Finally, the angle γ (which is needed for calculating θ_4) can now be determined by using Eqs. (3.2.11) and (3.2.12) to calculate $\sin \gamma = (L_3 \sin \beta)/L_4$ and $\cos \gamma = (L - L_3 \cos \beta)/L_4$ for each of the two possible solutions for β.

The angle θ_4 is then determined as:

$$\theta_4 = \pi - (\alpha + \gamma) \tag{3.2.17}$$

Example 3.2.1 Bar AB of the mechanism shown in either Fig. 3.2 (p. 224) or Fig. 3.5 (p. 230) is driven by a motor so that $\theta_2 = \pi/6 + 1.5\pi \sin(kt)$ rad, with $k = 2$ rad/s. The lengths of the bars are $L_2 = 0.45$ m, $L_3 = 0.8$ m, and $L_4 = 0.7$ m, while $L_1 = 1$ m. Determine, for the initial instant $t = 0$, and for $t = 5$ s, the angular velocity and the angular acceleration (magnitude and direction) of bars BC and DC.

■ Solution

Since only θ_2 is given, we need to calculate θ_3 and θ_4 to be able to determine the angular velocities and angular accelerations $\dot{\theta}_3, \dot{\theta}_4, \ddot{\theta}_3$, and $\ddot{\theta}_4$ of bars BC and DC. With $\theta_2(t) = \pi/6 + 1.5\pi \sin(2t)$, we have $\theta_2(0) = \pi/6$ rad $= 30°$, and $\theta_2(5) \approx -2.04$ rad $\approx -116.9°$. The results of the sequence of calculations presented in this section for these two values of θ_2 are shown in the following table.

θ_2 (degrees)	L(m)	$\sin\alpha$	$\cos\alpha$	$\cos\beta$	$\sin\beta$	$\sin\gamma$	$\cos\gamma$
30	0.6504	0.3459	0.9383	0.5507	±0.8347	±0.954	0.3
−116.9	1.269	-0.3163	0.9487	0.8669	±0.4986	±0.57	0.822

These results yield the following solutions for θ_3 and θ_4.

Solution 1 for $\theta_2 = 30°$:

$$\theta_3 = \beta - \alpha \approx 36.4° \qquad \theta_4 = 180 - \alpha - \gamma \approx 87.2°$$

Solution 2 for $\theta_2 = 30°$:

$$\theta_3 = \beta - \alpha \approx -76.8° \qquad \theta_4 = 180 - \alpha - \gamma \approx 232.3°$$

Solution 1 for $\theta_2 = -116.9°$:

$$\theta_3 = \beta - \alpha \approx 48.4° \qquad \theta_4 = 180 - \alpha - \gamma \approx 163.7°$$

Solution 2 for $\theta_2 = -116.9°$:

$$\theta_3 = \beta - \alpha \approx -11.5° \qquad \theta_4 = 180 - \alpha - \gamma \approx 233.2°$$

The angular velocity and angular acceleration corresponding to each of the two solutions can be calculated using Eqs. (3.2.5) and (3.2.8). With $\dot{\theta}_2(t) = 3\pi \cos(2t)$ rad/s and $\ddot{\theta}_2(t) = -6\pi \sin(2t)$ rad/s^2, this yields the following results.

Solution 1 for $\theta_2(0) = 30°$ [thus, $\dot{\theta}_2(0) = 3\pi$ rad/s and $\ddot{\theta}_2(0) = 0$], $\theta_3 \approx 36.4°$, $\theta_4 \approx 87.2°$:

$$\dot{\theta}_3 \approx -5.75 \text{ rad/s} \qquad \dot{\theta}_4 \approx -0.86 \text{ rad/s}$$
$$\ddot{\theta}_3 \approx 60.9 \text{ rad/s}^2 \qquad \ddot{\theta}_4 \approx 121.2 \text{ rad/s}^2$$

Solution 2 for $\theta_2(0) = 30°$ [thus, $\dot{\theta}_2(0) = 3\pi$ rad/s and $\ddot{\theta}_2(0) = 0$], $\theta_3 \approx -76.8°$, $\theta_4 \approx 232.3°$:

$$\dot{\theta}_3 \approx -2.6 \text{ rad/s} \qquad \dot{\theta}_4 \approx -7.48 \text{ rad/s}$$
$$\ddot{\theta}_3 \approx 117.2 \text{ rad/s}^2 \qquad \ddot{\theta}_4 \approx 56.9 \text{ rad/s}^2$$

Solution 1 for $\theta_2(5) = -116.9°$ [thus, $\dot{\theta}_2(5) \approx -7.91$ rad/s and $\ddot{\theta}_2(5) = 10.25$ rad/s²], $\theta_3 \approx 48.4°$, $\theta_4 \approx 163.7°$:

$$\dot{\theta}_3 \approx -4.84 \text{ rad/s} \qquad \dot{\theta}_4 \approx 1.43 \text{ rad/s}$$
$$\ddot{\theta}_3 \approx 0.35 \text{ rad/s}^2 \qquad \ddot{\theta}_4 \approx -14.3 \text{ rad/s}^2$$

Solution 2 for $\theta_2(5) = -116.9°$ [thus, $\dot{\theta}_2(5) \approx -7.91$ rad/s and $\ddot{\theta}_2(5) = 10.25$ rad/s²], $\theta_3 \approx -11.5°$, $\theta_4 \approx 233.2°$:

$$\dot{\theta}_3 \approx 0.85 \text{ rad/s} \qquad \dot{\theta}_4 \approx -5.4 \text{ rad/s}$$
$$\ddot{\theta}_3 \approx -10.6 \text{ rad/s}^2 \qquad \ddot{\theta}_4 \approx 4.02 \text{ rad/s}^2$$

The angular velocities $\dot{\theta}_3$ and $\dot{\theta}_4$ may also be calculated using the graphical construction that involves the instantaneous center of rotation of bar BC, as shown in Fig. 3.3. The use of the instantaneous center of rotation to do this is illustrated in Example 3.2.2.

Example 3.2.2 Bar AB of the mechanism shown in Fig. 3.6 is rotating clockwise with a constant angular velocity $\dot{\phi}_2 = -90$ rpm. The figure is a snapshot of the mechanism taken when $\phi_2 = 90°$. Bars AB, BC, and DC are rigid, and their lengths are $L_2 = 2$ m, $L_3 = 3$ m, and $L_4 = 2.5$ m. The distances D_1 and D_2 shown in the figure are also constants. Determine, for the instant under consideration:

a. The angular velocity and the angular acceleration of bars BC and DC. Indicate their magnitude and whether they are clockwise or counterclockwise.

b. The absolute velocity $\vec{v}_C$ and the absolute acceleration $\vec{a}_C$ of point C. Specify their magnitude and direction.

■ **Solution**

a. The sequence of calculations to obtain the angular velocity of bars BC and CD, using the instantaneous center of rotation of bar BC, is shown in Fig. 3.7. The instantaneous center of rotation of bar BC is point C_{BC}.

First, point C_{BC} is located at the intersection of the extension of bars AB and DC. The distance d_B, needed for calculating the angular velocity of bar BC, could have been measured from the figure, but it was determined analytically, using trigonometry, as shown in Fig. 3.7a. The vector $\vec{v}_B$, which is the absolute velocity of point B, is then added to the figure. It is perpendicular to bar AB and points to the right as shown in that figure. Because $\vec{v}_B$ is pointing to the right, bar BC has to be rotating counterclockwise, which means that the angular velocity ω_{BC} of bar BC is as indicated in the figure.

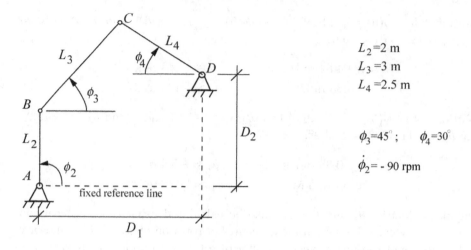

L_2=2 m
L_3=3 m
L_4=2.5 m

ϕ_3=45° ; ϕ_4=30°

$\dot{\phi}_2$= - 90 rpm

Figure 3.6: The mechanism for Example 3.2.2.

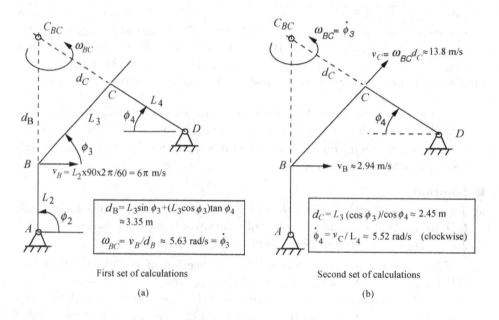

$d_B = L_3\sin\phi_3 + (L_3\cos\phi_3)\tan\phi_4$
≈ 3.35 m
$\omega_{BC} = v_B/d_B \approx 5.63$ rad/s $= \dot{\phi}_3$

First set of calculations

(a)

$d_C = L_3(\cos\phi_3)/\cos\phi_4 \approx 2.45$ m
$\dot{\phi}_4 = v_C/L_4 \approx 5.52$ rad/s   (clockwise)

Second set of calculations

(b)

Figure 3.7: Solution for the angular velocities $\omega_{BC} = \dot{\phi}_3$ and $\dot{\phi}_4$ for Example 3.2.2.

This, then, determines the direction of the absolute velocity $\vec{v}_C$ of point C, which is shown in Fig. 3.7b. Having the values for $v_B = |\vec{v}_B|$ and the distance d_B, the value of the angular velocity of bar BC, $\omega_{BC} = \dot{\phi}_3$, can then be determined as $\omega_{BC} = v_B/d_B \approx 5.63$ rad/s.

To determine the angular accelerations $\ddot{\phi}_3$ and $\ddot{\phi}_4$, we need to generate two equations that involve these two unknowns. The general approach to do this is to write two equations involving the angles ϕ_2, ϕ_3, and ϕ_4 shown in Fig. 3.6 in the same manner Eqs. (3.2.1) and (3.2.2) were generated for Fig. 3.2. This could be done, for this example, simply by determining the angles θ_2, θ_3, and θ_4 between line AD and bars AB, BC, and CD, respectively, in the same manner that such angles were defined in Fig. 3.2, and then use Eqs. (3.2.6) and (3.2.7) to calculate $\ddot{\phi}_3 = \ddot{\theta}_3$ and $\ddot{\phi}_4 = -\ddot{\theta}_4$. One also could look at Fig. 3.6 and write, by inspection, the two following equations by imagining that the figure was drawn with $\phi_2 < 90°$.

$$L_2 \cos \phi_2 + L_3 \cos \phi_3 + L_4 \cos \phi_4 = \text{constant} = D_1 \qquad (3.2.2a)$$
$$L_2 \sin \phi_2 + L_3 \sin \phi_3 - L_4 \sin \phi_4 = \text{constant} = D_2 \qquad (3.2.2b)$$

The left- and right-hand sides of Eq. (3.2.2a) represent two expressions for the same distance D_1, and the two sides of Eq. (3.2.2b) represent two expressions for the distance D_2 in Fig. 3.6.

The following equations involve the unknown angular velocities $\dot{\phi}_3$ and $\dot{\phi}_4$. They are obtained by taking the time derivative of Eqs. (3.2.2a) and (3.2.2b).

$$L_2 \dot{\phi}_2 \sin \phi_2 + L_3 \dot{\phi}_3 \sin \phi_3 + L_4 \dot{\phi}_4 \sin \phi_4 = 0 \qquad (3.2.2c)$$
$$L_2 \dot{\phi}_2 \cos \phi_2 + L_3 \dot{\phi}_3 \cos \phi_3 - L_4 \dot{\phi}_4 \cos \phi_4 = 0 \qquad (3.2.2d)$$

Equations (3.2.2c) and (3.2.2d) yield, of course, the same solutions for $\dot{\phi}_3$ and $\dot{\phi}_4$ that were obtained earlier using the instantaneous center of rotation. The next set of equations involve the angular accelerations. They are obtained simply by taking the time derivative of Eqs. (3.2.2c) and (3.2.2d).

$$L_2 \left(\ddot{\phi}_2 \sin \phi_2 + \dot{\phi}_2^2 \cos \phi_2 \right) + L_3 \left(\ddot{\phi}_3 \sin \phi_3 + \dot{\phi}_3^2 \cos \phi_3 \right)$$
$$+ L_4 \left(\ddot{\phi}_4 \sin \phi_4 + \dot{\phi}_4^2 \cos \phi_4 \right) = 0 \qquad (3.2.2e)$$
$$L_2 \left(\ddot{\phi}_2 \cos \phi_2 - \dot{\phi}_2^2 \sin \phi_2 \right) + L_3 \left(\ddot{\phi}_3 \cos \phi_3 - \dot{\phi}_3^2 \sin \phi_3 \right)$$
$$- L_4 \left(\ddot{\phi}_4 \cos \phi_4 - \dot{\phi}_4^2 \sin \phi_4 \right) = 0 \qquad (3.2.2f)$$

For the values of L_2, L_3, and L_4 shown in Fig. 3.6, Eqs. (3.2.2e) and (3.2.2f) yield the following results, with $\dot{\phi}_2 = -90$ rpm $= -3\pi$ rad/s and $\ddot{\phi}_2 = 0$.

$$\ddot{\phi}_3 \approx -4.13 \text{ rad/s}^2 \qquad\qquad \ddot{\phi}_4 \approx -99.6 \text{ rad/s}^2$$

b. The calculation of the absolute velocity $\vec{v}_C$ of point C is shown in Fig. 3.7b. Its direction is perpendicular to bar DC. Since it was determined that bar BC is rotating counterclockwise, the direction of $\vec{v}_C$ has to be as indicated in Fig. 3.7. Its magnitude is $v_C = \omega_{BC}d_C \approx 13.8$ m/s; the calculation of the distance d_C is shown in the figure. Now that v_C is known, the angular velocity of bar DC can then be determined. Because $\vec{v}_C$ is directed as shown in Fig. 3.7b, bar DC has to be rotating clockwise. Its angular velocity is $v_C/L_4 \approx 5.52$ rad/s. Therefore, $\dot{\phi}_4 \approx 5.52$ rad/s (compare Fig. 3.7b and Fig. 3.6 to see why the sign of $\dot{\phi}_4$ is positive).

Finally, we come to the calculation of the absolute acceleration $\vec{a}_C$ of point C. The simplest way to express $\vec{a}_C$ is in terms of unit vectors $\hat{r}_4$ and $\hat{\phi}_4$ that rotate with bar DC as shown in Fig. 3.8. However, this is a matter of personal choice since any other set of orthogonal unit vectors could be used. In terms of $\hat{r}_4$ and $\hat{\phi}_4$, the absolute acceleration $\vec{a}_C$ of point C is simply the following: (see Fig. 3.4, p. 228, if help is needed in understanding how to obtain such an expression)

$$\vec{a}_C = L_4\ddot{\phi}_4\hat{\phi}_4 - L_4\dot{\phi}_4^2\hat{r}_4 \approx -249\hat{\phi}_4 - 76\hat{r}_4 \text{ m/s}^2$$

The magnitude of $\vec{a}_C$ is $|\vec{a}_C| = \sqrt{249^2 + 76^2} \approx 260$ m/s^2. The vector $\vec{a}_C$ is directed to the left of bar DC in Fig. 3.8, at an angle equal to $\pi/2 + \arctan(76/249) \approx 1.87$ rad $\approx 107°$, measured counterclockwise from line DC. As an exercise, put it in the figure to visualize it, and check all the answers obtained in this example.

One other possible way to specify the absolute acceleration $\vec{a}_C$ is to express it, for example, in terms of the unit vectors $\hat{x}$ and $\hat{y}$ that are also shown in Fig. 3.8. Again, as mentioned earlier, this is a matter of personal choice. Such an expression is given here, with the specific answer $\vec{a}_C \approx -59\hat{x} - 253\hat{y}$ being for $\phi_4 = 30°$.

$$\begin{aligned} \vec{a}_C &= L_4\left(\ddot{\phi}_4\sin\phi_4 + \dot{\phi}_4^2\cos\phi_4\right)\hat{x} + L_4\left(\ddot{\phi}_4\cos\phi_4 - \dot{\phi}_4^2\sin\phi_4\right)\hat{y} \\ &\approx -59\hat{x} - 253\hat{y} \text{ m/s}^2 \end{aligned}$$

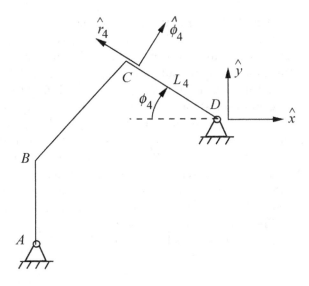

Figure 3.8: Unit vectors chosen for expressing the acceleration of point C for Example 3.2.2.

3.2.3   Automated Solution for the Four-Bar Mechanism

For practical applications dealing with investigations and design of four-bar mechanisms, the most efficient way to do all the calculations presented so far is to automate them in a computer.

⟹ Two similar programs named *fourbar_mechanism*, written as *functions*, are listed in Appendix E. They use another program named *fourbar_kinematics* (which implements all the calculations seen earlier), and animate the motion of the mechanism. One set of programs is for Scilab users (which Scilab saves as *fourbar_mechanism.sci* and *fourbar_kinematics.sci*, respectively), and the other is for Matlab users (which Matlab saves as *fourbar_mechanism.m* and *fourbar_kinematics.m*, respectively). The Scilab and Matlab syntax is almost the same, and specific instructions on how to use any of the programs listed in Appendix E are given at the beginning of that Appendix.

The "inputs" to these programs are the lengths L_1, L_2, L_3, and L_4, and the angle $\theta_2(t)$ and its first and second time derivatives $\dot{\theta}_2$ and $\ddot{\theta}_2$. Their calling syntax, which is the same for Scilab and Matlab, is (please see the *Note* that follows)

```
output=fourbar_mechanism(t, L1, L2, L3, L4, theta2, ...
```

dtheta2_dt, d2theta2_dt2, solution_number);

where t is either a scalar or a time array such as [t0, t1, ..., tfinal] or [t0:Δt:tfinal]. The time array t contains the instants the calculations are stored in the time array output, starting from an initial time t0 to the final time tfinal (in steps Δt if t is specified as [t0:Δt:tfinal]). The input solution_number must be either 1 or 2, and they correspond to "Solution number 1" and to "Solution number 2" indicated in Eqs. (3.2.14) and (3.2.15), respectively, for $\sin \beta = \pm\sqrt{1 - \cos^2 \beta}$. These solutions are implemented in the programs as $\sin \beta = (3 - 2 * \text{solution\_number} * \sqrt{1 - \cos^2 \beta})$. Any desired "name" can be given to the variable output. After such a call to the function fourbar_mechanism, the variable output is a ten-column array whose columns are $t, \theta_2, \theta_3, \theta_4, \dot{\theta}_2, \dot{\theta}_3, \dot{\theta}_4, \ddot{\theta}_2, \ddot{\theta}_3$, and $\ddot{\theta}4$. For a specific example of usage, see the first comment lines of the *fourbar_mechanism* programs (i.e., the comment lines that are seen before the first blank line in those programs).

Note: The three dots ... in a command line have been used by Scilab and Matlab for continuing a command that should be typed in a single line. However, as a precaution for the possibility that this feature may be discontinued on future versions of Scilab and/or Matlab, it is not used in any of the programs associated with this book and available for downloading at www.doverpublications.com. The three-dots feature is used in this book for illustrative purposes only.

Those who have Simulink and prefer to use it may want to construct the block diagram model shown in Fig. 3.9 for performing the same calculations mentioned earlier. Since the calculations do not involve integrations, choose *discrete (no continuous states)* in the solver part of the *Simulation → Parameters* menu. Except for round-off errors, the answers displayed in Fig. 3.9 correspond to the ones obtained for Solution number 1 in Example 3.2.1 (p. 231), with t = 5. To see an animation of the motion, choose a time array of the form $0 : \Delta t : tfinal$ and type animate_nbars(x_coordinates, y_coordinates) in the Matlab command window after the simulation stops. The Matlab program animate_nbars, listed in Appendix E, is needed for this.

All the steps presented in Section 3.2.2 for calculating θ_3 and θ_4 are embedded in the block labeled "*Four-bar mechanism*" seen in Fig. 3.9, and all the steps for calculating the angular velocities and angular accelerations are embedded in the block labeled "*ang_veloc, ang_accel*". Those blocks are the *subsystems*

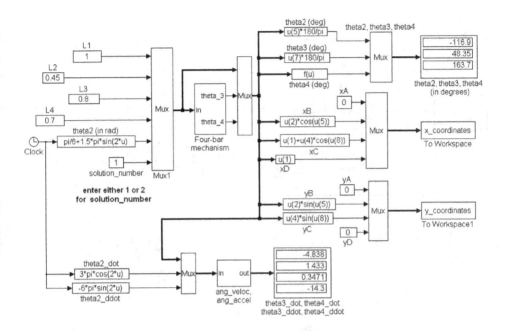

Figure 3.9: A Simulink block diagram model for the four-bar mechanism. For the model inputs shown, the time array [0:0.001:5] is suggested for the *Simulation →* *Parameters* settings.

shown in Fig. 3.10.[5] Subsystems are used for compacting a block diagram that might hardly fit on a computer screen (see the Simulink tutorial in Appendix A).

The internal forces acting on each bar of a mechanism also play a role in the design of mechanisms since some operational limits should not be exceeded. Section 5.6 in Chapter 5 is dedicated to the calculation of such forces.

---

[5]The expression in the Matlab Function block labeled "Solution of Eq. (3.2.8)" in Fig. 3.10b is [u(5)*sin(u(9)), -u(6)*sin(u(10)); u(5)*cos(u(9)), -u(6)*cos(u(10))]∧(-1)*[u(6)*u(2)∧2*cos(u(10)) -u(5)*u(1)∧2*cos(u(9))-u(4)*(u(12)*sin(u(7))+u(11)∧2*cos(u(7)));   -u(6)*u(2)∧2*sin(u(10)) +u(5)*u(1)∧2*sin(u(9))-u(4)*(u(12)*cos(u(7)) -u(11)∧2*sin(u(7)))]. It should be typed, *in a single line*, in the little window that appears on the computer screen after double-clicking that block. Spaces are not important. The expression in the Matlab Function block labeled "Solution of Eq. (3.2.5)" in the same figure is -u(2)*u(9)*[u(3)*sin(u(7)), -u(4)*sin(u(8)); u(3)*cos(u(7)), -u(4)*cos(u(8))]∧(-1)*[sin(u(5)); cos(u(5))].

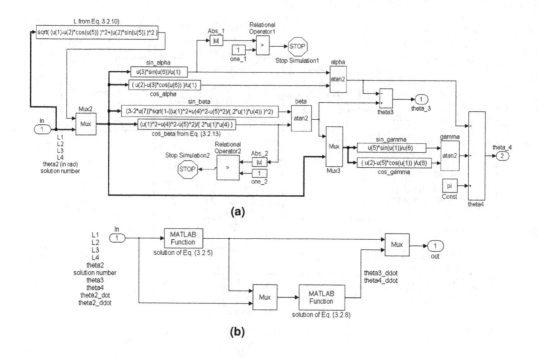

(a)

(b)

Figure 3.10: (a) The subsystem labeled *"Four-bar mechanism"*; (b) the subsystem labeled *"ang_veloc, ang_accel"* in Fig. 3.9, for Simulink users.

3.3   The Geneva Wheel Mechanism

This mechanism is used for creating intermittent motion. It consists of two wheels, one of which has a number of radial slots to engage a pin on the other wheel. Such a mechanism is used in mechanical clocks and watches, for example. A six-slot Geneva wheel is shown in Fig. 3.11.

Wheel 1 in Fig. 3.11 is the driver. It is driven by an external motor with a known angular velocity ω_1 (relative to the plane of the paper), and it has a pin B. The distance between the fixed pins A and C is such that B engages one of the slots of wheel 2 and then slides along that rotating slot without hitting the end of the slot. Thus, when B is engaged, the slotted wheel 2 is forced to rotate until B slides in and out of the slot as wheel 1 continues to turn. When B is disengaged from a slot on wheel 2, the slotted wheel will stop rotating until B engages the next slot and that wheel starts to rotate again. The slots on wheel 2 in Fig. 3.11 are equally spaced, and their openings should be in position to engage pin B.

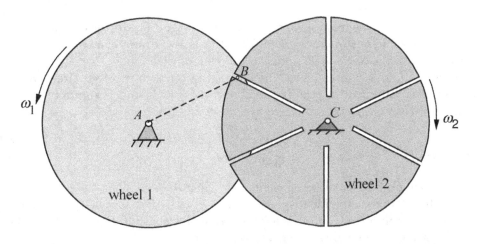

Figure 3.11: A six-slot Geneva wheel mechanism.

Given the angular velocity ω_1 of wheel 1 and the position of B at any instant t, the kinematic analysis of the Geneva wheel consists of determining the angular velocity and the angular acceleration of the slotted wheel and the velocity and acceleration of pin B relative to the slot in which it slides. The analysis is done with the aid of Fig. 3.12.

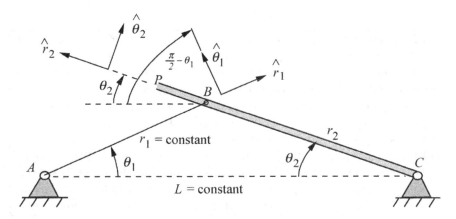

Figure 3.12: Diagram for analyzing the kinematics of a Geneva wheel mechanism.

Notice that the angular velocity ω_1 shown in Fig. 3.11, drawn with a counter-clockwise arrow, is the same as $\dot{\theta}_1$ because θ_1 in Fig. 3.12 is also drawn counterclockwise. Similarly, the angular velocity ω_2 shown in Fig. 3.11, drawn with a clockwise arrow, is the same as $\dot{\theta}_2$ because θ_2 in Fig. 3.12 is also drawn clockwise. Bar AB is driven to make the angle $\theta_1(t)$ a known function of time.

As in the four-bar mechanism, the kinematics problem is solved after two independent equations involving the two unknowns $r_2 = |\overrightarrow{CB}|$ and θ_2 shown in Fig. 3.12 are generated by inspection of the figure. Such an inspection yields

$$r_1 \cos\theta_1 + r_2 \cos\theta_2 \;=\; \text{constant} = L \tag{3.3.1}$$

$$r_1 \sin\theta_1 \;=\; r_2 \sin\theta_2 \tag{3.3.2}$$

These are simply the scalar components along line AC and along the perpendicular to line AC of the vector identity $\overrightarrow{AC} = \overrightarrow{AB} + \overrightarrow{BC}$. For simplicity, intermittent motion is not considered, and it is assumed that the length $|\overrightarrow{CP}|$ of bar CP is large enough for pin B to be always engaged on that bar.

The left- and right-hand sides of Eq. (3.3.1) represent two expressions for the same distance $|\overrightarrow{AC}|$, and the left- and right-hand sides of Eq. (3.3.2) represent two expressions for the perpendicular distance from point B to line AC.

Unlike Eqs. (3.2.1) and (3.2.2) for the four-bar mechanism, Eqs. (3.3.1) and (3.3.2) can be easily solved for r_2 and θ_2. The solution for r_2 is obtained by combining Eqs. (3.2.1) and (3.2.2) to yield Eq. (3.3.3):

$$r_2 = \sqrt{(L - r_1 \cos\theta_1)^2 + (r_1 \sin\theta_1)^2} = \sqrt{L^2 + r_1^2 - 2Lr_1 \cos\theta_1} \tag{3.3.3}$$

> With r_2 given by Eq. (3.3.3), the only unknown in Eqs. (3.3.1) and (3.3.2) is now the angle θ_2 and, for this mechanism, θ_2 can take any value in any quadrant. Therefore, one should resist the temptation to use only one of those equations to determine θ_2 based only on the value of either $\sin\theta_2$ or $\cos\theta_2$ (or even on $\theta_2 = \arctan\left[(\sin\theta_2)/(\cos\theta_2)\right] = \arctan\left[(r_1 \sin\theta_1)/(L - r_1 \cos\theta_1)\right]$, for example, as obtained by combining those equations), because the answer that is obtained from such a calculation is an angle that is only in the range $[-\pi/2, \pi/2]$ (i.e., only in two quadrants, instead of all four). A proper calculation of θ_2 for the mechanism should give a result in any of the four possible quadrants so that the actual position of bar CP of the mechanism can be determined.

Generally, having determined r_2, as given by Eq. (3.3.3), one then uses *both* Eqs. (3.3.1) and (3.3.2) to calculate $\sin\theta_2$ and $\cos\theta_2$ as

$$\sin\theta_2 = \frac{r_1 \sin\theta_1}{r_2} \quad \text{and} \quad \cos\theta_2 = \frac{L - r_1 \cos\theta_1}{r_2}$$

The angle θ_2 for bar CP is then determined based on the values of both $\sin\theta_2$ and $\cos\theta_2$. This and other calculations are illustrated in Example 3.3.1.

The problem of determining $\dot{r}_2$, which is the velocity of pin B as seen by an observer rotating with bar CP, and the angular velocity[6] $\dot{\theta}_2(t)$ of bar CP, is readily solved by using the following two equations that are obtained by simply taking the time derivative of Eqs. (3.3.1) and (3.3.2):

$$\dot{r}_2\cos\theta_2 - r_2\dot{\theta}_2\sin\theta_2 = r_1\dot{\theta}_1\sin\theta_1 \qquad (3.3.4)$$

$$\dot{r}_2\sin\theta_2 + r_2\dot{\theta}_2\cos\theta_2 = r_1\dot{\theta}_1\cos\theta_1 \qquad (3.3.5)$$

Equations (3.3.4) and (3.3.5) may be put in matrix form and readily solved for $\dot{r}_2$ and $\dot{\theta}_2$ by using a hand-held calculator that handles matrices. They can also be solved by hand, and their step-by-step solution is as follows.[7]

$$
\begin{bmatrix} \dot{r}_2 \\ \dot{\theta}_2 \end{bmatrix} = r_1\dot{\theta}_1 \begin{bmatrix} \cos\theta_2 & -r_2\sin\theta_2 \\ \sin\theta_2 & r_2\cos\theta_2 \end{bmatrix}^{-1} \begin{bmatrix} \sin\theta_1 \\ \cos\theta_1 \end{bmatrix}
$$

$$
= \frac{r_1\dot{\theta}_1}{r_2} \begin{bmatrix} r_2\cos\theta_2 & r_2\sin\theta_2 \\ -\sin\theta_2 & \cos\theta_2 \end{bmatrix} \begin{bmatrix} \sin\theta_1 \\ \cos\theta_1 \end{bmatrix}
$$

$$
= \begin{bmatrix} r_1\dot{\theta}_1\left(\sin\theta_1\cos\theta_2 + \sin\theta_2\cos\theta_1\right) \\ r_1\dot{\theta}_1\left(\cos\theta_1\cos\theta_2 - \sin\theta_1\sin\theta_2\right)/r_2 \end{bmatrix} \qquad (3.3.6)
$$

The velocity $\dot{r}_2$ may also be determined by simply taking the time derivative of Eq. (3.3.3) to obtain

$$\dot{r}_2 = \frac{Lr_1\dot{\theta}_1\sin\theta_1}{\sqrt{L^2 + r_1^2 - 2Lr_1\cos\theta_1}} \qquad (3.3.7)$$

Finally, equations that involve the accelerations $\ddot{r}_2$ (which is the acceleration of pin B as seen by an observer rotating with bar CP) and $\ddot{\theta}_2$ are obtained by taking the time derivative of Eqs. (3.3.4) and (3.3.5). The resulting equations that follow can be simultaneously solved for $\ddot{r}_2$ and $\ddot{\theta}_2$ to obtain numerical values for those quantities once r_2, θ_2, $\dot{r}_2$, and $\dot{\theta}_2$ have been determined.

$$\ddot{r}_2\cos\theta_2 - r_2\ddot{\theta}_2\sin\theta_2 = 2\dot{r}_2\dot{\theta}_2\sin\theta_2 + r_2\dot{\theta}_2^2\cos\theta_2 + r_1\left(\ddot{\theta}_1\sin\theta_1 + \dot{\theta}_1^2\cos\theta_1\right) \qquad (3.3.8)$$

$$\ddot{r}_2\sin\theta_2 + r_2\ddot{\theta}_2\cos\theta_2 = r_2\dot{\theta}_2^2\sin\theta_2 - 2\dot{r}_2\dot{\theta}_2\cos\theta_2 + r_1\left(\ddot{\theta}_1\cos\theta_1 - \dot{\theta}_1^2\sin\theta_1\right) \qquad (3.3.9)$$

---

[6]Notice that these scalar quantities are also being referred to as *velocity*, instead of *speed*, since they may be negative. Recall that the word *speed* is used for the magnitude of velocity.

[7]The analytical solution for $\dot{r}_2$ seen in Eq. (3.3.6) may also be obtained by multiplying Eq. (3.3.4) by $\cos\theta_2$, multiplying Eq. (3.3.5) by $\sin\theta_2$, adding the results, and then noticing that $\sin^2\theta_2 + \cos^2\theta_2 = 1$. Similarly, the analytical solution for $\dot{\theta}_2$ seen in Eq. (3.3.6) may also be obtained by multiplying Eq. (3.3.5) by $\cos\theta_2$, multiplying Eq. (3.3.4) by $\sin\theta_2$, and then subtracting the results.

An Alternative Way to Determine $\dot{r}_2$, $\dot{\theta}_2$, $\ddot{r}_2$, and $\ddot{\theta}_2$ The solution for the velocities $\dot{r}_2$ and $\dot{\theta}_2$ may also be obtained by making use of two different expressions for the absolute velocity $\vec{v}_B$ of point B, which are obtained by inspection of Fig. 3.12 (p. 241) as

$$\vec{v}_B = \frac{d}{dt}(r_1 \hat{r}_1) = r_1 \dot{\theta}_1 \hat{\theta}_1 \tag{3.3.10}$$

$$\vec{v}_B = \frac{d}{dt}(r_2 \hat{r}_2) = \dot{r}_2 \hat{r}_2 + r_2 \dot{\theta}_2 \hat{\theta}_2 \tag{3.3.11}$$

> The solutions for $\dot{r}_2$ and $\dot{\theta}_2$ in the resulting vector equation $r_1 \dot{\theta}_1 \hat{\theta}_1 = \dot{r}_2 \hat{r}_2 + r_2 \dot{\theta}_2 \hat{\theta}_2$ are simply obtained by projecting the known velocity $\vec{v}_B = r_1 \dot{\theta}_1 \hat{\theta}_1$ in the $\hat{r}_2$ and $\hat{\theta}_2$ directions (see such directions in Fig. 3.12, p. 241), and then equating the projections to $\dot{r}_2 \hat{r}_2$ and $r_2 \dot{\theta}_2 \hat{\theta}_2$, respectively, since those projections must be the ones given by Eq. (3.3.11).

To project $\vec{v}_B = r_1 \dot{\theta}_1 \hat{\theta}_1$ in the $\hat{r}_2$ and $\hat{\theta}_2$ directions, we need to express $\hat{\theta}_1$ in terms of $\hat{r}_2$ and $\hat{\theta}_2$, and this is done simply by inspection of Fig. 3.12. This yields the following expression:

$$\begin{aligned} \hat{\theta}_1 &= \hat{r}_2 \cos\left(\frac{\pi}{2} - \theta_1 - \theta_2\right) + \hat{\theta}_2 \sin\left(\frac{\pi}{2} - \theta_1 - \theta_2\right) \\ &= \hat{r}_2 \sin(\theta_1 + \theta_2) + \hat{\theta}_2 \cos(\theta_1 + \theta_2) \end{aligned} \tag{3.3.12}$$

Therefore, Eqs. (3.3.10) and (3.3.11) yield the following expressions for $\dot{r}_2$ and $\dot{\theta}_2$:

$$\dot{r}_2 = r_1 \dot{\theta}_1 \sin(\theta_1 + \theta_2) \tag{3.3.13}$$

$$r_2 \dot{\theta}_2 = r_1 \dot{\theta}_1 \cos(\theta_1 + \theta_2) \tag{3.3.14}$$

which are the same as those seen in Eq. (3.3.6) since $\sin(\theta_1 + \theta_2) = \sin\theta_1 \cos\theta_2 + \sin\theta_2 \cos\theta_1$, and that $\cos(\theta_1 + \theta_2) = \cos\theta_1 \cos\theta_2 - \sin\theta_1 \sin\theta_2$.

The solutions for the accelerations $\ddot{r}_2$ and $\ddot{\theta}_2$ may be obtained in the same manner by working with the following two expressions for the acceleration $\vec{a}_B$ (see Fig. 3.12):

$$\vec{a}_B = \frac{d}{dt}\left(r_1 \dot{\theta}_1 \hat{\theta}_1\right) = r_1 \left(\ddot{\theta}_1 \hat{\theta}_1 - \dot{\theta}_1^2 \hat{r}_1\right) \tag{3.3.15}$$

$$\vec{a}_B = (\ddot{r}_2 - r_2 \dot{\theta}_2^2)\hat{r}_2 + \left(r_2 \ddot{\theta}_2 + 2\dot{r}_2 \dot{\theta}_2\right)\hat{\theta}_2 \tag{3.3.16}$$

The solutions for $\ddot{r}_2$ and $\ddot{\theta}_2$ may be obtained by projecting the vector $r_1(\ddot{\theta}_1 \hat{\theta}_1 - \dot{\theta}_1^2 \hat{r}_1)$ along the $\hat{r}_2$ and $\hat{\theta}_2$ directions, and then equating those projections to $(\ddot{r}_2 - r_2 \dot{\theta}_2^2)\hat{r}_2$

and $(r_2\ddot{\theta}_2 + 2\dot{r}_2\dot{\theta}_2)\hat{\theta}_2$, respectively. The results obtained are the same (as they should be!) as those obtained from Eqs. (3.3.8) and (3.3.9).

> The use of any of the methodologies presented here is entirely a matter of personal preference. Some readers may naturally favor one of the methodologies over the other.

Example 3.3.1 For the mechanism shown in Fig. 3.12 (p. 241), $r_1 = 0.3$ m and $L = 1$ m. Bar AB is driven by a motor so that $\theta_1 = \pi/2 + 3\pi\sin(kt)$ rad, where $k = 1$ rad/s. For $t = 5$ s, determine the absolute velocity and absolute acceleration of point B, the angular velocity $\dot{\theta}_2$ and angular acceleration $\ddot{\theta}_2$ of bar PC, and the acceleration of point B relative to an observer fixed to the rotating bar PC.

■ **Solution**

We have

$$
\begin{aligned}
\theta_1(5) &= \pi/2 + 3\pi\sin(5) \approx -7.47 \text{ rad} & &(\approx -427.8^\circ \implies -67.8^\circ) \\
\dot{\theta}_1(t) &= 3k\pi\cos(kt) & \text{thus,} \quad & \dot{\theta}_1(5) \approx 2.67 \text{ rad/s} \\
\ddot{\theta}_1(t) &= -3k^2\pi\sin(kt) & \text{thus,} \quad & \ddot{\theta}_1(5) \approx 9.04 \text{ rad/s}^2 \\
r_2(t) &= \sqrt{L^2 + r_1^2 - 2Lr_1\cos\theta_1(t)} & \text{thus,} \quad & r_2(5) \approx 0.93 \text{ m} \\
\sin\theta_2 &= (r_1\sin\theta_1)/r_2 & \text{thus,} \quad & \sin\theta_2(5) \approx -0.299 \\
\cos\theta_2 &= (L - r_1\cos\theta_1)/r_2 & \text{thus,} \quad & \cos\theta_2(5) \approx 0.954
\end{aligned}
$$

These values for $\sin\theta_2(5)$ and $\cos\theta_2(5)$ then yield

$$\theta_2(5) \approx -17.4^\circ$$

The absolute velocity and acceleration of point B are simply

$$\vec{v}_B = r_1\dot{\theta}_1\hat{\theta}_1 \underset{\text{for } t=5}{\approx} 0.8\hat{\theta}_1 \text{ m/s}$$

$$\vec{a}_B = \frac{d}{dt}\vec{v}_B = r_1\ddot{\theta}_1\hat{\theta}_1 - r_1\dot{\theta}_1^2\hat{r}_1 \underset{\text{for } t=5}{\approx} 2.71\hat{\theta}_1 - 2.14\hat{r}_1 \text{ m/s}^2$$

The angular velocity $\dot{\theta}_2$ of bar PC and the angular acceleration $\ddot{\theta}_2$ may be calculated using any of the methodologies presented in this section. If we work with the projection of the velocity $\vec{v}_B$ along bar PC and along the perpendicular to bar PC, we obtain

$$\dot{r}_2 = r_1\dot{\theta}_1 \sin(\theta_1 + \theta_2) \qquad \text{which gives} \qquad \dot{r}_2 \underset{\text{for } t=5}{\approx} -0.8 \text{ m/s}$$

$$r_2\dot{\theta}_2 = r_1\dot{\theta}_1 \cos(\theta_1 + \theta_2) \qquad \text{which gives} \qquad \dot{\theta}_2 \underset{\text{for } t=5}{\approx} 0.07 \text{ rad/s}$$

Since $\dot{\theta}_2(5)$ is positive, and θ_2 is measured clockwise in Fig. 3.12, bar PC is rotating clockwise at $t = 5$ s.

The acceleration of point B as viewed by an observer fixed to the rotating bar PC is $\ddot{r}_2\hat{r}_2$. As mentioned earlier, one way to determine $\ddot{r}_2$ and the angular acceleration $\ddot{\theta}_2$ is by taking the time derivative of the given expressions that involve $\dot{r}_2$ and $r_2\dot{\theta}_2$. The results are

$$\ddot{r}_2 = r_1\left[\ddot{\theta}_1 \sin(\theta_1 + \theta_2) + \dot{\theta}_1\left(\dot{\theta}_1 + \dot{\theta}_2\right)\cos(\theta_1 + \theta_2)\right]$$

which gives

$$\ddot{r}_2 \underset{\text{for } t=5}{\approx} -2.52 \text{ m/s}^2$$

$$r_2\ddot{\theta}_2 + \dot{r}_2\dot{\theta}_2 = r_1\left[\ddot{\theta}_1 \cos(\theta_1 + \theta_2) - \dot{\theta}_1\left(\dot{\theta}_1 + \dot{\theta}_2\right)\sin(\theta_1 + \theta_2)\right]$$

which gives

$$\ddot{\theta}_2 \underset{\text{for } t=5}{\approx} 2.67 \text{ rad/s}^2 \qquad \text{i.e., } \dot{\theta}_2 \text{ is increasing}$$

Example 3.3.2 Bar AB of the mechanism shown in Fig. 3.12 (p. 241) is driven with a constant angular velocity $\dot{\theta}_1 = 2$ rad/s, and its motion is started with $\theta_1 = 0$. For $r_1 = 0.3$ m and $L = 1$ m, determine the first instant the angular velocity $\dot{\theta}_2$ of bar PC changes direction. Also determine, for that instant, the velocity and the acceleration of pin B as viewed by an observer fixed to bar PC, and the angular acceleration $\ddot{\theta}_2$ of bar PC.

■ Solution

With $\dot{\theta}_1(t) = \text{constant} = 2$ rad/s and $\theta_1(0) = 0$, we obtain $\theta_1(t) = 2t$ rad, with t in seconds. This determines $\theta_1(t)$ at all times.

At all times t we can also write the following expression either by inspecting Fig. 3.12 again or by using Eqs. (3.3.1) and (3.3.2), which were also obtained by inspection of that figure:

$$\frac{\sin\theta_2(t)}{\cos\theta_2(t)} = \frac{r_1 \sin\theta_1}{L - r_1 \cos\theta_1}$$

The rotation of bar PC starts changing direction at some time $t = t^*$ when its angular velocity $\dot{\theta}_2(t^*)$ is $\dot{\theta}_2(t^*) = 0$. With $\dot{\theta}_2(t^*) = 0$, we can also write from Eqs. (3.3.4) and (3.3.5)

$$\frac{\sin\theta_2(t^*)}{\cos\theta_2(t^*)} = \frac{\cos\theta_1(t^*)}{\sin\theta_1(t^*)}$$

Notice that this trigonometric result indicates that $\theta_1(t^*) + \theta_2(t^*) = \pm 90°$ and bars AB and PC are perpendicular to each other at $t = t^*$. This physical situation could also have been inferred by inspection of Fig. 3.12, where bar PC is long enough for it to be always in contact with bar AB. We then have

$$\frac{r_1\sin\theta_1(t^*)}{L - r_1\cos\theta_1(t^*)} = \frac{\cos\theta_1(t^*)}{\sin\theta_1(t^*)}$$

This equation may be rearranged to yield, after making use of the fact that $\sin^2\theta_1 + \cos^2\theta_1 = 1$,

$$\cos\theta_1(t^*) = \frac{r_1}{L}$$

A real solution for $\theta_1(t^*)$ exists only when $r_1 \leq L$. For $r_1/L = 0.3$, as specified in the problem statement, and $\theta_1 = 2t$ as already determined, the smallest value of t^* that satisfies this equation is $t^* \approx 0.633$ s. This is the time of the first change in the direction of the rotation of bar PC, and $\theta_1(t^*) = 2t^* = 1.266$ rad $\approx 72.5°$. Since it was concluded that bar PC in Fig. 3.12 is perpendicular to bar AB at $t = t^*$, and since $r_1/L < 1$ in this example, we must then have $\theta_2(t^*) = 90 - 72.5 = 17.5°$. With $r_2(t) = \sqrt{L^2 + r_1^2 - 2Lr_1\cos\theta_1}$, we then obtain $r_2(t^*) \approx 0.95$ m.

The rest of the calculation has been presented in detail in this section, and one could follow those same steps to obtain the answers to the questions that were asked in this problem. However, for the sake of diversity, let us make use of the fact that bars AB and PC are perpendicular to each other to solve the rest of this problem.

At the instant bars AB and PC are perpendicular to each other, we must have for $\dot{r}_2(t)$ [see Fig. 3.12 or use Eq. (3.3.13)]

$$\dot{r}_2(t) = r_1\dot{\theta}_1(t) \qquad \text{which gives} \qquad \dot{r}_2(t^*) = 2r_1 = 0.6 \text{ m/s}$$

The acceleration $\ddot{r}_2(t^*)$ and the angular acceleration $\ddot{\theta}_2(t^*)$ can be determined by using the following expressions for the acceleration $\vec{a}_B$ of point B in Fig. 3.12:

$$
\begin{aligned}
\vec{a}_B &= r_1\ddot{\theta}_1\hat{\theta}_1 - r_1\dot{\theta}_1^2\hat{r}_1 \\
\vec{a}_B &= (\ddot{r}_2 - r_2\dot{\theta}_2^2)\hat{r}_2 + (r_2\ddot{\theta}_2 + 2\dot{r}_2\dot{\theta}_2)\hat{\theta}_2
\end{aligned}
$$

These are Eqs. (3.3.15) and (3.3.16) whose basic derivations were first presented in Chapter 2 and are summarized in Fig. 3.4 (p. 228).

When bars AB and PC are perpendicular to each other, we must have (see Fig. 3.12) $\hat{\theta}_1 = \hat{r}_2$ and $\hat{r}_1 = \hat{\theta}_2$. Therefore, with $\dot{\theta}_2(t^*) = 0$, the two expressions for $\vec{a}_B$ yield the following results:

$$\ddot{r}_2(t^*) = r_1\ddot{\theta}_1 = 0$$
$$r_2(t^*)\ddot{\theta}_2(t^*) = -r_1\dot{\theta}_1^2 \qquad \text{which gives} \qquad \ddot{\theta}_2(t^*) \approx -1.26 \text{ m/s}^2$$

Notice that these two preceding expressions may be written right away by inspection of Fig. 3.12 (especially when bars AB and PC are perpendicular to each other) and with the knowledge of the results summarized in Fig. 3.4.

3.3.1   Automated Solution for the Geneva Wheel Mechanism

⟹   The calculations to determine r_2 and θ_2, $\dot{r}_2$ and $\dot{\theta}_2$, $\ddot{r}_2$ and $\ddot{\theta}_2$, based on Eqs. (3.3.1) to (3.3.9) for the mechanism shown in Fig. 3.12 (p. 241), under the assumption that the two bars AB and CP are engaged at all times, are implemented in two similar programs named *geneva_wheel* that are listed in Appendix E. One program is for Scilab users, which Scilab would save as *geneva_wheel.sci*, and the other is for Matlab users, which Matlab would save as *geneva_wheel.m*. As indicated in the comment lines at the beginning of each program (see their listing in Appendix E), they animate the motion from an initial time $t0$ to a final time $tfinal$, and generate an eight column matrix containing $t, \theta_1, \theta_2, r_2, \dot{r}_2, \dot{\theta}_2, \ddot{r}_2$, and $\ddot{\theta}_2$, which the user can use for generating plots, if desired. An example of usage is also indicated in those comment lines.

Those who have Simulink and prefer to use it may want to construct the block diagram model shown in Fig. 3.13 for performing the same calculations mentioned earlier.[8] Since the calculations do not involve integrations, choose *discrete (no continuous states)* in the solver part of the *Simulation → Parameters* menu. Except for round-off errors, the answers displayed in Fig. 3.13 correspond to the ones obtained in Example 3.3.1 (p. 245). To see an animation of the motion after the simulation stops, type `animate_nbars(x_coordinates, y_coordinates)` in the Matlab command window. The Matlab program `animate_nbars`, listed in Appendix E, is needed for this.

---

[8]The syntax of Eqs. (3.3.6) in the Matlab Function block that solves it is $[\cos(u(4)), -u(5) * \sin(u(4)); \sin(u(4)), u(5) * \cos(u(4))]^{\hat{}}(-1) * u(2) * u(6) * [\sin(u(3)); \cos(u(3))]$, which must be typed in a single line. The syntax of Eqs. (3.3.8) and (3.3.9) in the second Matlab Function block must be typed in a similar manner after double-clicking the block.

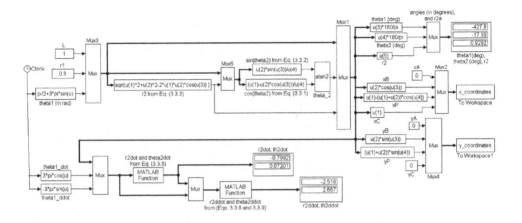

Figure 3.13: A Simulink block diagram model for two engaged bars of the Geneva wheel mechanism (see Fig. 3.12, p. 241).

Example 3.3.3 For the mechanism of Example 3.3.2 (see Fig. 3.12, p. 241), for which $r_1 = 0.3$ m, $L = 1$ m, and $\theta_1(t) = \pi/2 + 3\pi \sin(kt)$ rad, where $k = 1$ rad/s, determine the maximum and minimum values of the angular velocity $\dot{\theta}_2(t)$ of bar CP.

■ **Solution**

The maximum and minimum values of $\dot{\theta}_2(t)$ occur at the instants $t = t^*$ when $\left[d\dot{\theta}_2/dt\right]_{t=t^*} = \ddot{\theta}_2(t^*) = 0$; $\dot{\theta}_2(t^*)$ is a maximum if $[d^3\dot{\theta}_2/dt^3]_{t=t^*} < 0$ and minimum if $[d^3\dot{\theta}_2/dt^3]_{t=t^*} > 0$. If $[d^3\dot{\theta}_3/dt^3]_{t=t^*} = 0$, the sign of the next derivative $[d^4\dot{\theta}_2/dt^4]_{t=t^*}$ has to be checked; the process continues if $[d^4\dot{\theta}_2/dt^4]_{t=t^*}$ is also zero. Since the expression that is obtained for $\dot{\theta}_2(t)$, and even for $\theta_2(t)$, is somewhat complicated, a numerical solution for the maximum and minimum values of $\dot{\theta}_2$ is determined from a plot of $\dot{\theta}_2$ versus t, which is shown in Fig. 3.14. Either one of the programs mentioned earlier may be used for generating the data to create such a plot. The plot discloses that the function $\dot{\theta}_2(t)$ exhibits several maximum and minimum values. The *supremum* (i.e., the largest of the maximum values) and the *infimum* (i.e., the smallest of the minimum values) values for $\dot{\theta}_2(t)$ are approximately 4 rad/s and -4 rad/s, respectively.

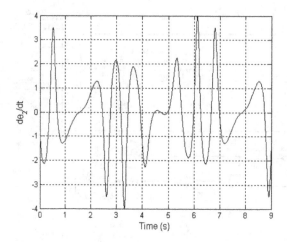

Figure 3.14: Plot of $\dot{\theta}_2(t)$ versus time t for Example 3.3.3.

3.4   The Slider Crank Mechanism

The slider crank mechanism is shown in Fig. 3.15. The small insert in the figure shows the mechanism in its more common configuration, for $D = 0$ in the larger drawing. The constant lengths L_1 and L_2 of links OA and AB and the design parameter D are known, and the "slider" at B is constrained to move along a straight line. In terms of the unit vectors $\hat{x}$ and $\hat{y}$ shown in the figure, the position vector of point B relative to the fixed point O is $\overrightarrow{OB} = x_B\hat{x} + D\hat{y}$; the constant D is positive for the configuration illustrated in the figure, but negative values may also be chosen.

In practice, the mechanism is constructed so that bars OA and AB and the slider at B move along different, but parallel, planes to avoid collisions. The pin at A, for example, would be long like a nail, but still perpendicular to the plane of the paper. If link OA is driven with a known angular velocity $\dot{\theta}_1(t)$, the objective of the kinematic analysis is to determine the angular velocity $\dot{\theta}_2$ and the angular acceleration $\ddot{\theta}_2$ of link AB and the velocity $\dot{x}_B\hat{x}$ and acceleration $\ddot{x}_B\hat{x}$ of the slider at B (see Fig. 3.15). Since the length L_2 of link AB is constant, the angular velocity $\dot{\theta}_2$ of that link may be determined by using its instantaneous center of rotation (which is point C_{AB} in Fig. 3.15), if one wishes to do so. This is illustrated in Fig. 3.15 and

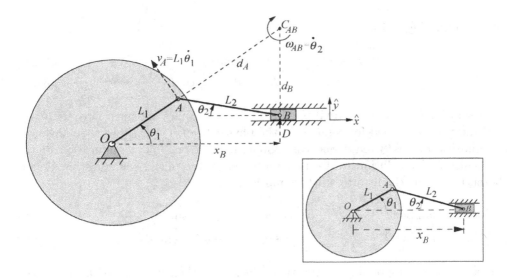

Figure 3.15: The slider crank mechanism.

will now be discussed.

Analysis 1: Using the Instantaneous Center of Rotation C_{AB} of Link AB

The location of the instantaneous center of rotation C_{AB} for link AB is determined as shown in Fig. 3.15. Since $v_A = L_1 \dot{\theta}_1$ is known, and since it is also equal to $\omega_{AB} d_A$, where d_A is determined from the figure (either analytically, as presented here, or by drawing the figure to scale and measuring it), ω_{AB} is then determined as $\omega_{AB} = L_1 \dot{\theta}_1 / d_A$. If ω_{AB} is positive, as illustrated in Fig. 3.15, the velocity $\dot{x}_B \hat{x}$ of point B will be directed along $-\hat{x}$ since $\dot{x}_B$ will be negative; its magnitude is found as $|\dot{x}_B| = \omega_{AB} d_B$, where d_B is also determined from the figure. The sequence of calculations is then (see Fig. 3.15)

$$
\begin{aligned}
L_1 \sin \theta_1 &= D + L_2 \sin \theta_2 & &\text{(this determines } \theta_2 \text{ in the figure) (3.4.1)} \\
d_A \cos \theta_1 &= L_2 \cos \theta_2 & &\text{(this determines } d_A) & &(3.4.2) \\
\omega_{AB} d_A &= L_1 \dot{\theta}_1 & &\text{(this determines } \omega_{AB} = \dot{\theta}_2) & &(3.4.3) \\
d_B + D &= (L_1 + d_A) \sin \theta_1 & &\text{(this determines } d_B) & &(3.4.4) \\
\dot{x}_B &= -\omega_{AB} d_B & &\text{(this determines } \vec{v}_B = \dot{x}_B \hat{x}) & &(3.4.5)
\end{aligned}
$$

Analysis 2: Starting with Two Basic Geometric Relations from Fig. 3.15

For this type of solution, start with the following relations, which are written simply by inspecting Fig. 3.15:

$$L_1 \sin \theta_1 = D + L_2 \sin \theta_2 \tag{3.4.6}$$
$$x_B = L_1 \cos \theta_1 + L_2 \cos \theta_2 \tag{3.4.7}$$

A solution for θ_2 exists only if $|L_1 \sin \theta_1 - D| \le L_2$, since one must have $|\sin \theta_2| \le 1$. If the value obtained for $\cos \theta_2$ is in the physical range $[-1, 1]$, we then have two possibilities for $\cos \theta_2$, each one corresponding to a possible configuration of the mechanism. The solution to use is determined by the initial configuration that is chosen for the mechanism to start its motion.

Solution number 1: $$\cos \theta_2 = \sqrt{1 - \sin^2 \theta_2} \tag{3.4.8}$$
Solution number 2: $$\cos \theta_2 = -\sqrt{1 - \sin^2 \theta_2} \tag{3.4.9}$$

The angle θ_2 corresponding to the two solutions is then determined by using the values found for $\sin \theta_2$ and $\cos \theta_2$.

The angular velocity $\dot{\theta}_2$ of bar AB and the velocity $\vec{v}_B = \dot{x}_B \hat{x}$ of point B are readily calculated from

$$L_1 \dot{\theta}_1 \cos \theta_1 = L_2 \dot{\theta}_2 \cos \theta_2 \tag{3.4.10}$$
$$\dot{x}_B = -L_1 \dot{\theta}_1 \sin \theta_1 - L_2 \dot{\theta}_2 \sin \theta_2 \tag{3.4.11}$$

These equations are obtained by taking the time derivative of Eqs. (3.4.6) and (3.4.7). It is a useful exercise to obtain such equations by inspecting Fig. 3.15 and using the vector identity $\vec{v}_B = \vec{v}_A + \vec{v}_{B/A}$, as explained at the beginning of this chapter.

Observe Eq. (3.4.10) and notice that $\dot{\theta}_2 = 0$ when $\theta_1 = \pm 90°$. Now look at Fig. 3.15 and notice that, when this happens, the instantaneous center of rotation of bar AB is at infinity. Also, notice, again by observing Eq. (3.4.10), that $\dot{\theta}_2 \to \infty$ as $\cos \theta_2 \to 0$, i.e., as $\theta_2 \to \pm 90°$. As indicated by Eq. (3.4.6), this situation can be avoided by having $|L_1 \sin \theta_1 - D| < L_2$. For $D = 0$, for example, a mechanism constructed with $L_1 < L_2$ will not encounter such a situation.

With the general approach used in this method of analysis, one can also immediately determine the angular acceleration $\ddot{\theta}_2$ of link AB, and the acceleration $\vec{a}_B = \ddot{x}_B \hat{x}$ of point B, from the two equations that are obtained by taking the time derivative of Eqs. (3.4.10) and (3.4.11). The resulting equations are:

$$L_1(\ddot{\theta}_1 \cos \theta_1 - \dot{\theta}_1^2 \sin \theta_1) = L_2(\ddot{\theta}_2 \cos \theta_2 - \dot{\theta}_2^2 \sin \theta_2) \tag{3.4.12}$$
$$\ddot{x}_B = -L_1(\ddot{\theta}_1 \sin \theta_1 + \dot{\theta}_1^2 \cos \theta_1) - L_2(\ddot{\theta}_2 \sin \theta_2 + \dot{\theta}_2^2 \cos \theta_2) \tag{3.4.13}$$

By using Eqs. (3.4.12) and (3.4.13), the numerical values for the accelerations $\ddot{\theta}_2$ and $\ddot{x}_B$ can be readily obtained.[9]

Example 3.4.1 For the slider crank mechanism shown in Fig. 3.15, $L_1 = 0.5$ m, $L_2 = 1$ m, and $D = 0.25$ m (i.e., since D is positive, the slider moves in a slot *above* point O as shown in the figure). At a certain instant, $\theta_2 = 0$, $\dot{\theta}_1 = 60$ rpm, and $\ddot{\theta}_1 = 0$ (see Fig. 3.15). Determine the angular velocity $\dot{\theta}_2$ of bar AB, and the velocity $\dot{x}_B$ of the slider at that instant. Also, determine the distance point B is from point O at that instant.

■ Solution
The steps leading to all the equations that are necessary for determining the angular velocity $\dot{\theta}_2$ and angular acceleration $\ddot{\theta}_2$ of bar AB, as well as the velocity $\dot{x}_B$ and acceleration $\ddot{x}_B$ of the slider for a general slider crank mechanism were presented in detail in this section.

At the instant under consideration, for which $\theta_2 = 0$, we have (see Fig. 3.15, and/or refer to the equations already formulated)

$$\sin \theta_1 = \frac{D}{L_1} = 0.5 \qquad \text{therefore,} \qquad \theta_1 = 30°$$

This yields $x_B = L_1 \cos \theta_1 + L_2 \approx 1.43$ m.

The distance from point O to point B at the instant under consideration is then

$$\sqrt{x_B^2 + D^2} \approx 1.45 \text{ m}$$

The angular velocity $\dot{\theta}_2$ of bar AB when $\theta_2 = 0$ is determined as

$$L_1 \dot{\theta}_1 \cos \theta_1 = L_2 \dot{\theta}_2 \qquad \text{which yields} \qquad \dot{\theta}_2 = 15\sqrt{3} \text{ rpm} \approx 26 \text{ rpm}$$

Notice that there is no need to convert revolutions per minute to radians per second for this calculation, and an answer in revolutions per minute is perfectly acceptable. With $\theta_2 = 0$, the preceding equation can be quickly written by inspecting Fig. 3.15 and using the vector identity $\vec{v}_B = \vec{v}_A + \vec{v}_{B/A}$, as suggested earlier. It is also obtained

---

[9]As a matter of minor importance, notice that one might prefer to express $\dot{\theta}_2$ as $\dot{\theta}_2 = \pm L_1 \dot{\theta}_1 (\cos \theta_1) / \sqrt{L_2^2 - (L_1 \sin \theta_1 - D)^2}$, which is Eq. (3.4.10) with $\cos \theta_2$ obtained from Eq. (3.4.6) as $\cos \theta_2 = \pm \sqrt{1 - [(L_1 \sin \theta_1 - D)/L_2]^2}$. The + sign should be used for Solution number 1, and the − sign for Solution number 2. This is a matter of personal choice. With such a choice, one might also prefer to determine $\ddot{\theta}_2$ using the expression that is obtained by taking the time derivative of this equation.

by using Eq. (3.4.10). Notice that the preceding equation is valid only when $\theta_2 = 0$ and, thus, one should never take its derivative to try to determine $\ddot{\theta}_2$, for example.

When $\theta_2 = 0$, the expression for the velocity $\dot{x}_B$ of point B given by Eq. (3.4.11) becomes

$$\dot{x}_B = -L_1\dot{\theta}_1 \sin\theta_1$$

With $\dot{\theta}_1 = 60$ rpm $= 60 \times 2\pi/60$ rad/s (so that a proper answer in m/s is obtained for $\dot{x}_B$), the preceding result yields $\dot{x}_B = -\pi/2 \approx -1.57$ m/s.

The angular velocity $\dot{\theta}_2$ and the velocity $\dot{x}_B$ may also be determined, if one prefers to do so, using the instantaneous center of rotation of bar AB. This is actually the quickest way to determine these quantities for this problem. The sequence of calculations in such a solution is shown here; it is based on the drawing shown in Fig. 3.15, with $\theta_2 = 0$.

Angle θ_1: $\theta_1 = \arctan(D/L_1) = 30°$

Distance d_B: $d_B = L_2 \tan\theta_1 = 1/\sqrt{3}$ m

Distance d_A: $d_A = \sqrt{L_2^2 + d_B^2} = 2/\sqrt{3}$ m

$$d_A\dot{\theta}_2 = L_1\dot{\theta}_1 \qquad \text{which yields} \qquad \dot{\theta}_2 = 15\sqrt{3} \text{ rpm} \approx 26 \text{ rpm}$$

$$\dot{x}_B = -d_B\omega_{AB} = -d_B\dot{\theta}_2 = -15 \times 2\pi/60 \approx -1.57 \text{ m/s}$$

Example 3.4.2 Point B of the mechanism shown in Fig. 3.16 is constrained to move along an inclined slot. The location of the slot is specified by the distance $L = 1$ m from point O and by the angle $\alpha = $ constant $= 30°$. For this mechanism, $L_1 = 1$ m, $L_2 = 2.5$ m, and bar OA is driven by a motor so that $\phi_1 = kt$ rad, where $k = 2\pi$ rad/s. Determine, for the instants of time when bar OA is parallel to the slot where B moves,

   a. The angular velocity $\dot{\phi}_2$ of bar AB and the velocity $\dot{r}_B$ of point B
   b. The angular acceleration $\ddot{\phi}_2$ of bar AB and the acceleration $\ddot{r}_B$ of point B

■ **Solution**

To solve this problem, it is better to look at this mechanism from the point of view shown in Fig. 3.17, which makes it clear that this is the same as the general slider crank mechanism of Fig. 3.15 (p. 251) that was already analyzed in this section. Thus, instead of working with the angles ϕ_1 and ϕ_2 and the variable r_B shown

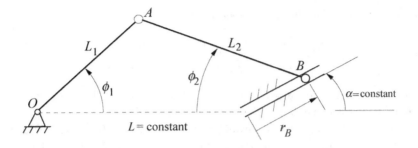

Figure 3.16: The slider crank mechanism for Example 3.4.2.

in Fig. 3.16, let us work with the angles θ_1 and θ_2 and the variable x_B shown in Fig. 3.17. Since $\phi_1 = \theta_1 + \alpha$, and $\phi_2 = \theta_2 - \alpha$, and α is constant, we then have: $\dot{\phi}_1 = \dot{\theta}_1$ and $\dot{\phi}_2 = \dot{\theta}_2$. Also, since x_B and r_B only differ by a constant, we have $\dot{r}_B = \dot{x}_B$.

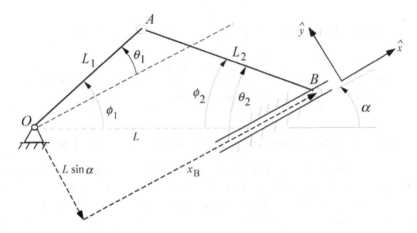

Figure 3.17: The slider crank mechanism for Example 3.4.2.

Comparing Figs. 3.17 and 3.15, we have

$$\overrightarrow{OB} = -(L\sin\alpha)\hat{y} + x_B\hat{x} = x_B\hat{x} + D\hat{y}$$

and, therefore, the variable D in the generic Fig. 3.15 is, for this problem,

$$D = -L\sin\alpha = -\sin(\pi/6) = -0.5 \text{ m}$$

The basic equations for the mechanism of Fig. 3.17 are

$$L_1 \sin\theta_1 = L_2 \sin\theta_2 - L\sin\alpha$$
$$x_B = L_1 \cos\theta_1 + L_2 \cos\theta_2$$

and these are, of course, the same as Eqs. (3.4.6) and (3.4.7). Therefore, there is no need to repeat the rest of the analysis here.

As seen in Fig. 3.17, bar OA is parallel to the slot where B moves when $\theta_1 = 0$ and $\theta_1 = 180°$. The results of the calculations of θ_2, x_B, $\dot{\theta}_2$, $\dot{x}_B$, $\ddot{\theta}_2$, and $\ddot{x}_B$, using the procedure that is presented starting on p. 251 are shown in the following table.

t (s)	Solution number	θ_1 (deg)	θ_2 (deg)	x_B (m)	$\dot{\theta}_2$ (rad/s)	$\dot{x}_B$ (m/s)	$\ddot{\theta}_2$ (rad/s²)	$\ddot{x}_B$ (m/s²)
1/12	1	0	11.5	3.45	2.56	-1.28	1.34	-56.3
	2	0	168.5	-1.45	-2.56	1.28	-1.34	-22.7
7/12	1	180	11.5	1.45	-2.56	1.28	1.34	22.7
	2	180	168.5	-3.45	2.56	-1.28	-1.34	56.3

Table 3.1: Results of the calculations for Example 3.4.2.

The entire analysis of the general slider crank mechanism shown in Fig. 3.15 (p. 251), based on the equations developed in this section, is automated in Section 3.4.1. This provides a convenient tool for analyzing such mechanisms.

3.4.1   Automated Solution for the Slider Crank Mechanism

⟹   The calculations presented in this section for the slider crank mechanism shown in Fig. 3.15 (p. 251) are implemented in two similar programs named *slider_crank* that are listed in Appendix E. One program is for Scilab users, which Scilab would save as *slider_crank.sci*, and the other is for Matlab users, which Matlab would save as *slider_crank.m.* As indicated in the comment lines at the beginning of each program (see their listing in Appendix E), they animate the motion from an initial time *t0* to a final time *tfinal*, and generate an eight column matrix containing $t, \theta_1, \theta_2, x_B, \dot{\theta}_2, \dot{x}_B, \ddot{\theta}_2$, and $\ddot{x}_B$, which the user can use for generating plots, if desired. Three examples of usage are also indicated in those comment lines.

Those who have Simulink and prefer to use it may want to construct the block diagram model shown in Fig. 3.18 for automating the same calculations. Since the calculations do not involve integrations, choose *discrete (no continuous states)* in the solver part of the *Simulation → Parameters* menu. Except for round-off errors, the answers displayed in Fig. 3.18 correspond to the ones obtained for *Solution 1* in

Example 3.4.2 with $tfinal = 1/12$. To see an animation of the motion after the simulation stops, type `animate_nbars(x_coordinates, y_coordinates)` in the Matlab command window. The Matlab program `animate_nbars`, listed in Appendix E, is needed for this.

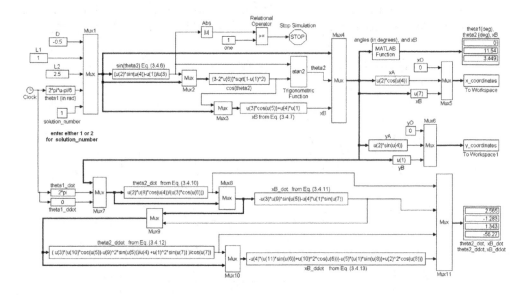

Figure 3.18: A Simulink block diagram model for the slider crank mechanism. The displayed answers are for a "Stop time" equal to $1/12$, chosen in the *Simulation* → *Parameters* menu.

Example 3.4.3 Bar OA of the mechanism shown in Fig. 3.15 (p. 251) is driven by a motor so that $\theta_1 = 2\sin(kt)$ rad, where $k = 1$ rad/s. For this mechanism, $D = 0$ and the lengths of the bars are $L_2 = 1$ m and $L_1 = 0.6$ m. Determine:

a. The maximum and minimum values of the angular velocity $\dot{\theta}_2$ of bar AB and the maximum and minimum values of the velocity $\dot{x}_B$ of point B

b. The maximum and minimum values of the angular acceleration $\ddot{\theta}_2(t)$ of bar AB

■ **Solution**

a. Instead of seeking an analytical solution for this problem, let us use plots of $\dot{\theta}_2(t)$, $\dot{x}_B(t)$, and $\ddot{\theta}_2$ generated from a numerical solution, and then extract the maximum and minimum values of those variables from such plots. Either one of the programs mentioned earlier may be used for generating the data to create such plots.

A plot of $\dot\theta_2(t)$ and $\dot x_B$ versus time t is shown in Fig. 3.19. The plot discloses that these variables exhibit several maximum and minimum values. Their supremum (i.e., the largest of the maximum values) and infimum (i.e., the smallest of the minimum values), obtained from Fig. 3.19, are

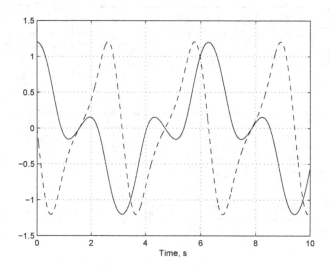

Figure 3.19: Plots of $\dot\theta_2(t)$ (solid line) in radians per second versus time t, and $\dot x_B(t)$ (dashed line) in meters per second for Example 3.4.3.

$$(\dot\theta_2)_{\min} \approx -1.2 \text{ rad/s} \qquad (\dot\theta_2)_{\max} \approx 1.2 \text{ rad/s}$$
$$(\dot x_B)_{\min} \approx -1.2 \text{ m/s} \qquad (\dot x_B)_{\max} \approx 1.2 \text{ m/s}$$

b. Figure 3.20 shows a plot of $\ddot\theta_2(t)$ versus t. The supremum and infimum values of the angular acceleration $\ddot\theta_2$ of bar AB, obtained from Fig. 3.20, are

$$(\ddot\theta_2)_{\min} \approx -1.9 \text{ rad/s}^2 \qquad (\ddot\theta_2)_{\max} \approx 1.9 \text{ rad/s}^2$$

3.5   The Scotch Yoke Mechanism

A scotch yoke mechanism is shown in Fig. 3.21. It is a very simple mechanism to analyze, and it is used for converting rotational motion into translational motion,

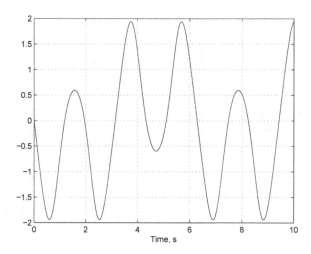

Figure 3.20: Plot of $\ddot{\theta}_2(t)$ in radians per second squared versus t for Example 3.4.3.

or vice versa.

The mechanism consists of a slotted body, which is constrained to move along a straight guide perpendicular to the slot, and of a rotating arm OP of length R, at the end of which there is a pin P that is constrained to move along the slot in the slotted body. By imparting a rotation to arm OP, the slotted body is forced to move along its guide, and the variable x_A that locates point A on the body relative to the fixed point O in Fig. 3.21 is simply $x_A = R\cos\theta$. If arm OP is driven by a motor with a constant angular velocity $\dot{\theta} = \omega$ [thus, $\theta = \omega t + \theta(0)$], this mechanism can be used, for example, to impart a simple harmonic motion to the slotted body, which could be a table shaker that can be used for vibration tests in a laboratory or in field experiments.

The objective of the kinematic analysis of this mechanism is to determine the velocity $\vec{v}_A = \dot{x}_A \hat{x}_1$ (see Fig. 3.21) and the acceleration $\vec{a}_A = \ddot{x}_A \hat{x}_1$ of the slotted body for any given angular motion $\theta(t)$ of arm OP. The only unknown in this kinematics problem is the variable x_A that tracks the motion of the slotted body, and the basic relationship for this mechanism is simply

$$x_A = R\cos\theta \tag{3.5.1}$$

Therefore, the velocity $\dot{x}_A$ and the acceleration $\ddot{x}_A$ are simply obtained as

$$\dot{x}_A = -R\dot{\theta}\sin\theta \tag{3.5.2}$$

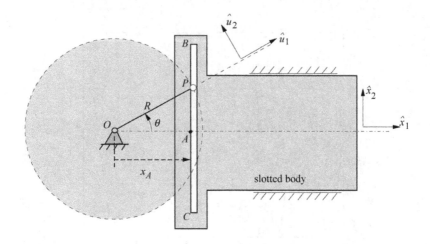

Figure 3.21: The scotch yoke mechanism.

$$\ddot{x}_A = -R(\ddot{\theta}\sin\theta + \dot{\theta}^2\cos\theta) \tag{3.5.3}$$

Note: The absolute velocity $\vec{v}_P$ and absolute acceleration $\vec{a}_P$ of pin P of bar OP, expressed in terms of the unit vectors $\hat{u}_1$ and $\hat{u}_2$ that rotate with bar OP (see Fig. 3.21), are

$$\vec{v}_P = R\dot{\theta}\hat{u}_2 \tag{3.5.4}$$

$$\vec{a}_P = \frac{d}{dt}\vec{v}_P = R\left(\ddot{\theta}\hat{u}_2 - \dot{\theta}^2\hat{u}_1\right) \tag{3.5.5}$$

Notice that the velocity $\vec{v}_A = \dot{x}_A\hat{x}_1$ and the acceleration $\vec{a}_A = \ddot{x}_A\hat{x}_1$ of point A of the slotted body are the same as the projection of the vectors $\vec{v}_P$ and $\vec{a}_P$, respectively, along the fixed direction $\hat{x}_1$. They are, respectively, the absolute velocity and absolute acceleration of point A.

$\implies$ The calculations presented in this section for the scotch yoke mechanism shown in Fig. 3.21 are implemented in two similar programs named *scotch_yoke* that are listed in Appendix E. As done earlier, one program is for Scilab users, which Scilab would save as *scotch_yoke.sci*, and the other is for Matlab users, which Matlab would save as *scotch_yoke.m*. As indicated in the comment lines at the beginning of each program (see their listing in Appendix E), they animate the motion and generate a four column matrix containing $t, \theta, \dot{x}_A$, and $\ddot{x}_A$, which the user can use for generating plots, if desired. An example of usage is given in those comment lines.

Those who have Simulink and prefer to use it may want to construct the block diagram model shown in Fig. 3.22 for automating the same calculations for this mechanism. Since the calculations do not involve integrations, choose *discrete (no continuous states)* in the solver part of the *Simulation → Parameters* menu. The answers displayed in Fig. 3.22 are for a final time $tfinal = 10$. To see an animation of the motion after the simulation stops, type `animate_nbars(x_coordinates, y_coordinates)` in the Matlab command window. For this, you need to have the Matlab program `animate_nbars`, listed in Appendix E, stored in your computer.

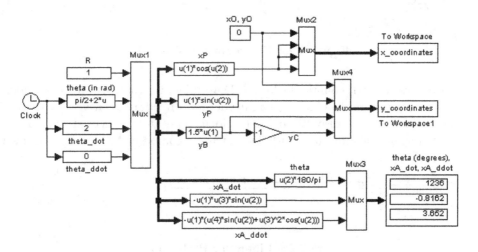

Figure 3.22: A Simulink block diagram model for the scotch yoke mechanism. For the model inputs shown, the time array $[0:0.002:10]$ is suggested for the *Simulation → Parameters* settings.

3.6 Problems

3.1 At a certain instant, bar AB of a four-bar mechanism is perpendicular to line AD and bar CD is along AD as shown in Fig. 3.23. The absolute angular velocity of bar AB at that time is 10 rad/s counterclockwise. Determine the absolute velocity of point C (magnitude and direction) and the absolute angular velocity (magnitude and direction) of bars BC and CD at that instant. Do not use a computer to solve this problem.

3.2 At a certain instant, the bars of a mechanism are aligned along a straight line as shown in Fig. 3.24. At that instant, the absolute angular velocity and angular acceleration of bar AB are $\omega_{AB} = 15$ rad/s and $\dot{\omega}_{AB} = 24$ rad/s^2,

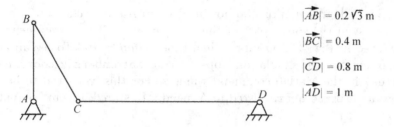

$|\overrightarrow{AB}| = 0.2\sqrt{3}$ m

$|\overrightarrow{BC}| = 0.4$ m

$|\overrightarrow{CD}| = 0.8$ m

$|\overrightarrow{AD}| = 1$ m

Figure 3.23: Figure for Problem 3.1.

respectively, both counterclockwise. Determine, for that instant, the velocity and acceleration of point B (magnitude and direction) relative to the slot it slides in, and the absolute angular velocity and angular acceleration of that slot (magnitude and direction). Do not use a computer to solve this problem.

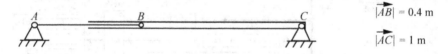

$|\overrightarrow{AB}| = 0.4$ m

$|\overrightarrow{AC}| = 1$ m

Figure 3.24: Figure for Problem 3.2.

3.3 At a certain instant, bar AB of a slider crank mechanism is perpendicular to the direction of the motion of the slider at C as shown in Fig. 3.25. That bar is rotating counterclockwise with constant absolute angular velocity $\omega_{AB} = 20$ rad/s. Determine, for that instant (without using a computer):

   a. The absolute velocity of point C (magnitude and direction) and the absolute angular velocity (magnitude and direction) of bar BC

   b. The absolute acceleration of point C (magnitude and direction) and the absolute angular acceleration (magnitude and direction) of bar BC

Figure 3.26 shows snapshots of a four-bar mechanism at different times. For all the snapshots, the distance L_1 between the fixed supports at A and D and the lengths of the rotating bars are indicated in the figure. At the instant represented by each snapshot, the absolute angular velocity and angular acceleration of bar AB are $\omega_{AB} = 100$ rpm and $\dot{\omega}_{AB} = 0.3$ rpm/s, respectively, both counterclockwise.

   For each problem 3.4 to 3.10, determine, without using the instantaneous center of rotation method, the absolute angular velocity of bars BC and CD

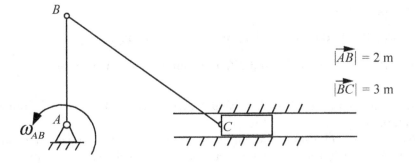

Figure 3.25: Figure for Problem 3.3.

and the absolute velocity of point C (specifying its magnitude and direction), at the instant represented by each snapshot in Fig. 3.26. Specify the magnitude of those angular velocities and indicate whether they are clockwise or counterclockwise.

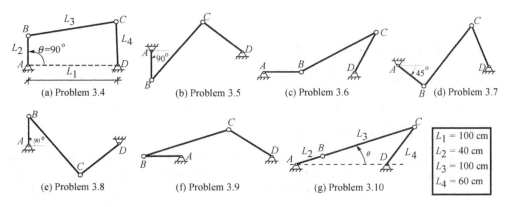

Figure 3.26: Figures for Problems 3.4 to 3.10.

3.4 See Fig. 3.26a.

3.5 See Fig. 3.26b.

3.6 See Fig. 3.26c. Bar AB is along line AD at that instant.

3.7 See Fig. 3.26d.

264 CHAPTER 3 *Kinematic Analysis of Planar Mechanisms*

3.8 See Fig. 3.26e.

3.9 See Fig. 3.26f. Bar AB is along line AD at that instant.

3.10 See Fig. 3.26g. Bars AB and BC are collinear at that instant.

For Problems 3.11 to 3.17, solve Problems 3.4 to 3.10 using instantaneous centers of rotation.
Suggestions: Determine the angles between bars AD and DC, and bars AD and BC, using the loop equations for those mechanisms; eliminate one of the unknown angles by combining those equations to obtain an equation that involves only one of the unknowns. When necessary, use a calculator to do a few numerical iterations to determine the unknown angle in the resulting equations, as you might do in a closed-book exam.

3.11 See Problem 3.4 and Fig. 3.26a.

3.12 See Problem 3.5 and Fig. 3.26b.

3.13 See Problem 3.6 and Fig. 3.26c.

3.14 See Problem 3.7 and Fig. 3.26d.

3.15 See Problem 3.8 and Fig. 3.26e.

3.16 See Problem 3.9 and Fig. 3.26f.

3.17 See Problem 3.10 and Fig. 3.26g.

For Problems 3.18 to 3.24, determine the absolute angular acceleration of bars BC and CD and the absolute acceleration of point C (specifying its magnitude and direction), at the instant represented by each snapshot in Fig. 3.26. Specify the magnitude of those angular accelerations, and indicate whether they are clockwise or counterclockwise.

3.18 See Fig. 3.26a.

3.19 See Fig. 3.26b.

3.20 See Fig. 3.26c. Bar AB is along line AD at that instant.

3.21 See Fig. 3.26d.

3.22 See Fig. 3.26e.

3.23 See Fig. 3.26f. Bar AB is along line AD at that instant.

3.24 See Fig. 3.26g. Bars AB and BC are collinear at that instant.

3.25 Determine the angular velocity $\dot{\theta}_2$ and the angular acceleration $\ddot{\theta}_2$ of bar CP for the mechanism shown in Fig. 3.12 (p. 241), at the instants when $\theta_1 = 45°$ and $\theta_1 = 90°$. Also, determine the velocity $\dot{r}_2$ and acceleration $\ddot{r}_2$ (i.e., the velocity and acceleration of point B as seen by an observer fixed to the rotating bar CP) at those same instants. When $\theta_1 = 45°$ (see Fig. 3.12), $\dot{\theta}_1 = -0.74$ rad/s and $\ddot{\theta}_1 = -0.55$ rad/s^2. When $\theta_1 = 90°$, $\dot{\theta}_1 = -1.05$ rad/s and $\ddot{\theta}_1 = 0$. For that mechanism, $|\overrightarrow{AB}| = 1.2$ m and $|\overrightarrow{AC}| = 1.5$ m. Do not use a computer to solve this problem.

3.26 Determine the angular velocity $\dot{\theta}_2$ and the angular acceleration $\ddot{\theta}_2$ of bar AB and the absolute velocity and acceleration of point B for the slider-crank mechanism shown in Fig. 3.15 (p. 251), at the instants when $\theta_1 = 45°$ and $\theta_1 = 90°$. When $\theta_1 = 45°$, $\dot{\theta}_1 = 0.74$ rad/s and $\ddot{\theta}_1 = 0.55$ rad/s^2. When $\theta_1 = 90°$, $\dot{\theta}_1 = 1.05$ rad/s and $\ddot{\theta}_1 = 0$. For that mechanism, $L_1 = 0.3$ m, $L_2 = 1.5$ m, and $D = 0$. Do not use a computer to solve this problem.

3.27 The mechanism shown in Fig. 3.27 consists of a bar AB, with a groove that is welded at its end B making a constant angle α with AB; a bar PC that is forced to pass through the groove that is welded to bar AB; and a bar CD. Points A and D are pinned to the ground, and bar AB is driven by a motor so that the angle $\theta_2(t)$ is a known function of time t. A pin at C allows PC to rotate relative to CD. For $L_1 = 1$ m, $\alpha = 30°$, $\theta_2(t) = \pi/2 + 3\pi \sin(kt)$ rad, where $k = 1$ rad/s, $L_2 = 0.4$ m, and $L_4 = 1.5$ m, determine the velocity $\dot{L}_3$ and the angular velocities $\dot{\theta}_3$ and $\dot{\theta}_4$, the acceleration $\ddot{L}_3$ (where $L_3 = |\overrightarrow{CB}|$), and the angular accelerations $\ddot{\theta}_3$ and $\ddot{\theta}_4$, when $t = 5$ and 10 s.

   Hint: Start, as done in Section 3.2.1, by writing the loop equations for this mechanism and the constraint equation between the appropriate angles labeled in Fig. 3.27 that guarantees that the angle α remains constant.

3.28 At a certain instant $t = t^*$, $\theta_2(t^*) = -45°$, $\dot{\theta}_2(t^*) = 100$ rpm, and $\ddot{\theta}_2(t^*) = 0.3$ rpm/s for a mechanism constructed in an identical manner to the one shown in Fig. 3.27 but driven with a different motor. Determine the angular velocity of bars CP and CD for that same instant.

3.29 What is the angular acceleration of bars CP and CD, at $t = t^*$, for the mechanism of Problem 3.28?

3.30 The distance x_A for the mechanism shown in Fig. 3.28 is changing with time as $x_A = 40 + 15\sin\left[\omega\,(t-5)\right]$ cm, where $\omega = 2$ rad/s. For that mechanism,

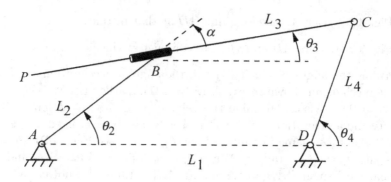

Figure 3.27: Figure for Problem 3.27.

$L_{BE} = 30$ cm, $L_{BC} = L_{ED} = 100$ cm, $L_{CD} = 60$ cm, and $L_{AB} = 50$ cm. Determine, for $t = 5$ s, the absolute angular velocity (magnitude and direction) of all the bars of the mechanism and the magnitude of the absolute velocity of points C and B. Treat points A and D, and the direction of line AED, as inertial.

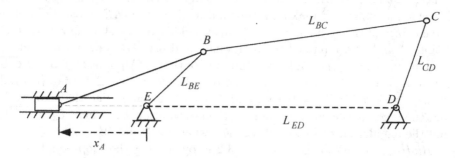

Figure 3.28: Figure for Problem 3.30.

3.31   For the mechanism in Problem 3.30, determine, for $t = 5$ s, the absolute angular acceleration (magnitude and direction) of all the bars of the mechanism and the magnitude of the absolute acceleration of points C and B.

3.32   If you have not done so yet, solve Problems 3.4 to 3.10 and 3.18 to 3.24 using either a computer program of your choosing or any of the ones mentioned in this book. Show hand calculation of the angle θ for the configuration in Fig. 3.26g.

3.33 Use either a computer program of your choosing or any of the ones mentioned in this book to solve Example 3.2.2 (p. 233). The mechanism is shown in Fig. 3.6 (p. 234).

3.34 It turns out that the mechanism of Example 3.2.2 (p. 233) is only able to operate without locking when ϕ_2 is in a certain range $\phi_{2\min} \leq \phi_2 \leq \phi_{2\max}$. Determine that range. Then, redesign the mechanism by changing the length of either AB, BC, or CD (without changing the location of the supports at A and D) to prevent locking, and then answer the questions in parts a and b of that example for your redesigned mechanism when $\phi_2 = \text{constant} = -90$ rpm. A computer program of your choosing or any of the ones mentioned in this book is a convenient tool to use to solve this problem.

3.35 Use either a computer program of your choosing or any of the ones mentioned in this book to solve Problem 3.25, and check the answers you obtained if you solved that problem earlier.

3.36 Use either a computer program of your choosing or any of the ones mentioned in this book to solve Problem 3.26, and check the answers you obtained if you solved that problem earlier.

3.37 Automate the solution of the equations you formulated for the mechanism in Problems 3.27, 3.28, and 3.29. Include an animation of the motion of that mechanism, and use your program (or model) to check the answers you obtained if you solved those problems earlier.

3.38 Automate the solution of the equations you formulated for the mechanism in Problems 3.30 and 3.31. Include an animation of the motion of that mechanism, and use your program (or model) to check the answers you obtained if you solved those problems earlier.

CHAPTER 4

System of Point Masses: Special Dynamical Properties

To analyze the motion of each particle in a system of N point masses, one starts with a free-body diagram for the particles, which is done by isolating them from their surroundings. From then on, the problem is no different than that presented in Chapter 2 for a single particle, where one of the forces acting on the particle is the gravitational force of attraction of the planet the particles are subjected to (say, the earth!).[1] For many applications, the mutual force of attraction between the particles, given by Newton's law of universal gravitation, may be neglected when compared to other forces acting on them (as in Example 2.7.10, p. 174).

The motion of a system of particles exhibits some special properties when the system is viewed as a whole. Such properties are of fundamental importance to the understanding of the motion and are essential to the formulation of the differential equations of motion of the system. They are essential to the analysis of the motion of continuum systems such as rigid bodies. Specifically, these properties deal with the *motion of the center of mass* of the system under the action of the external forces acting on the entire system, and with the relationship between the moment of the external forces acting on the system and the time rate of change of the absolute *angular momentum* associated with the motion of the system. The latter relationship provides the means for analyzing the rotational motion of the system, which is the motion associated with change in direction of a line that joins any two particles of the system. Such properties of the motion of system of particles are addressed and studied in detail in this chapter.

Two classical problems dealing with the motion of unconstrained particles are

---

[1] Actually, such a single-particle problem involves two particles, where the second particle represents the faraway center of mass C of the earth. If the arc length traveled by the particle is small compared to the radius of the earth, the direction of the gravitational force of attraction due to the earth remains essentially constant, and the earth acts as if it were flat. Examples of this were given in Section 2.5 of Chapter 2.

analyzed in Sections 4.4 and 4.6. These are the two-body problem of celestial mechanics (and of spacecraft dynamics) and the restricted three-body problem.

4.1   Motion of the Center of Mass of a System of Point Masses

Consider a system of N point masses P_i, each of mass m_i $(i = 1, 2, \ldots, N)$, as
shown in Fig. 4.1. The position vector of an arbitrary particle P_i, relative to a
reference point O, is the vector $\vec{r}_i = \overrightarrow{OP_i}$ shown in the figure.

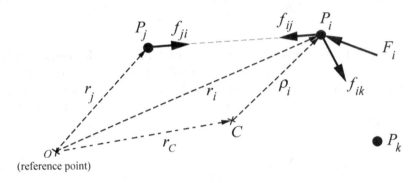

Figure 4.1: A system of N point masses (C is the *center of mass* of the system).

For convenience, the forces acting on each particle are separated into two types,
referred to as *internal* and *external* forces. The former are due to the mutual attraction between any two particles, and the latter constitute all the other forces
acting on the particle (such as gravitational forces, friction, and externally applied
forces). Each particle P_i in Fig. 4.1 is, thus, subjected to $N-1$ internal forces $\vec{f}_{ij}$
(which stands for the force of attraction on particle P_i due to particle P_j) and to
the resultant $\vec{F}_i$ of the external forces acting on that particle.

It will prove convenient in the development that follows, to define a point with
position vector $\vec{r}_C$ from O (point C in Fig. 4.1), determined as

$$\vec{r}_C \triangleq \frac{1}{m} \sum_{i=1}^{N} m_i \vec{r}_i \qquad (4.1.1)$$

where $m = \sum_{i=1}^{N} m_i$ is the total mass of the system. Point C is called the *center of
mass* of the system. Notice that if the position vector $\vec{r}_i$ of an arbitrary particle P_i

is decomposed as $\vec{r}_i = \vec{r}_C + \vec{\rho}_i$, where $\vec{\rho}_i = \overrightarrow{CP_i}$ is the position vector of P_i relative to C (see Fig. 4.1), Eq. (4.1.1) yields $\sum_{i=1}^{N} m_i \vec{\rho}_i = 0$. This is an alternative equation, which is equivalent to Eq. (4.1.1), for locating the center of mass C.

The center of mass plays a significant role in the dynamics of system of point masses, especially on the dynamics of bodies with finite dimensions (which is the case for all bodies in nature). As already indicated in Section 1.10 in Chapter 1, it is important not to confuse the concepts of *center of mass* and *center of gravity*.

By letting the reference point O in Fig. 4.1 be inertial (i.e., with no acceleration), application of Newton's second law to each particle P_i yields

$$\vec{F}_i + \sum_{\substack{j=1 \\ j \neq i}}^{N} \vec{f}_{ij} = m_i \frac{d}{dt} \vec{v}_i = m_i \frac{d^2}{dt^2} \vec{r}_i \qquad (i = 1, 2, ..., N) \qquad (4.1.2)$$

Equation (4.1.2) represents N vector equations (or $3N$ scalar equations if each particle is allowed to move in three-dimensional space). An important result is obtained by adding all the N vector equations obtained from Eq. (4.1.2). By making use of Eq. (4.1.1), which defines the position vector $\vec{r}_C$ of the center of mass C, Eq. (4.1.2) yields

$$\sum_{i=1}^{N} \vec{F}_i + \sum_{i=1}^{N} \sum_{\substack{j=1 \\ j \neq i}}^{N} \vec{f}_{ij} = \frac{d^2}{dt^2} \sum_{i=1}^{N} (m_i \vec{r}_i) = m \frac{d^2}{dt^2} \vec{r}_C \qquad (4.1.3)$$

By noticing that $\sum_{i=1}^{N} \vec{F}_i$ is the resultant $\vec{F}$ of all the external forces acting on the system of particles, and that the double summation in Eq. (4.1.3) vanishes since $\vec{f}_{ij} = -\vec{f}_{ji}$ (as a consequence of Newton's third law), Eq. (4.1.3) is reduced to

$$\boxed{\vec{F} \triangleq \sum_{i=1}^{N} \vec{F}_i = m \frac{d^2}{dt^2} \vec{r}_C = m \frac{d}{dt} \vec{v}_C} \qquad (4.1.4)$$

The equation $\sum_{i=1}^{N} [\vec{F}_i - m_i (d^2 \vec{r}_i / dt^2)] = 0$ is called *d'Alembert's principle*, and it is due to Jean Le Rond d'Alembert (1717-1783).

Equation (4.1.4) discloses that the motion of the center of mass of a system of N particles is governed by the differential equation of motion of a point C of mass equal to the total mass of the system and subjected to the resultant force $\vec{F}$ of all the external forces acting on the system. This is true whether or not the resultant $\vec{F}$ passes through the center of mass C. This is referred to as the *principle of the motion of the center of mass*, and it is one of the properties that were mentioned in the introduction to this chapter. The total absolute linear momentum of the system of particles, defined as $\vec{p} = \sum_{i=1}^{N} m_i \vec{v}_i$, is equal to the absolute linear momentum $m\vec{v}_C$ of the center of mass C. This conclusion is obtained from the time derivative of Eq. (4.1.1) that defines the position vector for C.

4.2 Time Rate of Change of Angular Momentum, and Moment

An important relation that is essential for analyzing the motion of a body, either rigid or flexible, is obtained by first defining a vector $\vec{H}_A$ for the system of particles as

$$\vec{H}_A \overset{\Delta}{=} \sum_{i=1}^{N} \vec{r}_{Ai} \times \frac{d}{dt}\left(m_i \vec{r}_i\right) \qquad (4.2.1)$$

where $\vec{r}_{Ai}$ is the vector from an arbitrary point A to particle P_i as shown in Fig. 4.2. The vector $\vec{H}_A$ is known as the *absolute angular momentum, about point A, of the system of particles*. It is the sum of the moment, about an arbitrary point A, of the absolute linear momentum for each particle of the system. The angular momentum $\vec{H}_A$, referred to an arbitrary point A, is a generalization, for a system of particles, of the quantity $\vec{H}_O$ that was introduced in Section 2.3 of Chapter 2 for a single point mass. The usefulness of the quantity $\vec{H}_A$, and the reason for using an arbitrary point A (instead of simply an inertial point O as done in Section 2.3 for a single point mass), are disclosed when we take its time derivative. By making use of Eq. (4.1.2) we obtain

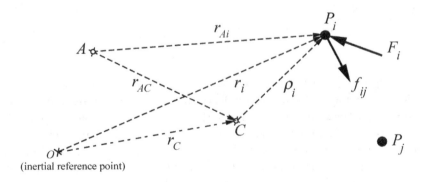

(inertial reference point)

Figure 4.2: A system of N point masses again.

$$\frac{d}{dt}\vec{H}_A = \sum_{i=1}^{N}\left(\frac{d}{dt}\vec{r}_{Ai}\right) \times \frac{d}{dt}(m_i\vec{r}_i) + \sum_{i=1}^{N}\vec{r}_{Ai} \times \left[\vec{F}_i + \sum_{\substack{j=1\\j\neq i}}^{N}\vec{f}_{ij}\right] \qquad (4.2.2)$$

With $\vec{f}_{ij} = -\vec{f}_{ji}$, the term involving the double summation in Eq. (4.2.2) vanishes. By expressing the position vectors $\vec{r}_{Ai}$ and $\vec{r}_i$ as (see Fig. 4.2)

$$\vec{r}_{Ai} = \vec{r}_{AC} + \vec{\rho}_i \qquad (4.2.3)$$
$$\vec{r}_i = \vec{r}_C + \vec{\rho}_i \qquad (4.2.4)$$

and by noticing that the second summation term in Eq. (4.2.2) is equal to the total moment $\vec{M}_A$ of the external forces $\vec{F}_i$, about point A, i.e., $\sum_{i=1}^{N}\vec{r}_{Ai} \times \vec{F}_i = \vec{M}_A$, Eq. (4.2.2) yields

$$\frac{d}{dt}\vec{H}_A = \sum_{i=1}^{N}\left(\frac{d}{dt}\vec{r}_{AC}\right) \times \frac{d}{dt}(m_i\vec{r}_C) + \vec{M}_A + \sum_{i=1}^{N}\left(\frac{d}{dt}\vec{r}_{AC}\right) \times \frac{d}{dt}(m_i\vec{\rho}_i)$$
$$+ \sum_{i=1}^{N}\left(m_i\frac{d}{dt}\vec{\rho}_i\right) \times \frac{d}{dt}\vec{r}_C \qquad (4.2.5)$$

The last two summation terms in Eq. (4.2.5) vanish because $\sum_{i=1}^{N}m_i\vec{\rho}_i = 0$. Therefore, Eq. (4.2.5) simply reduces to

$$\frac{d}{dt}\vec{H}_A = \vec{M}_A + m\left(\frac{d}{dt}\vec{r}_{AC}\right) \times \frac{d}{dt}\vec{r}_C \qquad (4.2.6)$$

In obtaining Eq. (4.2.6), no restrictions were imposed on the choice of point A about which the moment of the external forces and the absolute angular momentum of the system of particles were taken. Thus, that equation is completely general. The choice of working with an arbitrary point A now puts us in a privileged situation of having an equation that discloses that a proper choice of A can eliminate the cross product term in that equation, thus yielding a simpler relationship between the moment $\vec{M}_A$ and the time derivative of the absolute angular momentum $\vec{H}_A$. This can be accomplished by either choosing point A to be inertial, in which case $\vec{r}_{AC} = \vec{r}_C$, or to be the center of mass C, in which case $\vec{r}_{AC} = 0$. The following simpler relations, which still can be applied to any problem, are then obtained:

$$
\begin{array}{ll}
\dfrac{d}{dt}\vec{H}_O = \vec{M}_O & \text{when } O \text{ is inertial} \\[2em]
\text{and} & \qquad\qquad (4.2.7) \\[2em]
\dfrac{d}{dt}\vec{H}_C = \vec{M}_C & \text{when } C \text{ is the center of mass of the system}
\end{array}
$$

The second equation of (4.2.7) is particularly significant since it is valid even when the center of mass C of the system of particles is undergoing any arbitrary motion. When analyzing the motion of a particular system, it is more convenient, in general, to work with the moment/angular-momentum relationship relative to a point that is moving with the system rather than relative to an inertial point. In such cases, the simpler form of that relationship, given by the second equation of (4.2.7), evidences the advantage of working with the center of mass.

The equations of (4.2.7) constitute what was referred to as the second property of the motion of a system of point masses mentioned in the introduction to this chapter.

The material in Sections 4.1 and 4.2 leads, in a natural way, to the study of rigid body dynamics, which is started in Chapter 5.

4.3 An Integrated Form of Newton's Second Law and of the Moment Equation

When the resultant force $\vec{F}$ acting on a system mass particles is equal to zero, Eq. (4.1.4) discloses that the absolute velocity $\vec{v}_C$ of the center of mass of the system is constant. This means that point C moves along a straight line in space, with constant speed along that line. For this to happen, the system would have to be immune from the attracting force of any celestial body in the universe. What is more realistic, though, is that a component, along an inertial direction, of the resultant force $\vec{F}$ might be equal to zero. In such a case, the component, along that direction, of the absolute velocity $\vec{v}_C$ of the center of mass of the system of particles remains constant.

Since the position vector $\vec{r}_C$ of the center of mass C (see Fig. 4.2) is defined by Eq. (4.1.1), that equation yields $m\vec{v}_C = (m_1 + m_2 + \ldots + m_N)\vec{v}_C = m_1\vec{v}_1 + m_2\vec{v}_2 + \ldots + m_N\vec{v}_N$, where $\vec{v}_1, \vec{v}_2, \ldots$ are the absolute velocity of particle 1, particle 2, etc.

The result mentioned in the previous paragraph, i.e., $\vec{v}_C$ = constant when $\vec{F} = 0$, is known in dynamics as the *principle of conservation of linear momentum* for the system of N point masses.

Another conservation principle in dynamics is disclosed by either equation of (4.2.7), which relates the absolute angular momentum of the system of particles to the moment of the forces applied to the system, i.e., $d\vec{H}_O/dt = \vec{M}_O$ or $d\vec{H}_C/dt = \vec{M}_C$ where O is any inertial point. Either the absolute angular momentum $\vec{H}_O$ is constant if $\vec{M}_O = 0$ or the absolute angular momentum $\vec{H}_C$ is constant if $\vec{M}_C = 0$. This is known as the *principle of conservation of angular momentum*. The latter result (i.e., $\vec{H}_C$ = constant if $\vec{M}_C = 0$) is particularly useful because the center of mass of the system may be moving in any manner in space. For many problems, a component, along an inertial direction, of the moment $\vec{M}_C$ (or $\vec{M}_O$) might be equal to zero, instead of the moment itself being zero. In such a case, the component, along that direction, of the absolute angular momentum $\vec{H}_C$ (or $\vec{H}_O$) remains constant.

Example 4.3.1 A person of mass $m_1 = 70$ kg is standing on a stationary horizontal platform of mass $m_2 = 400$ kg that is on the smooth surface of a frozen lake, as illustrated in Fig. 4.3. Point C is the center of mass of that person. Neglecting friction between the platform and the frozen surface, determine the absolute velocity of the platform when the person starts to walk along the platform with a constant velocity $\dot{x}_1 = 4$ km/h relative to the platform.

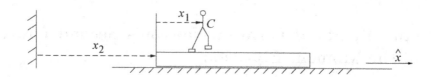

Figure 4.3: Motion on a platform that is free to slide (Example 4.3.1).

■ **Solution**

When walking, one holds a foot on the surface, lifts the other foot off the surface, and moves it forward. The same procedure is then repeated with the other foot. If there is no sliding, the friction force acting on the foot that is in contact with the surface is static.

The free-body diagrams for the person and platform are shown in Fig. 4.4; $\vec{f}_1$ and $\vec{f}_2$ are friction forces, with $\vec{f}_2 = 0$ since there is no friction between the platform and the stationary ground, which is approximated as being the earth when its translational and rotational motion are disregarded.[2] The absolute position vectors $x_2\hat{x}$ and $(x_1 + x_2)\hat{x}$ for the platform and for the center of mass C of the person are also shown in the figure; the unit vectors $\hat{x}$ and $\hat{y}$ are inertial (i.e., they maintain the same directions relative to inertial space).

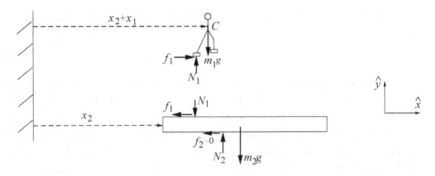

Figure 4.4: Free-body diagrams for Example 4.3.1.

Application of Newton's second law to the center of mass of the person and to the platform yield the following equations.

For the person:

$$N_1 = m_1 g \tag{4.3.1a}$$
$$f_1 = m_1(\ddot{x}_1 + \ddot{x}_2) \tag{4.3.1b}$$

For the platform:

$$N_2 = N_1 + m_2 g \tag{4.3.1c}$$
$$f_1 = -m_2 \ddot{x}_2 \tag{4.3.1d}$$

The following equation is obtained by combining Eqs. (4.3.1b) and (4.3.1d):

$$m_2 \ddot{x}_2 + m_1(\ddot{x}_1 + \ddot{x}_2) = 0$$

This equation can easily be integrated with respect to time. Since values for $\dot{x}_1$ and $\dot{x}_2$ are known just before the person starts to walk on the platform (and not

---

[2]The earth would be a third body on this problem (and other similar ones). However, because the mass of the earth is very large (essentially infinite) compared to the other masses involved and because of the small amount of time involved in such a motion, the earth behaves as an inertial body when its long time motion around the sun is not considered.

just after), let us integrate that equation from $t = 0^-$ (where $t = 0^-$ denotes the instant just before the person started walking) to an arbitrary value of t. This then yields

$$\int_{t=0^-}^{t} [m_2 \ddot{x}_2 + m_1 (\ddot{x}_1 + \ddot{x}_2)] \, dt = m_2 \dot{x}_2 + m_1 (\dot{x}_1 + \dot{x}_2) = \text{constant}$$

$$= [m_2 \dot{x}_2 + m_1 (\dot{x}_1 + \dot{x}_2)]_{t=0^-} \qquad (4.3.1e)$$

The result expressed by Eq. (4.3.1e) discloses that the linear momentum of the entire system, which is the quantity $[m_2 \dot{x}_2 + m_1 (\dot{x}_1 + \dot{x}_2)] \, \hat{x}$, is conserved (i.e., remains constant for all times). Such results, which are generally referred to as the principle of conservation of linear momentum, are obtained in a natural manner simply by application of Newton's second law to the appropriate parts of the system and by working with the resulting equations, as illustrated in this example.

Equation (4.3.1e) is easily solved for the desired unknown $\dot{x}_2$. With the given initial conditions, i.e., $\dot{x}_1(0^-) = \dot{x}_2(0^-) = 0$, we then obtain

$$\dot{x}_2 = -\frac{m_1}{m_1 + m_2} \dot{x}_1$$

Right after the person starts to walk along the platform (i.e., at $t = 0^+$) with $\dot{x}_1 = \text{constant} = 4$ km/h, the platform starts moving in the opposite direction and $\dot{x}_2 = \text{constant} \approx -0.596$ km/h ≈ -9.93 m/min.

Example 4.3.2 The *Two-Body* Problem of Celestial Mechanics (or of Spaceflight Dynamics)[3]

As indicated in the footnote, the system under consideration consists of two spherical bodies with masses m_1 and m_2, with geometric centers at P_1 and P_2, respectively, subjected only to their mutual forces of attraction. The free-body diagrams for the two bodies are shown in Fig. 4.5b, where $\vec{f}_{12}$ is the force on P_1 due to P_2 (in the sequel, the bodies will be referred to as P_1 and P_2), and $\vec{f}_{21} = -\vec{f}_{12}$ is the force on P_2 due to P_1. The vectors $\vec{r}_1$ and $\vec{r}_2$ shown in Fig. 4.5a are the position vectors of

---

[3]This is one of the classical problems in dynamics. It consists of determining the motion of two bodies subjected only to their mutual force of attraction. The solution to such a problem provides a first approximation for the motion of the center of mass of a satellite around a planet or the motion of any planet around the sun. In this classical problem, both bodies are treated as homogeneous spherical bodies. As shown in Section 1.7 of Chapter 1, the gravitational force of attraction due to a homogeneous sphere is the same as that of a particle with mass equal to the mass of the sphere, located at its center.

P_1 and P_2 relative to the center of mass C of the system while the vector $\vec{r} = \vec{r}_2 - \vec{r}_1$ is the position vector of P_2 relative to P_1. The x-y coordinate system shown in the figure is inertial (i.e., the orientation of the x- and y-axes do not change), and the angle θ is chosen for tracking the orientation of line P_1P_2. By working with the polar coordinates $r \overset{\Delta}{=} |\overrightarrow{P_1P_2}|$ (which is the distance from P_1 to P_2) and θ (see Fig. 4.5):

a. Obtain the differential equations that govern the motion of the system.
b. Verify that the differential equations of motion obtained in part a are of the same form as those for the central force problem analyzed in Example 2.7.11 in Chapter 2 (p. 179) and, thus, that all the conclusions and results obtained for that example are valid for the two-body problem under consideration here.

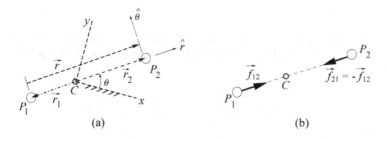

(a)   (b)

Figure 4.5: The two-body system and the free-body diagrams for Example 4.3.2.

■ Solution

a. Two important conclusions about the motion can be obtained right away.

■ Since the resultant of the forces acting on the system, $\vec{F} = \vec{f}_{12} + \vec{f}_{21}$, is equal to zero (see Fig. 4.5), Eq. (4.1.4) discloses that the center of mass C of the system is inertial (i.e., its absolute acceleration is zero). Therefore, the position vectors $\vec{r}_1$ and $\vec{r}_2$ shown in Fig. 4.5 are, respectively, the absolute position vectors for P_1 and P_2.

■ Since the moment about C of the forces $\vec{f}_{12}$ and $\vec{f}_{21}$ is equal to zero, the second equation of (4.2.7) discloses that the angular momentum $\vec{H}_C$ is constant (because $d\vec{H}_C/dt = \vec{M}_C = 0$). Thus, as concluded in Example 2.5.11 in Chapter 2, the motion takes place in a single plane in space. Therefore, only two variables are needed to describe the motion. With $\vec{r} \overset{\Delta}{=} r\hat{r}$ (see Fig. 4.5), the variables chosen here are the relative distance r between P_1 and P_2, and the angle θ, which is measured from a nonrotating reference line in the plane of the motion and tracks the direction of line P_1P_2 in that plane.[4]

Since Newton's second law requires the use of the absolute position vectors $\vec{r}_1$ for P_1, and $\vec{r}_2$ for P_2, and since we want to use the relative position vector $\vec{r} = r\hat{r}$, it is necessary to determine the expressions that relate $\vec{r}_1$ and $\vec{r}_2$ to $\vec{r}$. The first relationship, Eq. (4.3.2a), is simply obtained from the definition of the center of mass C of the system, and the second one, Eq. (4.3.2b), is obtained by inspecting Fig. 4.5.

$$m_1\vec{r}_1 + m_2\vec{r}_2 = 0 \tag{4.3.2a}$$
$$\vec{r} = \vec{r}_2 - \vec{r}_1 \tag{4.3.2b}$$

From Eqs. (4.3.2a) and (4.3.2b), the following expressions for the absolute position vectors $\vec{r}_1$ and $\vec{r}_2$ are then obtained in terms of the relative position vector $\vec{r}$.

$$\vec{r}_1 = -\frac{m_2}{m_1 + m_2}\vec{r} \qquad \vec{r}_2 = \frac{m_1}{m_1 + m_2}\vec{r} \tag{4.3.2c}$$

Notice that when $m_1 \gg m_2$, one obtains $\vec{r}_2 \approx \vec{r}$ and $\vec{r}_1 \approx 0$, which means that P_1 is essentially inertial. This was the central force problem considered in Examples 2.5.11 and 2.7.11 in Chapter 2.

Since we are dealing with relative motion, the same differential equations involving r and θ will be obtained if we consider either P_1 or P_2 for formulating those equations. By arbitrarily using P_2 we then have:

■ Absolute position vector for P_2:

$$\vec{r}_2 = \frac{m_1}{m_1 + m_2}\vec{r} = \frac{m_1}{m_1 + m_2}r\hat{r} \tag{4.3.2d}$$

■ Absolute velocity of P_2:

$$\vec{v}_2 = \frac{d\vec{r}_2}{dt} = \frac{m_1}{m_1 + m_2}\left[\dot{r}\hat{r} + r\frac{d\hat{r}}{dt}\right] = \frac{m_1}{m_1 + m_2}\left[\dot{r}\hat{r} + r\dot{\theta}\hat{\theta}\right] \tag{4.3.2e}$$

---

[4]In practical terms, it is more convenient to work with the relative distance r instead of either distance $r_1 = |\overrightarrow{CP_1}|$ or $r_2 = |\overrightarrow{CP_2}|$. The desirability of doing so is seen if we think of one body as being the earth, with a person on the earth observing the motion of the other body relative to the earth.

■ Absolute acceleration of P_2:

$$\vec{a}_2 = \frac{d\vec{v}_2}{dt} = \frac{m_1}{m_1 + m_2}\left[\ddot{r}\hat{r} + (r\ddot{\theta} + \dot{r}\dot{\theta})\hat{\theta} + \dot{r}\frac{d\hat{r}}{dt} + r\dot{\theta}\frac{d\hat{\theta}}{dt}\right]$$

$$= \frac{m_1}{m_1 + m_2}\left[(\ddot{r} - r\dot{\theta}^2)\hat{r} + (r\ddot{\theta} + 2\dot{r}\dot{\theta})\hat{\theta}\right] \qquad (4.3.2f)$$

■ Resultant force acting on P_2 due to the attraction of P_1:

$$\vec{f}_{21} = -\frac{Gm_1 m_2}{r^2}\hat{r} \qquad (4.3.2g)$$

Newton's second law, $\vec{f}_{21} = m_2\vec{a}_2$, then yields the following scalar differential equations of motion:

$$\ddot{r} - r\dot{\theta}^2 = -\frac{G(m_1 + m_2)}{r^2} \qquad (4.3.2h)$$

$$r\ddot{\theta} + 2\dot{r}\dot{\theta} = \frac{1}{r}\frac{d}{dt}(r^2\dot{\theta}) = 0 \quad \text{or} \quad r^2\dot{\theta} = \text{constant} \overset{\Delta}{=} h \qquad (4.3.2i)$$

b. Equations (4.3.2h) and (4.3.2i) are the differential equations that govern the relative motion between the two bodies (the trajectory of one body relative to the other is called an *orbit*). These differential equations are exactly of the same form as Eqs. (2.7.11a) and (2.7.11b) obtained for the central force problem analyzed in Example 2.7.11 in Chapter 2, with $GM_E = gR_E^2$ in those equations replaced by $G(m_1 + m_2)$. Therefore, all the conclusions and results obtained for the central force problem are valid for the two-body problem, for which both bodies are moving.

It is worth noticing that Eq. (4.3.2i) also follows directly from the fact that the angular momentum $\vec{H}_C$ is constant. The angular momentum $\vec{H}_C$ for this system is [see Fig. 4.5, and use the expressions for $\vec{r}_1$ and $\vec{r}_2$ given by Eqs. (4.3.2c)]:

$$\vec{H}_C = m_1\vec{r}_1 \times \frac{d}{dt}\vec{r}_1 + m_2\vec{r}_2 \times \frac{d}{dt}\vec{r}_2 = \frac{m_1 m_2}{m_1 + m_2}\vec{r} \times \frac{d}{dt}\vec{r}$$

$$= \frac{m_1 m_2}{m_1 + m_2}r\hat{r} \times (\dot{r}\hat{r} + r\dot{\theta}\hat{\theta}) = \frac{m_1 m_2}{m_1 + m_2}r^2\dot{\theta}\hat{r} \times \hat{\theta} = \text{constant} \qquad (4.3.2j)$$

The mass $m \overset{\Delta}{=} m_1 m_2/(m_1 + m_2)$ is called the *reduced mass* of the system. The quantity $h = r^2\dot{\theta} = [r^2\dot{\theta}]_{t=0}$ is called the *angular momentum per unit reduced mass* of the system. Such a quantity has a nice physical interpretation, which is arrived at with the aid of Fig. 4.6. The figure shows an infinitesimal portion of the trajectory of P_2 relative to P_1. The infinitesimal area dA swept by the position vector $\vec{r}$ during an infinitesimally small elapsed time dt is equal to half the area of a parallelogram of sides $r = |\vec{r}|$ and $r\,d\theta$. The area dA may be calculated as (refer to Fig. 4.5 when needed)

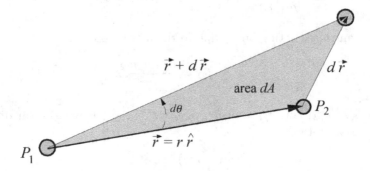

Figure 4.6: Infinitesimal area element swept by the position vector $\vec{r}$ for Example 4.3.2.

$$dA = \frac{1}{2}|\vec{r} \times d\vec{r}| = \frac{1}{2}\left|r\hat{r} \times \left[(dr)\hat{r} + r(d\theta)\,\hat{\theta}\right]\right| = \frac{1}{2}\left|r^2 d\theta\right| \qquad (4.3.2\text{k})$$

Equation (4.3.2k) discloses that $dA/dt = r^2|\dot{\theta}|/2$ remains constant during the motion since $r^2\dot{\theta}$ is constant. This is the same conclusion obtained by Kepler after many years of exhaustive analysis of the data gathered by Tycho Brahe concerning the motion of the planets. Kepler's second law states that *the motion of each planet is such that the radius vector from the sun to the planet sweeps equal areas in equal times.*

4.4　Detailed Analysis of the Two-Body Problem

Because of the importance of the two-body problem, a detailed investigation of the motion is presented in this section. As mentioned earlier, the solution to such a problem provides a first approximation for the motion of a satellite around a planet or for the motion of any planet around the sun. A more detailed model for the earth-satellite problem, for example, involves taking into account the oblateness of the earth. Such an effect, and other effects, are treated in advanced books on *celestial mechanics*.

When the expression for $\dot{\theta}$ obtained from Eq. (4.3.2i) is substituted into Eq. (4.3.2h), the resulting equation is of the form $\ddot{r} = f(r)$ and, thus, it can be integrated once with respect to r to obtain the following result. Such a result is the same as that given by Eq. (2.7.11d) of Example 2.7.11 in Chapter 2.

$$\frac{1}{2} \underbrace{\left[\dot{r}^2 + \frac{h^2}{r^2} \right]}_{=v^2} - \frac{\mu}{r} = \text{constant} \stackrel{\triangle}{=} E \qquad \text{where } \mu = G\,(m_1 + m_2) \qquad (4.4.1)$$

As indicated in Eq. (4.4.1), the quantity $\dot{r}^2 + h^2/r^2 = \dot{r}^2 + r^2\dot{\theta}^2$ is the square of the speed of P_2 relative to P_1. As indicated throughout the book, an expression of the form of Eq. (4.4.1), which only involves $\dot{r}$ and r, yields important information about the motion even without determining how $r(t)$ changes with time. Such an investigation is presented in Section 4.4.2. An analytical integration of Eq. (4.4.1) is presented in Section 4.4.3 with the objective of generating additional important information about the motion of P_2 relative to P_1. The knowledge gained from such analyses is of crucial importance in the planning of spacecraft missions. Before this is done, the physical interpretation of the terms that appear in Eq. (4.4.1) is presented in Section 4.4.1.

4.4.1 The Meaning of the Quantity E: Kinetic and Potential Energies of the Motion

The quantity E in Eq. (4.4.1) has a physical interpretation, which is arrived at if we apply the work-energy relationship $dW = d(T)$ to this problem. Such an approach is presented in Section 2.8 of Chapter 2. By referring to Fig. 4.5, and by using the expressions for $\vec{r}_1$ and $\vec{r}_2$ given by Eqs. (4.3.2c), the expressions for the infinitesimal work dW done by the forces acting on the system, and for the kinetic energy T of the motion, are developed as follows:

Infinitesimal work:

$$dW = \vec{f}_{21} \bullet d\vec{r}_2 + \vec{f}_{12} \bullet d\vec{r}_1 = \vec{f}_{21} \bullet d\vec{r}_2 - \vec{f}_{21} \bullet d\vec{r}_1 = \vec{f}_{21} \bullet d\,(\vec{r}_2 - \vec{r}_1)$$
$$= \vec{f}_{21} \bullet d\vec{r} = -\frac{Gm_1m_2}{r^2} dr \qquad (4.4.2)$$

Kinetic energy:

$$T = \frac{1}{2}m_1 \frac{d\vec{r}_1}{dt} \bullet \frac{d\vec{r}_1}{dt} + \frac{1}{2}m_2 \frac{d\vec{r}_2}{dt} \bullet \frac{d\vec{r}_2}{dt} = \frac{1}{2} \frac{m_1m_2}{m_1 + m_2} \frac{d\vec{r}}{dt} \bullet \frac{d\vec{r}}{dt}$$
$$= \frac{1}{2} \frac{m_1m_2}{m_1 + m_2} \left(\dot{r}^2 + r^2\dot{\theta}^2 \right) = \frac{1}{2} \frac{m_1m_2}{m_1 + m_2} \left(\dot{r}^2 + \frac{h^2}{r^2} \right) \qquad (4.4.3)$$

Equation (4.4.3) discloses that the quantity $\frac{1}{2}(\dot{r}^2 + h^2/r^2)$ is the *kinetic energy per unit reduced mass* of the system.

Since the expression for dW is an exact differential, integration of $dW = d(T)$ yields

$$T - W = \frac{m_1 m_2}{m_1 + m_2} \left\{ \frac{1}{2} \left[\dot{r}^2 + \frac{h^2}{r^2} \right] - \frac{\mu}{r} \right\} = \text{constant} \overset{\Delta}{=} \frac{m_1 m_2}{m_1 + m_2} E \qquad (4.4.4)$$

which is the total energy of the motion. The quantity $-\mu/r$ is called the *potential energy per unit reduced mass* of the system, and the quantity E is the *total energy per unit reduced mass* of the system. The potential energy of the system is the quantity U obtained by integrating $d(U) \overset{\Delta}{=} -dW = (Gm_1 m_2/r^2)dr$, where the minus sign is for conforming to universal convention. Thus, except for an arbitrary constant of integration that can be set to zero without any loss of generality, as discussed in Section 2.8 of Chapter 2, we then obtain $U = -Gm_1 m_2/r$.

Notice that, since $T + U = [m_1 m_2/(m_1 + m_2)](v^2/2 - \mu/r)$ always remains constant (i.e., a decrease in the kinetic energy is compensated by an increase in potential energy, and vice versa), the speed of P_2 relative to P_1 is maximum when P_2 is closest to P_1 and minimum when P_2 is farthest.

Note: The use of the work-energy equation $dW = d(T)$, together with the moment and angular momentum relationship $d\vec{H}_C/dt = \vec{M}_C$ yielded the integral of the motion given by Eq. (4.4.4) without generating Eq. (4.3.2h) (which required the use of acceleration to do so). This observation is made here to reinforce the notion that one actually has a choice on how to approach the formulation of a dynamics problem, and that different choices lead to different amounts of effort to arrive at the same place.

4.4.2   Determination of the Maximum and Minimum Values of $r(t)$

The maximum and minimum values of $r(t)$ are obtained by setting $\dot{r} = 0$ in Eq. (4.4.1). This yields the following equation, where r_m is being used for denoting such values.

$$\frac{h^2}{2r_m^2} - \frac{\mu}{r_m} - E = 0 \qquad (4.4.5)$$

Equation (4.4.5) is a quadratic equation in r_m. When $E \neq 0$, its roots are

$$r_m = -\frac{\mu}{2E}(1 \pm e) \qquad (4.4.6)$$

where

$$e = \sqrt{1 + \frac{2Eh^2}{\mu^2}} \tag{4.4.7}$$

The trajectory r versus θ of P_2 relative to P_1 is called an *orbit*, and the quantity e defined by Eq. (4.4.7) is called the *orbit eccentricity*. The following are all the different cases involving the possible values of e.

CASE 1 $0 \le e < 1$ (E **is negative**)

In this case both values given by Eq. (4.4.6) are positive and, thus, they are the maximum and minimum values of $r(t)$:

$$r_{\max} = -\frac{\mu}{2E}(1+e) \qquad r_{\min} = -\frac{\mu}{2E}(1-e) \tag{4.4.8}$$

The trajectory of P_2 relative to P_1 is then a closed curve in space. The points where the minimum and maximum values of $r(t)$ occur are called, respectively, the *pericenter* and the *apocenter* of the orbit.[5]

Notice that the analysis so far still did not reveal what kind of closed trajectory P_2 is executing around P_1. To determine that, it is necessary to integrate Eq. (4.4.1) analytically. This is done in Section 4.4.3. When $r_{\min} = r_{\max} \overset{\Delta}{=} r_{\text{circ}}$, which corresponds to $E = -\mu^2/(2h^2)$, the closed trajectory is a circle. The speed v_{circ} of a body P_2 in circular orbit of a radius r_{circ} around P_1 is obtained from Eq. (4.3.2h) and is given below. Notice that the constant value $r = r_{\text{circ}}$ is the same as the equilibrium solution to Eq. (4.3.2h). Equilibrium solutions were first presented in Chapter 1.

$$v_{\text{circ}} = \sqrt{r_{\text{circ}}^2 \dot{\theta}_{\text{circ}}^2} = \sqrt{\frac{\mu}{r_{\text{circ}}}} = \sqrt{\frac{G(m_1 + m_2)}{r_{\text{circ}}}}$$

The constant orbital angular velocity for a circular orbit is $\dot{\theta}_{\text{circ}} = v_{\text{circ}}/r_{\text{circ}} = \sqrt{\mu/r_{\text{circ}}^3}$, and the orbital period, i.e., the time it takes for P_2 to complete one revolution around P_1, is equal to $2\pi/\dot{\theta}_{\text{circ}}$.

Notice that to place a satellite in a circular orbit of radius r_{circ} around a planet, the satellite has to be "injected into orbit" so that its initial velocity relative to the attracting planet has no radial component, and its magnitude must be equal to v_{circ}.

---

[5]For an orbit around the earth, such points are called the *perigee* and *apogee*, respectively. For an orbit around the sun, they are called the *perihelion* and *apohelion*.

CASE 2 $e = 1$ $(E = 0)$

If the initial conditions are such that $E = 0$, the solutions to Eq. (4.4.5) are $r_m = h^2/(2\mu)$ and $r_m = \infty$. Since $r_m = h^2/(2\mu)$ is a positive quantity, we then have for this case

$$r_{\min} = \frac{h^2}{2\mu} \qquad\qquad r_{\max} = \infty \qquad\qquad (4.4.9)$$

This case represents a very interesting situation. If point P_1 represents the center of the earth, for example, and P_2 represents the center of mass of a spacecraft launched from the vicinity of the earth at a distance $r = r_0$ from P_1 and with a speed v_0 relative to the earth, the spacecraft would leave the vicinity of the earth and never return, unless it is equipped with a power system to do so. The required initial speed v_0 is the minimum launching speed for that to happen and, as indicated by Eq. (4.4.1), as r tends to infinity, the speed v of P_2 relative to P_1 tends to zero. The minimum launching speed, $v_0 \overset{\Delta}{=} v_{\mathrm{esc}}$, for P_2 to escape the gravitational field of P_1 is known either as the *escape speed* or *escape velocity*. For a launching at $r = r_0$, it is determined as

$$v_{\mathrm{esc}} = \sqrt{\frac{2\mu}{r_0}} = \sqrt{\frac{2G(m_1 + m_2)}{r_0}} \qquad\qquad (4.4.10)$$

Notice that the escape speed at $r = r_0$ is equal to $\sqrt{2}$ times the speed on a circular orbit of radius r_0.

The escape speed from the earth's surface for a spacecraft with mass $m_2 << m_1$ is about $40,270$ km/h ($\approx 25,000$ miles/hour). The escape speed from the surface of the moon for the same spacecraft is about 8550 km/h (≈ 5300 miles/hour).

CASE 3 $e > 1$ $(E$ **is positive**$)$

In this case only one of the solutions given by Eq. (4.4.5) is positive, and that solution is $r_m = \mu(e-1)/(2E) = (h^2/\mu)/(1+e)$. As in the case for $E < 0$, this solution corresponds to the minimum[6] value of $r(t)$. When $E > 0$, there is no other real and positive value of r for which $\dot{r} = 0$, which is also confirmed by observation of Eq. (4.4.1). Therefore, as in the case $E = 0$, the trajectory is also open and we then have

$$r_{\min} = \frac{h^2/\mu}{1+e} \qquad\qquad r_{\max} = \infty \qquad\qquad (4.4.11)$$

---

[6]This may also be verified by making use of Eq. (4.3.2h) to obtain, for $r = (h^2/\mu)/(1+e)$, $\ddot{r} = \mu^3 e(1+e)^3/h^4$. Since this value of $\ddot{r}$ is positive, the solution in question is a minimum.

According to Eq. (4.4.1), one obtains for the speed v, in the limit as r tends to infinity,

$$\lim_{r \to \infty} v = \sqrt{2E}$$

In this case, P_2 escapes the gravitational field of P_1 and, provided it does not come "close" to any other body in space, it continues to travel in space with a constant velocity (i.e., along a straight line and with constant speed).

Here is an important feature of the analysis presented in this section:

Given the initial conditions of the motion of P_2 relative to P_1, one can immediately determine whether the trajectory is closed or open simply by verifying whether the specific energy $E = \frac{1}{2}v^2 - \frac{\mu}{r}$ is negative or not.

4.4.3 Solution of the Orbital Differential Equation of Motion for r in Terms of θ

Equation (4.4.1) may be rearranged and then integrated analytically to determine the time t as a function of distance traveled along an orbit to yield the following.

$$\int_{t=0}^{t} dt = t = \left| \int_{r(0)}^{r} \frac{r\, dr}{\sqrt{2Er^2 + 2\mu r - h^2}} \right|$$

Although the integral that appears in this equation can be found in a table of integrals, the determination of $r(t)$ as a function of time is done more conveniently by numerical integration of Eq. (4.4.1). The absolute value in the preceding equation is introduced to ensure that the value found for t is positive.

It turns out that a variety of important information about the motion is obtained if Eq. (4.4.1) is integrated with respect to θ, which happens to be possible to do. This is done by using the chain rule of differentiation, i.e., $d(\ )/dt = [d(\ )/d\theta]d\theta/dt$ which yields the following expression for $\dot{r}$:

$$\dot{r} = \frac{dr}{d\theta}\dot{\theta} = \frac{h}{r^2}\frac{dr}{d\theta} = -h\frac{d}{d\theta}\left(\frac{1}{r}\right) \qquad (4.4.12)$$

By making use of this expression for $\dot{r}$, Eq. (4.4.1) can then be rearranged and integrated as

$$\int \frac{h\, d\,(1/r)}{\sqrt{2E + 2\mu/r - h^2/r^2}} = \pm \int d\theta$$

The integral that appears in this equation can be found in a table of integrals, and the integration yields the result shown in Eq. (4.4.13), where e is the same nondimensional parameter defined in Eq. (4.4.7).

$$\arcsin \frac{1/r - \mu/h^2}{\sqrt{(\mu/h^2)^2 + 2E/h^2}} = \pm\theta - \theta_0$$

or

$$\frac{1}{r} = \frac{\mu}{h^2} \left[1 + e \sin (\pm\theta - \theta_0) \right] \tag{4.4.13}$$

The quantity θ_0 in the solution for $1/r$ given by Eq. (4.4.13) is an arbitrary constant of integration. It may be set to any desired value without any loss of generality. The choice $\theta_0 = -\pi/2$ is a convenient one since it transforms the term $\sin(\pm\theta - \theta_0)$ into the more convenient form $\cos(\pm\theta) = \cos\theta$, thus eliminating the appearing of the $\pm$ signs. In practical terms, the choice $\theta_0 = -\pi/2$ implies that the tracking angle $\theta(t)$ (which is called the *true anomaly* of the orbit) shown in Fig. 4.5 is measured from a reference line that passes through the attracting body at P_1 and the point that corresponds to the minimum distance between P_1 and P_2. The resulting expression for $r(\theta)$, which is a form that is adopted universally, is

$$r(\theta) = \frac{h^2/\mu}{1 + e \cos\theta} \tag{4.4.14}$$

The solution given by Eq. (4.4.14) shows that for a closed trajectory (i.e., one for which $e < 1$), the maximum value of r, $r_{max} = (h^2/\mu)/(1 - e)$, occurs when $\theta = \pi$, i.e., at a point on the line joining P_1 and P_2 that is diametrically opposed to the point where $r_{min} = (h^2/\mu)/(1 + e)$ occurs.

Notice that if the values of r_{max} and r_{min} for a closed orbit are measured in practice by a *distance measuring equipment*, the expressions for r_{max} and r_{min} disclose that the orbit eccentricity can be readily determined as

$$e = \frac{r_{max} - r_{min}}{r_{max} + r_{min}} \tag{4.4.15}$$

In addition, when the value of r is measured when $\theta = 90°$, the orbit is completely determined since that value is equal to h^2/μ, as disclosed by Eq. (4.4.14).

It is known from calculus that Eq. (4.4.14) is the polar equation of a conic section (i.e., a circle, ellipse, parabola, or hyperbola) with the origin located at a focus of that conic. The value of the parameter e, which only depends on the initial conditions $r(0)$, $\dot{r}(0)$, and $\dot{\theta}(0)$, determines the type of conic. For $e < 1$, the trajectory is a closed curve, as it has been previously concluded even without having to find the solution for $r(\theta)$. As determined in this section, the closed curve is an ellipse (which becomes a circle when $e = 0$). The trajectory is open for $e \geq 1$, as also concluded earlier. It is a parabola if $e = 1$, and a hyperbola if $e > 1$. This is summarized in the following table.

Type of orbit	Total specific energy E	Eccentricity e
Circular	$E = -\frac{\mu^2}{2h^2}$	$e = 0$
Elliptic	$-\frac{\mu^2}{2h^2} < E < 0$	$0 < e < 1$
Parabolic	$E = 0$	$e = 1$
Hyperbolic	$E > 0$	$e > 1$

As previously concluded, if an artificial satellite P_2, of mass m_2, is to be launched so that it escapes the gravitational field of the attracting body P_1, of mass m_1, the initial conditions of the motion of P_2 relative to P_1 have to be such that the eccentricity e is no less than unity, with $e = 1$ accomplishing the desired task.

> These results show that the open orbit that requires the smallest energy for P_2 to escape the gravitational attraction of P_1 is a parabola.

Figure 4.7 illustrates the types of conic sections that are obtained when a right cone is cut by a plane, thus justifying their names. Figure 4.8 shows the types of orbits and some of their geometric properties. In each case, the point mass P_1 is located at the focus F_1 of any of the conics, as Kepler predicted in his second law. The focus F_2 of the conic section is known as the *vacant focus*. For an ellipse, the *semi-major axis*, of length a, and the *semi-minor axis*, of length b, shown in Fig. 4.8, are related to the orbital parameters h^2/μ and e as

$$a = \frac{1}{2}(r_{\min} + r_{\max}) = \frac{h^2/\mu}{1 - e^2} = \frac{r_{\min}}{1 - e} = \frac{r_{\max}}{1 + e} \qquad (4.4.16)$$

$$b = \sqrt{a^2 - (a - r_{\min})^2} = \sqrt{r_{\max} r_{\min}} = a\sqrt{1 - e^2} = r_{\min}\sqrt{\frac{1 + e}{1 - e}} \qquad (4.4.17)$$

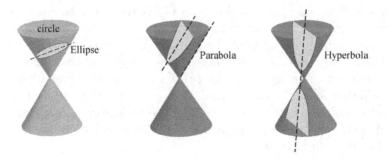

Figure 4.7: The three types of conic sections and how to obtain them.

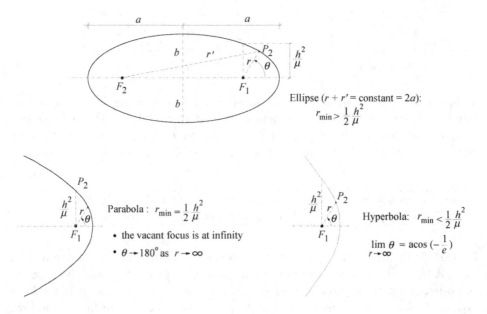

Figure 4.8: The three types of trajectories and some of their geometric properties.

Also, as disclosed by Eq. (4.4.14), the quantity h^2/μ is equal to the distance r when $\theta = 90°$. This provides a practical way for determining such a quantity from actual observation data involving a satellite.

The orbital period $t_{2\pi}$ for an elliptic orbit (i.e., the time it takes P_2 to complete one revolution around P_1) may be calculated by knowing, from calculus, that the area A of the ellipse is equal to $A = \pi ab$ (see Fig. 4.8), and by making use of Eq.

(4.3.2k), which yields

$$\frac{dA}{dt} = \frac{1}{2}|r^2\dot{\theta}| = \text{constant} = \frac{\text{area of the ellipse}}{t_{2\pi}} = \frac{\pi ab}{t_{2\pi}}$$

With $b = a\sqrt{1 - e^2}$, as per Eq. (4.4.17), and $|r^2\dot{\theta}| = |h| = \sqrt{\mu a\,(1 - e^2)}$, as per Eq. (4.4.16), the following result is obtained for the period $t_{2\pi}$:

$$t_{2\pi} = 2\pi\sqrt{\frac{a^3}{\mu}}$$

or

$$t_{2\pi} = 2\pi\sqrt{\frac{[(r_{\max} + r_{\min})/2]^3}{\mu}} \tag{4.4.18}$$

The result given by Eq. (4.4.18) agrees with Kepler's prediction stated in his third law, i.e., the square of the period of a planet in its motion around the sun is proportional to the cube of the semi-major axis of its elliptic orbit around the sun.

4.4.4 Solved Example Problems

Example 4.4.1 A detection equipment provided the following data at a time $t = t_0$ for the position and velocity, relative to the earth, of a small object traveling near the earth (the unit vectors $\hat{x}$, $\hat{y}$, and $\hat{z}$, with $\hat{z} = \hat{x} \times \hat{y}$, are inertial):

$$\vec{r}_0 = 5000(\hat{x} + \hat{y} + \hat{z}) \text{ km} \qquad \vec{v}_0 = 16,000(\hat{x} - \hat{y} + \hat{z}) \text{ km/h}$$

By approximating the earth as a homogeneous sphere of radius $R_E \approx 6378$ km, assuming that the object is unpowered, and disregarding the effect of the earth's atmosphere during the entire trajectory of the body:

a. Determine the specific angular momentum of the body.
b. Determine whether the body's trajectory is open or closed. What are the maximum and minimum values of the distance from the earth's center to the body?
c. Determine whether the body hits or misses the earth.

■ **Solution**

a. The specific angular momentum is simply

$$\vec{h} = \vec{r} \times \vec{v} = \text{constant} = \vec{r}_0 \times \vec{v}_0 = 16 \times 10^7 (\hat{x} - \hat{z}) \text{ km}^2/\text{h}$$

b. The specific energy $E = v^2/2 - \mu/r$ of a two-body system determines whether the trajectory is open or closed. The trajectory is open if $E \geq 0$ and closed otherwise. With $\mu = G(m_{\text{earth}} + m_{\text{object}}) \approx Gm_{\text{earth}} = gR_E^2$, we then obtain with $g \approx 9.81 \text{ m/s}^2 = 9.81 \times (3600^2/1000) \text{ km/h}^2$ and $R_E \approx 6378$ km:

$$E = \frac{v^2}{2} - \frac{\mu}{r} = \frac{\vec{v} \bullet \vec{v}}{2} - \frac{\mu}{|\vec{r}|} = \frac{\vec{v}_0 \bullet \vec{v}_0}{2} - \frac{\mu}{|\vec{r}_0|} \approx -2.132 \times 10^8 \text{ km}^2/\text{h}^2$$

Since $E < 0$, the trajectory is closed.

c. Since the energy during the motion is constant, and since $|\vec{h}| = r^2|\dot{\theta}|$ is also constant, we have

$$\frac{v^2}{2} - \frac{\mu}{r} = \frac{1}{2}\left(\dot{r}^2 + r^2\dot{\theta}^2\right) - \frac{\mu}{r} = \frac{1}{2}\left(\dot{r}^2 + \frac{h^2}{r^2}\right) - \frac{\mu}{r} = E$$

With $h^2 = \vec{h} \bullet \vec{h} = (16\sqrt{2} \times 10^7)^2 \text{ km}^4/\text{h}^2$, we then obtain for the values of r when $\dot{r} = 0$:

$$r_{\max} \approx 17{,}330 \text{ km} \qquad r_{\min} \approx 6929 \text{ km}$$

Since $r_{\min}$ is greater than the radius of the earth, the body will not hit the earth. The minimum height from the earth's surface will be $6929 - 6378 = 551$ km. The body that was observed traveling near the earth is then in orbit around the earth. We have already seen in this section that the orbit is an ellipse; its period is

$$t_{2\pi} = 2\pi\sqrt{\frac{[(r_{\max} + r_{\min})/2]^3}{\mu}} \approx 221 \text{ min}$$

Example 4.4.2 For the satellite considered in Example 4.4.1, determine the angle θ_p between the line of perigee (i.e., the line joining the center of the earth and the perigee) and the initial position vector $\vec{r}_0$. Also, determine the speed of the satellite at perigee and at apogee.

■ Solution

To determine θ_p one needs to use the orbital equation:

$$r = \frac{h^2/\mu}{1 + e\cos\theta}$$

With $e = (r_{\max} - r_{\min})/(r_{\max} + r_{\min}) \approx 0.429$, we then obtain, with $r = |\vec{r}_0| = 5000\sqrt{3}$ km,

$$\cos\theta \approx 0.334$$

Thus, θ is either $70.5°$ or $-70.5°$.

To determine the speed of the satellite at perigee and at apogee, notice that at both of those points the position vector $\vec{r}$ is normal to the velocity $\vec{v}$. Therefore, the speeds v_p (at perigee) and v_a (at apogee) may be obtained as

$$|\vec{h}| = v_p r_{\min} = v_a r_{\max} = 16\sqrt{2} \times 10^7 \ \text{km}^2/\text{h}$$

This yields

$$v_p \approx 32,656 \ \text{km/h} \qquad v_a \approx 13,057 \ \text{km/h}$$

Example 4.4.3 The *flight path angle* of a flying object is defined as the angle the velocity of the object with respect to the attracting planet makes with the local horizontal. Determine the flight path angle of the orbiting body in Example 4.4.1 at the instant it was sighted. Also determine, for that instant, the values of $\dot{r}$ and the absolute angular velocity of the line $P_1 P_2$.

■ Solution

To solve problems of this type start by drawing a figure such as Fig. 4.9. The *local horizontal* is, by definition, the line perpendicular to line $P_1 P_2$, and the flight path angle is the angle γ shown in the figure. To obtain the solution, observe Fig. 4.9 and notice the following items.

- ■ To "see" the flight path angle γ, the plane of the orbit needs to be determined. This is done by determining the direction of the unit vector $\hat{n}$ that is normal to that plane. Since $\vec{r}$ and $\vec{v}$ are known at a given instant, we then have $\hat{n} = (\vec{r} \times \vec{v})/|\vec{r} \times \vec{v}|$.

- ■ Having determined $\hat{n}$, which is the same as $\hat{r} \times \hat{\theta}$, the unit vector $\hat{\theta}$, parallel to the orbital plane and normal to $\hat{r}$, is then $\hat{\theta} = \hat{n} \times \hat{r}$. The flight path angle γ is then found from $\cos\gamma = \hat{\theta} \bullet (\vec{v}/|\vec{v}|)$.

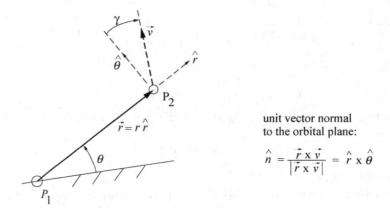

Figure 4.9: Flight path angle γ for Example 4.4.3.

At the instant the satellite was sighted, its position vector $\vec{r}_0$ and velocity $\vec{v}_0$ relative to the earth were given as $\vec{r} = 5000(\hat{x} + \hat{y} + \hat{z})$ km, and $\vec{v} = 16,000(\hat{x} - \hat{y} + \hat{z})$ km/h. This yields, at that instant,

$$\hat{r} = \frac{\vec{r}}{|\vec{r}|} = \frac{1}{\sqrt{3}}(\hat{x} + \hat{y} + \hat{z})$$

$$\hat{n} = \frac{\vec{r} \times \vec{v}}{|\vec{r} \times \vec{v}|} = \frac{(\hat{x} + \hat{y} + \hat{z}) \times (\hat{x} - \hat{y} + \hat{z})}{|(\hat{x} + \hat{y} + \hat{z}) \times (\hat{x} - \hat{y} + \hat{z})|} = \frac{\hat{x} - \hat{z}}{\sqrt{2}}$$

$$\hat{\theta} = \hat{n} \times \hat{r} = \frac{\hat{x} - 2\hat{y} + \hat{z}}{\sqrt{6}}$$

$$\cos\gamma = \frac{\vec{v}}{|\vec{v}|} \bullet \hat{\theta} = \frac{4}{\sqrt{18}} \quad \text{which yields } \gamma \approx \pm 19.47°$$

$$\dot{r} = \vec{v} \bullet \hat{r} \approx \frac{16,000}{\sqrt{3}} \approx 9237.6 \text{ km/h}$$

$$|\vec{r}|\dot{\theta} = \vec{v} \bullet \hat{\theta} \approx 26,128 \text{ km/h} \quad \text{which yields } \dot{\theta} \approx 3.017 \text{ rad/s}$$

Since the value of $\dot{r}$ is positive, inspection of Fig. 4.9 discloses that γ has to be in the range $0 < \gamma < \pi$. Therefore, the flight path angle (which, based on the value of $\cos\gamma$, was found to be $\pm 19.47°$) is $\gamma \approx 19.47°$.

4.5   Automated Solution for the Two-Body Problem

As it turned out, the nonlinear differential equations of motion for the two-body problem were in a form that made analytical investigations possible. Those investigations revealed a substantial amount of extremely useful information about the motion. Specifically, it allowed us to

- Determine whether the trajectory is open or closed simply by looking at the sign of the total energy E of the motion
- Determine the maximum and minimum values of $r(t)$ without having to solve the differential equation of motion for $r(t)$
- Obtain the expression for r as a function of the angle θ, which allowed for the identification of the different types of orbits
- Obtain the period of a closed orbit

This section presents the numerical determination of the solution of the differential equations of motion for the two-body system, which were formulated in Example 4.3.2. The data used for this are: the radius R_E of the attracting planet, the acceleration g of gravity at its surface, the quantity $m_2/m_1 + 1$ (where m_2 and m_1 are the masses of the spacecraft and the attracting planet, respectively), and the initial conditions $r(0)$, $v(0)$ (the magnitude of the initial velocity of the spacecraft relative to the attracting planet), and $\gamma(0)$ (the initial flight path angle) in degrees. The flight path angle of a trajectory is the angle shown in Fig. 4.9 (p. 292). The magnitude $v(0) = |\vec{v}(0)|$ and the initial flight path angle $\gamma(0)$ are used because these quantities are preferred in practice, instead of the scalar components $\dot{r}(0)$ and $r(0)\dot{\theta}(0)$. Notice that inspection of Fig. 4.9 discloses that $\dot{r}$ and $r\dot{\theta}$ are related to $v = |\vec{v}|$ and γ as $\dot{r} = v \sin\gamma$ and $r\dot{\theta} = v \cos\gamma$. The numerical solution is illustrated with Scilab, Matlab, and Simulink.

$\implies$ There are two similar programs named *two_body_simul* listed in Appendix E. One is for Scilab users and the other is for Matlab users. They integrate Eqs. (4.3.2h) and (4.3.2i) to obtain $r(t)$ and $\theta(t)$, stop the simulation if the spacecraft (represented by a small circle) crashes on the attracting planet, and animate the motion of the center of mass of the spacecraft. They also keep a circle on the computer screen, representing the attracting planet, a rotating radial line representing the planet's spin about its own axis, and record the day number on the animation window, assuming that it takes one 24-hour day for the planet to rotate 2π rad about its own axis. The day number, counting from the starting of the simulation, is automatically updated during the simulation every time that line rotates $360°$. Three examples of usage are given in

the first few comment lines at the beginning of those programs: one with data for a circular orbit, one with data for an elliptic orbit, and one for initial conditions that result in a crash on the surface of the attracting planet (neglecting atmospheric effects).[7] To use either of those programs, one types the command
`my_output=two_body_simul(t,Re,g, m2_over_m1_plus_1,r_0,v_0,gamma_0);`
either in the Scilab console or in the Matlab command window, with a given array for the time t, and values for the other parameters. For example, the command (try $\Delta t = 0.005$ for Scilab, and 0.001 for Matlab)
`my_output=two_body_simul([0:Δt:30],6378,9.81,1,6578,38040,0);`
yields an elliptic orbit. Decreasing (increasing) the step size of the time array will slow down (speed up) the animation. Figure 4.10 shows a snapshot of the animation of an elliptic orbit.

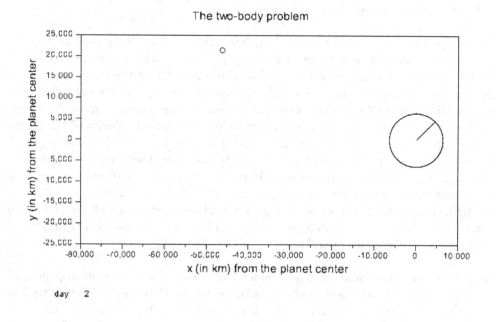

Figure 4.10: A snapshot of the animation of an elliptic orbit.

Those who have Simulink and prefer to use it may want to construct the block diagram model shown in Fig. 4.11 for simulating the motion of the spacecraft, setting the *Simulation* → *Parameters* menu for this model with time t of the form

---

[7]Reasonably realistic short-range or long-range ballistic problems could be addressed by including aerodynamic effects in the differential equations of motion.

[0:Δt:30], and using $\theta(0) = 0$. To see the animation of the motion after the simulation stops, without the sophistication shown in Fig. 4.10, type `animate_nbars(x, y)` in the Matlab command window. This assumes that you have the Matlab program `animate_nbars` stored on your computer. For a sophisticated display of the motion, use the Matlab function `two_body_simul` listed in Appendix E.

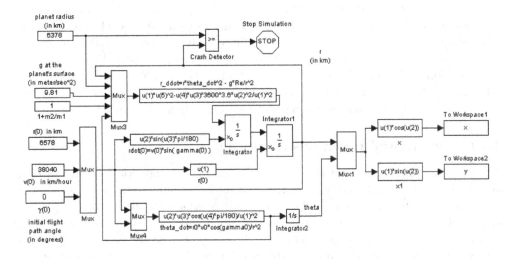

Figure 4.11: A Simulink block diagram model for the two-body problem.

The problem analyzed in Section 4.6 is another important classical problem in dynamics, and it has attracted the attention of scientists for a long time. Its analysis is the most involved one up to this point in the book, and the type of work involved is typical of what is encountered in professional practice. The entire analysis involves the formulation of the differential equations of motion, determination of the equilibrium solutions, linearization, and analysis of the stability of each equilibrium.

4.6 The Restricted Three-Body Problem

A spacecraft is subjected to the gravitational attraction of several celestial bodies, and relatively little can be done analytically for the general case involving any number N of bodies attracting each other. Numerical integration of the equations is the main recourse in such a case. The problem that is analyzed in this section involves two main attracting bodies and a third body whose mass is much smaller than either one of the other two so that its presence essentially does not affect their motions. The problem consists of investigating the motion of that smaller body when the only forces acting on it are the gravitational forces of attraction of the other two massive bodies. This problem is known as the *restricted three-body problem* of celestial mechanics (or of spacecraft dynamics).

This problem represents a first approximation to the motion of a spacecraft where the main external forces acting on it are the gravitational attraction of two celestial bodies (such as the earth and the moon, for example). As shown in Section 4.6.2, such a system exhibits several equilibrium points where the smaller body does not move relative to the other two bodies. Such equilibrium points have attracted the attention of the scientific community and might be used in the future for placing artificial satellites in their neighborhood for communication purposes, for example. The stability of those equilibrium points is also investigated in this section.

4.6.1   Differential Equations of Motion

The system under consideration consists of two massive spherical bodies in orbit around each other, both attracting a third body that is much smaller than the other two. Such a system is shown in Fig. 4.12a and b, where the two massive spherical bodies of masses m_1 and m_2 are represented by their geometric centers P_1 and P_2, respectively, and the smaller body, with center of mass at P_3 and mass m_3, is subjected to the forces $\vec{F}_{31}$ and $\vec{F}_{32}$ due to the gravitational attraction of P_1 and P_2, respectively. The mass m_3 of P_3 is assumed to be very small compared to m_1 and m_2 so that the forces acting on P_1 and P_2 are essentially due to their mutual attraction. Thus, their motion is not affected by the presence of the smaller body. Therefore, P_1 and P_2 constitute a two-body problem and, as such, they move in a plane, which is the orbital plane shown in Fig. 4.12.

Different coordinate systems, such as spherical, cylindrical, or rectangular, may be chosen to analyze the motion of P_3. Cylindrical coordinates r-θ-z, with r measured from P_1, are shown in Fig. 4.12a, while rectangular coordinates x-y-z, measured from the center of mass C of the system, are shown in Fig. 4.12b. Both of these coordinates have been used in the technical literature to investigate this system. Spherical coordinates would involve a distance (say, from either P_1 or P_2 to P_3) and two angles. No matter what choice is made, the differential equations of motion for the system will certainly not be very simple. They rarely are, either for the system under consideration or for other systems.

In Fig. 4.12a, $r \overset{\Delta}{=} |\overrightarrow{P_1 Q}|$, where $\overrightarrow{P_1 Q}$ is the projection of the vector $\overrightarrow{P_1 P_3}$ on the orbital plane of $P_1 P_2$ and z is the *elevation* of P_3 relative to that orbital plane. The mutually orthogonal x-, y-, z-axes shown in Fig. 4.12a and b with x and y in the orbital plane and rotating in space, are centered at the center of mass C of the P_1-P_2

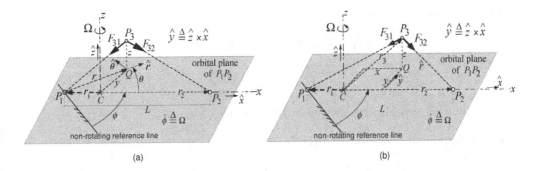

Figure 4.12: The three-body system and the free-body diagrams. (a) Using cylindrical (r, θ, z) coordinates to specify the position of P_3. (b) Using rectangular (x, y, z) coordinates to specify the position of P_3.

system. From the analysis of the two-body problem, it is already known that point C is inertial, i.e., that its absolute acceleration is equal to zero.

Several unit vectors, which will be used later, are also shown in Fig. 4.12a and b. The unit vector $\hat{z}$ is perpendicular to the orbital plane of $P_1 P_2$. At any time t the distance L between P_1 and P_2 is determined from the solution of the two-body problem. Line $P_1 P_2$ rotates in space with absolute angular velocity $\vec{\Omega} = \dot{\phi}\hat{z} \overset{\Delta}{=} \Omega \hat{z}$ where, as shown in Fig. 4.12, ϕ is the angle measured from a nonrotating reference line in the orbital plane to line $P_1 P_2$.

The analysis that follows is restricted to the case where P_1 and P_2 are in a circular orbit around C, which is known as the *restricted circular three-body problem*. In such a case, the distance $L = |\overrightarrow{P_1 P_2}|$ is constant. As determined from the analysis of the two-body problem, the angular velocity $\dot{\phi} = \Omega$ is also constant and is determined as

$$\Omega = \text{constant} = \sqrt{\frac{G\left(m_1 + m_2\right)}{L^3}} \tag{4.6.1}$$

where G, as before, is the universal gravitational constant.

The motion is analyzed below using the cylindrical coordinates r, θ, and z shown in Fig. 4.12a. The formulation of the differential equations involving the rectangular coordinates x, y, and z shown in Fig. 4.12b is left as an exercise (see Problem 4.13).

Differential Equations Using the Cylindrical Coordinates $r = |\overrightarrow{P_1 Q}|$, θ, and z:
Referring to Fig. 4.12, and to Eqs. (4.3.2c) (p. 278) for the two-body problem, the

absolute position vectors of P_1 and P_2 are

$$\vec{r}_1 \;=\; \overrightarrow{CP_1} = -\frac{m_2 L}{m_1 + m_2}\,\hat{x} \tag{4.6.2}$$

$$\vec{r}_2 \;=\; \overrightarrow{CP_2} = \frac{m_1 L}{m_1 + m_2}\,\hat{x} \tag{4.6.3}$$

The following can also be written after inspecting Fig. 4.12.

■ Absolute position vector of P_3:

$$\vec{r}_3 = \overrightarrow{CP_3} = \overrightarrow{CP_1} + \overrightarrow{P_1Q} + \overrightarrow{QP_3} = -\frac{m_2 L}{m_1 + m_2}\,\hat{x} + r\hat{r} + z\hat{z} \tag{4.6.4}$$

■ Absolute velocity of P_3:

$$\vec{v}_3 \;=\; \frac{d\vec{r}_3}{dt} = \underbrace{\dot{r}\hat{r} + \left[\left(\dot{\phi}+\dot{\theta}\right)\hat{z}\right]\times(r\hat{r})}_{\text{This is } \frac{d}{dt}(r\hat{r})} + \dot{z}\hat{z} - \frac{m_2 L}{m_1 + m_2}\underbrace{\left(\dot{\phi}\hat{z}\right)\times\hat{x}}_{\text{This is } \frac{d}{dt}\hat{x}}$$

$$= \;\dot{r}\hat{r} + r\left(\dot{\phi}+\dot{\theta}\right)\hat{\theta} + \dot{z}\hat{z} - \frac{m_2 L}{m_1 + m_2}\dot{\phi}\,\hat{y} \tag{4.6.5}$$

since $\hat{z}\times\hat{x}=\hat{y}$ and $\hat{z}\times\hat{r}=\hat{\theta}$ (see Fig. 4.12, and use the right-hand rule to do these cross products).

■ Absolute acceleration of P_3:

$$\vec{a}_3 = \frac{d\vec{v}_3}{dt} \;=\; \underbrace{\ddot{r}\hat{r} + \left[\left(\dot{\phi}+\dot{\theta}\right)\hat{z}\right]\times(\dot{r}\hat{r})}_{\text{This is } \frac{d}{dt}(\dot{r}\hat{r})}$$

$$+ \underbrace{\left[r\left(\ddot{\phi}+\ddot{\theta}\right)+\dot{r}\left(\dot{\phi}+\dot{\theta}\right)\right]\hat{\theta} + \left(\dot{\phi}+\dot{\theta}\right)\hat{z}\times\left[r\left(\dot{\phi}+\dot{\theta}\right)\hat{\theta}\right]}_{\text{This is } \frac{d}{dt}\left[r(\dot{\phi}+\dot{\theta})\hat{\theta}\right]}$$

$$+ \ddot{z}\hat{z} - \frac{m_2 L}{m_1 + m_2}\underbrace{\left[\ddot{\phi}\,\hat{y} + \left(\dot{\phi}\hat{z}\right)\times\left(\dot{\phi}\hat{y}\right)\right]}_{\text{This is } \frac{d}{dt}(\dot{\phi}\hat{y})}$$

$$= \;\left[\ddot{r} - r\left(\dot{\phi}+\dot{\theta}\right)^2\right]\hat{r} + \left[r\left(\ddot{\phi}+\ddot{\theta}\right)+2\dot{r}\left(\dot{\phi}+\dot{\theta}\right)\right]\hat{\theta}$$

$$+ \ddot{z}\hat{z} - \frac{m_2 L}{m_1 + m_2}\left[\ddot{\phi}\hat{y} - \dot{\phi}^2\hat{x}\right] \tag{4.6.6}$$

Since $\dot{\phi} = \text{constant} \overset{\Delta}{=} \Omega$, $\ddot{\phi} = 0$ in this expression for $\vec{a}_3$.

■ Resultant force $\vec{F}$ acting on P_3 due to the gravitational attractions of P_1 and P_2, as per Newton's law of universal gravitation:

$$\vec{F} = \vec{F}_{31} + \vec{F}_{32} = \frac{Gm_1m_3}{r_{31}^2}\frac{\overrightarrow{P_3P_1}}{|\overrightarrow{P_3P_1}|} + \frac{Gm_2m_3}{r_{32}^2}\frac{\overrightarrow{P_3P_2}}{|\overrightarrow{P_3P_2}|} \qquad (4.6.7)$$

where $r_{31} = |\overrightarrow{P_3P_1}|$ is the distance between P_3 and P_1, and $r_{32} = |\overrightarrow{P_3P_2}|$ is the distance between P_3 and P_2. The vectors $\overrightarrow{P_3P_1}/|\overrightarrow{P_3P_1}|$ and $\overrightarrow{P_3P_2}/|\overrightarrow{P_3P_2}|$ are the unit vectors from P_3 to P_1 and from P_3 to P_2, respectively.

By inspecting Fig. 4.12, the following expressions can be written for $\overrightarrow{P_3P_1}$ and $\overrightarrow{P_3P_2}$ in terms of the quantities r, θ, z, and L:

$$\overrightarrow{P_3P_1} = -z\hat{z} - r\hat{r} \qquad (4.6.8)$$

$$\overrightarrow{P_3P_2} = -z\hat{z} - r\hat{r} + L\hat{x} = -z\hat{z} - r\hat{r} + L\left(\hat{r}\cos\theta - \hat{\theta}\sin\theta\right) \qquad (4.6.9)$$

and this yields,

$$r_{31} = |\overrightarrow{P_3P_1}| = \sqrt{r^2 + z^2} \qquad (4.6.10)$$

$$r_{32} = |\overrightarrow{P_3P_2}| = \sqrt{L^2 + r^2 + z^2 - 2Lr\cos\theta} \qquad (4.6.11)$$

We now have all the ingredients to obtain the differential equations of motion for the system based on Newton's second law $\vec{F} = m_3\vec{a}_3$. With $\vec{a}_3$ given by Eq. (4.6.6) and $\vec{F}$ by Eq. (4.6.7), $\overrightarrow{P_3P_1}$ and $\overrightarrow{P_3P_2}$ given by Eqs. (4.6.8) and (4.6.9), and $\hat{x}$ and $\hat{y}$ that appear in Eq. (4.6.6) obtained by inspection of Fig. 4.12 as

$$\hat{x} = \hat{r}\cos\theta - \hat{\theta}\sin\theta \qquad \hat{y} = \hat{r}\sin\theta + \hat{\theta}\cos\theta \qquad (4.6.12)$$

you can verify that $\vec{F} = m_3\vec{a}_3$ yields the following scalar differential equations of motion:

$$\ddot{r} - r\left(\dot{\theta} + \Omega\right)^2 + \Omega^2\frac{m_2L}{m_1 + m_2}\cos\theta = \frac{Gm_2}{r_{32}^3}(L\cos\theta - r) - \frac{Gm_1}{r_{31}^3}r$$

$$= \frac{\Omega^2L^3}{m_1 + m_2}\left[\frac{m_2(L\cos\theta - r)}{r_{32}^3} - \frac{m_1r}{r_{31}^3}\right] \qquad (4.6.13)$$

$$r\ddot{\theta} + 2\dot{r}\left(\dot{\theta} + \Omega\right) - \Omega^2\frac{m_2L}{m_1 + m_2}\sin\theta = -\frac{Gm_2}{r_{32}^3}L\sin\theta = -\frac{\Omega^2L^3}{m_1 + m_2}\frac{m_2L\sin\theta}{r_{32}^3} \qquad (4.6.14)$$

$$\ddot{z} = -\left[\frac{Gm_1}{r_{31}^3} + \frac{Gm_2}{r_{32}^3}\right]z \qquad (4.6.15)$$

4.6.2 Equilibrium Solutions

The equilibrium solutions to Eqs. (4.6.13), (4.6.14), and (4.6.15) correspond to $r = \text{constant} \overset{\Delta}{=} r_e$, $\theta = \text{constant} \overset{\Delta}{=} \theta_e$, and $z = \text{constant} \overset{\Delta}{=} z_e$. Equation (4.6.15) immediately discloses that

$$z_e = 0 \tag{4.6.16}$$

Since $z_e = 0$, any equilibrium point lies in the orbital plane of $P_1 P_2$. Thus, at equilibrium, point P_3 in Fig. 4.12 is the same as Q, which means that the distance $r_{31} = \overrightarrow{P_3 P_1}$ is $r_{31} = r_e$ (which still needs to be determined). By inspecting Eq. (4.6.14), it follows that either

$$r_{32} = \text{constant} \overset{\Delta}{=} r_{32e} = L \qquad \text{or} \qquad \sin \theta_e = 0 \tag{4.6.17}$$

Both cases are considered. The notations r_{31e} and r_{32e} denote the equilibrium values of the distances r_{31} and r_{32}, respectively.

CASE 1 $\quad r_{32e} = L$

With $r_{31e} = r_e$, Eq. (4.6.13) yields

$$r_e = L \tag{4.6.18}$$

Since $r_{32e} = L$ and also $r_{31e} = r_e = L$, points P_1, P_2, and P_3 form an equilateral triangle for this solution. There are two such configurations, with $\theta_e = 60°$ and $\theta_e = -60°$, respectively. These two solutions are known as the L_4 and L_5 equilibrium points, respectively, and they were first found by Lagrange.

CASE 2 $\quad \sin \theta_e = 0$

There are two possibilities for this case, namely, $\theta_e = 0$ and $\theta_e = 180°$. For both cases, Eq. (4.6.13) yields

$$-(m_1 + m_2)r_e + m_2 L \cos \theta_e = L^3 \left[\frac{m_2(L \cos \theta_e - r_e)}{r_{32e}^3} - \frac{m_1}{r_e^2} \right] \tag{4.6.19}$$

where

$$r_{32e} = \sqrt{L^2 + r_e^2 - 2 r_e L \cos \theta_e} = \left\{ \begin{array}{ll} r_e + L & \text{when } \theta_e = 180° \\ |r_e - L| & \text{when } \theta_e = 0 \end{array} \right\} \tag{4.6.20}$$

Equation (4.6.19) is nonlinear in r_e. It was found many years ago that, by separately considering the two cases $\theta_e = 0$ and $\theta_e = 180°$, it yields a fifth degree polynomial equation in r_e/L when it is reduced to a single denominator for each one of these two cases. Numerical solutions for r_e/L can then be obtained for each value of either m_2/m_1 or m_1/m_2, with only the real and positive solutions being the acceptable ones. An alternative and much easier solution procedure is shown next.

Inspection of Eq. (4.6.19) discloses that it is certainly nonlinear in r_e, but it is linear in either m_2 or m_1. Therefore, why not solve it either for m_2/m_1 or m_1/m_2 in terms of r_e/L? A plot involving these variables can then be generated so that values of r_e/L versus either m_2/m_1 or m_1/m_2 are then taken from the plot. The solution for m_2/m_1 is

$$\frac{m_2}{m_1} = \frac{r_e/L - (L/r_e)^2}{(\cos\theta_e - r_e/L)\left\{1 - 1/[1 + (r_e/L)^2 - 2(r_e/L)\cos\theta_e]^{3/2}\right\}} \qquad (4.6.21)$$

Without any loss of generality, one can choose $m_2/m_1 \leq 1$. A plot of m_2/m_1 versus r_e/L (or vice versa, if preferred) can then be generated for any range of values of r_e/L using Eq. (4.6.21), and then limiting the range of the plot to the one that yields $m_2/m_1 \leq 1$. Such a plot is shown in Fig. 4.13b.

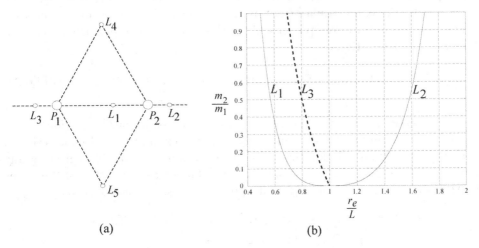

(a)   (b)

Figure 4.13: The equilibrium points L_1 to L_5 for the restricted circular three-body problem, and a plot of the nondimensional distances r_e/L for L_1, L_2, and L_3 (r_e is measured from P_1).

The three equilibrium points obtained from Eq. (4.6.21), which are along a straight line, are known as the L_1, L_2, and L_3 points. They were first found by

Euler. As an example, if P_1 represents the center of the earth and P_2 represents the center of the moon, the L_3 and L_2 equilibrium points are always behind the earth and the moon, respectively, and L_1 is between the earth and the moon. The five points $L_1, \ldots, L_5$ are known as the *libration points*, and that is what the L in their notation stands for. They are shown in Fig. 4.13a.

The question that needs to be answered now is whether a body P_3, with mass m_3 that is much smaller than m_1 and m_2, placed in the neighborhood of any of the equilibrium points $L_1, \ldots, L_5$ will remain in that neighborhood with the passage of time. The answer to such a question, which is important for space applications, involves the analysis of the stability of each equilibrium. Such an analysis is presented in Section 4.6.3.

4.6.3   Stability of the Equilibrium Solutions

If a spacecraft is placed in the vicinity of any of the equilibrium points $L_1, \ldots, L_5$, it is important to know if it stays near that equilibrium point for all times. If that happens, the equilibrium is said to be stable. Otherwise, it is unstable. As done in Section 1.15 of Chapter 1 and in other parts of the book (see, for example, the spring-pendulum on p. 162), stability of each equilibrium is analyzed here by linearizing the differential equations of motion about each equilibrium. To do this, the equilibrium solutions are "perturbed" by introducing small perturbations $r_s(t)$, $\theta_s(t)$, and $z_s(t)$, respectively, as

$$r(t) \overset{\Delta}{=} r_e + r_s(t) \qquad \theta(t) \overset{\Delta}{=} \theta_e + \theta_s(t) \qquad z(t) \overset{\Delta}{=} \underbrace{z_e}_{=0} + z_s(t) = z_s(t) \qquad (4.6.22)$$

These expressions are substituted into the differential equations of motion – Eqs. (4.6.13), (4.6.14), and (4.6.15) – and all terms are expanded in Taylor series in the small perturbations. Linearization of the equations is done by neglecting any product of small quantities in the Taylor series. The series expansion of the terms $r(\dot{\theta} + \Omega)^2$, $\sin \theta$, $\cos \theta$, r_{31}^{-3}, and r_{32}^{-3} that appear in those equations yields

$$
\begin{aligned}
r(\dot{\theta} + \Omega)^2 &= (r_e + r_s)(\dot{\theta}_s + \Omega)^2 = (r_e + r_s)(\Omega^2 + 2\Omega\dot{\theta}_s + \dot{\theta}_s^2) \\
&= \Omega^2 r_e + \Omega^2 r_s + 2\Omega r_e \dot{\theta}_s + \ldots
\end{aligned} \qquad (4.6.23)
$$

$$\sin \theta = \sin(\theta_e + \theta_s) = \sin \theta_e + \theta_s \cos \theta_e + \ldots \qquad (4.6.24)$$
$$\cos \theta = \cos(\theta_e + \theta_s) = \cos \theta_e - \theta_s \sin \theta_e + \ldots \qquad (4.6.25)$$

$$r_{31}^{-3} = \left[(r_e + r_s)^2 + z_s^2\right]^{-3/2} = r_e^{-3}\left(1 - 3\frac{r_s}{r_e}\right) + \ldots \qquad (4.6.26)$$

$$
\begin{aligned}
r_{32}^{-3} &= \left[L^2 + (r_e + r_s)^2 + z_s^2 - 2L \left(r_e + r_s \right) \cos \left(\theta_e + \theta_s \right) \right]^{-3/2} \\
&= r_{32e}^{-3} \left\{ 1 - \frac{3}{r_{32e}^2} \left[L r_e \left(\sin \theta_e \right) \theta_s + \left(r_e - L \cos \theta_e \right) r_s \right] \right\} + \dots \quad (4.6.27)
\end{aligned}
$$

where "..." denote all neglected products of small quantities.

By proceeding in this manner with the rest of the terms in the differential equations, and by noticing that the nonlinear term $\dot{r}\dot{\theta} = \dot{r}_s \dot{\theta}_s$ that appears in Eq. (4.6.14) is neglected in the linearization, the following linearized differential equations of motion are obtained.

$$
\begin{aligned}
\ddot{r}_s - 2\Omega r_e \dot{\theta}_s - \Omega^2 r_s - \frac{\Omega^2 m_2 L \sin \theta_e}{m_1 + m_2} \theta_s &= -\frac{\Omega^2 L^3 m_2}{(m_1 + m_2) r_{32e}^3} \left\{ r_s + L \left(\sin \theta_e \right) \theta_s \right. \\
&\left. + \frac{3 \left(L \cos \theta_e - r_e \right) \left[L r_e \left(\sin \theta_e \right) \theta_s + \left(r_e - L \cos \theta_e \right) r_s \right]}{r_{32e}^2} \right\} + \frac{2\Omega^2 L^3 m_1 r_s}{(m_1 + m_2) r_{32e}^3}
\end{aligned}
$$

$$(4.6.28)$$

$$
\begin{aligned}
r_e \ddot{\theta}_s + 2\Omega \dot{r}_s - \frac{\Omega^2 m_2 L \cos \theta_e}{m_1 + m_2} \theta_s &= -\frac{\Omega^2 L^4 m_2}{(m_1 + m_2) r_{32e}^3} \left\{ \theta_s \cos \theta_e \right. \\
&\left. - 3 \frac{L r_e \left(\sin \theta_e \right) \theta_s + \left(r_e - L \cos \theta_e \right) r_s}{r_{32e}^2} \sin \theta_e \right\}
\end{aligned}
$$

$$(4.6.29)$$

$$
\ddot{z}_s = -\left[\frac{Gm_1}{r_e^3} + \frac{Gm_2}{r_{32e}^3} \right] z_s \quad (4.6.30)
$$

These linearized differential equations are valid for perturbed motions about all five equilibrium solutions. Equation (4.6.30) immediately discloses that the out-of-plane z-motion is infinitesimally stable because the quantity $Gm_1/r_e^3 + Gm_2 r_{32e}^3$ is always positive. Therefore, the "linearized $z_s(t)$ motion" simply consists of an oscillation of frequency equal to $\sqrt{Gm_1/r_e^3 + Gm_2 r_{32e}^3}$. We now have to deal only with Eqs. (4.6.28) and (4.6.29), which are independent of $z_s(t)$. For convenience, the analysis is separated into two parts: the equilateral triangle and the straight-line configurations.

The Equilateral Triangle Configuration (Points L_4 and L_5 in Fig. 4.13)
For these equilibrium points, $r_e = r_{32e} = L$ and $\theta_e = \pm 60°$. In this case, $\cos \theta_e = 1/2$ and $\sin \theta_e = \pm\sqrt{3}/2$ for the L_4 and L_5 points, respectively, and you can verify

that Eqs. (4.6.28) and (4.6.29) simplify to the following, where $k \overset{\Delta}{=} m_2/m_1$:

$$\ddot{r}_s - 2\Omega L \dot{\theta}_s = \frac{3\left(1 + k/4\right) r_s \pm \frac{3\sqrt{3}}{4} k L \theta_s}{1 + k} \Omega^2 \qquad (4.6.31)$$

$$L\ddot{\theta}_s + 2\Omega \dot{r}_s = \frac{\Omega^2 k \left(9L\theta_s \pm 3\sqrt{3} r_s\right)}{4\left(1 + k\right)} \qquad (4.6.32)$$

As presented in Section 1.16 of Chapter 1 (p. 66), the general solutions to the linear differential equations (4.6.31) and (4.6.32) are exponentials of the form

$$r_s(t) = Ae^{st} \qquad\qquad \theta_s(t) = Be^{st}$$

where A, B, and s are constants. By substituting this general solution into Eqs. (4.6.31) and (4.6.32), the exponential function e^{st} will be common to all terms. This is the reason why the exponential function is the general solution to a set of "homogeneous" linear differential equations with constant coefficients. The following algebraic equations involving A, B, and s are then obtained:

$$\left[s^2 - \frac{3\Omega^2 \left(1 + k/4\right)}{1 + k} \right] A - \left[2s\Omega L \pm \frac{3\sqrt{3}L\Omega^2 k}{4\left(1 + k\right)} \right] B = 0 \qquad (4.6.33)$$

$$\left[2s\Omega \mp \frac{3\sqrt{3}\Omega^2 k}{4\left(1 + k\right)} \right] A + \left[Ls^2 - \frac{9\Omega^2 Lk}{4\left(1 + k\right)} \right] B = 0 \qquad (4.6.34)$$

By solving Eq. (4.6.33), say, for either A or B and by substituting the result into Eq. (4.6.34), the following equation is obtained:

$$\left[s^4 + \Omega^2 s^2 + \frac{27k\Omega^4}{4(1 + k)^2} \right] B = 0 \qquad (4.6.35)$$

By imposing the requirement that $B \neq 0$ (otherwise one would just have the obvious solution $r_s = \theta_s = 0$, i.e., the equilibrium is not perturbed!), the coefficient of B in Eq. (4.6.35) has to be equal to zero. As in Chapter 1, the resulting equation is called the *characteristic equation* associated with this equilibrium. The solution to the characteristic equation for s^2 is

$$s^2 = -\frac{\Omega^2}{2}\left[1 \pm \sqrt{1 - \frac{27k}{(1 + k)^2}} \right] \qquad (4.6.36)$$

If the roots of Eq. (4.6.36) are complex, the square root of such a root will also be complex, but with a positive real part.[8] The equilibrium is unstable in such a case.

---

[8]This is so because any complex number is of the form $C(\cos\alpha \pm i\sin\alpha) = Ce^{\pm i\alpha}$, where C and α are real numbers, with $C > 0$ and $0 \leq \alpha \leq \pi$, and $i = \sqrt{-1}$. The square root of this is then $\sqrt{C}e^{\pm i\alpha/2} = \sqrt{C}[\cos(\alpha/2) \pm i\sin(\alpha/2)]$, whose real part $\sqrt{C}\cos(\alpha/2)$ is never negative for any α.

Therefore, based on the linearized differential equations, the equilibrium is stable only if Eq. (4.6.36) yields two negative real roots for s^2, and this happens only if $(1+k)^2 - 27k > 0$. The roots of $(1+k)^2 - 27k = 0$ are $k = m_2/m_1 = (25 \pm 3\sqrt{69})/2 \approx$ 24.96 and also $k = 1/24.96$. Taking $m_2 \leq m_1$, without any loss of generality, the L_4 and L_5 equilibrium points are unstable if $m_2/m_1 > 1/24.96$, and infinitesimally stable otherwise. It can be easily verified that such an equilibrium is stable when $m_2/m_1 \leq 1/24.96$ by numerically integrating Eqs. (4.6.13), (4.6.14), and (4.6.15) with initial conditions near the equilibrium. When this is done with $m_1 = 80m_2$, $\dot{r}(0) = \dot{\theta}(0) = \dot{z}(0) = 0$, $\theta(0) = 59°$, and $z(0)/L = 0.01$, for example, the maximum values obtained for r/L are approximately 1.01 and 1.006 when $r(0)/L = 1.001$ and $r(0)/L = 1.0005$, respectively.

For the earth-moon system $m_2/m_1 \approx 1/81.3$, and for the sun-earth system $m_2/m_1 \approx 3 \times 10^{-6}$. The L_4 and L_5 points for both of these systems are stable.

$\Longrightarrow$ For Scilab users, the program *restricted_3body.sci* listed in Appendix E integrates Eqs. (4.6.13), (4.6.14), and (4.6.15) to obtain $r(t)/L$, $\theta(t)$, and $z(t)/L$, plots these variables versus time t and also y/L versus x/L, and animates the motion. For Matlab users, the equivalent program *restricted_3body.m*, also listed in Appendix E, does the same. An example of their use [with the initial conditions $r(0)/L = 1.001$, $\dot{r}(0) = 0$, $\theta(0) = 59\pi/180$, $\dot{\theta}(0) = 0$, $z(0)/L = 0.01$, and $\dot{z}(0) = 0$] is shown in the comment lines at the beginning of both programs.

Those who have Simulink and prefer to use it may want to construct the block diagram model shown in Fig. 4.14 for integrating the same differential equations, and generating the same plots, setting the *Simulation* $\rightarrow$ *Parameters* menu for this model with time t of the form $[0:\Delta t:100]$, and using the same initial conditions mentioned in the previous paragraph. To see an animation of the motion after the simulation stops, type

```
animate_nbars(r_over_L.*cos(theta), r_over_L.*sin(theta))
```

in the Matlab command window. You need to have the Matlab program `animate_nbars` stored on your computer (please see the first paragraph of Appendix E).

The Straight-Line Configuration (Points L_1, L_2, and L_3 in Fig. 4.13) For this configuration $\theta_e = 0$ for the equilibrium points L_1 and L_2, and $\theta_e = 180°$ for the L_3 point (see Fig. 4.13, p. 301). For these cases, $r_{32e} = |r_e \pm L|$, as per Eq. (4.6.20) (p. 300). The term $r_e - L\cos\theta_e$ that appears in Eqs. (4.6.28) and (4.6.29) becomes

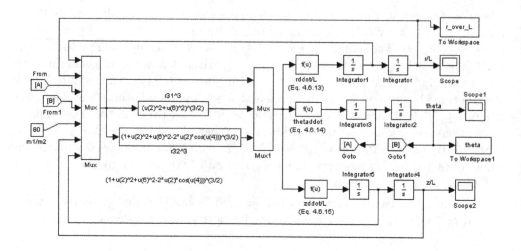

Figure 4.14: A Simulink block diagram model for the restricted three-body problem.

$|r_e \pm L|$, and those linearized differential equations reduce to

$$\ddot{r}_s - 2\Omega r_e \dot{\theta}_s = \Omega^2 \underbrace{\left[1 + \frac{2m_1 L^3}{(m_1 + m_2)r_e^3} + \frac{2m_2 L^3}{(m_1 + m_2)r_{32e}^3}\right]}_{\text{Call this } f_1} r_s \overset{\Delta}{=} \Omega^2 f_1 r_s \qquad (4.6.37)$$

$$r_e \ddot{\theta}_s + 2\Omega \dot{r}_s = \Omega^2 \underbrace{\frac{m_2 L \cos \theta_e}{m_1 + m_2} \left[1 - \left(\frac{L}{r_{32e}}\right)^3\right]}_{\text{Call this } f_2} \theta_s \overset{\Delta}{=} \Omega^2 f_2 \theta_s \qquad (4.6.38)$$

The characteristic equation in this case is readily found to be

$$r_e s^4 + \Omega^2 \left(4r_e - f_2 - r_e f_1\right) s^2 + \Omega^4 f_1 f_2 = 0 \qquad (4.6.39)$$

This equation is of the form $s^4 + a_2 s^2 + a_0 = 0$, which yields the roots

$$s^2 = \frac{-a_2 \pm \sqrt{a_2^2 - 4a_0}}{2} \qquad (4.6.40)$$

Two different roots for s^2 that are real and negative are obtained only when $a_2 > 0$, $a_0 > 0$, and $a_2^2 - 4a_0 > 0$. Thus, all three conditions must be satisfied for the equilibrium to be infinitesimally stable. The equilibrium is unstable if $a_2 < 0$ *or* $a_0 < 0$ *or* $a_2^2 - 4a_0 < 0$.

Looking at Eq. (4.6.39), the quickest way to perform the stability test for this case is to start examining the term $f_1 f_2$. Since $f_1 > 0$, as seen from its definition in Eq. (4.6.37), one of the conditions for an infinitesimally stable equilibrium is for f_2 to be positive. From the definition of f_2 in Eq. (4.6.38), this requires $[1 - (L/r_e)^3] \cos \theta_e$ to be positive, where, as per Eq. (4.6.20) (p. 300), $r_{32e} = r_e + L$ when $\theta_e = 180°$, and $r_{32e} = |r_e - L|$ when $\theta_e = 0$. However, for both cases, inspection of Fig. 4.13 (p. 301) discloses that $[1 - (L/r_e)^3] \cos \theta_e$ is negative. Therefore, all three equilibrium points L_1, L_2, and L_3 are unstable. In practice, a propulsion device activated by a control system is required if a spacecraft is to be maintained near any of these three equilibrium points.

4.7 Problems

Some of the problems have to be solved numerically since the equations for them cannot be solved analytically.

4.1 A geosynchronous satellite is one whose orbital period is equal to the period of the rotation of the earth about its own axis, and whose orbit is on the equatorial plane. The satellite is geostationary (i.e., appears to be at the same point on the sky when it is viewed from the earth) if the orbit is a perfect circle. Small errors in orbit injection will cause small drifts in the satellite position relative to the earth. Determine the height above the surface of the earth a satellite should be placed for it to be geostationary.

4.2 An astronomical observation revealed that a small moon of a distant planet is in a nearly circular orbit about that planet. The diameter of the orbit, and the orbital period (measured in earth time), are $80,000$ km and 10 hours 35 min., respectively. Calculate the mass of the planet, in terms of the mass of the earth, by treating the planet and its moon as a two-body system.

4.3 If the maximum and minimum distances between an inertial center of attraction O and the center of mass C of a spacecraft in an elliptic orbit around O are related as $r_{max}/r_{min} = k$, show that the maximum and minimum magnitudes of the absolute velocity of C are $v_{max} = \sqrt{2(\mu/r_{min})k/(k+1)}$ and $v_{min} = v_{max}/k$.

4.4 The center of mass C of a spacecraft is in a circular orbit 300 km above the earth's surface, which is well above the earth's atmosphere. The mass of the spacecraft is $m = 2500$ kg. The spacecraft is equipped with an engine that, when ignited, generates a constant thrust $\vec{F}$ of magnitude $F = 2000$ N. To change the orbit to an elliptical orbit, the spacecraft is oriented with small thrusters so that $\vec{F}$ is directed along the local horizontal. The small thrusters are then turned off, and the engine is fired for a short time Δt s. Estimate the value of Δt so that the minimum and maximum heights of the new orbit are 300 and 450 km, respectively. Should the initial speed of C be increased of decreased by the engine firing to achieve the new orbit? Approximate the earth as a homogeneous sphere of radius $R_e = 6378$ km. *Hint:* See the note on p. 96 in Chapter 2; use the relationship given in Problem 4.3.

4.5   The spacecraft in Problem 4.4 is in an elliptical orbit of minimum and maximum heights of 300 and 450 km, respectively, above the earth's surface after the reorientation and firing maneuver described in that problem. It is now desired to fire the engine a second time, for a short time Δt_2 s, when the spacecraft reaches its apogee, to transfer the spacecraft to a circular orbit 450 km above the surface of the earth. Estimate the value of Δt_2 to accomplish the task. Should the speed of C at apogee be increased of decreased to achieve the circular orbit? Approximate the earth as a homogeneous sphere of radius $R_E = 6378$ km. See the hint in Problem 4.4.

4.6   A space shuttle was in a circular orbit at a height of 300 km above the earth's surface when it released a satellite with a velocity relative to the shuttle of 20 km/h, and a flight path angle equal to $10°$. Determine the maximum and minimum values of the subsequent distance $r(t)$ between the earth's center and the center of mass of the satellite, and the eccentricity of the resulting orbit.

4.7   A small space probe approaches a distant planet P_1 as shown in Fig. 4.15. The distant planet is Uranus, and its mass is 14.536 times the mass of the earth, and its radius is $R \approx 4.007R_E$, where $R_E \approx 6378$ km is the radius of the earth. When the probe is far away from P_1, it is traveling essentially along a straight line with velocity $v_0 = 60,000$ km/h relative to P_1, and the perpendicular distance from the center of P_1 to the asymptote of the flight path is $D = 1.8R$. By treating the probe and Uranus as a two-body system, determine the minimum value $r_{\min}$ of the distance $r(t)$ from the center of the planet to the probe. Does the probe crash on P_1?

4.8   For Problem 4.7, plot $r_{\min}$ versus the distance D. What is the minimum distance D for the probe not to crash on the planet?

4.9   A small body P of mass m is launched, along the equator, from a height of 1 km above the earth's surface (treating the earth as done in this chapter) with a speed $v_0 = 20,000$ km/h and a flight path angle $\gamma_0 = 30°$. Determine the range R of its trajectory, i.e., the distance, measured along the earth's surface,

Figure 4.15: Figure for Problem 4.7.

from the launching point to the point it hits the earth, and the total time of flight. Consider only the mutual force of attraction between the earth and P. Estimate the error in the answers when the earth's rotation about its own axis is taken into account. For simplicity, an equatorial trajectory is assumed, so that the plane of the trajectory is perpendicular to the axis of rotation of the earth. Use $R_E = 6378$ km for the radius of the earth. What would the range be if the constant gravitational force approximation were used?

4.10 For Problem 4.9, plot the range R versus γ_0 [with $20° \leq \gamma(0) \leq 60°$] for $v_0 = 18,000$ km/h and $v_0 = 20,000$ km/h, preferably superimposing the plots on the same graph (such types of plots are prepared, for a number of other problems, for engineering manuals). For $v_0 = 20,000$ km/h, what value of γ_0 maximizes the range R, and what should be the values of γ_0 so that $R = 3500$ km?

4.11 In addition to the gravitational force of attraction due to the earth, a satellite of mass m is subjected to a force $\vec{f_1} = -c_1 \vec{v}$, and to a drag force modeled as $\vec{f_2} = -c_2 v^2 (\vec{v}/|\vec{v}|)$, where $\vec{v} = \dot{r}\hat{r} + r\dot{\theta}\hat{\theta}$ is the velocity of the center of mass C of the satellite relative to the earth, and c_1 and c_2 are constants.

   a. What are the differential equations that govern the motion of the center of mass of the satellite? Approximate the earth as an inertial homogeneous sphere of radius $R_E \approx 6378$ km.

   b. Numerically integrate the differential equations of motion, animate the motion, and record the total flight time. Automatically stop the simulation if $r \leq R_E$. The satellite was put in space at an altitude of 102 km above the surface of the earth, with a velocity of 29,500 km/h directed along the local horizontal. For $c_1 = 0$, $c_1/m = 0.0001$ h^{-1}, and $c_1/m = 0.001$ h^{-1}, record the flight path and observe what happens during the first 72 h of flight if $c_2/m = 5 \times 10^{-9}$ km^{-1}, and write a few paragraphs about your observations. Also, indicate what happens when $c_1 = c_2 = 0$. *Suggestion:* Either the Scilab or Matlab program *two_body_simul* listed in Appendix E may be modified to solve this problem.

Caution: The flight times are long and can be sensitive to the step size Δt of the numerical integration, and also to the integration method. Therefore, one may need to experiment with smaller values of Δt to verify that the answers are independent of it.

4.12 A common theoretical model for the earth's atmosphere is one for which the air density changes exponentially with height H from the earth's surface as $\rho = \rho_0 e^{-kH}$. Such a model is an approximation for the *standard atmosphere*, for which you can find details on the World Wide Web from the sites www.digitaldutch.com/atmoscalc/ and www.grc.nasa.gov/www/k-12/airplane/atmosmet.html, for example. For altitudes up to about 90 km or so, a fair approximation for the air density from the standard atmosphere model is obtained with $\rho_0 = 1.5$ kg/m^3 and $k = 0.14$ km^{-1}.

It is extremely difficult to accurately account for the effect of air drag on satellites when they enter the atmosphere since the drag depends on the shape of the vehicle and also on changes of atmospheric conditions. The purpose of this problem is to estimate the effect of air drag on the motion of the center of mass of a spacecraft using the exponential model.

    a. Obtain the differential equations that govern the motion of the center of mass C of a satellite of mass m taking into account the drag force modeled as $\vec{F}_{\text{drag}} = -\frac{1}{2}C_D\rho A v^2 \hat{v}$, where $\vec{v} = \dot{r}\hat{r} + r\dot{\theta}\hat{\theta}$ is the velocity of C relative to the earth, and $\hat{v} = \vec{v}/|\vec{v}|$. The nondimensional quantity C_D is called the *drag coefficient*, and A is the area of the satellite that is exposed to the airflow. Approximate the earth as an inertial homogeneous sphere of radius $R_E \approx 6378$ km, treat C_D and A as constants, and assume $\vec{F}_{\text{drag}}$ acts on the satellite all the time .

    b. At a certain time t_0, the speed of a satellite circling the earth is $v_0 = 29,500$ km/h, its height from the earth surface is 102 km, and its flight path angle is $\gamma = 0$. With $C_D A/m = 5 \times 10^{-3}$m^2/kg, numerically integrate the differential equations of motion and then determine how much time it takes for the satellite to crash on the earth's surface (assuming it survives the heat generated due to travel in the atmosphere). What are the values of the crash times when $C_D A/m$ is changed by ± 10 percent? Write a short paragraph commenting on your observations of the motion. *Note:* See the suggestion in Problem 4.11.

4.13 Show that, in terms of the x, y, and z rectangular coordinates shown in Fig. 4.12 (p. 297), the differential equations of motion for the restricted three-body problem investigated in Section 4.6 are

$$\ddot{x} - 2\Omega\dot{y} - \Omega^2 x = -\left[\frac{Gm_1}{r_{31}^3} + \frac{Gm_2}{r_{32}^3}\right]x \tag{4.7.1}$$

$$\ddot{y} + 2\Omega\dot{x} - \Omega^2 y = -\left[\frac{Gm_1}{r_{31}^3} + \frac{Gm_2}{r_{32}^3}\right]y - \frac{Gm_1 m_2 L}{m_1 + m_2}\left[\frac{1}{r_{31}^3} - \frac{1}{r_{32}^3}\right] \tag{4.7.2}$$

$$\ddot{z} = -\left[\frac{Gm_1}{r_{31}^3} + \frac{Gm_2}{r_{32}^3}\right]z \tag{4.7.3}$$

4.14 Body AB shown in Fig. 4.16 consists of two mass particles, each of mass $m/2$, connected by a rigid bar of negligible mass and length L. Body AB is able to rotate freely, in the plane of the figure, about its center of mass C, which

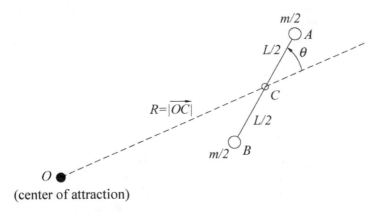

Figure 4.16: Figure for Problem 4.14.

is connected to a fixed pin. The system is attracted by an inertial center of attraction at O that exerts at any point P of mass M a force whose magnitude is equal to $\mu M/|\overrightarrow{OP}|^2$, where μ is a constant. Show that the system exhibits the equilibrium orientations $\theta = \text{constant} \overset{\Delta}{=} \theta_e$, where $\theta_e = 0$ and $\theta_e = 90°$. Which one of these is stable? For $R = 3$ m, $L = 2$ m, and $\mu = 1.8$ m^3/s^2, determine the frequency of the small oscillations about the stable equilibrium orientation of the system. *Hint:* The moment $\vec{M}_C$ was formulated in Example 1.10.1 (p. 27).

4.15 For the system shown in Fig. 4.16, point O represents the center of the earth, and the earth is approximated as a homogeneous sphere of radius $R_E \approx 6378$ km. Instead of being stationary as in Problem 4.14, point C is in a circular orbit of radius R about O, above the earth's atmosphere. For $L \ll R_E$, show, either analytically or numerically, that the frequency ω of the small oscillations of the system about the equilibrium $\theta = \theta_e = 0$ is $\omega = \sqrt{3}\Omega$, where $\Omega = \sqrt{gR_E^2/R^3}$ (with g being the acceleration of gravity at the surface of the earth) is the absolute orbital angular velocity of the circular orbit. *Hint:* The unit vectors $\hat{x}$ and $\hat{y}$ shown in Fig. 1.9 (p. 27) are no longer inertial. Introduce an angle ϕ, where $\dot{\phi} = \Omega$, to keep track of their rotation.

4.16 For the system in Problem 4.15, show that the linearized differential equation of motion about the equilibrium $\theta_e = 0$ is $\ddot{\theta} + 3\Omega^2\theta = 0$, where Ω is the orbital angular velocity of the circular orbit.

CHAPTER 5

Dynamics of Rigid Bodies in "Simpler" Planar Motion

The motion of each particle of a system of N point masses is investigated simply by applying Newton's second law to each particle. As shown in Chapter 4, the resulting vector equations can also be manipulated to yield a vector equation that governs the translational motion of the center of mass of the system, and also a vector equation that involves the time rate of change of the *absolute angular momentum* of the system. These vector equations, which yield six scalar equations for a general motion in three-dimensional space, provide the transition from a system composed of discrete mass particles to a continuum of matter (which is achieved in the limit as $N \to \infty$), which includes either rigid or flexible bodies.

A *rigid body* is a continuum of matter for which the distance between any two of its points is constant. A rigid body may be viewed as a very large, in fact infinite, number of particles satisfying such a distance constraint. The fundamental vector equations for analyzing the motion of a rigid body consist of the two equations previously mentioned, namely, the equation that governs the translational motion of the center of mass of the system and a vector equation that involves the time rate of change of the absolute angular momentum of the system. As shown in this chapter, the latter equation yields a set of scalar equations that govern the *rotational motion* (i.e., motion that involves the change of orientation of a line) of the rigid body. Internal *action and reaction forces* between individual particles of the body do not appear in the differential equations of motion for the system. Those sets of equations are the fundamental equations that govern the most general motion of a rigid body.

This chapter starts with a section showing that a rigid body that is able to move in three-dimensional space has six degrees of freedom; i.e., six independent coordinates are needed to specify the position and orientation of the body in three-dimensional space. The rest of the chapter is restricted to *"simpler" planar motion* (defined at the end of Section 5.1) of rigid bodies. Section 5.6 is dedicated to the calculation of the reaction forces acting on the members of mechanisms.

5.1   Degrees of Freedom of a Rigid Body

A particle P_1 that is able to move in three-dimensional space has three degrees of freedom since three time-dependent variables, or motion *coordinates*, are necessary and sufficient to locate the particle in that space and, thus, to describe its motion. Similarly, two unconstrained particles P_1 and P_2 have six degrees of freedom, while such a number is reduced to five if the distance between the particles is constrained to be constant. This is so because if six variables were used, say six rectangular coordinates $x_1, y_1, z_1, x_2, y_2, z_2$, those variables would be related by one equation of constraint, namely, $(x_2-x_1)^2+(y_2-y_1)^2+(z_2-z_1)^2 = \text{constant} = L^2$, where L is the constant distance between the two particles. Thus, only five of those six variables would be independent of each other. This is illustrated in Fig. 5.1. The addition of one more particle P_3 to the constrained P_1-P_2 system increases the number of degrees of freedom of the entire system from five to eight. However, if the distances between P_3 and the other two particles are constant, so that the particles form a rigid body, the number of degrees of freedom of such a system is reduced from eight to six. This happens because each one of the two additional rigid connections that are needed to transform the system into a rigid body introduces one constraint between the three rectangular coordinates of each particle.

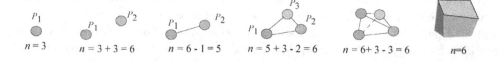

Figure 5.1: From a single particle to a rigid body in three-dimensional motion (n is the number of degrees of freedom).

As illustrated in Fig. 5.1, further addition of any number of particles to a rigid three-particle system, in such a way that the resulting N-particle system is rigid, will not change the number of degrees of freedom of the resulting system. In the limit, as $N \to \infty$, the system becomes a continuum body.

As the number of particles is increased, three new variables are added for each particle that is added to the system. However, three new constraint equations are also introduced when the distance constraints are imposed, and the number of degrees of freedom of the system remains equal to six. Thus, although a rigid body has an infinite number of particles, its number of degrees of freedom is still $n = 6$ for motion in three-dimensional space. This is an important result, and it implies that six scalar differential equations are necessary and sufficient to analyze the most general motion of a rigid body in three-dimensional space. If the motion of the body is constrained to a single plane in space, the number of degrees of freedom for such motion is equal to $n = 3$. In such cases, the motion of the body is governed by only three scalar differential equations.

In this chapter, only "simpler" planar motion of rigid bodies is considered. A *simpler* (but not necessarily "simple" by all means!) planar motion is defined in this book as follows: For thin bodies that may be approximated as being planar, the motion is restricted to a plane that is parallel to the plane of the body; for general three-dimensional bodies, only bodies whose points move on parallel planes are considered (thus, if such bodies are rotating, the rotation is about an axis perpendicular to such planes). Such cases still encompass a large class of problems that are encountered in engineering practice. They do not involve *products of inertia*, which are quantities that appear in Chapter 6 in connection with the general motion of rigid bodies. That chapter includes the basic material that is necessary for dealing with the general case of three-dimensional motion for more advanced studies involving such motions (such as studies involving satellite dynamics, aircraft and ship dynamics, and gyroscopic instruments). The material presented in this chapter provides a transition to the more general material presented in Chapter 6.

5.2 Types of Rigid Body Displacements

A rigid body can undergo two types of basic motion: *translation* and *rotation*. In translation, the orientation of any line fixed to the body does not change as the body moves. This is illustrated in Fig. 5.2a where a top is simply moved from one location to another without spinning it or without changing its orientation. In *rotation*, there is at least one line painted on the body that changes orientation as the body moves. The rotation of a line is illustrated in Fig. 5.2b. If the top in Fig. 5.2a were rotating, line AB would change direction during its motion.

The most general motion of a rigid body can be "decomposed" into a translation of an arbitrarily chosen point A of the body and a rotation about an axis that passes through A. Each one of these two basic motions accounts for three of the six degrees of freedom of the motion of a rigid body. The choice of a point A is arbitrary. Such a decomposition is illustrated by the planar motion of the pendulum shown in Fig. 5.3. The motion may be thought of as either one of the following:

■ A rotation θ about an axis that passes through O and is perpendicular to the plane of the paper, as shown in Fig. 5.3a.

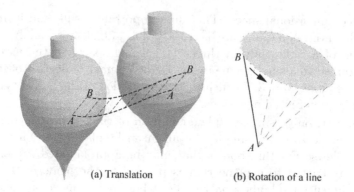

Figure 5.2: The two types of motion of a rigid body (translation and rotation).

■ A translation of an arbitrary point A on the body (i.e., no rotation is involved in this step, as shown in Fig. 5.3b), followed by a rotation θ about an axis that passes through A and is also perpendicular to the plane of the paper (as shown in Fig. 5.3c). Such a rotation brings point O to its actual position at all times in space, without changing the position of point A (which is already in the right place after the translation shown in Fig. 5.3b occurs).

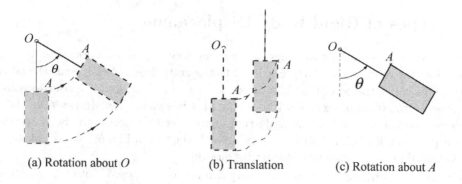

Figure 5.3: Decomposition of a motion (a) into a translation of a point (b), and a rotation (c).

The rotation shown in Fig. 5.3c brings the body shown in Fig. 5.3b to its correct orientation in space as shown in Fig. 5.3a. Notice that the angle the body rotates (namely, the angle θ) is independent of the choice of point A. Thus, the rotation θ is the same for any line drawn on the body shown in Fig. 5.3.

The motion of a rigid body that is only translating in space is similar to that of a point mass since each point of the body is moving in parallel paths. Thus, for such cases, the motion of any point of the body describes the motion of the entire body.

> The motion of a rigid body will be purely translational only if the forces acting on the body are equivalent to a single resultant force $\vec{F}$ that passes through its center of mass C (i.e., the moment $\vec{M}_C$ about C of all forces acting on the body is equal to zero), *and* if the motion is started with zero angular velocity.

If the moment $\vec{M}_C$, about the center of mass C, of the forces acting on a body is not zero, such a moment will cause the body to rotate.

The formulation of the equations that are needed for analyzing the motion of a rigid body in simpler planar motion is presented in Section 5.3, followed by several solved examples. Section 5.4 deals with rolling of rigid bodies.

5.3 Dynamics of Simpler Planar Motion of a Rigid Body

If a rigid body is viewed as a system of a large number of point masses connected by massless bars of constant lengths, it is seen that the results obtained in Sections 4.1 and 4.2 of Chapter 4 are applicable to such bodies. For a continuum body, the summations in the equations in that chapter are replaced by an integration that spans the volume of the entire body.

Figure 5.4 shows a planar rigid body subjected to a set of forces whose resultant is $\vec{F}$. Some of the applied forces may appear as a couple (i.e., two forces that are equal in magnitude and opposite in direction), and their effect is a moment applied to the body (such as the moment M_{bar} in Example 2.7.6, p. 154). In general, the body will be translating and rotating. Specific details about Fig. 5.4 are listed here.

- Point P (whose absolute position vector is $\vec{r} = \overrightarrow{OP}$, where O is an inertial point) represents an infinitesimal mass element of the body, with mass dm. The total mass of the body is $m = \int_{\text{body}} dm$, where the integration spans the entire body.

- Point C is the center of mass of the body, and its absolute position vector is $\vec{r}_C = \overrightarrow{OC}$. The absolute position vector $\vec{r}_C$ was defined in Chapter 4 for a system of point masses. For continuum bodies, either rigid or flexible, it is

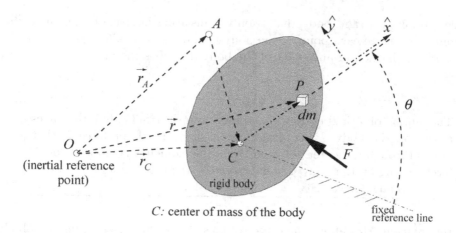

C: center of mass of the body

Figure 5.4: A rigid body subjected to a resultant force $\vec{F}$.

determined as

$$\vec{r}_C = \frac{1}{m} \int_{body} \vec{r}\, dm \tag{5.3.1}$$

Notice that the center of mass of the body may not coincide with any point on the body itself, as in the case of a ring, for example. In such cases, the center of mass may be simply viewed as belonging to an extension of the body.

■ The angle θ is the variable that is used for keeping track of the rotational motion of the body in inertial space. The angle measured from a line that is fixed in space to *any* line that is fixed to the body will accomplish such a purpose. Since θ is measured from a nonrotating reference line, the absolute angular velocity of the body is, according to the right-hand rule, $\vec{\omega} = \dot{\theta}\hat{z}$, where $\hat{z} = \hat{x} \times \hat{y}$ is a unit vector perpendicular to the plane of Fig. 5.4, and pointing toward you, the reader.

■ Point A is an arbitrary point in space. Its absolute position vector is $\vec{r}_A = \overrightarrow{OA}$.

It was shown in Chapter 4 that the motion of the center of mass C of a system of particles is governed by Newton's second law as if the entire mass of the system were concentrated at C. The same result is applicable, of course, to a rigid body. If the body is subjected to several forces (say, N forces $\vec{F}_1, \vec{F}_2, \dots, \vec{F}_N$), the motion of its center of mass is then governed by Newton's second law as [notice that m is assumed to be constant so that $d(m\vec{v}_C)/dt = m\, d\vec{v}_C/dt$]:

$$\boxed{\vec{F} \overset{\Delta}{=} \sum_{i=1}^{N} \vec{F}_i = m\frac{d\vec{v}_C}{dt} = m\frac{d^2\vec{r}_C}{dt^2}} \tag{5.3.2}$$

The motion governed by this equation was considered in detail in Chapter 2.

We now need to analyze the rotational motion of the rigid body and, for this, it is necessary to look at the *absolute angular momentum* of the body. Returning to Fig. 5.4, let us look at the absolute angular momentum referred to an arbitrary point A, which is denoted by $\vec{H}_A$. For a continuum body, $\vec{H}_A$ is still defined by Eq. (4.2.1) in Chapter 4, with the summation replaced by an integral that should span every infinitesimal mass element of the body. Thus, $\vec{H}_A$, which involves the moment about A of the absolute linear momentum $(d\vec{r}/dt)dm$, is defined as being the following quantity:

$$\vec{H}_A \overset{\Delta}{=} \int_{\text{body}} \left[\overrightarrow{AP} \times \frac{d\vec{r}}{dt} \right] dm \tag{5.3.3}$$

The rotational motion of the body is governed by Eq. (4.2.6) in Chapter 4. That equation is repeated here.

$$\frac{d}{dt}\vec{H}_A = \vec{M}_A + m\left(\frac{d}{dt}\overrightarrow{AC} \right) \times \frac{d\vec{r}_C}{dt} \tag{5.3.4}$$

The most convenient way to "read" Eq. (5.3.4) is, "The moment $\vec{M}_A$, about a point A, of the resultant force $\vec{F}$ acting on a body is equal to the time rate of change of the absolute angular momentum $\vec{H}_A$, plus a correction." The "correction" is zero for two particularly convenient choices of point A, as indicated in the next paragraph (which is essentially repeated from Chapter 4). When read as indicated, Eq. (5.3.4) is the "counterpart" to the fundamental equation $\vec{F} = m\, d\vec{v}_C/dt$. It is also a fundamental equation in dynamics, an equation whose form is simple and not difficult to remember.[1] This simpler form of Eq. (5.3.4) is the one that is used in the solution to problems in the book.

Equation (5.3.4) is completely general since no restrictions were imposed on the choice of point A about which the moment of the external forces and the absolute angular momentum of the system of particles were taken. However, by proper choice of a point A (as discussed in Section 4.2 of Chapter 4), the cross product term in Eq. (5.3.4) can be eliminated, thus yielding a simpler relationship (that still can be

---

[1] The similarity between the two equations is seen by replacing the force $\vec{F}$ in $\vec{F} = m\, d\vec{v}_C/dt$ by the moment $\vec{M}_C$, and the linear momentum $m\vec{v}_C$ by the angular momentum $\vec{H}_C$.

applied to any problem) between the moment $\vec{M}_A$ and the time derivative of the angular momentum $\vec{H}_A$. This can be accomplished by either choosing point A to be inertial, in which case $\overrightarrow{AC} = \vec{r}_C$, or to be the center of mass C of the body, in which case $\overrightarrow{AC} = 0$. The following simpler relations are then obtained. These equations were originally developed by Euler, and they are appropriately called *Euler's vector equations*, or simply Euler's equations.

$$\frac{d}{dt}\vec{H}_O = \vec{M}_O \qquad \text{(when } O \text{ is inertial)}$$

$$\frac{d}{dt}\vec{H}_C = \vec{M}_C \qquad \text{(when } C \text{ is the center of mass)}$$

$$(5.3.5)$$

The second equation of (5.3.5) is particularly significant since it is valid even when the center of mass C is not inertial, and, thus, may be moving in an arbitrary manner in space. The first equation of (5.3.5) is useful in problems when a body is rotating about a fixed point O that is either a point on the body or on a "rigid extension" of the body.

To apply the second equation of (5.3.5) to specific problems, an expression for the angular momentum $\vec{H}_C$ is needed. With $\vec{r} = \vec{r}_C + \overrightarrow{CP}$ (see Fig. 5.4), such an expression is developed as follows:

$$\vec{H}_C = \int_{\text{body}} \left[\overrightarrow{CP} \times \frac{d}{dt}\vec{r} \right] dm = \int_{\text{body}} \left[\overrightarrow{CP} \times \frac{d}{dt}\vec{r}_C \right] dm + \int_{\text{body}} \left[\overrightarrow{CP} \times \frac{d}{dt}\overrightarrow{CP} \right] dm$$

$$= \underbrace{\left[\int_{\text{body}} \overrightarrow{CP}\, dm \right]}_{=0} \times \frac{d}{dt}\vec{r}_C + \int_{\text{body}} \left[\overrightarrow{CP} \times \frac{d}{dt}\overrightarrow{CP} \right] dm$$

The first integral in the right-hand side of this expression is zero because that is how the location of the center of mass C is defined. For the second integral, it is convenient to express the vector $\overrightarrow{CP}$ in terms of the unit vector $\hat{x}$ fixed to the rotating rigid body as $\overrightarrow{CP} = |\overrightarrow{CP}|\hat{x}$ (see Fig. 5.4), where $|\overrightarrow{CP}| = $ constant because the body is rigid. Therefore, since $d\hat{x}/dt = \dot{\theta}\hat{y}$ (again, see Fig. 5.4), it follows that

$$\frac{d}{dt}\overrightarrow{CP} = |\overrightarrow{CP}|\dot{\theta}\hat{y}$$

This immediately yields

$$\vec{H}_C = I_C\dot{\theta}\hat{z}$$

$$(5.3.6)$$

where

$$I_C \overset{\Delta}{=} \int_{\text{body}} |\overrightarrow{CP}|^2 \, dm \tag{5.3.7}$$

is called the *mass moment of inertia* (or simply, the *moment of inertia*) of the rigid body about the line that passes through the center of mass C and is perpendicular to the plane of Fig. 5.4. Notice that the integrand in the expression for I_C is simply the square of the distance from point C to a mass element dm of the body and, thus, I_C is never a negative quantity. The unit vector $\hat{z} = \hat{x} \times \hat{y}$ is perpendicular to the plane of the motion in Fig. 5.4, and the quantity $\vec{\omega} \overset{\Delta}{=} \dot{\theta}\hat{z}$ that appears in Eq. (5.3.6) is the absolute angular velocity of the body in planar motion.

> The definition of moment of inertia about a line that passes through the center of mass, given by Eq. (5.3.7), is extended for any other line in space, say a line AA, as $I_A = \int_{\text{body}} |\overrightarrow{AP}|^2 \, dm$, where $|\overrightarrow{AP}|$ is the perpendicular distance from the mass element dm of the body to line AA. This general definition appears in the analysis of three-dimensional motion of rigid bodies.

The moments of inertia of a body about two parallel lines, one of which passes through the center of mass of the body, are related to each other in a relatively simple manner. Such a relation, which is presented next, is known as the *parallel axis theorem*. When the moment of inertia I_C about an axis that passes through C is known, use of the parallel axis theorem provides a simple means for calculating the moment of inertia I_A about an axis that passes through A and is parallel to the one that passes through C.

The Parallel Axis Theorem
By making use of Fig. 5.4, and of the definition of the moment of inertia, it can be shown that:

$$I_A = I_C + m|\overrightarrow{AC}|^2 \tag{5.3.8}$$

This relationship between the moments of inertia about two parallel lines, one of which passes through the center of mass of the body, is known as the *parallel axis theorem*.

The relationship between I_A and I_C given by Eq. (5.3.8) is readily developed by referring to Fig. 5.4. From the definition of the moment of inertia, we have for I_A and for I_C:

$$I_A = \int_{body} |\overrightarrow{AP}|^2 dm \qquad I_C = \int_{body} |\overrightarrow{CP}|^2 dm$$

Since $\overrightarrow{AP} = \overrightarrow{AC} + \overrightarrow{CP}$ (see Fig. 5.4), we can then write

$$|\overrightarrow{AP}|^2 = \overrightarrow{AP} \bullet \overrightarrow{AP} = \overrightarrow{AC} \bullet \overrightarrow{AC} + 2\overrightarrow{AC} \bullet \overrightarrow{CP} + \overrightarrow{CP} \bullet \overrightarrow{CP} = |\overrightarrow{AC}|^2 + 2\overrightarrow{AC} \bullet \overrightarrow{CP} + |\overrightarrow{CP}|^2$$

and this gives

$$I_A = \int_{body} |\overrightarrow{AC}|^2 dm + 2\underbrace{\int_{body} \overrightarrow{AC} \bullet \overrightarrow{CP}\, dm}_{= \overrightarrow{AC}\, \bullet\, \int_{body} \overrightarrow{CP}\, dm = 0} + \int_{body} |\overrightarrow{CP}|^2 dm = I_C + m|\overrightarrow{AC}|^2$$

The second integral in this expression is zero because of the definition of the center of mass C.

Notes

1. The absolute angular momentum $\vec{H}_A$ about an arbitrary point A is related to the absolute angular momentum $\vec{H}_C$ about the center of mass C as (see Problem 5.4)

$$\vec{H}_A = \vec{H}_C + m\overrightarrow{AC} \times \vec{v}_C$$

2. In general, the relationship between the absolute angular momentum $\vec{H}_A$ about an arbitrary point A, and the body's absolute angular velocity $\vec{\omega} = \dot{\theta}\hat{z}$, is not as simple as Eq. (5.3.6) (with I_C replaced by I_A). One would have $\vec{H}_A = I_A\dot{\theta}\hat{z}$ only when point A is inertial and the distance between A and the center of mass C of the body is constant (see Problem 5.5).

Several examples of application of the material that was presented so far are given next. Examples 5.3.1, 5.3.2, and 5.3.3 illustrate the calculations involved in finding the location of the center of mass of a body, and Examples 5.3.4 through 5.3.8 illustrate the calculation of mass moments of inertia. Examples 5.3.9 through 5.3.13 involve the dynamics of rigid bodies undergoing the plane motions addressed in this chapter.

Example 5.3.1 Three wires AB, BD, and AD, of lengths $L_1 = 3$ m, $L_2 = 4$ m, and $L_3 = \sqrt{3^2 + 4^2} = 5$ m, respectively, are welded together to form a rectangular triangle as shown in Fig. 5.5a. The wires are made of the same material (i.e., they have the same mass density ρ kg/m). Determine the location of the center of mass of the wire triangle.

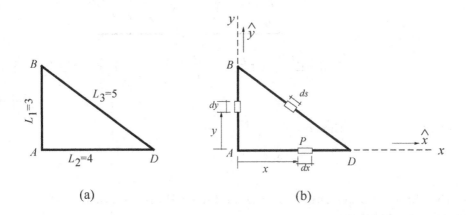

Figure 5.5: The wire triangle and its infinitesimal mass elements for Example 5.3.1.

■ **Solution**

The additional details needed for the calculations are shown in Fig. 5.5b. They are:

■ A rectangular coordinate system (x, y) that is arbitrarily chosen to be centered at A.

■ Two orthogonal unit vectors $\hat{x}$ and $\hat{y}$ that will allow us to keep track of directions.

■ Three infinitesimal line segments along the wires AD, AB, and BD. The lengths of the first two are dx and dy, respectively. Since the equation for the line that passes through points B and D is $y = L_1 - (L_1/L_2)x$, the length of the infinitesimal line segment along BD is $ds = \sqrt{(dx)^2 + [(L_1/L_2)\,dx]^2} = (L_3/L_2)\,dx$.

Solution 1: Direct Integration

Since the mass of the wires AD, AB, and BD are $\rho\,dx$, $\rho\,dy$, and $\rho\,ds$, respectively, the position vector $\overrightarrow{AC}$ for the center of mass of the bent wire is then

$$\underbrace{\rho(L_1 + L_2 + L_3)}_{\text{total mass}} \overrightarrow{AC} = \underbrace{\int\limits_{x=0}^{L_2} (x\hat{x})\, \rho\, dx}_{\text{for wire } AD} + \underbrace{\int\limits_{y=0}^{L_1} (y\hat{y})\, \rho\, dy}_{\text{for wire } AB}$$

$$+ \underbrace{\int\limits_{x=0}^{L_2} \left[x\hat{x} + \left(L_1 - \frac{L_1}{L_2}x \right) \hat{y} \right] \rho \frac{L_3}{L_2} dx}_{\text{for wire } BD}$$

$$= \frac{\rho L_2^2}{2}\hat{x} + \frac{\rho L_1^2}{2}\hat{y} + \frac{\rho L_3}{L_2} \left[\frac{L_2^2}{2}\hat{x} + \left(L_1 L_2 - \frac{L_1}{L_2}\frac{L_2^2}{2} \right) \hat{y} \right]$$

$$(5.3.1a)$$

From this equation we then obtain

$$\overrightarrow{AC} = 1.5\hat{x} + \hat{y} \text{ m}$$

Solution 2: Reducing the Problem to Three Point Masses (Indirect Integration)

Since each wire has a constant mass density, the center of mass of each wire is located at its geometric center (as also obtained by direct integration). With this observation, the problem is then reduced to finding the center of mass of a system of three mass particles: a particle of mass ρL_1 located in the middle of wire AB, a particle of mass ρL_2 located in the middle of wire AD, and a particle of mass ρL_3 located in the middle of wire BD. The position vector $\overrightarrow{AC}$ of the center of mass of this system is then given by

$$(\rho L_1 + \rho L_2 + \rho L_3)\overrightarrow{AC} = \rho L_1 \frac{L_1}{2}\hat{y} + \rho L_2 \frac{L_2}{2}\hat{x} + \rho L_3 \left(\frac{L_2}{2}\hat{x} + \frac{L_1}{2}\hat{y} \right) \quad (5.3.1b)$$

which is the same as the expression obtained in Solution 1.

Example 5.3.2 Determine the location of the center of mass C of the homogeneous triangular plate ABD shown in Fig. 5.6a. The mass of the plate is m, and its thickness is small enough (when compared to either L_1 or L_2) to be neglected in the calculations.

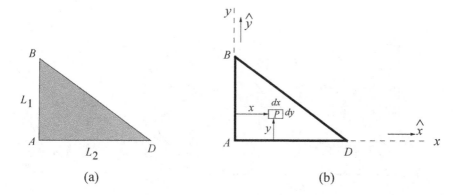

Figure 5.6: The triangular plate and the infinitesimal mass element for Example 5.3.2.

■ Solution

As shown in Fig. 5.6b, let $\overrightarrow{AP} = x\hat{x} + y\hat{y}$ be the position vector from the arbitrarily chosen reference point A to an infinitesimal mass element whose mass is equal to $dm = \rho\, dx\, dy$, where $\rho = 2m/(ab)$ is the mass density of the plate. By noticing that the boundary BD of the plate is the straight line $y = L_1 - (L_1/L_2)x$, we then obtain for the position vector $\overrightarrow{AC}$ of the center of mass of the plate

$$m\overrightarrow{AC} = \frac{2m}{L_1 L_2} \int\limits_{x=0}^{L_2} \int\limits_{y=0}^{L_1 - \frac{L_1}{L_2}x} (x\hat{x} + y\hat{y})\, dy\, dx \qquad (5.3.2a)$$

This gives

$$\overrightarrow{AC} = \frac{L_2}{3}\hat{x} + \frac{L_1}{3}\hat{y}$$

Example 5.3.3 A Composite Body

Determine the location of the center of mass C of the *composite* body (i.e., a relatively complex body that consists of several simple bodies joined to each other) shown in Fig. 5.7, based on the x-y rectangular coordinate system (centered at point A) that is also shown in the figure. If needed, use the $\hat{x}$ and $\hat{y}$ unit vectors shown in the figure. The body consists of a homogeneous triangular plate ABD welded to a homogeneous rectangular plate $DEFG$. The mass of the entire body is m kg, and its thickness is small enough to be neglected in the calculations. All dimensions shown in Fig. 5.7 are in meters.

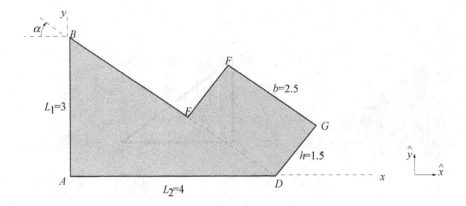

Figure 5.7: The composite body for Example 5.3.3.

■ **Solution**

The easiest way to calculate the coordinates of the center of mass for a composite body is to reduce the problem to locating the center of mass of a system of point masses (as done in Solution 2 of Example 5.3.1). For this, one needs to determine the location of the center of mass of each individual body that makes up the composite body. Here, the composite body can be divided into a homogeneous triangular plate and a homogeneous rectangular plate.

The x- and y-coordinates of the center of mass C_1 of the homogeneous triangular plate ABD are (see Example 5.3.2)

$$x_{C_1} = \frac{L_2}{3} = \frac{4}{3} \text{ m}$$

$$y_{C_1} = \frac{L_1}{3} = 1 \text{ m}$$

The center of mass C_2 of the homogeneous rectangular plate $DEFG$ is located at the middle of the plate. Since the angle α shown in Fig. 5.7 is $\alpha = \arctan \frac{3}{4}$, the x- and y-coordinates of C_2 are then (see Fig. 5.7 to understand the calculations)

$$x_{C_2} = L_2 - \frac{b}{2} \cos\alpha + \frac{h}{2} \sin\alpha = 4 - \frac{2.5}{2} \times \frac{4}{\sqrt{3^2 + 4^2}} + \frac{1.5}{2} \times \frac{3}{\sqrt{3^2 + 4^2}} = 3.45 \text{ m}$$

$$y_{C_2} = \frac{h}{2} \cos\alpha + \frac{b}{2} \sin\alpha = \frac{1.5}{2} \times \frac{4}{\sqrt{3^2 + 4^2}} + \frac{2.5}{2} \times \frac{3}{\sqrt{3^2 + 4^2}} = 1.35 \text{ m}$$

The area of the triangular plate is $A_1 = L_1 L_2/2 = 6$ m^2, and the area of the rectangular plate is $A_2 = bh = 3.75$ m^2.

The problem is now reduced to finding the center of mass of two mass particles: a particle of mass $m_1 = mA_1/(A_1 + A_2)$, located at C_1, and a particle of mass $m_2 = mA_2/(A_1 + A_2)$, located at C_2. The position vector $\overrightarrow{AC}$ of the center of mass of the composite plate is then given by

$$m\overrightarrow{AC} = m_1\overrightarrow{AC_1} + m_2\overrightarrow{AC_2} \tag{5.3.3a}$$

This gives

$$m\overrightarrow{AC} = \frac{6}{9.75}m\left(\frac{4}{3}\hat{x} + \hat{y}\right) + \frac{3.75}{9.75}m\left(3.45\hat{x} + 1.35\hat{y}\right)$$

or

$$\overrightarrow{AC} \approx 2.15\hat{x} + 1.13\hat{y} \text{ m}$$

The x- and y-coordinates of the center of mass of the composite plate are then $x_C \approx 2.15$ m and $y_C \approx 1.13$ m.

Example 5.3.4 For the three-wire triangle of Example 5.3.1, determine the moments of inertia I_A and I_C about the axes that are perpendicular to the plane of Fig. 5.5 and that pass through point A and the center of mass C of the system, respectively.

■ Solution

Solution 1
By definition, the moment of inertia I_A is given by the following expression (see the boxed text on p. 321):

$$I_A = \int_{\text{body}} \left(x^2 + y^2\right) dm = \int_{\text{body}\,AB} \left(x^2 + y^2\right) dm + \int_{\text{body}\,AD} \left(x^2 + y^2\right) dm$$

$$+ \int_{\text{body}\,BD} \left(x^2 + y^2\right) dm \tag{5.3.4a}$$

Therefore, referring to Fig. 5.5, we obtain for I_A [where $ds = (L_3/L_2)\,dx)$]

$$I_A = \int_{y=0}^{L_1} y^2\rho\,dy + \int_{x=0}^{L_2} x^2\rho\,dx + \int_{x=0}^{L_2}\left[x^2 + \left(L_1 - \frac{L_1}{L_2}x\right)^2\right]\rho\,ds$$

$$= \frac{\rho}{3}\left[L_1^3 + L_2^3 + L_3\left(L_1^2 + L_2^2\right)\right] = 72\rho \text{ kg·m}^2$$

The moment of inertia I_C may be readily calculated by using the parallel axis theorem as

$$I_C = I_A - m|\overrightarrow{AC}|^2 \qquad (5.3.4b)$$

We then obtain

$$I_C = 72\rho - (3 + 4 + 5)\rho \left(1.5^2 + 1\right) = 33\rho \text{ kg·m}^2$$

Solution 2: Using the moment of inertia about the center of mass of each bar and the parallel axis theorem

The moment of inertia $(I_C)_{\text{bar}}$ of a homogeneous bar of mass density ρ kg/m and length L is $(I_C)_{\text{bar}} = \int_{-L/2}^{L/2} \rho x^2 dx = \rho L^3/12 = mL^2/12$ (where $m = \rho L$ is the mass of the bar). Therefore, the moment of inertia I_A for the three-wire triangle may be calculated as

$$I_A = \underbrace{\frac{\rho L_1^3}{12} + (\rho L_1)\left(\frac{L_1}{2}\right)^2}_{\text{for bar } AB} + \underbrace{\frac{\rho L_2^3}{12} + (\rho L_2)\left(\frac{L_2}{2}\right)^2}_{\text{for bar } AD}$$

$$+ \underbrace{\frac{\rho L_3^3}{12} + (\rho L_3)\left[\left(\frac{L_1}{2}\right)^2 + \left(\frac{L_2}{2}\right)^2\right]}_{\text{for bar } BD} = 72\rho \text{ kg·m}^2$$

Notice that the first two underbraced terms in this result, which reduce to $\rho L_1^3/3$ and $\rho L_2^3/3$, are the moments of inertia about A of bars AB and AD, respectively. The third underbraced term is the sum of the moment of inertia $\rho L_3^3/12$ of bar BD about its center of mass, with a term that is equal to the product of the mass ρL_3 of bar BD and the square of the distance from A to the center of mass of BD.

Example 5.3.5 For the triangular plate of Example 5.3.2, determine the moments of inertia I_A and I_C about the axes that are perpendicular to the plane of the figure and that pass through point A and the center of mass of the plate, respectively.

■ **Solution**

Referring to Fig. 5.6 and to the definition of the moment of inertia, we obtain for I_A

$$I_A = \int_{\text{body}} \left(x^2 + y^2\right) dm$$

$$= \frac{m}{L_1 L_2/2}\left[\int_{x=0}^{L_2}\int_{y=0}^{L_1 - \frac{L_1}{L_2}x} x^2 dy\, dx + \int_{x=0}^{L_2}\int_{y=0}^{L_1 - \frac{L_1}{L_2}x} y^2 dy\, dx\right] = \frac{m}{6}\left(L_1^2 + L_2^2\right)$$

The moment of inertia I_C may be calculated by using the parallel axis theorem as

$$I_C = I_A - m|\overrightarrow{AC}|^2 = I_A - m\left[\left(\frac{L_1}{3}\right)^2 + \left(\frac{L_2}{3}\right)^2\right] = \frac{m}{18}\left(L_1^2 + L_2^2\right)$$

Example 5.3.6 A thin rectangular plate with sides $a = 4$ m and $b = 3$ m is welded to a thin triangular plate with sides $b = 3$ m and $L = 2$ m as shown in Fig. 5.8. Both plates are made of steel, and have the same thickness. The total mass of the system is $m = 115.5$ kg. Points C_1 and C_2 shown in the figure are the centers of mass of the two plates. Determine the following quantities for the entire plate.

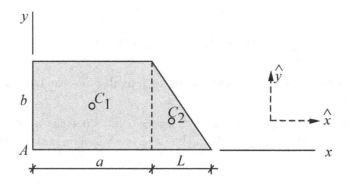

Figure 5.8: The composite plate for Example 5.3.6.

a. The x- and y-coordinates of its center of mass using the rectangular coordinate system shown in Fig. 5.8, which is centered at point A. If needed, use the two unit vectors $\hat{x}$ and $\hat{y}$ that are shown in the figure.

b. The moment of inertia I_A, about point A (more precisely, about the line that is perpendicular to the plane of the figure and passes through point A). The moment of inertia of a rectangular plate of mass m_1 and sides a and b, about its center of mass, is $m_1(a^2 + b^2)/12$.

■ **Solution**

a. The location of the center of mass of a homogeneous triangular plate was determined in Example 5.3.2, and the center of mass of a homogeneous rectangular plate is at the middle of the plate. By choosing point A as the reference point (notice that such a choice is arbitrary), and also choosing to work with the unit vectors $\hat{x}$ and $\hat{y}$ shown in Fig. 5.8, the position vectors for C_1 and C_2 relative to point A are then

$$\vec{r}_{C_1} = \overrightarrow{AC_1} = \frac{a}{2}\hat{x} + \frac{b}{2}\hat{y} = 2\hat{x} + 1.5\hat{y} \text{ m}$$

$$\vec{r}_{C_2} = \overrightarrow{AC_2} = \left(a + \frac{L}{3}\right)\hat{x} + \frac{b}{3}\hat{y} \approx 4.67\hat{x} + \hat{y} \text{ m}$$

The position vector $\vec{r}_C$ of the center of mass C of the entire system, relative to point A, is determined by using the equation that defines a center of mass, i.e.,

$$(m_1 + m_2)\vec{r}_C = m_1\vec{r}_{C_1} + m_2\vec{r}_{C_2} \tag{5.3.6a}$$

where m_1 and m_2 are the masses of the rectangular and the triangular plates, respectively.

Since both plates have the same thickness, and since the total mass of the system is $m = m_1 + m_2 = 115.5$ kg, the mass of each plate is then proportional to the plate's area:

$$m_1 = \frac{ab}{ab + bL/2}m = 92.4 \text{ kg} \qquad m_2 = \frac{bL/2}{ab + bL/2}m = 23.1 \text{ kg}$$

This then gives for $\vec{r}_C$

$$\vec{r}_C \approx 2.53\hat{x} + 1.4\hat{y} \text{ m}$$

b. As determined in Example 5.3.5, the moment of inertia of the triangular plate about its center of mass is $m_2(b^2 + L^2)/18$. Since the moment of inertia of the plate about its center of mass is already given as $m_1(a^2 + b^2)/12$ (as an exercise for yourself, do the appropriate integrations to obtain this result), we can use the parallel axis theorem to determine the moment of inertia I_A. We then have

$$I_A = [I_A]_{\text{rectangle}} + [I_A]_{\text{triangle}} \tag{5.3.6b}$$

where (see Fig. 5.8)

$$[I_A]_{\text{rectangle}} = [I_{C_1}]_{\text{rectangle}} + m_1|\overrightarrow{AC_1}|^2$$

$$= [I_{C_1}]_{\text{rectangle}} + m_1\left[\left(\frac{a}{2}\right)^2 + \left(\frac{b}{2}\right)^2\right]$$

$$[I_A]_{\text{triangle}} = [I_{C_2}]_{\text{triangle}} + m_2|\overrightarrow{AC_2}|^2$$

$$= [I_{C_2}]_{\text{triangle}} + m_2\left[\left(a + \frac{L}{3}\right)^2 + \left(\frac{b}{3}\right)^2\right]$$

With $[I_{C_2}]_{\text{triangle}} = m_2(b^2 + L^2)/18$, as determined in Example 5.3.5, and $[I_{C_1}]_{\text{rectangle}} = m_1(a^2 + b^2)/12$, we obtain

$$I_A \approx 1313 \text{ kg·m}^2$$

Example 5.3.7 A homogeneous bar of mass m is bent as shown in Fig. 5.9 (the unit vectors $\hat{x}$ and $\hat{y}$ shown in the figure are used in the calculation of the location of the center of mass C). The thickness of the cross section of the bar is very small and may be neglected.

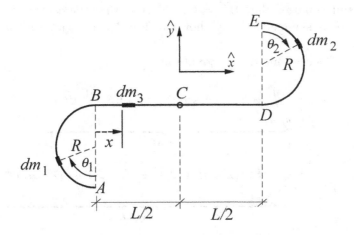

Figure 5.9: The bent bar for Example 5.3.7.

a. Verify that the center of mass of the bar is at its geometric center C shown in the figure.

b. Determine the moments of inertia I_B and I_C about the axes that are perpendicular to the plane of the figure and pass through points B and C, respectively.

■ **Solution**

a. Since the bar is homogeneous, its density (mass per unit length) is equal to $\rho = m/(L+2\pi R)$. Therefore, by dividing the bar into the curved segments AB and DE, and the straight segment BD (see Fig. 5.9), the masses of the infinitesimal mass elements for each segment are equal to $dm_1 = \rho R\, d\theta_1$, $dm_2 = \rho R\, d\theta_2$, and $dm_3 = \rho\, dx$, respectively.

By arbitrarily taking point B as the reference point, we obtain from the definition of the center of mass C

$$
m\overrightarrow{BC} = \underbrace{\int_0^\pi \left[(-R\sin\theta_1)\,\hat{x} - (R + R\cos\theta_1)\,\hat{y}\right]\rho R\, d\theta_1}_{\text{for wire } AB} + \underbrace{\int_0^L (x\hat{x})\,\rho\, dx}_{\text{for wire } BD}
$$

$$
+ \underbrace{\int_0^\pi \left[(L + R\sin\theta_2)\,\hat{x} + (R + R\cos\theta_2)\,\hat{y}\right]\rho R\, d\theta_2}_{\text{for wire } DE}
$$

$$
= \rho\,(L + 2\pi R)\frac{L}{2}\hat{x} = m\frac{L}{2}\hat{x}
$$

This result confirms that $\overrightarrow{BC} = (L/2)\hat{x}$ (i.e., that the center of mass of the homogeneous symmetric body shown in Fig. 5.9 is located at its geometric center).

b. For the moment of inertia I_B, we obtain

$$
I_B = \underbrace{\int_{\theta_1=0}^\pi \left[(R + R\cos\theta_1)^2 + (R\sin\theta_1)^2\right]\rho R\, d\theta_1}_{\text{for } AB}
$$

$$
+ \underbrace{\int_{\theta_2=0}^\pi \left[(L + R\sin\theta_2)^2 + (R + R\cos\theta_2)^2\right]\rho R\, d\theta_2}_{\text{for } DE}
$$

$$
+ \underbrace{\int_{x=0}^L x^2\rho\, dx}_{\text{for } BD} = \left[\pi R\left(L^2 + 4R^2\right) + 4LR^2 + \frac{L^3}{3}\right]\rho
$$

$$= \frac{m}{2\pi R + L}\left[\pi R\left(L^2 + 4R^2\right) + 4LR^2 + \frac{L^3}{3}\right]$$

For the moment of inertia I_C, since $I_B = I_C + m(L/2)^2$ (as per the parallel axis theorem), we then have

$$I_C = I_B - m\left(\frac{L}{2}\right)^2$$

Example 5.3.8 Figure 5.10a shows a hollow cylinder of mass m, inner radius R_1 and outer radius R_2, and length L. The cylinder is homogeneous, and its mass is m kg. Determine its mass moment of inertia about the axial axis that passes through the center C of the cylinder cross section, as shown in the figure.

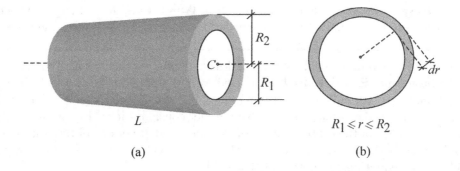

Figure 5.10: The hollow cylinder for Example 5.3.8.

■ **Solution**
A cross section of the cylinder, perpendicular to the axial axis, is shown in Fig. 5.10. By definition, the moment of inertia I_C is equal to the integral, over the entire volume of the cylinder (i.e., a triple integral), of the product of the infinitesimal mass dm and the square of the perpendicular distance from the axial axis to dm, i.e.,

$$I_C = \int_{\text{cylinder}} r^2\, dm \qquad (5.3.8a)$$

Since the cylinder is homogeneous, the triple integral that spans the entire cylinder can be replaced by a single integral if we work with a larger mass element of radius r, width dr, and length L. The side view of this mass element is shown in Fig. 5.10b. The volume dV of such a larger element is $dV = (2\pi r\, dr)\, L$, and its mass density (i.e., its mass per unit volume) is $\rho = m/(\pi R_2^2 L - \pi R_1^2 L)$. Therefore, since $dm = \rho\, dV$, the moment of inertia I_C can be obtained as

$$I_C = \frac{m}{(\pi R_2^2 L - \pi R_1^2 L)} \int_{r=R_1}^{R_2} r^2 \times (2\pi r L\, dr) = \frac{m}{2(R_2^2 - R_1^2)}\left(R_2^4 - R_1^4\right) = \frac{m}{2}\left(R_2^2 + R_1^2\right)$$

Notice that two particularly interesting cases can be extracted from this result, namely, a ring and a massive cylinder.

> *A ring or a cylindrical shell.* For this case $R_1 = R_2$, but the mass of the body is still m (i.e., the mass density is the mass per unit length, which is m/L). Thus, $I_C = mR_2^2$. This result should be expected since the perpendicular distance between any mass element of the ring and the axis through C is equal to R_2.
>
> *A massive cylinder.* For this case, $R_1 = 0$ and, thus, $I_C = mR_2^2/2$ (which happens to be one half of the moment of inertia of a ring with the same mass).

The following examples deal with dynamics of simpler planar motion of rigid bodies. The basic equations that need to be used were developed in Section 5.3. They are Eq. (5.3.2) (p. 319) and either one of the equations of (5.3.5) (p. 320) (whichever is simpler to apply to any particular problem under consideration). Equation (5.3.2) governs the motion of the center of mass of the body and either one of the equations of (5.3.5) governs the rotational motion of the body. Together, those equations yield the differential equations of motion of the body.

Example 5.3.9 The system shown in Fig. 5.11 consists of a rigid body of mass m kg, pinned to a fixed point O. Motion takes place on a vertical plane, and the body is free to rotate about O, with no friction. The distance between O and the center of mass C of the body is r_C m. The body is subjected to an external force $P(t)$ that is always perpendicular to line OC, applied at a distance D from O. The motion is started with given initial conditions $\theta(0) = \theta_0$ and $\dot{\theta}(0) = \dot{\theta}_0$.

    a. Obtain the equations for determining the support reaction forces at O, along OC and perpendicular to OC, and the differential equation that governs the resulting motion for the system. Describe, in a few words, the necessary steps for calculating the support reactions as a function of time.

    b. Is it possible to predict how the component of the support reaction perpendicular to OC changes with the angle θ? If so, determine how.

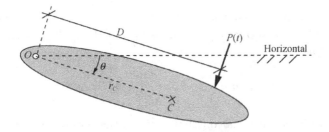

Figure 5.11: A pinned rigid body subjected to a force $P(t)$ for Example 5.3.9.

■ Solution

a. The free-body diagram for the system, using reaction force components along line OC and along the perpendicular to OC, is shown in Fig. 5.12. Two rotating unit vectors, $\hat{r}$ and $\hat{\theta}$, are also shown in the figure. In terms of these unit vectors, the expression for the absolute acceleration of the center of mass C is $\vec{a}_C = r_C(\ddot{\theta}\hat{\theta} - \dot{\theta}^2\hat{r})$.

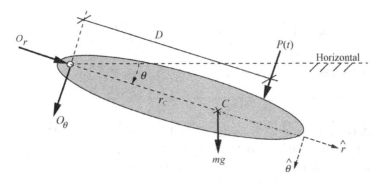

Figure 5.12: Free-body diagram for Example 5.3.9.

From Newton's second law $\vec{F} = m\vec{a}_C$, the following equations are obtained (see Fig. 5.12):

$$O_r + mg\sin\theta = -mr_C\dot{\theta}^2 \tag{5.3.9a}$$
$$O_\theta + mg\cos\theta + P = mr_C\ddot{\theta} \tag{5.3.9b}$$

These equations involve three unknowns, namely, $\theta(t), O_r(t)$, and $O_\theta(t)$. Therefore, one more equation is needed. The third equation is the moment equation.

By taking moments about the inertial point O, the following equation that does not involve the support reactions is obtained (see Fig. 5.12, and use the right-hand rule).

$$PD + mgr_C \cos\theta = \frac{d}{dt}\left(I_O \dot\theta\right) = I_O \ddot\theta \qquad (5.3.9c)$$

This is the differential equation that governs the motion of the system. The quantity I_O that appears in this equation is the moment of inertia of the body about an axis that passes through point O and is perpendicular to the plane of Fig. 5.12. For a homogeneous thin bar of length L, for example, $I_O = mL^2/3$.

To calculate the support reactions in terms of time t, Eq. (5.3.9c) needs to be integrated with the known initial conditions to determine $\theta(t)$. This is done numerically since that equation is nonlinear in $\theta(t)$. Equations (5.3.9a) and (5.3.9b) are then used with the numerical integration for calculating O_r and O_θ.

b. By combining Eqs. (5.3.9b) and (5.3.9c), the following expression is obtained for the component O_θ of the support reaction.

$$O_\theta = \left(\frac{mr_C^2}{I_O} - 1\right) mg \cos\theta + \left(\frac{mDr_C}{I_O} - 1\right) P \qquad (5.3.9d)$$

Equation (5.3.9d) discloses that O_θ changes with θ as $\cos\theta$.

Example 5.3.10 The system considered in Example 2.7.6 (p. 154) consisted of a particle of mass m attached to the end B of a massless bar AB. What is the differential equation of motion of the system if the massless bar and the point mass are replaced by a thin homogeneous bent bar ABD, of mass m and total length $L = L_1 + L_2$, as shown in Fig. 5.13? The support point A is given a known displacement $x(t)$ m as indicated in the figure.

■ **Solution**

Two free-body diagrams for the bent bar are shown in Fig. 5.14. They are equivalent to each other, and one can choose to work with either one of them. Two rotating unit vectors, $\hat{r}$ and $\hat\theta$, and a nonrotating unit vector $\hat{x}$ are also shown in the figure.

Point C in Fig. 5.14 is the center of mass of the bent bar. With

$$m_1 = \frac{mL_1}{L_1 + L_2} \qquad\qquad m_2 = \frac{mL_2}{L_1 + L_2} \qquad (5.3.10a)$$

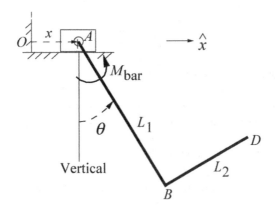

Figure 5.13: A pendulum on a moving support for Example 5.3.10.

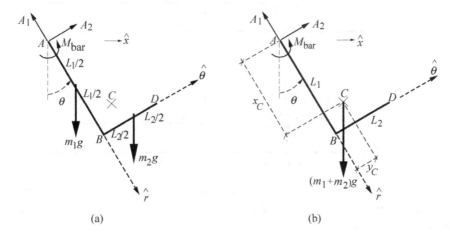

Figure 5.14: Two equivalent free-body diagrams for Example 5.3.10 (point C is the center of mass of the system).

the location of point C is determined as follows:

$$m\overrightarrow{AC} = m_1 \frac{L_1}{2}\hat{r} + m_2(L_1\hat{r} + \frac{L_2}{2}\hat{\theta}) \triangleq m(x_C\hat{r} + y_C\hat{\theta})$$

This gives for the distances x_C and y_C shown in Fig. 5.14

$$x_C = \frac{L_1}{L_1 + L_2}\left(L_2 + \frac{L_1}{2}\right) \qquad (5.3.10b)$$

$$y_C = \frac{L_2^2}{2(L_1 + L_2)} \qquad (5.3.10c)$$

The support point A shown in Fig. 5.14 is not inertial and, thus, $\vec{M}_A \neq d\vec{H}_A/dt$. If we take moments about the center of mass C, and use the basic equation $\vec{M}_C = d\vec{H}_C/dt$, where $\vec{H}_C = I_C\dot{\theta}\,\hat{r}\times\hat{\theta}$, we obtain the following equation (see Fig. 5.14, and use the right-hand rule to obtain the correct signs as they appear in the equation):

$$I_C\ddot{\theta} = M_{\text{bar}} - A_1 y_C - A_2 x_C \qquad (5.3.10d)$$

The expression for the moment of inertia I_C is obtained as follows:

$$I_C = \underbrace{m_2\frac{L_2^2}{12}}_{\substack{\text{moment of inertia of bar } BD \\ \text{about its center of mass}}} + \underbrace{m_2\left[(L_1 - x_C)^2 + \left(\frac{L_2}{2} - y_C\right)^2\right]}_{\substack{m_2 \text{ times the square of the distance from } C \\ \text{to the center of mass of line } BD}}$$

The sum of these two terms is I_C for bar BD

$$+ \underbrace{m_1\frac{L_1^2}{12}}_{\substack{\text{moment of inertia of bar } AB \\ \text{about its center of mass}}} + \underbrace{m_1\left[\left(x_C - \frac{L_1}{2}\right)^2 + y_C^2\right]}_{\substack{m_1 \text{ times the square of the distance from } C \\ \text{to the center of mass of line } AB}}$$

The sum of these two terms is I_C for bar AB

where use is made of the parallel axis theorem for bars BD and AB in each step of the calculation. The expressions $m_1 L_1^2/12$ and $m_2 L_2^2/12$ were first obtained in Solution 2 of Example 5.3.4 (p. 327).

This expression is good enough for generating a numerical value for I_C, but a much simpler expression for I_C is obtained if the expressions for m_1 (which is the mass of bar AB) and m_2 (which is the mass of bar BD) given by the two equations in (5.3.10a) are substituted in this equation. The following simpler expression for I_C is obtained after some algebraic manipulations:

$$I_C = \frac{m}{12}(L_1 + L_2)^2 - \frac{m}{2}\left(\frac{L_1 L_2}{L_1 + L_2}\right)^2 \qquad (5.3.10e)$$

Now, use Newton's second law for the motion of the center of mass, $F = m\vec{a}_C = md\vec{v}_C/dt$, where $\vec{v}_C = \dot{x}\hat{x} + d(x_C\hat{r} + y_C\hat{\theta})/dt$ as seen from Fig. 5.14. With $\hat{z} \overset{\Delta}{=} \hat{r} \times \hat{\theta}$, and noticing that $\hat{z} \times \hat{r} = \hat{\theta}$ and $\hat{z} \times \hat{\theta} = -\hat{r}$, the following expressions are then obtained for $\vec{v}_C$ and $\vec{a}_C$:

$$
\begin{aligned}
\vec{v}_C &= \dot{x}\hat{x} + \frac{d}{dt}\left(x_C\hat{r} + y_C\hat{\theta}\right) = \dot{x}\hat{x} + \dot{\theta}\hat{z} \times \left(x_C\hat{r} + y_C\hat{\theta}\right) \\
&= \dot{x}\hat{x} + \dot{\theta}\left(x_C\hat{\theta} - y_C\hat{r}\right) \qquad (5.3.10\mathrm{f})
\end{aligned}
$$

$$
\begin{aligned}
\vec{a}_C &= \frac{d}{dt}\vec{v}_C = \ddot{x}\hat{x} + \ddot{\theta}\left(x_C\hat{\theta} - y_C\hat{r}\right) + \dot{\theta}\hat{z} \times \left(x_C\dot{\theta}\hat{\theta} - y_C\dot{\theta}\hat{r}\right) \\
&= \ddot{x}\hat{x} + \left(x_C\ddot{\theta} - y_C\dot{\theta}^2\right)\hat{\theta} - \left(y_C\ddot{\theta} + x_C\dot{\theta}^2\right)\hat{r} \qquad (5.3.10\mathrm{g})
\end{aligned}
$$

Since $\hat{x} = \hat{r}\sin\theta + \hat{\theta}\cos\theta$ (as disclosed by Fig. 5.14), $\vec{F} = m\vec{a}_C$ yields the following additional equations.

$$
m\left(\ddot{x}\sin\theta - y_C\ddot{\theta} - x_C\dot{\theta}^2\right) = mg\cos\theta - A_1 \qquad (5.3.10\mathrm{h})
$$

$$
m\left(\ddot{x}\cos\theta + x_C\ddot{\theta} - y_C\dot{\theta}^2\right) = A_2 - mg\sin\theta \qquad (5.3.10\mathrm{i})
$$

Finally, by combining Eqs. (5.3.10d), (5.3.10h), and (5.3.10i), the following differential equation that governs the motion of the bent bar is obtained (where $\ddot{x}$ is a prescribed function, such as $\ddot{x} = \text{constant} \stackrel{\Delta}{=} -a_A$ of any known function of time t).

$$
\begin{aligned}
\left[I_C + m\left(x_C^2 + y_C^2\right)\right]\ddot{\theta} &= M_{\text{bar}} + m\left(y_C\sin\theta - x_C\cos\theta\right)\ddot{x} \\
&\quad - mg\left(x_C\sin\theta + y_C\cos\theta\right) \qquad (5.3.10\mathrm{j})
\end{aligned}
$$

Equation (5.3.10j), which only involves θ and the applied moment M_{bar}, is the differential equation of motion of the system. The analysis of the motion is done in Example 5.3.11.

Example 5.3.11 The support point A of the system considered in Example 5.3.10 is given an acceleration $\ddot{x} = \text{constant} = -2g$ m/s^2. For $M_{\text{bar}} = 0$ and $L_2 = 2L_1$:

a. Plot the angular velocity $\dot{\theta}$ versus θ and describe, by observing the plot, the types of motion the system may exhibit.

b. Determine the maximum and minimum values of the angle $\theta(t)$ if the motion is started with the initial conditions $\theta(0) = 90°$ and $\dot{\theta}(0) = 0$.

c. Determine the equilibrium value (or values if there is more than one equilibrium solution) for the angle θ.

d. Determine the frequency of the small oscillations of the system about the stable equilibrium values of the angle θ.

■ **Solution**

a. From Eqs. (5.3.10b), (5.3.10c), and (5.3.10e) developed in Example 5.3.10 we obtain

$$x_C = \frac{5}{6}L_1 \qquad y_C = \frac{2}{3}L_1 \qquad I_C = \frac{19}{36}mL_1^2$$

The differential equation that governs the motion of the system, Eq. (5.3.10j), then becomes

$$\frac{5}{3}\ddot{\theta} + \frac{g}{L_1}\left(\frac{13}{6}\sin\theta - \cos\theta\right) = M_{\text{bar}} = 0 \qquad (5.3.11a)$$

Since Eq. (5.3.11a) is of the form $\ddot{\theta} + f(\theta) = 0$, it can be integrated analytically once (as shown in Chapter 1 and in several examples in Chapter 2) to yield

$$\dot{\theta}^2 - \frac{6g}{5L_1}\left(\frac{13}{6}\cos\theta + \sin\theta\right) = \text{constant} \overset{\Delta}{=} \frac{g}{L_1}H$$

$$= \left[\dot{\theta}^2 - \frac{6g}{5L_1}\left(\frac{13}{6}\cos\theta + \sin\theta\right)\right]_{t=0} \qquad (5.3.11b)$$

For the initial conditions $\dot{\theta} = 0$ and $\theta(0) = 90°$, the constant H in Eq. (5.3.11b) is equal to $-6/5$. The plot of the nondimensional angular velocity $d\theta/\tau$ (where $\tau = t\sqrt{g/L_1}$ is a nondimensional time) versus the angle θ is shown in Fig. 5.15. The corresponding values for the constant H associated with each one of the four curves shown in the plot are indicated in the figure.

b. Observation of Fig. 5.15 discloses that the system may exhibit two types of motion depending on the initial conditions of the motion:

1. One type of motion, which corresponds to the closed curves in Fig. 5.15, is an oscillation about the point on the θ axis marked with a •. That point, which corresponds to $\theta = \text{constant} \overset{\Delta}{=} \theta_e \approx 24.77°$, is an equilibrium solution, and that equilibrium is, therefore, stable (because a motion that starts in its neighborhood stays in that neighborhood). Other stable equilibrium points are seen to be $24.77 \pm 360n$ degrees, where n=1,2, ..., but these new points correspond to the same physical orientation of line AB in Fig. 5.13.

   The maximum and minimum values of the angle θ for the motion just described correspond to $\dot{\theta} = 0$ in Eq. (5.3.11b). For the initial conditions given in the problem statement, which correspond to $H = -\frac{6}{5}$

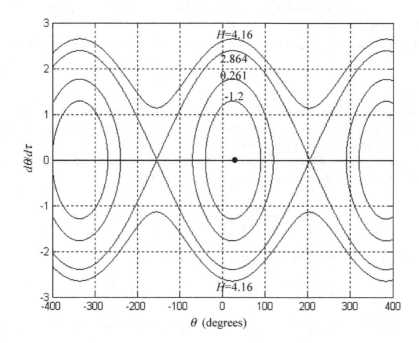

Figure 5.15: Plot of $d\theta/d\tau$ versus θ, where $\tau = t\sqrt{g/L_1}$ is a nondimensional time for Example 5.3.11.

as obtained earlier, that equation yields $\theta_{max} = \pi/2$ rad $= 90°$ and $\theta_{min} \approx -0.71$ rad $\approx -40.7°$. Reasonably accurate values for θ_{max} and θ_{min} could, of course, be extracted from the plot in Fig. 5.15. That figure also discloses that $\theta \approx 24.77 + 180 = 204.77°$ is an equilibrium point, and that equilibrium is unstable.

2. The other type of motion the system can exhibit corresponds to open curves in Fig. 5.15. Two such curves (for which $H = 4.16$) are shown in that figure. These motions occur when the system is given a big enough initial angular velocity $\dot{\theta}$, either clockwise or counterclockwise, causing the system to rotate about the support point A shown in Fig. 5.13.

c. The equilibrium solutions for the angle θ are obtained by setting $\theta = $ constant $\overset{\Delta}{=} \theta_e$ in the differential equation of motion, Eq. (5.3.11a). They correspond to

$$\frac{13}{6}\sin\theta_e - \cos\theta_e = 0 \quad \text{or} \quad \tan\theta_e = \frac{6}{13}$$

This gives two physically distinct solutions $\theta_e \approx 24.77°$ and $\theta_e \approx 204.77°$, which are the ones mentioned earlier.

d. To determine the frequency of the small oscillations about the stable equilibrium, let

$$\theta(t) \stackrel{\Delta}{=} \theta_e + \theta_s(t)$$

and expand in Taylor series, for small θ_s, the nonlinear terms $\sin\theta = \sin(\theta_e + \theta_s)$ and $\cos\theta = \cos(\theta_e + \theta_s)$ that appear in the differential equation of motion, Eq. (5.3.11a), and disregard powers in the small quantities that appear in the expansion. The resulting approximate differential equation is the linearized differential equation of motion:

$$\frac{5}{3}\ddot{\theta}_s + \frac{g}{L_1}\left(\frac{13}{6}\cos\theta_e + \sin\theta_e\right)\theta_s = 0$$

Specifically, we then have for each equilibrium:

for $\theta_e = 24.77°$:
$$\frac{5}{3}\ddot{\theta}_s + 2.39\frac{g}{L_1}\theta_s = 0$$

for $\theta_e = 204.77°$:
$$\frac{5}{3}\ddot{\theta}_s - 2.39\frac{g}{L_1}\theta_s = 0$$

The linearized differential equation discloses that the equilibrium $\theta_e = 204.77°$ is unstable (as already concluded from Fig. 5.15), and that the equilibrium $\theta_e = 24.77°$ is infinitesimally stable (it would be asymptotically stable if there were damping in the system). Actually, inspection of the $\dot{\theta}$ versus θ plot shown in Fig. 5.15 already disclosed that the equilibrium $\theta_e = 24.77°$ is stable, instead of simply infinitesimally stable.

As disclosed by the linearized differential equation $\frac{5}{3}\ddot{\theta}_s + 2.39\frac{g}{L_1}\theta_s = 0$, the frequency of the small oscillations about the stable equilibrium $\theta_e = 24.77°$ is equal to

$$\omega = \sqrt{\frac{3 \times 2.39}{5}\frac{g}{L_1}} \approx 1.2\sqrt{\frac{g}{L_1}} \text{ rad/s}$$

For $L_1 = 1$ m, for example, the frequency is $\omega \approx 3.75$ rad/s.

The motion of the bent bar can be simulated and animated by numerically integrating Eq. (5.3.10j) (p. 339). By making use of the expressions for x_C, y_C, and I_C given by Eqs. (5.3.10b), (5.3.10c), and (5.3.10e), respectively, and introducing, for generality, a linear viscous damping moment $-c\dot{\theta}\hat{z}$ (where $c > 0$ is the viscous damping coefficient) applied to the pin at point

A (see Fig. 5.13), Eq. (5.3.10j) can be rearranged and written in terms of the nondimensional time $\tau = t\sqrt{g/L_1}$ as:

$$\left[1 + 3\frac{L_2}{L_1} + \left(\frac{L_2}{L_1}\right)^3\right]\theta''$$

$$+ \frac{3}{2}\left\{\left[1 + 2\frac{L_2}{L_1} - \frac{\ddot{x}}{g}\left(\frac{L_2}{L_1}\right)^2\right]\sin\theta + \left[\left(1 + 2\frac{L_2}{L_1}\right)\frac{\ddot{x}}{g} + \left(\frac{L_2}{L_1}\right)^2\right]\cos\theta\right\}$$

$$= 3\left(1 + \frac{L_2}{L_1}\right)\left(\frac{M_{\text{bar}}}{mgL_1} - c^*\theta'\right) \tag{5.3.11c}$$

where $c^* = c/(mL_1\sqrt{gL_1})$ and primes denote differentiation with respect to τ.

$\Longrightarrow$ For Scilab users, the program *sim_model_ex5_3_11.sci* listed in Appendix E integrates Eq. (5.3.11c), plots θ versus τ, and animates the motion as viewed by an observer fixed to the moving support at point A. For Matlab users, the similar program *sim_model_ex5_3_11.m* that is also listed in Appendix E will do the same. An example of their use with $L_2/L_1 = 2$, $\ddot{x}/g = -2$, $M_{\text{bar}}/(mgL_1) = 0$, $c^* = 0$, and the initial conditions $\theta(0) = \pi/2$ and $[d\theta/d\tau]_{\tau=0} = 0$, as given in the comment lines at the beginning of each program, is
In Scilab: `my_output=sim_model_ex5_3_11([0:0.005:10],2,-2,0,0,%pi/2,0);`
In Matlab: `my_output=sim_model_ex5_3_11([0:0.002:10],2,-2,0,0,pi/2,0);`
Such a call generates a three-column matrix `my_output` whose columns are the nondimensional time values specified in the first element of the call to the function *sim_model_ex5_3_11*, θ, and $d\theta/d\tau$, which may then be used for creating plots as desired.

Those who have Simulink and prefer to use it may want to construct the block diagram model shown in Fig. 5.16 and use the suggestion indicated for the Solver settings in the *Simulation* $\rightarrow$ *Parameters* menu. To see an animation of the motion after the simulation stops, type `animate_nbars(x_coordinates, y_coordinates)` in the Matlab command window (assuming you have stored the Matlab program `animate_nbars` in your computer).

Example 5.3.12 A thin homogeneous rectangular plate of mass m_2 kg, with dimensions b and h m, and a thin homogeneous bar of mass m_1 kg and length L m, are welded at the center of mass C_2 of the plate. The other side of the bar is connected to a fixed bearing at point O as shown in Fig. 5.17. Friction in the bearing is of the linear viscous type with friction coefficient c_1 N·m/(rad/s). The mass moments of inertia for the plate and the bar are given in the figure. The motion of the system

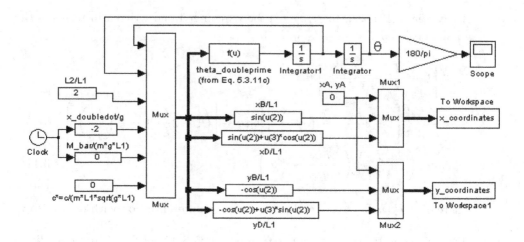

Figure 5.16: A Simulink model for integrating Eq. (5.3.11c) for Example 5.3.11. For the model inputs shown, the time array [0:0.002:10] is suggested for the *Simulation* → *Parameters* settings.

takes place on a vertical plane. The angle ϕ, measured from a fixed vertical line, is used for describing the orientation of the system. The angle θ is constant since the plate and the bar are welded to each other.

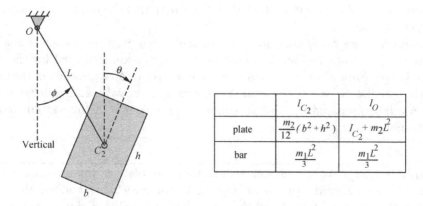

	I_{C_2}	I_O
plate	$\frac{m_2}{12}(b^2+h^2)$	$I_{C_2}+m_2 L^2$
bar	$\frac{m_1 L^2}{3}$	$\frac{m_1 L^2}{3}$

Figure 5.17: Figure for Example 5.3.12, and mass moments of inertia.

a. Obtain the expressions for the angular momentum of the entire system about C_2 and about O.

b. What is the differential equation of motion for the system?

■ Solution

a. Let $\hat{z}$ be a unit vector perpendicular to the plane of Fig. 5.17, and coming out of the paper toward you, the reader. The absolute angular velocity of the rigid system is then

$$\vec{\omega} = \dot{\phi}\hat{z} \qquad (5.3.12a)$$

Therefore, by noticing that $(I_O)_{\text{bar}} = (I_{C_2})_{\text{bar}}$ (because both C_2 and O are endpoints of the bar), the angular momentum vectors $\vec{H}_{C_2}$ and $\vec{H}_O$ for the entire rigid system are calculated as follows:

$$\vec{H}_{C_2} = (I_{C_2})_{\text{plate}}\,\vec{\omega} + (I_{C_2})_{\text{bar}}\,\vec{\omega} = \left[\frac{m_2(b^2 + h^2)}{12} + \frac{m_1 L^2}{3}\right]\dot{\phi}\hat{z} \qquad (5.3.12b)$$

$$\vec{H}_O = (I_O)_{\text{plate}}\,\vec{\omega} + (I_O)_{\text{bar}}\,\vec{\omega} = \left[\underbrace{\frac{m_2(b^2 + h^2)}{12} + m_2 L^2}_{\substack{\text{as per the parallel} \\ \text{axis theorem}}} + \frac{m_1 L^2}{3}\right]\dot{\phi}\hat{z}$$

$$(5.3.12c)$$

b. The differential equation of motion for the system may be obtained either from $\vec{M}_O = d\vec{H}_O/dt$ or from $\vec{M}_{C_2} = d\vec{H}_{C_2}/dt$. The support reactions at O consist of a force $\vec{F}_O$ and of the moment $-c_1\dot{\phi}\hat{z}$ due to viscous friction in the bearing at O. Since the moment about C_2 of force $\vec{F}_O$ is not zero, the equation $\vec{M}_{C_2} = d\vec{H}_{C_2}/dt$ will involve three unknowns, namely, the angle ϕ and two components of the reaction force $\vec{F}_O$. Thus, using the equation $\vec{M}_{C_2} = d\vec{H}_{C_2}/dt$ will also require the use of Newton's second law for the center of mass of the system to obtain two additional equations involving those same unknowns. Therefore, to obtain the differential equation of motion for the system, it is much simpler to use, instead, the equation $\vec{M}_O = d\vec{H}_O/dt$ since this will only involve the desired unknown $\phi(t)$.

The moment $\vec{M}_O$ (which consists of the reaction moment at the bearings, and of the moment due to the gravitational forces) and the differential equation of motion that is immediately obtained from $\vec{M}_O = d\vec{H}_O/dt$ are

$$\vec{M}_O = -c_1\dot{\phi}\hat{z} - m_1 g\frac{L}{2}(\sin\phi)\,\hat{z} - m_2 g L(\sin\phi)\,\hat{z} \qquad (5.3.12d)$$

$$\left[\frac{m_2(b^2 + h^2)}{12} + m_2 L^2 + \frac{m_1 L^2}{3}\right]\ddot{\phi} = -c_1\dot{\phi} - \left(m_1 g\frac{L}{2} + m_2 g L\right)\sin\phi$$

$$(5.3.12e)$$

If the system were a simple pendulum consisting of a particle of mass m_2 connected at the end C_2 of a massless bar of length L, the differential equation of motion would simply be

$$m_2 L^2 \ddot{\phi} + c_1\dot{\phi} + m_2 g L\sin\phi = 0 \qquad (5.3.12f)$$

The effect of having a finite body of mass m_2 connected at C_2, and the bar OC_2 having mass was simply to change the coefficients of $\ddot{\phi}$ and $\sin\phi$ in Eq. (5.3.12f). The coefficient of $\dot{\phi}$ was not changed because the term $c_1\dot{\phi}$ is dependent only on the viscous damper, and not on the physical characteristics of any of the bodies.

Suggested problem: Choose your own values for the system parameters (plate dimensions, masses, etc.) and determine the frequencies of the damped and undamped oscillations about the stable equilibrium $\phi = 0$ (we already know it is stable because of similar problems in Chapter 2). Compare the answers with those you would have if the plate were to be replaced by a point mass at C_2.

Example 5.3.13 Instead of being welded to bar OC_2, the plate of Example 5.3.12 is now connected at C_2 by a bearing, and friction in that bearing is also of the linear viscous type, with friction coefficient c_2 N·m/(rad/s). The motion of the system is started with the initial conditions $\phi(0) = 20°$, $\dot{\phi}(0) = \dot{\phi}_0$, $\theta(0) = 32°$, and $\dot{\theta}(0) = \dot{\theta}_0$. Obtain the differential equations of motion for the system, and:

a. Determine the angle $\theta(t)$ for the case when $c_2 = 0$ and $\dot{\theta}_0 = 0$.
b. Verify that $\phi = 0$ is a stable equilibrium.
c. Determine the frequency Ω of the undamped linearized oscillations of the system about the equilibrium $\phi = 0$, in terms of m_1, m_2, L, b, and h.
d. Rewrite the differential equations of motion in terms of a nondimensional time $\tau \stackrel{\Delta}{=} \Omega t$, with Ω being the frequency obtained in part c. The damping coefficients c_1 and c_2 should appear in the resulting equations as the nondimensional quantities $c_1^* = \Omega c_1/(m_1 g L/2 + m_2 g L)$ and $c_2^* = \Omega c_2/(m_1 g L/2 + m_2 g L)$, respectively. Integrate the resulting equations numerically, and plot the angles $\phi(\tau)$ and $\theta(\tau)$ versus τ, for $0 \leq \tau \leq 50$. To do this, use $(b^2 + h^2)/L^2 = 7/6$,

$m_1/m_2 = 1/2$, $c_1^* = 0.1$, $c_2^* = 0.1$, and the initial conditions $\phi(0) = 20°$, $\theta(0) = 32°$, $[d\phi/d\tau]_{\tau=0} = 0$, and $[d\theta/d\tau]_{\tau=0} = 50$.

■ **Solution**

a. Unlike Problem 5.3.12, the system is no longer a rigid body. The free-body diagram for each rigid body component of the system is shown in Fig. 5.18. The figure shows a convenient unit vector triad for expressing the absolute velocity and acceleration of point C_2, and the absolute angular velocity of each body. The moments M_1 and M_2 shown in the free-body diagrams are the reaction moments due to the friction in the system. The friction moment $\vec{M}_2$ that is acting on the plate is proportional to the angular velocity of the plate relative to the bar. Thus, $\vec{M}_2 = -c_2[\dot\theta\hat z - (-\dot\phi\hat z)] = -c_2(\dot\theta + \dot\phi)\hat z$, which accounts for the way it is shown in Fig. 5.18a.

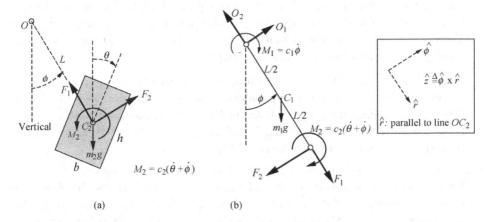

(a)   (b)

Figure 5.18: Free-body diagrams for the system of Example 5.3.13.

Notice that the two free-body diagrams involve six unknowns, namely, ϕ, θ, F_1, F_2, O_1, and O_2, and six scalar equations can be generated by using both free-body diagrams (i.e., each diagram yields one moment equation and two equations obtained from $\vec{F} = m\vec{a}$ for the center of mass of the body).

The following equations are obtained for the free-body diagram of the plate shown in Fig. 5.18a.

$$\frac{m_2}{12}\left(b^2 + h^2\right)\ddot\theta = -M_2 = -c_2\left(\dot\theta + \dot\phi\right) \tag{5.3.13a}$$

$$F_1 - m_2 g\cos\phi = m_2 L\dot\phi^2 \tag{5.3.13b}$$

$$F_2 - m_2 g\sin\phi = m_2 L\ddot\phi \tag{5.3.13c}$$

Since the absolute angular velocity of the plate is $\dot{\theta}\hat{z}$ (see Fig. 5.18), the moment equation for the plate is $\vec{M}_{C_2} = d[(I_{C_2})_{\text{plate}}\dot{\theta}\hat{z}]/dt$, where $\vec{M}_{C_2} = -c_2(\dot{\theta} + \dot{\phi})\hat{z}$ and $(I_{C_2})_{\text{plate}} = m_2(b^2 + h^2)/12$, as given in Example 5.3.12.

Equation (5.3.13a) involves only the unknown coordinates θ and ϕ and, thus, is one of the two differential equations of motion for the system. It can be integrated analytically once to yield

$$\frac{m_2}{12}\left(b^2 + h^2\right)\dot{\theta} + c_2\left(\theta + \phi\right) = \text{constant} = \left[\frac{m_2}{12}\left(b^2 + h^2\right)\dot{\theta} + c_2\left(\theta + \phi\right)\right]_{t=0}$$

(5.3.13d)

Equation (5.3.13d) immediately discloses that the absolute angular momentum of the plate, $(I_{C_2})_{\text{plate}}\dot{\theta}$, is conserved [i.e., $(I_{C_2})_{\text{plate}}\dot{\theta}(t) = \text{constant} = (I_{C_2})_{\text{plate}}\dot{\theta}(0)$] if there is no damping in the pin at point C_2 (i.e., if $c_2 = 0$). Thus, the absolute angular velocity of the plate is constant in such a case and equal to its initial absolute angular velocity.

For $\dot{\theta} = 0$ (i.e., if the initial absolute angular velocity of the plate is zero), as specified in part a, the plate maintains a fixed direction in space for all times if $c_2 = 0$ [i.e., the angle θ shown in Fig. 5.18 remains constant and equal to $\theta(0) = 32°$ during the entire motion of the system]. Notice that the solution to part a, with $\dot{\theta}(0) = 0$, only involved Eq. (5.3.13a).

b. Another differential equation is needed since the problem involves two unknown coordinates, i.e., θ and ϕ . The easiest and quickest way to obtain the second differential equation of motion is to use the free-body diagram for the bar shown in Fig. 5.18b, take moments about point O to obtain an expression for F_2, and then use the resulting expression in Eq. (5.3.13c).

The moment equation for the bar is $\vec{M}_O = d[-(I_O)_{\text{bar}}\dot{\phi}\hat{z}]/dt$, where $(I_O)_{\text{bar}} = m_1 L^2/3$ as determined in Example 5.3.4 and also given in Example 5.3.12, and $\vec{M}_O$ is given as follows:

$$\vec{M}_O = \left[c_1\dot{\phi} + c_2\left(\dot{\theta} + \dot{\phi}\right) + m_1 g\frac{L}{2}\sin\phi + LF_2\right]\hat{z}$$

(5.3.13e)

Such an expression is obtained by inspecting Fig. 5.18b. Therefore, the approach just described yields

$$\frac{m_1 L^2}{3}\ddot{\phi} = -c_1\dot{\phi} - c_2\left(\dot{\theta} + \dot{\phi}\right) - m_1 g\frac{L}{2}\sin\phi - LF_2$$

(5.3.13f)

By solving Eq. (5.3.13f) for F_2 and substituting the result into Eq. (5.3.13c), the following differential equation is obtained:

$$\left(\frac{m_1 L^2}{3} + m_2 L^2\right)\ddot{\phi} + \left(m_1 g \frac{L}{2} + m_2 g L\right)\sin\phi + c_1\dot{\phi} + c_2\left(\dot{\theta} + \dot{\phi}\right) = 0$$

$$(5.3.13g)$$

Equations (5.3.13a) and (5.3.13g) [or (5.3.13d) and (5.3.13g)] are the differential equations that govern the motion of the system in terms of the two unknowns $\phi(t)$ and $\theta(t)$. Notice that there was no need to apply Newton's second law to the free-body diagram for bar OC_1 to answer the questions that were asked in this problem. However, we would need to do so if we wanted to find the support reactions O_1 and O_2 at point O.

c. For the undamped system $c_1 = c_2 = 0$, and Eq. (5.3.13g) reduces to

$$\left(\frac{m_1 L^2}{3} + m_2 L^2\right)\ddot{\phi} + \left(m_1 g \frac{L}{2} + m_2 g L\right)\sin\phi = 0 \qquad (5.3.13h)$$

while $\dot{\theta} = \text{constant} = \dot{\theta}_0$, as indicated by Eq. (5.3.13d). The equilibrium solutions to Eq. (5.3.13h) are $\phi = \text{constant} \triangleq \phi_e$, with $\phi_e = 0$ and $\phi_e = \pi$. Only the equilibrium $\phi_e = 0$ is stable, and the frequency Ω of the undamped (i.e., $c_1 = c_2 = 0$) linearized ϕ-oscillation is

$$\Omega = \sqrt{\frac{m_2 + m_1/2}{m_2 + m_1/3}\frac{g}{L}} \qquad (5.3.13i)$$

Notice that as $m_1/m_2 \to \infty$ this expression yields $\Omega \to \sqrt{3g/(2L)}$, which agrees with the frequency of a pendulum that consists of a homogeneous bar of length L. On the other hand, as $m_2/m_1 \to \infty$, $\Omega \to \sqrt{g/L}$, which is the frequency of a pendulum that consists of a massless bar of length L with a mass particle at the end of the bar.

d. Equations (5.3.13a) and (5.3.13g) [or (5.3.13d) and (5.3.13g)] for the motion coordinates $\phi(t)$ and $\theta(t)$ involve five dimensional coefficients $m_1 L^2/3$, $(m_2 + m_1/2)gL$, c_1, c_2, and $m_2/(b^2 + h^2)$. As indicated in Chapter 2, the number of coefficients in the equations can be reduced if nondimensional variables are used (notice, however, that this is an optional step in any analysis). Since the variables ϕ and θ are already nondimensional, time is the only variable to be nondimensionalized. The frequency Ω of the undamped oscillation is a convenient nondimensionalization factor to use.

With $\tau \overset{\Delta}{=} \Omega t$, the time derivatives $\dot{\phi}$, $\ddot{\phi}$, $\dot{\theta}$, and $\ddot{\theta}$ are transformed as follows (using the chain rule):

$$\dot{\phi} = \frac{d\phi}{dt} = \frac{d\phi}{d\tau}\frac{d\tau}{dt} = \Omega\frac{d\phi}{d\tau} \qquad \ddot{\phi} = \frac{d\dot{\phi}}{dt} = \left[\frac{d}{d\tau}\left(\Omega\frac{d\phi}{d\tau}\right)\right]\frac{d\tau}{dt} = \Omega^2\frac{d^2\phi}{d\tau^2}$$

$$\dot{\theta} = \Omega\frac{d\theta}{d\tau} \qquad \ddot{\theta} = \Omega^2\frac{d^2\theta}{d\tau^2}$$

By arbitrarily choosing to work with Eq. (5.3.13a), instead of Eq. (5.3.13d), Eqs. (5.3.13g) and (5.3.13a) become, after minor manipulations that can be readily checked,

$$\frac{d^2\phi}{d\tau^2} + \sin\phi + c_1^*\frac{d\phi}{d\tau} + c_2^*\left(\frac{d\theta}{d\tau} + \frac{d\phi}{d\tau}\right) = 0 \qquad (5.3.13j)$$

$$\frac{d^2\theta}{d\tau^2} + \frac{12\left[1 + m_1/(3m_2)\right]}{(b/L)^2 + (h/L)^2}c_2^*\left(\frac{d\theta}{d\tau} + \frac{d\phi}{d\tau}\right) = 0 \qquad (5.3.13k)$$

In these equations c_1^* and c_2^* are the nondimensional damping coefficients as follows:

$$c_1^* = \frac{\Omega c_1}{(m_2 + m_1/2)gL} = \frac{c_1}{L\sqrt{gL(m_2 + m_1/2)(m_2 + m_1/3)}} \qquad (5.3.13l)$$

$$c_2^* = \frac{\Omega c_2}{(m_2 + +m_1/2)gL} = \frac{c_2}{L\sqrt{gL(m_2 + m_1/2)(m_2 + m_1/3)}} \qquad (5.3.13m)$$

Notice that the new equations involve only three different nondimensional coefficients instead of the five dimensional ones that appeared before.

$\Longrightarrow$ For Scilab users, the program *sim_model_ex5_3_13.sci* listed in Appendix E integrates Eqs. (5.3.13j) and (5.3.13k), plots ϕ and θ versus the nondimensional time τ, and animates the motion. For Matlab users, the similar program *sim_model_ex5_3_13.m* that is also listed in Appendix E will do the same. An example of their use as given in the comment lines at the beginning of each program is (the ... is used either in Scilab or in Matlab for continuation of a command if it is not typed in a single line):

```
my_output = sim_model_ex5_3_13([0:Δt:50], 12, 0.1, 0.1, ...
     20*_pi_/180, 0, 32*_pi_/180, 50);
```

where _pi_ should be typed as %pi if Scilab is used, and pi if Matlab is used. Try $\Delta t = 0.005$ for Scilab and 0.002 for Matlab, and change it if the animation is either too fast or too slow in your computer. After the simulation stops, the variable my_output is a five-column matrix whose columns are the time values in a given array τ, ϕ, $d\phi/d\tau$, θ, and $d\theta/d\tau$, which may then be used for creating plots as desired.

Those who have Simulink and prefer to use it may want to construct the block diagram model shown in Fig. 5.19 and use the suggestion shown for the Solver settings in the *Simulation* $\rightarrow$ *Parameters* menu. To see an animation of the motion after the simulation stops, type

`animate_nbars(x_coordinates, y_coordinates)`

in the Matlab command window. The Matlab program `animate_nbars`, listed in Appendix E, is needed for this. The output seen in Fig. 5.19 is essentially the equilibrium solution for ϕ when $\theta(0) = 32°$.

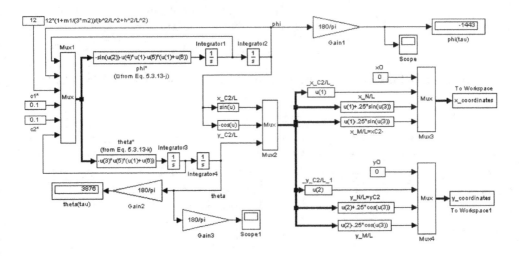

Figure 5.19: A Simulink model for Example 5.3.13. For the model inputs shown, the time array [0:0.002:50] is suggested for the *Simulation* $\rightarrow$ *Parameters* settings.

Figure 5.20 shows the expressions for the x- and y-coordinates of points O, C_2, N, and M. Instead of a plate, a bar MN of length $h = 0.25L$ is used for the animation.

Figure 5.21 shows plots of $\theta(\tau)$ and $\phi(\tau)$ versus τ. As seen in the plots, the angle θ practically reaches its equilibrium when $\tau \approx 50$ while it takes a little longer for the angle ϕ to reach its equilibrium.

Comments on the Observed Motion

When the simulation is started with the system parameters and initial conditions specified in part d, the system responds as described in the following.

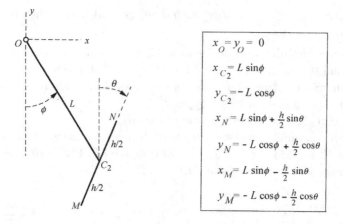

$$x_O = y_O = 0$$

$$x_{C_2} = L \sin\phi$$

$$y_{C_2} = -L \cos\phi$$

$$x_N = L \sin\phi + \frac{h}{2} \sin\theta$$

$$y_N = -L \cos\phi + \frac{h}{2} \cos\theta$$

$$x_M = L \sin\phi - \frac{h}{2} \sin\theta$$

$$y_M = -L \cos\phi - \frac{h}{2} \cos\theta$$

Figure 5.20: Details for the figure used in the animation for Example 5.3.13.

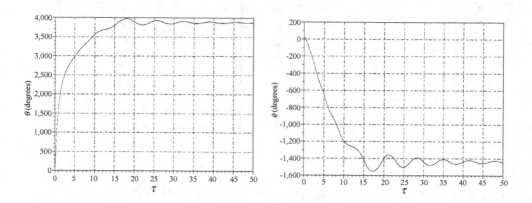

Figure 5.21: Plot of $\theta(\tau)$ versus $\tau = t\sqrt{g/L}$, and $\phi(\tau)$ versus τ, in degrees, for Example 5.3.13, with $\phi(0) = 20°$, $[d\phi/d\tau]_{\tau=0} = 0$, $\theta(0) = 32°$, and $[d\theta/d\tau]_{\tau=0} = 50$.

The plate starts to spin, and bar OC_2 starts to revolve around point O. As time progresses, the damping in both pins (at O and at C_2) slows down the motion until bar OC_2 and the plate stop spinning and then oscillate. The oscillations are damped and eventually become unnoticeable; i.e., the system essentially reached an equilibrium state $\phi = \text{constant} \overset{\Delta}{=} \phi_e$ and $\theta = \text{constant} \overset{\Delta}{=} \theta_e$, where $\phi_e = 0$ (which is the expected stable equilibrium). Bar MN that imitates the plate in the simulation ends up inclined with the horizontal. Here is a curious observation. Why does that happen? To investigate this further, change the initial conditions of the motion and see what happens. If the motion is started with $\theta(0) = 0$ and the other initial conditions are kept the same, bar MN ends up in a different orientation. The results of these two observations are shown in Fig. 5.22, which consists of two snapshots of the animation.

The snapshot in Fig. 5.22a is for $\theta(0) = 32°$, and that in Fig. 5.22b is for $\theta(0) = 0$. The values of the angle θ_e for both cases are listed in the following table.

$\theta(0)$	θ_e
32°	3879°, which is equivalent to $3879 - 10 \times 360 = 279°$
0	3847°, which is equivalent to $3847 - 10 \times 360 = 247°$

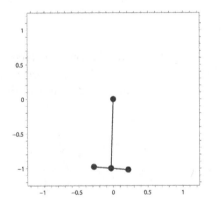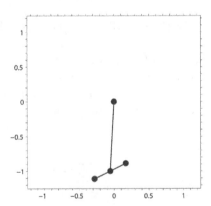

Figure 5.22: The equilibrium orientation of bars OC_2 and MN for Example 5.3.13: (a) with $\theta(0) = 32°$ and (b) with $\theta(0) = 0$.

To understand what is happening, look at the differential equations of motion and determine their equilibrium solutions. With $\theta = \text{constant} \overset{\Delta}{=} \theta_e$ and $\phi = \text{constant} \overset{\Delta}{=} \phi_e$, Eq. (5.3.13a) is automatically satisfied, and Eq. (5.3.13g) yields $\sin \phi_e = 0$. Thus, $\phi_e = 0$ and $\phi_e = 180°$ are the two physically different solutions, with the solution $\phi_e = 0$ being the stable equilibrium as seen in Chapters 1 and 2. This still does not explain why there should be an equilibrium value for θ, as observed in the simulation. The explanation lies in the fact that Eq. (5.3.13a) can be integrated once to yield Eq. (5.3.13d), and that equation explains why the angle θ tends to an equilibrium value that depends on the initial condition θ_0. Using Eq. (5.3.13k) with $\phi_e = 0$, which is the nondimensional form of Eq. (5.3.13a), the equilibrium angle θ_e is obtained as follows:

$$\theta_e = \theta(0) + \phi(0) + \frac{1}{12c_2^* \left[1 + m_1/(3m_2)\right]} \left[\left(\frac{b}{L}\right)^2 + \left(\frac{h}{L}\right)^2\right] \left[\frac{d\theta}{d\tau}\right]_{\tau=0} \quad (5.3.13\text{n})$$

Equation (5.3.13n) yields values of θ_e that are equivalent to the ones shown in the table presented earlier. For example, for $\phi(0) = 20°$, $\theta(0) = 32°$, $\phi_e = 0$, $\left(b^2 + h^2\right)/L^2 = 7/6$, $m_1/m_2 = 1/2$, $c_2^* = 0.1$, and $[d\theta/d\tau]_{\tau=0} = 50$, Eq. (5.3.13n) yields $\theta_e \approx 2439°$. Such a value is equivalent to $\theta_e = 2439 - 6 \times 360 = 279°$, which is shown in the table presented earlier.

5.4   Rolling of a Rigid Body

A special case of rigid body motion consists of a cylinder rolling on a surface, with friction between the cylinder and the surface modeled as Coulomb friction. Two typical cases are shown in Fig. 5.23, namely, rolling on a flat surface and rolling on the inside of another cylinder.

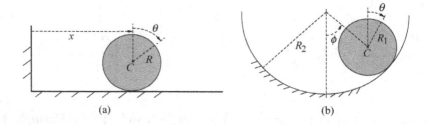

(a)   (b)

Figure 5.23: A cylinder rolling on a flat surface, and on a cylinder of radius R_2.

The analysis of the motion of the cylinder in either case shown in Fig. 5.23 involves four unknowns, namely, two motion coordinates and two unknown reactions. The unknown reactions are a normal force and the friction force of contact between the surfaces. The motion coordinates are x and θ for Fig. 5.23a (i.e., for the cylinder rolling on a flat surface), and θ and ϕ for Fig. 5.23b. However, the equations of dynamics [i.e., $\vec{F} = m\vec{a}$ for the motion of the center of mass C, and the moment equation $\vec{M}_C = d\vec{H}_C/dt$] yield only three scalar equations. Thus, one more equation is needed since four unknowns are involved. The fourth equation in this type of problem has to do with whether the cylinder rolls with or without slipping, and it is obtained as described in the following.

Rolling is said to be without slipping when the length of the infinitesimal path traveled by the center of the rolling cylinder during an infinitesimal time dt is equal to the length of the infinitesimal path traveled along the surface of contact during the same time. Thus, for each one of the two cases shown in Fig. 5.23, rolling occurs without slipping only when the condition indicated for each case is satisfied. The variables x and θ shown in Fig. 5.23a and ϕ and θ shown in Fig. 5.23b, are related to each other in these cases.

Case 1. For a cylinder of radius R, rolling without slipping on a flat surface: $dx = R\,d\theta$. This implies

$$\dot{x} = R\dot{\theta} \tag{5.4.1}$$

Case 2. For a cylinder of radius R_1, rolling without slipping on a cylinder of radius $R_2 > R_1$: $(R_2 - R_1)d\phi = R_1 d\theta$. This implies

$$(R_2 - R_1)\,\dot{\phi} = R_1\dot{\theta} \tag{5.4.2}$$

When the cylinder rolls without slipping, the fourth equation needed for determining all the unknowns is either Eq. (5.4.1) for rolling on a flat surface or Eq. (5.4.2) for rolling on another cylinder. In such a case, the friction force is determined from one of the equations obtained from $\vec{F} = m\vec{a}$. Its magnitude is smaller than the maximum possible value $\mu_s|N|$, where μ_s is the static coefficient of friction between the surfaces in contact and $|N|$ is the magnitude of the normal force of contact. If the magnitude of the friction force happens to be equal to $\mu_s|N|$, then slipping is impending. The magnitude of the friction force becomes smaller than $\mu_s|N|$ when slipping occurs, and such a motion is treated separately.

Rolling occurs with slipping if the no-slipping condition given either by Eq. (5.4.1) or Eq. (5.4.2) is violated. In such a case, the magnitude of the Coulomb friction force acting on the body is equal to $\mu_k|N|$, where μ_k is the kinetic coefficient of friction between the surfaces in contact. In this case, this is the fourth equation that is needed for solving the problem. The variables x and θ shown in Fig. 5.23a, and ϕ and θ shown in Fig. 5.23b, are independent of each other in this case.

Problems involving dry friction are solved by guessing whether the motion occurs with or without slipping and then checking whether the guess is true or false at the end of the calculations. Examples 5.4.1 through 5.4.3 illustrate the calculations that are typical of such problems. For the sake of completeness, details of the calculations are repeated in the solution to each example.

Example 5.4.1 A homogeneous cylinder of mass m and radius R rolls on a fixed inclined plane that makes an angle α with the horizontal. The static and kinetic coefficients of Coulomb friction between the cylinder and the plane are μ_s and μ_k, respectively.

a. What are the differential equations that govern the motion of the cylinder? For this, consider the cases when the cylinder rolls without slipping and the case when it slips.

b. Determine, in terms of μ_s, the maximum value of the angle α, $\alpha \overset{\Delta}{=} \alpha_{\text{slip}}$, for the cylinder to roll on the plane without slipping. What is the direction of the friction force acting on the cylinder? What is its magnitude? Sketch the magnitude of the friction force versus α (this is a good way to understand whether slipping occurs or not).

c. What are the values of α_{slip} (in terms of μ_s) in the case where the cylinder is a ring and the case where it is a solid homogeneous body?

■ Solution

a. Figure 5.24 shows the cylinder on the inclined plane and the free-body diagram for the cylinder. Since the cylinder is homogeneous, its center of mass coincides with its geometric center C. The absolute position vector $\overrightarrow{OC}$ of the center of mass C, and the angle θ that was chosen to track the rotation of the cylinder, are also shown in the figure.

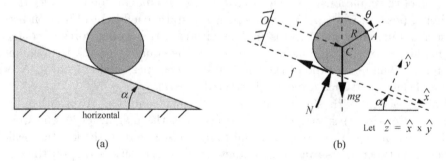

(a)   (b)

Figure 5.24: Planar rolling motion of a rigid body for Example 5.4.1.

The forces acting on the cylinder are the gravitational force, the normal force (which is perpendicular to the inclined plane), and the friction force (which is tangent to the inclined plane). The direction of these forces, indicated by the arrows, were arbitrarily drawn as shown in Fig. 5.24. If those directions are correct, the calculations will yield positive values for N and for f. Otherwise, they will yield a negative value for either f or N or both. Two mutually orthogonal unit vectors that were conveniently chosen for expressing the vector quantities that will appear in the formulation presented in the following are also shown in the figure.

In terms of the unit vector triad $\{\hat{x}, \hat{y}, \hat{z}\}$ shown in Fig. 5.24, we then have

Resultant force acting on the cylinder: $\vec{F} = (mg \sin \alpha - f)\,\hat{x} + (N - mg \cos \alpha)\,\hat{y}$

Absolute position vector for the center
of mass C of the cylinder: $\qquad\qquad\qquad \vec{r}_C = \overrightarrow{OC} = r\hat{x}$

Absolute velocity of C: $\qquad\qquad\qquad\qquad\quad \vec{v}_C = \frac{d}{dt}\vec{r}_C = \dot{r}\hat{x}$

Absolute acceleration of C: $\qquad\qquad\qquad\quad \vec{a}_C = \frac{d}{dt}\vec{v}_C = \ddot{r}\hat{x}$

$\vec{F} = m\vec{a}_C$ then yields the following scalar equations

$$m\ddot{r} = mg \sin \alpha - f \qquad\qquad (5.4.1a)$$
$$N = mg \cos \alpha \qquad\qquad (5.4.1b)$$

Next, take moments about the center of mass C of the cylinder to obtain the third equation that governs the motion. With $\vec{M}_C = d\vec{H}_C/dt$, where $\vec{M}_C = -fR\hat{z}$, and $\vec{H}_C = -I_C\dot{\theta}\hat{z}$, with I_C being the mass moment of inertia of the cylinder about the axis that passes through C and is perpendicular to the plane of the figure, we then obtain

$$I_C\ddot{\theta} = fR \qquad\qquad (5.4.1c)$$

The three equations (5.4.1a), (5.4.1b), and (5.4.1c) involve four unknowns, namely, r, θ, f, and N, which, as per Eq. (5.4.1b), is simply $N = mg \cos \alpha$. Therefore, one more equation is needed to determine the remaining unknowns. The difficulty we are now facing is that all the equations from dynamics, namely $\vec{F} = m\vec{a}_C$ and $\vec{M}_C = d\vec{H}_C/dt$, have already been used in the formulation, and those vector equations already yielded all the scalar equations they can yield. A fourth equation must then be generated by some other means. Such a difficulty is typical of problems that involve friction.

To generate a fourth equation for such a problem with dry friction, we must make a guess about the motion, and then check whether the guess is true or false at the end of the calculations. The guess to be made is that the body rolls either with or without slipping. It is simpler to assume that the cylinder rolls without slipping, and then check whether such an assumption is valid or not based on the magnitude of the friction force.

If the body rolls without slipping, the displacement dr of the center C of the cylinder must be equal to the displacement $Rd\theta$ of the point of contact of the cylinder with the inclined plane. The no-slipping condition then relates the variables $r(t)$ and $\theta(t)$ as

$$dr = Rd\theta$$

which yields

$$\dot{r} = R\dot{\theta} \tag{5.4.1d}$$

By combining Eqs. (5.4.1a), (5.4.1c), and (5.4.1d), one obtains for the angular acceleration $\ddot{\theta}$ and for the friction force f

$$\left(I_C + mR^2\right)\ddot{\theta} \;=\; mgR\sin\alpha \tag{5.4.1e}$$

$$f \;=\; \frac{mg\sin\alpha}{1 + mR^2/I_C} \tag{5.4.1f}$$

If the cylinder is not slipping, its motion is governed only by one of the two variables: $\theta(t)$ or $r(t) = R\theta(t)$. In such a case, Eq. (5.4.1e) is the differential equation of motion in terms of $\theta(t)$. That equation, which discloses that $\ddot{\theta}$ is constant, can be easily integrated. For the differential equation of motion when there is slipping, see the note at the end of this problem.

Since the scalar quantity f given by Eq. (5.4.1f) is a positive quantity, it can now be said with certainty that the direction of the friction force is indeed as shown in Fig. 5.24.

b. It is now necessary to check whether these solutions are valid or not, which then ascertains whether the cylinder is slipping or not. The way to do this is to determine whether the solution for the friction force f given by Eq. (5.4.1f) is valid or not. This is done by comparing the magnitude of f with the maximum possible magnitude of the friction force. If μ_s is the static coefficient of friction between the cylinder and the plane, the maximum magnitude of the friction force, $|f_{\max}|$, must be equal to

$$|f_{\max}| = \mu_s|N|$$

Or, since $N = mg\cos\alpha$ is a positive quantity, and f given by Eq. (5.4.1f) is also positive, we can then write

$$f_{\max} = \mu_s mg \cos\alpha \qquad (5.4.1g)$$

The body rolls without slipping if the friction force calculated by using Eq. (5.4.1f) is at most equal to $f_{\max}$ as given by Eq. (5.4.1g). Figure 5.25 shows a sketch of $f/(mg)$ versus the angle α. Such a sketch is a good way to understand whether slipping occurs or not.

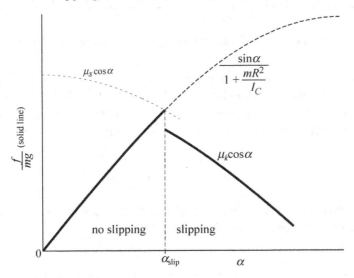

Figure 5.25: Nondimensional friction force $f/(mg)$ versus the angle α for a rolling rigid body.

As α is increased from $\alpha = 0$, the friction force follows the sinusoidal path shown in Fig. 5.25, as given by Eq. (5.4.1f), while the maximum value of f follows a cosine path that starts at $f = mg$ when $\alpha = 0$. When α reaches a value $\alpha = \alpha_{\text{slip}}$, the two paths intersect each other and the friction force becomes equal to $f_{\max} = \mu_s|N| = \mu_s mg \cos\alpha_{\text{slip}}$ when $\alpha = \alpha_{\text{slip}}$. If α is increased further, slipping occurs and the friction force becomes $f = \mu_k|N| = \mu_k mg \cos\alpha$. The value of α_{slip}, above which slipping occurs, is determined as follows:

$$\frac{\sin\alpha_{\text{slip}}}{1 + mR^2/I_C} = \mu_s \cos\alpha_{\text{slip}} \quad \text{or} \quad \tan\alpha_{\text{slip}} = \mu_s\left(1 + mR^2/I_C\right) \qquad (5.4.1h)$$

c. Table 5.1 shows the relationship between $\tan\alpha_{\text{slip}}$ and the static coefficient of friction for a ring or a cylindrical shell, and for a solid homogeneous cylinder. For a hollow cylinder of mass m, inner radius R_1, and outer radius R_2, the moment of inertia I_C is $I_C = m(R_2^2 + R_1^2)/2$ (see Example 5.3.8, p. 333), and the value of the angle of slip is between $\arctan(2\mu_s)$ and $\arctan(3\mu_s)$.

Type of body	I_C	$\tan\alpha_{\text{slip}}$
ring or cylindrical shell of radius R	mR^2	$2\mu_s$
solid homogeneous cylinder of radius R	$mR^2/2$	$3\mu_s$

Table 5.1: Values of $\tan\alpha_{\text{slip}}$ for two rolling bodies.

Note: For $\alpha > \alpha_{\text{slip}}$, the equations that govern the motion are Eqs. (5.4.1a) and (5.4.1c), with the friction force f given by Eq. (5.4.1g) with μ_s replaced by μ_k, i.e., $f = \mu_k N = \mu_k mg\cos\alpha$. For such a case, those equations then yield the following differential equations for the independent variables $r(t)$ and $\theta(t)$:

$$\ddot{r} = g(\sin\alpha - \mu_k\cos\alpha) \tag{5.4.1i}$$
$$I_C\ddot{\theta} = \mu_k mgR\cos\alpha \tag{5.4.1j}$$

The right-hand sides of Eqs. (5.4.1i) and (5.4.1j) are constants, and thus, they can be easily integrated to obtain $r(t)$ and $\theta(t)$.

Example 5.4.2 A wheel, modeled as a solid homogeneous cylinder of mass m and radius R, is able to roll on a horizontal surface. The wheel is subjected to an applied clockwise moment $M(t)$ that is equal to $M(t) = kmgRt$ N·m (where $k = 0.1$ s^{-1}) for $0 \le t \le 6$ s. The power source that produces the moment is disengaged from the wheel right after 6 s so that $M = 0$ for $t > 6$ s. Friction between the wheel and the horizontal surface is of the Coulomb type; the static and kinetic coefficients of friction are, respectively, $\mu_s = 0.3$ and $\mu_k = 0.25$. The wheel is initially at rest on the horizontal surface.

a. Obtain the differential equations of motion for the wheel for each phase of the motion, namely, when the wheel does not slip and when it slips.

b. Integrate, analytically, the differential equations of motion for each phase of the motion, and determine the times when the wheel starts and stops to slip. How far does the center of mass of the wheel travel after 10 s has elapsed since the application of the moment $M(t)$?

c. Plot the velocity of the center of mass of the wheel versus time t for $0 \le t \le$ 10 s, and indicate, in the plot, the time intervals when the wheel slips and when it does not.

■ Solution

a. The free-body diagram for the wheel is shown in Fig. 5.26. The figure also shows the absolute position vector $\overrightarrow{OC}$ for the center of mass C of the wheel, and an angle θ that is introduced for tracking the rotation of the wheel.

   The steps that should be followed to formulate the equations that govern the motion of the wheel are the same as those presented in Example 5.4.1. With the absolute acceleration of point C written as $\vec{a}_C = \ddot{r}\hat{x}$, and the resultant force acting on the wheel being equal to $\vec{F} = -f\hat{x} + (N - mg)\hat{y}$, Newton's second law $\vec{F} = m\vec{a}_C$ then yields

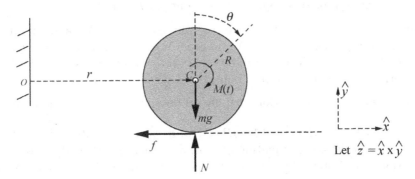

Figure 5.26: A driven wheel rolling on a horizontal surface for Example 5.4.2.

$$m\ddot{r} \;=\; -f \tag{5.4.2a}$$
$$N \;=\; mg \tag{5.4.2b}$$

Also, by taking moments about the center of mass C, the equation $\vec{M}_C = d\vec{H}_C/dt$ yields, with $\vec{M}_C = -(M + fR)\hat{z}$ (see Fig. 5.26) and $\vec{H}_C = -I_C\dot{\theta}\hat{z}$, where $I_C = mR^2/2$:

$$I_C\ddot{\theta} = M(t) + fR \tag{5.4.2c}$$

There are two possibilities to investigate: rolling without slipping, and with slipping.

1. *Rolling with slipping.* For rolling with slipping, r and θ are independent of each other. Thus, the motion is governed by two differential equations, and they are Eqs. (5.4.2a) and (5.4.2c). The magnitude of the friction force for this type of motion is $|f| = \mu_k |N| = \mu_k mg$.

2. *Rolling without slipping.* For rolling without slipping, we have $\dot{r} = R\dot{\theta}$. In this case, Eqs. (5.4.2a) and (5.4.2c) can be combined to yield, with $M(t) = 0.1mgRt$ and $I_C = mR^2/2$,

$$\ddot{\theta} = \frac{\ddot{r}}{R} = \frac{M(t)}{I_C + mR^2} = \frac{2M(t)}{3mR^2} = \frac{0.2gt}{3R} \qquad (5.4.2\text{d})$$

$$f = -\frac{mRM(t)}{I_C + mR^2} = -\frac{2M(t)}{3R} = -\frac{0.2mgt}{3} \qquad (5.4.2\text{e})$$

Notice that, since $f = -0.2mgt/3$ is a negative quantity, the direction of the friction force is opposite to the one shown in Fig. 5.26.

b. The wheel will roll without slipping as long as $|f| < \mu_s |N|$, and slipping will be impending when $|f| = \mu_s |N|$. Since $N = mg$, this then occurs when $0.2mgt/3 = \mu_s mg$. With $\mu_s = 0.3$, this gives $t = 4.5$ s. Thus, the wheel starts its motion by rolling without slipping.

The analysis of the motion now has to be separated into several parts.

$0 \le t \le 4.5$ s: $M(t) = 0.1mgRt$ N, *and the wheel rolls without slipping.* The differential equation of motion is then Eq. (5.4.2d), and the friction force is given by Eq. (5.4.2e). Since $\ddot{r} = R\ddot{\theta}$ during this time interval, successive integrations yield

$$\int_{\dot{r}=0}^{\dot{r}} \ddot{r}\, dt = \int_{t=0}^{t} \frac{0.2gt}{3}\, dt \qquad \text{i.e.,} \qquad \dot{r}(t) = \frac{0.1gt^2}{3}$$

$$\int_{r=r(0)}^{r} \dot{r}\, dt = \int_{t=0}^{t} \frac{0.1gt^2}{3}\, dt \qquad \text{i.e.,} \qquad r(t) = r(0) + \frac{0.1gt^3}{9}$$

At the end of 4.5 s, the velocity $\dot{r}$ of the center of mass C of the wheel, and the distance $r(4.5) - r(0)$ traveled by C, are then

$$\dot{r}(4.5) = \frac{0.1g \times 4.5^2}{3} \approx 6.6 \text{ m/s}$$

$$r(4.5) - r(0) = \frac{0.1g \times 4.5^3}{9} \approx 9.9 \text{ m}$$

$4.5 \leq t \leq 6$ s: $M(t) = 0.1mgRt$ N, *and slipping occurs.* The friction force f was negative for $0 \leq t \leq 4.5$ s, and continues to be negative after that. However, due to slipping, it now becomes $f = -\mu_k |N| = -\mu_k mg = -0.25mg$. The variables $r(t)$ and $\theta(t)$ are now independent of each other, and the differential equations that govern the motion are now Eqs. (5.4.2a) and (5.4.2c). With $I_C = mR^2/2$, and $M = 0.1mgRt$, successive integrations now yield

$$\int_{\dot{r}=\dot{r}(4.5)}^{\dot{r}} \ddot{r} \, dt \;=\; 0.25g \int_{t=4.5}^{t} dt \quad \text{i.e.,} \quad \dot{r}(t) = \dot{r}(4.5) + 0.25g \, (t - 4.5)$$

$$\int_{r=r(4.5)}^{r} \dot{r} \, dt \;=\; \int_{t=4.5}^{t} [\dot{r}(4.5) + 0.25g \, (t - 4.5)] \, dt$$

$$\text{i.e.,} \qquad r(t) = r(4.5) + [\dot{r}(4.5)] \, (t - 4.5) + 0.25g \frac{(t - 4.5)^2}{2}$$

$$\int_{\dot{\theta}=\dot{\theta}(4.5)}^{\dot{\theta}} \ddot{\theta} \, dt \;=\; \int_{t=4.5}^{t} \left[\frac{0.1mgRt - 0.25mgR}{mR^2/2} \right] dt$$

$$\text{i.e.,} \qquad \dot{\theta}(t) = \frac{\dot{r}(4.5)}{R} + \frac{0.1g}{R} \, (t^2 - 4.5^2) - \frac{0.5g}{R} \, (t - 4.5)$$

At the end of 6 s, these solutions give for the velocity $\dot{r}$ of the center of mass C of the wheel, and for the distance $r(6) - r(0)$ traveled by C,

$$\dot{r}(6) \;\approx\; 10.3 \text{ m/s}$$
$$r(6) - r(0) \;\approx\; 22.6 \text{ m}$$

$t > 6$ s *until slipping stops:* $M(t) = 0$. Since neither the angular velocity of the wheel nor the velocity of its center of mass can change instantaneously,[2] the wheel continues to roll and slip after $t = 6$ s. Since slipping continues, the subsequent motion is still governed by Eqs. (5.4.2a) and (5.4.2c), but with $M(t) = 0$. Integration of the differential equations of

---

[2]This can be shown as follows. Since $\vec{F} = m \, d\vec{v}_C / dt$, an instantaneous change of $\vec{v}_C$ implies that $d\vec{v}_C / dt$ is infinite, which requires an abrupt, infinite change in $\vec{F}$ for that to happen. By the same argument, since $\vec{M}_C = I_C d\vec{\omega}/dt$, an instantaneous change of the angular velocity $\vec{\omega}$ of the wheel requires an abrupt, infinite change in the moment $\vec{M}_C$.

motion now yields

$$\int_{\dot{r}=\dot{r}(6)}^{\dot{r}} \ddot{r}\, dt = 0.25g \int_{t=6}^{t} dt \qquad \text{i.e.,} \qquad \dot{r}(t) = \dot{r}(6) + 0.25g\,(t-6)$$

$$\int_{r=r(6)}^{r} \dot{r}\, dt = \int_{r=r(6)}^{r} [\dot{r}(6) + 0.25g\,(t-6)]\, dt$$

i.e., $$r(t) = r(6) + [\dot{r}(6)]\,(t-6) + 0.25g\frac{(t-6)^2}{2}$$

$$\int_{\dot{\theta}=\dot{\theta}(6)}^{\dot{\theta}} \ddot{\theta}\, dt = \int_{t=6}^{t} \left[\frac{-0.25mgR}{mR^2/2}\right] dt \quad \text{i.e.,} \quad \dot{\theta}(t) = \frac{\dot{r}(6)}{R} - \frac{0.5g}{R}(t-6)$$

After $t = 6$ s, slipping stops when $\dot{r} = R\dot{\theta}$, and this occurs when

$$\dot{r}(6) + 0.25g\,(t-6) = R\dot{\theta}(6) - 0.5g\,(t-6) \qquad \text{which gives} \qquad t = 6.6\text{ s}$$

As obtained from these equations, the velocity of the center of mass C of the wheel and the distance traveled by C at the end of 6.6 s are

$$\dot{r}(6.6) \approx 11.8 \text{ m/s}$$
$$r(6.6) - r(0) \approx 29.2 \text{ m}$$

$t > 6.6$ s: $M(t) = 0$, *and there is no slipping.* During this last phase of the motion, $\dot{r} = R\dot{\theta}$. The motion is governed by Eq. (5.4.2d) again, and the friction force is given by Eq. (5.4.2e), with $M(t) = 0$. In this case, both the angular acceleration $\ddot{\theta}$ and the friction force are zero. Since $\dot{r} = R\dot{\theta}$, and $\dot{\theta} = $ constant (because $\ddot{\theta} = 0$), we then have

$$\dot{r} = \text{constant} = \dot{r}(6.6) = 11.8 \text{ m/s}$$

$$\int_{r=r(6.6)}^{r} \dot{r}\, dt = r(t) - r(6.6) = \int_{t=6.6}^{t} 11.8\, dt = 11.8(t - 6.6)$$

Since $r(6.6) = r(0) + 29.2$ m, the total distance traveled by the center of mass C after 10 s is then

$$r_{\text{total}} = r(10) - r(0) = 69.3 \text{ m}$$

c. Figure 5.27 shows the plot of $\dot{r}(t)$ versus time t for $0 \le t \le 10$ s, created by piecing together the solutions that were obtained for $\dot{r}$ at the various time intervals the problem had to be divided into. The time interval where the wheel slips, and those where it does not, are indicated in the plot.

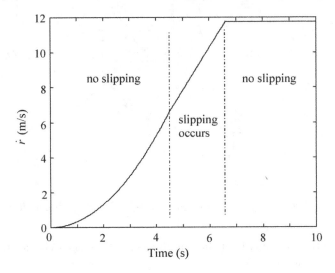

Figure 5.27: Velocity of point C (in meters per second) in Fig. 5.26 versus time t for Example 5.4.2.

Example 5.4.3 A solid cylinder of mass m and radius $R_1 = 8$ cm rolls on a vertical semicircle of radius $R_2 = 24$ cm as shown in the free-body diagram in Fig. 5.28 (the unit vectors $\hat{r}$, $\hat{\phi}$, and $\hat{z} \stackrel{\Delta}{=} \hat{\phi} \times \hat{r}$ included in the figure are used in the analysis of the problem). The angles ϕ and θ are measured from a fixed vertical line. Friction between the surfaces that are in contact is of the Coulomb type, and the static and kinetic coefficients of friction are, respectively, $\mu_s = 0.3$ and $\mu_k = 0.25$. The cylinder is released from rest at an angle $\phi = \phi_0$ so that it rolls without slipping.

a. Obtain the differential equation of motion for the cylinder in terms of the angle ϕ. How does the magnitude and the direction of the friction force change with ϕ?

b. Show that the differential equation of motion predicts that $\phi(t)$ is an oscillatory motion about the equilibrium $\phi = 0$. What is the frequency of the small (linearized) oscillation about that equilibrium?

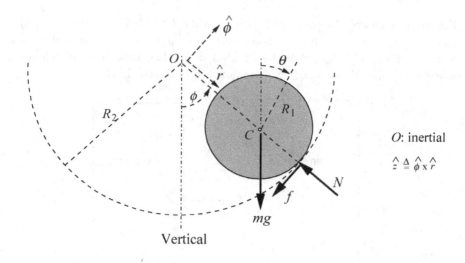

Figure 5.28: Free-body diagram for Example 5.4.3 (point O is inertial).

c. Determine the magnitude of the angular velocity of the cylinder in terms of ϕ_0, at the time when $\phi = 0$.

d. Determine the maximum value ϕ_0 of the angle ϕ for the cylinder to roll without slipping.

■ **Solution**

a. In terms of the unit vectors $\hat{r}$ and $\hat{\phi}$ shown in Fig. 5.28, the absolute position vector for the center of mass C of the cylinder is

$$\vec{r}_C = \overrightarrow{OC} = (R_2 - R_1)\,\hat{r}$$

Therefore, the following expressions are obtained for the absolute velocity $\vec{v}_C$ and for the absolute acceleration $\vec{a}_C$ of point C.

$$\vec{v}_C = \frac{d}{dt}\vec{r}_C = (R_2 - R_1)\frac{d}{dt}\hat{r} = (R_2 - R_1)\,\dot{\phi}\,\hat{\phi}$$

$$\vec{a}_C = \frac{d}{dt}\vec{v}_C = (R_2 - R_1)\,\ddot{\phi}\,\hat{\phi} + (R_2 - R_1)\,\dot{\phi}\frac{d}{dt}\hat{\phi}$$

$$= (R_2 - R_1)\,\ddot{\phi}\,\hat{\phi} - (R_2 - R_1)\,\dot{\phi}^2\hat{r}$$

The resultant force acting on the cylinder is

$$\vec{F} = (mg\cos\phi - N)\,\hat{r} - (f + mg\sin\phi)\,\hat{\phi}$$

Therefore, $\vec{F} = m\vec{a}_C$ yields the following scalar equations:

$$m\left(R_2 - R_1\right)\ddot{\phi} = -mg\sin\phi - f \tag{5.4.3a}$$

$$N - mg\cos\phi = m\left(R_2 - R_1\right)\dot{\phi}^2 \tag{5.4.3b}$$

As indicated by the equations of (5.3.5) (p. 320), the third equation that governs the motion is obtained either from $\vec{M}_C = d\vec{H}_C/dt$ or, since point O is inertial, from $\vec{M}_O = d\vec{H}_O/dt$, where $\vec{M}_C$ and $\vec{M}_O$ are the moment of the resultant $\vec{F}$ about the center of mass C and the inertial point O, respectively. Since the absolute angular velocity of the cylinder is $\vec{\omega} = \dot{\theta}\hat{z}$ (see Fig. 5.28 and use the right-hand rule), the absolute angular momentum $\vec{H}_C$ is $\vec{H}_C = I_C\dot{\theta}\hat{z} = (mR_1^2/2)\dot{\theta}\hat{z}$.

By making use of the equation $\vec{M}_C = d\vec{H}_C/dt$, the following scalar equation is obtained:

$$\frac{mR_1^2}{2}\ddot{\theta} = fR_1 \tag{5.4.3c}$$

We now have three equations, namely, Eqs. (5.4.3a), (5.4.3b), and (5.4.3c), but they involve the four unknowns θ, ϕ, N, and f. Another equation is needed, and it is the constraint relation between $\dot{\theta}$ and $\dot{\phi}$ which, as shown earlier in this section, is

$$\left(R_2 - R_1\right)\dot{\phi} = R_1\dot{\theta} \tag{5.4.3d}$$

By combining Eqs. (5.4.3a), (5.4.3c), and (5.4.3d), the following differential equation for the angle ϕ is obtained:

$$\ddot{\phi} + \frac{2g}{3\left(R_2 - R_1\right)}\sin\phi = 0 \tag{5.4.3e}$$

Equation (5.4.3e) is the differential equation that governs the motion of the cylinder. The friction force f can now be obtained by combining Eqs. (5.4.3a) and (5.4.3e) to yield

$$f = -\frac{mg}{3}\sin\phi \tag{5.4.3f}$$

This solution for f shows that the direction of the friction force is opposite to that shown in Fig. 5.28 when $\phi > 0$. The friction force is equal to zero when $\phi = 0$ (i.e., when C crosses the vertical line), and it will point to the left in Fig. 5.28 when $\phi < 0$.

b. Equation (5.4.3e) is a nonlinear differential equation that is of the same form as the differential equation for a pendulum, as considered in Example 1.13.2 in Chapter 1 (p. 50). Its equilibrium solutions $\phi = $ constant $\overset{\Delta}{=} \phi_e$ satisfy the equation $\sin \phi_e = 0$, and the only solution of interest here is $\phi_e = 0$. As shown in Example 1.13.2, the equilibrium $\phi_e = 0$ is stable. Here, the motion of the cylinder is started with $\dot{\phi} = 0$, and the subsequent motion is an oscillation about the equilibrium $\phi = 0$ (see Fig. 1.13, p. 53).

To determine the frequency of the small oscillations of the system, linearize Eq. (5.4.3e) about the stable equilibrium $\phi = 0$. For small values of ϕ, $\sin \phi \approx \phi$ and linearization of Eq. (5.4.3e) yields

$$\ddot{\phi} + \frac{2g}{3\left(R_2 - R_1\right)}\phi = 0$$

Thus, the frequency of the linearized oscillations is $\sqrt{2g/[3(R_2 - R_1)]}$, which, for $R_1 = 8$ cm and $R_2 = 24$ cm, is approximately 6.4 rad/s.

c. The nonlinear differential equation (5.4.3e) is also of the same form as Eq. (1.13.14) (p. 48), and it can be integrated once with respect to ϕ by using the identity $\ddot{\phi} = [d\dot{\phi}/d\phi]\dot{\phi}$, and the given initial conditions, as

$$\int\limits_{\dot{\phi}=0}^{\dot{\phi}} \frac{d\dot{\phi}}{d\phi}\dot{\phi}\,d\phi + \int\limits_{\phi=\phi_0}^{\phi} \frac{2g\sin\phi}{3\left(R_2 - R_1\right)}d\phi = 0$$

This gives

$$\frac{1}{2}\dot{\phi}^2 - \frac{2g\cos\phi}{3\left(R_2 - R_1\right)} = -\frac{2g\cos\phi_0}{3\left(R_2 - R_1\right)} \tag{5.4.3g}$$

or

$$\dot{\phi} = \pm\sqrt{\frac{4g\left(\cos\phi - \cos\phi_0\right)}{3\left(R_2 - R_1\right)}}$$

The angular velocity of the cylinder is $\vec{\omega} = \dot{\theta}\hat{z}$, and its magnitude, $|\dot{\theta}| = (R_2 - R_1)|\dot{\phi}|/R_1$, can now be readily obtained by using the solution found for ϕ. With $R_1 = 8$ cm and $R_2 = 24$ cm, we obtain when $\phi = 0$,

$$|\dot{\theta}|_{\phi=0} \approx 18.1\sqrt{1 - \cos\phi_0}\ \text{rad/s}$$

d. The last question that was asked was to determine the maximum value of ϕ_0 for the cylinder to roll without slipping. This is done by setting the magnitude of the friction force, $|f|$, to its maximum possible value $\mu_s|N|$.

By combining Eqs. (5.4.3b) and (5.4.3g), the normal force N is obtained as

$$N = \frac{mg}{3}\left(7\cos\phi - 4\cos\phi_0\right) \tag{5.4.3h}$$

With f given by Eq. (5.4.3f), the condition $|f| = \mu_s|N|$ yields

$$|\sin\phi| = \mu_s|7\cos\phi - 4\cos\phi_0|$$

By setting $\phi = \phi_0$ in this equation, it follows that $|\tan\phi_0| = 3\mu_s$. For $\mu_s = 0.3$, this gives $|\phi_0| \approx 42°$. For the cylinder to be able to roll without slipping when the motion is started from rest, we must have $-42° \le \phi_0 \le 42°$.

5.5   A Work-Energy Approach for Rigid Body Dynamics

A work-energy approach involving the relationship $dW = d(T)$ for a system of point masses was presented in Section 2.8 of Chapter 2. The same relationship and all the results presented in that section are also valid for a rigid body. The infinitesimal work dW done by a set of N forces $\vec{F}_1, \vec{F}_2, \ldots, \vec{F}_N$ acting on the rigid body is still given as

$$dW \triangleq \sum_{i=1}^{N} \vec{F}_i \bullet d\vec{r}_i \tag{5.5.1}$$

where $\vec{r}_i$ is the absolute position vector of the point on the body where the force $\vec{F}_i$ is applied. The dot product $\vec{F}_i \bullet d\vec{r}_i$ is a simple operation to perform. It is even simpler when it involves a pair of forces that are equal in magnitude and opposite in direction, which is called a *couple*. The infinitesimal work done by a couple is equal to the dot product of the moment of the couple and the infinitesimal rotation $\vec{\omega}\,dt$ of the body, where $\vec{\omega}$ is the absolute angular velocity of the body.

For the kinetic energy associated with the motion of the rigid body, one simply needs to replace the summation in Eq. (2.8.5), p. 189, in Chapter 2 by an integral that spans the entire body. Therefore, the kinetic energy of the motion of a rigid body is defined as

$$T \triangleq \frac{1}{2} \int_{\text{body}} \vec{v}_P \bullet \vec{v}_P \, dm \tag{5.5.2}$$

where $\vec{v}_P$ is the absolute velocity of an infinitesimal mass element of the body, represented by point P in Fig. 5.4 (p. 318).

Let us now take a closer look at the expression for the kinetic energy T, and manipulate it to obtain a more useful expression for calculating T. Referring to Fig. 5.4, by writing the absolute position vector of P as $\vec{r} = \vec{r}_C + \overrightarrow{CP}$, where $\vec{r}_C$ is the absolute position vector of the body's center of mass, the expression for the kinetic energy T becomes

$$T = \frac{1}{2} \int_{\text{body}} \left(\vec{v}_C + \frac{d}{dt}\overrightarrow{CP} \right) \bullet \left(\vec{v}_C + \frac{d}{dt}\overrightarrow{CP} \right) dm$$

$$= \frac{1}{2} \int_{\text{body}} \vec{v}_C \bullet \vec{v}_C \, dm + \int_{\text{body}} \vec{v}_C \bullet \left(\frac{d}{dt}\overrightarrow{CP} \right) dm + \frac{1}{2} \int_{\text{body}} \left(\frac{d}{dt}\overrightarrow{CP} \right) \bullet \left(\frac{d}{dt}\overrightarrow{CP} \right) dm$$

$$= \frac{1}{2} m \vec{v}_C \bullet \vec{v}_C + \vec{v}_C \bullet \frac{d}{dt} \underbrace{\left(\int_{\text{body}} \overrightarrow{CP} \, dm \right)}_{=0} + \frac{1}{2} \int_{\text{body}} \left(\frac{d}{dt}\overrightarrow{CP} \right) \bullet \left(\frac{d}{dt}\overrightarrow{CP} \right) dm$$

The first integral in this expression is zero because of the definition of the center of mass C. Since the magnitude of $\overrightarrow{CP}$ is constant because the body is rigid, and thus, $|d\,\overrightarrow{CP}/dt| = |\dot{\theta}\,\overrightarrow{CP}|$ (see Fig. 5.4), the expression for T then becomes

$$T = \frac{1}{2} m \vec{v}_C \bullet \vec{v}_C + \frac{1}{2} I_C \dot{\theta}^2 \qquad (5.5.3)$$

where

$$I_C = \int_{\text{body}} |\overrightarrow{CP}|^2 \, dm$$

is recognized as the body's moment of inertia about the axis that passes through the center of mass C and is perpendicular to the plane of Fig. 5.4.

Notice that, as indicated by Eq. (5.5.3), the kinetic energy of the motion of a rigid body is simply the sum of two simple terms that actually have a physical interpretation. The first term in that equation is the kinetic energy associated only with the translational motion of the center of mass of the body. The second term is the kinetic energy due only to rotation of the body about an axis that passes through the body's center of mass.

As indicated in Chapter 2, when dW is an exact differential, the equation $T - W = $ constant does not involve any time derivative higher than the first derivative (i.e., velocities, but not accelerations) of any variable. In such cases, $dW = d(T)$ yields $T - W = $ constant, which is a first-order differential equation. This is the differential equation of motion if the motion is governed by only one variable (i.e., a one degree of freedom system). Its time derivative, $d(T - W)/dt = 0$, gives the same second-order differential equation that one would obtain by direct use of a combination of Newton's second law $\vec{F} = m\vec{a}_C$ and the moment equation $\vec{M}_C = d\vec{H}_C/dt$ (or $\vec{M}_O = d\vec{H}_O/dt$). The advantage of the work-energy approach, as described earlier, is that the differential equation of motion is obtained with much less effort. When dW is not an exact differential, one could still obtain a second-order differential equation via $dW/dt = d(T)/dt$.

As indicated in Chapter 2, the work-energy approach is actually the basis for another approach to dynamics that is called *analytical dynamics*, which is mostly based on work developed by Lagrange. In such an approach, there is no need to formulate expressions for acceleration to obtain the differential equations of motion of dynamical systems. A number of other great masters, such as Euler, Hamilton, and Jacobi, were also creators of very elegant and powerful theories that are part of analytical dynamics. The basics of analytical dynamics are presented in Chapter 7.

5.6   Forces on Mechanisms

This section is devoted to the calculation of the reaction forces acting on each member of a mechanism. These calculations are typically very lengthy and require the use of the free-body diagram for each member of the mechanism. However, the formulation of the required equations is trivial and it only involves the use of the basic equation $\vec{F} = m\vec{a}_C$, for the motion of the center of mass C of a rigid body, and either one of the moment equations $\vec{M}_O = d\vec{H}_O/dt$ (if point O is inertial) or $\vec{M}_C = d\vec{H}_C/dt$ for the rotational motion of the body. The kinematic analysis of mechanisms was presented in detail in Chapter 3, and the results obtained in that chapter are needed here. A four-bar mechanism is used for illustrating the calculations involved in determining the forces acting on each bar of the mechanism, but the procedure shown here is the same for other mechanisms.

A four-bar mechanism is shown in Fig. 5.29. It is assumed that bar AB is driven by an external motor so that the angle $\theta_2(t)$ is known at all times. Points A and D are treated as being inertial. The reaction forces acting on each bar of the mechanism will be determined in terms of the mass, length, angular position, angular velocity, and angular acceleration of each of its members.

Figure 5.30 shows free-body diagrams for each bar of the four-bar mechanism for

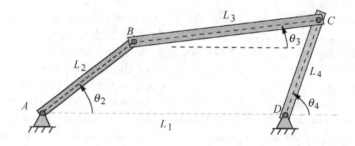

Figure 5.29: A four-bar mechanism (the angles θ_2, θ_3, and θ_4 are changing with time).

the simpler case where the mechanism is either on a horizontal plane or the effect of the gravitational forces acting on each member is disregarded. The reaction forces in the plane perpendicular to the plane of the motion of the mechanism are not shown since they do not affect the solution to the problem. The moment M_{ext} is applied to bar AB by a motor to make $\theta_2(t)$ a desired function of time. Points C_2, C_3, and C_4 shown in Fig. 5.30 are the centers of mass of bars AB, BC, and CD, respectively. Three sets of orthogonal unit vectors, which will be used in the formulation of the equations for obtaining the reaction forces, are also shown in the figure.

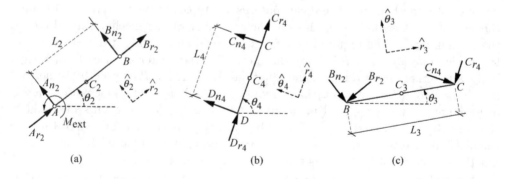

Figure 5.30: Free-body diagram for each bar of the four-bar mechanism of Fig. 5.29.

In terms of the unit vector triad $\{\hat{r}_2, \hat{\theta}_2, \hat{z} \overset{\Delta}{=} \hat{r}_2 \times \hat{\theta}_2\}$ that rotates with bar AB

shown in Fig. 5.30, we then have

Absolute position vector for the center of mass C_2 of bar AB:

$$\vec{r}_{C_2} = \overrightarrow{AC_2} = \frac{L_2}{2}\hat{r}_2$$

Absolute velocity of C_2:

$$\vec{v}_{C_2} = \frac{d}{dt}\vec{r}_{C_2} = \dot{\theta}_2\hat{z} \times \left(\frac{L_2}{2}\hat{r}_2\right) = \frac{L_2}{2}\dot{\theta}_2\hat{\theta}_2$$

Absolute acceleration of C_2:

$$\vec{a}_{C_2} = \frac{d}{dt}\vec{v}_{C_2} = \frac{L_2}{2}\ddot{\theta}_2\hat{\theta}_2 + \dot{\theta}_2\hat{z} \times \vec{v}_{C_2} = \frac{L_2}{2}\left(\ddot{\theta}_2\hat{\theta}_2 - \dot{\theta}_2^2\hat{r}_2\right)$$

Resultant force acting on bar AB:

$$\vec{F}_{AB} = (A_{r_2} + B_{r_2})\hat{r}_2 + (A_{n_2} + B_{n_2})\hat{\theta}_2$$

The motion of the center of mass of the bar is governed by Newton's second law $\vec{F} = m_2\vec{a}_{C_2}$, and this yields the following scalar equations:

$$A_{r_2} + B_{r_2} = -m_2\frac{L_2}{2}\dot{\theta}_2^2 \tag{5.6.1}$$

$$A_{n_2} + B_{n_2} = m_2\frac{L_2}{2}\ddot{\theta}_2 \tag{5.6.2}$$

Since point A is inertial, let us take moments about A and use the simple moment equation $\vec{M}_A = d\vec{H}_A/dt$, where $\vec{H}_A = I_A\dot{\theta}_2\hat{z}$ and $I_A = m_2L_2^2/3$. This moment equation yields the following scalar equation:

$$M_{\text{ext}} + L_2B_{n_2} = \frac{1}{3}m_2L_2^2\ddot{\theta}_2 \tag{5.6.3}$$

So far we have three equations, but they involve four unknown reaction forces and one unknown moment M_{ext}. To be able to solve the problem, more equations have to be generated until the number of equations and unknowns are the same.

Before proceeding, notice that the form of Eqs. (5.6.1) to (5.6.3) are relatively simple because force components A_{r_2}, B_{r_2}, A_{n_2}, and B_{n_2} in the directions $\hat{r}_2$ along bar AB, and $\hat{\theta}_2$ normal to AB, are being used in the analysis. Because of such a choice, the coefficients of these force components in Eqs. (5.6.1) and (5.6.2) are equal to one, and the expressions for the acceleration components of $\vec{a}_{C_2}$ are simply $(L_2/2)\ddot{\theta}_2\hat{\theta}_2$ and $-(L_2/2)\dot{\theta}_2^2\hat{r}_2$.

Therefore, to obtain more equations that are as simple as Eqs. (5.6.1) to (5.6.3), consider bar CD next, for which point D is inertial. Treat that bar the same way bar AB was treated, but using the unit vectors $\hat{r}_4$ and $\hat{\theta}_4$ that rotate with bar CD. With the absolute position vector of the center of mass C_4 of bar CD being $\overrightarrow{DC_4} = (L_4/2)\hat{r}_4$, the following equations are then obtained for that bar:

$$C_{r_4} + D_{r_4} = -m_4 \frac{L_4}{2} \dot{\theta}_4^2 \tag{5.6.4}$$

$$C_{n_4} + D_{n_4} = m_4 \frac{L_4}{2} \ddot{\theta}_4 \tag{5.6.5}$$

$$L_4 C_{n_4} = \frac{1}{3} m_4 L_4^2 \ddot{\theta}_4 \tag{5.6.6}$$

Finally, consider the last bar, which is bar BC. The reaction forces acting on that bar are B_{r_2}, B_{n_2}, C_{r_4}, and C_{n_4} and they appear as shown in the free-body diagram in Fig. 5.30c. Since these forces are not in the directions along the bar and perpendicular to the bar, and since neither point B nor point C is inertial, the equations in which the unknown reaction forces B_{r_2}, B_{n_2}, C_{r_4}, and C_{n_4} will appear will be a little more complicated than the equations that were obtained for the other two bars. The formulation of the equations involve obtaining expressions for the absolute acceleration of the center of mass of bar BC and the resultant force acting on the bar.

Absolute position vector for the center of mass C_3 of bar BC:

$$\vec{r}_{C_3} = \overrightarrow{AC_3} = \overrightarrow{AB} + \frac{L_3}{2} \hat{r}_3$$

Absolute velocity of C_3:

$$\vec{v}_{C_3} = \frac{d}{dt} \vec{r}_{C_3} = \underbrace{\frac{d}{dt} \overrightarrow{AB}}_{=L_2 \dot{\theta}_2 \hat{\theta}_2} + \dot{\theta}_3 \hat{z} \times \left(\frac{L_3}{2} \hat{r}_3 \right) = L_2 \dot{\theta}_2 \hat{\theta}_2 + \frac{L_3}{2} \dot{\theta}_3 \hat{\theta}_3$$

Absolute acceleration of C_3:

$$\begin{aligned}
\vec{a}_{C_3} &= \frac{d}{dt} \vec{v}_{C_3} = L_2 \ddot{\theta}_2 \hat{\theta}_2 + \dot{\theta}_2 \hat{z} \times \left(L_2 \dot{\theta}_2 \hat{\theta}_2 \right) + \frac{L_3}{2} \ddot{\theta}_3 \hat{\theta}_3 + \dot{\theta}_3 \hat{z} \times \left(\frac{L_3}{2} \dot{\theta}_3 \hat{\theta}_3 \right) \\
&= L_2 \left(\ddot{\theta}_2 \hat{\theta}_2 - \dot{\theta}_2^2 \hat{r}_2 \right) + \frac{L_3}{2} \left(\ddot{\theta}_3 \hat{\theta}_3 - \dot{\theta}_3^2 \hat{r}_3 \right)
\end{aligned}$$

Resultant force acting on bar BC:

$$\vec{F}_{BC} = -B_{r_2} \hat{r}_2 - B_{n_2} \hat{\theta}_2 - C_{r_4} \hat{r}_4 - C_{n_4} \hat{\theta}_4$$

Newton's second law is now applied to bar BC, i.e., $\vec{F}_{BC} = m_3 \vec{a}_{C_3}$. To do this, first express $\vec{F}_{BC}$ and $\vec{a}_{C_3}$ in the same orthogonal set of unit vectors. By arbitrarily choosing to work with vector components along $\hat{r}_2$ and $\hat{\theta}_2$, the transformations that relate $\{\hat{r}_3, \hat{\theta}_3\}$ (which appear in the expression for $\vec{a}_{C_3}$) and $\{\hat{r}_4, \hat{\theta}_4\}$ (which appear in the expression for $\vec{F}_{BC}$) to $\{\hat{r}_2, \hat{\theta}_2\}$ are obtained by inspection of Fig. 5.31. These relations are as follows:

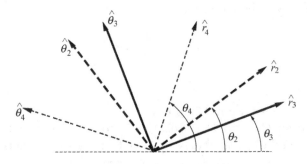

Figure 5.31: The three unit vector systems used with Fig. 5.29.

$$\hat{r}_3 = \hat{r}_2 \cos(\theta_2 - \theta_3) - \hat{\theta}_2 \sin(\theta_2 - \theta_3) \qquad \hat{\theta}_3 = \hat{r}_2 \sin(\theta_2 - \theta_3) + \hat{\theta}_2 \cos(\theta_2 - \theta_3)$$
$$\hat{r}_4 = \hat{r}_2 \cos(\theta_4 - \theta_2) + \hat{\theta}_2 \sin(\theta_4 - \theta_2) \qquad \hat{\theta}_4 = \hat{\theta}_2 \cos(\theta_4 - \theta_2) - \hat{r}_2 \sin(\theta_4 - \theta_2)$$

With $\hat{r}_3$, $\hat{\theta}_3$, $\hat{r}_4$, and $\hat{\theta}_4$ expressed in these equations, the following scalar equations are obtained from the vector equation $\vec{F}_{BC} = m_3 \vec{a}_{C_3}$:

$$B_{r_2} + C_{r_4} \cos(\theta_4 - \theta_2) - C_{n_4} \sin(\theta_4 - \theta_2)$$
$$= m_3 \frac{L_3}{2} \left[\dot{\theta}_3^2 \cos(\theta_2 - \theta_3) - \ddot{\theta}_3 \sin(\theta_2 - \theta_3) \right] + m_3 L_2 \dot{\theta}_2^2 \qquad (5.6.7)$$

$$B_{n_2} + C_{r_4} \sin(\theta_4 - \theta_2) + C_{n_4} \cos(\theta_4 - \theta_2)$$
$$= -m_3 \frac{L_3}{2} \left[\dot{\theta}_3^2 \sin(\theta_2 - \theta_3) + \ddot{\theta}_3 \cos(\theta_2 - \theta_3) \right] - m_3 L_2 \ddot{\theta}_2 \qquad (5.6.8)$$

Since neither point B nor point C is inertial, let us take moments about the center of mass C_3 of bar BC and use the moment equation $\vec{M}_{C_3} = d\vec{H}_{C_3}/dt$, where $\vec{H}_{C_3} = I_{C_3} \dot{\theta}_3 \hat{z}$ and $I_{C_3} = m_3 L_3^2 / 12$. This moment equation yields the following scalar equation, where the moment $\vec{M}_{C_3}$ is obtained by inspecting Fig. 5.30c.

$$\frac{L_3}{2} \left[B_{r_2} \sin(\theta_2 - \theta_3) + B_{n_2} \cos(\theta_2 - \theta_3) - C_{r_4} \sin(\theta_4 - \theta_3) - C_{n_4} \cos(\theta_4 - \theta_3) \right]$$
$$= I_{C_3} \ddot{\theta}_3 \qquad (5.6.9)$$

Equations (5.6.1) to (5.6.9) constitute a system of nine equations involving the nine unknowns M_{ext}, A_{r_2}, A_{n_2}, B_{r_2}, B_{n_2}, C_{r_4}, C_{n_4}, D_{r_4}, and D_{n_4}. They are linear in these variables and can be easily solved for them. They could, for example, be put in matrix form and then solved by inverting a 9×9 matrix, as done in Chapter 3 for a 2×2 matrix. However, they can be solved in a simpler manner.

Equations (5.6.5) and (5.6.6) can immediately be solved for the reaction forces C_{n_4} and D_{n_4} to obtain the following results:

$$C_{n_4} = \frac{1}{3} m_4 L_4 \ddot{\theta}_4 \qquad (5.6.10)$$

$$D_{n_4} = \frac{1}{2} C_{n_4} \qquad (5.6.11)$$

with $\ddot{\theta}_4$ determined as shown in Chapter 3.

Having solved for C_{n_4} and D_{n_4}, Eqs. (5.6.7) to (5.6.9) now only involve the unknowns B_{r_2}, B_{n_2}, and C_{r_4}. They may be solved with a simple matrix inversion and a matrix multiplication as follows, and numerical values may be readily obtained either by hand or with a calculator.

$$
\begin{bmatrix} B_{r_2} \\ B_{n_2} \\ C_{r_4} \end{bmatrix} =
\begin{bmatrix}
1 & 0 & \cos(\theta_4 - \theta_2) \\
0 & 1 & \sin(\theta_4 - \theta_2) \\
\sin(\theta_2 - \theta_3) & \cos(\theta_2 - \theta_3) & -\sin(\theta_4 - \theta_3)
\end{bmatrix}^{-1} \times
$$

$$
\begin{bmatrix}
\frac{1}{2} m_3 L_3 \left(\dot{\theta}_3^2 \cos(\theta_2 - \theta_3) - \ddot{\theta}_3 \sin(\theta_2 - \theta_3) \right) + m_3 L_2 \dot{\theta}_2^2 + C_{n_4} \sin(\theta_4 - \theta_2) \\
-\frac{1}{2} m_3 L_3 \left(\dot{\theta}_3^2 \sin(\theta_2 - \theta_3) + \ddot{\theta}_3 \cos(\theta_2 - \theta_3) \right) - m_3 L_2 \ddot{\theta}_2 - C_{n_4} \cos(\theta_4 - \theta_2) \\
\frac{1}{6} m_3 L_3 \ddot{\theta}_3 + C_{n_4} \cos(\theta_4 - \theta_3)
\end{bmatrix}
$$

$$(5.6.12)$$

The remaining unknowns A_{r_2}, A_{n_2}, M_{ext}, and D_{r_4} are readily obtained from Eqs. (5.6.1) to (5.6.4).

$\Longrightarrow$ The calculation of all the reactions, as presented here, are implemented, in the form of functions, in the similar programs *fourbarforces.sci* and *fourbarforces.m* listed in Appendix E. The former is for Scilab users and the latter is for Matlab users. An example of their use, as given in the comment lines at the beginning of each program, is

```
t=[0:0.1:5]; theta2=_pi_/6+1.5*_pi_*sin(2*t);
dtheta2_dt=3*_pi_*cos(2*t); d2theta2_dt2=-6*_pi_*sin(2*t);
my_output=fourbarforces(t, 1, 0.45, 0.8, 0.7, theta2, ...
          dtheta2_dt, d2theta2_dt2, 1, 0.15, 0.225, 0.1875);
```

where `_pi_` should be typed as `%pi` for Scilab users, and `pi` for Matlab users.

Such a call results in a 51 by 10 matrix whose columns are t, M_{ext}, A_{r_2}, A_{n_2}, D_{r_4}, B_{r_2}, B_{n_2}, C_{r_4}, D_{n_4}, and C_{n_4}. Any of these can then be plotted by using the `plot` command, which is the same for Scilab and Matlab.

Those who have Simulink and prefer to use it may want to construct the block diagram model shown in Fig. 5.32 for automating the same calculations pre-

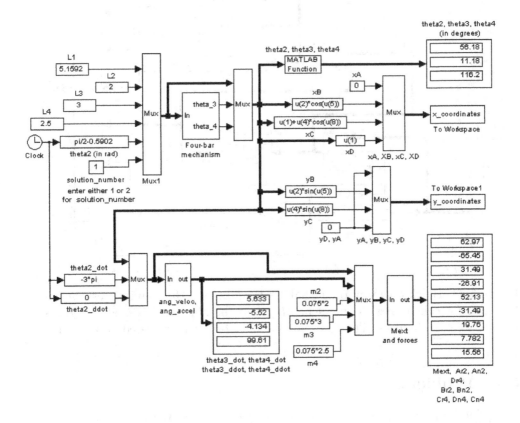

Figure 5.32: A Simulink model for calculating the joint reactions in a four-bar mechanism. This illustration uses the input data for Example 3.2.2 in Chapter 3, with the stop time equal to $t = 0$.

sented here. The model shown in Fig. 5.32 is based on the *fourbar_mechanism* model shown in Fig. 3.9 (p. 239). The only difference between the two models is the addition of the blocks seen to the right of the block that displays the angular velocities $\dot{\theta}_3$ and $\dot{\theta}_4$ and angular accelerations $\ddot{\theta}_3$ and $\ddot{\theta}_4$.[3]

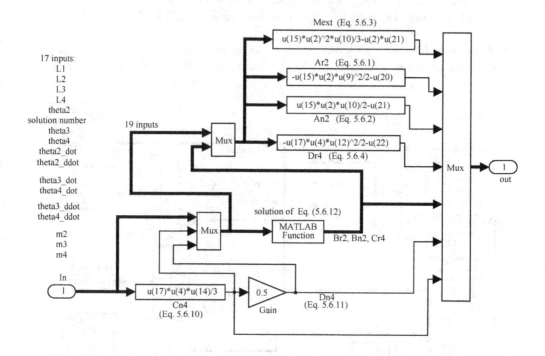

Figure 5.33: Block diagram for the subsystem labeled "Mext and forces" shown in Fig. 5.32 for calculating the moment M_{ext} and the reaction forces at the joints of the four-bar mechanism. The 17 inputs to the subsystem are listed on the left.

5.7 Problems

5.1 The body in Fig. 5.34 is a thin homogeneous plate with dimensions $b = 100$ cm, $h = 50$ cm, and $L = 20$ cm. The plate would be rectangular if it were not for the cutout. Its mass density is $\rho = 15.4$ kg/m^2, which is for a steel sheet

---

[3]All the steps presented for calculating the applied moment M_{ext} and the reaction forces on each bar of the mechanism are embedded in the subsystem block labeled "Mext and forces" that is seen in Fig. 5.32. A subsystem is created as shown in the tutorial in Appendix A. One is used in Fig. 5.32 with the sole purpose of compacting the entire block diagram which, otherwise, might hardly fit on the computer screen. The block diagram that is grouped in that subsystem is shown in Fig. 5.33. The calculation of the reactions B_{r2}, B_{n_2}, and C_{r_4} using Eq. (5.6.12) is done in the *Matlab Function* block that is seen in Fig. 5.33, typed as explained in Appendix A.

of approximately 2 mm of thickness. Determine the location of its center of mass C relative to the corner at point O shown in the figure. Also, determine the moments of inertia I_O and I_C about the axes that are perpendicular to the plate and pass through O and through the center of mass C of the plate, respectively. Disregard the thickness in the calculations. *Suggestion:* Solve the problem for a plate without the cutout, and then subtract the cutout (which is the same as treating the cutout as if its mass were negative).

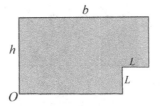

Figure 5.34: Figure for Problem 5.1.

5.2 The body shown in Fig. 5.35 consists of a homogeneous solid ring of inner and outer radius $R_1 = 90$ cm and $R_2 = 100$ cm, respectively. To reinforce the ring, three crossbars with rectangular cross section are symmetrically welded to its inside. The ring and the crossbars are made of steel, and their thickness is 2 mm, which is very small. The density of steel is 7700 kg/m^3. Determine the body's moment of inertia I_C about the axis that is perpendicular to the plane of the ring and passes through its center of mass C. What percentage do the crossbars contribute to the answer? Disregard the thickness in the calculations.

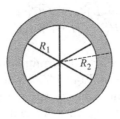

Figure 5.35: Figure for Problem 5.2.

5.3 The body in Fig. 5.36 is a thin homogeneous triangular plate with a square cutout and $L = 100$ cm. Its mass density is $\rho = 15.4$ kg/m^2, which is for a steel sheet of approximately 2 mm of thickness. Determine the location of its center of mass C relative to point O shown in the figure, which is in the middle of the bottom part of the plate. Also, determine the moments of inertia I_O and I_C about the axes that are perpendicular to the plate and pass through O and through its center of mass C, respectively. Disregard the thickness in the calculations.

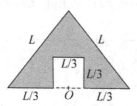

Figure 5.36: Figure for Problem 5.3.

5.4 Show that the absolute angular momentum $\vec{H}_A$ of a body, about an arbitrary point A, is related to the absolute angular momentum $\vec{H}_C$ about the center of mass C of the body as $\vec{H}_A = \vec{H}_C + m\overrightarrow{AC} \times \vec{v}_C$, where $\vec{v}_C$ is the absolute velocity of C.

5.5 A rigid body of mass m, such as the one in Fig. 5.4 (p. 318), is rotating, in a plane, with absolute angular velocity $\dot{\theta}\hat{z}$, where $\hat{z}$ is a unit vector normal to the plane of rotation. The rotation is about an axis that passes through the inertial point O shown in that figure, and line OC is a massless bar rigidly attached to the body. Show that the absolute angular momentum $\vec{H}_O$ is equal to $\vec{H}_O = I_O\dot{\theta}\hat{z}$, where $I_O = I_C + m|\overrightarrow{OC}|^2$. *Suggestion:* Start by using the result of Problem 5.4, i.e., $\vec{H}_O = \vec{H}_C + m\overrightarrow{OC} \times \vec{v}_C$.

5.6 An engineering company needs to determine the location of the center of mass C of the thin plate shown in Fig. 5.37. The moment of inertia I_C of the plate, about an axis perpendicular to the plate and passing through C, also needs to be determined. The problem is that the plate is inhomogeneous, in addition to its shape being somewhat complicated. Experiments were then devised to determine those quantities. First, two points A and B were marked on the plate, with the distance between them equal to $D = 1.1$ m, as shown in Fig. 5.37. Thin pins were attached at those points so that the plate could be suspended from them in such a way that it remained on a vertical plane. The plate was also weighted, and its mass was found to be equal to 30.5 kg.

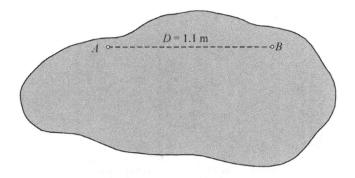

Figure 5.37: Figure for Problem 5.6.

a. The plate was hung from A, and line AB remained at $19°$ with the vertical when the plate was in static equilibrium (i.e., no motion). It was then given a small angular displacement from that equilibrium, and the period of the resulting oscillation was found to be 3.3 s. The engineers made sure that friction in the system was small enough to be disregarded in their subsequent calculations.

b. The plate was also hung from B and, in static equilibrium, line AB remained at $48°$ with the vertical.

Is the data gathered sufficient to perform the task at hand? If so, determine the location of C, and the moment of inertia I_C. What would be the period of the resulting small oscillations if the plate were suspended from B?

5.7 Figure 5.38 is a side view of a solid homogeneous cylinder of radius R and mass m, rolling on a plane that is inclined α degrees with the horizontal. The center of mass C of the cylinder is on its axial axis. That axis is connected to two identical linear springs (one on each face of the cylinder) of unstressed length L_u. The other end of each spring is connected to fixed points on the inclined plane. The total stiffness of the parallel springs is k N/m. Friction between the cylinder and the plane is of the Coulomb type, and the static and kinetic friction coefficients are μ_s and μ_k, respectively. The cylinder rolls without slipping on the inclined plane.

a. Obtain the differential equation of motion for the system in terms of the variable $x(t)$ shown in Fig. 5.38. Also, determine the equilibrium $x = \text{constant} \overset{\Delta}{=} x_e$.

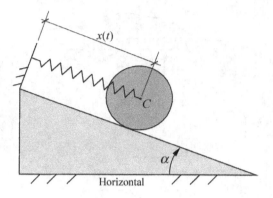

Figure 5.38: Figure for Problem 5.7.

b. Is the differential equation obtained in part a linear or nonlinear? If it is nonlinear, linearize it about $x = x_e$ and then obtain the frequency of the small oscillations of the system about each stable equilibrium. If it is linear, obtain the frequency of the oscillations of the system about its equilibrium. The system constants are $k/m = 54\,\mathrm{s}^{-2}$, $\mu_s = 0.3$, $\mu_k = 0.25$, $L_u = 50$ cm, and $\alpha = 30°$.

5.8 The motion of the system described in Problem 5.7, with its parameter values given in part b of that problem, is started with the initial conditions $x(0) = x_0$ and $\dot{x}(0) = 0$. Physically one would expect that the cylinder will roll with slipping if $x(0)$ is "too big." Determine the range of x_0 for the cylinder to roll without slipping.

5.9 The system shown in Fig. 5.39 consists of a homogeneous circular plate of radius R and a homogeneous rectangular plate of length R and $h = (\sqrt{3}/2)R$. The two plates are glued to each other, are of small thickness, and their mass density is ρ kg/m^2. Point A is the center of mass of the circular plate, and line AB is parallel to one of the sides of the rectangular plate. The plate rolls without slipping on a horizontal surface, and the system is constrained to move on a vertical plane by two horizontal frictionless guides that are not shown in the figure. Friction between the plate and the horizontal surface shown is of the Coulomb type, and the static and kinetic coefficients of friction are μ_s and μ_k, respectively.

a. Obtain the differential equation of motion for the system in terms of the angle $\theta(t)$ shown in Fig. 5.39, and all the equilibrium solutions $\theta =$

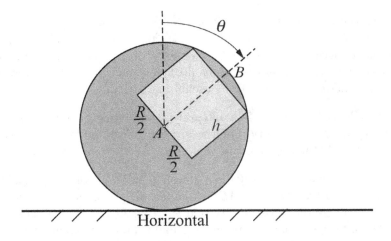

Figure 5.39: Figure for Problem 5.9.

constant $\overset{\Delta}{=} \theta_e$.

  b. Analyze the stability of each equilibrium solution, and then determine
     the frequency of the small oscillations of the system about each stable
     equilibrium for $R = 15$ cm.

5.10 It is desired to determine if the wheel described in Problem 5.9 slips during
      its motion. This is to be done by numerically solving the differential equation
      of motion of the system and then calculating the friction and normal forces
      acting on the wheel. Obtain the differential equation of motion of the system
      and integrate it numerically to solve the problem. First, verify that no slipping
      occurs if the motion is started from rest and using $\theta(0) = 0.01$ rad. Then, by
      numerical experimentation, determine the maximum value of $|\dot{\theta}(0)/\sqrt{g/R}|$ to
      prevent slipping. The static and kinetic coefficients of Coulomb friction are
      $\mu_s = 0.3$ and $\mu_k = 0.25$, respectively. *Suggestion:* See Examples 5.4.1 and
      5.4.2.

5.11 Consider the system of Examples 5.3.10 (p. 336) and 5.3.11 (p. 339), with the
      value of L_2/L_1 being such that the equilibrium value of the angle θ is $\theta_e = 30°$
      when $M_{\text{bar}} = -c\dot{\theta}$, and the support point A is given a constant acceleration
      $\ddot{x} = -7$ m/s² (see Fig. 5.13, p 337). For $L_1 = 1$ m and $c/m = 4.5$ m²/s,
      determine the damped and undamped frequencies of the small oscillations of
      the system about its equilibrium.

5.12 Consider the system of Examples 5.3.10 (p. 336) and 5.3.11 (p. 339), with $L_1 = 1$ m, $L_2 = 0.5$ m, $M_{\text{bar}} = 0$, and when the support point A is given a constant acceleration $\ddot{x} = -7$ m/s² (see Fig. 5.13, p. 337). If the motion $\theta(t)$ is started with the initial conditions $\theta(0) = 0$ and $\dot{\theta}(0) = \dot{\theta}_0$, determine the range of values of $\dot{\theta}_0$ for $\theta(t)$ to be oscillatory.

5.13 Solve Problems 5.11 and 5.12 by integrating numerically the differential equation that governs the motion of the system. For Problem 5.11, use $L_2 = 0.42L_1$, which is the approximate answer for L_2 for that problem.

5.14 The system shown in Fig. 5.40a is a double pendulum constructed of two homogeneous thin bars OA and AB of constant length L_1 and L_2, respectively, and having the same mass density ρ kg/m. Bar OA is pivoted to an inertial point O, and the two bars are pinned at A and can rotate about that point relative to each other. The motion of the entire system takes place on a vertical plane, and friction in the system can be neglected.

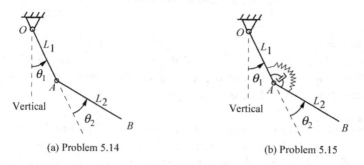

(a) Problem 5.14   (b) Problem 5.15

Figure 5.40: Figure for Problems 5.14 and 5.15.

a. Obtain the differential equations of motion for the system in terms of the angles θ_1 and θ_2 shown in Fig. 5.40a.

b. Use the energy method $dW = d(T)$ to show that $dW = -d(U)$ and that the mechanical energy $T+U$ of this system remains constant for all times.

c. Numerically integrate the differential equations of motion. For this, use the nondimensional time $\tau = t\sqrt{g/L_1}$. Do the integration with $L_2/L_1 = 2$, and with the initial conditions $\theta_1(0) = 30°$, $\theta_2(0) = 0$, $\dot{\theta}_1(0) = \dot{\theta}_2(0) = 0$, and plot $\theta_1(\tau)$ and $\theta_2(\tau)$ versus τ. Also, numerically verify that $T + U = \text{constant}$. Do some numerical experiments by changing the initial conditions of the motion, and then write a few paragraphs describing the resulting motion.

5.15 The double pendulum of Problem 5.14 is modified as in Fig. 5.40b. The mod-
      ification consists of adding the torsional spring and torsional viscous damper
      at the connection between the two bars. The spring and damper are linear,
      with coefficients equal to k N·m/rad and c N·m/(rad/s), respectively. The
      spring is unstressed when $\theta_2 = 0$. Obtain the differential equations of mo-
      tion for this system, and integrate them numerically using the nondimensional
      time $\tau = t\sqrt{g/L_1}$. Do the integration with $L_2/L_1 = 2$, and with the initial
      conditions $\theta_1(0) = 30°$, $\theta_2(0) = 0$, $\dot{\theta}_1(0) = \dot{\theta}_2(0) = 0$, and plot $\theta_1(\tau)$ and $\theta_2(\tau)$
      versus τ. Choose your own non-zero values for the nondimensional spring and
      damping coefficients $k^* = k/(m_1 g L_1)$ and $c^* = c/(m_1 L_1\sqrt{gL_1})$. Animate
      the motion to observe how the two bodies move. Do some numerical experi-
      ments by changing the initial conditions of the motion, and then write a few
      paragraphs describing the resulting motion.

5.16 Determine the maximum values of the nondimensional force reactions $F_1/(m_2 g)$
      and $F_2/(m_2 g)$ acting on the plate of Example 5.3.13 (p. 346). Use the data and
      the initial conditions given in part d of that example, and plot those reactions
      versus the normalized time τ that is also defined in that example. What are
      the maximum values of those nondimensional reaction forces if the nondimen-
      sional damping coefficients are changed to $c_1^* = c_2^* = 2$? Write a brief comment
      on your observations about those reaction forces, including your conclusion
      about what their steady state values should be. *Suggestion:* The differential
      equations have to be integrated numerically since they are nonlinear. Any
      appropriate numerical software can be used for doing this; if you use either
      Scilab or Matlab, either one of the programs named *sim_model_ex5_3_13* listed
      in Appendix E could be modified for doing this problem.

5.17 For the four-bar mechanism shown in Fig. 3.2 (p. 224), $\dot{\theta}_2 = \text{constant} = -100$
      rpm. Friction in the system can be neglected. The motion of the mechanism
      takes place on a horizontal plane. Each bar is homogeneous and is approx-
      imated as a slender line with mass density equal to 0.075 kg/m. For that
      mechanism, $L_2 = 0.4$ m, $L_3 = 1$ m, $L_4 = 0.6$ m, and $L_1 = 1$ m. Determine
      the magnitude of the reaction forces acting at the ends of each bar of the
      mechanism, and on the plane of the motion, when the mechanism is in the
      position labeled "Problem 3.6" in Fig. 3.26 (p. 263). Also one needs to buy
      the right motor to drive the mechanism and, for that, one needs to know the
      maximum torque to be supplied by the motor. Plot the torque M_{ext} versus
      time and then specify the maximum torque the motor must supply. Make sure
      that your answers have appropriate units, and clearly specify those units for
      each answer. *Suggestion:* If you use either Scilab or Matlab, either one of the

programs named *fourbarforces* listed in Appendix E could be modified to solve this problem.

5.18 Repeat Problem 5.17 if motion takes place on a vertical plane. Use the equations you have developed for this problem to verify that the answers to Problem 5.17 may be obtained by simply setting the value of the acceleration of gravity in that problem to $g = 0$. If that does not happen, you can conclude that there is an error in your calculations.

5.19 The four-bar mechanism shown in Fig. 5.41 moves on a vertical plane, and its motion is started when $\theta_2 = 0$ and point C is above line AD. Friction in the pins can be neglected. Bar AB is driven by a motor so that $\dot{\theta}_2 = $ constant $ = 100$ rpm. Each bar (including the bent bar $BCPQ$) is homogeneous, with mass density equal to 0.075 kg/m, and may be approximated as a line. For that mechanism, $L_2 = 0.25$ m, $L_{31} = 0.6$ m, $L_{32} = 0.26$ m, $L_{33} = 0.14$ m, $L_4 = 0.7$ m, $L_1 = 1$ m, and $\alpha = $ constant $ = 30°$. Determine the reaction forces acting on each bar of the mechanism (clearly indicating them in a free-body diagram), and the external moment (magnitude and direction) applied to bar AB, when θ_2 reaches 45° after bar AB turns two revolutions. While performing your calculations, approximate each bar as a slender line. Make sure that your answers have appropriate units, and clearly specify those units for each answer.

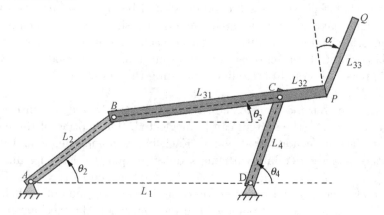

Figure 5.41: Figure for Problem 5.19.

5.20 If you have not done so yet, solve Problem 5.19 numerically and plot the applied moment versus time t while bar AB turns only three revolutions. What is the maximum value of the magnitude of that moment?

5.21 Bar AB of the mechanism shown in Fig. 3.12 (p. 241) is driven by a motor that applies to it a moment M_{ext} so that $\theta_1 = \pi/6 + kt$ rad, where $k = -8\pi$ rad/s. Friction in the pins and in the slotted bar can be neglected. The motion of the mechanism takes place on a horizontal plane. Each bar is homogeneous, with mass density equal to 0.075 kg/m, and may be approximated as a slender line. For that mechanism, $r_1 = 0.4$ m, $L = 1$ m, and the length of the slotted bar is equal to $L + 1.2r_1$.

Determine the reaction forces acting on each bar of the mechanism and on the plane of the motion, the reaction moments (if any), and the external moment M_{ext} (magnitude and direction) applied to bar AB (clearly indicating them in a free-body diagram and showing the details of each step of the calculations) immediately after bar AB turns four complete revolutions.

5.22 For the slider-crank mechanism shown in Fig. 3.15 (p. 251), $D = 0$, and $\theta_1 = kt - \pi/6$ rad, where $k = 2\pi$ rad/s. The motion of the mechanism takes place on a horizontal plane, and friction in the system can be neglected. Each bar of the mechanism is homogeneous, with mass density equal to 0.075 kg/m, and may be approximated as a slender line. For that mechanism, $L_1 = 0.4$ m, $L_2 = 1$ m, and the mass of the block at B is 1 kg. The frictionless pin that connects bar AB to the block at B passes through the center of mass of the block.

Determine the reaction forces acting on each bar of the mechanism and on the plane of the motion, the reaction moments (if any), and the external moment M_{ext} (magnitude and direction) applied to bar OA (clearly indicating them in a free-body diagram and showing the details of each step of the calculations) immediately after bar OA turns two complete revolutions. *Note:* The problem has two solutions, one for $x_B > 0$ and another for $x_B < 0$.

5.23 Repeat Problem 5.21 for the case when the mechanism is vertical.

5.24 Repeat Problem 5.22 for the case when the mechanism is vertical. Automate the solution of your equations, display all your answers, generate a plot of the external moment M_{ext} during the entire time interval, and animate the motion of the mechanism. The simulation should stop automatically at the end of the specified time. What is the maximum value of the magnitude of the moment M_{ext}? Make sure that your answers have appropriate units, and clearly specify those units for each answer.

5.25 Bar AB of the mechanism in Problem 3.27, shown in Fig. 3.27 (p. 266), is driven by a motor that applies to it a moment M_{ext} so that $\theta_2(t) = \pi/2 + k_1 \sin(k_2 t)$ rad, where $k_1 = 3\pi$ rad and $k_2 = 1$ rad/s. For that mechanism,

$L_1 = 1$ m, $L_2 = 0.4$ m, $|\overrightarrow{CP}| = 3$ m, $L_4 = |\overrightarrow{CD}| = 1.5$ m, and $\alpha = 30°$. Motion takes place on a horizontal plane, and friction can be neglected. Each bar is uniform and may be approximated as a line with density $\rho = 0.075$ kg/m. Determine the reaction forces acting on each bar of the mechanism and on the plane of the motion, the reaction moments (if any), and the external moment M_{ext} when $t = 20$ s.

5.26   Repeat Problem 5.25 when the mechanism is on a vertical plane.

5.27   In the system shown in Fig. 5.42, a thin homogeneous bar of mass m_{bar} and length L is able to rotate about a pin located at a fixed point O. Two bodies (approximated as point masses), each of mass m kg, are attached to the ends of the bar, and those ends are connected by linear springs, each of stiffness k N/m and unstressed length L_u m, to the fixed points P and Q. The springs are unstretched when $\theta = 0$. A motor applies a moment M_{motor} to the bar, and there is viscous friction at the pin contact at O, with friction coefficient c N·m/(rad/s). The motion of the system takes place on a horizontal plane, and the moment M_{motor} is a known function of time.

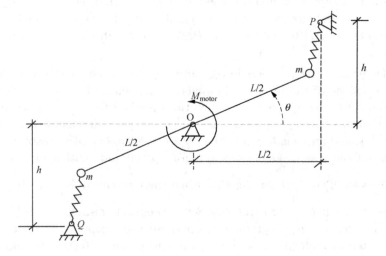

Figure 5.42: Figure for Problem 5.27.

   a. Obtain the differential equation of motion for the system using the angle θ to describe the motion. Notice that the equations may be obtained either by the vectorial formulation of Newton and Euler or by the energy method. Use whatever one you choose.

  b. With $M_{motor} = 0$, determine *all* the equilibrium values θ_e of the angle θ
     for $h = L$.

  c. Analyze the stability of each equilibrium obtained in part b by linearizing
     the differential equation of motion about each equilibrium. What is the
     frequency of the small undamped oscillations of the system about each
     stable equilibrium?

5.28 Repeat Problem 5.27 if the plane of Fig. 5.42 is vertical and the motion takes
      place on that vertical plane. Disregard friction in the system. Does the me-
      chanical energy $T + U$ of this system remain constant for all times? Explain.
      What is the expression for U? In addition, consider the case when the motor
      moment is a function of θ, instead of being a known function of time, i.e.,
      $M_{motor} = Kf(\theta)$ where $K = $ constant. What is the expression for U in this
      case, and what can be said about the change of $T + U$ with time?

5.29 The system shown in Fig. 5.43 consists of a rigid homogeneous T-beam of
      dimensions L, $L/4$, and $L/4$, where $L = 70$ cm, with a pin support at point A
      of the T. Motion takes place on a horizontal plane. An external force is

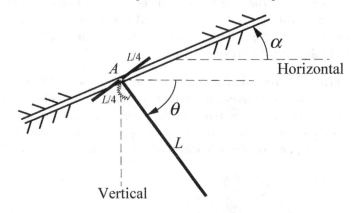

Figure 5.43: Figure for Problems 5.29 to 5.32.

applied at A to make that point move with a constant absolute velocity of
magnitude $|\vec{v}_A| = 2$ m/s, along a straight track on the horizontal plane. The
torsional spring that is attached to the pin at A and to the beam is linear with
stiffness $k = 3$ N·m/rad. The spring is unstressed when $\theta = 0$. Determine
the frequency of the undamped oscillations of the beam about its equilibrium
position by approximating its two parts as thin bars of density $\rho = 0.075$
kg/m.

5.30 Repeat Problem 5.29 if the smaller part of the T-beam is a homogeneous thin bar of constant density $\rho_1 = 0.075$ kg/m and the larger part of length L has variable density $\rho_2 = 0.05 + 0.03r/L$ kg/m, with r being distance, measured from point A, along that part of the beam.

5.31 The system shown in Fig. 5.43 consists of a rigid homogeneous T-beam of dimensions L, $L/4$, and $L/4$, where $L = 70$ cm, with a pin support at point A of the T. An external force is applied at A to make that point move with a constant absolute velocity of magnitude $|\vec{v}_A| = 2$ m/s, along a straight track that is inclined $\alpha = 30°$ with the horizontal. Motion of the beam takes place on a vertical plane. The torsional spring that is attached to the pin at A and to the beam is linear with stiffness k N·m/rad, and k is such that the beam is in equilibrium when $\theta \triangleq \theta_e = 30°$. The spring is unstressed when $\theta \triangleq \theta_u = 0$. Determine the frequency of the small oscillations of the beam about its equilibrium position by approximating its two parts as thin bars of density ρ kg/m. Neglect friction in the support at A.

5.32 Repeat Problem 5.31 if the smaller part of the T-beam is a homogeneous thin bar of constant density $\rho_1 = 0.075$ kg/m, and the larger part of length L has variable density $\rho_2 = 0.05 + 0.03r/L$ kg/m, with r being distance, measured from point A, along that part of the beam.

5.33 A thin homogeneous rectangular plate of dimensions $a = 30$ cm and $b = 2a$ is welded to a thin homogeneous triangular plate of dimensions a, b, and $c = \sqrt{a^2 + b^2}$. Both plates are made of the same material and have the same thickness. The system is pinned to a fixed point A as shown in Fig. 5.44 and motion takes place on the vertical plane, which is the plane of the figure.

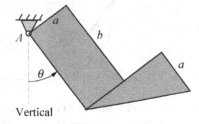

Figure 5.44: Figure for Problem 5.33.

Neglecting friction, obtain the differential equation of motion of the system in terms of the angle θ, the equilibrium value (in degrees) for θ, and the frequency of the small oscillations about the stable equilibrium of the system. Disregard the thickness of the plates in your calculations.

5.34 A thin bar AB of length $L = 1.8$ m and mass m_1 is welded to a thin triangular plate of dimensions $a = 30$ cm, $b = 45$ cm, $c = \sqrt{a^2 + b^2}$, and mass $m_2 = 3m_1$, and the system is pinned to a fixed point A as shown in Fig. 5.45. The motion of the system takes place on a vertical plane, which is the plane of the figure. Neglecting friction, obtain the differential equation of motion of the system in terms of the angle θ, the equilibrium solutions (in degrees) for θ, and the frequency of the small oscillations about the stable equilibrium of the system.

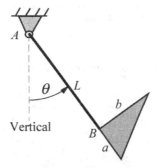

Figure 5.45: Figure for Problem 5.34.

5.35 The system shown in Fig. 5.46 consists of a homogeneous disk of mass m_1 and radius R and a homogeneous rectangular plate of dimensions a and $b > a$ that is pinned, with a frictionless pin, at the center of one of its sides to the center C of the disk. The disk rolls without slipping on a plane that is inclined at an angle α with the horizontal, and motion of the system takes place on a vertical plane. For $m_1 = m_2$, $b = 2a$, and $\alpha = 30°$, determine:

a. The equilibrium values $\theta = \text{constant} \overset{\Delta}{=} \theta_e$, in degrees, and the corresponding acceleration of the center of mass C of the disk.

b. The frequency of the small oscillations of the plate about its stable equilibrium θ_e for $b = 35$ cm and $b = 25$ cm. Does the frequency increase or decrease with the dimension b?

5.36 The system shown in Fig. 5.47 consists of a thin block of mass m, with a pin through its center of mass C that is able to slide along a vertical circular groove of radius R. The system also has a spring that is attached to a fixed point O and to C so that the distance from O to C is equal to $2R$ when line

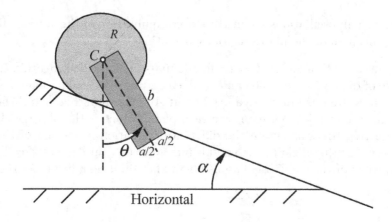

Figure 5.46: Figure for Problem 5.35.

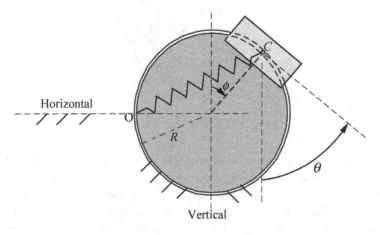

Figure 5.47: Figure for Problem 5.36.

OC is horizontal. The spring is linear with stiffness k N/m, and is unstressed when $\phi = 0$ (i.e., C is along the vertical). The block can rotate about C, on a vertical plane, as indicated by the angle θ shown in the figure. Neglect friction in the system.

a. Obtain the differential equations of motion of the system and, based on such equations, explain what $\theta(t)$ does if the motion is started with the initial conditions $\theta(0) = \theta_0$ (arbitrary), $\dot{\theta}(0) = 20$ rpm, and with any $\phi(0)$

and any $\dot{\phi}(0)$.

b. Verify that $\phi = 0$ and $\phi = 180°$ are two of the equilibrium solutions. There is a third equilibrium solution $\phi = $ constant $\overset{\Delta}{=} \phi_e$. Plot it versus $kR/(mg)$ (where $g = 9.81$ m/s^2 is the acceleration of gravity) for ϕ_e in the range $-60 \le \phi_e \le 60°$.

c. When $kR/(mg) = \sqrt{3} + \sqrt{2}$, $\phi_e = 30°$ is one equilibrium of the system. Determine all the other equilibrium solutions ϕ_e and analyze the stability of all of them (including the equilibrium $\phi_e = 30°$), specifying whether they are stable, asymptotically stable, or unstable.

d. For $kR/(mg) = \sqrt{3} + \sqrt{2}$, plot $d\phi/d\tau$ versus $\phi(\tau)$, where $\tau = t\sqrt{g/R}$ is nondimensional time, and identify in your plot the region of $\phi(0)$ and $[d\phi/d\tau]_{\tau=0}$ that result in an oscillatory motion $\phi(t)$ about $\phi_e = 0$. *Suggestion:* It is simpler to use the approach $dW = d(T)$, instead of $\vec{F} = m\vec{a}_C$.

5.37   The system shown in Fig. 5.48 consists of a thin block of mass m, with a pin through its center of mass C that is able to slide along a vertical circular plate

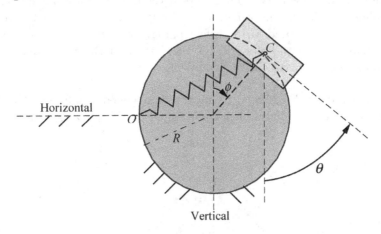

Figure 5.48: Figure for Problem 5.37.

of radius R. The block may leave the surface of the plate because no provision is made to keep it there at all times. A spring is attached from a fixed point O to C, and the distance from O to C is equal to $2R$ when line OC is horizontal. The spring is linear with stiffness k N/m, is unstressed when $\phi = 0$ (i.e., when C is along the vertical), and contacts are approximated as frictionless. The block can rotate about C, as indicated by the angle θ shown in the figure.

Motion takes place on the vertical plane

a. With g being the acceleration of gravity, show that the normal reaction force N applied by the plate to the block is (the direction of the force is left for you to determine)

$$\frac{N}{mg} = 3\cos\phi - 2H + \frac{kR}{mg}\left[5 + 3\sin\phi - 5\sqrt{1 + \sin\phi}\right]$$

where (with $\tau = t\sqrt{g/R}$),

$$H = \frac{1}{2}\left(\frac{d\phi}{d\tau}\right)^2 + \cos\phi + \frac{kR}{mg}\left(\sqrt{1 + \sin\phi} - 1\right)^2 = \text{constant} = [H]_{\tau=0}$$

b. With $kR/(mg) = 5.5$, $\phi(0) = 30^o$, and $\dot{\phi}(0) = 0$, integrate numerically the differential equation of motion stopping the integration automatically if the block starts to leave the plate, and plot $\phi(\tau)$ and $N/(mg)$ versus τ and $d\phi/d\tau$ versus $\phi(\tau)$. Then, repeat the integration using slightly larger values of $\phi(0)$ in the range $30^o < \phi(0) \le 40^o$. Observe the results of the integration and write a few paragraphs with your comments and conclusions. If the block leaves the surface of the plate, determine the value of $\phi(0)$ for which that happens.

5.38   The body shown in Fig. 5.49 consists of a homogeneous half-cylinder of radius R, mass m, and negligible thickness that can roll on a horizontal surface.

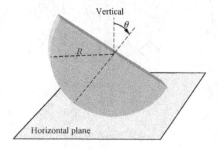

Figure 5.49: Figure for Problem 5.38.

Motion takes place on a vertical plane (i.e., the body neither tips over nor rotates laterally) and without slipping. Show that the center of mass C of the body is located at a distance equal to $4R/(3\pi)$ from the flat portion of the body and that its moment of inertia I_C about the axis that passes through C and is perpendicular to the plane of the body is equal to $I_C = \left[1/2 - 16/ \left(9\pi^2 \right) \right] mR^2$. Using these results, obtain the differential equation of motion for the body in terms of the angle θ shown in Fig. 5.49. Verify that the differential equation exhibits the equilibrium $\theta = \text{constant} \overset{\triangle}{=} \theta_e = 0$ and that the equilibrium is stable. For $R = 1$ m, what is the frequency of the small oscillations of the system about its equilibrium?

CHAPTER 6

Dynamics of Rigid Bodies in General Motion

The general motion of a rigid body consists of the translation of a chosen reference point on the body, and a rotation of the body about an axis that passes through that reference point. In general, the direction of that axis is also changing with time, and, for that reason, such an axis is called an *instantaneous axis of rotation*. The simpler case when the motion is in two-dimensional space was dealt with in Chapter 5. Two-dimensional motion involves only three degrees of freedom, namely, two for the motion of any reference point on the body (normally taken to be the center of mass of the body) and one that describes the rotational motion (i.e., change in angular orientation) of the body. Three-dimensional motion involves six degrees of freedom; i.e., six variables are necessary to completely describe the position and orientation of the body in space. Three such variables locate the position of a reference point of the body (again, normally taken to be the center of mass of the body), and the other three specify the angular orientation of the body in space. The basic vector equations that govern the motion are the same as Eqs. (5.3.2) and (5.3.5) in Chapter 5. Those equations govern the motion of the center of mass of the body and the rotational motion, respectively.

For rotational motion in three-dimensional space, the absolute angular momentum and absolute angular velocity of the body are not necessarily parallel to each other. For such a motion, the expression for the angular momentum involves three *mass moments of inertia* and three *mass products of inertia* for the body. These are essential quantities in dynamics, and they appear in the formulation of the differential equations that govern the most general rotational motion of a body.

Section 6.1 introduces a mathematical operation with vectors that was not presented in Chapter 1. The operation is the *juxtaposition* of two vectors side by side to form what is called a *dyadic* (or second-order *tensor*). The matrix representation of a tensor is also presented and used in this chapter. The use of dyadics is optional (since one could work, instead, with the matrix representation of a dyadic), but they are convenient quantities that naturally appear when the explicit expression for the angular momentum of a body is formulated. The studying of Section 6.1 may be skipped for now since the use of dyadics appears for the first time in the book in Eq. (6.3.7) (p. 404).

Several examples involving kinematics and dynamics of rigid bodies in three dimensions are included in this chapter. Also included are several classical problems involving rotational motion in three-dimensional space.

6.1 Juxtaposition of Two Vectors: Dyadics

To see how dyadics appear naturally, consider two parallel vectors $\vec{v}$ and $\vec{c}$. Since the vectors are parallel to each other, they are related by some scalar quantity k as $\vec{v} = k\vec{c}$. Now, express k as a dot product of two other vectors, say $\vec{a}$ and $\vec{b}$, i.e., $k = \vec{a} \bullet \vec{b}$. Thus, we can always write $\vec{v} = (\vec{a} \bullet \vec{b})\vec{c}$. The right-hand side of this expression simply reads, "a scalar times $\vec{c}$," which results in a vector, of course. If the parentheses are left out, such an expression is written as $\vec{v} = \vec{a} \bullet \vec{b}\vec{c}$, and we may now stare at the quantity $\vec{b}\vec{c}$ and ask, "What is that?" The quantity $\vec{b}\vec{c}$, which is just the juxtaposition of two vectors, i.e., two vectors written side by side, is called a *dyad*. If we have, for example, the vectors $2\vec{v}_1$ and $3\vec{v}_2$, their juxtaposition is, by definition, $6\vec{v}_1\vec{v}_2$, i.e., the scalars associated with them are, by definition of the juxtaposition operation, multiplied by each other.

It is seen that the dot product of a vector with a dyad, or vice versa, yields a vector. Notice that $\vec{a} \bullet (\vec{b}\vec{c}) \neq (\vec{b}\vec{c}) \bullet \vec{a}$ in general. Also notice that $\vec{b}\vec{c} \neq \vec{c}\vec{b}$. The use of dyads is especially convenient in rigid body kinematics for transforming the inertia matrix of a body from a set of axis into another (see Example 6.3.5, p. 415).

A *dyadic* (which is also called a *second-order tensor*) is a quantity defined as a sum of dyads, such as $\vec{a}_1\vec{a}_2 + \vec{b}_1\vec{b}_2 + \dots$. In this book, dyadics are denoted by a letter with two arrows over it, such as $\overset{\leftrightarrow}{c}$.

Since any vector $\vec{a}$ may be written in terms of three components in any triad $\{\hat{e}_1, \hat{e}_2, \hat{e}_3\}$ as $\vec{a} = a_{e_1}\hat{e}_1 + a_{e_2}\hat{e}_2 + a_{e_3}\hat{e}_3$, and since $a_{e_i} = \vec{a} \bullet \hat{e}_i$ for $i = 1, 2, 3$, it follows that

$$
\begin{aligned}
\vec{a} &= (\vec{a} \bullet \hat{e}_1)\hat{e}_1 + (\vec{a} \bullet \hat{e}_2)\hat{e}_2 + (\vec{a} \bullet \hat{e}_3)\hat{e}_3 \\
&= \vec{a} \bullet (\hat{e}_1\hat{e}_1 + \hat{e}_2\hat{e}_2 + \hat{e}_3\hat{e}_3) \overset{\Delta}{=} \vec{a} \bullet \hat{U}
\end{aligned}
\tag{6.1.1}
$$

The quantity $\hat{U} \overset{\Delta}{=} \hat{e}_1\hat{e}_1 + \hat{e}_2\hat{e}_2 + \hat{e}_3\hat{e}_3$ is called the *unit dyadic*. Notice that $\vec{a} \bullet \hat{U} = \hat{U} \bullet \vec{a} = \vec{a}$. Also, if $\vec{b}$ is any other vector expressed in the same $\{\hat{e}_1, \hat{e}_2, \hat{e}_3\}$ triad as $\vec{a}$, i.e., $\vec{b} = b_{e_1}\hat{e}_1 + b_{e_2}\hat{e}_2 + b_{e_3}\hat{e}_3$, it is seen that any dyadic $\vec{a}\vec{b}$ has nine elements, as follows:

$$\begin{aligned}
\vec{a}\vec{b} =&(a_{e_1}\hat{e}_1 + a_{e_2}\hat{e}_2 + a_{e_3}\hat{e}_3)(b_{e_1}\hat{e}_1 + b_{e_2}\hat{e}_2 + b_{e_3}\hat{e}_3)\\
=&a_{e_1}b_{e_1}\hat{e}_1\hat{e}_1 + a_{e_1}b_{e_2}\hat{e}_1\hat{e}_2 + a_{e_1}b_{e_3}\hat{e}_1\hat{e}_3\\
&+a_{e_2}b_{e_1}\hat{e}_2\hat{e}_1 + a_{e_2}b_{e_2}\hat{e}_2\hat{e}_2 + a_{e_2}b_{e_3}\hat{e}_2\hat{e}_3\\
&+a_{e_3}b_{e_1}\hat{e}_3\hat{e}_1 + a_{e_3}b_{e_2}\hat{e}_3\hat{e}_2 + a_{e_3}b_{e_3}\hat{e}_3\hat{e}_3
\end{aligned} \tag{6.1.2}$$

The dot product of a dyadic $\vec{\vec{I}} = I_{e_1e_1}\hat{e}_1\hat{e}_1 + \ldots + I_{e_3e_3}\hat{e}_3\hat{e}_3$ with a vector $\vec{a} = a_{e_1}\hat{e}_1 + a_{e_2}\hat{e}_2 + a_{e_3}\hat{e}_3$ is another vector $\vec{b} = \vec{\vec{I}} \bullet \vec{a}$ that is obtained as follows:

$$\begin{aligned}
\vec{b} = \vec{\vec{I}} \bullet \vec{a} =& (I_{e_1e_1}a_{e_1} + I_{e_1e_2}a_{e_2} + I_{e_1e_3}a_{e_3})\hat{e}_1\\
&+ (I_{e_2e_1}a_{e_1} + I_{e_2e_2}a_{e_2} + I_{e_2e_3}a_{e_3})\hat{e}_2\\
&+ (I_{e_3e_1}a_{e_1} + I_{e_3e_2}a_{e_2} + I_{e_3e_3}a_{e_3})\hat{e}_3
\end{aligned} \tag{6.1.3}$$

As suggested by Eqs. (6.1.2) and (6.1.3), all nine elements of the dyadic $\vec{\vec{I}}$ may be represented by a matrix $[I]$ as

$$[I] = \begin{bmatrix} I_{e_1e_1} & I_{e_1e_2} & I_{e_1e_3}\\ I_{e_2e_1} & I_{e_2e_2} & I_{e_2e_3}\\ I_{e_3e_1} & I_{e_3e_2} & I_{e_3e_3} \end{bmatrix} \tag{6.1.4}$$

Thus, in matrix notation, the representation of the dot product $\vec{b} = \vec{\vec{I}} \bullet \vec{a}$ given by Eq. (6.1.3), with both $\vec{a}$ and $\vec{b}$ expressed in a same unit vector triad $\{e\} = \{\hat{e}_1, \hat{e}_2, \hat{e}_3\}$, is simply obtained by a matrix multiplication as follows:

$$\underline{b} \stackrel{\Delta}{=} \begin{bmatrix} b_{e_1}\\ b_{e_2}\\ b_{e_3} \end{bmatrix} = [I]\,\underline{a} = \begin{bmatrix} I_{e_1e_1} & I_{e_1e_2} & I_{e_1e_3}\\ I_{e_2e_1} & I_{e_2e_2} & I_{e_2e_3}\\ I_{e_3e_1} & I_{e_3e_2} & I_{e_3e_3} \end{bmatrix} \begin{bmatrix} a_{e_1}\\ a_{e_2}\\ a_{e_3} \end{bmatrix} \tag{6.1.5}$$

Notice that the *matrix representation of a vector* is a column matrix whose elements are the scalar components of the vector. A column matrix is represented in this book with an underlined symbol, such as

$$\underline{v} = \begin{bmatrix} v_1\\ v_2\\ v_3 \end{bmatrix} = [v_1 \; v_2 \; v_3]^T$$

with the superscript T denoting the transpose operation.

A dyadic $\vec{\vec{I}}$ is said to be symmetric if its matrix representation $[I]$ is symmetric, i.e., if $I_{ij} = I_{ji}$.

6.2   The Two Basic Vector Equations of Rigid Body Dynamics

Figure 6.1 illustrates an arbitrary rigid body subjected to a set of forces $\vec{F}_1, \vec{F}_2, \ldots$. Point C is the center of mass of the body, O is an inertial point, and A is a point that may move in any manner in space.

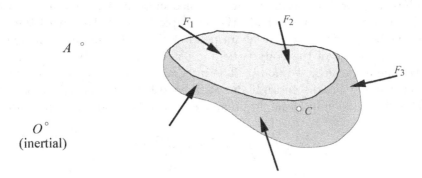

Figure 6.1: A rigid body subjected to a set of forces (C is the center of mass of the body).

The body may or may not be given an initial motion, and the basic dynamics problem consists of determining the subsequent motion of the body, i.e., its position and its angular orientation in space at all times. As mentioned in the introduction to this chapter, motion in three-dimensional space involves six degrees of freedom; i.e., six variables are needed to describe the motion. Three such variables describe the translation of a reference point that is either on the body or on a rigid extension of the body (such as the case of a doughnut, for example). The other three variables describe the rotational motion of the body, i.e., motion that is characterized by the change of orientation of a set of axes fixed to the body.

The basic vector equations that describe the general motion of a rigid body are the same as those already obtained in Chapter 4 and used in Chapter 5. They are the following.

■ For the motion of the center of mass (see Fig. 6.1):

$$\vec{F} \triangleq \sum_{i=1}^{N} \vec{F}_i = m \frac{d}{dt} (\vec{v}_C) = m \frac{d^2}{dt^2} \overrightarrow{OC} \tag{6.2.1}$$

■ For the rotational motion of the body:

$$\frac{d}{dt}\vec{H}_A = \vec{M}_A + m\left(\frac{d}{dt}\overrightarrow{AC}\right) \times \left(\frac{d}{dt}\overrightarrow{OC}\right) \tag{6.2.2}$$

where $\vec{M}_A$ is the moment, about point A (see Fig. 6.1) of all forces applied to the body, and $\vec{H}_A$ is the absolute angular momentum (referred to point A) of the body. The absolute angular momentum $\vec{H}_A$ was defined in Chapters 4 and 5. For a rigid body it is the integral over the body (i.e., a triple integral that spans the entire volume of the body) of the moment about point A of the absolute linear momentum of a mass element dm of the body (where dm is equal to the body mass density times a volume element of the body). The integral over the body assures that the mass element dm spans the entire body. Mathematically, $\vec{H}_A$ is defined as follows (where P is the location of an infinitesimal element of the body, with mass dm):

$$\vec{H}_A \triangleq \int_{\text{body}} \left[\overrightarrow{AP} \times \frac{d}{dt}\overrightarrow{OP}\right] dm \tag{6.2.3}$$

Equation (6.2.2) reduces to the simpler forms as follows, when point A is either an inertial point O or the center of mass C of the body.

$$\frac{d}{dt}\vec{H}_O = \vec{M}_O \qquad \text{(when } O \text{ is inertial)} \tag{6.2.4}$$

$$\frac{d}{dt}\vec{H}_C = \vec{M}_C \qquad \text{(when } C \text{ is the center of mass)} \tag{6.2.5}$$

These forms are used in this book. When point A is an inertial point O, $\overrightarrow{AP}$ is, of course, $\overrightarrow{OP}$ in Eq. (6.2.3). Similarly, $\overrightarrow{AP} = \overrightarrow{CP}$ when point A is the center of mass C.

To apply either Eq. (6.2.5) or Eq. (6.2.4) to specific problems, an explicit expression for the angular momentum $\vec{H}_C$ (or $\vec{H}_O$) needs to be developed. This is done in Section 6.3. Several problems are then presented afterward.

6.3   Rotational Kinematics of Rigid Bodies in General Motion

As will be seen in this section, the absolute angular momentum of a rigid body depends on the angular velocity $\vec{\omega}^B$ of the body with respect to inertial space, and on a set of quantities that are called moments of inertia and products of inertia of the body. It will also be seen that the angular momentum vector and the angular velocity vector are not necessarily parallel to each other, as was the case for the simpler type of motion considered in Chapter 5. To analyze the rotational motion of the body, it is necessary to determine the relationship between the absolute angular momentum and the absolute angular velocity of the body before Eq. (6.2.5) or Eq. (6.2.4) can be used for solving problems. This is done in Section 6.3.1.

6.3.1 Absolute Angular Momentum

Consider a rigid body undergoing general motion. As illustrated in Fig. 6.2, its center of mass C is translating in space, and the body is also rotating with absolute (i.e., relative to inertial space) angular velocity $\vec{\omega}^B(t)$. Both the magnitude and the direction of $\vec{\omega}^B(t)$ may be changing with time. The x-y-z axes shown in that figure are centered at C, but are not necessarily fixed to the body. The unit vectors $\hat{x}$, $\hat{y}$, and $\hat{z}$ are parallel to the x-, y-, and z-axes, respectively. If one chooses the x-y-z axes not to be fixed to the rigid body, their absolute angular velocity is different from $\vec{\omega}^B$ and the coordinates of any vector $\overrightarrow{CP}$ change with time (but the magnitude of $\overrightarrow{CP}$ still remains constant because the body is rigid).

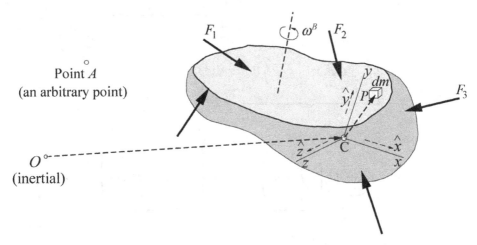

Figure 6.2: A rigid body able to move in space with six degrees of freedom.

The absolute angular momentum $\vec{H}_A$ of the body, referred to any point A, is given by Eq. (6.2.3). By letting point A in Eq. (6.2.3) be the center of mass C, that equation yields the following expression for the absolute angular momentum $\vec{H}_C$:

$$\vec{H}_C = \int_{body} \left[\overrightarrow{CP} \times \frac{d}{dt}\overrightarrow{OP} \right] dm$$

$$= \underbrace{\left[\int_{body} \overrightarrow{CP}\, dm\right]}_{=0} \times \frac{d}{dt}\overrightarrow{OC} + \int_{body}\left[\overrightarrow{CP} \times \frac{d}{dt}\overrightarrow{CP}\right] dm \qquad (6.3.1)$$

The first integral in Eq. (6.3.1) is equal to zero because of the definition of the center of mass.

Note: For any other point A, the angular momentum $\vec{H}_A$ is the sum of the angular momentum $\vec{H}_C$ and the moment, about A, of the linear momentum $m\vec{v}_C$ (where $\vec{v}_C$ is the absolute velocity of C) of the body, i.e.,

$$\vec{H}_A = \vec{H}_C + \overrightarrow{AC} \times (m\vec{v}_C)$$

This relationship between $\vec{H}_A$ and $\vec{H}_C$ is obtained as follows:

$$\vec{H}_A = \int_{body}\left(\overrightarrow{AP} \times \frac{d}{dt}\overrightarrow{OP}\right) dm = \int_{body}\left(\overrightarrow{AC} \times \frac{d}{dt}\overrightarrow{OP}\right) dm$$

$$= \underbrace{\int_{body}\left(\overrightarrow{AC}\times\frac{d}{dt}\overrightarrow{OC}\right)dm + \int_{body}\left(\overrightarrow{AC}\times\frac{d}{dt}\overrightarrow{CP}\right)dm}$$

$$+ \underbrace{\int_{body}\left(\overrightarrow{CP} \times \frac{d}{dt}\overrightarrow{OP}\right) dm}_{=\vec{H}_C}$$

$$= \overrightarrow{AC} \times \underbrace{\frac{d}{dt}\overrightarrow{OC}}_{=\vec{v}_C}\underbrace{\left(\int_{body} dm\right)}_{=m} + \overrightarrow{AC} \times \left(\frac{d}{dt}\underbrace{\int_{body}\overrightarrow{CP}\, dm}_{=0}\right) + \vec{H}_C$$

$$= \vec{H}_C + \overrightarrow{AC} \times (m\vec{v}_C)$$

The vector identities $\overrightarrow{AP} = \overrightarrow{AC} + \overrightarrow{CP}$ and $\overrightarrow{OP} = \overrightarrow{OC} + \overrightarrow{CP}$ used in this equation are obtained by inspection of Fig. 6.2. Also, the last integral that appears in the equation is equal to zero because of the definition of the center of mass C.

To develop a more useful expression for the angular momentum $\vec{H}_C$ from Eq. (6.3.1), it is convenient to express the vector $\overrightarrow{CP}$ and the absolute angular velocity $\vec{\omega}^B$ of the body in terms of orthogonal unit vectors $\hat{x}$, $\hat{y}$, and $\hat{z}$ that are parallel to the x-, y-, and z-axes, respectively, centered at C as shown in Fig. 6.2. At this point of this development, those axes are not necessarily fixed to the body, and they may be chosen to be rotating in any manner whatsoever. It will be seen later on, however, that it is advantageous to choose such axes to be fixed to the rotating rigid body. In general, the expressions for $\overrightarrow{CP}$ and $\vec{\omega}^B$ are of the following forms:

$$\overrightarrow{CP} \triangleq x\hat{x} + y\hat{y} + z\hat{z} \tag{6.3.2}$$

$$\vec{\omega}^B \triangleq \omega_x\hat{x} + \omega_y\hat{y} + \omega_z\hat{z} \tag{6.3.3}$$

Since the magnitude of the vector $\overrightarrow{CP}$ from C to an element of mass dm (located at point P in Fig. 6.2) is constant, the time derivative of the vector $\overrightarrow{CP}$ that appears in Eq. (6.3.1) for the angular momentum $\vec{H}_C$ is then $d\overrightarrow{CP}/dt = \vec{\omega}^B \times \overrightarrow{CP}$ (because the angular velocity $\vec{\omega}^B$ is what causes the direction of $\overrightarrow{CP}$ to change).

The cross product $\overrightarrow{CP} \times (\vec{\omega}^B \times \overrightarrow{CP})$ is then

$$\overrightarrow{CP} \times \left(\vec{\omega}^B \times \overrightarrow{CP}\right) = \overrightarrow{CP} \times \begin{vmatrix} \hat{x} & \hat{y} & \hat{z} \\ \omega_x & \omega_y & \omega_z \\ x & y & z \end{vmatrix}$$

$$= \begin{vmatrix} \hat{x} & \hat{y} & \hat{z} \\ x & y & z \\ \omega_y z - \omega_z y & \omega_z x - \omega_x z & \omega_x y - \omega_y x \end{vmatrix}$$

$$= \left[\left(y^2 + z^2\right)\omega_x - xy\omega_y - xz\omega_z\right]\hat{x}$$
$$+ \left[-yx\omega_x + \left(x^2 + z^2\right)\omega_y - yz\omega_z\right]\hat{y}$$
$$+ \left[-zx\omega_x - zy\omega_y + \left(x^2 + y^2\right)\omega_z\right]\hat{z} \tag{6.3.4}$$

By substituting this expression for $\overrightarrow{CP} \times (\vec{\omega}^B \times \overrightarrow{CP})$ into Eq. (6.3.1), the following expression is then obtained for the angular momentum $\vec{H}_C$:

$$\vec{H}_C = \left(I_{xx}\omega_x + I_{xy}\omega_y + I_{xz}\omega_z\right)\hat{x}$$
$$+ \left(I_{yx}\omega_x + I_{yy}\omega_y + I_{yz}\omega_z\right)\hat{y}$$
$$+ \left(I_{zx}\omega_x + I_{zy}\omega_y + I_{zz}\omega_z\right)\hat{z}$$
$$\triangleq H_x\hat{x} + H_y\hat{y} + H_z\hat{z} \tag{6.3.5}$$

where

$$I_{xx} = \int_{\text{body}} \left(y^2 + z^2\right) dm \qquad I_{yy} = \int_{\text{body}} \left(x^2 + z^2\right) dm \qquad I_{zz} = \int_{\text{body}} \left(x^2 + y^2\right) dm$$

$$I_{xy} = I_{yx} = -\int_{\text{body}} xy\, dm \qquad I_{xz} = I_{zx} = -\int_{\text{body}} xz\, dm \qquad I_{yz} = I_{zy} = -\int_{\text{body}} yz\, dm$$

$$\tag{6.3.6}$$

The expression for the angular momentum given in Eq. (6.3.5) may be written in the following form, which resembles the one that was obtained in Chapter 5 for the case of simpler planar motion of rigid bodies.

$$\boxed{\vec{H}_C = \vec{\vec{I}}_C \bullet \vec{\omega}^B} \qquad (6.3.7)$$

In this equation, the quantity $\vec{\vec{I}}_C$ is called the *inertia dyadic*, or *inertia tensor* of the body, and it is equal to the following nine-element dyadic:

$$\begin{aligned}
\vec{\vec{I}}_C = \quad & I_{xx}\hat{x}\hat{x} + I_{xy}\hat{x}\hat{y} + I_{xz}\hat{x}\hat{z} \\
+ \ & I_{yx}\hat{y}\hat{x} + I_{yy}\hat{y}\hat{y} + I_{yz}\hat{y}\hat{z} \\
+ \ & I_{zx}\hat{z}\hat{x} + I_{zy}\hat{z}\hat{y} + I_{zz}\hat{z}\hat{z}
\end{aligned} \qquad (6.3.8)$$

The relationship between the angular momentum $\vec{H}_C$ and absolute angular velocity $\vec{\omega}^B$ of the body may also be represented in matrix form as follows:

$$\begin{bmatrix} H_x \\ H_y \\ H_z \end{bmatrix} = [I_C] \begin{bmatrix} \omega_x \\ \omega_y \\ \omega_z \end{bmatrix}$$

or, in short notation,

$$\boxed{\underline{H}_C = [I_C]\underline{\omega}^B} \qquad (6.3.9)$$

where $\underline{H}_C \triangleq [H_x\ H_y\ H_z]^T$, $\underline{\omega}^B = [\omega_x\ \omega_y\ \omega_z]^T$, and $[I_C]$ is the *symmetric* 3×3 matrix,

$$[I_C] = \begin{bmatrix} I_{xx} & I_{xy} & I_{xz} \\ I_{yx} & I_{yy} & I_{yz} \\ I_{zx} & I_{zy} & I_{zz} \end{bmatrix} \qquad (6.3.10)$$

which is called the *inertia matrix* of the body, referred to the center of mass C.

> The inertia matrix $[I_C]$ is the *matrix representation* of the inertia dyadic $\vec{\vec{I}}_C$. Whether one uses dyadics or matrices and whether one writes the angular momentum as either $\vec{H}_C = \vec{\vec{I}}_C \bullet \vec{\omega}^B$ or $\underline{H}_C = [I_C]\underline{\omega}^B$ is entirely a matter of individual preference. But one should never equate $\vec{H}_C$ to $\underline{H}_C$ because the former is a vector $\vec{H}_C = H_x\hat{x} + H_y\hat{y} + H_z\hat{z}$ and the latter is the column matrix $\underline{H}_C = [H_x\ H_y\ H_z]^T$ (which is the *matrix representation* of the vector $\vec{H}_C$).

■ The elements I_{xx}, I_{yy}, and I_{zz} that appear in the diagonal of the inertia matrix are called the *mass moments of inertia* (or, for short, the *moments of inertia*) of the body about the x-, y-, and z-axes, respectively.

■ The quantities I_{xy}, I_{xz}, and I_{yz} defined by the last three equations of (6.3.6) are called the *products of inertia* of the body with respect to the x-y-z axes.

■ Notice that each one of the quantities $y^2 + z^2$, $x^2 + z^2$, and $x^2 + y^2$ that appear in the definitions of the moments of inertia are simply the squares of the perpendicular distances from the mass element dm to the x-, y-, and z-axes, respectively. Although the value of each one of the moments of inertia of the body depends on the particular orientation of the x-y-z axes, it is of interest to notice that the sum $I_{xx} + I_{yy} + I_{zz}$ does not since, as disclosed by the equations of (6.3.6),

$$I_{xx} + I_{yy} + I_{zz} = 2 \int_{\text{body}} \left(x^2 + y^2 + z^2 \right) dm = 2 \int_{\text{body}} \overrightarrow{CP} \bullet \overrightarrow{CP} \, dm$$

$$= 2 \int_{\text{body}} |\overrightarrow{CP}|^2 dm \qquad (6.3.11)$$

where $|\overrightarrow{CP}|$ is the constant distance from C to the mass element dm. This is one of the properties of any inertia matrix.

Inspection of the equations of (6.3.6), which define the moments and products of inertia, also reveals the following two properties of the inertia matrix.

■ The equations of (6.3.6) also disclose that any of the three moments of inertia of a body is never greater than the sum of the other two. This is an important property of inertia matrices. For example, $I_{zz} \leq I_{xx} + I_{yy}$ (since $I_{xx} + I_{yy} = I_{zz} + 2 \int_{\text{body}} z^2 dm$ and the integral term is never negative), with the equal sign holding for a flat body for which $z = 0$.

■ Also notice that $I_{xx} \pm 2I_{yz} = \int_{\text{body}} \left(y^2 + z^2 \mp 2yz \right) dm = \int_{\text{body}} \left(y \mp z \right)^2 dm$ is never negative. The other two similar combinations involve I_{yy} and I_{xz}, and also I_{zz} and I_{xy}. Therefore, the elements of any inertia matrix must also satisfy the following conditions:

$$I_{xx} \pm 2I_{yz} \geq 0 \qquad I_{yy} \pm 2I_{xz} \geq 0 \qquad I_{zz} \pm 2I_{xy} \geq 0$$

The other properties of the inertia matrix are given in Appendix C.

As mentioned earlier, the quantities I_{xy}, I_{xz}, and I_{yz} defined by the last three equations of (6.3.6) are called the *products of inertia* of the body with respect to the chosen x-, y-, and z-axes. Depending on the orientation of such axes, each one of those quantities may be positive, negative, or equal to zero.

Note: Notice that the negative signs seen in Eq. (6.3.4) were just embedded in the definitions of products of inertia given by the last three equations of (6.3.6). Many authors (perhaps most) define products of inertia in this manner, and several others do not embed that negative sign in their definitions. Actually, there is no universally adopted sign convention for the definition of products of inertia, and this can cause confusion in engineering practice. If one defines I_{xy} as $\int_{\text{body}} xy \, dm$ (with similar sign definitions for I_{xz} and I_{yz}), the off-diagonal elements of the matrix $[I_C]$ in Eq. (6.3.9) for the angular momentum will be the negative of such products of inertia, instead of being the products of inertia themselves. However, regardless of the manner in which they are defined, the actual values for products of inertia may be positive, negative, or zero, while none of the moments of inertia of a body can be negative.

If the x-y-z axes are not fixed to the rigid body, the moments and products of inertia of the body referred to those axes will, in general, change with time as the body rotates, which is inconvenient. To formulate and analyze dynamics problems, it is convenient to work with axes for which the moments and products of inertia are constants.

It is shown in Section 6.3.4 that it is always possible to find an orientation for any set of orthogonal x-y-z axes, centered at any point on the body, so that the three products of inertia of the body are zero when referred to such axes. Such axes are called the *principal axes of inertia* of the body. Thus, if the x-y-z axes are principal axes of inertia, the expression for the angular momentum $\vec{H}_C$ given by Eq. (6.3.5) reduces to the simpler expression $\vec{H}_C = I_{xx}\omega_x\hat{x} + I_{yy}\omega_y\hat{y} + I_{zz}\omega_z\hat{z}$, which is another advantage of working with such axes once they are located. The moments of inertia of a body about its principal axes are called the *principal moments of inertia* (or, often, the *central principal moments of inertia* if the axes are centered at the center of mass C).

6.3.2   The Parallel Axis Theorem

Some applications require calculation of the moments and products of inertia of a body about two sets of orthogonal axes, one of which is centered at the center of mass of the body, and with each axis of one set being parallel to a corresponding axis of the other set. Once the moments and products of inertia of the body relative to one such set of axes are known, the moments and products of inertia relative to the other set of axes can be calculated in a simple manner. The relationships between the inertias of the body about those two sets of orthogonal axes are called the *parallel axis theorem*. This is presented in this section; for an application, see Example 6.3.2.

Figure 6.3 shows a rigid body and two sets of parallel axes; the x-y-z axes are centered at the center of mass C of the body, and the x_1-y_1-z_1 axes are centered at

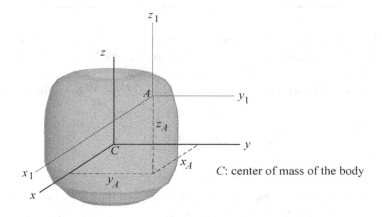

Figure 6.3: Figure for calculations involving the parallel axis theorem.

some other point A, whose rectangular coordinates relative to C are x_A, y_A, and z_A, respectively. The elements of the inertia matrix $[I_C]$ about the axes centered at the center of mass C are calculated by performing the integrations shown in the equations of (6.3.6). The elements $I_{x_1x_1}, I_{x_1y_1}$, etc., of the inertia matrix $[I_A]$ about the axes centered at the origin A of the x_1-y_1-z_1 axes are calculated in the same manner, with x, y, and z replaced by x_1, y_1, and z_1, respectively. The steps involved in the calculation of $I_{x_1x_1}$ and $I_{x_1y_1}$ are presented in the following. The calculation of the remaining moments and products of inertia follows the same procedure, and the results are similar to the ones obtained for $I_{x_1x_1}$ and $I_{x_1y_1}$.

By letting $x = x_A + x_1$, $y = y_A + y_1$, and $z = z_A + z_1$ (see Fig. 6.3) we obtain:

$$I_{x_1x_1} = \int_{\text{body}} \left(y_1^2 + z_1^2\right) dm = \int_{\text{body}} \left[(y - y_A)^2 + (z - z_A)^2\right] dm$$

$$= \underbrace{\int\limits_{\text{body}} \left(y^2 + z^2\right) dm}_{=I_{xx}} + m\left(y_A^2 + z_A^2\right) - 2y_A \underbrace{\int\limits_{\text{body}} y\, dm}_{=0} - 2z_A \underbrace{\int\limits_{\text{body}} z\, dm}_{=0}$$

$$I_{x_1 y_1} = I_{y_1 x_1} = -\int\limits_{\text{body}} x_1 y_1\, dm = -\int\limits_{\text{body}} (x - x_A)(y - y_A)\, dm$$

$$= \underbrace{-\int\limits_{\text{body}} xy\, dm}_{=I_{xy}} - mx_A y_A + y_A \underbrace{\int\limits_{\text{body}} x\, dm}_{=0} + x_A \underbrace{\int\limits_{\text{body}} y\, dm}_{=0}$$

The last two integrals in the expressions for $I_{x_1 x_1}$ and $I_{x_1 y_1}$ are zero because of the definition of the center of mass C. They would not be zero if the x-y-z axes were not centered at the center of mass of the body.

These results can be summarized, with similar expressions involving z as follows[1]:

$$I_{x_1 x_1} = I_{xx} + m\left(y_A^2 + z_A^2\right) \qquad (6.3.12)$$

$$I_{x_1 y_1} = I_{xy} - mx_A y_A \qquad (6.3.13)$$

The relationship given by Eq. (6.3.12) between the moment of inertia I_{xx} about an x-axis centered at the center of mass, and the moment of inertia $I_{x_1 x_1}$ about an x_1-axis parallel to the x-axis, is of the same form as Eq. (5.3.8) (p. 321). The quantity $m(y_A^2 + z_A^2)$ is simply the product of the mass of the body and the square of the perpendicular distance between the two parallel axes x and x_1.

Equation (6.3.13) is the parallel axis theorem for products of inertia. The quantity $mx_A y_A$ that appears in the expression for $I_{x_1 y_1}$ is simply the product of the mass of the body and the x-y rectangular coordinates of the new origin A relative to the center of mass C. Thus, for $I_{y_1 z_1}$, for example, one would have $I_{y_1 z_1} = I_{yz} - my_A z_A$.

6.3.3   Examples Involving Moments and Products of Inertia

Example 6.3.1 A homogeneous bar of mass m is bent as shown in Fig. 6.4. The dimensions of the cross section of the bar are very small and may be neglected. For

---

[1]Equations (6.3.12) and (6.3.13) would also have been obtained if one used, instead, $x = x_A - x_1$, $y = y_A - y_1$, and $z = z_A - z_1$.

the x-y-z axes shown, which are centered at the center of mass C of the bar and with the z-axis being perpendicular to the plane of the figure, determine all the elements of the inertia matrix $[I_C]$ for the bent bar.

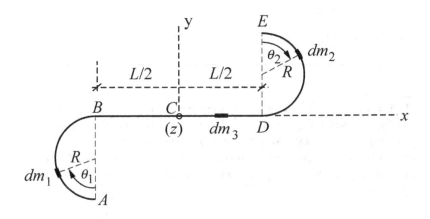

Figure 6.4: The bar for Example 6.3.1.

■ Solution

Since the bar is homogeneous, its density (i.e., its mass per unit length) is equal to $\rho = m/(L + 2\pi R)$. Therefore, by dividing the bar into the segments AB, BD, and DE shown in Fig. 6.4, the masses of the infinitesimal mass elements for each segment are equal to $dm_1 = \rho R\, d\theta_1$, $dm_2 = \rho R\, d\theta_2$, and $dm_3 = \rho\, dx$, respectively. With $z = 0$, we then obtain

$$
\begin{aligned}
I_{xx} &= \int y^2\, dm = \int_{\theta_1=0}^{\pi} (R + R\cos\theta_1)^2 \rho R\, d\theta_1 + \int_{\theta_2=0}^{\pi} (R + R\cos\theta_2)^2 \rho R\, d\theta_2 \\
&= 2\int_{\theta=0}^{\pi} (R + R\cos\theta)^2 \rho R\, d\theta = 3\pi\rho R^3
\end{aligned}
$$

$$
\begin{aligned}
I_{yy} &= \int x^2\, dm = 2\int_{\theta=0}^{\pi} \left(\frac{L}{2} + R\sin\theta\right)^2 \rho R\, d\theta + \int_{x=-L/2}^{L/2} x^2 \rho\, dx \\
&= \rho R\left[\left(R^2 + \frac{L^2}{2}\right)\pi + 4LR\right] + \rho\frac{L^3}{12}
\end{aligned}
$$

$$I_{zz} = I_{xx} + I_{yy}$$

$$I_{xy} = -\int xy\, dm = -2 \int_{\theta=0}^{\pi} \left(\frac{L}{2} + R\sin\theta\right)(R + R\cos\theta)\,\rho R\, d\theta$$

$$= -\rho R^2 (\pi L + 4R)$$

$$I_{xz} = I_{yz} = 0$$

Example 6.3.2 Figure 6.5 shows a homogeneous steel triangular plate ABD of mass $m = 52$ kg and sides $L_1 = 3$ m and $L_2 = 4.5$ m. Determine the moments and products of inertia of the plate about the x-y-z axes and the x_1-y_1-z_1 axes shown in the figure. The x-y-z axes are centered at the center of mass C of the plate. The thickness of the plate, which is 2 mm, is small enough to be neglected in the calculations.

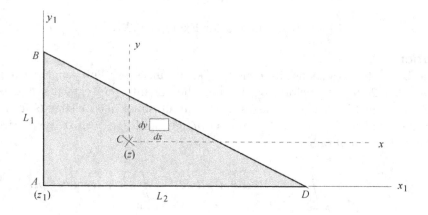

Figure 6.5: The triangular plate for example 6.3.2.

■ **Solution**

As shown in Example 5.3.2 (p. 324), the center of mass of this triangular plate is located at $x_1 = L_2/3 = 1.5$ m and $y_1 = L_1/3 = 1$ m relative to the corner A. To calculate the inertias, let us use the infinitesimal mass element with mass $dm = [2m/(L_1L_2)]dx\, dy = [2m/(L_1L_2)]dx_1\, dy_1$, shown in Fig. 6.5. Since the boundary BD of the plate is the straight line $y_1 = L_1 - L_1x_1/L_2$, and $z_1 = z = 0$, we then obtain for the inertias relative to the x_1-y_1-z_1 axes:

$$I_{x_1x_1} = \frac{2m}{L_1L_2} \int_{x_1=0}^{L_2} \int_{y_1=0}^{L_1-L_1x_1/L_2} y_1^2 dy_1\, dx_1 = \frac{1}{6}mL_1^2 = 78 \text{ kg·m}^2$$

$$I_{y_1y_1} = \frac{2m}{L_1L_2} \int_{x_1=0}^{L_2} \int_{y_1=0}^{L_1-L_1x_1/L_2} x_1^2 dy_1\, dx_1 = \frac{1}{6}mL_2^2 = 175.5 \text{ kg·m}^2$$

$$I_{z_1z_1} = \frac{2m}{L_1L_2} \int_{x_1=0}^{L_2} \int_{y_1=0}^{L_1-L_1x_1/L_2} (x_1^2 + y_1^2)\, dy_1\, dx_1 = I_{x_1x_1} + I_{y_1y_1}$$

$$I_{x_1y_1} = -\frac{2m}{L_1L_2} \int_{x_1=0}^{L_2} \int_{y_1=0}^{L_1-L_1x_1/L_2} x_1y_1 dy_1\, dx_1 = -\frac{1}{12}mL_1L_2 = -58.5 \text{ kg·m}^2$$

$$I_{x_1z_1} = I_{z_1x_1} = 0 \qquad I_{y_1z_1} = I_{z_1y_1} = 0 \qquad (\text{because } z_1 = 0)$$

The moments and products of inertia referred to the x-y-z axes centered at the center of mass C may be calculated in one of the following manners to obtain the results shown.

■ Let $x_1 = L_2/3 + x$ and $y_1 = L_1/3 + y$ in the earlier integrals and proceed as done in Section 6.3.2.

■ Simply notice that the rectangular coordinates of A relative to C are $x_A = -L_2/3$, $y_A = -L_1/3$, $z_A = 0$, and use the simple and general results obtained in Section 6.3.2.

$$I_{x_1x_1} = I_{xx} + my_A^2 \qquad \therefore \qquad I_{xx} = \frac{1}{18}mL_1^2 = 26 \text{ kg·m}^2$$

$$I_{y_1y_1} = I_{yy} + mx_A^2 \qquad \therefore \qquad I_{yy} = \frac{1}{18}mL_2^2 = 58.5 \text{ kg·m}^2$$

$$I_{z_1z_1} = I_{zz} + m\left(x_A^2 + y_A^2\right) \qquad \therefore \qquad I_{zz} = \frac{1}{18}m(L_1^2 + L_2^2) = 84.5 \text{ kg·m}^2$$

$$I_{x_1y_1} = I_{xy} - mx_Ay_A \qquad \therefore \qquad I_{xy} = \frac{1}{36}mL_1L_2 = 19.5 \text{ kg·m}^2$$

$$I_{x_1z_1} = I_{z_1x_1} = I_{xz} - mx_Az_A = 0$$

$$I_{y_1z_1} = I_{z_1y_1} = I_{yz} - my_Az_A = 0$$

Example 6.3.3 Figure 6.6 shows a homogeneous rectangular plate of mass m and dimensions a, b, and L. Determine the moments and products of inertia of the plate about the x-y-z axes that are parallel to the sides of the plate and are centered at its center of mass C.

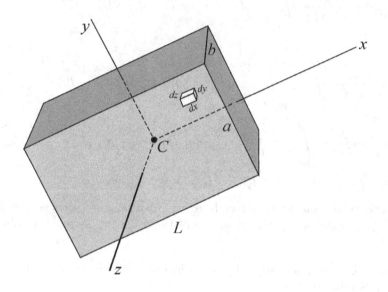

Figure 6.6: The rectangular plate for Example 6.3.3.

■ **Solution**

Since the plate is homogeneous, its mass density (mass per unit volume) is equal to $\rho = m/(abL)$. Therefore, the mass of the infinitesimal volume element with sides dx, dy, and dz, shown in Fig. 6.6, is $\rho\, dx\, dy\, dz$. The moment of inertia I_{xx} and the product of inertia I_{xy} are then determined as follows:

$$
I_{xx} = \int_{z=-b/2}^{b/2} \int_{y=-a/2}^{a/2} \int_{x=-L/2}^{L/2} (y^2 + z^2)\,\rho\, dx\, dy\, dz = Lb\left[\frac{y^3}{3}\right]_{y=-a/2}^{a/2} + La\left[\frac{z^3}{3}\right]_{z=-b/2}^{b/2}
$$

$$
= \underbrace{\frac{\rho Lab}{12}}_{=m/12}\left(a^2 + b^2\right) = \frac{m}{12}\left(a^2 + b^2\right)
$$

$$I_{xy} = -\int_{z=-b/2}^{b/2} \int_{y=-a/2}^{a/2} \int_{x=-L/2}^{L/2} \rho xy \, dx \, dy \, dz = 0$$

By inspecting Fig. 6.6 and these results, the following can be inferred:

$$I_{yy} = \frac{m}{12} \left(L^2 + b^2 \right)$$

$$I_{zz} = \frac{m}{12} \left(L^2 + a^2 \right)$$

$$I_{xz} = I_{yz} = 0$$

Examples 6.3.4 and 6.3.5 show how the elements of the inertia matrix are affected when the axes are rotated and how the inertias referred to the rotated axes are readily calculated using dyadics.

Example 6.3.4 The inertia matrix $[I]_{x_1,x_2,x_3}$ of a body, referred to a set of orthogonal axes x_1, x_2, x_3, is given below. Determine the inertia matrix of the body referred to a set of orthogonal axes y_1, y_2, y_3 that are rotated θ degrees about the x_3-axis.

$$[I]_{x_1,x_2,x_3} = \begin{bmatrix} 4 & -3 & 0 \\ -3 & 5 & 0 \\ 0 & 0 & 9 \end{bmatrix} \text{ kg·m}^2$$

■ **Solution**
As shown in the following, this type of calculation is readily done if dyadics and the transformations of triads presented in Section 1.2 of Chapter 1 are used. Two sets of unit vectors, one parallel to the x_1-x_2-x_3 axes and the other parallel to the y_1-y_2-y_3 axes, are shown in Fig. 6.7. Using those unit vectors, the dyadic representation $\vec{\vec{I}}$ of the given inertia matrix is expressed as follows:

$$\begin{aligned} \vec{\vec{I}} &= I_{x_1x_1}\hat{x}_1\hat{x}_1 + I_{x_1x_2}\hat{x}_1\hat{x}_2 + I_{x_1x_3}\hat{x}_1\hat{x}_3 + I_{x_2x_1}\hat{x}_2\hat{x}_1 + I_{x_2x_2}\hat{x}_2\hat{x}_2 + I_{x_2x_3}\hat{x}_2\hat{x}_3 \\ &\quad + I_{x_3x_1}\hat{x}_3\hat{x}_1 + I_{x_3x_2}\hat{x}_3\hat{x}_2 + I_{x_3x_3}\hat{x}_3\hat{x}_3 \\ &= 4\hat{x}_1\hat{x}_1 - 3\hat{x}_1\hat{x}_2 - 3\hat{x}_2\hat{x}_1 + 5\hat{x}_2\hat{x}_2 + 9\hat{x}_3\hat{x}_3 \end{aligned}$$

Inspection of Fig. 6.7 discloses the following relations between the unit vectors $\hat{x}_1$, $\hat{x}_2$, $\hat{x}_3$ and $\hat{y}_1$, $\hat{y}_2$, $\hat{y}_3$:

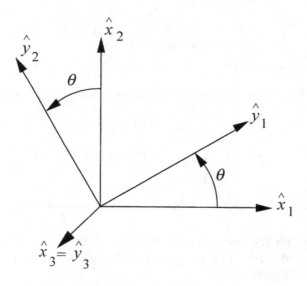

Figure 6.7: The triads $\{\hat{x}_1, \hat{x}_2, \hat{x}_3\}$ and $\{\hat{y}_1, \hat{y}_2, \hat{y}_3\}$ for Example 6.3.4.

$$\hat{x}_1 = \hat{y}_1 \cos\theta - \hat{y}_2 \sin\theta$$
$$\hat{x}_2 = \hat{y}_1 \sin\theta + \hat{y}_2 \cos\theta$$
$$\hat{x}_3 = \hat{y}_3$$

By making use of these relations, the inertia dyadic $\overset{\leftrightarrow}{I}$ becomes

$$\overset{\leftrightarrow}{I} = 4\,(\hat{y}_1 \cos\theta - \hat{y}_2 \sin\theta)\,(\hat{y}_1 \cos\theta - \hat{y}_2 \sin\theta)$$
$$-3\,(\hat{y}_1 \cos\theta - \hat{y}_2 \sin\theta)\,(\hat{y}_1 \sin\theta + \hat{y}_2 \cos\theta)$$
$$-3\,(\hat{y}_1 \sin\theta + \hat{y}_2 \cos\theta)\,(\hat{y}_1 \cos\theta - \hat{y}_2 \sin\theta)$$
$$+5\,(\hat{y}_1 \sin\theta + \hat{y}_2 \cos\theta)\,(\hat{y}_1 \sin\theta + \hat{y}_2 \cos\theta) + 9\hat{y}_3\hat{y}_3$$

Or, by carrying out the multiplications such as $(\cos\theta)\cos\theta$, $(\cos\theta)\sin\theta$, etc., we then obtain

$$\overset{\leftrightarrow}{I} = \left[4\cos^2\theta - 6(\cos\theta)\sin\theta + 5\sin^2\theta\right]\hat{y}_1\hat{y}_1 + \left[3\sin^2\theta - 3\cos^2\theta + (\cos\theta)\sin\theta\right]\hat{y}_1\hat{y}_2$$
$$+ \left[3\sin^2\theta - 3\cos^2\theta + (\cos\theta)\sin\theta\right]\hat{y}_2\hat{y}_1$$
$$+ \left[4\sin^2\theta + 6(\cos\theta)\sin\theta + 5\cos^2\theta\right]\hat{y}_2\hat{y}_2 + 9\hat{y}_3\hat{y}_3$$

This expression discloses that the elements of the inertia matrix of the body, referred to the y_1-y_2-y_3 axes, are

$$I_{y_1 y_1} = 4\cos^2\theta - 6(\cos\theta)\sin\theta + 5\sin^2\theta \quad I_{y_1 y_2} = 3\sin^2\theta - 3\cos^2\theta + (\cos\theta)\sin\theta$$
$$I_{y_1 y_3} = 0 \quad I_{y_2 y_2} = 4\sin^2\theta + 6(\cos\theta)\sin\theta + 5\cos^2\theta \quad I_{y_2 y_3} = 0 \quad I_{y_3 y_3} = 9$$

Example 6.3.5 A rigid body has the following moments and products of inertia (in kg·m^2) about a set of orthogonal axes x, y, z that are centered at a point A of the body and are fixed to the body:

$$I_{xx} = 3 \quad I_{xy} = \sqrt{2} \quad I_{xz} = -\sqrt{1.5}$$
$$I_{yy} = 4 \quad I_{yz} = -1 \quad I_{zz} = 6$$

The body is rigidly mounted on a horizontal shaft as shown in Fig. 6.8a. Point A

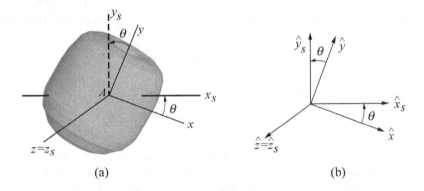

(a) (b)

Figure 6.8: A simple transformation of triads for Example 6.3.5.

is in the middle of the shaft and is inertial, and the angle θ between x and x_s (and between y and y_s) is constant. The x_s axis shown in the figure is aligned with the shaft, while the y_s and z_s axes (with $z_s = z$) are perpendicular to the shaft.

a. Determine the inertia matrix of the body about the x_s-y_s-z_s axes for $\theta = 30°$.

b. Determine the angular momentum $\vec{H}_A$ of the body if the shaft is spun counterclockwise with an angular velocity of 120 rpm about the x_s axis. Express the answer in terms of the unit vectors $\hat{x}_s$, $\hat{y}_s$, and $\hat{z}_s$ shown in Fig. 6.8b.

c. What is the time derivative of $\vec{H}_A$ if the only moment applied to the body is an internal moment $\vec{M} = M_{xs}\hat{x}_s + M_{ys}\hat{y}_s + M_{zs}\hat{z}_s$ N·m due to the contact with the shaft?

■ **Solution**

Figure 6.8b shows a set of unit vectors that will be used in the following calculations.

a. As shown in Example 6.3.4, this type of calculation is readily done if we use dyadics and the transformations of triads presented in Section 1.2 of Chapter 1. Start by writing the following expression for the inertia dyadic $\overset{\rightrightarrows}{I}_A$ for the body, based on the given information:

$$
\begin{aligned}
\overset{\rightrightarrows}{I}_A &= I_{xx}\hat{x}\hat{x} + I_{xy}\hat{x}\hat{y} + I_{xz}\hat{x}\hat{z} + I_{yx}\hat{y}\hat{x} + I_{yy}\hat{y}\hat{y} + I_{yz}\hat{y}\hat{z} + I_{zx}\hat{z}\hat{x} + I_{zy}\hat{z}\hat{y} + I_{zz}\hat{z}\hat{z} \\
&= 3\hat{x}\hat{x} + \sqrt{2}\hat{x}\hat{y} - \sqrt{1.5}\hat{x}\hat{z} + \sqrt{2}\hat{y}\hat{x} + 4\hat{y}\hat{y} - \hat{y}\hat{z} - \sqrt{1.5}\hat{z}\hat{x} - \hat{z}\hat{y} + 6\hat{z}\hat{z}
\end{aligned}
$$

Now, express the unit vectors $\hat{x}$, $\hat{y}$, and $\hat{z}$ in terms of the unit vectors $\hat{x}_s$, $\hat{y}_s$, and $\hat{z}_s$ that are parallel to the x_s-y_s-z_s axes. The relationships between these two sets of unit vectors are obtained by inspecting Fig. 6.8b, and they are:

$$
\begin{aligned}
\hat{x} &= \hat{x}_s \cos\theta - \hat{y}_s \sin\theta = \frac{\sqrt{3}}{2}\hat{x}_s - \frac{1}{2}\hat{y}_s \\
\hat{y} &= \hat{x}_s \sin\theta + \hat{y}_s \cos\theta = \frac{1}{2}\hat{x}_s + \frac{\sqrt{3}}{2}\hat{y}_s \\
\hat{z} &= \hat{z}_s
\end{aligned}
$$

When these relationships are substituted into the expression for $\overset{\rightrightarrows}{I}_A$ and all the multiplications are performed [such as $3\hat{x}\hat{x} = 3(\hat{x}_s \cos\theta - \hat{y}_s \sin\theta)(\hat{x}_s \cos\theta - \hat{y}_s \sin\theta) = (3\cos^2\theta)\hat{x}_s\hat{x}_s - 3(\sin\theta)(\cos\theta)\hat{x}_s\hat{y}_s - 3(\sin\theta)(\cos\theta)\hat{y}_s\hat{x}_s + (3\sin^2\theta)\hat{y}_s\hat{y}_s$, etc.], an expression of the following type is obtained for the inertia dyadic $\overset{\rightrightarrows}{I}_A$ in terms of the unit vectors $\hat{x}_s$, $\hat{y}_s$, and $\hat{z}_s$:

$$
\begin{aligned}
\overset{\rightrightarrows}{I}_A &= I_{x_s x_s}\hat{x}_s\hat{x}_s + I_{x_s y_s}\hat{x}_s\hat{y}_s + I_{x_s z_s}\hat{x}_s\hat{z}_s + I_{y_s x_s}\hat{y}_s\hat{x}_s + I_{y_s y_s}\hat{y}_s\hat{y}_s + I_{y_s z_s}\hat{y}_s\hat{z}_s \\
&\quad + I_{z_s x_s}\hat{z}_s\hat{x}_s + I_{z_s y_s}\hat{z}_s\hat{y}_s + I_{z_s z_s}\hat{z}_s\hat{z}_s
\end{aligned}
$$

The expressions that are obtained for all the moments and products of inertia $I_{x_s x_s}$, $I_{x_s y_s}$, etc., and their values for $\theta = 30°$, are as follows:

$$
\begin{aligned}
I_{x_s x_s} &= I_{xx}\cos^2\theta + 2I_{xy}(\sin\theta)\cos\theta + I_{yy}\sin^2\theta \approx 4.47 \text{ kg·m}^2 \\
I_{x_s y_s} &= (I_{yy} - I_{xx})(\sin\theta)\cos\theta + I_{xy}(\cos^2\theta - \sin^2\theta) \approx 1.14 \text{ kg·m}^2
\end{aligned}
$$

$$I_{x_s z_s} = I_{xz} \cos\theta + I_{yz} \sin\theta \approx -1.56 \text{ kg·m}^2$$

$$I_{y_s x_s} = I_{x_s y_s}$$

$$I_{y_s y_s} = I_{xx} \sin^2\theta - 2I_{xy}(\sin\theta)\cos\theta + I_{yy}\cos^2\theta \approx 2.53 \text{ kg·m}^2$$

$$I_{y_s z_s} = I_{yz} \cos\theta - I_{xz} \sin\theta \approx -0.25 \text{ kg·m}^2$$

$$I_{z_s x_s} = I_{x_s z_s}$$

$$I_{z_s y_s} = I_{y_s z_s}$$

$$I_{z_s z_s} = I_{zz} = 6 \text{ kg·m}^2$$

The algebraic steps that lead to these expressions are left for you, the reader, to do.

b. The angular velocity $\vec{\omega}^B$ of the body is the same as the angular velocity of the shaft because the two are welded to each other. Therefore, $\vec{\omega}^B = 120\hat{x}_s$ rpm $= 120 \times (2\pi/60)\hat{x}_s$ rad/s $= 4\pi\hat{x}_s$ rad/s. In terms of the unit vectors $\hat{x}_s$, $\hat{y}_s$, and $\hat{z}_s$, the angular momentum $\vec{H}_A$ is then

$$\begin{aligned}\vec{H}_A = \vec{\overline{I}}_A \bullet \vec{\omega}^B &= (I_{x_s x_s}\hat{x}_s\hat{x}_s + I_{x_s y_s}\hat{x}_s\hat{y}_s + I_{x_s z_s}\hat{x}_s\hat{z}_s + I_{y_s x_s}\hat{y}_s\hat{x}_s + I_{y_s y_s}\hat{y}_s\hat{y}_s \\ &\quad + I_{y_s z_s}\hat{y}_s\hat{z}_s + I_{z_s x_s}\hat{z}_s\hat{x}_s + I_{z_s y_s}\hat{z}_s\hat{y}_s + I_{z_s z_s}\hat{z}_s\hat{z}_s) \bullet (4\pi\hat{x}_s) \\ &= 4\pi\left(I_{x_s x_s}\hat{x}_s + I_{y_s x_s}\hat{y}_s + I_{z_s x_s}\hat{z}_s\right) \\ &\approx 56.23\hat{x}_s + 14.33\hat{y}_s - 19.61\hat{z}_s \text{ kg·m}^2/\text{s}\end{aligned}$$

c. Since point A is inertial, it follows that (see Chapter 5 or Section 6.2) $d\vec{H}_A/dt = \vec{M}_A = \vec{M}$. Therefore, no calculations are needed to answer the question in this item!

6.3.4 Principal Axes and Principal Moments of Inertia

It was mentioned in Section 6.3.1 that it is always possible to find the orientation for any set of orthogonal x-y-z axes so that the three products of inertia of the body are zero when referred to such axes. Such axes are called the *principal axes of inertia* of the body.

The principal axes of inertia of a body are, in general, fixed to the body. As will be seen, the principal axes need not be fixed to the body when the body exhibits symmetries, as in the case of a homogeneous cylinder or a sphere, for example. For the purpose of calculating the six inertias of a body, the x-y-z axes may be centered anywhere on the body. When they are centered at the center of mass of the body, the corresponding principal axes are called the *central principal axes of inertia*. Frequently, however, they are simply referred to as the *principal axes* of a body, with the word *central* added at times for more clarity.

To locate the central principal axes of a body, use is made of Eq. (6.3.7) (p. 404), which relates the angular momentum $\vec{H}_C$ to the absolute angular velocity $\vec{\omega}^B$ of the body, i.e., $\vec{H}_C = \vec{\vec{I}}_C \bullet \vec{\omega}^B$. However, the calculations that will follow are more conveniently done if one works with the matrix form given by Eq. (6.3.9), i.e., $\underline{H}_C = [I_C]\underline{\omega}^B$, where $[I_C]$ is the symmetric 3×3 inertia matrix of the body (referred to the orthogonal x-y-z axes centered at C) as in Eq. (6.3.10). Recall that $\underline{H}_C$ is the 3×1 column matrix of the components of $\vec{H}_C$, i.e., $\underline{H}_C = [H_x \ H_y \ H_z]^T$, and that $\underline{\omega}^B$ is the 3×1 column matrix of the components of $\vec{\omega}^B$, i.e., $\underline{\omega}^B = [\omega_x \ \omega_y \ \omega_z]^T$.

In general, the angular momentum $\vec{H}_C$ and the angular velocity of the body, $\vec{\omega}^B$, are not parallel to each other since they are not necessarily proportional to each other. If the x-y-z axes were the principal axes of the body, the angular momentum $\vec{H}_C$ would be simply equal to $\vec{H}_C = I_{xx}\omega_x\hat{x} + I_{yy}\omega_y\hat{y} + I_{zz}\omega_z\hat{z}$. In such a case, if the body were given a spin about any of its principal axes, say the x-axis, the resulting angular momentum would be a vector along that axis. In other words, if $\omega_y = \omega_z = 0$ the angular momentum would simply be a vector parallel to $\omega_x\hat{x}$ and equal to $\vec{H}_C = I_{xx}\omega_x\hat{x}$. This observation is the key for the mathematical determination of the direction of the principal axes of the body with respect to an arbitrarily chosen x-y-z axes.

To determine the direction of the central principal axes of a body, choose a set of orthogonal axes centered at the center of mass C of the body and calculate all the moments and products of inertia of the body relative to such axes. Next, consider the following question: If the body is given an angular velocity $\vec{\omega} = \omega_x\hat{x} + \omega_y\hat{y} + \omega_z\hat{z}$, what should be the direction of $\vec{\omega}$ so that the angular momentum of the body is parallel to $\vec{\omega}$? The answer to such a question gives the direction of a principal axis of the body. Since a vector parallel to $\vec{\omega}$ must be equal to $\lambda\vec{\omega}$, where λ is a scalar, the determination of the direction of any of the principal axes of a body is accomplished by finding the solution to the matrix equation $[I_C]\underline{\omega} = \lambda\underline{\omega}$, i.e.,

$$\begin{bmatrix} I_{xx} - \lambda & I_{xy} & I_{xz} \\ I_{yx} & I_{yy} - \lambda & I_{yz} \\ I_{zx} & I_{zy} & I_{zz} - \lambda \end{bmatrix} \begin{bmatrix} \omega_x \\ \omega_y \\ \omega_z \end{bmatrix} \overset{\Delta}{=} [B] \begin{bmatrix} \omega_x \\ \omega_y \\ \omega_z \end{bmatrix} = \begin{bmatrix} 0 \\ 0 \\ 0 \end{bmatrix} \quad (6.3.14)$$

This is a classical eigenvalue-eigenvector problem of matrix algebra.

If the matrix $[B]$ has an inverse, the only solution to Eq. (6.3.14) is $\omega_x = \omega_y = \omega_z = 0$. For Eq. (6.3.14) to have a nonzero solution, the determinant of the matrix $[B]$ has to be equal to zero. This condition yields the following cubic polynomial equation for λ, which is called the *characteristic equation* for the matrix $[B]$:

$$\begin{vmatrix} I_{xx} - \lambda & I_{xy} & I_{xz} \\ I_{yx} & I_{yy} - \lambda & I_{yz} \\ I_{zx} & I_{zy} & I_{zz} - \lambda \end{vmatrix} \overset{\Delta}{=} P(\lambda) = -\left(\lambda^3 - a_2\lambda^2 + a_1\lambda - a_0\right) = 0 \quad (6.3.15)$$

In this equation,

$$a_2 = I_{xx} + I_{yy} + I_{zz} \tag{6.3.16}$$

$$a_1 = \begin{vmatrix} I_{xx} & I_{xy} \\ I_{yx} & I_{yy} \end{vmatrix} + \begin{vmatrix} I_{xx} & I_{xz} \\ I_{zx} & I_{zz} \end{vmatrix} + \begin{vmatrix} I_{yy} & I_{yz} \\ I_{zy} & I_{zz} \end{vmatrix} \tag{6.3.17}$$

$$a_0 = \begin{vmatrix} I_{xx} & I_{xy} & I_{xz} \\ I_{yx} & I_{yy} & I_{yz} \\ I_{zx} & I_{zy} & I_{zz} \end{vmatrix} \tag{6.3.18}$$

Equation (6.3.15) is a cubic polynomial equation and, thus, admits three solutions $\lambda = \lambda_1$, $\lambda = \lambda_2$, and $\lambda = \lambda_3$. These solutions, which are called the *eigenvalues* of the inertia matrix $[I_C]$, are the central principal moments of inertia of the body.

When the solutions of the characteristic equation are found, each individual eigenvalue λ_i should be substituted into Eq. (6.3.14) to determine the components ω_x, ω_y, and ω_z of the vector $\vec{\omega}$ that are associated with each λ_i. The column matrix $[\omega_x \ \omega_y \ \omega_z]^T$ is an *eigenvector* of the inertia matrix $[I_C]$. However, as a consequence of Eq. (6.3.15), at most two of the three equations involving the quantities ω_x, ω_y, and ω_z obtained for each λ_i ($i = 1, 2, 3$) will be independent of each other. Thus, since we are left with more unknowns than equations to determine the components of $\vec{\omega}$, none of the eigenvectors $[\omega_x \ \omega_y \ \omega_z]^T$ are uniquely determined. In other words, what we are left with is an equation of a straight line in terms of the originally chosen x-y-z axes. But these straight lines, which are the central principal axes of the body, are precisely what we want and need. These straight lines may be specified by arbitrarily choosing a non-zero value for any of the components of $\vec{\omega}$, say ω_x, and solving for the other unknowns ω_y and ω_z in terms of ω_x. Any column matrix $[\omega_x \ \omega_y \ \omega_z]^T$ representing a vector $\omega_x \hat{x} + \omega_y \hat{y} + \omega_x \hat{z}$ along such a straight line is, of course, an eigenvector of the inertia matrix $[I_C]$. Although not necessary, these eigenvectors may be "normalized," if desired, so that their magnitude $\sqrt{\omega_x{}^2 + \omega_y{}^2 + \omega_z{}^2}$ is equal to 1. If this is done, the values obtained for ω_x, ω_y, and ω_z are the direction cosines of the principal axes of the body relative to the originally chosen x-y-z axes.

It turns out that the eigenvectors of a symmetric matrix, corresponding to distinct eigenvalues, are orthogonal to each other. The proof of this and of other properties of the inertia matrix are presented in Appendix C. Thus, since the inertia matrix is symmetric, the principal axes of any body having three distinct principal moments of inertia are orthogonal to each other. In contrast, the angle between the principal directions corresponding to equal principal inertias is arbitrary. For example, for a homogeneous cylinder, the central principal axes are the axial axis and any transverse axis that passes through the center of mass of the cylinder. For a homogeneous sphere, any axis through the center of the sphere is a central principal axis. Also, if a homogeneous body has a plane of symmetry, any axis through its center of mass and perpendicular to that plane of symmetry is a central principal axis.

Several examples of determination of principal axes are given in Section 6.3.5.

6.3.5 Examples of Determination of Principal Axes and Principal Inertias

---

Example 6.3.6 Determine the central principal moments of inertia and the direction of the principal axes of the triangular plate of Example 6.3.2 (p. 410).

■ **Solution**
The elements I_{xx}, I_{xy}, etc., of the inertia matrix for that plate were calculated in Example 6.3.2, and the inertia matrix $[I_C]$, relative to the center of mass C of the plate, was found to be the following:

$$[I_C] = \begin{bmatrix} I_{xx} & I_{xy} & I_{xz} \\ I_{yx} & I_{yy} & I_{yz} \\ I_{zx} & I_{zy} & I_{zz} \end{bmatrix} = \begin{bmatrix} 26 & 19.5 & 0 \\ 19.5 & 58.5 & 0 \\ 0 & 0 & 84.5 \end{bmatrix} \tag{6.3.6a}$$

The principal moments of inertia of the plate and the corresponding principal axes are determined by solving the eigenvalue-eigenvector problem given by the following matrix equation:

$$\begin{bmatrix} 26 - \lambda & 19.5 & 0 \\ 19.5 & 58.5 - \lambda & 0 \\ 0 & 0 & 84.5 - \lambda \end{bmatrix} \begin{bmatrix} \omega_x \\ \omega_y \\ \omega_z \end{bmatrix} = \begin{bmatrix} 0 \\ 0 \\ 0 \end{bmatrix} \tag{6.3.6b}$$

The characteristic equation for Eq. (6.3.6b) is

$$(\lambda - 84.5)(\lambda^2 - 84.5\lambda + 1140.75) = 0$$

whose roots λ_1, λ_2, and λ_3 are

$$\lambda_1 = 84.5 \text{ kg·m}^2$$

$$\lambda_2 = \frac{84.5 + \sqrt{84.5^2 - 4 \times 1140.75}}{2} \approx 67.63 \text{ kg·m}^2$$

$$\lambda_3 = \frac{84.5 - \sqrt{84.5^2 - 4 \times 1140.75}}{2} \approx 16.87 \text{ kg·m}^2$$

Notice that the z-axis is a principal axis of the plate because $I_{xz} = I_{yz} = 0$. The two eigenvectors in the x-y plane, corresponding to the eigenvalues $\lambda = \lambda_2$ and $\lambda = \lambda_3$, are determined by using one of the two equations obtained from Eq. (6.3.6b), i.e., either from $(I_{xx} - \lambda)\omega_x + I_{xy}\omega_y = 0$ or from $(I_{yy} - \lambda)\omega_y + I_{yx}\omega_x = 0$. These two equations are actually the same since the determinant of the matrix shown in Eq. (6.3.6b) is equal to zero. For example, for $\lambda = 67.63$, they give $-41.63\omega_x + 19.5\omega_y = 0$ and $19.5\omega_x - 9.13\omega_y = 0$. Except for the round-off error in $\lambda \approx 67.63$, these equations are the same as $\omega_y - 2.135\omega_x = 0$.

By arbitrarily working with the first row of Eq. (6.3.6b), we obtain the following eigenvectors $\underline{\omega}_2$ and $\underline{\omega}_3$ corresponding to the eigenvalues λ_2 and λ_3, respectively:

$$\text{for } \lambda = \lambda_2 = 67.63: \qquad \underline{\omega}_2 = \omega_x \begin{bmatrix} 1 \\ 2.135 \end{bmatrix}$$

$$\text{for } \lambda = \lambda_3 = 16.87: \qquad \underline{\omega}_3 = \omega_x \begin{bmatrix} 1 \\ -1/2.135 \end{bmatrix}$$

The angle between the x-axis shown in Fig. 6.5 (p. 410) and the principal axis for which the moment of inertia is equal to $\lambda_2 = 67.63$ kg·m^2 is $\arctan(2.135) \approx 64.9°$. The angle between that same x-axis and the third principal axis (for which the moment of inertia is $\lambda_3 = 16.87$ kg·m^2) is $\arctan(-1/2.135) \approx -25.1°$.

Notice that the three principal axes of the triangular plate, which are associated with three different principal moments of inertia, are perpendicular to each other. As mentioned earlier, such a property of the inertia matrix is proved in Appendix C.

Example 6.3.7 Determine the central principal moments of inertia of the homogeneous solid rotor of mass m, radius R, and height h shown in Fig. 6.9. Because of symmetry, the center of mass C of the rotor is located at its geometric center. The x-y-z axes shown in the figure, which are centered at C, are the central principal axes of the rotor. The figure also shows a mass element of the rotor (conveniently using cylindrical coordinates) to be used in the calculations.

■ Solution

Because of the symmetry, the z-axis shown in Fig. 6.9 is a principal axis, which means that $I_{zx} = I_{zy} = 0$. Also, any axis through the center of mass C, which is the geometric center of the rotor, and perpendicular to the z-axis is a principal axis. This is confirmed by the results of the following calculations.

With the mass density of the rotor being equal to $\rho = m/(\pi R^2 h)$, the mass of the infinitesimal volume element shown in Fig. 6.9 is equal to $dm = \rho(r\,d\theta)dr\,dz$.

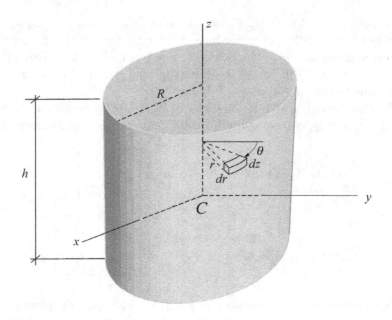

Figure 6.9: The solid rotor of Example 6.3.7.

Therefore, with $x = r \sin \theta$ and $y = r \cos \theta$, the principal moments of inertia of the rotor are obtained as follows:

$$I_{zz} = \frac{m}{\pi R^2 h} \int_{r=0}^{R} \int_{\theta=0}^{2\pi} \int_{z=-h/2}^{h/2} r^2 \left(r \, dz \, d\theta \, dr \right) = m \frac{R^2}{2}$$

$$I_{yy} = \frac{m}{\pi R^2 h} \int_{r=0}^{R} \int_{\theta=0}^{2\pi} \int_{z=-h/2}^{h/2} \left(r^2 \sin^2 \theta + z^2 \right) r \, dz \, d\theta \, dr$$

$$= \frac{m}{\pi R^2 h} \int_{r=0}^{R} r^3 dr \int_{\theta=0}^{2\pi} \frac{h}{2} (1 - \cos 2\theta) \, d\theta + \frac{mh^2}{4} = m \left(\frac{R^2}{4} + \frac{h^2}{12} \right)$$

Because of the symmetry in the x-y plane, the x-axis is really indistinguishable from the y-axis and, thus, $I_{xx} = I_{yy}$.

For the product of inertia I_{xz} (which is equal to I_{yz} because the x-axis is indistinguishable from the y-axis) we have, with $x = r \sin \theta$,

$$I_{xz} = -\frac{m}{\pi R^2 h} \int_{r=0}^{R} \int_{z=-h/2}^{h/2} \int_{\theta=0}^{2\pi} (z\, r \sin \theta)\, r\, d\theta\, dz\, dr = 0$$

For the product of inertia I_{xy} we have, with $xy = (r \sin \theta) r \cos \theta$,

$$I_{xy} = -\frac{m}{\pi R^2 h} \left[\int_{r=0}^{R} \int_{z=-h/2}^{h/2} r^3 dr\, dz \right] \times \underbrace{\int_{\theta=0}^{2\pi} (\sin \theta)(\cos \theta)\, d\theta}_{=\left[\frac{1}{4}\left(\sin^2 \theta - \cos^2 \theta\right)\right]_{\theta=0}^{2\pi} = 0} = 0$$

Example 6.3.8 A rigid body has the following moments and products of inertia (in kg·m^2) about a set of orthogonal x-y-z axes centered at its center of mass:

$$\begin{array}{ccc} I_{xx} = 3 & I_{yy} = 4 & I_{zz} = 6 \\ I_{xy} = \sqrt{2} & I_{xz} = -\sqrt{1.5} & I_{yz} = -1 \end{array}$$

Determine the central principal moments of inertia of the body and the direction cosines of its central principal axes.

■ Solution

Let λ be the moment of inertia of the body about one of its principal axes, whose orientation is still to be determined. Imagine now that we spin the body about that axis, and let ω_x, ω_y, and ω_z denote, respectively, the components of the angular velocity $\vec{\omega}_B$ of the body along unit vectors $\hat{x}$, $\hat{y}$, and $\hat{z}$ that are parallel to those axes. If the spin axis is a principal axis, the angular momentum of the body, $\vec{\vec{I}}_C \bullet \vec{\omega}_B$ (or $[I_C]\underline{\omega}^B$ in matrix form) must be parallel to its angular velocity, which means that it must be equal to $\lambda \vec{\omega}^B$ (or $\lambda \underline{\omega}^B$ in matrix notation). Therefore, to determine λ and the direction of a principal axis, we must solve the eigenvalue-eigenvector problem given by the following matrix equation:

$$\begin{bmatrix} 3 - \lambda & \sqrt{2} & -\sqrt{1.5} \\ \sqrt{2} & 4 - \lambda & -1 \\ -\sqrt{1.5} & -1 & 6 - \lambda \end{bmatrix} \begin{bmatrix} \omega_x \\ \omega_y \\ \omega_z \end{bmatrix} = \begin{bmatrix} 0 \\ 0 \\ 0 \end{bmatrix} \tag{6.3.8a}$$

To have a non-zero solution for the angular velocity $\vec{\omega}^B$, the determinant of the matrix shown in Eq. (6.3.8a) must be equal to zero. This gives the following polynomial equation for λ:

$$\lambda^3 - 13\lambda^2 + 49.5\lambda - \left(51 + 2\sqrt{3}\right) = 0 \tag{6.3.8b}$$

The following roots of Eq. (6.3.8b) can be obtained by a trial-and-error process done with a pocket calculator. They may also be obtained with either Scilab or Matlab using the command `roots([1, -13, 49.5, -(51 + 2 * sqrt(3))])`.

$$\lambda_1 \approx 7.126 \qquad \lambda_2 \approx 3.928 \qquad \lambda_3 \approx 1.946$$

These roots are the central principal moments of inertia of the body. Notice that $\lambda_1 + \lambda_2 + \lambda_3 = 13$, which is the *trace* of the inertia matrix (i.e., the sum of its diagonal elements). This provides a spot check on the validity of the numbers obtained for the three roots. One might also use, of course, approximations as $\lambda_1 \approx 7.13$, $\lambda_2 \approx 3.93$, and $\lambda_3 \approx 1.95$, which yields $\lambda_1 + \lambda_2 + \lambda_3 = 13.01$ instead of 13. The difference is simply due to round-off errors.

Let us denote the three central principal axes of the body by x_1 (corresponding to the moment of inertia λ_1), x_2 (corresponding to the moment of inertia λ_2), and x_3 (corresponding to the moment of inertia λ_3). To determine the direction of any of these axes, Eq. (6.3.8a) needs to be solved for ω_y and ω_z, say, in terms of ω_x for each of the eigenvalues λ. Since the direction cosines of a line are the scalar components of a unit vector parallel to that line, it follows that the quantities ω_x, ω_y, and ω_z are the direction cosines of a principal axis if they are normalized such that $\sqrt{\omega_x^2 + \omega_y^2 + \omega_z^2} = 1$.

By arbitrarily choosing to work with the first two rows of the matrix shown in Eq. (6.3.8a), the following eigenvector $\underline{\omega}_1 = [\omega_{1x} \; \omega_{1y} \; \omega_{1z}]^T$ corresponding to the eigenvalue $\lambda_1 \approx 7.126$ is obtained:

$$\underline{\omega}_1 \approx \omega_x \begin{bmatrix} 1 \\ 1.117 \\ -2.078 \end{bmatrix}$$

Or, after normalization so that $(1 + 1.117^2 + 2.078^2)\omega_x^2 = 1$,

$$\underline{\omega}_1 \approx \begin{bmatrix} 0.390 \\ 0.436 \\ -0.811 \end{bmatrix} \triangleq \begin{bmatrix} \cos\theta_{1x} \\ \cos\theta_{1y} \\ \cos\theta_{1z} \end{bmatrix}$$

The angle this principal axis makes with the x-, y-, and z-axes are, respectively, $\theta_{1x} \approx 67^\circ$, $\theta_{1y} \approx 64.2^\circ$, and $\theta_{1z} \approx 144.2^\circ$. Notice that the components obtained for the eigenvector $\underline{\omega}_1$ satisfy, within the accuracy of the numerical computations, the equation obtained from the third row of the matrix in Eq. (6.3.8a).

By repeating this procedure for the other two eigenvalues, the following normalized eigenvectors are obtained.

For the eigenvalue $\lambda = \lambda_2 \approx 3.928$:

$$\underline{\omega}_2 \approx \begin{bmatrix} 0.366 \\ 0.735 \\ 0.571 \end{bmatrix} \triangleq \begin{bmatrix} \cos\theta_{2x} \\ \cos\theta_{2y} \\ \cos\theta_{2z} \end{bmatrix}$$

For the eigenvalue $\lambda = \lambda_3 \approx 1.946$:

$$\underline{\omega}_3 \approx \begin{bmatrix} 0.845 \\ -0.520 \\ 0.126 \end{bmatrix} \triangleq \begin{bmatrix} \cos\theta_{3x} \\ \cos\theta_{3y} \\ \cos\theta_{3z} \end{bmatrix}$$

For the principal axis x_2 we obtain $\theta_{2x} \approx 68.5°$, $\theta_{2y} \approx 42.7°$, and $\theta_{2z} \approx 55.2°$, and for the principal axis x_3, $\theta_{3x} \approx 32.3°$, $\theta_{3y} \approx 121.3°$, and $\theta_{3z} \approx 82.8°$. Notice that the products $\underline{\omega}_1^T \underline{\omega}_2$, $\underline{\omega}_1^T \underline{\omega}_3$, and $\underline{\omega}_2^T \underline{\omega}_3$, are all zero, which indicates that the three principal axes are orthogonal to each other, as expected.

6.4 Euler's Differential Equations of Rotational Motion

It was seen in Section 6.2 that the rotational motion of a rigid body is governed by the vector equation

$$\frac{d}{dt}\vec{H}_C = \vec{M}_C \tag{6.4.1}$$

where the expression for the angular momentum $\vec{H}_C$ is given by Eq. (6.3.5). For convenience, that expression for $\vec{H}_C$ is repeated here.

$$\begin{aligned} \vec{H}_C &= (I_{xx}\omega_x + I_{xy}\omega_y + I_{xz}\omega_z)\,\hat{x} \\ &\quad + (I_{yx}\omega_x + I_{yy}\omega_y + I_{yz}\omega_z)\,\hat{y} \\ &\quad + (I_{zx}\omega_x + I_{zy}\omega_y + I_{zz}\omega_z)\,\hat{z} \\ &\triangleq H_x\hat{x} + H_y\hat{y} + H_z\hat{z} \end{aligned} \tag{6.4.2}$$

If the x-y-z axes shown in Fig. 6.2 (p. 401) are not fixed to the rigid body, the moments and products of inertia I_{xx}, I_{xy}, etc., change with time. In such a case, Eq. (6.4.1) yields the following result:

$$\frac{d}{dt}\vec{H}_C = \dot{H}_x\hat{x} + \dot{H}_y\hat{y} + \dot{H}_z\hat{z} + \vec{\omega} \times \vec{H}_C = \vec{M}_C \tag{6.4.3}$$

where

$$
\begin{aligned}
\dot{H}_x &= \dot{I}_{xx}\omega_x + \dot{I}_{xy}\omega_y + \dot{I}_{xz}\omega_z + I_{xx}\dot{\omega}_x + I_{xy}\dot{\omega}_y + I_{xz}\dot{\omega}_z \\
\dot{H}_y &= \dot{I}_{yx}\omega_x + \dot{I}_{yy}\omega_y + \dot{I}_{yz}\omega_z + I_{yx}\dot{\omega}_x + I_{yy}\dot{\omega}_y + I_{yz}\dot{\omega}_z \\
\dot{H}_z &= \dot{I}_{zx}\omega_x + \dot{I}_{zy}\omega_y + \dot{I}_{zz}\omega_z + I_{zx}\dot{\omega}_x + I_{zy}\dot{\omega}_y + I_{zz}\dot{\omega}_z
\end{aligned}
\qquad (6.4.4)
$$

It is convenient to avoid the additional complication associated with time-varying inertias by choosing to work with a set of axes that are fixed to the rigid body. In such a case, the absolute angular velocity $\vec{\omega}$ of the unit vector triad $\{\hat{x}, \hat{y}, \hat{z}\}$ is equal to the absolute angular velocity $\vec{\omega}^B \triangleq \omega_x\hat{x} + \omega_y\hat{y} + \omega_z\hat{z}$ of the rigid body. By expressing the moment $\vec{M}_C$, acting on the body, in terms of the unit vectors $\{\hat{x}, \hat{y}, \hat{z}\}$, an expression of the following form is obtained for each particular problem under consideration.

$$
\vec{M}_C = M_x\hat{x} + M_y\hat{y} + M_z\hat{z}
\qquad (6.4.5)
$$

In such a case, Eq. (6.4.3) yields the following scalar differential equations for the body-fixed angular velocity components ω_x, ω_y, and ω_z:

$$
\begin{aligned}
I_{xx}\dot{\omega}_x + I_{xy}\dot{\omega}_y + I_{xz}\dot{\omega}_z + \omega_y H_z - \omega_z H_y &= M_x \qquad (6.4.6) \\
I_{yx}\dot{\omega}_x + I_{yy}\dot{\omega}_y + I_{yz}\dot{\omega}_z + \omega_z H_x - \omega_x H_z &= M_y \qquad (6.4.7) \\
I_{zx}\dot{\omega}_x + I_{zy}\dot{\omega}_y + I_{zz}\dot{\omega}_z + \omega_x H_y - \omega_y H_x &= M_z \qquad (6.4.8)
\end{aligned}
$$

where H_x, H_y, and H_z are the $\hat{x}$, $\hat{y}$, and $\hat{z}$ components, respectively, of the angular momentum $\vec{H}_C$, defined by Eq. (6.4.2). Equations (6.4.6) to (6.4.8) are called *Euler's equations for nonprincipal axes*. They are used in some applications such as aircraft and ship dynamics where the moments due to the aerodynamic forces and the hydrodynamic forces, as the case may be, are more conveniently expressed in terms of some nonprincipal axes that are chosen as a trade-off in complexity between the left-hand sides of Euler's equations and the expressions for the moment components that appear on their right-hand sides.

In numerous applications, working with nonprincipal axes means an unnecessary complication of the differential equations that govern the rotational motion of the body. In principal axes, one simply has $H_x = I_{xx}\omega_x$, $H_y = I_{yy}\omega_y$, and $H_z = I_{zz}\omega_z$. Thus, when all the products of inertia are equal to zero, the following simpler (but not simple!) differential equations are obtained from Eqs. (6.4.6) to (6.4.8). In such cases, it is more convenient to use a single subscript for denoting the three principal moments of inertia, i.e., to use I_x instead of I_{xx}, I_y instead of I_{yy}, and I_z instead of I_{zz}. Such a convention will be adopted, from this point on, when working with principal axes.

$$I_x \dot{\omega}_x + (I_z - I_y)\,\omega_y \omega_z \;\; = \;\; M_x \qquad (6.4.9)$$

$$I_y \dot{\omega}_y + (I_x - I_z)\,\omega_x \omega_z \;\; = \;\; M_y \qquad (6.4.10)$$

$$I_z \dot{\omega}_z + (I_y - I_x)\,\omega_x \omega_y \;\; = \;\; M_z \qquad (6.4.11)$$

These three equations (which are nonlinear and coupled) are called *Euler's equations for principal axes*. Notice the symmetry that appears in Eqs. (6.4.9) to (6.4.11), which is evidenced when each equation is divided by the moment of inertia that multiplies the time derivative of the respective component of the angular velocity: the moments of inertia in the second term in each equation appear in a counterclockwise x-y-z sequence (with $\hat{z} = \hat{x} \times \hat{y}$), such as $(I_z - I_y)/I_x$ in the first equation.

When the components of the moment $\vec{M}_C$ are not functions of the angular orientation of the body, integration of Eqs. (6.4.9) to (6.4.11) yields the components ω_x, ω_y, and ω_z of the absolute angular velocity $\vec{\omega}^B$ of the body as a function of time. Otherwise, the left-hand side of those equations would also have to be expressed in terms of the variables that are chosen to describe the orientation of the body in space, and this results in a set of three nonlinear second-order differential equations for those variables.

Before dealing with several examples of the direct application of Euler's equations for investigating rotational motion, the work-energy approach is presented in Section 6.5. Then several important classical problems in dynamics, such as the stability of a body that is spun about one of its principal axes, dynamic balancing of a spinning rotor with rigid bodies attached to it, and the motion of a gyroscope (or a top) are presented in a separate section (Section 6.6) due to their special significance.

6.5   Kinetic Energy and the Work-Energy Approach

The determination of the kinetic energy for a rigid body undergoing a simpler plane motion as defined in Section 5.1 was presented in Section 5.5 of Chapter 5. Here, the material in that section is extended to rigid bodies that can undergo any type of three-dimensional motion. The calculations involve the same steps seen in that chapter. However, the final expression for the kinetic energy for a general motion involves other elements of the inertia matrix of the body, instead of only one for the case of the type of simpler planar motion that was dealt with in Chapter 5.

Referring to Fig. 6.2 (p. 401), recall that the kinetic energy of the motion of a rigid body is defined as

$$T \stackrel{\Delta}{=} \frac{1}{2} \int_{\text{body}} \vec{v}_P \bullet \vec{v}_P \, dm \qquad (6.5.1)$$

where $\vec{v}_P = d\overrightarrow{OP}/dt$ is the absolute velocity of an infinitesimal mass element dm of the body, which is represented by point P in Fig. 6.2. It is important to remember that time derivatives such as $d\overrightarrow{OP}/dt$ have to take into account any change of orientation of $\overrightarrow{OP}$ relative to an inertial reference frame, and that point O has to be an inertial reference point. Such a fact is embedded in the phrase "where $\vec{v}_P$ is the absolute velocity . . . ."

With $\overrightarrow{OP} = \overrightarrow{OC} + \overrightarrow{CP}$, where C is the center of mass of the body (see Fig. 6.2), it was seen in Chapter 5 that Eq. (6.5.1) reduces to

$$T = \frac{1}{2} \int_{\text{body}} \left(\vec{v}_C + \frac{d}{dt}\overrightarrow{CP}\right) \bullet \left(\vec{v}_C + \frac{d}{dt}\overrightarrow{CP}\right) dm$$

$$= \frac{1}{2}m\vec{v}_C \bullet \vec{v}_C + \frac{1}{2} \int_{\text{body}} \left(\frac{d}{dt}\overrightarrow{CP}\right) \bullet \left(\frac{d}{dt}\overrightarrow{CP}\right) dm \qquad (6.5.2)$$

where $\vec{v}_C = d\overrightarrow{OC}/dt$ is the absolute velocity of the center of mass C of the body.

It is at this point that the development of the final expression for the kinetic energy T for a body moving in three-dimensional space starts to be a little more involved than the planar motion case considered in Chapter 5. Basically, however, the only thing that needs to be done is to take the time derivative of the rotating vector $\overrightarrow{CP}$ (as already done in Section 6.3.1, p. 401), use the result in Eq. (6.5.2), and perform the integration to come up with the final expression for the kinetic energy T. By expressing $\overrightarrow{CP}$, and also the absolute angular velocity $\vec{\omega}^B$, as given in Eqs. (6.3.2) and (6.3.3) (p. 403), the following result is obtained for the kinetic energy T, where $\left(\underline{\omega}^B\right)^T$ is the transpose of the 3×1 column matrix $\underline{\omega}^B$.

$$T = \frac{1}{2}m\vec{v}_C \bullet \vec{v}_C + \underbrace{\frac{1}{2}\left[I_{xx}\omega_x^2 + I_{yy}\omega_y^2 + I_{zz}\omega_z^2 + 2I_{xy}\omega_x\omega_y + 2I_{xz}\omega_x\omega_z + 2I_{yz}\omega_y\omega_z\right]}$$

$$\text{same as } \tfrac{1}{2}\vec{\omega}^B \bullet \vec{H}_C = \tfrac{1}{2}\vec{\omega}^B \bullet \overleftrightarrow{I}_C \bullet \vec{\omega}^B \quad \text{or} \quad \tfrac{1}{2}(\underline{\omega}^B)^T[I_C]\underline{\omega}^B$$

In summary, the kinetic energy is determined as

$$T = \frac{1}{2}m\vec{v}_C \bullet \vec{v}_C + \frac{1}{2}\vec{\omega}^B \bullet \vec{H}_C = \frac{1}{2}m\vec{v}_C \bullet \vec{v}_C + \frac{1}{2}\vec{\omega}^B \bullet \overleftrightarrow{I}_C \bullet \vec{\omega}^B \qquad (6.5.3)$$

or, in matrix notation for the last term,

$$T = \frac{1}{2}m\vec{v}_C \bullet \vec{v}_C + \frac{1}{2}\left(\underline{\omega}^B\right)^T [I_C]\,\underline{\omega}^B \qquad (6.5.4)$$

The first term in these expressions is the kinetic energy due to translation of the center of mass C of the body, and the second term is the kinetic energy due to rotation of the body about an axis that passes through the center of mass of the body. Notice that it is not inconsistent to "mix" the vector and matrix representations in the expression for T because each one of the two terms in that expression is a scalar.

Notes

1. When principal axes of inertia are used, the kinetic energy due to rotation about the center of mass reduces to the simpler expression

$$T_{\text{rot}} = \frac{1}{2} \left[I_x \omega_x^2 + I_y \omega_y^2 + I_z \omega_z^2 \right]$$

2. If the rigid body is rotating about a fixed point O so that the distance from O to any point of the body is constant, the kinetic energy given by Eq. (6.5.4) reduces to the simpler form

$$T = \frac{1}{2} \vec{\omega}^B \bullet \vec{I}_O \bullet \vec{\omega}^B \qquad \text{or} \qquad T = \frac{1}{2} \left(\underline{\omega}^B \right)^T [I_O] \underline{\omega}^B$$

In such an expression, $\vec{I}_O$ is the inertia dyadic of the body, referred to point O, and $[I_O]$ is its matrix representation, i.e., the inertia matrix of the body about orthogonal axes that pass through O.

The work-energy approach involving the relationship $dW = d(T)$ for a rigid body is the same as presented in Section 5.5 of Chapter 5. Referring again to Fig. 6.2 (p. 401), recall that the infinitesimal work dW done by a set of N forces $\vec{F}_1, \vec{F}_2, \ldots, \vec{F}_N$ acting on the rigid body is determined by a relatively simple operation involving dot products of vectors, i.e.,

$$dW \overset{\Delta}{=} \sum_{i=1}^{N} \vec{F}_i \bullet d\vec{r}_i \tag{6.5.5}$$

In Eq. (6.5.5), $\vec{r}_i$ is the absolute position vector of the point on the body where the force $\vec{F}_i$ is applied.

As indicated in Chapters 2 and 5, when dW is an exact differential, Eq. (6.5.5) can be readily integrated to yield $T - W = $ constant, which is a first-order differential equation. In other words, such a differential equation only involves the first derivative of any motion variable. As mentioned in Chapter 5, that is the differential equation of motion if the motion is governed by only one variable (i.e., a one-degree-of-freedom system). Generally, for motion involving more than one variable, additional work is needed to make full use of the work-energy approach. Such work has been developed by Lagrange and other great masters of dynamics and mathematics. That branch of dynamics is known as *Analytical Dynamics*, or *Lagrangian Dynamics*, or *Lagrangian and Hamiltonian Dynamics*, and is presented in Chapter 7. That material is an extremely powerful, versatile, and convenient methodology for formulating the differential equations of motion of dynamical systems and for searching for integrals of motion.

6.6 Classical Problems

The problems in this section are classical problems that have repercussions in a number of important practical applications, and they involve direct application of Euler's differential equations of motion.

6.6.1 Stability of a Body Spun About One of Its Principal Axes

> This problem has repercussions in a number of important applications involving rotating bodies, such as balancing of rotors and satellite dynamics.

Problem Description   A rigid body of mass m, and principal moments of inertia I_x, I_y, and I_z, is thrown upward with an angular velocity Ω about one of its principal axes, as shown in Fig. 6.10a. The angle α shown in that figure is supposedly equal to zero, but it is essentially impossible for that to happen in practice. Therefore, α is interpreted to be a very small "pointing error" due to the practical inability to make it exactly equal to zero. All forces acting on the body other than the force due to gravity are neglected.

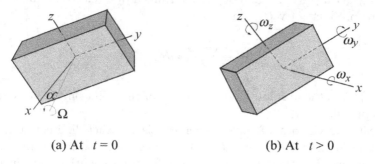

(a) At $t = 0$ (b) At $t > 0$

Figure 6.10: Motion of a body given a spin that is nominally about a principal axis.

■ What are the differential equations of motion for the system in terms of the angular velocity components ω_x, ω_y, and ω_z, of the subsequent motion, shown in Fig. 6.10b?

■ The differential equations of motion admit a constant solution that corresponds to the body spinning about one of its principal axes of inertia. Linearize the differential equations of motion about that solution, and then investigate the stability of the motion using the linearized differential equations.

Solution   Since the resultant of the gravitational force acting on the body passes through the center of mass C of the body, the moment $\vec{M}_C$ of that force is equal to zero. The rotational motion of the body is then governed by Eqs. (6.4.9) to (6.4.11), with $M_x = M_y = M_z = 0$.

With $M_x = M_y = M_z = 0$, Eqs. (6.4.9) to (6.4.11) admit the following three particular solutions. They also admit the solution $\omega_x = \omega_y = \omega_z = 0$, but such a solution is of no interest here because it only corresponds to translation of the body, with no rotation.

Solution 1: $\qquad\qquad \omega_x = \text{constant} \overset{\Delta}{=} \Omega \qquad \omega_y = \omega_z = 0$

Solution 2: $\qquad\qquad \omega_y = \text{constant} \overset{\Delta}{=} \Omega \qquad \omega_x = \omega_z = 0$

Solution 3: $\qquad\qquad \omega_z = \text{constant} \overset{\Delta}{=} \Omega \qquad \omega_x = \omega_y = 0$

All three solutions have the same physical meaning, namely, a body spinning with a constant angular velocity about one of its three principal axes.

Thus, if the body is given a constant spin about any of its x, y, or z principal axis of inertia, it continues to spin about that axis with the same angular velocity it started with. The question that arises now is the following:

What happens if that motion is perturbed, i.e., if the spin is about an axis that makes a slight angle α (as illustrated in Fig. 6.10a) with the principal axis we were supposed to spin the body about?

To answer this question, we can take, without any loss of generality, the particular solution

$$\omega_x = \text{constant} \overset{\Delta}{=} \Omega \qquad\qquad \omega_y = \omega_z = 0 \qquad\qquad (6.6.1a)$$

and, then, introduce a small perturbation $\omega_{xs}(t)$ as

$$\omega_x(t) = \Omega + \omega_{xs}(t) \qquad\qquad (6.6.1b)$$

In terms of the variables $\omega_{xs}(t), \omega_y(t)$, and $\omega_z(t)$, the differential equations of motion become

$$I_x\dot{\omega}_{xs} + (I_z - I_y)\,\omega_y\omega_z \;=\; 0 \qquad\qquad (6.6.1\text{c})$$
$$I_y\dot{\omega}_y + (I_x - I_z)\,(\Omega + \omega_{xs})\,\omega_z \;=\; 0 \qquad\qquad (6.6.1\text{d})$$
$$I_z\dot{\omega}_z + (I_y - I_x)\,(\Omega + \omega_{xs})\,\omega_y \;=\; 0 \qquad\qquad (6.6.1\text{e})$$

To analyze the stability of the motion, we linearize these differential equations. This is done by simply neglecting products of small quantities to obtain the following equations:

$$I_x\dot{\omega}_{xs} \;\approx\; 0 \qquad\qquad (6.6.1\text{f})$$
$$I_y\dot{\omega}_y + (I_x - I_z)\,\Omega\omega_z \;\approx\; 0 \qquad\qquad (6.6.1\text{g})$$
$$I_z\dot{\omega}_z + (I_y - I_x)\,\Omega\omega_y \;\approx\; 0 \qquad\qquad (6.6.1\text{h})$$

The first one of these equations discloses that $\omega_{xs} = $ constant and, without loss of generality, the perturbation ω_{xs} may be equated to zero.

Notice that Eqs. (6.6.1f) to (6.6.1h) are also obtained in the situation where an external control moment $\vec{M} = M_x\hat{x}$ is applied to the body to maintain the x component of its angular velocity constant for all times, i.e., $\omega_x = $ constant $\overset{\Delta}{=}\Omega$. This would be the case if the body were rigidly attached to a fixed spinning shaft, for example.

By taking the first time derivative of Eq. (6.6.1g), and combining the result with Eq. (6.6.1h), the following second-order linear differential equation with constant coefficients is obtained for ω_y (the same differential equation is also obtained for ω_z):

$$\ddot{\omega}_y + \frac{(I_x - I_z)(I_x - I_y)}{I_yI_z}\Omega^2\omega_y = 0 \qquad\qquad (6.6.1\text{i})$$

As shown in Section 1.13 of Chapter 1, the particular solution corresponding to $\omega_x = $ constant $\overset{\Delta}{=}\Omega$ is unstable if the coefficient of ω_y in Eq. (6.6.1i) is negative (because, in such a case, the solution for ω_y grows with time). This occurs when $I_x > I_z$ and $I_x < I_y$ or when $I_x < I_z$ and $I_x > I_y$, i.e., if the axis about which the body is spun is the axis of intermediate inertia. This is an extremely important result in practice, and it is emphasized in the following boxed text.

> The rotational motion of a moment-free rigid body is unstable if the body is spun about its axis of intermediate inertia.

For an infinitesimally stable particular solution $\omega_x = $ constant, the coefficients of $\ddot{\omega}_y$ and ω_y must have the same sign. Therefore, for an infinitesimally stable particular solution, we must have one of the following:

- $I_x > I_z$ and $I_x > I_y$, which means that the body is spun about its axis of maximum inertia

- $I_x < I_z$ and $I_x < I_y$, which means that the body is spun about its axis of minimum inertia

It can be easily verified by numerical integration of Eqs. (6.6.1c) to (6.6.1e) that the particular solution under consideration is stable when either one of these conditions is satisfied. In such a case, the frequency of the small oscillations governed by Eq. (6.6.1i) is equal to $\Omega\sqrt{(I_x/I_z - 1)(I_x/I_y - 1)}$.

> **An Interesting Experiment** The conclusions about the stability of a body spun about one of its principal axes are very easy to verify in practice with a simple experiment with a solid rectangular plate (such as one made of wood) or with a book that is solidly wrapped with a few rubber bands. The principal axes of inertia of such a body are easily identified. Just toss the rectangle into space (say, along the vertical) by trying to spin it about one of its principal axes, and observe the resulting motion. Do this in three separate instances, by spinning the body about a principal axis in each case. You will see that the body will always tumble when the spin is about its intermediate axis of inertia. For the other two cases (i.e., a spin about either the axis of maximum or minimum moment of inertia), the body continues to spin about that axis, with an unnoticeable deviation of the direction of that axis as the body moves in space.

It is of interest to notice that two integrals of motion can be immediately obtained from Eqs. (6.4.9) to (6.4.11), and they yield important information about the rotational motion of the body — information that is not obtained from the linearized differential equations. A well-known visual description of the rotational motion of the moment-free body, based on such integrals, was developed by Louis Poinsot (1777-1859). One of those integrals, $\vec{H}_C = $ constant, which is obtained from Euler's vector equation $\vec{M}_C = d\vec{H}_C/dt$ with $\vec{M}_C = 0$, is

$$\vec{H}_C = I_x\omega_x\hat{x} + I_y\omega_y\hat{y} + I_z\omega_z\hat{z} = \text{constant} \qquad (6.6.1j)$$

This means that the direction of the vector $\vec{H}_C$ remains fixed in inertial space, and the magnitude of $\vec{H}_C$, which is $\sqrt{(I_x\omega_x)^2 + (I_y\omega_y)^2 + (I_z\omega_z)^2}$, is also constant. Using the square of the magnitude of $\vec{H}_C$, we then have

$$(I_x\omega_x)^2 + (I_y\omega_y)^2 + (I_z\omega_z)^2 = \text{constant} \triangleq H^2 \qquad (6.6.1\text{k})$$

Since Eqs. (6.4.9) to (6.4.11) for the rotational motion are uncoupled from the equations that govern the motion of the center of mass C of the body, the work-energy equation $dW = d(T)$ applied to the rotational motion implies that the kinetic energy of rotation remains constant during the motion, i.e.,[2]

$$I_x\omega_x^2 + I_y\omega_y^2 + I_z\omega_z^2 = \text{constant} = 2T \qquad (6.6.1\text{l})$$

Equations (6.6.1k) and (6.6.1l) represent two ellipsoids in the ω_x-ω_y-ω_z space. They are known as the *momentum ellipsoid* and the *energy ellipsoid*[3], respectively. The intersection of these two ellipsoids is a curve called the *polhode*. For a body that is given a spin about an axis that is nearly coincident with either its maximum or minimum axis of inertia, the polhode is a geometric representation of the stable motion analyzed earlier using linearized differential equations, but without restricting them to infinitesimally small motions about the equilibrium solution $\vec{\omega} = \omega_x\hat{x}$, say. For a spin about an axis near the intermediate axis of inertia of the body, the polhode corresponds to the unstable motion indicated earlier, where the body wobbles in three-dimensional space.

Poinsot's visualization is completed by looking at the gradient of the function $T = \left(I_x\omega_x^2 + I_y\omega_y^2 + I_z\omega_z^2\right)/2$, which is

$$\overrightarrow{\text{grad } T} = \frac{\partial T}{\partial \omega_x}\hat{x} + \frac{\partial T}{\partial \omega_y}\hat{y} + \frac{\partial T}{\partial \omega_z}\hat{z} = I_x\omega_x\hat{x} + I_y\omega_y\hat{y} + I_z\omega_z\hat{z} = \vec{H}_C \qquad (6.6.1\text{m})$$

The gradient at any point P of a function $f(x_1, x_2, \ldots, x_n)$ is normal to the tangent plane of f at P. Since $\overrightarrow{\text{grad } T} = \vec{H}_C$, and the angular momentum $\vec{H}_C$ is a vector that maintains a fixed direction in space, the plane that is normal to $\overrightarrow{\text{grad } T}$ also maintains a fixed direction in space. Such a plane is known as the *invariable*

---

[2]Eq. (6.6.1k) may also be obtained by multiplying Eq. (6.4.9) by I_x, Eq. (6.4.10) by I_y, Eq. (6.4.11) by I_z, adding the results and integrating the resulting equation. Such steps are, of course, avoided by using the work and energy equation.

[3]By comparing Eqs. (6.6.1l) and (C.15) in Appendix C (p. 633) for principal axes, it is seen that the energy ellipsoid is similar to the inertia ellipsoid presented in that Appendix.

plane. Now, letting $\overrightarrow{CP} = \vec{\omega}$ be a vector from the center of mass C to any point P on the surface of the energy ellipsoid, it follows that the perpendicular distance from C to the invariable plane, which is the dot product of $\vec{\omega}$ with the unit vector parallel to $\vec{H}_C$, i.e., $\vec{\omega} \bullet \vec{H}_C/|\vec{H}_C| = 2T/|\vec{H}_C|$, is constant. Therefore, the rotational motion of a moment-free body can be visualized as the rolling, without slipping, of the energy ellipsoid on the invariable plane. The curve traced by the contact point P on the invariable plane is called the *herpolhode*. Notice that if the body has an axis of symmetry, point P moves on a circle and, thus, the angular velocity vector describes, in this case, a circular cone around the direction defined by the angular momentum vector.[4]

6.6.2 Dynamic Balancing of a Rotating Part of a Machine

Problem Description   A rigid body of mass m is welded to a horizontal symmetric shaft of mass m_{shaft} as shown in Fig. 6.11.   Point O is the center of mass

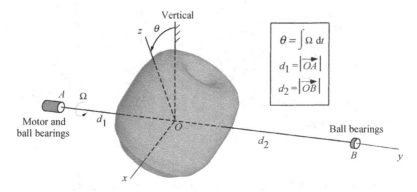

Figure 6.11: A body welded to a spinning shaft (balancing of rotors).

of the body and is on the centerline of the shaft, and line AB is inertial. The x-y-z axes shown in the figure are fixed to the body, with the y-axis coinciding with the centerline of the horizontal shaft. Thus, the x- and z-axes are on a vertical plane. The orthogonal x-y-z axes that are fixed to the body are arbitrarily chosen and, therefore, are not necessarily the principal axes of that body. The shaft is forced by a motor to rotate with a known angular velocity $\dot{\theta} \overset{\Delta}{=} \Omega(t)$, where t denotes time and

---

[4]To see this, let z, say, be the symmetry axis so that $I_x = I_y$. Since Euler's equation $I_z\dot{\omega}_z + (I_y - I_x)\omega_y\omega_x = 0$ yields, in this case, $\omega_z = $ constant, the rotational energy integral then becomes $T = I_x(\omega_x^2 + \omega_y^2) + I_z\omega_z^2 = $ constant. This yields $\omega_x^2 + \omega_y^2 = $ constant, which is the equation of a circle in the ω_x-ω_y space.

θ is the angle shown in Fig. 6.11. The quantities d_1 and d_2 that are also shown in the figure are the distances from O to A and O to B, respectively, i.e., $d_1 = |\overrightarrow{OA}|$ and $d_2 = |\overrightarrow{OB}|$.

The following need to be determined:

■ The moment $\vec{M}_{\text{motor}}$ applied to the shaft by the motor

■ The support reactions at the bearings at A and B

It is also necessary to know whether it is possible to weld the body to the shaft so that the support reactions at A and B are not affected by the angular velocity $\Omega(t)$. If that is possible to do, how should the body be mounted on the shaft for that to happen?

Solution   The free-body diagram in Fig. 6.12 shows the details that are necessary

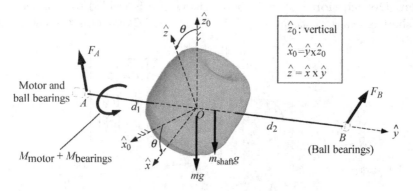

Figure 6.12: Details needed for finding the bearing reactions and the moment M_{motor} applied by the motor.

for solving the problem. The diagram is for the rigid system composed of the shaft and the body that is welded to it. Such a system is rigid because of the weld connection. We could also have split the system into two parts, namely the shaft and the body, to work with the free-body diagram for each part, thus generating a set of equations for the dynamics of both parts. However, that would introduce additional unknowns into the problem and would lead to additional unnecessary work to eliminate them by combining the resulting equations.

Note the following from Fig. 6.12:

■ $\vec{F}_A$ and $\vec{F}_B$ are the bearing reactions at A and B, respectively, and m_{shaft} is the mass of the shaft. The gravitational force with magnitude $m_{\text{shaft}}g$ acts at the middle of the homogeneous shaft.

■ $\vec{M}_{\text{motor}} \overset{\Delta}{=} M_{\text{motor}} \hat{y}$ is the moment applied to the system by the motor. The moment $\vec{M}_{\text{motor}}$ is a vector directed along the shaft, but it is indicated in the figure by a curved arrow to distinguish it from a force. The counterclockwise sense shown in the figure was arbitrarily chosen.

■ $\vec{M}_{\text{bearings}} \overset{\Delta}{=} M_{\text{bearings}} \hat{y}$ is a reaction moment due to friction in the two bearings. It is also a vector directed along the shaft, but it is indicated in the figure as part of the same curved arrow mentioned previously. Again, the counterclockwise sense shown in the figure was arbitrarily chosen. $\vec{M}_{\text{bearings}} = 0$ if the bearings are frictionless. If friction in the bearings is of the linear viscous type, with coefficients c_A and c_B at the bearings at A and B, respectively, then $M_{\text{bearings}} = -(c_A + c_B)\Omega$.

For convenience, five unit vectors are also included in Fig. 6.12, namely $\hat{z}_0$; three orthogonal unit vectors $\hat{x}$, $\hat{y}$, and $\hat{z} \overset{\Delta}{=} \hat{x} \times \hat{y}$; and $\hat{x}_0 \overset{\Delta}{=} \hat{y} \times \hat{z}_0$. The unit vectors $\hat{z}_0$ and $\hat{x}_0$, with $\hat{z}_0$ vertical, do not change direction, while the unit vectors $\hat{x}$, $\hat{y}$, and $\hat{z}$ are parallel to the x-y-z axes that are fixed to the rotating body. The unit vector $\hat{y}$ is horizontal and, therefore, does not change direction either.

Let us now start the mathematical formulation of the problem to answer the questions that were asked. Three scalar equations for the system are obtained from Euler's equation for the rotational motion, and three more from Newton's second law for the motion of the center of mass C of the entire system. Point C, which is not shown in Fig. 6.12, lies on the centerline of the shaft. Since point O is inertial, and it is also fixed to the rotating system, let us use the expression $\vec{M}_O = d\vec{H}_O/dt$.

The inertia dyadic $\overset{\Rightarrow}{I}_O$ of the system is

$$\overset{\Rightarrow}{I}_O = \left(\overset{\Rightarrow}{I}_O\right)_{\text{body}} + \left(\overset{\Rightarrow}{I}_O\right)_{\text{shaft}} \tag{6.6.2a}$$

For the body that is welded to the shaft,

$$\left(\overset{\Rightarrow}{I}_O\right)_{\text{body}} = I_{xx}\hat{x}\hat{x} + I_{xy}\hat{x}\hat{y} + I_{xz}\hat{x}\hat{z} + I_{yx}\hat{y}\hat{x} + I_{yy}\hat{y}\hat{y} + I_{yz}\hat{y}\hat{z}$$
$$+ I_{zx}\hat{z}\hat{x} + I_{zy}\hat{z}\hat{y} + I_{zz}\hat{z}\hat{z} \tag{6.6.2b}$$

and for the symmetric shaft (for which all the products of inertia referred to the x-y-z axes are zero),

$$\left(\overset{\Rightarrow}{I}_O\right)_{\text{shaft}} = (I_{xx})_{\text{shaft}} \left(\hat{x}\hat{x} + \hat{z}\hat{z}\right) + (I_{yy})_{\text{shaft}} \hat{y}\hat{y} \tag{6.6.2c}$$

Referring to Fig. 6.12, the absolute angular velocity of the system is simply

$$\vec{\omega} = \Omega \hat{y} \tag{6.2.2d}$$

Since the x-y-z axes are not principal axes of the system, the scalar equations that are obtained from $d\vec{H}_O/dt$ are Eqs. (6.4.6) to (6.4.8) that were developed in Section 6.4, with $\omega_x = \omega_z = 0$ and $\omega_y = \Omega$, as determined earlier. One could then formulate the expressions for the x-y-z components of the moment $\vec{M}_O$ and directly apply those equations to the problem at hand. One could also reformulate the equations for the rotational motion starting from the basic equation $\vec{M}_O = d\vec{H}_O/dt$, without having to consult the reasonably complicated Eqs. (6.4.6) to (6.4.8), as shown in the following.

With the inertia dyadic and the absolute angular velocity of the body given as determined earlier, the absolute angular momentum $\vec{H}_O$ and its time derivative are

$$\vec{H}_O = \overleftrightarrow{I}_O \bullet \vec{\omega} = I_{xy}\Omega\hat{x} + \left[I_{yy} + (I_{yy})_{\text{shaft}}\right]\Omega\hat{y} + I_{zy}\Omega\hat{z} \tag{6.2.2e}$$

$$\frac{d}{dt}\vec{H}_O = I_{xy}\dot{\Omega}\hat{x} + \left[I_{yy} + (I_{yy})_{\text{shaft}}\right]\dot{\Omega}\hat{y} + I_{zy}\dot{\Omega}\hat{z} + \vec{\omega} \times \vec{H}_O$$

$$= \left[I_{xy}\dot{\Omega} + \Omega^2 I_{zy}\right]\hat{x} + \left[I_{yy} + (I_{yy})_{\text{shaft}}\right]\dot{\Omega}\hat{y} + \left[I_{zy}\dot{\Omega} - \Omega^2 I_{xy}\right]\hat{z} \tag{6.2.2f}$$

Referring again to Fig. 6.12, the moment $\vec{M}_O$ that is applied to the system is calculated as follows:

$$\vec{M}_O = d_2\hat{y} \times \vec{F}_B - d_1\hat{y} \times \vec{F}_A$$

$$+ \underbrace{\left(\frac{d_1 + d_2}{2} - d_1\right)\hat{y} \times (-m_{\text{shaft}}g\hat{z}_0)}_{= -\left(\frac{d_2 - d_1}{2}m_{\text{shaft}}g\right)\hat{x}_0} + \vec{M}_{\text{motor}} + \vec{M}_{\text{bearings}} \tag{6.2.2g}$$

The expression for the reaction forces $\vec{F}_A$ and $\vec{F}_B$ at the bearings are of the following form:

$$\vec{F}_A = F_{A_x}\hat{x} + F_{A_y}\hat{y} + F_{A_z}\hat{z} \tag{6.2.2h}$$

$$\vec{F}_B = F_{B_x}\hat{x} + F_{B_y}\hat{y} + F_{B_z}\hat{z} \tag{6.2.2i}$$

Therefore, since $\hat{x}_0 = \hat{x}\cos\theta + \hat{z}\sin\theta$ (see Fig. 6.12), Eq. (6.2.2g) yields for the moment $\vec{M}_O$:

$$\vec{M}_O = \left(d_2 F_{B_z} - d_1 F_{A_z} - \frac{d_2 - d_1}{2}m_{\text{shaft}}g\cos\theta\right)\hat{x} + (M_{\text{motor}} + M_{\text{bearings}})\,\hat{y}$$

$$+ \left(d_1 F_{A_x} - d_2 F_{B_x} - \frac{d_2 - d_1}{2}m_{\text{shaft}}g\sin\theta\right)\hat{z} \tag{6.2.2j}$$

With $\vec{M}_O$ given by Eq. (6.6.2j), and $d\vec{H}_O/dt$ by Eq. (6.6.2f), the following scalar equations are then obtained from Euler's equation $d\vec{H}_O/dt = \vec{M}_O$:

$$d_2 F_{Bz} - d_1 F_{Az} - \frac{d_2 - d_1}{2} m_{\text{shaft}} g \cos\theta = I_{xy}\dot{\Omega} + \Omega^2 I_{zy} \tag{6.6.2k}$$

$$M_{\text{motor}} + M_{\text{bearings}} = \left[I_{yy} + (I_{yy})_{\text{shaft}}\right]\dot{\Omega} \tag{6.6.2l}$$

$$d_1 F_{Ax} - d_2 F_{Bx} - \frac{d_2 - d_1}{2} m_{\text{shaft}} g \sin\theta = I_{zy}\dot{\Omega} - \Omega^2 I_{xy} \tag{6.6.2m}$$

It can be readily verified that Eqs. (6.6.2k), (6.6.2l), and (6.6.2m) are the same as Eqs. (6.4.6), (6.4.7), and (6.4.8), respectively.

Equation (6.6.2l) gives the moment M_{motor} that the motor has to apply to the shaft to drive the system with a desired angular velocity $\Omega(t)$. Notice that if the angular velocity is to be maintained constant (which is achieved by a motor that is called a synchronous motor), the motor only needs to apply a moment to counteract the friction in the bearings, if any.

Equations (6.6.2k) and (6.6.2m) involve four unknowns, namely, F_{Ax}, F_{Az}, F_{Bx}, and F_{Bz}. Therefore, two more equations involving those same unknowns are needed to determine them. The additional equations are obtained by applying Newton's second law to the motion of the center of mass of the system. Since the centerline of the shaft is inertial, and the center of mass C of the system is in that centerline, it follows that the absolute acceleration $\vec{a}_C$ of C is equal to zero. Newton's second law governing the motion of C then reduces to $\vec{F} = 0$, where (see the free-body diagram in Fig. 6.12)

$$
\begin{aligned}
\vec{F} &= \vec{F}_A + \vec{F}_B - (m + m_{\text{shaft}})g\hat{z}_0 \\
&= \left(F_{Ax} + F_{Bx}\right)\hat{x} + \left(F_{Ay} + F_{By}\right)\hat{y} + \left(F_{Az} + F_{Bz}\right)\hat{z} - (m + m_{\text{shaft}})g\hat{z}_0 = 0
\end{aligned}
\tag{6.6.2n}
$$

Since (again, see Fig. 6.12)

$$\hat{z}_0 = \hat{z}\cos\theta - \hat{x}\sin\theta \tag{6.6.2o}$$

Equation (6.6.2n) yields the following scalar equations involving the force components F_{Ax}, F_{Ay}, F_{Az}, F_{Bx}, F_{By}, and F_{Bz}:

$$F_{Ax} + F_{Bx} = -(m + m_{\text{shaft}})g\sin\theta \tag{6.6.2p}$$

$$F_{Ay} + F_{By} = 0 \tag{6.6.2q}$$

$$F_{Az} + F_{Bz} = (m + m_{\text{shaft}})g\cos\theta \tag{6.6.2r}$$

The four equations (6.6.2k), (6.6.2m), (6.6.2p), and (6.6.2r) can now be solved for F_{A_x}, F_{A_z}, F_{B_x}, and F_{B_z}. The components F_{A_y} and F_{B_y} have opposite signs, and that is all that can be said about them. The following expressions are then obtained for the support reactions $\vec{F}_A = F_{A_x}\hat{x} + F_{A_y}\hat{y} + F_{A_z}\hat{z}$ and $\vec{F}_B = F_{B_x}\hat{x} + F_{B_y}\hat{y} + F_{B_z}\hat{z}$ at the bearings:

$$\vec{F}_A = F_{A_y}\hat{y} + \left(\frac{md_2}{d_1 + d_2} + \frac{1}{2}m_{\text{shaft}}\right)g\underbrace{\left(\hat{z}\cos\theta - \hat{x}\sin\theta\right)}_{=\hat{z}_0}$$

$$+ \frac{\left(\dot{\Omega}I_{zy} - \Omega^2 I_{xy}\right)\hat{x} - \left(\dot{\Omega}I_{xy} + \Omega^2 I_{zy}\right)\hat{z}}{d_1 + d_2} \tag{6.6.2s}$$

$$\vec{F}_B = -F_{A_y}\hat{y} + \left(\frac{md_1}{d_1 + d_2} + \frac{1}{2}m_{\text{shaft}}\right)g\underbrace{\left(\hat{z}\cos\theta - \hat{x}\sin\theta\right)}_{=\hat{z}_0}$$

$$- \frac{\left(\dot{\Omega}I_{zy} - \Omega^2 I_{xy}\right)\hat{x} - \left(\dot{\Omega}I_{xy} + \Omega^2 I_{zy}\right)\hat{z}}{d_1 + d_2} \tag{6.6.2t}$$

As disclosed by Eqs. (6.6.2s) and (6.6.2t), the bearing reactions $\vec{F}_A$ and $\vec{F}_B$ consist of constant vertical forces (i.e., in the $\hat{z}_0$ direction shown in Fig. 6.12) whose sum simply counteracts the gravitational force $-(m + m_{\text{shaft}})g\hat{z}_0$ acting on the system, of the forces $F_{A_y}\hat{y}$ and $F_{B_y}\hat{y} = -F_{A_y}\hat{y}$ along the shaft, and of a pair of forces $\vec{F}'_A$ and $\vec{F}'_B = -\vec{F}'_A$, acting at A and at B, respectively, where

$$\vec{F}'_A \triangleq \frac{\left(\dot{\Omega}I_{zy} - \Omega^2 I_{xy}\right)\hat{x} - \left(\dot{\Omega}I_{xy} + \Omega^2 I_{zy}\right)\hat{z}}{d_1 + d_2} \tag{6.6.2u}$$

The forces $\vec{F}'_A$ and $\vec{F}'_B$ rotate with the shaft with angular velocity $\vec{\omega} = \Omega\hat{y}$ (since both $\hat{x}$ and $\hat{y}$ rotate with that angular velocity). The magnitude of these rotating "imbalance forces" is

$$|\vec{F}'_A| = |\vec{F}'_B| = \frac{\sqrt{I_{xy}^2 + I_{zy}^2}}{d_1 + d_2}\sqrt{\Omega^4 + \dot{\Omega}^2} \tag{6.6.2v}$$

Notice that the magnitude of the rotating imbalance forces $\vec{F}'_A$ and $F'_B = -\vec{F}'_A$ is proportional to the quantity $\sqrt{\Omega^4 + \dot{\Omega}^2}$. Such a quantity is greater than or equal to the square of the shaft angular velocity Ω, and $|\vec{F}'_A|$ is proportional to Ω^2 if Ω is constant. These imbalance forces will cause, in practice, major damage to the bearings. Fortunately, such undesirable pair of forces may be eliminated by welding the body to the shaft so that I_{xy} and I_{zy} are equal to zero. Such a condition means that the axis about which the body is rotated (which is the y-axis in Fig. 6.12) is a principal axis of the body. When this is done, the rotating system is said to be *dynamically balanced.*

> In summary, to minimize damage to the bearings due to the rotating imbalance forces, it is important that the rotor be *dynamically balanced.* This is accomplished by assuring that the axis of rotation is a principal axis of the system composed of the rotor and of the body (or bodies) attached to it.

In practice, one should ensure that the system is rotated about an axis that is as close as possible to one of its principal axis to make the imbalance rotating forces acting on the bearings as close to zero as possible for any angular velocity $\Omega(t)$.

As indicated, careful alignment is required to keep the magnitude of the imbalance forces $\vec{F}'_A$ and $\vec{F}'_B = -F'_A$ below acceptable levels. The manufacturer of a part that is meant to be spun (such as a vehicle tire) tries to make the body symmetric about one of its principal axes, but small deviations from symmetry are always bound to occur in practice. To counteract such small unavoidable deviations, small masses are placed at strategic locations on the body to achieve dynamic balancing. In the balancing process, the imbalance forces $\vec{F}'_A$ and $\vec{F}'_B$ are measured by sensors that are part of a "balancing machine."

A Note on Balancing of Rotating Parts of Machines: In addition to dynamic balancing a system consisting of a rotating shaft attached to one or more bodies, the system should also be *statically balanced.* Static balancing involves only the balancing of the gravitational forces acting on the shaft so that the shaft does not turn when it is not driven. This is achieved by connecting the bodies on the shaft so that their centers of mass coincide with the axis of rotation of the shaft. Undesirable rotating bearing reactions will also appear if the perpendicular distance e between the center of mass of any of such bodies and the centerline of the shaft in Fig. 6.11 is not equal to zero (see Problem 6.13).

6.6.3 Motion of a Symmetric Top (or of a Gyroscope)

Problem Description   A system consisting of a homogeneous rotor of radius R, height h, and mass m is welded to an arm of negligible mass that coincides with the axial axis of the rotor. The whole assembly is connected by a ball-and-socket joint to a point O as shown in Fig. 6.13, where $L = |\overrightarrow{OC}|$ is the distance between O and the center of mass C of the rotor. The system is a model for a top and for a gyroscope (in which case, $L = 0$). The unit vectors $= \hat{z}$, $\hat{n}$, and $\hat{\phi} \triangleq \hat{z} \times \hat{n}$ that are included in the figure will be used in the analysis that follows. Point O is inertial,

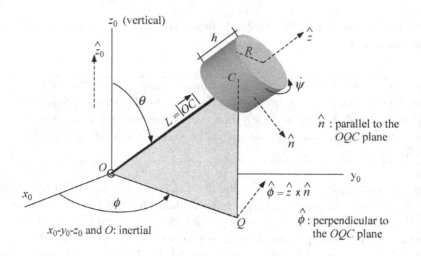

Figure 6.13: A symmetric top (or a gyroscope) subjected to a gravitational force that creates a moment $\vec{M}_O$.

the x_0-y_0-z_0 axes are fixed in inertial space, and friction in the system is neglected. The angles $\theta(t)$ and $\phi(t)$ locate arm OC in inertial space. The rotor is given an initial spin $[\dot{\psi}(t)]_{t=0} = \dot{\psi}(0)$ rad/s while arm OC is held still with $[\theta(t)]_{t=0} = \theta(0)$ and $[\phi(t)]_{t=0} = \phi(0)$. The system is then released with $[\dot{\theta}(t)]_{t=0} = \dot{\theta}(0)$ rad/s and $[\dot{\phi}(t)]_{t=0} = \dot{\phi}(0)$ rad/s.

  a. Obtain the differential equations that govern the rotational motion of the system in terms of the angles $\theta(t)$, $\phi(t)$, and, if needed, an angle $\psi(t) \triangleq \int \dot{\psi}\, dt$, which is an angle measured from a nonspinning line parallel to the plane of the flat part of the rotor.

  b. Obtain all the equilibrium solutions for the system. Show that one equilibrium corresponds to $\theta = \text{constant} \triangleq \theta_e$, with $\theta_e \neq 0$ and $\theta_e \neq \pi$, and that for such an equilibrium to exist the rotor must be spun about its axial axis with an angular velocity greater than a certain minimum value. Determine the expression for that minimum. For $R = 6$ cm and $R = 10$ cm, what is that minimum angular velocity so that $\theta_e = 30°$ when $L/R = 1$ and $h/R = 1/3$?

Note: In addition to answering these questions (which could be typical of a problem in a formal course), the solution shown here includes a number of additional instructive comments and observations about the motion.

Solution   To formulate the differential equations of motion, let us work with the rotating unit vector triad $\{\hat{z}, \hat{n}, \hat{\phi} \stackrel{\triangle}{=} \hat{z} \times \hat{n}\}$ shown in Fig. 6.13, with $\hat{n}$ being normal to arm OC, and in the plane defined by OC and the vertical z_0 axis (i.e., plane OQC shown in the figure). Both $\hat{n}$ and $\hat{\phi}$ are perpendicular to the axial axis of the rotor. Although the axes that are centered at the center of mass C of the rotor and are parallel to the unit vectors $\hat{n}$ and $\hat{\phi}$ are not fixed to the spinning rotor, they are still principal axes of the system because the rotor is symmetric about its axial axis.

After the system is released, the only external moment acting on it is due to the gravitational force $-mg\hat{z}_0$ (see Fig. 6.13). The moment $\vec{M}_O$, about point O, of this force is

$$\vec{M}_O = \overrightarrow{OC} \times (-mg\hat{z}_0) = \overrightarrow{OQ} \times (-mg\hat{z}_0) + \underbrace{\overrightarrow{QC} \times (-mg\hat{z}_0)}_{=0} = mgL\,(\sin\theta)\,\hat{\phi}$$

$$(6.6.3a)$$

The differential equations that govern the motion of the system, which are needed for investigating the motions, are obtained from $\vec{M}_O = d\vec{H}_O/dt$. There is no need to consider the reaction forces at O because they do not contribute to $\vec{M}_O$. Notice that the inertial $\hat{z}_0$ component of the angular momentum $\vec{H}_O$ is constant because the component of the moment $\vec{M}_O$ in that inertial direction is equal to zero.

Before proceeding any further, notice that the equation $d\vec{H}_O/dt = \vec{M}_O$ indicates that the infinitesimal change of the angular momentum of the body is in the same direction of the applied moment $\vec{M}_O$. This discloses that if the motion is started with $\theta(0) \neq 0$ and $\theta(0) \neq \pi$ (i.e., the initial angular momentum is not along the vertical direction), line OC should rotate about the vertical z_0 axis (which is something anyone who played with tops has observed). But, as it will be seen, there is more to the resulting motion, as the analysis of the differential equations of motion (and experiments) reveal.

The expression for the absolute angular momentum $\vec{H}_O$ is developed next. With the axial and transverse principal moments of inertia of the rotor taken from Example 6.3.7 (p. 421), the inertia dyadic of the rotor, referred to its center of mass C, is

$$\vec{\vec{I}}_C = \frac{mR^2}{2}\hat{z}\hat{z} + m\left(\frac{R^2}{4} + \frac{h^2}{12}\right)\left(\hat{n}\hat{n} + \hat{\phi}\hat{\phi}\right) \qquad (6.6.3b)$$

Using the parallel axis theorem, the inertia dyadic $\vec{\vec{I}}_O$ is then

$$
\begin{aligned}
\vec{\vec{I}}_O &= \vec{\vec{I}}_C + mL^2\left(\hat{n}\hat{n} + \hat{\phi}\hat{\phi}\right) \\
&= m\frac{R^2}{2}\hat{z}\hat{z} + m\left(\frac{R^2}{4} + \frac{h^2}{12} + L^2\right)\left(\hat{n}\hat{n} + \hat{\phi}\hat{\phi}\right) \stackrel{\triangle}{=} I_z\hat{z}\hat{z} + I_n\left(\hat{n}\hat{n} + \hat{\phi}\hat{\phi}\right)
\end{aligned}
$$

$$(6.6.3c)$$

This expression for $\vec{\vec{I}}_O$ is the inertia dyadic for the entire system since the bar that is welded to the rotor is being treated as massless.

Inspection of Fig. 6.13 discloses that the absolute angular velocity $\vec{\omega}_{\text{rotor}}$ of the rotor (which is the same as the absolute angular velocity of the entire system because of the rigid connection between the rotor and the bar) is

$$\vec{\omega}_{\text{rotor}} = \dot{\phi}\hat{z}_0 + \dot{\theta}\hat{\phi} + \dot{\psi}\hat{z} = \dot{\phi}\left(\hat{z}\cos\theta - \hat{n}\sin\theta\right) + \dot{\theta}\hat{\phi} + \dot{\psi}\hat{z} \qquad (6.6.3\text{d})$$

Note: Having the three components of the absolute angular velocity $\vec{\omega}_{\text{rotor}}$ and the three principal inertias I_z, I_n, and I_n again, it is natural to ask whether they could be used directly in Euler's equations [i.e., Eqs. (6.4.9) to (6.4.11), p. 427] to obtain the differential equations of motion for the system. It is important to notice that doing so would yield wrong equations. The reason is that those equations are for axes that are fixed to the rigid body, which is not the case for the axes that are being used here. If the body were not symmetric (i.e., if the three principal moments of inertia were different), then one would work with body-fixed axes and those equations could be used directly. One could also use body-fixed axes for the symmetric body, of course, and that would require introducing an angle $\psi = \int \dot{\psi}\,dt$ in Fig. 6.13. It is suggested that, in addition to using the formulation that follows, you also derive the equations using body fixed-axes (see Problem 6.16 in Section 6.10). You will then see that the angle ψ can be eliminated by combining the resulting equations to yield those that are formulated in the following (with a little less effort!).

Using the expression for $\vec{\omega}_{\text{rotor}}$ given by Eq. (6.6.3d), the absolute angular momentum $\vec{H}_O$ of the system is then obtained as follows:

$$\vec{H}_O = \vec{\vec{I}}_O \bullet \vec{\omega}_{\text{rotor}} = I_z(\dot{\psi} + \dot{\phi}\cos\theta)\hat{z} + I_n\left[\dot{\theta}\hat{\phi} - \left(\dot{\phi}\sin\theta\right)\hat{n}\right] \overset{\Delta}{=} H_z\hat{z} + H_\phi\hat{\phi} + H_n\hat{n}$$
$$(6.6.3\text{e})$$

The time derivative of $\vec{H}_O$ is obtained, where $(\ )^\bullet$ denotes $d(\ )/dt$, as:

$$\frac{d}{dt}\vec{H}_O = I_z\left(\dot{\psi} + \dot{\phi}\cos\theta\right)^\bullet \hat{z} + I_n\left[\ddot{\theta}\hat{\phi} - \left(\dot{\phi}\sin\theta\right)^\bullet \hat{n}\right] + \vec{\omega}\times\vec{H}_O \qquad (6.6.3\text{f})$$

In this expression, $\vec{\omega}$ is the absolute angular velocity of the unit vector triad $\{\hat{n}, \hat{z}, \hat{\phi} = \hat{z}\times\hat{n}\}$ in which the absolute angular momentum $\vec{H}_O$ has been expressed. This is a fine point to pay attention to since one could inadvertently make a mistake here! Since that unit vector triad is not fixed to the spinning rotor, its absolute angular velocity $\vec{\omega}$ is not equal to $\vec{\omega}_{\text{rotor}}$, but rather

$$\vec{\omega} = \dot{\phi}\hat{z}_0 + \dot{\theta}\hat{\phi} = \dot{\phi}\left(\hat{z}\cos\theta - \hat{n}\sin\theta\right) + \dot{\theta}\hat{\phi} \qquad (6.6.3\text{g})$$

This yields the following expression for $\vec{\omega} \times \vec{H}_O$:

$$\vec{\omega} \times \vec{H}_O = \left[I_z \left(\dot{\psi} + \dot{\phi} \cos\theta \right) - I_n \dot{\phi} \cos\theta \right] \left[\dot{\theta}\, \hat{n} + \left(\dot{\phi} \sin\theta \right) \hat{\phi} \right] \qquad (6.6.3h)$$

Finally, the equation $\vec{M}_O = d\vec{H}_O/dt$ yields the following differential equations involving the variables ψ, θ, and ϕ:

$$I_z \left(\dot{\psi} + \dot{\phi} \cos\theta \right)^{\bullet} = 0 \quad \text{or} \quad \dot{\psi} + \dot{\phi} \cos\theta = \text{constant} \overset{\Delta}{=} \omega_z \qquad (6.6.3i)$$

$$I_n \left(\dot{\phi} \sin\theta \right)^{\bullet} + \left[I_n \dot{\phi} \cos\theta - I_z \omega_z \right] \dot{\theta} = 0 \qquad (6.6.3j)$$

$$I_n \ddot{\theta} + \left[I_z \omega_z - I_n \dot{\phi} \cos\theta \right] \dot{\phi} \sin\theta = mgL \sin\theta \qquad (6.6.3k)$$

The constant $\omega_z = \dot{\psi} + \dot{\phi} \cos\theta = [\dot{\psi} + \dot{\phi} \cos\theta]_{t=0}$ is the axial $\hat{z}$ component (see Fig. 6.13) of the absolute angular velocity $\vec{\omega}_{\text{rotor}}$ of the system.

The two equations (6.6.3j) and (6.6.3k) are the differential equations that govern the motion of the system. They involve only the two unknown angles θ and ϕ, and the constant angular velocity ω_z, which is determined by the initial conditions $\dot{\psi}(0)$, $\dot{\phi}(0)$, and $\theta(0)$ of the motion.

Notice that Eq. (6.6.3j) can be integrated once to obtain another first integral of the motion. An alternative way that yields such an integral is to notice that the moment $\vec{M}_O$ given by Eq. (6.6.3a) has no component along the inertial $\hat{z}_0$ direction shown in Fig. 6.13. Therefore, since $d\vec{H}_O/dt = \vec{M}_O$, the $\hat{z}_0$ component of $\vec{H}_O$ is constant, i.e., $\vec{H}_O \bullet \hat{z}_0 = \text{constant} \overset{\Delta}{=} H_{z_0}$. By making use of the expression for $\vec{H}_O = H_z \hat{z} + H_\phi \hat{\phi} + H_n \hat{n}$ given by Eq. (6.6.3e), and by inspecting Fig. 6.13, the following is obtained[5]:

$$I_n \dot{\phi} \sin^2\theta + I_z \omega_z \cos\theta = \text{constant} = \overset{\Delta}{=} H_{z_0} \qquad (6.6.3l)$$

Notice that the time derivative of this equation yields Eq. (6.6.3j).

Equilibrium solutions Equations (6.6.3j) and (6.6.3k) admit several equilibrium solutions $\theta = \text{constant} \overset{\Delta}{=} \theta_e$. One such equilibrium corresponds to $\sin\theta_e = 0$, which has two physically distinct solutions, namely, $\theta_e = 0$ and $\theta_e = \pi$. For these equilibrium solutions, the top is vertical and spinning about that axis with a constant angular velocity equal to ω_z. The top is called a *sleeping top* when it is spinning with $\theta = \theta_e = 0$.

---

[5]One could also obtain Eq. (6.6.3l) by multiplying Eq. (6.6.3j) by $\sin\theta$, which allows the terms involving $\dot{\phi}$ to be combined as $d(\dot{\phi}\sin^2\theta)/dt$, and then integrating the result with respect to time.

Another type of equilibrium with $\theta_e = $ constant $\neq 0$ corresponds to

$$(I_z\omega_z - I_n\dot\phi\cos\theta_e)\dot\phi = mgL \qquad (6.6.3\text{m})$$

with $\dot\phi = $ constant $\overset{\Delta}{=} \dot\phi_e$. This also implies that $\dot\psi = $ constant $\overset{\Delta}{=} \dot\psi_e$, as per Eq. (6.6.3i). In this case, Eq. (6.6.3m) yields the following relation for calculating the angle θ_e in terms of the angular velocities ω_z and $\dot\phi_e$:

$$\cos\theta_e = \frac{1}{I_n\dot\phi_e}\left(I_z\omega_z - \frac{mgL}{\dot\phi_e}\right) \qquad (6.6.3\text{n})$$

For the equilibrium solution given by Eq. (6.6.3n), the top is spinning about its axial axis with a constant angular velocity ω_z. While doing so, that axis is inclined at a constant angle θ_e with the vertical and is turning about the vertical with a constant angular velocity $\dot\phi_e$. Such a motion is called *precession*.

Equation (6.6.3m) is also a quadratic equation in $\dot\phi_e$, i.e.,

$$I_n(\cos\theta_e)\dot\phi_e^2 - I_z\omega_z\dot\phi_e + mgL = 0 \qquad (6.6.3\text{o})$$

This quadratic equation in $\dot\phi_e$ has two solutions when $\cos\theta_e \neq 0$ (i.e., when the spin axis of the top is not horizontal), which are

$$\dot\phi_e = \frac{I_z\omega_z}{2I_n\cos\theta_e}\left[1 \pm \sqrt{1 - 4\frac{mgL}{I_n}\left(\frac{I_n}{I_z}\right)^2\frac{\cos\theta_e}{\omega_z^2}}\right] \qquad (6.6.3\text{p})$$

With $I_z = mR^2/2$ and $I_n = m(R^2/4 + h^2/12 + L^2)$, as given by Eq. (6.6.3c), we have $I_z/I_n = (1/2)/[1/4 + (h/R)^2/12 + (L/R)^2]$ and $mgL/I_n = (gL/R^2)/[1/4 + (h/R)^2/12 + (L/R)^2]$.

Figure 6.14 shows a plot of the smallest root of Eq. (6.6.3o), versus θ_e, for $L/R = 1$, $h/R = 1/3$, with $R = 10$ cm and $\omega_z = 50$ rad/s (see Observation 2, which follows).

The top exhibits an equilibrium with $\theta = $ constant $\overset{\Delta}{=} \theta_e$ only if the solutions to Eq. (6.6.3p) are real. This happens only if

$$|\omega_z| \geq \frac{\sqrt{4mgLI_n\cos\theta_e}}{I_z} \qquad (6.6.3\text{q})$$

The top has to be given a minimum spin rate $|\omega_{z\min}| = \sqrt{4mgLI_n\cos\theta_e}/I_z$ for it to exhibit a motion with $\theta = $ constant $\overset{\Delta}{=} \theta_e$. For $\theta_e = 30°$, $L/R = 1$, and $h/R = 1/3$, it follows that $|\omega_{z\min}| \approx 53.4$ rad/s (≈ 510 rpm) for $R = 6$ cm, and $|\omega_{z\min}| \approx 41.4$ rad/s (≈ 395 rpm) for $R = 10$ cm. In practice, when friction is present, one should

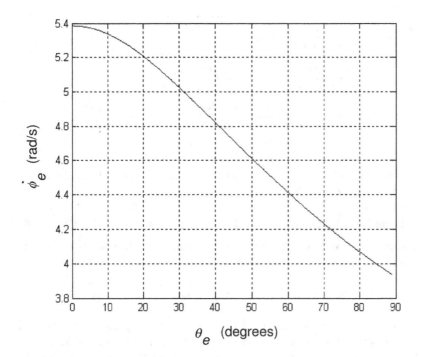

Figure 6.14: Plot of $\dot{\phi}_e$ versus θ_e for $L/R = 1$, $h/R = 1/3$, with $R = 10$ cm, $mgL/I_n = 196.2/2.52$, and $\omega_z = 50$ rad/s.

spin the rotor with an angular speed $|\omega_z|$ much greater than its minimum value. Friction would cause the value of $|\omega_z|$ to decrease, and eventually the top would fall down as $|\omega_z|$ becomes smaller than $\omega_{z\min} = \sqrt{4mgLI_n \cos\theta_e}/I_z$.

Observation 1. When $\cos\theta_e = 0$, i.e., when the spin axis of the top is horizontal, the solution to Eq. (6.6.3o) is

$$\dot{\phi}_e = \frac{mgL}{I_z \omega_z}$$

which again shows that the higher the value of the spin rate $|\omega_z|$, the slower the top rotates around the vertical.

Observation 2. When $I_z|\omega_z| \gg \sqrt{4mgLI_n \cos\theta_e}$, the roots of Eq. (6.6.3o) are

$$
\begin{aligned}
\dot\phi_e &= \frac{I_z\omega_z}{2I_n\cos\theta_e}\left[1\pm\sqrt{1-\frac{4mgLI_n\cos\theta_e}{(I_z\omega_z)^2}}\right]\\[2mm]
&\approx \frac{I_z\omega_z}{2I_n\cos\theta_e}\left[1\pm\left(1-\frac{2mgLI_n\cos\theta_e}{(I_z\omega_z)^2}\right)\right]\\[2mm]
&= \frac{I_z\omega_z}{I_n\cos\theta_e}-\frac{mgL}{I_z\omega_z}\approx\frac{I_z\omega_z}{I_n\cos\theta_e}\quad\text{and}\quad\frac{mgL}{I_z\omega_z}
\end{aligned}
$$

Of the two roots, the one that matches the result for $\cos\theta_e = 0$ is the smallest one $\dot\phi_e = mgL/(I_z\omega_z)$. Generally, the actual value of $\dot\phi_e$ exhibited by a top is the smallest root of Eq. (6.6.3o).

Stability Analysis by Linearization   To linearize the differential Eqs. (6.6.3j) and (6.6.3k) about $\theta = \text{constant} \triangleq \theta_e$ (with $\theta_e \neq 0$) and $\dot\phi = \text{constant} \triangleq \dot\phi_e$, start by introducing small perturbations $\theta_s(t)$ and $\dot\phi_s(t)$ by letting

$$\theta(t) = \theta_e + \theta_s(t) \qquad \dot\phi(t) = \dot\phi_e + \dot\phi_s(t)$$

By substituting these expressions into Eq. (6.6.3j), and expanding the resulting equation in Taylor series for small θ_s and $\dot\phi_s$, the following linearized differential equation is obtained after products of small quantities in the expansion are disregarded:

$$I_n\ddot\phi_s\sin\theta_e + (2I_n\dot\phi_e\cos\theta_e - I_z\omega_z)\dot\theta_s = 0 \tag{6.6.3r}$$

Following the same procedure with Eq. (6.6.3k), the following linearized equation is obtained:

$$
I_n\ddot\theta_s + \overbrace{\left[I_z\omega_z\dot\phi_e\cos\theta_e - I_n\dot\phi_e^2(\cos^2\theta_e - \sin^2\theta_e) - mgL\cos\theta_e\right]}^{=I_n\dot\phi_e^2\sin^2\theta_e\text{ since }I_z\omega_z\dot\phi_e-mgL-I_n\dot\phi_e^2\cos\theta_e=0\text{ as per Eq. }(6.6.3m)}\theta_s
$$

$$
+\left[I_z\omega_z\sin\theta_e - 2I_n\dot\phi_e(\sin\theta_e)\cos\theta_e\right]\dot\phi_s = 0 \tag{6.6.3s}
$$

By expressing the solution to Eqs. (6.6.3r) and (6.6.3s) as $\theta_s = Ae^{st}$ and $\dot\phi_s = Be^{st}$, the following characteristic equation is obtained:

$$
s\left[s^2I_n^2 + I_n^2\dot\phi_e^2\sin^2\theta_e + \left(I_z\omega_z - 2I_n\dot\phi_e\cos\theta_e\right)^2\right] = 0
$$

This discloses that the perturbed motion will exhibit an oscillation with frequency Ω, where

$$\Omega = \sqrt{\dot{\phi}_e^2 \sin^2 \theta_e + (I_z \omega_z / I_n - 2\dot{\phi}_e \cos \theta_e)^2}$$

This oscillation of the spin axis of a top or of a gyroscope is called *nutation*.

Initial Conditions for Preventing Nutation   In practice, the nutational motion is eliminated with no major difficulty by inserting a damper in the system. Here, as a curiosity, we show how it can be prevented by proper choice of the initial conditions $\dot{\phi}(0)$ and $\dot{\psi}(0)$ for a desired value of the equilibrium θ_e. With $\theta(0) = \theta_e$, $\dot{\phi}(0) = \dot{\phi}_e$, Eq. (6.6.3o), with $\omega_z = \dot{\psi}(0) + \dot{\phi}(0) \cos\theta(0)$, yields the following solutions for $\dot{\phi}(0)$ (with $I_n \neq I_z$):

$$\dot{\phi}(0) = \frac{I_z \dot{\psi}(0) \pm \sqrt{I_z^2 \dot{\psi}^2(0) - 4(I_n - I_z)mgL \cos\theta(0)}}{2(I_n - I_z)\cos\theta(0)}$$

A real value for $\dot{\phi}(0)$ exists only if $I_z \dot{\psi}(0) \geq \sqrt{4(I_n - I_z)mgL \cos(\theta(0)}$ with, of course, $I_n > I_z$.

Additional Comments

■ It can be easily verified by numerical integration of the nonlinear differential equations of motion that the equilibrium $\theta = \text{constant} \stackrel{\Delta}{=} \theta_e$ and $\dot{\phi} = \text{constant} \stackrel{\Delta}{=} \dot{\phi}_e$, when it exists, is stable.

■ The precession rate of a top may be readily obtained from a very simple approximate calculation when the spin rate $|\omega_z|$ is very large, so that the absolute angular momentum $\vec{H}_O = I_z \omega_z \hat{z} + I_n \dot{\theta} \hat{\phi} - I_n(\dot{\phi} \sin\theta)\hat{n}$ may be approximated as $H_O \approx I_z \omega_z \hat{z}$. In such a case, application of the equation $d\vec{H}_O/dt = \vec{M}_O = (mgL \sin\theta_e) \hat{\phi}$ is readily visualized as shown in Fig. 6.15. The figure shows the large angular momentum $\vec{H}_O \approx I_z \omega_z \hat{z}$ and its change $d\vec{H}_O$ after an infinitesimal time dt is elapsed while $\theta = \text{constant} = \theta_e$. That change is in the same direction of the moment $\vec{M}_O$. Thus, it occurs on a horizontal plane. It is equal to $(I_z \omega_z \sin\theta_e d\phi) \hat{\phi}$, and it is represented by the arrow labeled "$M_O dt = mgL \sin\theta_e dt$" in the figure. As shown in the insert of the figure, this yields $\dot{\phi}_e = mgL/(I_z \omega_z)$, which is the smallest root of Eq. (6.6.3o) as explained earlier.

■ The top spins about its axial axis OC with a constant angular velocity ω_z. When in its equilibrium state $\theta = \text{constant} = \theta_e$, that axis rotates about the

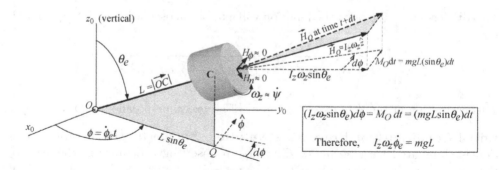

Figure 6.15: Visualization of $d\vec{H}_O/dt = \vec{M}_O$ when $\vec{H}_O = I_z\omega_z\hat{z} + I_n\dot{\theta}\hat{\phi} - I_n(\dot{\phi}\sin\theta)\hat{n} \approx I_z\omega_z\hat{z}$.

vertical with a constant angular velocity $\dot{\phi}_e$ causing points on the axial axis of the top to describe a vertical cone. When ω_z is counterclockwise (i.e., $\omega_z > 0$), the cone is traced in a counterclockwise manner, and clockwise otherwise. It was seen that higher values of the spin rate $|\omega_z|$ yield smaller values of the angular velocity $\dot{\phi}_e$; i.e., it takes longer for the spin axis of the top to change direction. That is the type of behavior one wants in a gyroscope for, in such a case, its spin axis would serve as a nearly fixed reference direction in space. As long as Eq. (6.6.3p) yields a real solution for $\dot{\phi}_e$, the motion with θ = constant occurs when it is started with the initial conditions $\theta(0) = \theta_e$, $\dot{\theta}(0) = 0$, and $\dot{\phi}(0) = \dot{\phi}_e$ as determined earlier. The stability analysis of the equilibrium disclosed that when such an equilibrium exists (i.e., when $|\omega_z| \geq \sqrt{4mgLI_n\cos\theta_e}/I_z$), it is stable, and $\theta(t)$ exhibits a small harmonic oscillation.

■ The type of motion where the spin axis of the top revolves around the vertical is called *precession*. When θ = constant $\stackrel{\triangle}{=} \theta_e$, the precession is said to be *steady*. If the angle θ_e is perturbed (say, by hitting the spin axis a little bit or by adding a little additional mass to the tip of the top), the axial axis of the top still revolves around the vertical but in a manner that neither $\theta(t)$ nor $\dot{\phi}(t)$ remain constant. The precession is unsteady in such a situation.

■ The oscillation of line OC about the equilibrium $\theta = \theta_e$ is called *nutation*. Precession and nutation are two important responses exhibited by spinning bodies such as a top or a gyroscope. For a gyroscope, one wants to have $L = 0$ to eliminate the gravitational moment. In other words, in a gyroscope the spinning body is attached to a set of gimbals that allow for the direction of the spin axis to change, as illustrated in Fig. 6.16.

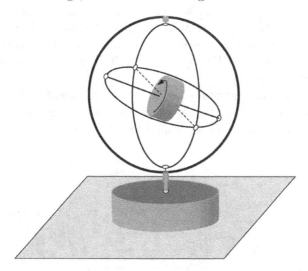

Figure 6.16: A typical gyroscope, showing its gimbals.

An ideal gyroscope is one that is not subjected to a moment $\vec{M}_C$, and, thus, its angular momentum would be constant. In such a case, the magnitude of $\vec{H}_C$ would be constant, and $\vec{H}_C$ would maintain the same direction in inertial space. In practice, there is always a drift of the spin axis due to small disturbing moments that could be caused by small mass asymmetry for example. High-precision gyroscopes are widely used for inertial navigation, such as in commercial aircraft instrumentation, and in many other applications.

$\Longrightarrow$ **Numerical Integration of the Differential Equations of Motion**
For Scilab or Matlab users, the respective programs *top_dynamics* listed in Appendix E solve Eqs. (6.6.3j) and (6.6.3k) for ϕ and θ, plot ϕ and θ versus time, and animate the motion of the spin axis of the top. One uses them by simply typing, either in the Scilab console or in the Matlab command window (the ... is for continuing a command if it is not typed in a single line),
`my_output=top_dynamics(t, IzoverIn, mgLoverIn, phidot0, psidot0, ...`
`theta0, thetadot0);`
with desired values for `IzoverIn`=I_z/I_n, `mgLoverIn`=mgL/I_n, for the initial conditions `phidot0`=$\dot{\phi}(0)$, `psidot0`=$\dot{\psi}(0)$, `theta0`=$\theta(0)$, `thetadot0`=$\dot{\theta}(0)$, and for a time array `t` of the form [`t0` : Δ`t` : `tfinal`]. A specific example of their use is shown in the comment lines at the beginning of both programs.

Those who have Simulink and prefer to use it may want to construct the block diagram model shown in Fig. 6.17 for integrating the same differential equations.[6] A suggestion for the Solver settings in the *Simulation → Parameters* menu is indicated in the figure caption. See the Simulink tutorial in Appendix A for the suggested settings of the *To Workspace* block. To see an animation of the motion after the simulation stops, type

`animate3d_nbars(x_coordinates, y_coordinates, z_coordinates)`

in the Matlab command window (assuming you have stored in your computer the program `animate3d_nbars` listed in Appendix E).

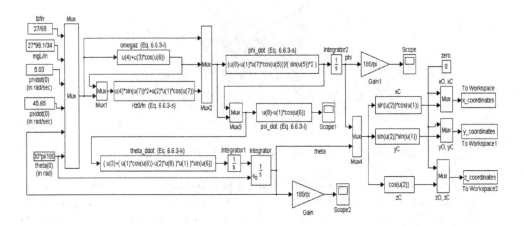

Figure 6.17: A Simulink model for the top dynamics. For the model inputs shown, the time array [0:0.001:10] is suggested for the *Simulation → Parameters* settings. Also, one may set $\phi(0) = 0$ and $\dot{\theta}(0) = 0$ by double-clicking on the integrators labeled "Integrator2" and "Integrator1," respectively.

For a cylinder with an arm, such as the rotor shown in Fig. 6.13, $I_z = mR^2/2$ and $I_n = m(3R^2 + h^2)/12 + mL^2$, as indicated in Eq. (6.6.3c). The values $I_z/I_n = 27/68$ and $mgL/I_n = 27*98.1/34$ correspond to $L/R = 1$ and $h/R = 1/3$, with $R = 10$ cm. The initial conditions $\dot{\phi}(0) = 5.03$ rad/s, and $\dot{\psi}(0) = 45.65$ rad/s yield $\omega_z \approx 50$ rad/s for $\theta_e = 30°$ (which is the value for $\theta(0)$ shown in Fig. 6.17). The value $\dot{\phi}_e \approx 5.03$ rad/s is obtained from Eq. (6.6.3p) with $\theta_e = 30°$. It is also interesting to do the numerical integration with a smaller value of either $\dot{\phi}(0)$ or $\dot{\psi}(0)$, including values

---

[6]The fifth user defined input in the block diagram model shown in Fig. 6.17 is the initial value $\theta(0)$ for the integrator whose output is the angle $\theta(t)$. That initial condition is supplied "externally" to that integrator only for convenience.

that yield $\omega_z < \omega_{z_{\min}}$, and watch the animation to see what happens to the top.

Figure 6.18 shows the response $\theta(t)$ obtained with either the Scilab or the Matlab program mentioned earlier, with the initial conditions and parameter values

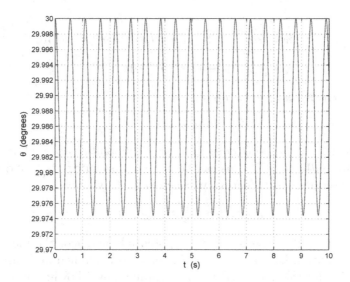

Figure 6.18: Response $\theta(t)$ obtained with the initial conditions and parameter values mentioned in the text.

also mentioned earlier, using the calling command shown in a comment line at the beginning of the program. The values mentioned for $\dot{\phi}(0)$ and $\dot{\psi}(0)$ were calculated to prevent nutation with $\theta(0) = 30°$. However, due to round-off errors in the calculations, the response shown in Fig. 6.18 is not constant, but the amplitude of the resulting oscillation is very small, as expected. Also its frequency matches the value $\Omega \approx 11.5$ rad/s obtained from the expression for Ω given on p. 449.

6.6.4 On the Nutational Motion of a Symmetric Top

Further insight into the motion of the top can be obtained by pushing analytical techniques as far as possible, and it turns out that this can be done if we solve Eq. (6.6.3l) for $\dot{\phi}$ in terms of θ and use the solution to eliminate $\dot{\phi}$ from Eq. (6.6.3k) to obtain a single differential equation whose only unknown is the angle $\theta(t)$. The final equation is of the general form $I_n \ddot{\theta} + f(\theta) = 0$, which can be integrated once to obtain an equation that involves $\dot{\theta}$ and θ, as shown in Chapter 1 using the chain rule. However, it turns out that such an integration is much easily done if we make use of the basic work-energy equation $dW = d(T)$, where T is the kinetic energy of the motion and dW is the infinitesimal work done by all forces acting on the system.

The infinitesimal work done by the gravitational force $-mg\hat{z}_0$ (see Fig. 6.13) is simply

$$
\begin{aligned}
dW &= -mg\hat{z}_0 \bullet \vec{v}_C \, dt = -mg\hat{z}_0 \bullet \left[(L\dot{\theta}\,dt)\hat{n} + (L\sin\theta)(\dot{\phi}\,dt)\hat{\phi} \right] \\
&= -mg\hat{z}_0 \bullet \left[(L\,d\theta)\hat{n} + (L\sin\theta)(d\phi)\hat{\phi} \right] \tag{6.6.4a}
\end{aligned}
$$

and since $\hat{z}_0 = \hat{z}\cos\theta - \hat{n}\sin\theta$, we obtain:

$$
dW = mgL\sin\theta\,d\theta \tag{6.6.4b}
$$

The expression $dW = mgL\sin\theta\,d\theta$ is the exact differential $d(-mgl\cos\theta)$, plus a constant, and since $dW = d(T)$, this can be immediately integrated (and this is the reason why we used the work-energy approach here) to yield

$$
T + mgL\cos\theta = \text{constant} \overset{\Delta}{=} E \tag{6.6.4c}
$$

The quantity $mgL\cos\theta$ happens to be the potential energy of the motion, and it could have been written by inspection of Fig. 6.13, as explained in Chapter 2.

The expression for the kinetic energy T is obtained by following the steps shown in the boxed text on p. 429 in Section 6.5. Using the expressions for the absolute angular velocity of the rotor and for the inertia dyadic $\vec{\vec{I}}_O$, we obtain

$$
T = \frac{1}{2}\vec{\omega}_{\text{rotor}} \bullet \vec{\vec{I}}_O \bullet \vec{\omega}_{\text{rotor}} = \frac{1}{2}I_n\dot{\theta}^2 + \frac{1}{2}I_n\dot{\phi}^2\sin^2\theta + \frac{1}{2}I_z\left(\dot{\psi} + \dot{\phi}\cos\theta\right)^2 \tag{6.6.4d}
$$

Since $\dot{\psi} + \dot{\phi}\cos\theta$ was found to be constant, $T + mgL\cos\theta = \text{constant}$ then yields, also using the expression for $\dot{\phi}$ obtained from Eq. (6.6.31),

$$
\frac{1}{2}I_n\left[\dot{\theta}^2 + \dot{\phi}^2\sin^2\theta\right] + mgL\cos\theta = \frac{1}{2}I_n\left[\dot{\theta}^2 + \frac{(H_{z0} - I_z\omega_z\cos\theta)^2}{I_n^2\sin^2\theta}\right] + mgL\cos\theta
$$

$$
= \text{constant} \overset{\Delta}{=} C \tag{6.6.4e}
$$

This equation looks more complicated than it actually is! By premultiplying it by $\sin^2\theta$, because the quantity $\dot{\theta}^2\sin^2\theta$ is the same as $[d(\cos\theta)/dt]^2$, we can give preference to working with $\cos\theta$ to reduce it to

$$
\left(\frac{d}{dt}\cos\theta\right)^2 = \frac{2C}{I_n}\left(1 - \cos^2\theta\right) - \left(\frac{H_{z0}}{I_n} - \frac{I_z}{I_n}\omega_z\cos\theta\right)^2 - \frac{2mgL}{I_n}\left(\cos\theta\right)\left(1 - \cos^2\theta\right)
$$

$$
\tag{6.6.4f}
$$

The right-hand side of Eq. (6.6.4f) is a cubic function of $\cos\theta$ and, thus, can easily be plotted versus $\cos\theta$ for given values of the system parameters and initial conditions of the motion. A sketch of a plot of $[d(\cos\theta)/dt]^2$ versus $\cos\theta$ is shown in Fig. 6.19.

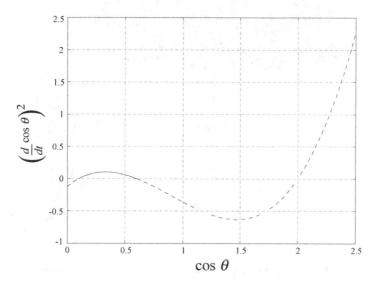

Figure 6.19: A sketch of $[d(\cos\theta)/dt]^2$ versus $\cos\theta$ for Eq. (6.6.4f).

Since $[d(\cos\theta)/dt]^2$ is never negative, and $|\cos\theta| \leq 1$, the plot is valid only in the region defined by the solid part of the curve. Since $[d(\cos\theta)/dt]^2 = \dot\theta^2\sin^2\theta = 0$ when $\dot\theta = 0$ (i.e., when $\theta(t)$ reaches a maximum or minimum value), the sketch in Fig. 6.19 is disclosing that the spin axis of the top oscillates between a minimum and a maximum value of $\theta(t)$. The minimum and maximum values of $\theta(t)$ can coalesce into a single value for certain initial conditions. When that happens, the equilibrium solution determined in Section 6.6.3 is then observed. The solid part of the curve in Fig. 6.19 is shifted down until $[d(\cos\theta)/dt]^2 = 0$ in such a case. A further shift down of that curve implies that the spin rate $\dot\psi_e$ of the top is too small and the top keeps "falling" as it precesses.

6.7 Solved Example Problems

Example 6.7.1 A homogeneous rectangular plate of mass m and dimensions L, a, and b is hinged to a rotating horizontal circular platform as shown in Fig. 6.20.

The center O of the platform is inertial. The plate is connected to the platform by

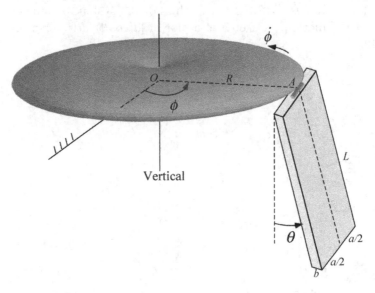

Figure 6.20: A plate on a rotating platform for Example 6.7.1.

a hinge and, thus, is forced to turn with the platform. The hinge allows the angle θ shown in Fig. 6.20 to change with time. The platform is driven by a motor with a known absolute angular velocity $\dot{\phi}$ (i.e., ϕ is measured from a line fixed in space and in the plane of rotation of line OA of the platform). Friction in the hinge may be neglected.

  a. Obtain the differential equation for the plate in terms of the angle θ.

  b. For $\dot{\phi} = \text{constant} \overset{\Delta}{=} \Omega$, with $\Omega^2 L/g = 1.5$, $L = 3R$, $a = L/4$, and b so small that it can be neglected:

     1. Determine the equilibrium solutions $\theta = \text{constant} \overset{\Delta}{=} \theta_e$, and analyze the stability of each equilibrium by linearizing the differential equation obtained in part a. What is the frequency of the small oscillations of the plate about its stable equilibrium?

     2. Plot the magnitude of the reaction force and the magnitude of the reaction moment at the hinge versus nondimensional time $\tau = t\sqrt{g/L}$ and then determine, from the plots, their maximum values when the motion is started with the initial conditions $\theta(0) = 30°$ and $\dot{\theta}(0) = 0$.

■ Solution

a. Figure 6.21 shows the free-body diagram for the plate in Fig. 6.20. A unit
vector triad $\{\hat{x}, \hat{y}, \hat{z}\}$, parallel to the principal x-y-z axes of the plate, and a
vertical unit vector $\hat{z}_0 = \hat{z}\sin\theta - \hat{x}\cos\theta$, are also superimposed on the diagram.
The principal x-y-z axes of the plate are centered at its center of mass C. The
vectors labeled A_1, A_2, and A_3 are the $\hat{x}$, $\hat{y}$, and $\hat{z}$ components, respectively, of
the reaction force applied at point A to the plate. The curved arrows labeled
M_1 and M_3 represent the reaction moments $\vec{M}_1 = M_1\hat{x}$ and $\vec{M}_3 = M_3\hat{z}$ applied
by the hinge to the plate. Those reaction moments are nonzero because the
hinge, by its construction, prevents the plate from rotating along axes parallel
to $\hat{x}$ and $\hat{z}$, respectively. The component of the reaction moment parallel to $\hat{y}$
is equal to zero because the hinge is frictionless.

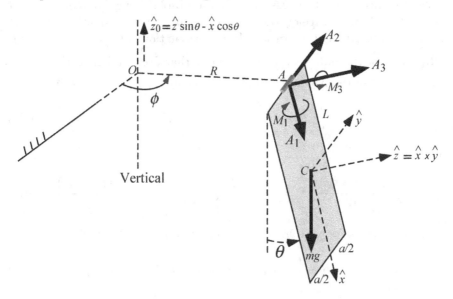

Figure 6.21: Free-body diagram (also showing a unit vector triad $\{\hat{x}, \hat{y}, \hat{z}\}$) for the
plate in Fig. 6.20.

Six scalar equations involving the six unknowns θ, A_1, A_2, A_3, M_1, and M_3
are obtained from the equation $\vec{F} = m\vec{a}_C$, where $\vec{a}_C$ is the absolute acceleration
of the center of mass C of the plate, and from $\vec{M}_C = d\vec{H}_C/dt$, where $\vec{H}_C$ is the
absolute angular momentum of the plate. Of those six equations, only one of

them is a differential equation of motion since θ is the only unknown motion variable.

The resultant $\vec{F}$ of the forces acting on the plate, and the moment $\vec{M}_C$, about C, of those forces are (see Fig. 6.21)

$$
\begin{aligned}
\vec{F} &= A_1\hat{x} + A_2\hat{y} + A_3\hat{z} - mg\hat{z}_0 \\
&= A_1\hat{x} + A_2\hat{y} + A_3\hat{z} - mg\left(\hat{z}\sin\theta - \hat{x}\cos\theta\right) \qquad (6.7.1a)
\end{aligned}
$$

$$
\begin{aligned}
\vec{M}_C &= M_1\hat{x} + M_3\hat{z} - \frac{L}{2}\hat{x} \times (A_1\hat{x} + A_2\hat{y} + A_3\hat{z}) \\
&= M_1\hat{x} + \frac{L}{2}A_3\hat{y} + \left(M_3 - \frac{L}{2}A_2\right)\hat{z} \overset{\Delta}{=} M_x\hat{x} + M_y\hat{y} + M_z\hat{z} \quad (6.7.1b)
\end{aligned}
$$

We now need to get expressions for the absolute angular momentum $\vec{H}_C$ and for the absolute acceleration $\vec{a}_C$ to use them in the equations $\vec{F} = m\vec{a}_C$ and $\vec{M}_C = d\vec{H}_C/dt$. Let us start with the formulation of the expression for $\vec{H}_C$.

In terms of the unit vectors $\hat{x}$, $\hat{y}$, and $\hat{z}$ that are parallel to the principal axes of inertia of the plate (see Fig. 6.21), the inertia dyadic $\vec{\vec{I}}_C$ of the plate is

$$
\vec{\vec{I}}_C = I_x\hat{x}\hat{x} + I_y\hat{y}\hat{y} + I_z\hat{z}\hat{z} \qquad (6.7.1c)
$$

where

$$
I_x = \frac{m\left(a^2 + b^2\right)}{12} \qquad I_y = \frac{m\left(L^2 + b^2\right)}{12} \qquad I_z = \frac{m\left(L^2 + a^2\right)}{12} \qquad (6.7.1d)
$$

The absolute angular velocity $\vec{\omega}$ of the plate is (see Fig. 6.21)

$$
\vec{\omega} = \dot{\phi}\hat{z}_0 - \dot{\theta}\hat{y} = \dot{\phi}\left(\hat{z}\sin\theta - \hat{x}\cos\theta\right) - \dot{\theta}\hat{y} \overset{\Delta}{=} \omega_x\hat{x} + \omega_y\hat{y} + \omega_z\hat{z} \quad (6.7.1e)
$$

Euler's equations for principal axes, which are Eqs. (6.4.9) to (6.4.11) in Section 6.4, then yield the following equations for this problem:

$$
-I_x\left(\ddot{\phi}\cos\theta - \dot{\phi}\dot{\theta}\sin\theta\right) - (I_z - I_y)\dot{\phi}\dot{\theta}\sin\theta = M_1 \qquad (6.7.1f)
$$

$$
-I_y\ddot{\theta} - (I_x - I_z)\dot{\phi}^2\sin\theta\cos\theta = \frac{L}{2}A_3 \qquad (6.7.1g)
$$

$$
I_z\left(\ddot{\phi}\sin\theta + \dot{\phi}\dot{\theta}\cos\theta\right) + (I_y - I_x)\dot{\phi}\dot{\theta}\cos\theta = M_3 - \frac{L}{2}A_2 \quad (6.7.1h)
$$

Equations (6.7.1f), (6.7.1g), and (6.7.1h) may also be obtained directly from $d\vec{H}_C/dt = \vec{M}_C$ without either "consulting" or remembering Eqs. (6.4.9), (6.4.10), and (6.4.11). The expressions for $\vec{H}_C$ and $d\vec{H}_C/dt$ for doing so are as follows:

$$\vec{H}_C = \overleftrightarrow{I}_C \bullet \vec{\omega} = I_z \left(\dot{\phi} \sin\theta \right) \hat{z} - I_y \dot{\theta}\hat{y} - I_x \left(\dot{\phi}\cos\theta \right) \hat{x} \qquad (6.7.1i)$$

$$\frac{d}{dt}\vec{H}_C = I_z \left(\ddot{\phi}\sin\theta + \dot{\phi}\dot{\theta}\cos\theta \right) \hat{z} - I_y\ddot{\theta}\hat{y} - I_x \left(\ddot{\phi}\cos\theta - \dot{\phi}\dot{\theta}\sin\theta \right) \hat{x} + \vec{\omega} \times \vec{H}_C$$

$$= \left[(I_y - I_z)\dot{\phi}\dot{\theta}\sin\theta - I_x \left(\ddot{\phi}\cos\theta - \dot{\phi}\dot{\theta}\sin\theta \right) \right] \hat{x}$$

$$+ \left[(I_z - I_x)\dot{\phi}^2 \sin\theta\cos\theta - I_y\ddot{\theta} \right] \hat{y}$$

$$+ \left[(I_y - I_x)\dot{\phi}\dot{\theta}\cos\theta + I_z \left(\ddot{\phi}\sin\theta + \dot{\phi}\dot{\theta}\cos\theta \right) \right] \hat{z} \qquad (6.7.1j)$$

The expression for the absolute acceleration $\vec{a}_C = d^2\overrightarrow{OC}/dt^2$ of C is developed next, starting from the expression for $\overrightarrow{OC}$ as obtained by inspection of Fig. 6.21.

■ Absolute position vector for the center of mass C of the plate:

$$\vec{r}_C = \overrightarrow{OC} = \overrightarrow{OA} + \overrightarrow{AC} = R\left(\hat{x}\sin\theta + \hat{z}\cos\theta \right) + \frac{L}{2}\hat{x} \qquad (6.7.1k)$$

■ Absolute velocity of C:

$$\vec{v}_C = \frac{d\vec{r}_C}{dt} = R\dot{\theta}\left(\hat{x}\cos\theta - \hat{z}\sin\theta \right) + \vec{\omega} \times \vec{r}_C = \left(R + \frac{L}{2}\sin\theta \right)\dot{\phi}\,\hat{y} + \frac{L}{2}\dot{\theta}\hat{z}$$

$$\qquad (6.7.1l)$$

■ Absolute acceleration of C:

$$\vec{a}_C = \frac{d\vec{v}_C}{dt} = \left[\left(R + \frac{L}{2}\sin\theta \right)\ddot{\phi} + \frac{L}{2}\dot{\phi}\dot{\theta}\cos\theta \right]\hat{y} + \frac{L}{2}\ddot{\theta}\hat{z} + \vec{\omega} \times \vec{v}_C$$

$$= -\left[\frac{L}{2}\dot{\theta}^2 + \left(R + \frac{L}{2}\sin\theta \right)\dot{\phi}^2\sin\theta \right]\hat{x} + \left[\left(R + \frac{L}{2}\sin\theta \right)\ddot{\phi} + L\dot{\phi}\dot{\theta}\cos\theta \right]\hat{y}$$

$$+ \left[\frac{L}{2}\ddot{\theta} - \left(R + \frac{L}{2}\sin\theta \right)\dot{\phi}^2\cos\theta \right]\hat{z} \qquad (6.7.1m)$$

With $\vec{F}$ given by Eq. (6.7.1a) and $\vec{a}_C$ by Eq. (6.7.1m), the following scalar equations are then obtained from $\vec{F} = m\vec{a}_C$:

$$A_1 + mg\cos\theta = -m\left[\frac{L}{2}\dot{\theta}^2 + \left(R + \frac{L}{2}\sin\theta \right)\dot{\phi}^2\sin\theta \right] \qquad (6.7.1n)$$

$$A_2 = m\left[\left(R + \frac{L}{2}\sin\theta\right)\ddot{\phi} + L\dot{\phi}\dot{\theta}\cos\theta\right] \qquad (6.7.1\text{o})$$

$$A_3 - mg\sin\theta = m\left[\frac{L}{2}\ddot{\theta} - \left(R + \frac{L}{2}\sin\theta\right)\dot{\phi}^2\cos\theta\right] \qquad (6.7.1\text{p})$$

The six unknowns θ, A_1, A_2, A_3, M_1, and M_3 can now be determined by solving the six equations (6.7.1f), (6.7.1g), (6.7.1h), (6.7.1n), (6.7.1o), and (6.7.1p). Equations (6.7.1g) and (6.7.1p) can be combined to yield

$$\left(\frac{L^2}{3} + \frac{b^2}{12}\right)\ddot{\theta} - \left[\frac{RL}{2} + \left(\frac{L^2}{3} - \frac{b^2}{12}\right)\sin\theta\right]\dot{\phi}^2\cos\theta + g\frac{L}{2}\sin\theta = 0 \;\;(6.7.1\text{q})$$

where use was made of the expressions for the moments of inertia I_x, I_y, and I_z shown in the equations of (6.7.1d). Equation (6.7.1q), whose only unknown is the angle θ when $\dot{\phi}$ is prescribed, is the differential equation of motion for the system.

Notice that the motion $\theta(t)$ does not depend on the lateral dimension a of the plate. Also notice that Eq. (6.7.1q) is the only equation that is needed if only the angle $\theta(t)$, and not reactions, are to be determined. In such a case the equations that involve the hinge reactions can be disregarded.

b1. When $\dot{\phi} = \text{constant} \overset{\Delta}{=} \Omega$, Eq. (6.7.1q) admits the solution $\theta = \text{constant} \overset{\Delta}{=} \theta_e$. For b small enough to be neglected (i.e., $b << 2L$), the equilibrium θ_e is obtained by solving the following equation, which is obtained from Eq. (6.7.1q):

$$\sin\theta_e - \frac{\Omega^2 L}{g}\left(\frac{R}{L} + \frac{2}{3}\sin\theta_e\right)\cos\theta_e = 0 \qquad (6.7.1\text{r})$$

The values of θ_e that satisfy Eq. (6.7.1r) may be obtained by plotting the left-hand side of that equation versus θ_e. Such a plot is shown in Fig. 6.22 for $\Omega^2 L/g = 1.5$ and $L = 3R$, and the solutions for θ_e are found to be

$$\theta_e \approx 52.2^\circ \qquad \text{and} \qquad \theta_e \approx 194.2^\circ$$

To linearize the differential equation of motion, Eq. 6.7.1q, let

$$\theta(t) \overset{\Delta}{=} \theta_e + \theta_s(t)$$

For very small perturbations $\theta_s(t)$ about the equilibrium, linearization of Eq. (6.7.1q) in θ_s, as presented in Chapter 1, yields the following linear constant coefficient differential equation when $b << 2L$:

$$\ddot{\theta}_s + \frac{g}{L}\left[\frac{\Omega^2 L}{g}\left(\frac{3R}{2L}\sin\theta_e - \cos^2\theta_e + \sin^2\theta_e\right) + \frac{3}{2}\cos\theta_e\right]\theta_s = 0 \quad (6.7.1\text{s})$$

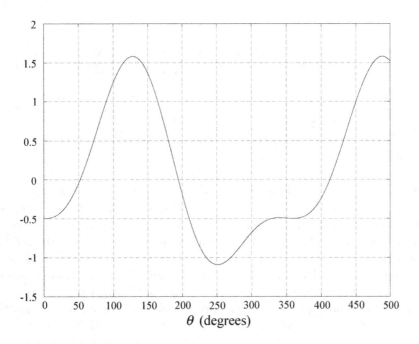

Figure 6.22: Plot of $f \overset{\Delta}{=} \sin\theta - (\Omega^2 L/g)(R/L + \frac{2}{3}\sin\theta)\cos\theta$ versus θ, for $\Omega^2 L/g = 1.5$ and $L = 3R$ for Example 6.7.1.

We then have for each equilibrium when $\Omega^2 L/g = 1.5$ and $L = 3R$:

$$\text{For } \theta_e = 52.2°: \qquad \ddot\theta_s + 1.89\frac{g}{L}\theta_s = 0 \qquad\qquad (6.7.1t)$$

$$\text{For } \theta_e = 194.2°: \qquad \ddot\theta_s - 2.96\frac{g}{L}\theta_s = 0 \qquad\qquad (6.7.1u)$$

These linearized differential equations disclose that the equilibrium solution $\theta_e = 52.2°$ is infinitesimally stable and the one with $\theta_e = 194.2°$ is unstable. The frequency of the small oscillations of the plate about its stable equilibrium $\theta_e = 52.2°$ is equal to $\sqrt{1.89g/L} \approx 1.375\sqrt{g/L}$. The period associated with this frequency is $2\pi/(1.375\sqrt{g/L}) \approx 4.57\sqrt{L/g}$. With $g = 9.81$ m/s^2 and $L = 1$ m, for example, the frequency of the small oscillations about $\theta_e \approx 52.2°$ is approximately 4.3 rad/s (for a period of $2\pi/4.3 \approx 1.46$ s).

b2. The following normalized components of the reactions are obtained from Eqs. (6.7.1n), (6.7.1o), and (6.7.1p) combined with Eqs. (6.7.1q), (6.7.1f), and

(6.7.1h):

$$\frac{A_1}{mg} = -\cos\theta - \frac{1}{2}\left(\frac{d\theta}{d\tau}\right)^2 - \frac{\Omega^2 L}{g}\left(\frac{R}{L} + \frac{1}{2}\sin\theta\right)\sin\theta$$

$$\frac{A_2}{mg} = \left(\Omega\sqrt{\frac{L}{g}}\right)\frac{d\theta}{d\tau}\cos\theta \qquad \frac{A_3}{mg} = \frac{1}{4}\left(\sin\theta - \frac{\Omega^2 R}{g}\cos\theta\right)$$

$$\frac{M_1}{mgL} = 0 \qquad\qquad \frac{M_3}{mgL} = \frac{2}{3}\frac{A_2}{mg}$$

$\Longrightarrow$ The plots of θ, and of the magnitude of the reaction force (normalized by mg) and of the reaction moment (normalized by mgL) at the hinge, versus the nondimensional time $\tau = t\sqrt{g/L}$ are shown in Fig. 6.23 for the case when $\Omega^2 L/g = 1.5$, $L = 3R$, and $b = 0$, when the motion is started with the initial conditions $\theta(0) = 30°$ and $\dot\theta(0) = 0$. Scilab users can generate these plots using the program *example6_7_1.sci* listed in Appendix E, and Matlab users can do the same by using the similar program *example6_7_1.m* that is also listed in Appendix E. The support reactions are calculated after a numerical solution for $\theta(\tau)$ is obtained by integrating the normalized form of Eq. (6.7.1q). Both programs are written as a function *example6_7_1*, and an example of their use, as shown in the comment lines at the beginning of both programs, is
my_output = example6_7_1([0:0.1:10], 1.5, 1/3, 30*_pi_/180, 0);
where _pi_ should be typed as %pi if Scilab is used, and pi if Matlab is used.

Suggestion: Investigate the effect of nonzero values of b/L on the motion and on the support reactions.

Notice that the response $\theta(\tau)$ in Fig. 6.23, which was started with the initial conditions $\theta(0) = 30°$ and $\dot\theta(0) = 0$, has a period of approximately $4.72\sqrt{L/g}$. This is slightly higher than the period of the very small oscillations about the equilibrium $\theta_e = 52.2°$, which was found to be $4.57\sqrt{L/g}$. This increase in the period with the amplitude of the motion is expected of a pendulum, and it was first presented in Chapter 1 (see Example 1.13.3, p. 52). The maximum value of the magnitude of the reaction force $A_1\hat{x}_1 + A_2\hat{x}_2 + A_3\hat{x}_3$ at the frictionless hinge is approximately $1.64mg$, and that of the reaction moment $M_1\hat{x}_1 + M_3\hat{x}_3$ is $0.26mgL$ (see Fig. 6.23).

Example 6.7.2 A homogeneous rectangular plate of mass m and dimensions L, a, and b is connected to a horizontal shaft by a gimbal as shown in Fig. 6.24. The shaft is driven by a motor with a prescribed angular velocity $\Omega(t)$. Point O is the center

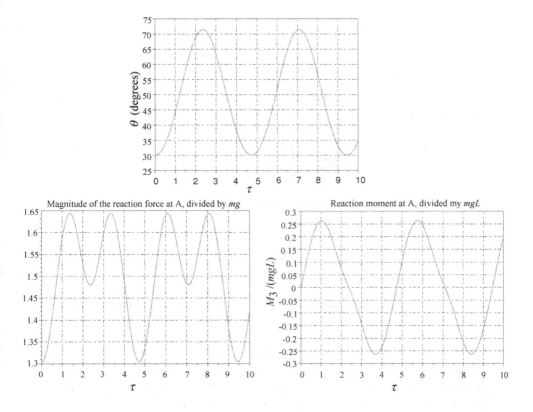

Figure 6.23: Angle θ, and the normalized force and moment reactions at A (see Fig. 6.21) versus nondimensional time $\tau = t\sqrt{g/L}$ for Example 6.7.1.

of mass of the plate, and is on the midline of the shaft. The gimbal allows the plate to rotate relative to the shaft, as indicated by the angle $\theta(t)$ in the figure. Friction in the hinges at A and B can be neglected. The orthogonal x-, y-, and z-axes shown in the figure are principal axes of the rotating plate. The unit vectors $\hat{x}$ and $\hat{y}$ are parallel to those axes, and $\hat{z} \overset{\Delta}{=} \hat{x} \times \hat{y}$. The rotation of the plate relative to the shaft is about the z-axis of the plate. The unit vector $\hat{x}_0$ is parallel to the shaft midline and is inertial (i.e., its direction does not change).

a. Obtain the differential equation for the angle θ, and verify that the system exhibits several equilibrium solutions $\theta = \text{constant} \overset{\Delta}{=} \theta_e$, but only $\theta_e = 0$ and

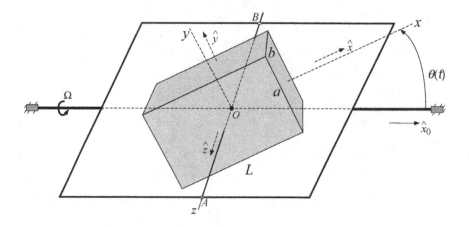

Figure 6.24: The plate and shaft system for Example 6.7.2.

$\theta_e = \pi/2$ correspond to physically different orientations of the plate relative to the shaft.

b. For $\Omega = $ constant, analyze the stability of the equilibrium solutions $\theta_e = 0$ and $\theta_e = \pi/2$, and show that, for the equilibrium $\theta_e = 0$ to be stable, the dimensions of the plate have to be such that $a > b$. Do the stability analysis by linearizing the differential equation obtained in part a.

c. Plot $\dot{\theta}$ versus θ and, from the plot, specify the region where the pair of initial conditions $\theta(0)$ and $\dot{\theta}(0)$ have to lie for the motion (even large motions) $\theta(t)$ to be oscillatory about the equilibrium $\theta_e = 0$.

■ Solution

a. The rotational motion of the plate is governed by the basic equation $d\vec{H}_O/dt = \vec{M}_O$, which yields Euler's three scalar equations. If we work with the principal axes of the plate, which are centered at point O, such equations are Eqs. (6.4.9), (6.4.10), and (6.4.11) (p. 427). The formulation is repeated here for this problem.

Inspection of Fig. 6.24 discloses that the absolute angular velocity $\vec{\omega}_{\text{plate}}$ of the plate is

$$\vec{\omega}_{\text{plate}} = \Omega \hat{x}_0 + \dot{\theta} \hat{z} = \Omega \left(\hat{x} \cos \theta - \hat{y} \sin \theta \right) + \dot{\theta} \hat{z} \overset{\Delta}{=} \omega_x \hat{x} + \omega_y \hat{y} + \omega_z \hat{z} \quad (6.7.2a)$$

Since the inertia dyadic of the plate is simply

$$\vec{\vec{I}}_{\text{plate}} = I_x \hat{x}\hat{x} + I_y \hat{y}\hat{y} + I_z \hat{z}\hat{z} \tag{6.7.2b}$$

where

$$I_x = \frac{m\left(a^2 + b^2\right)}{12} \qquad I_y = \frac{m\left(L^2 + b^2\right)}{12} \qquad I_z = \frac{m\left(L^2 + a^2\right)}{12} \tag{6.7.2c}$$

the absolute angular momentum $(\vec{H}_O)_{\text{plate}}$ of the rotating plate is then

$$(\vec{H}_O)_{\text{plate}} = \Omega\left(\hat{x} I_x \cos\theta - \hat{y} I_y \sin\theta\right) + I_z \dot{\theta}\hat{z} \tag{6.7.2d}$$

Therefore, as shown earlier, $d(\vec{H}_O)_{\text{plate}}/dt$ is obtained as follows:

$$
\begin{aligned}
\frac{d}{dt}(\vec{H}_O)_{\text{plate}} &= I_x \left[\dot{\Omega}\cos\theta - \Omega\dot{\theta}\sin\theta\right]\hat{x} - I_y\left[\dot{\Omega}\sin\theta + \Omega\dot{\theta}\cos\theta\right]\hat{y} \\
&\quad + I_z\ddot{\theta}\hat{z} + \vec{\omega}_{\text{plate}} \times \vec{H}_O \\
&= \left[I_x\left(\dot{\Omega}\cos\theta - \Omega\dot{\theta}\sin\theta\right) + (I_y - I_z)\Omega\dot{\theta}\sin\theta\right]\hat{x} \\
&\quad - \left[I_y\left(\dot{\Omega}\sin\theta + \Omega\dot{\theta}\cos\theta\right) + (I_z - I_x)\Omega\dot{\theta}\cos\theta\right]\hat{y} \\
&\quad + \left[I_z\ddot{\theta} + (I_x - I_y)\Omega^2(\sin\theta)\cos\theta\right]\hat{z} \tag{6.7.2e}
\end{aligned}
$$

The plate is subjected to a moment $\vec{M}_O$ that is applied to it by the shaft. Such a moment is of the form

$$\vec{M}_O = M_x\hat{x} + M_y\hat{y} + M_z\hat{z} \tag{6.7.2f}$$

where the component M_z is only due to friction in the hinges at A and B (see Fig. 6.24). For frictionless hinges, $M_z = 0$. The $\hat{z}$ component of Eq. (6.7.2e) then yields the following equation.

$$I_z\ddot{\theta} + (I_x - I_y)\Omega^2(\sin\theta)\cos\theta = 0 \tag{6.7.2g}$$

Equation (6.7.2g), which only involves the unknown variable θ, is the differential equation of motion. For $\Omega = $ constant, the equation has constant coefficients, and this is the case considered here. The steps that led to Eq. (6.7.2g) could have been omitted, of course, and Eq. (6.7.2g) could have been obtained directly from Eq. (6.4.11), (p. 427), with ω_x, ω_y, and ω_z given by Eq. (6.7.2a) for the problem at hand.

Equation (6.7.2g) admits the equilibrium solution $\theta = \text{constant} \overset{\Delta}{=} \theta_e$, where either $\sin\theta_e = 0$ or $\cos\theta_e = 0$. The equilibrium solutions corresponding to $\sin\theta_e = 0$ are $\theta_e = 0$ and integer multiples of $\theta_e = \pi$, and those corresponding to $\cos\theta_e = 0$ are $\pi/2, 3\pi/2, 5\pi/2, \ldots$. Of these equilibrium solutions, only $\theta_e = 0$ and $\theta_e = \pi/2$ correspond to physically different orientations of the plate relative to the shaft.

b. One way to analyze the stability of the equilibrium of the system is to let $\theta(t) \overset{\Delta}{=} \theta_e + \theta_s(t)$ and then linearize the differential equation for small θ_s. The following linearized differential equations are obtained for the perturbed motion about $\theta_e = 0$ and about $\theta_e = \pi/2$:

Linearization about $\theta_e = 0$: $\qquad\qquad I_z\ddot{\theta}_s + (I_x - I_y)\Omega^2\theta_s = 0$

Linearization about $\theta_e = \pi/2$: $\qquad\quad I_z\ddot{\theta}_s - (I_x - I_y)\Omega^2\theta_s = 0$

These linearized differential equations disclose that when one equilibrium is infinitesimally stable, the other is unstable (because of the opposite signs of the θ_s terms in the differential equations). The equilibrium $\theta_e = 0$ is infinitesimally stable if $I_x > I_y$ and unstable if $I_x < I_y$. Other problems whose motion are governed by a differential equation similar to Eq. (6.7.2g) were analyzed in this book, and those analyses disclosed that the equilibrium $\theta_e = 0$ is stable (instead of just infinitesimally stable) when $I_x > I_y$. Thus, as per Eq. (6.7.2g) and the equations of (6.7.2c), the equilibrium $\theta_e = 0$ is stable if $a > L$ in Fig. 6.24 and unstable if $a < L$.

c. A plot of $d\theta/dt$ versus θ may be done by integrating Eq. (6.7.2g) numerically and then using the result of the numerical integration to generate the plot. However, there is no need to do the numerical integration at all since advantage may be taken of the fact that Eq. (6.7.2g) may be integrated once with respect to θ, as first presented in Chapter 1, to yield

$$\frac{1}{2}\dot{\theta}^2 - \frac{\Omega^2}{4}\frac{I_x - I_y}{I_z}\left(\cos^2\theta - \sin^2\theta\right) = \text{constant} \overset{\Delta}{=} C$$

$$= \left[\frac{1}{2}\dot{\theta}^2 - \frac{\Omega^2}{4}\frac{I_x - I_y}{I_z}\left(\cos^2\theta - \sin^2\theta\right)\right]_{t=0} \qquad (6.7.2\text{h})$$

Figure 6.25 shows a plot of the angular velocity of the plate relative to the shaft, versus θ, for the case when $I_x > I_y$. The plot shows the curves that surround the stable equilibrium point $\theta_e = 0$ and that pass through the unstable equilibrium points $\theta_e = -\pi/2$ and $\pi/2$.

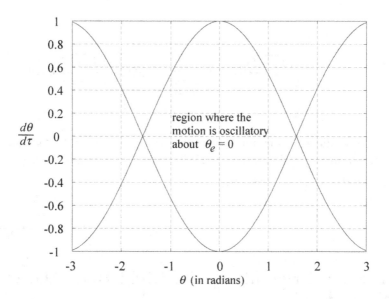

Figure 6.25: Plot of $d\theta/d\tau = \pm \cos\theta$ for Example 6.7.2, where $\tau = \Omega t\sqrt{(I_x - I_y)/I_z}$, for $I_x > I_z$.

To avoid drawing a plot for a specific value of $\Omega^2(I_x - I_y)/I_z$, the normalized time $\tau = \Omega t\sqrt{(I_x - I_y)/I_z}$ was used and $d\theta/d\tau$, instead of $d\theta/dt$, was plotted versus θ. Such a plot is valid for any positive value one wishes to choose for the quantity $\Omega\sqrt{(I_x - I_y)/I_z}$.

For an oscillatory motion about the equilibrium $\theta_e = 0$, the pair of values $\theta(0)$-$[d\theta/d\tau]_{\tau=0}$ must lie inside the region shown in Fig. 6.25. For the curves shown in Fig. 6.25 the value of the constant C in Eq. (6.7.2h) is obtained simply by setting $\dot{\theta} = 0$ and $\theta = \pi/2$ in that equation. That value is

$$C = \frac{\Omega^2}{4}\frac{I_x - I_y}{I_z}$$

Notice that this value of C yields the following relationship between $\dot{\theta}$ and θ for the curves shown in Fig. 6.25:

$$\frac{1}{2}\dot{\theta}^2 = \frac{\Omega^2}{4}\frac{I_x - I_y}{I_z}\left(\cos^2\theta - \sin^2\theta + 1\right) = \frac{\Omega^2}{2}\frac{I_x - I_y}{I_z}\cos^2\theta$$

or

$$\dot{\theta} = \pm\,\Omega\sqrt{\frac{I_x - I_y}{I_z}}\,|\cos\theta| \underbrace{=}_{\text{for the plate}} \pm\,\Omega\sqrt{\frac{a^2 - L^2}{a^2 + L^2}}\,\cos\theta \qquad (6.7.2\text{i})$$

As a specific example, for a plate with $a = 2L$ and $-\pi/2 < \theta(0) < \pi/2$, this equation and inspection of Fig. 6.25 discloses that the motion is oscillatory as long as $|\dot{\theta}(0)| < \sqrt{\frac{3}{5}}\,|\Omega\cos\theta(0)|$.

6.8   Angular Orientation of a Rigid Body: Orientation Angles

It was seen that a rigid body rotating about a fixed point has three degrees of freedom, while a rigid body translating and rotating in space has six. Three of the six degrees of freedom of the body describe the translation of a point of the body (or of a rigid extension of it), which is normally chosen to be the center of mass C of the body (for which one has $\vec{F} = m\vec{a}_C$). The other three degrees of freedom describe the rotational motion of the body, for which one has $\vec{M}_C = d\vec{H}_C/dt$.

The rotational motion of a body is described by its instantaneous orientation (also called the *attitude*) relative to some reference frame that is chosen to suit the particular application that is being investigated. For example, the convenient reference frame used for investigating the motion of an aircraft or a ship relative to the earth is a set of axes fixed to the earth. The orientation of an earth-orbiting satellite, for example, is normally referred to a set of earth-orbiting reference axes that consist of an axis that points to the center of the earth, a second axis that is perpendicular to the orbital plane of the center of mass of the satellite, and a third axis that is normal to the other two. As another example, the reference frame for investigating the trajectory of unwanted particles that may have entered a jet engine (and thus, are likely to cause erosion of the engine's rotating blades) might be a set of axes fixed to a rotating blade; in such a case, a set of coordinates chosen in that reference frame would be desirable to use so that the investigation of the particle's trajectories might disclose regions of the blade where erosion caused by those particles might occur.

> However, notice that when applying Newton's second law $\vec{F} = m\vec{a}_C$ and the moment equation (Euler's equation) $\vec{M}_C = d\vec{H}_C/dt$ to any problem, one ultimately needs to work with position vectors relative to an inertial point, with angles measured from nonrotating reference lines, and the absolute acceleration of the center of mass C. Also, the angular acceleration of the bodies must be *absolute*. In other words, after introducing convenient intermediate reference frames for a particular problem, one must also account for the motion of the reference frame relative to inertial space.

One widely used way to specify the orientation of a body relative to a desired reference frame was introduced by Euler. It consists of describing the orientation of a set of body-fixed axes, say x-y-z, relative to a chosen set of reference axes x_0-y_0-z_0, by first aligning x-y-z with x_0-y_0-z_0 and then performing three successive rotations $\theta_1, \theta_2, \theta_3$ of the body axes to bring them into alignment with their actual orientation relative to the chosen reference frame. Such angles are called *orientation angles*, and there are 24 possible sets of them, 12 of which involve rotations about an updated orientation of the same axis (as originally introduced by Euler), as will be seen in the following. They are also often referred to as *Euler angles*, although the original Euler angles consist only of some of the 12 sets of orientation angles just mentioned. For some problems, a particular sequence of orientation angles is preferred over the others when it leads to equations that are simpler (but generally never simple!) to analyze. It is fair to say, however, that the use of a particular sequence over any other is often a matter of personal preference, and this can create some difficulties in comparing results of attitude dynamics problems done by different persons with different orientation angle sequences for the same problem. Figure 6.26a shows a possible orientation angle sequence θ_1, θ_2, and θ_3.

The sequence shown in Fig. 6.26 describes the orientation of a set orthogonal body-fixed axes x-y-z (represented by the unit vectors $\hat{x}, \hat{y}$, and $\hat{z}$, respectively, that are parallel to those axes) with respect to the orthogonal reference axes x_0-y_0-z_0 (also represented by the unit vectors $\hat{x}_0, \hat{y}_0$, and $\hat{z}_0$, respectively). Other unit vectors that are parallel to the orientation of the axes after they are rotated are also shown. Referring to Fig. 6.26a, start by aligning x-y-z with x_0-y_0-z_0, i.e., start with $\hat{x} = \hat{x}_0$, $\hat{y} = \hat{y}_0$, and $\hat{z} = \hat{z}_0$. Then, perform the following successive rotations:

First rotation. A counterclockwise (counterclockwise simply by choice; it could be clockwise) rotation θ_1 about x_0 to bring the body axes to their first intermediate orientation x_1-y_1-z_1 (with $\hat{x}_1 = \hat{x}_0$). The angular velocity of the

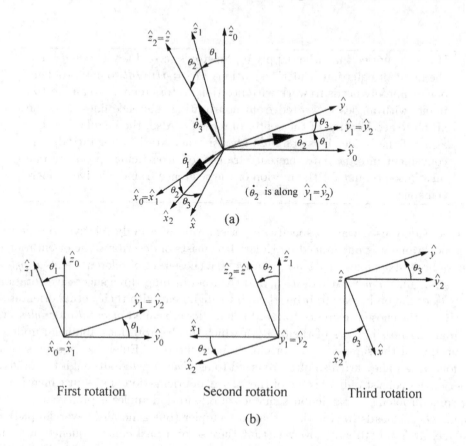

Figure 6.26: A three-axis orientation sequence to describe the orientation of a body.

x_1-y_1-z_1 orthogonal axis system relative to the x_0-y_0-z_0 reference system, associated with this rotation, is simply $\dot{\theta}_1 \hat{x}_0$, and it is shown by a bold arrow in Fig. 6.26a.

Second rotation. A rotation θ_2 about the new orientation y_1 of the y-axis to bring the body axes to their second intermediate orientation x_2-y_2-z_2 (with $\hat{y}_2 = \hat{y}_1$). The angular velocity of the x_2-y_2-z_2 system relative to the x_1-y_1-z_1 system, associated with this second rotation, is simply $\dot{\theta}_2 \hat{y}_1$, and it is also shown by a bold arrow in Fig. 6.26a.

Third and final rotation. A rotation θ_3 about the new and final orientation of the z-axis to bring the body axes to their final orientation x-y-z (with $\hat{z} = \hat{z}_2$).

The angular velocity of the body-fixed x-y-z axis system relative to the x_2-y_2-z_2 system, associated with this third rotation, is simply $\dot{\theta}_3 \hat{z}$, and it is also shown by a bold arrow in Fig. 6.26a.

The rotation sequence shown in Fig. 6.26 is widely used in aircraft and satellite dynamics, and in many applications in mechanical engineering and in structural dynamics. All three axes of the body were used in these rotations and, for this reason, they are known as *three-axis body-fixed* angles. There are six possible rotation sequences of this type, with two different rotations performed after the first rotation.[7]

Note: Instead of using rotations about the updated orientations of the body-fixed axes, one could also use rotations that are always about the x_0, y_0, and z_0 axes. There are also $2 + 2 + 2 = 6$ possible combinations when all three axes are used, but such rotations are generally not used in practice. Such orientation angles are called *three-axis space-fixed angles*.

Once an angle sequence is chosen to describe the orientation of the body, the next task is to obtain the expression for the absolute angular velocity $\vec{\omega}^B$ of the body. Having the expression for $\vec{\omega}^B$ one can then obtain the expression for the absolute angular momentum of the motion and, from there, the differential equations for investigating the rotational motion of the body.

The angular velocity $\vec{\omega}$ of the body, relative to the reference frame x_0-y_0-z_0, is simply the sum of the individual angular velocities $\dot{\theta}_1 \hat{x}_0$, $\dot{\theta}_2 \hat{y}_1$, and $\dot{\theta}_3 \hat{z}$ that are associated with each rotation. If the x_0-y_0-z_0 axes are inertial, $\vec{\omega} = \dot{\theta}_1 \hat{x}_0 + \dot{\theta}_2 \hat{y}_1 + \dot{\theta}_3 \hat{z}$ is the absolute angular velocity $\vec{\omega}^B$ of the body. If the x_0-y_0-z_0 directions are not inertial, the absolute angular velocity of the body is simply obtained by adding to $\vec{\omega}$ the absolute angular velocity of the x_0-y_0-z_0 frame.

The expression $\vec{\omega} = \dot{\theta}_1 \hat{x}_0 + \dot{\theta}_2 \hat{y}_1 + \dot{\theta}_3 \hat{z}$, obtained by inspection of Fig. 6.26, involves three nonorthogonal unit vectors. One now needs to express it in terms of an orthogonal triad, and such a transformation, first presented in Chapter 1, is simply obtained with the aid of the snapshots shown in Fig. 6.26b. In terms of the $\hat{x}, \hat{y}$, and $\hat{z}$ unit vectors that are parallel to the body-fixed x-y-z axes, the following expressions for $\hat{x}_0$ and $\hat{y}_1$ (which appear in the expression for $\vec{\omega}$) are written by inspection of Fig. 6.26b:

$$\hat{x}_0 = \hat{x}_1 = \hat{x}_2 \cos\theta_2 + \hat{z}\sin\theta_2 = (\hat{x}\cos\theta_3 - \hat{y}\sin\theta_3)\cos\theta_2 + \hat{z}\sin\theta_2$$
$$\hat{y}_1 = \hat{y}_2 = \hat{y}\cos\theta_3 + \hat{x}\sin\theta_3 \qquad (6.8.1)$$

---

[7]For the sequence starting with a rotation about x_0, the second rotation could have been about z_1 (instead of y_1), and the third rotation about y. This accounts for the two possible sequences that start with θ_1 about x_0. Two other possible sequences would start with a rotation about y_0. Similarly, the other two would start with a rotation about z_0. This brings the total number of such rotation sequences to the six that were mentioned.

This then gives the following expression for the angular velocity $\vec{\omega}$ of the body, relative to the x_0-y_0-z_0 reference frame:

$$
\begin{aligned}
\vec{\omega} &= \dot{\theta}_1 \hat{x}_0 + \dot{\theta}_2 \hat{y}_1 + \dot{\theta}_3 \hat{z} \\
&= \left(\dot{\theta}_1 \cos\theta_2 \cos\theta_3 + \dot{\theta}_2 \sin\theta_3 \right) \hat{x} + \left(\dot{\theta}_2 \cos\theta_3 - \dot{\theta}_1 \cos\theta_2 \sin\theta_3 \right) \hat{y} \\
&\quad + \left(\dot{\theta}_3 + \dot{\theta}_1 \sin\theta_2 \right) \hat{z} \overset{\Delta}{=} \omega_x \hat{x} + \omega_y \hat{y} + \omega_z \hat{z}
\end{aligned}
\tag{6.8.2}
$$

If the x_0-y_0-z_0 axes are nonrotating (i.e., inertial), these expressions for the angular velocity components ω_x, ω_y, and ω_z are the ones that go into Euler's Eqs. (6.4.9) to (6.4.11), developed in Section 6.4.

Figure 6.26b can also be used for quickly developing expressions that relate any of the unit vectors in that figure. For example, the following relations involving the orientations of the unit vectors of a triad before and after a rotation are readily obtained. They are written here in matrix form:

$$
\begin{bmatrix} \hat{x}_0 \\ \hat{y}_0 \\ \hat{z}_0 \end{bmatrix} = \begin{bmatrix} 1 & 0 & 0 \\ 0 & \cos\theta_1 & -\sin\theta_1 \\ 0 & \sin\theta_1 & \cos\theta_1 \end{bmatrix} \begin{bmatrix} \hat{x}_1 \\ \hat{y}_1 \\ \hat{z}_1 \end{bmatrix} \overset{\Delta}{=} [A_1] \begin{bmatrix} \hat{x}_1 \\ \hat{y}_1 \\ \hat{z}_1 \end{bmatrix}
\tag{6.8.3}
$$

$$
\begin{bmatrix} \hat{x}_1 \\ \hat{y}_1 \\ \hat{z}_1 \end{bmatrix} = \begin{bmatrix} \cos\theta_2 & 0 & \sin\theta_2 \\ 0 & 1 & 0 \\ -\sin\theta_2 & 0 & \cos\theta_2 \end{bmatrix} \begin{bmatrix} \hat{x}_2 \\ \hat{y}_2 \\ \hat{z}_2 \end{bmatrix} \overset{\Delta}{=} [A_2] \begin{bmatrix} \hat{x}_2 \\ \hat{y}_2 \\ \hat{z}_2 \end{bmatrix}
\tag{6.8.4}
$$

$$
\begin{bmatrix} \hat{x}_2 \\ \hat{y}_2 \\ \hat{z}_2 \end{bmatrix} = \begin{bmatrix} \cos\theta_3 & -\sin\theta_3 & 0 \\ \sin\theta_3 & \cos\theta_3 & 0 \\ 0 & 0 & 1 \end{bmatrix} \begin{bmatrix} \hat{x} \\ \hat{y} \\ \hat{z} \end{bmatrix} \overset{\Delta}{=} [A_3] \begin{bmatrix} \hat{x} \\ \hat{y} \\ \hat{z} \end{bmatrix}
\tag{6.8.5}
$$

Each of these matrices, $[A_1]$, $[A_2]$, and $[A_3]$, is a simple *transformation matrix* that involves only one rotation. These matrices were first introduced in Chapter 1. Their determinant is always equal to 1, and their inverse is always equal to their transpose.

From Eqs. (6.8.3) to (6.8.5), the transformation from $\{\hat{x}, \hat{y}, \hat{z}\}$ to $\{\hat{x}_0, \hat{y}_0, \hat{z}_0\}$ (i.e., $\{\hat{x}_0, \hat{y}_0, \hat{z}_0\}$ written in terms of $\{\hat{x}, \hat{y}, \hat{z}\}$) is seen to be given as follows:

$$
\begin{bmatrix} \hat{x}_0 \\ \hat{y}_0 \\ \hat{z}_0 \end{bmatrix} \overset{\Delta}{=} [A] \begin{bmatrix} \hat{x} \\ \hat{y} \\ \hat{z} \end{bmatrix}
\tag{6.8.6}
$$

where the transformation matrix $[A]$ is

$$
[A] = [A_1][A_2][A_3]
\tag{6.8.7}
$$

The elements A_{ij} of the matrix $[A]$ are:

$$A_{11} = \cos\theta_2 \cos\theta_3 \qquad A_{12} = -\cos\theta_2 \sin\theta_3 \qquad A_{13} = \sin\theta_2$$
$$A_{21} = \sin\theta_1 \sin\theta_2 \cos\theta_3 + \cos\theta_1 \sin\theta_3$$
$$A_{22} = -\sin\theta_1 \sin\theta_2 \sin\theta_3 + \cos\theta_1 \cos\theta_3 \qquad A_{23} = -\sin\theta_1 \cos\theta_2$$
$$A_{31} = -\cos\theta_1 \sin\theta_2 \cos\theta_3 + \sin\theta_1 \sin\theta_3$$
$$A_{32} = \cos\theta_1 \sin\theta_2 \sin\theta_3 + \sin\theta_1 \cos\theta_3 \qquad A_{33} = \cos\theta_1 \cos\theta_2$$

Notice that each column and row of a transformation matrix is composed of the components of a unit vector. Thus, each element of the matrix $[A]$ is a direction cosine between two unit vectors. The sum of the squares of each column, and of each row, of a transformation matrix is always equal to 1. Also, the inverse of the transformation matrix $[A] = [A_1][A_2][A_3]$ is equal to its transpose, which is a nice feature to use in any numerical computation that involves finding the inverse of such matrices.

Example 6.8.1 For the rotation sequence shown in Fig. 6.26, with $\theta_1 = 20°$, $\theta_2 = 30°$, and $\theta_3 = 15°$, determine the angle between the unit vectors $\hat{y}_0$ and $\hat{y}$, and also between $\hat{z}_1$ and $\hat{y}$.

■ **Solution**

As indicated in Chapter 1, the angle θ between two unit vectors $\hat{a}$ and $\hat{b}$ is simply obtained from the dot product $\hat{a} \bullet \hat{b} = |\hat{a}|\,|\hat{b}|\cos\theta = \cos\theta$. From Eq. (6.8.6), with $[A]$ given by Eq. (6.8.7), we then obtain

$$\hat{y}_0 \bullet \hat{y} = \cos\theta_1 \cos\theta_3 - \sin\theta_1 \sin\theta_2 \sin\theta_3 \approx 0.863 \overset{\Delta}{=} \cos\alpha$$

This gives $\alpha \approx \pm30.3°$.

The angle (which we will call β) between $\hat{z}_1$ and $\hat{y}$ is obtained from the 3-2 element of the product $[A_2][A_3]$ of the $[A_2]$ and $[A_3]$ matrices given by Eqs. (6.8.4) and (6.8.5), respectively. Therefore:

$$\hat{z}_1 \bullet \hat{y} = \sin\theta_2 \sin\theta_3 \approx 0.129 \overset{\Delta}{=} \cos\beta$$

This gives $\beta \approx \pm82.6°$.

For some applications, such as the investigation of the motion of a top or a gyroscope, *two-axis body-fixed angles* (the original Euler angles) are more "natural"

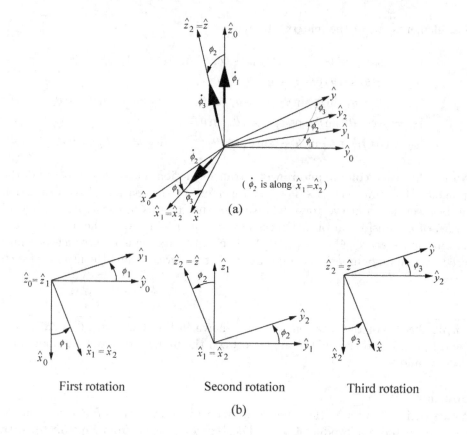

Figure 6.27: A two-axis orientation sequence to describe the orientation of a body.

to use. A rotation sequence, which we will call ϕ_1, ϕ_2, ϕ_3 here,[8] is illustrated in Fig. 6.27a. Figure 6.27b shows snapshots of the individual rotations.

The following sequence of rotations is performed in Fig. 6.27 after aligning $\hat{x}$ with $\hat{x}_0$, $\hat{y}$ with $\hat{y}_0$, and $\hat{z}$ with $\hat{z}_0$:

First rotation. A counterclockwise (again, counterclockwise simply by choice; it could be clockwise) rotation ϕ_1 about z_0 to bring the body axes to their first

---

[8]If one is investigating the same problem using two different sequences of orientation angles, it is good practice to use different symbols for the angles of both sequences. Such a practice is advisable because the corresponding angles in two different sequences are different variables, and they will not be equal to each other. Using the same symbols for the angles of the two different sequences will cause confusion and is prone to lead to erroneous conclusions about the motion.

intermediate orientation x_1-y_1-z_1 (with $\hat{z}_1 = \hat{z}_0$). The angular velocity of the x_1-y_1-z_1 axis system relative to the x_0-y_0-z_0 reference system, associated with this rotation, is simply $\dot{\phi}_1 \hat{z}_0$, and it is shown by a bold arrow in Fig. 6.27a.

Second rotation. A rotation ϕ_2 about the new orientation x_1 of the x-axis to bring the body axes to their second intermediate orientation x_2-y_2-z_2 (with $\hat{x}_2 = \hat{x}_1$). The angular velocity of the x_2-y_2-z_2 system relative to the x_1-y_1-z_1 system, associated with this second rotation, is simply $\dot{\phi}_2 \hat{x}_1$, and it is also shown by a bold arrow in Fig. 6.27a.

Third and final rotation. A rotation ϕ_3 about the new and final orientation of the z-axis, again, to bring the body axes to their final orientation x-y-z (with $\hat{z} = \hat{z}_2$). The angular velocity of the body-fixed x-y-z axis system relative to the x_2-y_2-z_2 system, associated with this third rotation, is simply $\dot{\phi}_3 \hat{z}$, and it is also shown by a bold arrow in Fig. 6.27a.

For two-axes orientation angles, the third and final rotation is performed about the updated position of the axis about which the first rotation was made.[9] As for the case of three-axis body-fixed orientation angles, there are also six possible rotation sequences for two-axis orientation angles, with two different ones following the first rotation, as indicated earlier (which brings the total number of combinations to twelve). In addition, there are also six possible combinations for two-axis space-fixed orientation angles (where the rotations are always about two of the x_0, y_0, and z_0 axes), thus bringing the total number of combinations to 24. Space-fixed orientation angles are generally not used in practice.

With the two-axis body-fixed rotation sequence just illustrated, the angular velocity $\vec{\omega}$ of the body, relative to the reference frame x_0-y_0-z_0, is $\vec{\omega} = \dot{\phi}_1 \hat{z}_0 + \dot{\phi}_2 \hat{x}_1 + \dot{\phi}_3 \hat{z}$. Such an expression is converted to the triad $\{\hat{x}, \hat{y}, \hat{z}\}$ that is parallel to the body-fixed x-y-z axes with the following transformations that are obtained by inspection of the three snapshots in Fig. 6.27b.

$$\hat{z}_0 = \hat{z}_1 = \hat{z}\cos\phi_2 + \hat{y}_2\sin\phi_2 = \hat{z}\cos\phi_2 + (\hat{y}\cos\phi_3 + \hat{x}\sin\phi_3)\sin\phi_2$$
$$\hat{x}_1 = \hat{x}_2 = \hat{x}\cos\phi_3 - \hat{y}\sin\phi_3 \tag{6.8.8}$$

This then gives the following expression for the angular velocity $\vec{\omega}$ of the body relative to the x_0-y_0-z_0 reference frame:

$$\begin{aligned}
\vec{\omega} &= \dot{\phi}_1 \hat{z}_0 + \dot{\phi}_2 \hat{x}_1 + \dot{\phi}_3 \hat{z} \\
&= \left(\dot{\phi}_1 \sin\phi_2 \sin\phi_3 + \dot{\phi}_2 \cos\phi_3\right)\hat{x} + \left(\dot{\phi}_1 \sin\phi_2 \cos\phi_3 - \dot{\phi}_2 \sin\phi_3\right)\hat{y}
\end{aligned}$$

---

[9] The angles shown in Fig. 6.13 (p. 442) are two-axis body-fixed angles for which the sequence of rotations is ϕ, θ, ψ.

$$+ \left(\dot{\phi}_3 + \dot{\phi}_1 \cos \phi_2 \right) \hat{z} \triangleq \omega_x \hat{x} + \omega_y \hat{y} + \omega_z \hat{z} \tag{6.8.9}$$

The individual transformations associated with each one of these rotations are given as

$$\begin{bmatrix} \hat{x}_0 \\ \hat{y}_0 \\ \hat{z}_0 \end{bmatrix} = \begin{bmatrix} \cos \phi_1 & -\sin \phi_1 & 0 \\ \sin \phi_1 & \cos \phi_1 & 0 \\ 0 & 0 & 1 \end{bmatrix} \begin{bmatrix} \hat{x}_1 \\ \hat{y}_1 \\ \hat{z}_1 \end{bmatrix} \triangleq [B_1] \begin{bmatrix} \hat{x}_1 \\ \hat{y}_1 \\ \hat{z}_1 \end{bmatrix} \tag{6.8.10}$$

$$\begin{bmatrix} \hat{x}_1 \\ \hat{y}_1 \\ \hat{z}_1 \end{bmatrix} = \begin{bmatrix} 1 & 0 & 0 \\ 0 & \cos \phi_2 & -\sin \phi_2 \\ 0 & \sin \phi_2 & \cos \phi_2 \end{bmatrix} \begin{bmatrix} \hat{x}_2 \\ \hat{y}_2 \\ \hat{z}_2 \end{bmatrix} \triangleq [B_2] \begin{bmatrix} \hat{x}_2 \\ \hat{y}_2 \\ \hat{z}_2 \end{bmatrix} \tag{6.8.11}$$

$$\begin{bmatrix} \hat{x}_2 \\ \hat{y}_2 \\ \hat{z}_2 \end{bmatrix} = \begin{bmatrix} \cos \phi_3 & -\sin \phi_3 & 0 \\ \sin \phi_3 & \cos \phi_3 & 0 \\ 0 & 0 & 1 \end{bmatrix} \begin{bmatrix} \hat{x} \\ \hat{y} \\ \hat{z} \end{bmatrix} \triangleq [B_3] \begin{bmatrix} \hat{x} \\ \hat{y} \\ \hat{z} \end{bmatrix} \tag{6.8.12}$$

and this yields

$$\begin{bmatrix} \hat{x}_0 \\ \hat{y}_0 \\ \hat{z}_0 \end{bmatrix} \triangleq [B] \begin{bmatrix} \hat{x} \\ \hat{y} \\ \hat{z} \end{bmatrix} \tag{6.8.13}$$

where

$$[B] = [B_1][B_2][B_3] \tag{6.8.14}$$

The elements B_{ij} of the matrix $[B]$ are:

$$B_{11} = \cos \phi_1 \cos \phi_3 - \sin \phi_1 \cos \phi_2 \sin \phi_3$$

$$B_{12} = -\cos \phi_1 \sin \phi_3 - \sin \phi_1 \cos \phi_2 \cos \phi_3 \qquad B_{13} = \sin \phi_1 \sin \phi_2$$

$$B_{21} = \sin \phi_1 \cos \phi_3 + \cos \phi_1 \cos \phi_2 \sin \phi_3$$

$$B_{22} = -\sin \phi_1 \sin \phi_3 + \cos \phi_1 \cos \phi_2 \cos \phi_3 \qquad B_{23} = -\cos \phi_1 \sin \phi_2$$

$$B_{31} = \sin \phi_2 \sin \phi_3 \qquad B_{32} = \sin \phi_2 \cos \phi_3 \qquad B_{33} = \cos \phi_2$$

Clearly, the transformation matrix $[B] = [B_1][B_2][B_3]$ obtained with the given rotation sequence is different from the $[A]$ matrix given by Eq. (6.8.7).

Example 6.8.2 For the two-axis body-fixed rotation sequence shown in Fig. 6.27, with $\phi_1 = 20°$, $\phi_2 = 30°$, and $\phi_3 = 15°$, determine the angle between the unit vectors $\hat{y}_0$ and $\hat{y}$, and also between $\hat{z}_1$ and $\hat{y}$.

■ Solution

For this example we have for the angle between $\hat{y}_0$ and $\hat{y}$ (which we will call α_1),

$$\hat{y}_0 \bullet \hat{y} = \cos \phi_1 \cos \phi_2 \cos \phi_3 - \sin \phi_1 \sin \phi_3 \approx 0.697 \overset{\Delta}{=} \cos \alpha_1$$

This gives $\alpha_1 \approx \pm 45.8°$.

The angle (which we will call β_1) between $\hat{z}_1$ and $\hat{y}$ is obtained from the 3-2 element of the product $[B_2][B_3]$, with the matrices $[B_2]$ and $[B_3]$ as given previously. Therefore:

$$\hat{z}_1 \bullet \hat{y} = \sin \phi_2 \cos \phi_3 \approx 0.483 \overset{\Delta}{=} \cos \beta_1$$

This gives $\beta_1 \approx \pm 61.1°$.

Example 6.8.3 Gravitational Moment on a Spacecraft

There are many cases where the external forces acting on a body depend on the orientation of the body. Such is the case for the aerodynamic and hydrodynamic forces acting, respectively, on aircraft and ships, for example. The gravitational force acting on an orbiting satellite is another example (see Problem 4.15, p. 311, which is based on Example 1.10.1, p. 27).

In this example, the expression for the gravitational moment $\vec{M}_C$ about the center of mass C of a rigid body of mass m in circular orbit around a planet is developed. The moment is certainly very small, but it cannot be neglected since it causes the orientation of orbiting bodies to slowly change with time. Such a moment depends on the spatial orientation of the body, and orientation angles can be used for describing the rotational dynamics of the body relative to a convenient "orbiting reference frame."

The situation is illustrated in Fig. 6.28, where O is the center of the attracting planet. The planet (say, the earth) is modeled as a homogeneous sphere of mass M_E and radius R_E, as done in Chapters 1 and 4. The scaling of the figure is highly exaggerated since only bodies with maximum dimension that is much smaller than $|\overrightarrow{OC}|$ are considered. This "restriction" is well justified in actual practical situations.

The convenient orbital reference frame previously mentioned is the set of orthogonal axes centered at C and parallel to the reference unit vectors $\hat{x}_0$ directed along the "local vertical" OC, $\hat{y}_0$ perpendicular to OC and in the orbital plane,

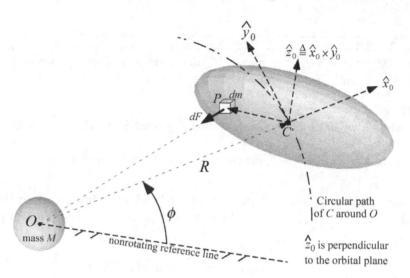

Figure 6.28: A rigid body subjected to the gravitational attraction of a planet, with scaling highly exaggerated for sake of clarity.

and $\hat{z}_0 \overset{\Delta}{=} \hat{x}_0 \times \hat{y}_0$, which is perpendicular to the orbital plane. The radius of the circular orbit is $R = |\overrightarrow{OC}|$ and, as seen in Chapter 4, the orbital angular velocity is $\dot{\phi} \overset{\Delta}{=} \Omega = \sqrt{GM_E/R^3}$, where G is the universal gravitational constant. In terms of the mass M_E and radius R_E of the earth, and of the acceleration g of gravity at the earth's surface, $GM_E = gR_E^2$.

Figure 6.28 shows an infinitesimal element of the body, with mass dm, and the infinitesimal gravitational force $d\vec{F}$ acting on it. The determination of the moment $\vec{M}_C$ is done along the same steps followed in Example 1.10.1 (p. 27), with the exception that an integration that spans the entire volume of the rigid body needs to be done. The gravitational force $d\vec{F}$ acting on dm is calculated using Newton's law of universal gravitation (see Fig. 6.28):

$$d\vec{F} = -\frac{GM_E \, dm}{|\overrightarrow{OP}|^2} \frac{\overrightarrow{OP}}{|\overrightarrow{OP}|} \qquad (6.8.3a)$$

Therefore, the moment $\vec{M}_C$ is obtained by integrating $\overrightarrow{CP} \times d\vec{F}$ over the entire rigid body. With $\overrightarrow{OP} = \overrightarrow{OC} + \overrightarrow{CP} = R\hat{x}_0 + \overrightarrow{CP}$ (see Fig. 6.28), we then obtain

$$\vec{M}_C = \int_{\text{body}} \overrightarrow{CP} \times d\vec{F} = \int_{\text{body}} \overrightarrow{CP} \times \left(-\frac{GM_E \, dm}{|\overrightarrow{OP}|^2} \frac{\overrightarrow{OP}}{|\overrightarrow{OP}|} \right)$$

$$= -\int_{\text{body}} \frac{GM_E}{|\overrightarrow{OP}|^3}\overrightarrow{CP} \times \left(R\hat{x}_0 + \overrightarrow{CP}\right) dm = -\int_{\text{body}} \frac{GM_E R\, dm}{|\overrightarrow{OP}|^3}\left(\overrightarrow{CP} \times \hat{x}_0\right) dm$$

(6.8.3b)

But (again, see Fig. 6.28),

$$|\overrightarrow{OP}|^2 = \overrightarrow{OP}\bullet\overrightarrow{OP} = \left(R\hat{x}_0 + \overrightarrow{CP}\right)\bullet\left(R\hat{x}_0 + \overrightarrow{CP}\right) = R^2 + 2R\hat{x}_0 \bullet \overrightarrow{CP} + |\overrightarrow{CP}|^2$$

$$= R^2\left[1 + 2\hat{x}_0 \bullet \frac{\overrightarrow{CP}}{R} + \underbrace{\left(\frac{|\overrightarrow{CP}|}{R}\right)^2}_{\text{negligible}}\right] \approx R^2\left[1 + 2\hat{x}_0 \bullet \frac{\overrightarrow{CP}}{R}\right] \qquad (6.8.3c)$$

where, for bodies whose maximum dimension is much smaller than R, the quadratic term $(|\overrightarrow{CP}|/R)^2$ is neglected when compared to the term $|\overrightarrow{CP}|/R$. In Eq. (6.8.3c), $|\overrightarrow{CP}|/R$ is certainly much smaller than 1, but if it is neglected, its small (but important) effect on the moment $\vec{M}_C$ is lost. Therefore, it must be kept.

By substituting in Eq. (6.8.3b) the expression for $|\overrightarrow{OP}|$ obtained from Eq. (6.8.3c), and expanding the term $|\overrightarrow{OP}|^{-3}$ in Taylor series for small $|\overrightarrow{CP}|/R$, the following approximate expression for $\vec{M}_C$ is obtained:

$$\vec{M}_C = -\int_{\text{body}} \frac{GM_E}{R^2}\left[1 - 3\hat{x}_0 \bullet \frac{|\overrightarrow{CP}|}{R}\right]\left(\overrightarrow{CP} \times \hat{x}_0\right) dm$$

$$= \frac{3GM_E}{R^3}\int_{\text{body}} \left(\hat{x}_0 \bullet \overrightarrow{CP}\right)\left(\overrightarrow{CP} \times \hat{x}_0\right) dm \qquad (6.8.3d)$$

In this equation, $\int_{\text{body}} \overrightarrow{CP}\, dm = 0$ because C is the center of mass of the body. Also, notice that terms that do not affect the integration in dm can be (and were) put outside the integrand.

To proceed, the integral in Eq. (6.8.3d) needs to be evaluated. Since the integrand involves products of the components of the vector $\overrightarrow{CP}$, moments and products of inertia are expected to appear in the final result. To avoid products of inertia, and also moments of inertia that are time dependent, let us then proceed by using the central principal axes of the body. Let us denote such axes as x-y-z, and introduce a set of orientation variables to describe their orientation relative to x_0-y_0-z_0 orbital reference axes that are parallel to the $\hat{x}_0$-$\hat{y}_0$-$\hat{z}_0$ unit vectors shown in Fig. 6.28. With $\hat{x}$-$\hat{y}$-$\hat{z}$ being unit vectors parallel to x-y-z, respectively, the orientation of the body may be described by first aligning $\hat{x}$-$\hat{y}$-$\hat{z}$ with $\hat{x}_0$-$\hat{y}_0$-$\hat{z}_0$ and then performing three successive rotations as shown earlier in this section. It was seen that the transformation relating $\hat{x}_0$-$\hat{y}_0$-$\hat{z}_0$ and $\hat{x}$-$\hat{y}$-$\hat{z}$ is of the form

$$
\begin{bmatrix} \hat{x}_0 \\ \hat{y}_0 \\ \hat{z}_0 \end{bmatrix} = \begin{bmatrix} A_{11} & A_{12} & A_{13} \\ \cdots & \cdots & \cdots \\ \cdots & \cdots & \cdots \end{bmatrix} \begin{bmatrix} \hat{x} \\ \hat{y} \\ \hat{z} \end{bmatrix} \triangleq [A] \begin{bmatrix} \hat{x} \\ \hat{y} \\ \hat{z} \end{bmatrix} \tag{6.8.3e}
$$

The elements of the matrix $= [A]$ when three-axis and two-axis body-fixed angles are used are given in Eqs. (6.8.7) (p. 472) and (6.8.14) (p. 476), respectively.

With $\hat{x}_0$ and $\overrightarrow{CP}$ expressed as

$$
\hat{x}_0 = A_{11}\hat{x} + A_{12}\hat{y} + A_{13}\hat{z} \tag{6.8.3f}
$$
$$
\overrightarrow{CP} = x\hat{x} + y\hat{y} + z\hat{z} \tag{6.8.3g}
$$

the integrand in Eq. (6.8.3d) becomes

$$
(\hat{x}_0 \bullet \overrightarrow{CP})(\overrightarrow{CP} \times \hat{x}_0) = (A_{11}x + A_{12}y + A_{13}z)\Big[(yA_{13} - zA_{12})\,\hat{x}
$$
$$
+ (zA_{11} - xA_{13})\,\hat{y} + (xA_{12} - yA_{11})\,\hat{x}\Big] \tag{6.8.3h}
$$

and, when principal axes are used, this integrates to the following

$$
\vec{M}_C = 3\frac{GM_E}{R^3}\Big[(I_z - I_y)\,A_{12}A_{13}\hat{x} + (I_x - I_z)\,A_{11}A_{13}\hat{y}
$$
$$
+ (I_y - I_x)\,A_{11}A_{12}\hat{z}\Big] \tag{6.8.3i}
$$

where I_x, I_y, and I_z are the three principal moments of inertia of the body, i.e.,

$$
I_x = \int_{body} \left(y^2 + z^2\right) dm \qquad I_y = \int_{body} \left(x^2 + z^2\right) dm \qquad I_z = \int_{body} \left(x^2 + y^2\right) dm
$$

$$
\tag{6.8.3j}
$$

The expressions for A_{11}, A_{12}, and A_{13} for the three-axis body-fixed angles are simpler than those for the two-axis body-fixed angles, as disclosed by Eqs. (6.8.7) (p. 472) and (6.8.14) (p. 476), respectively. The three-axis body-fixed angles presented in this section are the ones that appear widely in the technical literature dealing with investigations of rotational motions of satellites using Eq. (6.8.3i) for the gravitational moment. In terms of the angles θ_1, θ_2, and θ_3 shown in Fig. 6.26, the expression for the moment $\vec{M}_C$ is then

$$
\vec{M}_C = 3\frac{GM_E}{R^3}\Big[- (I_z - I_y)\,(\sin\theta_2 \cos\theta_2 \sin\theta_3)\,\hat{x}
$$
$$
+ (I_x - I_z)\,(\sin\theta_2 \cos\theta_2 \cos\theta_3)\,\hat{y}
$$
$$
- (I_y - I_x)\,(\cos^2\theta_2 \sin\theta_3 \cos\theta_3)\,\hat{z}\Big] \triangleq M_x\hat{x} + M_y\hat{y} + M_z\hat{z} \tag{6.8.3k}
$$

> To analyze the rotational motion of a satellite, the expressions given for the moment components M_x, M_y, and M_z are the ones that are used in Eqs. (6.4.9), (6.4.10), and (6.4.11), respectively, that were developed in Section 6.4. Such an analysis, which is an important problem in spacecraft dynamics, is left as a series of exercises for you, the reader (see Problems 6.21 to 6.24 in Section 6.10).

For a circular orbit, the term GM_E/R^3 in Eq. (6.8.3i) is constant and equal to the square of the orbital angular velocity Ω. As seen from the result given by Eq. (6.8.3i), the gravitational moment $\vec{M}_C$ acting on an orbiting body is proportional to the square of the orbital angular velocity $\Omega = \sqrt{GM_E/R^3}$, and it also depends on the principal inertia differences $I_x - I_y$, $I_x - I_z$, and $I_y - I_z$. Since Ω is very small, the magnitude of $\vec{M}_C$ is also very small. Physically, such a moment is due to the change in the gravitational force of attraction with distance from the attracting planet. Such a moment is called the *gravity gradient moment*.

As an illustration of the orders of magnitude of the moment $\vec{M}_C$, consider a homogeneous thin bar of mass m and length L in a circular orbit 200 km above the surface of the earth. With the earth approximated as a sphere of radius $R_E = 6378$ km, and with $R = 6578$ km and $g = 9.81$ m/s^2, the orbital angular velocity in such a case is $\Omega = \sqrt{gR_E^2/R^3} \approx 0.0012$ rad/s (which gives $t_{2\pi} = 2\pi/\Omega \approx 87$ min for the orbital period). For motion in the orbital plane, for example, $\theta_1 = \theta_2 = 0$ in Fig. 6.26, and this yields

$$\vec{M}_C = -3\Omega^2 \frac{mL^2}{12} (\sin\theta_3)(\cos\theta_3)\,\hat{z}$$

The maximum magnitude of $\vec{M}_C$ occurs for $\theta_3 = 45°$ in this case, and it is approximately equal to $1.75 \times 10^{-7} mL^2$ N·m, with m in kg and L in meters.

For the dumbbell of Example 1.10.1 (p. 27) and of Problem 4.15 (p. 311), with $I_x = 0$, and $I_y = I_z = mL^2/4$, one obtains $|\vec{M}_C|_{\max} = 5.26 \times 10^{-7} mL^2$ N·m under the same conditions stated for the thin bar. Although such a moment is very small, it causes the orbiting dumbbell to oscillate about the local vertical with a long period of about 51 min for a circular orbit 200 km above the earth's surface (see Problem 4.15, p. 311, and also Problem 4.16).

A detailed analysis of the rotational dynamics of a spin-stabilized satellite, using the results developed in this section, is presented in Example 6.8.4. Such an analysis is another important problem in spacecraft dynamics. The entire analysis involves the formulation of the differential equations of motion, and determination of the equilibrium solutions and their stability. Here, the stability analysis is done based on the linearized differential equations of motion about their equilibrium.

Example 6.8.4 Stability of a Spin-Stabilized Symmetric Satellite

Let the satellite illustrated in Fig. 6.28 (p. 478) be symmetric about its principal z-axis so that $I_x = I_y$, as for a cylinder, for example. The three-axis body-fixed angles θ_1, θ_2, and θ_3 shown in Fig. 6.26 (p. 470) will be used for describing the rotational motion of the satellite relative to the orbital reference frame $\hat{x}_0$-$\hat{y}_0$-$\hat{z}_0$ seen in Fig. 6.28. The principal x-, y-, and z-axes of the satellite are aligned, at all times, with the $\hat{x}$, $\hat{y}$, and $\hat{z}$ unit vectors, respectively, which are shown in Fig. 6.26. In the following, the differential equations for the rotational motion of a spin-stabilized satellite are developed. This is followed by the determination of the equilibrium solutions, and by the stability analysis.

Differential Equations of Motion

In terms of the angles θ_1, θ_2, and θ_3 shown in Fig. 6.26, the angular velocity of the satellite relative to the orbital reference frame $\hat{x}_0$-$\hat{y}_0$-$\hat{z}_0$ shown in Figs. 6.26 and 6.28 is given by Eq. (6.8.2) (p. 472). Since the absolute angular velocity of the $\{\hat{x}_0, \hat{y}_0, \hat{z}_0\}$ triad is $\dot{\phi}\hat{z}_0 \overset{\Delta}{=} \Omega\hat{z}_0$ (see Fig. 6.28, p. 478), the absolute angular velocity of the satellite is then

$$\vec{\omega} = (\dot{\theta}_1 \cos\theta_2 \cos\theta_3 + \dot{\theta}_2 \sin\theta_3)\hat{x} + (\dot{\theta}_2 \cos\theta_3 - \dot{\theta}_1 \cos\theta_2 \sin\theta_3)\hat{y}$$
$$+(\dot{\theta}_3 + \dot{\theta}_1 \sin\theta_2)\hat{z} + \Omega\hat{z}_0 \overset{\Delta}{=} \omega_x\hat{x} + \omega_y\hat{y} + \omega_z\hat{z} \qquad (6.8.4a)$$

From either one of the three snapshots shown in Fig. 6.26 (p. 470) or from Eq. (6.8.6) (p. 472), the expression for $\hat{z}_0$ in terms of $\hat{x}$, $\hat{y}$, and $\hat{z}$ is [see Eq. (6.8.6)]:

$$\hat{z}_0 = (\sin\theta_1 \sin\theta_3 - \cos\theta_1 \sin\theta_2 \cos\theta_3)\,\hat{x}$$
$$+ (\sin\theta_1 \cos\theta_3 + \cos\theta_1 \sin\theta_2 \sin\theta_3)\,\hat{y} + (\cos\theta_1 \cos\theta_2)\,\hat{z} \qquad (6.8.4b)$$

This gives the following expressions for the components ω_x, ω_y, and ω_z of the absolute angular velocity $\vec{\omega}$ of the satellite:

$$\omega_x = \left(\dot{\theta}_1 \cos\theta_2 - \Omega\cos\theta_1 \sin\theta_2\right)\cos\theta_3 + \left(\dot{\theta}_2 + \Omega\sin\theta_1\right)\sin\theta_3 \qquad (6.8.4c)$$

$$\omega_y = \left(\dot{\theta}_2 + \Omega\sin\theta_1\right)\cos\theta_3 - \left(\dot{\theta}_1 \cos\theta_2 - \Omega\cos\theta_1 \sin\theta_2\right)\sin\theta_3 \qquad (6.8.4d)$$

$$\omega_z = \dot{\theta}_3 + \dot{\theta}_1 \sin\theta_2 + \Omega\cos\theta_1 \cos\theta_2 \qquad (6.8.4e)$$

These expressions are then used in Euler's differential equations, which are Eqs. (6.4.9) to (6.4.11) developed in Section 6.4. For $I_x = I_y$, and with $\vec{M}_C$ given by

Eq. (6.8.3k) (p. 480), those equations yield the following differential equations for the rotational motion of the satellite

$$\dot{\omega}_x + k_1 \left(\omega_y\omega_z + 3\Omega^2 \sin\theta_2 \cos\theta_2 \sin\theta_3\right) = 0 \qquad (6.8.4\text{f})$$

$$\dot{\omega}_y - k_1 \left(\omega_x\omega_z - 3\Omega^2 \sin\theta_2 \cos\theta_2 \cos\theta_3\right) = 0 \qquad (6.8.4\text{g})$$

$$\dot{\omega}_z = 0 \qquad (6.8.4\text{h})$$

where

$$k_1 \triangleq \frac{I_z - I_y}{I_x} = \frac{I_z}{I_x} - 1 \qquad (6.8.4\text{i})$$

Notice that, using the definitions of moments of inertia given by the first three of Eqs. (6.3.6) (p. 403), it is seen that $k_1 = [\int_{\text{body}} \left(y^2 - z^2\right) dm]/(\int_{\text{body}} \left(y^2 + z^2\right) dm$. Therefore, for any rigid body, the parameter k_1 is in the range $-1 \le k_1 \le 1$.

Equation (6.8.4h) immediately discloses that the $\hat{z}$ component of the absolute angular velocity of the satellite (i.e., its spin about its axis of symmetry) is constant, i.e.,

$$\omega_z = \dot{\theta}_3 + \dot{\theta}_1 \sin\theta_2 + \Omega\cos\theta_1 \cos\theta_2 = \text{constant} \qquad (6.8.4\text{j})$$

Equilibrium Solutions

Equations (6.8.4f) and (6.8.4g) admit the equilibrium solutions $\theta_1 = \text{constant} \triangleq \theta_{1_e}$ and $\theta_2 = \text{constant} \triangleq \theta_{2_e}$, where $\theta_{1_e} = 0$ and $\theta_{1_e} = \pm\pi$, and $\theta_{2_e} = 0$ and $\theta_{2_e} = \pm\pi$. These equilibrium solutions represent the same physical situation, namely, a satellite with its symmetry axis perpendicular to the orbital plane and spinning relative to inertial space with a constant spin rate ω_z about that axis. The stability of such a state is analyzed next. The satellite is like a gyroscope in such a situation.

Stability Analysis of the Spin-Stabilized Satellite

The stability of each equilibrium is analyzed here by linearizing the differential equations of motion about the equilibrium solution $\theta_{1_e} = \theta_{2_e} = 0$, as just determined. Since $\theta_{1_e} = \theta_{2_e} = 0$, there is no need to introduce new symbols for the perturbed variables about the equilibrium. Thus, by letting $\theta_1(t)$ and $\theta_2(t)$ be small, the following linearized differential equations of motion are obtained. The algebraic steps that lead to Eqs. (6.8.4k) and (6.8.4l) are left as an exercise (see Problem 6.21, p. 495). Notice that no approximations should be made for θ_3 since that angle is continuously changing with time.

$$\ddot{\theta}_1 - \Omega\dot{\theta}_2 + [k_1 + \beta\left(1 + k_1\right)]\Omega\left(\dot{\theta}_2 + \Omega\theta_1\right) = 0 \qquad (6.8.4\text{k})$$

$$\ddot{\theta}_2 + \Omega\dot{\theta}_1 - [k_1 + \beta\left(1 + k_1\right)]\Omega\left(\dot{\theta}_1 - \Omega\theta_2\right) + 3\Omega^2 k_1\theta_2 = 0 \qquad (6.8.4\text{l})$$

where β is a constant "spin parameter" defined as

$$\beta = \frac{\omega_z - \Omega}{\Omega} = \frac{\omega_z}{\Omega} - 1 \tag{6.8.4m}$$

The general solutions to these equations are of the form $\theta_1 = Ae^{st}$ and $\theta_2 = Be^{st}$. As explained in Chapter 1, this yields the following algebraic equation for the exponent s, which is the characteristic equation for this system:

$$s^4 + a_2 \Omega^2 s^2 + a_0 \Omega^4 = 0 \tag{6.8.4n}$$

where

$$a_2 = 1 + 3k_1 + k_1^2 + (1 + k_1)\left[2k_1 + (1 + k_1)\beta\right]\beta \tag{6.8.4o}$$

$$a_0 = \left[k_1 + (1 + k_1)\beta\right]\left[4k_1 + (1 + k_1)\beta\right] \tag{6.8.4p}$$

As indicated on p. 306 in Chapter 4, the conditions for the equilibrium to be infinitesimally stable are $a_2 > 0$, $a_0 > 0$, and $a_2^2 - 4a_0 > 0$. All of these three conditions must be satisfied for the equilibrium to be infinitesimally stable. The equilibrium is unstable if either $a_2 < 0$ or $a_0 < 0$ or $a_2^2 - 4a_0 < 0$. Numerical checks to verify stability by integrating the nonlinear differential equations of motion are left as an exercise (see Problem 6.21, p. 495).

Figure 6.29a shows the "stability chart" β versus k_1, for any spinning rigid symmetric satellite, disclosing the regions where the equilibrium is stable, and where it is unstable. Figure 6.29b illustrates, using a cylinder, how the parameter k_1 is affected by the shape of the body. For a homogeneous cylinder of radius a and length L, $k_1 = (3a^2 - L^2)/(3a^2 + L^2)$.

Looking at Fig. 6.29, it is seen that the equilibrium of a spinning satellite with $k_1 \approx 1$, for example, is unstable when $-\Omega < \omega_z < \Omega/2$. When $\omega_z = -\Omega$, the satellite is not spinning relative to inertial space. High spin rates, either clockwise (i.e., $\omega_z \gg 0$) or counterclockwise (i.e., $\omega_z \ll 0$), result in an infinitesimally stable equilibrium as indicated by the chart in Fig. 6.29. The satellite is like a fast spinning gyroscope in such cases.

The lines that separate the infinitesimally stable from the unstable regions in Fig. 6.29 are obtained from the equations $a_2 = 0$, $a_0 = 0$, and $a_2^2 - 4a_0 = 0$. The equation $a_2 = 0$ yields $\beta = (-k_1 \pm \sqrt{-1 - 3k_1})/(1 + k_1)$. The equation $a_0 = 0$ yields $\beta = -k_1/(1 + k_1)$ and $\beta = -4k_1/(1 + k_1)$. The equation $a_2^2 - 4a_0 = 0$ yields the following polynomial:

$$\beta^4 (1 + k_1)^4 + 4k_1 (1 + k_1)^3 \beta^3 + 2 (1 + k_1)^2 \left(3k_1^2 + 3k_1 - 1\right)\beta^2$$
$$+ 4k_1 (1 + k_1)\left(k_1^2 + 3k_1 - 4\right)\beta + \left(k_1^4 + 6k_1^3 - 5k_1^2 + 6k_1 + 1\right) = 0$$

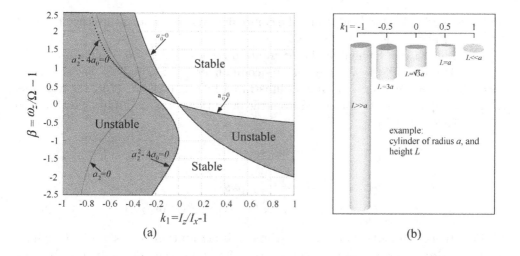

Figure 6.29: Stability chart for spinning satellites in circular orbit (dark region: unstable; white region: infinitesimally stable).

whose real roots then yield additional lines in the β versus k_1 parameter space. The lines corresponding to $a_2 = 0$, $a_0 = 0$, and to $a_2^2 - 4a_0 = 0$ are shown in Fig. 6.29 to disclose how the stability chart was generated. To determine which region separated by those lines correspond to an infinitesimally stable and to an unstable equilibrium, one only needs to test the signs of a_0, a_2, and $a_2^2 - 4a_0$ for any point on either "side" of those lines.

6.9   Angular Velocity and the Orientation Angles: Notes

The rotational motion of a body is governed by the vector differential equation $d\vec{H}_C/dt = \vec{M}_C$, where $\vec{H}_C$ is related to the absolute angular velocity of the body by Eq. (6.3.7) (p. 404). When the body is able to move unconstrained in three-dimensional space, this gives the three scalar differential equations that govern the rotational motion of the body. When the moment $\vec{M}_C$ depends on the orientation of the body (as in the case of airplanes, ships, satellites, and artillery shells, for example), it is necessary to express the components of its absolute angular velocity in terms of orientation variables and their time derivatives to be able to integrate the equations for determining the orientation of the body at any time. Assuming, for simplicity, that the orientation angles of the body are measured from a nonrotating reference frame x_0-y_0-z_0, the absolute angular velocity of the body is the one presented in Section 6.8 [and, thus, the components ω_x, ω_y, and ω_z are the ones that "go into" Euler's Eqs. (6.4.9) to (6.4.11)].

The time rate of change of the orientation angles θ_1, θ_2, and θ_3 may always be expressed in terms of the angular velocity components ω_x, ω_y, and ω_z. For the three-axis orientation angles θ_1, θ_2, and θ_3 illustrated in Fig. 6.26 (p. 470), for example, the following relations are obtained from the definitions of ω_x, ω_y, and ω_z introduced in Eq. (6.8.2) (p. 472):

$$\dot{\theta}_1 = \frac{\omega_x \cos \theta_3 - \omega_y \sin \theta_3}{\cos \theta_2}$$

$$\dot{\theta}_2 = \omega_x \sin \theta_3 + \omega_y \cos \theta_3$$

$$\dot{\theta}_3 = \omega_z - (\omega_x \cos \theta_3 - \omega_y \sin \theta_3) \tan \theta_2 \qquad (6.9.1)$$

Equations (6.9.1) disclose that the three-axis orientation angles of Fig. 6.26 have a "singularity" when $\cos \theta_2 = 0$, i.e., at $\theta_2 = \pm \pi/2$ rad (i.e., denominators become equal to zero). Actually, an inspection of Fig. 6.26 reveals that such a singularity occurs because the rotations θ_1 and θ_3 become indistinguishable from each other when $\theta_2 = \pm \pi/2$ (i.e., the rotations θ_1 and θ_3 are about the same axis in such a case).

For the two-axis orientation angles of Fig. 6.27, the expressions that relate $\dot{\phi}_1, \dot{\phi}_2$, and $\dot{\phi}_3$ to the new expressions of the components ω_x, ω_y, and ω_z defined by Eq. (6.8.9) (p. 476) are

$$\dot{\phi}_1 = \frac{\omega_x \sin \phi_3 + \omega_y \cos \phi_3}{\sin \phi_2}$$

$$\dot{\phi}_2 = \omega_x \cos \phi_3 - \omega_y \sin \phi_3$$

$$\dot{\phi}_3 = \omega_z - \frac{\omega_x \sin \phi_3 + \omega_y \cos \phi_3}{\tan \phi_2} \qquad (6.9.2)$$

As indicated by Eqs. (6.9.2), the singularity for these two-axis orientation angles occurs when $\sin \phi_2 = 0$, i.e., when $\phi_2 = 0$ or $\phi_2 = \pi$ rad. By observing Fig. 6.27, the rotations ϕ_1 and ϕ_3 become indistinguishable from each other when either $\phi_2 = 0$ or $\phi_2 = \pi$ rad, and this is the cause of the singularity. Other orientation variables, such as *Euler parameters*, can be used to avoid singularities. Interested readers can find details in Reference 13 in Appendix G.

6.10   Problems

6.1 Determine the central principal moments of inertia and the direction of the three principal axes of the bent bar of Example 6.3.1 (p. 409) (see Fig. 6.4).

For this, take $L = 0.5$ m, $R = 0.15$ m, and the mass density to be 0.063 kg/m, which is the approximate specific mass of a copper wire of 3 mm in diameter. Neglect the small diameter of the bar in your calculations.

6.2 Determine the central principal moments of inertia and the direction of the three principal axes of the triangular plate of Example 6.3.2 (p. 410) (see Fig. 6.5).

6.3 Two rectangular blocks are rigidly attached to each other as shown in Fig. 6.30, where $L = 1.2$ m, $a = 50$ cm, and $b = 40$ cm. The mass density of the larger block is $\rho_1 = 7700$ kg/m^3 and that of the smaller block is $\rho_2 = 2600$ kg/m^3. Determine, for the entire system:

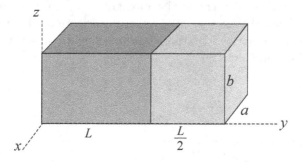

Figure 6.30: Figure for Problem 6.3.

a. The location of its center of mass

b. The central principal moments of inertia

c. The direction of the principal axes relative to the rectangular x-y-z coordinate system shown in the figure

6.4 Based on the definitions of moments and products of inertia, which of the following are inertia matrices? Why?

$$
\begin{bmatrix} 1 & -1 & \sqrt{3} \\ -1 & 2.5 & 0 \\ \sqrt{3} & 0 & 4 \end{bmatrix}
\qquad
\begin{bmatrix} 3 & -2 & 3 \\ -2 & 9 & 2 \\ 3 & 2 & 7 \end{bmatrix}
\qquad
\begin{bmatrix} 10 & 2 & -3 \\ 2 & 20 & -1 \\ -3 & -1 & 12 \end{bmatrix}
$$

6.5   The inertia matrix of a body, referred to an orthogonal set of axes centered at the center of mass of the body, is

$$
\begin{bmatrix} 5 & 0 & 3 \\ 0 & 7 & 0 \\ 3 & 0 & 4 \end{bmatrix} \text{ kg·m}^2
$$

What are the principal moments of inertia of the body?

6.6   A rigid body has the following moments and products of inertia about a set of orthogonal axes x-y-z that are centered at the center of mass C of the body and are fixed to the body (all quantities are in kg·m^2): $I_{xx} = 3, I_{yy} = 7, I_{zz} = 8, I_{xy} = 2, I_{xz} = 1, I_{yz} = -1$.

   a. Using dyadics, determine the moments and products of inertia of the body about the axes x_s-y_s-z_s whose orientation with respect to the x-y-z axes is described by the rotation sequence shown in Fig. 6.31, with $\theta_1 = 30°$ and $\theta_2 = 45°$.

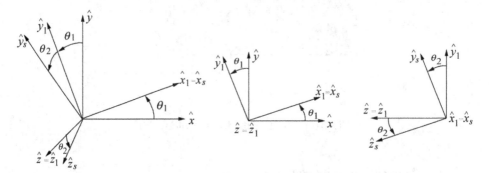

Figure 6.31: Figure for Problem 6.6.

   b. Show that the solution to part a may also be obtained by matrix algebra as $[R]^T[I_C][R]$, where $[I_C]$ is the inertia matrix whose elements are as given previously and $[R]$ is the transformation matrix that relates the unit vectors $\hat{x}$-$\hat{y}$-$\hat{z}$ and $\hat{x}_s$-$\hat{y}_s$-$\hat{z}_s$ as

$$
\begin{bmatrix} \hat{x} \\ \hat{y} \\ \hat{z} \end{bmatrix} = [R] \begin{bmatrix} \hat{x}_s \\ \hat{y}_s \\ \hat{z}_s \end{bmatrix}
$$

Determine the nine elements of $[R]$.

6.7 A rigid body is translating and rotating in space with absolute angular velocity $\vec{\omega} = \omega_x \hat{x} + \omega_y \hat{y} + \omega_z \hat{z}$, where $\{\hat{x}, \hat{y}, \hat{z}\}$ are orthogonal unit vectors parallel to the x-y-z principal axes of the body. While rotating, the body is subjected to a moment $\vec{M}_C = M_x \hat{x} + M_y \hat{y} + M_z \hat{z}$ about its center of mass C. The components $\omega_x(t), \omega_y(t)$, and $\omega_z(t)$ are measured by sensors, and the moment $\vec{M}_C$ is applied by actuators driven by a control system that guarantees that $M_x \omega_x + M_y \omega_y + M_z \omega_z = k \cos t$, where k is constant. Determine how the kinetic energy of the rotational motion of the body changes with time t for the cases when $k \neq 0$ and when $k = 0$. *Hint*: Use Euler's equations for principal axes.

6.8 A solid steel rotor (see Fig. 6.9, p. 422) of radius $R = 30$ cm and height $h = 30$ cm is given an initial absolute angular velocity $\vec{\omega} = 200 \hat{z}$ rpm, where $\hat{z}$ is a unit vector parallel to the axial z-axis shown in Fig. 6.9. The rotor is subjected to a moment $\vec{M}_C = k_1 t \hat{x} + k_2 \hat{y}$, where $k_1 = 4$ N·m/s and $k_2 = 2$ N·m, and $\hat{x}$ and $\hat{y} = \hat{z} \times \hat{x}$ are unit vectors parallel to the x- and y-axes shown in Fig. 6.9, which are fixed to the rotor. Determine the components $\omega_x(t), \omega_y(t)$, and $\omega_z(t)$ of the absolute angular velocity $\vec{\omega} = \omega_x \hat{x} + \omega_y \hat{y} + \omega_z \hat{z}$ of the rotor. The density of steel is $\rho = 7700$ kg/m^3, and the principal moments of inertia of a solid rotor are given in Example 6.3.7 (p. 421).

6.9 A solid steel rotor (see Fig. 6.9, p. 422) of radius $R = 30$ cm and height $h = 30$ cm is spinning about its axial z-axis with a constant absolute angular velocity equal to 6000 rpm when a small disturbing moment of magnitude 0.1 N·m is applied to it in the direction perpendicular to the axial axis. *Estimate* the small change of orientation of the spin axis, in degrees, after 30 min have elapsed. The density of steel is $\rho = 7700$ kg/m^3, and the principal moments of inertia of a solid rotor are given in Example 6.3.7 (p. 421).

6.10 A thin homogeneous rectangular plate of mass m and dimensions L and b is hinged to a rotating horizontal circular platform of radius $|\overrightarrow{OA}| = R$ as shown in Fig. 6.32. The center O of the platform is inertial, and the hinge at A only allows the plate to rotate about line OA relative to the platform. That rotation is tracked by the angle $\theta(t)$ shown in the figure. The absolute angular velocity of the platform is constant, i.e., $\dot{\phi} = \text{constant} \overset{\Delta}{=} \Omega$.

    a. Obtain the differential equation of motion for the system, all the equilibrium orientations $\theta = \text{constant} \overset{\Delta}{=} \theta_e$ in terms of Ω, L, and b, and analyze their stability. Indicate whether they are stable, unstable, or asymptotically stable.

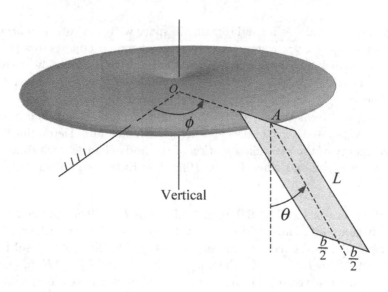

Figure 6.32: Figure for Problem 6.10.

  b. Determine the maximum and minimum values of $\theta(t)$ when $\theta(0) = 5°$
     and $\dot{\theta}(0) = 0.08\sqrt{g/L}$, for $\Omega^2 L/g = 1$ and 2.

  c. Verify the stability of each equilibrium solution for the system by sketch-
     ing phase plane plots (i.e., $\dot{\theta}$ versus θ) for $\Omega^2 L/g = 1$ and 2.

6.11 For the system described in Problem 6.10, determine the value (or values) of
      θ that yield the maximum and minimum values of the component of the force
      reaction at A along the perpendicular to the plate when $\Omega^2 L/g = 1$ and when
      $\Omega^2 L/g = 3$. The motion is started with the initial conditions $\theta(0) = 30°$ and
      $\dot{\theta}(0) = 0$.

6.12 A thin homogeneous rectangular plate of mass M and dimensions a and b
      is welded to a shaft, in the plane of the plate, as shown in Fig. 6.33 (C is
      the center of mass of the plate). One of the sides of the plate is inclined
      by α degrees with the shaft, and the shaft is driven at a constant absolute
      angular velocity ω. It is desired to dynamically balance the system with four
      small masses (modeled as point masses), each with mass $m = M/48$, placed
      as shown in the figure. For $L = a/2$ and $b = 1.9L$, determine the angle β to
      achieve dynamic balancing.

6.13 A homogeneous steel rod PQ of mass m and length L is welded to a horizontal

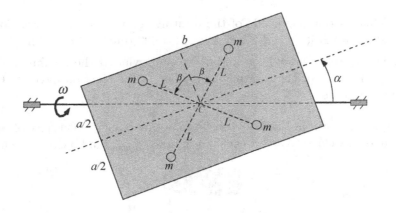

Figure 6.33: Figure for Problem 6.12.

shaft as shown in Fig. 6.34, where $d_1 = |\overrightarrow{OA}|$ and $d_2 = |\overrightarrow{OB}|$. The bar is perpendicular to the shaft, but its center of mass C is at a distance e (called

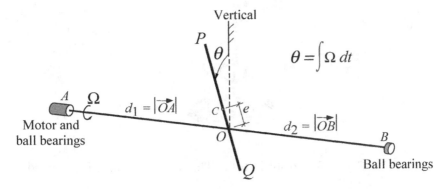

Figure 6.34: Figure for Problem 6.13.

the *offset distance*) from the shaft midline. The shaft is driven at a constant absolute angular velocity Ω. For simplicity, neglect the mass of the shaft and approximate the bar as a thin line.

a. Determine the bearing reaction forces at A and B. Show that there is a component of those reaction forces that is proportional to $e\Omega^2$ and is perpendicular to the shaft, and rotates with the same angular velocity as the shaft.

   b. What is the magnitude of the rotating component of the reaction force at A and B if $d_1 = d_2$, $m = 1$ kg, $\Omega = 600$ rpm, and $e = 6$ mm?

   c. For $d_1 = d_2$, $m = 1$ kg, and $e = 6$ mm, what is the maximum value of Ω (in rpm) for the rotating component of the reaction force at A and B not to exceed 2 N?

6.14 The system shown in Fig. 6.35 consists of a thin homogeneous disc of radius R and mass m welded to a vertical shaft of mass m_{shaft} and length L. The plane

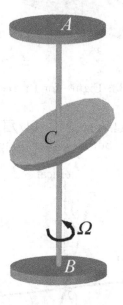

Figure 6.35: Figure for Problem 6.14. The bases at A and B are fixed.

of the disc is inclined by α degrees with the perpendicular to the shaft, and its center of mass is on the centerline of the shaft, with $\overrightarrow{AC} = L/2$. The shaft is rotating with a constant absolute angular velocity Ω and is supported by ball-and-socket joints at A and B.

   a. Determine the components of the reaction forces at A and B that are perpendicular to the shaft. Express the answers in terms of the quantities m, R, L, α, and Ω.

   b. For $m = 0.5$ kg, $R = 30$ cm, $L = 2$ m, and $\alpha = 30°$, determine the maximum value of Ω (in rpm) so that the rotating component of the reaction force at A indicated in part a does not exceed 2 N.

6.15 The system shown in Fig. 6.36 consists of a thin homogeneous disc of radius R and mass m welded to a bent shaft ABC, with lengths L_1 and L_2, respectively,

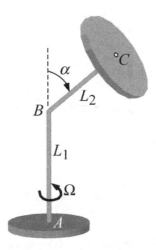

Figure 6.36: Figure for Problem 6.15. The angle α is constant, and the base at A is fixed.

and with C coinciding with the center of mass of the disc. Branch AB of the shaft is vertical, and the plane of the disc is perpendicular to branch BC of the shaft. The shaft is rotating with a constant absolute angular velocity Ω. Friction in the system, and the mass of the shaft, can be neglected.

a. Determine the force and moment reactions at A.

b. You should find a component of the reaction force and reaction moment that varies as Ω^2. Determine the maximum magnitude of those components in terms of m, R, L_1, L_2, and α.

c. For $L_2 = 25$ cm and $\alpha = 30°$, what should be the maximum value of Ω so that the magnitude of the reaction force that is perpendicular to AB does not exceed mg?

6.16 The angular orientation of a body may be described by a set of three orientation angles, as presented in Section 6.8. As indicated in the note on page 444 in Section 6.6.3, the differential equations of motion for the top analyzed in that section could have been formulated using the orientation angles for axes fixed to the rotating top. Such an orientation sequence is shown in Fig. 6.37, where the unit vectors $\hat{x}$, $\hat{y}$, and $\hat{z} \overset{\Delta}{=} \hat{x} \times \hat{y}$ are parallel to principal x-y-z axes fixed to the rotating body.

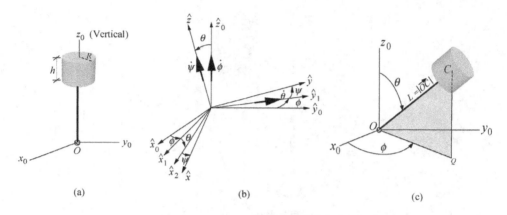

Figure 6.37: Figure for Problem 6.16. The axes x_0-y_0-z_0 are inertial.

The orientation of the system shown in Fig. 6.37 is described by first aligning the x-y-z body-fixed axes with the x_0-y_0-z_0 reference axes, as shown in Fig. 6.37a. The actual orientation of x-y-z relative to x_0-y_0-z_0 is then described by the rotation sequence ϕ-θ-ψ shown in Fig. 6.37b. This puts the system in its actual orientation shown in Fig. 6.37c. The moments of inertia of the system about the x-y-z axes that are centered at the fixed point O are I_x, $I_y = I_x$, and I_z. Using Euler's equations for principal axes, which are Eqs. (6.4.9) to (6.4.11) on p. 427:

a. Obtain the differential equations of motion for the top in terms of the angles ϕ, θ, and ψ.

b. Combine the appropriate equations obtained in part a and show that they yield Eqs. (6.6.3i) to (6.6.3k) that were formulated in Section 6.6.3, where I_n in those equations is the same as I_x in this problem.

6.17 For the rotation sequence shown in Fig. 6.26 (p. 470), $\theta_1 = 10°$, $\theta_2 = 30°$, and $\theta_3 = 20°$. Determine the angles between the unit vector $\hat{y}$ and the unit vectors $\hat{x}_0$, $\hat{y}_0$, and $\hat{z}_0$.

6.18 For the rotation sequence shown in Fig. 6.26 (p. 470), $\theta_1 = 30°$, $\theta_2 = -30°$, and $\theta_3 = 45°$. Determine the direction cosines of the unit vectors $\hat{x}$, $\hat{y}$, and $\hat{z}$, relative to the unit vectors $\hat{x}_0$, $\hat{y}_0$, and $\hat{z}_0$.

6.19 For the rotation sequence shown in Fig. 6.27 (p. 474), $\phi_1 = 10°$, $\phi_2 = 30°$, and $\phi_3 = 20°$. Determine the angles between the unit vector $\hat{y}$ and the unit vectors $\hat{x}_0$, $\hat{y}_0$, and $\hat{z}_0$.

6.20 For the rotation sequence shown in Fig. 6.27 (p. 474), $\phi_1 = 30°$, $\phi_2 = -30°$, and $\phi_3 = 45°$. Determine the direction cosines of the unit vectors $\hat{x}$, $\hat{y}$, and $\hat{z}$, relative to the unit vectors $\hat{x}_0$, $\hat{y}_0$, and $\hat{z}_0$.

6.21 Show that Eqs. (6.8.4f) and (6.8.4g) for a spin-stabilized symmetric satellite in circular orbit can be combined to yield the following two equations, which only involve θ_1 and θ_2:

$$\dot{\omega}_1 + \omega_2\dot{\theta}_3 + k_1\omega_z\omega_2 = 0$$
$$\dot{\omega}_2 - \omega_1\dot{\theta}_3 - k_1\omega_z\omega_1 + 3k_1\Omega^2 \sin\theta_2 \cos\theta_2 = 0$$

In these equations, $\omega_1 = \dot{\theta}_1 \cos\theta_2 - \Omega\cos\theta_1 \sin\theta_2$, $\omega_2 = \dot{\theta}_2 + \Omega\sin\theta_1$, $\dot{\theta}_3 = \omega_z - \dot{\theta}_1 \sin\theta_2 - \Omega\cos\theta_1 \cos\theta_2$, as obtained from Eqs. (6.8.4c), (6.8.4d), and (6.8.4e) on p. 482, and $\omega_z = \Omega(\beta+1)$, as given by Eq. (6.8.4m). Then, linearize the given equations for small θ_1 and θ_2 to verify the validity of Eqs. (6.8.4k) and (6.8.4l) shown on p. 483.

Using a few points (k_1, β) in the stable regions of the stability chart in Fig. 6.29, numerically integrate the nonlinear differential equations of motion (i.e., either Eqs. (6.8.4f) and (6.8.4g) on p. 483 or the equations given in this problem), after writing them as second-order differential equations for θ_1 and θ_2, to verify the results of the stability analysis presented in Example 6.8.4. Use the nondimensional time $\tau = \Omega t$ in the numerical integration.

6.22 The center of mass C of a rigid body of central principal moments of inertia I_x, I_y, and I_z is in a circular orbit of radius R around the earth, as indicated in Fig. 6.28 (p. 478). The principal axes of the body, x-y-z, are aligned with the unit vectors $\hat{x}$-$\hat{y}$-$\hat{z}$ shown in Fig. 6.26 (p. 470).

   a. What is the expression for the absolute angular velocity $\vec{\omega}$ of the body in terms of the three-axis body-fixed orientation angles shown in Fig. 6.26, and of their time derivatives? Express the answer in the $\hat{x}$-$\hat{y}$-$\hat{z}$ triad.

   b. Obtain the differential equations for the rotational motion of the body in terms of the three-axis body-fixed orientation angles shown in Fig. 6.26. Approximate the earth as a solid homogeneous sphere of radius R_E, and treat its center as inertial.

c. Show that the differential equations admit the equilibrium solution $\theta_{1_e} = \theta_{2_e} = \theta_{3_e} = 0$, which physically means that the three principal axes of the body coincide with the orbital reference axes.

d. Linearize the differential equations obtained in part a about the equilibrium $\theta_{1_e} = \theta_{2_e} = \theta_{3_e} = 0$, and show that the linearized differential equations of motion are

$$\ddot{\theta}_1 + (k_1 - 1)\,\Omega\dot{\theta}_2 + k_1\Omega^2\theta_1 = 0$$
$$\ddot{\theta}_2 + (k_2 + 1)\,\Omega\dot{\theta}_1 - 4\Omega^2 k_2\theta_2 = 0$$
$$\ddot{\theta}_3 + 3k_3\Omega^2\theta_3 = 0$$

where

$$k_1 = \frac{I_z - I_y}{I_x} \qquad k_2 = \frac{I_x - I_z}{I_y} \qquad k_3 = \frac{I_y - I_x}{I_z}$$

e. Using the definitions of moments of inertia, show that $k_3 = -(k_1 + k_2)/(1 + k_1k_2)$ and that $-1 \le k_1 \le 1$ and $-1 \le k_2 \le 1$. The parameters k_1 and k_2 are widely used in satellite dynamics studies found in the technical literature.

Suggestion: For part b, use Euler's equations for principal axes, with $\vec{M}_C$ given by Eq. (6.8.3i) (p. 480). For part e, use the definition of moments of inertia.

6.23 The homogeneous rectangular plate shown in Fig. 6.6 (p. 412), with dimensions $L = 3b$ and $a = 1.5b$, is put in orbit around the earth. The center of mass of the plate is in a circular orbit of radius 300 km above the surface of the earth. Using the linearized differential equations given in Problem 6.22, determine, for each one of the following cases, whether the equilibrium $\theta_{1_e} = \theta_{2_e} = \theta_{3_e} = 0$ indicated in that problem is infinitesimally stable, asymptotically stable, or unstable. In real situations, one tries to put the body in orbit in an orientation that is as close as possible to a stable equilibrium orientation relative to the orbiting reference frame shown in the book. Try to visualize each one of the given situations before you start solving the problem. Notice that each of the following statements defines how the parameters k_1 and k_2 should be calculated.

a1. At equilibrium, the plate is oriented with its axis of minimum inertia aligned with the local vertical and its axis of maximum inertia perpendicular to the orbital plane.

a2. At equilibrium, the plate is oriented with its axis of minimum inertia aligned with the local vertical and its axis of maximum inertia tangent to the orbit (i.e., the same as the velocity of the center of mass of the plate).

b1. At equilibrium, the plate is oriented with its axis of minimum inertia tangent to the orbit and its axis of maximum inertia perpendicular to the orbital plane.

b2. At equilibrium, the plate is oriented with its axis of minimum inertia tangent to the orbit and its axis of maximum inertia aligned with the local vertical.

c1. At equilibrium, the plate is oriented with its axis of minimum inertia perpendicular to the orbital plane and its axis of maximum inertia tangent to the orbit.

c2. At equilibrium, the plate is oriented with its axis of minimum inertia perpendicular to the orbital plane and its axis of maximum inertia aligned with the local vertical.

From the results you obtain, specify in which orientation (or orientations, if that is the case) the plate should be put in orbit. Does the radius of the circular orbit affect the answer?

6.24 Figure 6.38 shows the stability chart for gravity-gradient stabilized satellites in circular orbit, in the absence of damping, in terms of the parameters k_1 and k_2 defined in Problem 6.22. Four distinct regions are shown in the figure. Using the linearized differential equations presented in Problem 6.22, analyze the stability of the equilibrium $\theta_{1_e} = \theta_{2_e} = \theta_{3_e} = 0$ and then generate that stability chart identifying the regions where the equilibrium is infinitesimally stable and those where it is unstable. *Hint:* You should get a characteristic equation of the form $(s^2 + b)(s^4 + a_2 s^2 + a_0) = 0$.

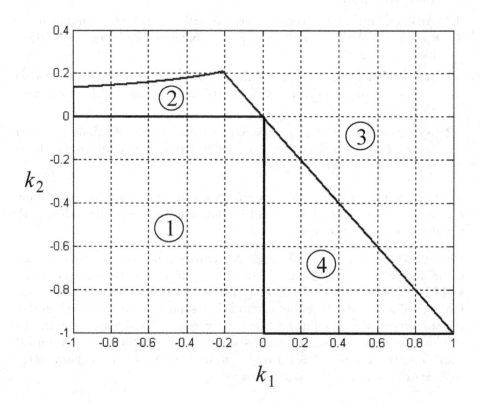

Figure 6.38: Stability chart for gravity-gradient stabilized satellites.

CHAPTER 7

Analytical Dynamics

The use of Newton's second law $\vec{F} = d(m\vec{v})/dt = m\vec{a}$ for formulating the differential equations of motion of a single particle, and the use of its extension to the motion of the center of mass of a system of particles and continuum bodies, requires the inclusion of all forces applied to the individual particles (or bodies, for the case of multiple bodies) of the system. This includes even reaction forces that do not appear in the differential equations of motion, as they are eventually eliminated from the formulation by proper manipulation of the scalar equations that are obtained from $\vec{F} = m\vec{a}$. Many examples of this appear in the book, starting in Chapter 2. For instance, in Example 2.5.9 (p. 125), the normal reaction forces N_1 and N_2 appear in the free-body diagrams in Fig. 2.17 (p. 126), but they do not appear in the differential equations that are eventually obtained for the system. In Example 2.7.6 (p. 154), the reaction forces B_1 and B_2 appear in Eqs. (2.7.6a) and (2.7.6b) that are obtained from $\vec{F} = m\vec{a}$, but they do not appear in the differential equation that governs the motion of the system, which is Eq. (2.7.6d) (p. 156). Therefore, one natural question that comes to mind is: "Is there a better way to obtain the differential equations of motion of dynamical systems, by eliminating such forces right at the beginning of the formulation?" The answer is a definitive "yes," and the approach that makes that possible is a part of *analytical dynamics*. It is a powerful approach and surprisingly easy to use. Analytical dynamics is based on a few fundamental principles, starting with d'Alembert's principle, due to Jean Le Rond d'Alembert (1717-1783), and is based on work that was mostly developed by Joseph Louis Lagrange (1736-1813), by William Rowan Hamilton (1805-1865), and by Karl Gustav Jacobi (1804-1851). In addition to his far-reaching work on rigid body dynamics, Leonhard Euler (1707-1783) also did pioneering work on calculus of variations, which was contemporaneously advanced by Lagrange as applied to dynamics.

Analytical dynamics is a work-energy approach and, as seen in this chapter, its fundamental quantities for obtaining the differential equations of motion of dynamical systems are the *virtual work* (defined in Section 7.2) done by the forces acting on the entire system and the *kinetic energy* of the motion of the entire system. Thus, formulation of expressions for acceleration is also avoided. Analytical dynamics views the system in its entirety, instead of looking at free-body diagrams of individual parts. Just like analytic geometry deals with geometry problems using calculus, *analytical dynamics* deals with dynamics of both simple and complex systems in a more mathematical manner than the vectorial approach.

7.1   Virtual Variations and Virtual Displacements

In basic calculus, the concepts of a function and the infinitesimal change and differential of a function are familiar. In analytical dynamics, the concept of *virtual variation* plays a significant role and *functionals* also appear. These concepts, which are from the *calculus of variations*, are compared, in the following, with their analogies from calculus.

Analogies between calculus and calculus of variations:

1. In calculus, for a *function* $y(x)$, a value of x yields a value of $y(x)$. The *argument* x of a function $y(x)$ is the *independent variable*. For example, $y_1(x) = \sin(x)$ and $y_2(x) = \sqrt{1+x}$ are functions of x since a value of the independent variable x yields a value for both $y_1(x)$ and $y_2(x)$.

   In calculus of variations, $f(y(x))$ is called a *functional* when a function $y(x)$ of an independent variable x yields a value for f. Different functions $y(x)$ generally yield a different value for the functional f. For example, $f = \int_0^6 [(dy/dx)^2 - y^2 + x^2]\, dx$, where $y(x)$ is a function of x, is a functional. The value of f depends, of course, on the function $y(x)$ in the integrand. It turns out (as it will be seen in Section 7.4.2) that functionals of the form $\int_{x_1}^{x_2} g(dy/dx, y, x)\, dx$, where y is an unknown function of the independent variable x, are the ones that appear in dynamics of point masses and rigid bodies.

2. In calculus, if x denotes an independent variable, a change in x causes, of course, a change in a function $y(x)$. The quantity dx is called the *infinitesimal change of x*, which is equal to the difference between two infinitesimally close values of x. The corresponding infinitesimal change of a function $y(x)$ is denoted by dy and is called the *differential of the function*. For a function $y(x)$ of only one variable x, for example, one has $dy = (dy/dx)dx$. The infinitesimal changes dx and dy are illustrated in Fig. 7.1a.

   In calculus of variations, an infinitesimal change of the function $y(x)$ in the functional $f(y(x))$ to another function $y_{\text{new}}(x)$ is called the *virtual variation*

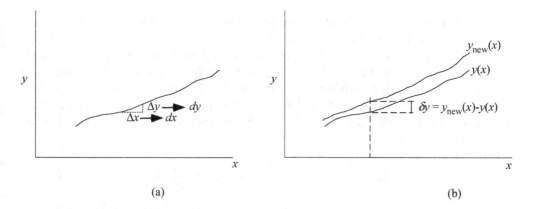

Figure 7.1: (a) Infinitesimal changes dx and dy of a function $y(x)$ and (b) virtual change δy of a function $y(x)$ that may appear in a functional $f(y(x))$.

of $y(x)$ and is denoted by δy (and pronounced "del y" instead of "delta y"). Notice that, in calculus, the change of a function $y(x)$ involves the change of x. In contrast, the change of a functional in calculus of variations involves the change of the argument $y(x)$ without changing the independent variable x. The virtual variation $\delta y = y_{\text{new}}(x) - y(x)$ of a function $y(x)$ that may appear in a functional $f(y(x))$ is illustrated in Fig. 7.1b.

3. In calculus, the values of x for which $dy(x) = 0$ are called the values that make y stationary. When there are no restrictions on x, the points where $dy = 0$ are either maximum or minimum or inflection points of $y(x)$.

   In calculus of variations, the functions $y(x)$ for which the variation $\delta f(y(x))$ of a functional f is zero are called the functions that make f stationary.

Functionals   As indicated, the value of a functional $f(y(x))$ depends on a function $y(x)$, i.e., a functional associates a number to a specific function. For example, the integral $I = \int_{x_1}^{x_2} \sqrt{1 + (dy/dx)^2}\, dx$ is a functional whose meaning is the length of a curve $y(x)$ defined between the points $x = x_1$ and $x = x_2$. One might want to find the curve of minimum length, for example, that connects two given points $(x = x_1, y = y_1)$ and $(x = x_2, y = y_2)$ on a plane. Such a curve is the one that yields $\delta I = 0$, and it happens to be a straight line. The integral $\int_0^6 [(dy/dx)^2 - y^2 + x^2]\, dx$ is also a functional, which depends on the function $y(x)$ and on its derivative dy/dx; it evaluates to different values for different functions $y(x)$.

Total differential df **and virtual variation** δf Notice the difference between the infinitesimal change df of a function $f(q_1, q_2, \ldots, q_n; x)$ that depends on n functions $q_1(x)$, $q_2(x), \ldots, q_n(x)$ (and maybe also depend explicitly on the independent variable x) and the virtual variation δf (in dynamics of point masses and rigid bodies, the independent variable x is the time t). The total differential df is the actual infinitesimal change of f, and it is due to changes in $q_1, q_2, \ldots, q_n$ and also to a change in the independent variable x. As reviewed in Section 1.12 of Chapter 1 (p. 38):

$$df(q_1, q_2, \ldots, q_n; x) = \frac{\partial f}{\partial q_1}dq_1 + \frac{\partial f}{\partial q_2}dq_2 + \ldots + \frac{\partial f}{\partial q_n}dq_n + \frac{\partial f}{\partial x}dx \qquad (7.1.1)$$

The coefficient of dq_1, for example, in Eq. (7.1.1) is the partial derivative of f with respect to q_1 (i.e., the derivative obtained when all the other arguments $q_2, \ldots, q_n$ and the independent variable x are held constant).

On the other hand, the virtual variation of $f(q_1, \ldots, q_n; x)$ is a variation that is obtained by holding the independent variable constant, i.e., "freezing" it, but changing the argument functions $q_1, \ldots, q_n$. The new symbol δ is used for such variations, and the virtual variation of f (which is due to virtual variations in $q_1, \ldots, q_n$) is then

$$\delta f(q_1, \ldots, q_n, \dot{q}_1, \ldots, \dot{q}_n; t) = \frac{\partial f}{\partial q_1}\delta q_1 + \ldots + \frac{\partial f}{\partial q_n}\delta q_n \qquad (7.1.2)$$

In dynamics one encounters the kinetic energy T of the motion, which depends on n functions $q_1(t)$, $q_2(t), \ldots, q_n(t)$, where t is time, and on the time derivatives $\dot{q}_1(t)$, $\dot{q}_2(t), \ldots, \dot{q}_n(t)$, and maybe also depends explicitly on the independent variable t, i.e., $T(q_1, \ldots, q_n, \dot{q}_1, \ldots \dot{q}_n; t)$. The total differential dT and the virtual variation δT of such a function is obtained as follows:

$$d(T(q_1, \ldots, q_n, \dot{q}_1, \ldots, \dot{q}_n; t)) = \frac{\partial T}{\partial q_1}dq_1 + \ldots + \frac{\partial T}{\partial q_n}dq_n + \frac{\partial T}{\partial \dot{q}_1}d\dot{q}_1$$

$$+ \ldots + \frac{\partial T}{\partial \dot{q}_n}d\dot{q}_n + \frac{\partial T}{\partial t}dt \qquad (7.1.3)$$

$$\delta(T(q_1, \ldots, q_n, \dot{q}_1, \ldots, \dot{q}_n; t)) = \frac{\partial T}{\partial q_1}\delta q_1 + \ldots + \frac{\partial T}{\partial q_n}\delta q_n + \frac{\partial T}{\partial \dot{q}_1}\delta \dot{q}_1$$

$$+ \ldots + \frac{\partial T}{\partial \dot{q}_n}\delta \dot{q}_n \qquad (7.1.4)$$

The variables $q_1, \ldots, q_n$ are the ones that are chosen for describing the motion of any particular dynamical system under investigation. They may be the coordinates of a point or the angle a rotating line makes with a known reference line. The virtual variations of such variables are called *virtual displacements*, and the virtual variations of their time derivatives (i.e, $\delta \dot{q}_1, \ldots, \delta \dot{q}_n$) are called *virtual velocities*. Such variables become a known function of the independent variable (time) only by integrating the differential equations that are obtained for them at the end of the *formulation phase* of a problem (i.e., the phase that consists of formulating the differential equations of motion for the system), after which the *analysis of the motion* phase starts. The virtual variations of such unknown variables are then arbitrary; i.e., they can be anything one desires, as long as they are infinitesimally small variations and do not violate any constraint relation that might exist between them (such as $q_1^2 + q_2^2 = $ constant, or any other relation). Often one is able to choose a set of n independent (i.e., not restricted by any constraint relation) variables $q_1, \ldots, q_n$ to work with in a dynamics problem. The unknown variables $q_1, \ldots, q_n$ that are necessary for describing the motion of a system are called the *generalized coordinates* of the motion. Generally, generalized coordinates are independent of each other. However, there are cases when one is forced to work with constrained coordinates, as in Example 7.9.2 (p. 567).

If a variable were known a priori, say $q_1 = \sin t$ (where, as before, t is the independent variable time), then one would have $dq_1 = (\cos t)dt$ and $\delta q_1 = (\cos t)\delta t = 0$ because $\delta t = 0$ by definition of a virtual variation. In other words, the virtual variation of any known (i.e., specified a priori) function of the independent variable is zero.

Relations between $d(\delta y)$ and $\delta(dy)$, and between $d(\delta y)/dt$ and $\delta(dy/dt)$
The virtual variation of y is defined as (see Fig. 7.1b):

$$\delta y \overset{\Delta}{=} y_{\text{new}} - y \tag{7.1.5}$$

In a similar manner, the virtual variations of dy and of dy/dt are defined as

$$\delta(dy) \overset{\Delta}{=} dy_{\text{new}} - dy \quad ; \quad \delta\left(\frac{dy}{dt}\right) \overset{\Delta}{=} \frac{dy_{\text{new}}}{dt} - \frac{dy}{dt} \tag{7.1.6}$$

These definitions then yield the following results involving "a del of a d" and "a d of a del":

$$\delta(dy) \overset{\Delta}{=} dy_{\text{new}} - dy = d(y_{\text{new}} - y) = d(\delta y) \tag{7.1.7}$$

$$\delta\left(\frac{dy}{dt}\right) \overset{\Delta}{=} \frac{d}{dt}y_{\text{new}} - \frac{d}{dt}y = \frac{d}{dt}(y_{\text{new}} - y) = \frac{d}{dt}(\delta y) \tag{7.1.8}$$

Conclusion: The operators d and δ are exchangeable.

The use of this conclusion will be crucial in Section 7.4.2.

7.2 Virtual Work and Generalized Forces, and Constraint Forces

> The material presented here, concerning generalized forces and potential energy, parallels that presented in Section 2.8 of Chapter 2. Therefore, reference will be made to that section when appropriate.

If $\vec{r}$ denotes the absolute position vector of a point P that is moving in space, i.e., the position vector of P relative to an inertial point O, the infinitesimal displacement of P relative to O is the vector $d\vec{r} \overset{\Delta}{=} \vec{r}(t + dt) - \vec{r}(t)$ and the virtual displacement (vector) of P is the vector $\delta\vec{r}$. By definition of a virtual displacement $\delta\vec{r} \overset{\Delta}{=} \vec{r}_{\text{new}}(t) - \vec{r}(t)$ is an arbitrary infinitesimal change of the position of P that one makes at will, without the passage of the independent variable time. If a force $\vec{F}$ is acting at that point P, the virtual work done by that force, denoted by δW, is the scalar quantity defined as the dot product:

$$\delta W \overset{\Delta}{=} \vec{F} \bullet \delta\vec{r} \tag{7.2.1}$$

For a single particle P moving along a given surface in space, one of the forces acting on the particle is a reaction force $\vec{N}$ that is perpendicular to that surface at the point of contact with the particle. That reaction force is the one that is responsible for the motion of the particle to be constrained to that given surface, which is called the *surface of constraint*. The infinitesimal work $\vec{N} \bullet (d\vec{r})$ done by that constraint force is zero if the surface is not moving in space. However, $\vec{N} \bullet (d\vec{r})$ is not zero if the surface itself is given a known motion in any direction in space since the actual infinitesimal displacement $d\vec{r}$ of point P in this latter case is not necessarily perpendicular to the constraint force $\vec{N}$, which is still a force normal to the moving surface of constraint where the motion of particle P takes place. If, instead of working with the actual infinitesimal displacement $d\vec{r}$, one works with the virtual displacement $\delta\vec{r}$, the virtual work $\vec{N} \bullet (\delta\vec{r})$ done by the reaction $\vec{N}$ is equal to zero because as one freezes the independent variable t the virtual displacement $\delta\vec{r}$, which must not violate any constraint, must be along the surface where P moves on. In other words, $N \bullet (\delta\vec{r}) = 0$ because $\delta\vec{r}$ is normal to $\vec{N}$. By working with virtual displacements, one then eliminates that constraint force from the formulation.

In this example, there must also be an external force applied to the surface of constraint to make it move in the desired manner. Again, the infinitesimal work dW done by such a force is not zero in general, but the virtual work done by the same force is zero because time is frozen in any virtual displacement. In other words, that virtual displacement would be a virtual displacement of a point whose motion is known, which is equal to zero by definition.

In Newtonian dynamics, forces are classified into internal and external forces. Some internal forces (such as the normal reaction force due to contact between two bodies, the forces acting on any two particles whose distance between them is constant, and the tension in an inextensible bar) do not appear in the differential equations of motion of a body. However, they start showing up early in the formulation of those equations via Newton's second law $\vec{F} = m\vec{a}$.

In analytical dynamics, forces are classified into constraint and nonconstraint forces. Analytical dynamics considers the system in its entirety, without breaking it into parts and without dealing with individual free-body diagrams as done in Newtonian dynamics. Although the virtual work done by a specific constraint force in a free-body diagram may not be zero, the virtual work done by all constraint forces acting on the entire system (i.e., the sum of the virtual work done by every constraint force in the system) is equal to zero. The virtual work done by any force that needs to be applied to a point of the system to impose a *known* motion to that point is also equal to zero. This was illustrated in the example just discussed regarding a particle moving on a given surface. The double pendulum example shown in Fig. 7.2 also illustrates this point.

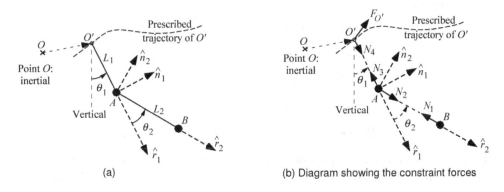

(a)   (b) Diagram showing the constraint forces

Figure 7.2: A double pendulum and free-body diagrams (also showing several unit vectors).

In Fig. 7.2, two mass particles A and B, of masses m_1 and m_2, respectively, are connected by a rigid bar forming a double pendulum that moves on a vertical plane. The support point O' of the pendulum is given a known motion. Fig. 7.2a shows the system, and two sets $\{\hat{r}_1, \hat{n}_1\}$ and $\{\hat{r}_2, \hat{n}_2\}$ of orthogonal unit vectors that will be used in our analysis. Figure 7.2b shows the internal constraint forces $\vec{N}_1$, $\vec{N}_2 = -\vec{N}_1$ (as per Newton's third law), $\vec{N}_3$, and $\vec{N}_4 = -\vec{N}_3$ (also as per Newton's third law). It also shows a force $\vec{F}_{O'}$ that is applied at point O' to make that point move in a desired manner. The virtual work done by all the forces shown in Fig. 7.2 is

$$\delta W = (\vec{F}_{O'} + \vec{N}_4) \bullet \underbrace{\delta(\overrightarrow{OO'})}_{= \, 0}$$

$$+ \underbrace{(\vec{N}_3)}_{= \, -N_3 \hat{r}_1} \bullet \delta(\overrightarrow{OA}) + (\vec{N}_2) \bullet \delta(\overrightarrow{OA}) + (\vec{N}_1) \bullet \delta(\overrightarrow{OB})$$

In this equation, $\delta(\overrightarrow{OO'}) = 0$ by definition of virtual variation, i.e., because the motion $\overrightarrow{OO'}$ of O' is prescribed. Notice that the infinitesimal displacement of O', which is $d(\overrightarrow{OO'}) = [d(\overrightarrow{OO'})/dt]\,dt$ is not zero and, therefore, the infinitesimal work $(\vec{F}_{O'} + \vec{N}_4) \bullet d(\overrightarrow{OO'})$ is not zero.

Since (see Fig. 7.2) $\overrightarrow{OA} = \overrightarrow{OO'} + \overrightarrow{O'A}$, $\overrightarrow{O'A} = L_1 \hat{r}_1$, and $d(\overrightarrow{O'A})/dt = L_1 \dot{\theta}_1 \hat{n}_1$ [which discloses that $\delta(\overrightarrow{O'A}) = (L_1 \delta\theta_1)\hat{n}_1$], we then obtain

$$\delta(\overrightarrow{OA}) = \delta(\overrightarrow{OO'}) + \delta(\overrightarrow{O'A}) = (L_1 \delta\theta_1)\hat{n}_1$$

Also, since for the vectors $\vec{N}_2$ and $\vec{N}_1$, $\vec{N}_2 = -\vec{N}_1$ (see Fig. 7.2), the expression for δW then reduces to

$$\delta W = \underbrace{(-N_3 \hat{r}_1) \bullet [(L_1 \delta\theta_1)\hat{n}_1]}_{= \, 0 \text{ because } \hat{r}_1 \, \bullet \, \hat{n}_1 \, = \, 0} + \vec{N}_1 \bullet [\delta(\overrightarrow{OB}) - \delta(\overrightarrow{OA})]$$

$$= \vec{N}_1 \bullet \underbrace{\delta(\overrightarrow{OB} - \overrightarrow{OA})}_{= \, \overrightarrow{AB} \text{ (see Fig. 7.2)}} = \vec{N}_1 \bullet \delta(\overrightarrow{AB})$$

$$= -N_1 \hat{r}_2 \bullet [L_2(\delta\theta_1 + \delta\theta_2)\hat{n}_2] = 0 \qquad \text{since } \hat{r}_2 \bullet \hat{n}_2 = 0$$

These calculations and inspection of Fig. 7.2 also allow for the following explanations of why the virtual work of all the forces shown in that figure is equal to zero.

- For the pair of forces $\vec{N}_1$ and $\vec{N}_2 = -\vec{N}_1$, the virtual work is $\vec{N}_1 \bullet \delta(\overrightarrow{OB}) + \vec{N}_2 \bullet \delta(\overrightarrow{OA}) = \vec{N}_1 \bullet \delta(\overrightarrow{OB} - \overrightarrow{OA})$. But $\delta(\overrightarrow{OB} - \overrightarrow{OA})$ is the virtual displacement $\delta(\overrightarrow{AB})$ of point B relative to point A without violating the constraint $|\overrightarrow{AB}| = \text{constant} = L_2$. Thus, such a displacement takes place along a circle of radius L_2 centered at A. If the pendulum were moving in three-dimensional space, that displacement would be along a sphere of radius L_2 centered at A. That surface is the *surface of constraint* (which is reduced to a circle for the two-dimensional case) for the particle located at point B. The constraint forces $\vec{N}_1$ and $\vec{N}_2 = -\vec{N}_1$ are always perpendicular to that surface of constraint and, therefore, the virtual work $\vec{N}_1 \bullet \delta(\overrightarrow{AB})$ has to be equal to zero.

- The same explanation applies to the virtual work done by the forces $\vec{N}_3$ and $\vec{N}_4 = -\vec{N}_3$, which is $\vec{N}_3 \bullet \delta(\overrightarrow{O'A})$. If the motion were three-dimensional, the surface of constraint in this case would be the sphere of radius L_1 centered at point O' (again, for the two-dimensional case, that surface of constraint reduces to a circle).

- The virtual work $\vec{F}_{O'} \bullet \delta(\overrightarrow{OO'})$ done by the force $\vec{F}_{O'}$ is equal to zero for a reason that is entirely independent of the trajectory of O'. It is zero because the motion of point O' is prescribed (i.e., it is a known desired motion) and, thus, $\delta(\overrightarrow{OO'}) = 0$. Therefore, $\vec{F}_{O'} \bullet \delta(\overrightarrow{OO'}) = 0$ no matter the direction of $\vec{F}_{O'}$.

For a system composed of N particles (or, in general, a system composed of discrete particles and of rigid bodies), the expression for the virtual work δW done by a set of forces $\vec{F}_1, \vec{F}_2, \ldots, \vec{F}_N$ acting on such a system is defined as the following sum of dot products:

$$\delta W \overset{\Delta}{=} \vec{F}_1 \bullet \delta\vec{r}_1 + \vec{F}_2 \bullet \delta\vec{r}_2 + \ldots + \vec{F}_N \bullet \delta\vec{r}_N \qquad (7.2.2)$$

The motion of such a system is described by a set of n variables $q_1(t)$, $q_2(t)$, $\ldots$, $q_n(t)$, each of which must be single valued for any specific value of t. Frequently, such variables are either distances or angles, but generalized coordinates are not restricted to such choices. They may also be chosen to be something else, such as the sine of an angle (if that is more advantageous for a specific problem) or the amplitudes in a Fourier expansion of some variable. As mentioned earlier, they are called the *generalized coordinates* of the motion.

The absolute position vector $\vec{r}_i$ of any particle P_i is a function of the generalized coordinates $q_1, \ldots, q_n$ and may also be an explicit function of the time t, i.e.,

$$\vec{r}_i = \vec{r}_i(q_1, q_2, \ldots, q_n; t) \qquad (7.2.3)$$

Therefore, after each dot product is performed in Eq. (7.2.2) for any particular problem, the expression that is obtained for the virtual work δW will be of the following form:

$$\delta W = \sum_{i=1}^{N} \vec{F}_i \bullet \delta\vec{r}_i = Q_{q_1}(q_1, q_2, \ldots, q_n, \dot{q}_1, \dot{q}_2, \ldots, \dot{q}_n; t)\,\delta q_1$$

$$+ \quad Q_{q_2}(q_1, q_2, \ldots, q_n, \dot{q}_1, \dot{q}_2, \ldots, \dot{q}_n; t)\,\delta q_2$$

$$+ \quad \ldots$$

$$+ \quad Q_{q_n}(q_1, q_2, \ldots, q_n, \dot{q}_1, \dot{q}_2, \ldots, \dot{q}_n; t)\,\delta q_n \qquad (7.2.4)$$

which is the same form of Eq. (2.8.2) (p. 188) without the term in dt and with virtual displacements $\delta q_1, \ldots, \delta q_n$ instead of the actual infinitesimal displacements $dq_1, \ldots, dq_n$.

The coefficients $Q_{q_1}, Q_{q_2}, \ldots, Q_{q_n}$ of each virtual displacement $\delta q_1, \delta q_2, \ldots,$ δq_n are called the *generalized forces* associated with the generalized coordinates $q_1, q_2, \ldots$. A generalized force Q_{q_k} has the unit of force if q_k is a linear displacement and the unit of a moment if q_k is an angular displacement. Generally, it is a relatively simple matter to obtain the generalized forces $Q_{q_1}, Q_{q_2}, \ldots, Q_{q_n}$, since all one has to do is dot multiply all the forces by the respective virtual displacements of the point to which they are applied.

> The generalized forces are one of the two fundamental quantities that are needed to obtain the differential equations of motion of a dynamical system using the Lagrangian formulation of dynamics. As we will see shortly, the other fundamental quantity is the kinetic energy of the motion of the system.

In general, δW is not the "del" of (i.e., the virtual variation of) a function W. However, the expression for δW can always be separated into the sum of two parts: one part that *is* the del of a function U (to be determined) and is conventionally written with a minus sign as $-\delta(U)$, and another part that is not the del of anything but is simply whatever is left out of the expression for δW. Accordingly, one can then write[1]

$$\delta W \;=\; \underbrace{-\delta(U) + \text{whatever is left out of the expression of }\delta W}_{\triangleq (\delta W)_{nc}} \qquad (7.2.5)$$

Notice that $(\delta W)_{nc}$ consists of all the terms in δW that cannot be obtained from the "del of a function." The minus sign in the term $-\delta(U)$ in Eq. (7.2.5) is a convention that is adopted universally. With such a sign convention, the function U, which can only be a function of the generalized coordinates $q_1, \ldots, q_n$, and may also be an explicit function of time t (but never depend on the generalized velocities $\dot{q}_1, \ldots, \dot{q}_n$ because the expression for δW does not involve any $\delta\dot{q}$), is called the *potential energy* of the motion.

The remaining part of δW in Eq. (7.2.5), which is denoted by $(\delta W)_{nc}$ and is of the form

$$(\delta W)_{nc} \triangleq (Q_{q_1})_{nc}\delta q_1 + \ldots + (Q_{q_n})_{nc}\delta q_n$$

is the nonpotential part of δW. The coefficients of $\delta q_1, \ldots, \delta q_n$ in $(\delta W)_{nc}$, i.e., $(Q_{q_1})_{nc}, \ldots, (Q_{q_n})_{nc}$, are called the *generalized nonconservative forces* acting on the system.

---

[1]$\delta(U)$, with the parentheses to make it clear that this is the *del* of a function, is read as "del of U," where U is a function of $q_1, q_2, \ldots, q_n$ and may also be an explicit function of time t.

The expression of $\delta(U)$ for any particular problem is of the form

$$\delta(U) = \frac{\partial U}{\partial q_1}\delta q_1 + \ldots + \frac{\partial U}{\partial q_n}\delta q_n$$

and the quantities $-\partial U/\partial q_1, \ldots, -\partial U/\partial q_n$ are called the *generalized conservative forces* acting on the system. Since the potential energy U is introduced through its partial derivatives $\partial U/\partial q_1, \ldots, \partial U/\partial q_n$, the expression for U is, then, obtained by integrations. Therefore, its expression will always contain an arbitrary constant of integration (which, actually, may also be an arbitrary function of time t). However, since only changes in potential energy are important, because motion is always accompanied by changes, the constant of integration is irrelevant. In this book, it will always be set to zero for convenience.

Several examples involving the development of expressions for potential energy were presented at the end of Section 2.8 (p. 191).[2] Additional examples are presented here.

Example 7.2.1 The double pendulum shown in Fig. 7.3 consists of two rigid massless bars $O'A$ and AB of lengths L_1 and L_2, respectively. A particle of mass M_1 is placed at A and another of mass M_2 is placed at the end B of bar AB. Point O' represents a hinge that is forced to move (by an external force) with a known motion represented by the dashed path in Fig. 7.3, and the motion of the system takes place on a vertical plane. The torsional spring connected to bars $O'A$ and AB is linear with stiffness k N·m/rad and is unstressed when $\theta_2 = \theta_u$ rad. Obtain the expression for the generalized forces acting on the system and for the potential energy of the motion.

■ Solution
The only forces and moments acting on this system for which the virtual work is not zero are the gravitational forces whose magnitudes are $M_1 g$ and $M_2 g$ (treating the gravitational field as uniform so that the gravitational forces are independent of distance), which are acting at points A and B, respectively, and the moment due to the torsional spring. Since they are only a function of position, their virtual work can be obtained from a potential energy U. From the presentation in Section 2.8 of Chapter 2, one could obtain the following expression for the potential energy simply by inspection of Fig. 7.3:

$$U = -M_1 g L_1 \cos\theta_1 - M_2 g[L_1 \cos\theta_1 + L_2 \cos(\theta_1 + \theta_2)] + \frac{k}{2}(\theta_2 - \theta_u)^2 \quad (7.2.1a)$$

---

[2]Although in Chapter 2, the concept of virtual variation was not introduced at that level of presentation, the generation of the expression for U using Eq. (2.8.6) (p. 191) is fully consistent with the material presented here. The expression for δW corresponds to the one that is obtained by replacing d by δ in Eq. (2.8.6) and setting $\delta t = 0$ in the resulting equation.

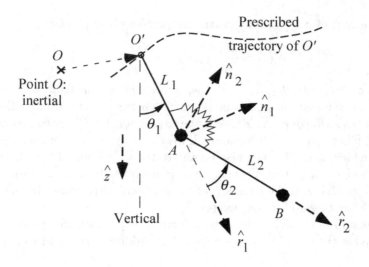

Figure 7.3: A double pendulum and several convenient unit vectors for Example 7.2.1.

from which the generalized forces Q_{θ_1} associated with $\delta\theta_1$, and Q_{θ_2} associated with $\delta\theta_2$, are

$$Q_{\theta_1} = -\frac{\partial U}{\partial \theta_1} = -(M_1 + M_2)gL_1 \sin\theta_1 - M_2 g L_2 \sin(\theta_1 + \theta_2) \quad (7.2.1\text{b})$$

$$Q_{\theta_2} = -\frac{\partial U}{\partial \theta_2} = -M_2 g L_2 \sin(\theta_1 + \theta_2) - k(\theta_2 - \theta_u) \quad (7.2.1\text{c})$$

Instead of obtaining the generalized potential forces from the expression of potential energy U when U can be written by inspection of a figure (which is actually not possible to do for examples that are either not familiar or, more likely, more complicated), one would formulate the expression for the virtual work from its definition. This, then, would yield all the generalized forces acting on the system, and also the expression for potential energy (if needed). This is done as follows. Referring to Fig. 7.3, the following expression for δW is written, where the third term is the virtual work done by the spring force.

$$\delta W = M_1 g \hat{z} \bullet \delta \vec{r}_1 + M_2 g \hat{z} \bullet \delta \vec{r}_2 - k(\theta_2 - \theta_u)\delta\theta_2 \quad (7.2.1\text{d})$$

But $\vec{r}_1 = L_1 \hat{r}_1$ and $\vec{v}_1 = d\vec{r}_1/dt = L_1 \dot{\theta}_1 \hat{n}_1$, and either one of these expressions gives $\delta \vec{r}_1 = (L_1 \delta\theta_1)\hat{n}_1$. Also, $\vec{r}_2 = \vec{r}_1 + L_2 \hat{r}_2$ and $\vec{v}_2 = d\vec{r}_2/dt = d\vec{r}_1/dt + L_2(\dot{\theta}_1 + \dot{\theta}_2)\hat{n}_2$, and either one of these expressions gives $\delta\vec{r}_2 = \delta\vec{r}_1 + L_2(\delta\theta_1 + \delta\theta_2)\hat{n}_2$. Therefore, Eq. (7.2.1d) yields

$$
\delta W = (M_1 + M_2)gL_1(\delta\theta_1)\underbrace{\hat{z}\bullet\hat{n}_1}_{=-\sin\theta_1} + M_2gL_2(\delta\theta_1 + \delta\theta_2)\underbrace{\hat{z}\bullet\hat{n}_2}_{=-\sin(\theta_1+\theta_2)} -k(\theta_2-\theta_u)\delta\theta_2
$$

$$
\triangleq Q_{\theta_1}\delta\theta_1 + Q_{\theta_2}\delta\theta_2 \tag{7.2.1e}
$$

As indicated in this equation, the generalized forces Q_{θ_1} and Q_{θ_2} are, by definition, the coefficients of $\delta\theta_1$ and $\delta\theta_2$ in the expression for δW.

The search for the expression for the potential energy involves testing the expressions for the generalized forces, as done here. For this example, we have

$$
\frac{\partial Q_{\theta_1}}{\partial\theta_2} = -M_2gL_2\cos(\theta_1 + \theta_2) \tag{7.2.1f}
$$

$$
\frac{\partial Q_{\theta_2}}{\partial\theta_1} = -M_2gL_2\cos(\theta_1 + \theta_2) \tag{7.2.1g}
$$

Since $\partial Q_{\theta_1}/\partial\theta_2 = \partial Q_{\theta_2}/\partial\theta_1$, the entire expression for δW is of the form $\delta W = -\delta(U) = -[(\partial U/\partial\theta_1)\delta\theta_1 + (\partial U/\partial\theta_2)\delta\theta_2]$, where

$$
\frac{\partial U}{\partial\theta_1} = -Q_{\theta_1} = (M_1 + M_2)gL_1\sin\theta_1 + M_2gL_2\sin(\theta_1 + \theta_2) \tag{7.2.1h}
$$

$$
\frac{\partial U}{\partial\theta_2} = -Q_{\theta_2} = M_2gL_2\sin(\theta_1 + \theta_2) + k(\theta_2 - \theta_u) \tag{7.2.1i}
$$

To find U, start by integrating either $\partial U/\partial\theta_1$ with respect to θ_1 or $\partial U/\partial\theta_2$ with respect to θ_2. Integration of $\partial U/\partial\theta_2$ with respect to θ_2 yields, except for an arbitrary constant of integration that is set to zero for convenience,

$$
U = -M_2gL_2\cos(\theta_1 + \theta_2) + \frac{k}{2}(\theta_2 - \theta_u)^2 + f(\theta_1) \tag{7.2.1j}
$$

where $f(\theta_1)$ is to be determined so that $\partial U/\partial\theta_1$ matches the expression in Eq. (7.2.1h), i.e.,

$$
\frac{\partial U}{\partial\theta_1} = M_2gL_2\sin(\theta_1 + \theta_2) + \frac{df}{d\theta_1} = (M_1 + M_2)gL_1\sin\theta_1 + M_2gL_2\sin(\theta_1 + \theta_2)
$$

$$
\tag{7.2.1k}
$$

This then yields

$$
\frac{df}{d\theta_1} = (M_1 + M_2)gL_1\sin\theta_1
$$

or
$$f(\theta_1) = -(M_1 + M_2)gL_1 \cos\theta_1$$

plus an arbitrary constant of integration that, for convenience, is again set to zero. With $f(\theta_1)$ as determined here, the expression for U given by Eq. (7.2.1j) is the same as the one in Eq. (7.2.1a) that was written by inspection of Fig. 7.3.

Note: Labeling the first approach to find U as an *inspection* process is actually a misnomer since there is only one way to obtain, in general cases, the expression of the potential energy for any problem, i.e., by starting with δW and then proceeding to find which part of δW, if any, can be obtained from a $-\delta(U)$ term.

For a more elaborate version of this example, see Problem 7.7 at the end of this chapter.

Example 7.2.2 In terms of two generalized coordinates q_1 and q_2, the force acting on a particle P is given as

$$\vec{F} = (q_1 q_2^3 + q_1 + q_2^2)\hat{x}_1 + (\tfrac{3}{2}q_1^2 q_2^2 + q_2 + q_1^2)\hat{x}_2$$

while the absolute velocity of P, $\vec{v} = d\vec{r}/dt$, is

$$\vec{v} = (\dot{q}_1 + k\sin t)\hat{x}_1 + \dot{q}_2\hat{x}_2 \ \ \text{m/s}$$

where k is constant and t is nondimensional time. Determine whether $\vec{F}$ is conservative or nonconservative, and obtain the expression for the potential energy of the motion of P.

■ **Solution**
From the expression for the velocity $\vec{v} = d\vec{r}/dt$ we can write

$$d\vec{r} = [dq_1 + (k\sin t)dt]\hat{x}_1 + (dq_2)\hat{x}_2 \tag{7.2.2a}$$

and, therefore, the virtual displacement $\delta\vec{r}$ is simply

$$\delta\vec{r} = (\delta q_1)\hat{x}_1 + (\delta q_2)\hat{x}_2 \tag{7.2.2b}$$

The virtual work done by $\vec{F}$ during a virtual displacement $\delta\vec{r}$ is then

$$\delta W = \vec{F} \bullet \delta\vec{r} = \left(q_1 q_2^3 + q_1 + q_2^2\right)\delta q_1 + \left(\tfrac{3}{2}q_1^2 q_2^2 + q_2 + q_1^2\right)\delta q_2$$

$$\overset{\Delta}{=} Q_{q_1}\delta q_1 + Q_{q_2}\delta q_2 \tag{7.2.2c}$$

The coefficients Q_{q_1} and Q_{q_2} of δq_1 and δq_2, respectively, in the expression for δW are the generalized forces associated with the generalized coordinates q_1 and q_2.

Since $\partial Q_{q_1}/\partial q_2 \neq \partial Q_{q_2}/\partial q_1$, δW is not the del of anything; i.e., there is no function W for which δW is of the form $\delta(W)$. Therefore, the force $\vec{F}$ is classified as nonconservative. The only parts $Q^*_{q_1}$ and $Q^*_{q_2}$ of Q_{q_1} and Q_{q_2}, respectively, for which $\partial Q^*_{q_1}/\partial q_2 = \partial Q^*_{q_2}/\partial q_1$ are $Q^*_{q_1} = q_1 q_2^3 + q_1$ and $Q^*_{q_2} = \frac{3}{2} q_1^2 q_2^2 + q_2$. The virtual work δW^* associated with these parts can be written in terms of a potential energy function $U(q_1, q_2)$ as

$$
\begin{aligned}
\delta W^* &= \left(q_1 q_2^3 + q_1 \right) \delta q_1 + \left(\frac{3}{2} q_1^2 q_2^2 + q_2 \right) \delta q_2 \\
&= - \left(\frac{\partial U}{\partial q_1} \delta q_1 + \frac{\partial U}{\partial q_2} \delta q_2 \right)
\end{aligned}
\tag{7.2.2d}
$$

The expression for the potential U is obtained by integrating the expressions for $\partial U/\partial q_1$ and $\partial U/\partial q_2$ as follows. For $\partial U/\partial q_1$, we have from Eq. (7.2.2d),

$$
\frac{\partial U}{\partial q_1} = -q_1 q_2^3 - q_1
\tag{7.2.2e}
$$

and this gives, by integrating with respect to q_1,

$$
U = -\frac{1}{2} q_1^2 q_2^3 - \frac{1}{2} q_1^2 + f(q_2)
\tag{7.2.2f}
$$

where $f(q_2)$ is to be determined so that $\partial U/\partial q_2$ matches the expression in Eq. (7.2.2g), which follows.

For $\partial U/\partial q_2$, we have from Eq. (7.2.2d), and also from Eq. (7.2.2f),

$$
\frac{\partial U}{\partial q_2} = -\frac{3}{2} q_1^2 q_2^2 - q_2 \underbrace{=}_{\text{from Eq. (7.2.2f)}} -\frac{3}{2} q_1^2 q_2^2 + \frac{df}{dq_2}
\tag{7.2.2g}
$$

and since this gives $df/dq_2 = -q_2$, we then obtain for $f(q_2)$, with the constant of integration set to zero as done previously,

$$
f = -\frac{1}{2} q_2^2
\tag{7.2.2h}
$$

The potential energy U is then, except for an arbitrary constant of integration,

$$
U = -\frac{1}{2} \left(q_1^2 q_2^3 + q_1^2 + q_2^2 \right)
\tag{7.2.2i}
$$

Example 7.2.3 A rigid body is subjected to a nondimensionalized force $\vec{F}$ that is related to three nondimensional generalized coordinates q_1, q_2, and q_3 as $\vec{F} = q_1\hat{x}_1 + 3(\cos q_2)(\sin t)\hat{x}_2 + (q_1\sin q_2)\hat{x}_3$, where t is nondimensional time. The force $\vec{F}$ is applied at a point P whose nondimensional absolute position vector $\vec{r}$ is $\vec{r} = q_1\hat{x}_1 + (2\cos t)\hat{x}_2$. The unit vectors form an orthogonal triad, with $\hat{x}_3 = \hat{x}_1 \times \hat{x}_2$, and that triad rotates in space with nondimensional absolute angular velocity $\vec{\omega} = (5\cos t)\hat{x}_1 + \dot{q}_2(3\hat{x}_2 + 2\hat{x}_3)$.

    a. Formulate the expression for the infinitesimal work and for the virtual work done by $\vec{F}$. What are the expressions for the generalized forces acting on the body?

    b. Is $\vec{F}$ a conservative or nonconservative force? What is the expression for the potential energy of the motion?

    c. If $\vec{F}$ is a nonconservative force, what are the expressions for the nonconservative generalized forces?

■ **Solution**

    a. The absolute velocity $\vec{v}$ of point P is

$$
\begin{aligned}
\vec{v} = \frac{d}{dt}\vec{r} = \frac{d}{dt}&[q_1\hat{x}_1 + (2\cos t)\hat{x}_2] = [\dot{q}_1\hat{x}_1 - (2\sin t)\hat{x}_2] \\
&+ \vec{\omega} \times [q_1\hat{x}_1 + (2\cos t)\hat{x}_2] = (\dot{q}_1 - 4\dot{q}_2\cos t)\,\hat{x}_1 \\
&+ (2q_1\dot{q}_2 - 2\sin t)\,\hat{x}_2 + (10\cos^2 t - 3q_1\dot{q}_2)\,\hat{x}_3 \qquad (7.2.3a)
\end{aligned}
$$

and, therefore, the infinitesimal displacement $d\vec{r} = \vec{v}\,dt$ is

$$
d\vec{r} = [dq_1 - 4(\cos t)dq_2]\,\hat{x}_1 + [2q_1 dq_2 - 2(\sin t)dt]\,\hat{x}_2 + [10(\cos^2 t)dt - 3q_1 dq_2]\,\hat{x}_3
$$

This then yields the following expression for the virtual displacement $\delta\vec{r}$:

$$
\delta\vec{r} = [\delta q_1 - 4(\cos t)\delta q_2]\,\hat{x}_1 + (2q_1\delta q_2)\,\hat{x}_2 - (3q_1\delta q_2)\,\hat{x}_3 \qquad (7.2.3b)
$$

The infinitesimal work dW done by the force $\vec{F}$ is then

$$
\begin{aligned}
dW \;\overset{\Delta}{=}\; \vec{F}\bullet(d\vec{r}) &= q_1\,dq_1 + [6q_1(\cos q_2)\sin t - 4q_1\cos t - 3q_1^2\sin q_2]dq_2 \\
&\quad + [10q_1(\sin q_2)\cos^2 t - 6(\cos q_2)\sin^2 t]dt \\
&\overset{\Delta}{=}\; Q_{q_1}dq_1 + Q_{q_2}dq_2 + a_0 dt
\end{aligned}
$$

while the virtual work δW done by $\vec{F}$ is

$$\delta W \stackrel{\triangle}{=} \vec{F} \bullet (\delta \vec{r}) = q_1 \delta q_1 + \left[6q_1(\cos q_2) \sin t - 4q_1 \cos t - 3q_1^2 \sin q_2\right] \delta q_2$$
$$= Q_{q_1} \delta q_1 + Q_{q_2} \delta q_2 \qquad (7.2.3\text{c})$$

The coefficients of δq_1 and δq_2 seen in the expression for the virtual work δW (and also in the expression for dW) are the generalized forces Q_{q_1} and Q_{q_2} associated with the virtual displacements δq_1 and δq_2, respectively.

b. Inspection of Eq. (7.2.3c) discloses that

$$\frac{\partial Q_{q_1}}{\partial q_2} = 0 \qquad \frac{\partial Q_{q_2}}{\partial q_1} = 6(\cos q_2) \sin t - 4 \cos t - 6q_1 \sin q_2 \qquad (7.2.3\text{d})$$

Since $\partial Q_{q_1}/\partial q_2 \neq \partial Q_{q_2}/\partial q_1$, the virtual work δW is not the del of anything. Therefore, the force $\vec{F}$ is nonconservative.

The potential energy U is obtained from a part of δW (if any) that allows one to write

$$\delta W \stackrel{\triangle}{=} - \underbrace{\delta(U)} \quad + \quad \text{whatever is left out of the expression of } \delta W$$
$$= \tfrac{\partial U}{\partial q_1} \delta q_1 + \tfrac{\partial U}{\partial q_2} \delta q_2$$

Since $\partial Q_{q_1}/\partial q_2 = 0$, with $Q_{q_1} = q_1$, as seen from Eq. (7.2.3c), the term $Q_{q_1} \delta q_1 = q_1 \delta q_1$ can be expressed as

$$q_1 \delta q_1 = \delta \left(\frac{1}{2}q_1^2 + C\right) \qquad (7.2.3\text{e})$$

where C is any desired constant (or even a specified function of time only), which will be set to zero since there is no need to keep it. Therefore, it is seen that a part of δW can be obtained from the potential energy

$$U = -\frac{1}{2}q_1^2 \qquad (7.2.3\text{f})$$

d. The rest of the expression for δW is obtained from the following nonconservative generalized force associated with q_2:

$$(Q_{q_2})_{nc} = 6q_1(\cos q_2) \sin t - 4q_1 \cos t - 3q_1^2 \sin q_2 \qquad (7.2.3\text{g})$$

7.3   The Principles of d'Alembert and of Virtual Work

Consider a system of N particles (which may also form any number of bodies of finite dimension) P_i of mass m_i $(i = 1, 2, \ldots, N)$, subjected to a set of forces. For convenience, let us group the forces acting on particle P_i into two types: the resultant $\vec{F}_{i\_c}$ of the *constraint forces*, if any, and the resultant $\vec{F}_i$ of all other forces. Now, letting $\vec{r}_i$ denote the absolute position vector for the ith particle P_i, as shown in Fig. 7.4, and letting $\vec{v}_i = d\vec{r}_i/dt$ be the absolute velocity of P_i, one can then write according to Newton's second law:

$$\vec{F}_i + \vec{F}_{i\_c} = \frac{d}{dt}(m_i \vec{v}_i) \tag{7.3.1}$$

Notice that, in the most general situation, Eq. (7.3.1) represents $3N$ scalar equations if each particle P_i is able to move in three-dimensional space.

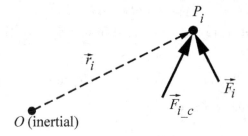

Figure 7.4: Forces acting on a particle.

Although the virtual work of the force of constraint $\vec{F}_{i\_c}$ acting on particle P_i may be nonzero, the virtual work done by all forces of constraint acting together on the entire system is zero. This was illustrated in an example using Fig. 7.2 (p. 505). One can then take advantage of such a fact to eliminate the constraint forces right at the beginning of the formulation of the differential equations of motion for a particular system under investigation. Unless one wants to determine the constraint forces, this is better than carrying them throughout the formulation, which is what happens when free-body diagrams for each part of the system are used as required in the Newtonian approach. Their elimination is accomplished by dot multiplying both sides of Eq. (7.3.1) by the virtual displacement $\delta\vec{r}_i$ of particle P_i and then adding all the resulting vector equations as

$$\sum_{i=1}^{N} \vec{F}_i \bullet \delta\vec{r}_i + \sum_{i=1}^{N} \vec{F}_{i\_c} \bullet \delta\vec{r}_i = \sum_{i=1}^{N} \left[\frac{d}{dt}(m_i \vec{v}_i) \right] \bullet \delta\vec{r}_i$$

The second summation on the left-hand side of this equation is the virtual work done by all the constraint forces acting on the system and, therefore, it is equal to zero. The first summation term is, by definition, the virtual work δW done by all the nonconstraint forces such as the externally applied forces that are not associated with a specified motion, friction forces, internal elastic forces due to springs, etc. Therefore, this equation reduces to

$$\delta W \triangleq \sum_{i=1}^{N} \vec{F}_i \bullet \delta \vec{r}_i = \sum_{i=1}^{N} \left[\frac{d}{dt}(m_i \vec{v}_i) \right] \bullet \delta \vec{r}_i \qquad (7.3.2)$$

For continuous bodies, the mass m_i is replaced by an infinitesimal mass element $dm = \rho \, dV$ of the body, and the summation in Eq. (7.3.2) is replaced by an integral that should span the entire volume of the body. Here, ρ and dV are the volumetric material density and the volume of a mass element of the body, respectively.

Equation (7.3.2) is known as *Lagrange's form of d'Alembert's principle*. For $d(m_i \vec{v}_i)/dt = 0$, i.e., for problems for which $\vec{v}_i = \text{constant} \neq 0$ or for those for which $\vec{v}_i = 0$ (i.e., statics problems), Eq. (7.3.2) reduces to $\delta W = 0$ and, in this case, it is known as the *principle of virtual work*.

As will be shown in the subsequent sections, Eq. (7.3.2) has a far-reaching consequence since it leads to *Hamilton's principle*, from which Lagrange's equation and the differential equations of motion of discrete systems can be readily obtained in a relatively simple manner.[3]

7.4 Hamilton's Principle and Lagrange's Equation

Let us look at Eq. (7.3.2) and the two ways of obtaining the virtual work δW done by the forces $\vec{F}_i$ during a virtual displacement $\delta \vec{r}_i$. By definition, δW is equal to the sum of the dot products of each $\vec{F}_i$ with the corresponding virtual displacement $\delta \vec{r}_i$. As already shown, this simply yields an expression of the form given by Eq. (7.2.4) (p. 507), from which the generalized forces, as defined by that same equation, are extracted. But making use of Newton's second law, Eq. (7.3.2) will also yield another expression for δW. As shown in the following, manipulation of the expression $\sum_{i=1}^{N}[d(m_i \vec{v}_i)/dt] \bullet \delta \vec{r}_i$, and a subsequent integration in time, yields *Hamilton's principle*, which is a fundamental principle in dynamics. Hamilton's principle, in turn, yields *Lagrange's equation* in terms of the n generalized coordinates $q_1, q_2, \ldots, q_n$

---

[3]Another beauty, so to speak, of Hamilton's principle is that it also yields the partial differential equations of motion and boundary condition equations for structural dynamics problems. The principle is widely used in dynamics and also in structural dynamics. The reader who is interested in such applications has much to gain by delving into such an approach.

for any *discrete system,* which is a system of points masses and/or rigid bodies (i.e., systems whose motion are governed by ordinary differential equations).

7.4.1   Hamilton's Principle

To manipulate Eq. (7.3.2), one starts by making use of the general form of the expression for the absolute position vector $\vec{r}_i$ of any point P_i. That expression, which is Eq. (7.2.3) (p. 507), is repeated here for convenience.

$$\vec{r}_i = \vec{r}_i(q_1, q_2, \ldots, q_n; t)$$

The absolute velocity $\vec{v}_i = d\vec{r}_i/dt$ of a particle P_i is then of the following form:

$$\vec{v}_i = \frac{d}{dt}\vec{r}_i = \frac{\partial \vec{r}_i}{\partial q_1}\dot{q}_1 + \frac{\partial \vec{r}_i}{\partial q_2}\dot{q}_2 + \ldots + \frac{\partial \vec{r}_i}{\partial q_n}\dot{q}_n + \frac{\partial \vec{r}_i}{\partial t} = \sum_{k=1}^{n} \frac{\partial \vec{r}_i}{\partial q_k}\dot{q}_k + \frac{\partial \vec{r}_i}{\partial t} \quad (7.4.1)$$

The rest of the derivations in this section will soon involve multiple summations. The matrix notation, introduced next, will be used because it is less cumbersome to write and read.

■ An underlined letter is used for denoting a column matrix, such as

$$\underline{f} = \left[\begin{array}{c} f_1 \\ f_2 \end{array} \right]$$

The transpose of such a column matrix is denoted by $\underline{f}^T$, i.e., $\underline{f}^T = [f_1\ f_2]$.

■ As per the notation just introduced, the n generalized coordinates $q_1, q_2, \ldots, q_n$ are denoted by the elements of an $n \times 1$ column matrix $\underline{q}$ as

$$\underline{q} = \left[\begin{array}{c} q_1 \\ q_2 \\ \vdots \\ q_n \end{array} \right]$$

The transpose of $\underline{q}$ is $\underline{q}^T = [q_1\ q_2\ \cdots\ q_n]$. The explicit dependence of a function $\underline{f}$ on n generalized coordinates $q_1, \ldots, q_n$ (which implicitly depend on time t) and also on t is denoted by $\underline{f}(\underline{q}; t)$.

■ The three scalar components of the absolute position vector $\vec{r}_i \overset{\Delta}{=} x_i\hat{x} + y_i\hat{y} + z_i\hat{z}$ expressed in a triad $\{\hat{x}, \hat{y}, \hat{z} \overset{\Delta}{=} \hat{x} \times \hat{y}\}$ are denoted by

$$\underline{r}_i = \left[\begin{array}{c} x_i \\ y_i \\ z_i \end{array} \right]$$

The transpose of $\underline{r}_i$ is $\underline{r}_i^T = [x_i \ y_i \ z_i]$, where $x_i = x_i(\underline{q}; t)$, $y_i = y_i(\underline{q}; t)$, and $z_i = z_i(\underline{q}; t)$.

■ Similarly, the three scalar components of the absolute velocity vector $\vec{v}_i = d\vec{r}_i/dt \overset{\Delta}{=} v_{ix}\hat{x} + v_{iy}\hat{y} + v_{iz}\hat{z}$ expressed in a triad $\{\hat{x}, \hat{y}, \hat{z} \overset{\Delta}{=} \hat{x} \times \hat{y}\}$ are denoted by

$$\underline{v}_i = \begin{bmatrix} v_{ix} \\ v_{iy} \\ v_{iz} \end{bmatrix}$$

The transpose of $\underline{v}_i$ is $\underline{v}_i^T = [v_{ix} \ v_{iy} \ v_{iz}]$.

Notice that if the unit vector triad $\{\hat{x}, \hat{y}, \hat{z}\}$ is not rotating, $v_{ix} = \dot{x}_i$, $v_{iy} = \dot{y}_i$, and $v_{iz} = \dot{z}_i$, whose matrix representation is simply $\underline{v}_i = d\underline{r}_i/dt$. Otherwise, $\vec{v}_i = \dot{x}_i\hat{x} + \dot{y}_i\hat{y} + \dot{z}_i\hat{z} + \vec{\omega} \times \vec{r}_i$, where $\vec{\omega}$ is the absolute angular velocity of the $\{\hat{x}, \hat{y}, \hat{z}\}$ triad. With $\vec{\omega}$ expressed as $\vec{\omega} = \omega_x\hat{x} + \omega_y\hat{y} + \omega_z\hat{z}$, this would yield $\vec{v}_i = (\dot{x}_i + \omega_y z_i - \omega_z y_i)\hat{x} + (\dot{y}_i + \omega_z x_i - \omega_x z_i)\hat{y} + (\dot{z}_i + \omega_x y_i - \omega_y x_i)\hat{z}$, and its matrix representation has to be

$$\underline{v}_i = \dot{\underline{r}}_i + \begin{bmatrix} 0 & -\omega_z & \omega_y \\ \omega_z & 0 & -\omega_x \\ -\omega_y & \omega_x & 0 \end{bmatrix} \underline{r}_i$$

In view of the differences in the representation of the derivative of a vector shown in this item, a nonrotating unit vector triad $\{\hat{x}, \hat{y}, \hat{z}\}$ will be used to simplify the derivations that will follow. Notice that this choice is made only for these derivations. For solving individual problems using the results obtained from these derivations, the use of rotating unit vectors is, in general, more convenient as it has already been seen in several examples.

■ Using the compact matrix notation just described, the components of the linear momentum of a particle P_i with mass m_i may be written as

$$\begin{bmatrix} m_i v_{ix} \\ m_i v_{iy} \\ m_i v_{iz} \end{bmatrix} = \begin{bmatrix} m_i & 0 & 0 \\ 0 & m_i & 0 \\ 0 & 0 & m_i \end{bmatrix} \begin{bmatrix} v_{ix} \\ v_{iy} \\ v_{iz} \end{bmatrix}$$

We are now ready to operate on Eq. (7.3.2), which is repeated here for convenience, using the compact matrix notation and also using $\vec{F}_i = d(m_i\vec{v}_i)/dt$, as per Newton's second law.

$$\delta W \overset{\Delta}{=} \sum_{i=1}^{N} \vec{F}_i \bullet \delta\vec{r}_i = \sum_{i=1}^{N} \left[\frac{d}{dt}(m_i\vec{v}_i) \right] \bullet \delta\vec{r}_i$$

As shown here, the matrix representation of the first summation term in this equation (which is simply the definition of δW) is $\underline{Q}^T \delta \underline{q}$, i.e.,:

$$\delta W \triangleq \sum_{i=1}^{N} \vec{F}_i \bullet \delta \vec{r}_i = Q_{q_1}\delta q_1 + Q_{q_2}\delta q_2 + \ldots + Q_{q_n}\delta q_n$$

$$= [Q_{q_1} \ Q_{q_2} \ \ldots Q_{q_n}]\delta \underline{q} \triangleq \underline{Q}^T \delta \underline{q} \tag{7.4.2}$$

where

$$\underline{Q} \triangleq \begin{bmatrix} Q_{q_1} \\ Q_{q_2} \\ \vdots \\ Q_{q_n} \end{bmatrix}$$

is the generalized force column matrix.

The second summation term, $\sum_{i=1}^{N}[d(m_i\vec{v}_i)/dt]\bullet\delta\vec{r}_i$, can be represented in matrix notation as

$$\sum_{i=1}^{N}\left[\frac{d}{dt}(m_i\vec{v}_i)\right] \bullet \delta\vec{r}_i = \left[\frac{d}{dt}(M\underline{v})\right]^T \delta\underline{r} \tag{7.4.3}$$

with $\underline{r} = \underline{r}(\underline{q}\,;t)$ and $\underline{v}(\underline{q},\ \dot{\underline{q}}\,;t) = d\underline{r}/dt$ being the following $3N \times 1$ column matrices:

$$\underline{r} = [x_1(\underline{q};t)\ y_1(\underline{q};t)\ z_1(\underline{q};t) \vdots \ldots \vdots x_N(\underline{q};t)\ y_N(\underline{q};t)\ z_N(\underline{q};t)]^T \tag{7.4.4}$$

$$\underline{v} = \frac{d\underline{r}}{dt} = \underbrace{\begin{bmatrix} \frac{\partial x_1}{\partial q_1} & \frac{\partial x_1}{\partial q_2} & \cdots & \frac{\partial x_1}{\partial q_n} \\ \frac{\partial y_1}{\partial q_1} & \frac{\partial y_1}{\partial q_2} & \cdots & \frac{\partial y_1}{\partial q_n} \\ & & \vdots & \\ \frac{\partial z_N}{\partial q_1} & \frac{\partial z_N}{\partial q_2} & \cdots & \frac{\partial z_N}{\partial q_n} \end{bmatrix}}_{\text{Call this matrix } \frac{\partial \underline{r}}{\partial \underline{q}}} \dot{\underline{q}} + \underbrace{\begin{bmatrix} \frac{\partial x_1}{\partial t} \\ \frac{\partial y_1}{\partial t} \\ \vdots \\ \frac{\partial z_N}{\partial t} \end{bmatrix}}_{\text{This is } \frac{\partial \underline{r}}{\partial t}} = \frac{\partial \underline{r}}{\partial \underline{q}}\dot{\underline{q}} + \frac{\partial \underline{r}}{\partial t} \tag{7.4.5}$$

and M being the $3N \times 3N$ symmetric mass matrix defined as

$$M = \begin{bmatrix} m_1 & 0 & 0 & \vdots & & & & & & \\ 0 & m_1 & 0 & \vdots & & & & & & \\ 0 & 0 & m_1 & \vdots & & & & & & \\ \cdots & \cdots & \cdots & & \ddots & & & & & \\ & & & & & \ddots & & \text{zeros} & & \\ & \text{zeros} & & & & & \ddots & & & \\ & & & & & & & \cdots & \cdots & \cdots \\ & & & & & & \vdots & m_N & 0 & 0 \\ & & & & & & \vdots & 0 & m_N & 0 \\ & & & & & & \vdots & 0 & 0 & m_N \end{bmatrix}$$

Notice that Eq. (7.4.5) is the matrix representation of Eq. (7.4.1) for all particles $P_1, P_2, \ldots, P_N$.

Now work on Eq. (7.3.2), performing the transpose operation[4] and rearranging the intermediate results as follows:

$$\delta W = \left[\frac{d}{dt}(M\underline{v}) \right]^T \delta\underline{r} = \left[\frac{d}{dt}(\underline{v}^T M) \right] \delta\underline{r} = \frac{d}{dt}(\underline{v}^T M \delta\underline{r}) - \underline{v}^T M \underbrace{\frac{d}{dt}(\delta\underline{r})}_{\text{same as } \delta\frac{dr}{dt}}$$

$$= \frac{d}{dt}(\underline{v}^T M \delta\underline{r}) - \underline{v}^T M \underbrace{\delta\frac{dr}{dt}}_{\text{same as } \underline{v}^T M \delta\underline{v}} = \frac{d}{dt}(\underline{v}^T M \delta\underline{r}) - \delta(T) \qquad (7.4.6)$$

where

$$T = \frac{1}{2}\underline{v}^T M \underline{v} \qquad (7.4.7)$$

is the kinetic energy of the motion of the system.

By rearranging Eq. (7.4.6) and integrating in time the resulting equation between two arbitrary instants $t = t_1$ and $t = t_2$, and by noticing that

$$\int_{t=t_1}^{t=t_2} \frac{d}{dt}(\underline{v}^T M \delta\underline{r}) \, dt = \left(\underline{v}^T M \delta\underline{r} \right)_{t=t_1}^{t=t_2}$$

---

[4]Recall from matrix algebra that the transpose $(AB)^T$ of a product AB of two matrices A and B is equal to $B^T A^T$.

the following result is then obtained:

$$\boxed{\int\limits_{t=t_1}^{t=t_2} [\delta W + \delta(T)]\, dt = \left(\underline{v}^T M \delta \underline{r}\right)_{t=t_1}^{t=t_2}} \qquad (7.4.8)$$

Equation (7.4.8) is called *Hamilton's extended principle*, and it is a fundamental principle in dynamics. The general form of the kinetic energy T for any dynamics problem is examined in Section 7.4.2, and the formulation is then continued in Section 7.4.3. The material presented in Section 7.4.3 will greatly reduce the mathematical operations that are needed in the search for an integral of the motion of a system. A number of solved examples using the theory presented in this section and in Sections 7.4.2 and 7.4.3 are included in Section 7.7.

Note: It is common to set the right-hand side of Eq. (7.4.8) to zero by considering variations for which $(\delta \vec{r})_{t=t_1} = 0$ and $(\delta \vec{r})_{t=t_2} = 0$, i.e., by considering variations for which $\vec{r}$ is known at two times t_1 and t_2. However, this is generally not the case since, in dynamics problems, conditions at $t = t_1$ (i.e., initial conditions) for $\vec{r}$ and $\vec{v}$ are what is specified, thus making $\delta \vec{r}$ unknown at any time $t = t_2$ and, consequently, $(\delta \vec{r})_{t=t_2}$ is not zero before the motion becomes known. It turns out, though, that integration of $\delta(T)$ with respect to time yields another term that happens to cancel out the term in the right-hand side of Eq. (7.4.8). This is presented in Section 7.4.2, where Lagrange's equation is obtained from Eq. (7.4.8). Lagrange's equation is a fundamental equation in dynamics. Its use will be illustrated with a number of examples presented at the end of these derivations.

7.4.2 Lagrange's Equation (Discrete Systems) Obtained from Hamilton's Principle

Equation (7.4.8) yields an equation, called *Lagrange's equation*, that allows one to obtain the differential equations of motion of a system in terms of the generalized coordinates $q_1(t), \ldots, q_n(t)$. Lagrange's equation is obtained by making use of the general expression for δW in Eq. (7.4.8) as $\delta W = \underline{Q}^T \delta \underline{q}$ [see Eq. (7.4.2), p. 520], taking the variation $\delta(T)$ of the kinetic energy $T(q_1, \ldots, q_n, \dot{q}_1, \ldots, \dot{q}_n; t)$, which is

$$
\begin{aligned}
\delta(T) &= \frac{\partial T}{\partial q_1}\delta q_1 + \ldots \frac{\partial T}{\partial q_n}\delta q_n + \frac{\partial T}{\partial \dot{q}_1}\delta \dot{q}_1 + \ldots + \frac{\partial T}{\partial \dot{q}_n}\delta \dot{q}_n \\
&= \left(\frac{\partial T}{\partial q_1} \cdots \frac{\partial T}{\partial q_n}\right)\delta \underline{q} + \left(\frac{\partial T}{\partial \dot{q}_1} \cdots \frac{\partial T}{\partial \dot{q}_n}\right)\delta \dot{\underline{q}} \overset{\Delta}{=} \frac{\partial T}{\partial \underline{q}}\delta \underline{q} + \frac{\partial T}{\partial \dot{\underline{q}}}\delta \dot{\underline{q}} \quad (7.4.9)
\end{aligned}
$$

and then proceeding as follows:

$$\int\limits_{t=t_1}^{t=t_2} \left(\underline{Q}^T \delta \underline{q} + \frac{\partial T}{\partial \underline{q}}\delta \underline{q} + \frac{\partial T}{\partial \dot{\underline{q}}}\delta \dot{\underline{q}}\right) dt - \left(\underline{v}^T M \frac{\partial \underline{r}}{\partial \underline{q}}\delta \underline{q}\right)_{t=t_1}^{t=t_2} = 0 \qquad (7.4.10)$$

Since

$$(\delta\dot{\underline{q}})dt = \left[\delta\left(\frac{d\underline{q}}{dt}\right)\right]dt = \left[\frac{d}{dt}(\delta\underline{q})\right]dt = d(\delta\underline{q}) \qquad (7.4.11)$$

an integration by parts in Eq. (7.4.10) yields

$$\int_{t=t_1}^{t=t_2}\left(\underline{Q}^T + \frac{\partial T}{\partial\underline{q}} - \frac{d}{dt}\frac{\partial T}{\partial\dot{\underline{q}}}\right)\delta\underline{q}\,dt + \left[\left(\frac{\partial T}{\partial\dot{\underline{q}}} - \underline{v}^T M\frac{\partial\underline{r}}{\partial\underline{q}}\right)\delta\underline{q}\right]_{t=t_1}^{t=t_2} = 0 \qquad (7.4.12)$$

Since $T = \frac{1}{2}\underline{v}^T M\underline{v}$, it follows that

$$\frac{\partial T}{\partial\dot{\underline{q}}} = \frac{1}{2}\underline{v}^T(M + M^T)\frac{\partial\underline{v}}{\partial\dot{\underline{q}}} \underbrace{=}_{\text{since } M^T = M} \underline{v}^T M\frac{\partial\underline{v}}{\partial\dot{\underline{q}}} \qquad (7.4.13)$$

and also, from Eq. (7.4.5), that[5]

$$\frac{\partial\underline{v}}{\partial\dot{\underline{q}}} = \frac{\partial\underline{r}}{\partial\underline{q}} \qquad (7.4.14)$$

Therefore, as a consequence of Eqs. (7.4.13) and (7.4.14), the boundary terms in Eq. (7.4.12), i.e., the terms that are evaluated at $t = t_1$ and at $t = t_2$, are identically equal to zero, and Eq. (7.4.12) reduces to

$$\int_{t=t_1}^{t=t_2}\left(\underline{Q}^T + \frac{\partial T}{\partial\underline{q}} - \frac{d}{dt}\frac{\partial T}{\partial\dot{\underline{q}}}\right)\delta\underline{q}\,dt = 0 \qquad (7.4.15)$$

---

[5] More explicitly, Eq. (7.4.14) means

$$\begin{bmatrix} \frac{\partial v_{1x}}{\partial\dot{q}_1} & \frac{\partial v_{1x}}{\partial\dot{q}_2} & \cdots & \frac{\partial v_{1x}}{\partial\dot{q}_n} \\ \frac{\partial v_{1y}}{\partial\dot{q}_1} & \frac{\partial v_{1y}}{\partial\dot{q}_2} & \cdots & \frac{\partial v_{1y}}{\partial\dot{q}_n} \\ \frac{\partial v_{1z}}{\partial\dot{q}_1} & \frac{\partial v_{1z}}{\partial\dot{q}_2} & \cdots & \frac{\partial v_{1z}}{\partial\dot{q}_n} \\ \vdots & \vdots & & \vdots \\ \frac{\partial v_{Nx}}{\partial\dot{q}_1} & \frac{\partial v_{Nx}}{\partial\dot{q}_2} & \cdots & \frac{\partial v_{Nx}}{\partial\dot{q}_n} \\ \frac{\partial v_{Ny}}{\partial\dot{q}_1} & \frac{\partial v_{Ny}}{\partial\dot{q}_2} & \cdots & \frac{\partial v_{Ny}}{\partial\dot{q}_n} \\ \frac{\partial v_{Nz}}{\partial\dot{q}_1} & \frac{\partial v_{Nz}}{\partial\dot{q}_2} & \cdots & \frac{\partial v_{Nz}}{\partial\dot{q}_n} \end{bmatrix} = \begin{bmatrix} \frac{\partial r_{1x}}{\partial q_1} & \frac{\partial r_{1x}}{\partial q_2} & \cdots & \frac{\partial r_{1x}}{\partial q_n} \\ \frac{\partial r_{1y}}{\partial q_1} & \frac{\partial r_{1y}}{\partial q_2} & \cdots & \frac{\partial r_{1y}}{\partial q_n} \\ \frac{\partial r_{1z}}{\partial q_1} & \frac{\partial r_{1z}}{\partial q_2} & \cdots & \frac{\partial r_{1z}}{\partial q_n} \\ \vdots & \vdots & & \vdots \\ \frac{\partial r_{Nx}}{\partial q_1} & \frac{\partial r_{Nx}}{\partial q_2} & \cdots & \frac{\partial r_{Nx}}{\partial q_n} \\ \frac{\partial r_{Ny}}{\partial q_1} & \frac{\partial r_{Ny}}{\partial q_2} & \cdots & \frac{\partial r_{Ny}}{\partial q_n} \\ \frac{\partial r_{Nz}}{\partial q_1} & \frac{\partial r_{Nz}}{\partial q_2} & \cdots & \frac{\partial r_{Nz}}{\partial q_n} \end{bmatrix}$$

Equation (7.4.15) is satisfied only if the integrand is zero, i.e., if

$$
\left(\frac{d}{dt}\frac{\partial T}{\partial \dot{q}_1} - \frac{\partial T}{\partial q_1} - Q_{q_1}\right)\delta q_1 + \left(\frac{d}{dt}\frac{\partial T}{\partial \dot{q}_2} - \frac{\partial T}{\partial q_2} - Q_{q_2}\right)\delta q_2 + \dots
$$
$$
+ \left(\frac{d}{dt}\frac{\partial T}{\partial \dot{q}_n} - \frac{\partial T}{\partial q_n} - Q_{q_n}\right)\delta q_n = 0
$$

(7.4.16)

Equation (7.4.16) is valid in the most general case, even if some of the variables $q_1, \dots, q_n$ are related to each other. Such a case is dealt with in Section 7.8.

For the case when there are no constraints between any of the variables $q_1, \dots, q_n$, the virtual displacements $\delta q_1, \dots, \delta q_n$ are independent of each other. In this case, since each virtual displacement is arbitrary, each one of the bracketed terms in Eq. (7.4.16) must then be set to zero to satisfy that equation. This then yields the following equation that must be satisfied by each one of the n variables $q_1, \dots, q_n$ *when they are independent of each other:*

$$
\frac{d}{dt}\frac{\partial T}{\partial \dot{q}_k} - \frac{\partial T}{\partial q_k} = Q_{q_k} \qquad \text{for } k = 1, 2, \dots, n
$$

(7.4.17)

Equation (7.4.17) is known as *Lagrange's equation.* Since the kinetic energy T is a quadratic function of the generalized velocities $\dot{q}_1, \dot{q}_2, \dots, \dot{q}_n$ (as shown in Section 7.4.3), Lagrange's equation represents n second-order differential equations for the generalized coordinates $q_1, \dots, q_n$. In general, the resulting differential equations are nonlinear and coupled.

The fundamental quantities that are needed for formulating the second-order differential equations of motion of a dynamical system using Lagrange's equation are the kinetic energy T of the entire system and the generalized forces $Q_{q_1}, \dots, Q_{q_n}$. The expressions for the generalized forces are simply the coefficients of $\delta q_1, \dots, \delta q_n$ that appear in the expression for the virtual work δW. The general form of the expression for the kinetic energy for any dynamics problem is presented and discussed in Section 7.4.3.

Several examples using this formulation are presented in Section 7.7.

7.4.3   General Form of the Kinetic Energy T

The velocity $\vec{v}$ is always an explicit linear function of the generalized velocities $\dot{q}_1, \ldots, \dot{q}_n$, and it may also explicitly depend (in general, in a nonlinear manner) on the generalized coordinates $q_1, \ldots, q_n$ and on the time t. The general form of the velocity $\vec{v}$ is given in matrix form in Eq. (7.4.5).

Since $T = \frac{1}{2}\underline{v}^T M \underline{v}$, use of Eq. (7.4.5) yields the following general result, which is valid for *all* dynamics problems:

$$
\begin{aligned}
T &= \frac{1}{2}\underline{v}^T M \underline{v} = \frac{1}{2}\left[\frac{\partial \underline{r}}{\partial \underline{q}}\dot{\underline{q}} + \frac{\partial \underline{r}}{\partial t}\right]^T M \left[\frac{\partial \underline{r}}{\partial \underline{q}}\dot{\underline{q}} + \frac{\partial \underline{r}}{\partial t}\right] \\
&= \underbrace{\frac{1}{2}\dot{\underline{q}}^T \left[\left(\frac{\partial \underline{r}}{\partial \underline{q}}\right)^T M \frac{\partial \underline{r}}{\partial \underline{q}}\right]\dot{\underline{q}}}_{\text{Call this part } T_2} + \underbrace{\left(\frac{\partial \underline{r}}{\partial t}\right)^T M \frac{\partial \underline{r}}{\partial \underline{q}}\dot{\underline{q}}}_{\text{Call this part } T_1} + \underbrace{\frac{1}{2}\left(\frac{\partial \underline{r}}{\partial t}\right)^T M \left(\frac{\partial \underline{r}}{\partial t}\right)}_{\text{Call this part } T_0} \quad (7.4.18)
\end{aligned}
$$

or, summarizing,

$$
\boxed{T = T_2 + T_1 + T_0} \quad (7.4.19)
$$

As disclosed by Eq. (7.4.18), the general form of the kinetic energy is always a sum of three distinct parts:

- A part, denoted by $T_2 \overset{\Delta}{=} \frac{1}{2}\dot{\underline{q}}^T A(\underline{q}; t)\dot{\underline{q}}$, where $A = (\partial \underline{r}/\partial \underline{q})^T M(\partial \underline{r}/\partial \underline{q})$ is a symmetric matrix (because its transpose is equal to itself), that is a purely quadratic function of the generalized velocities $\dot{q}_1, \dot{q}_2, \ldots, \dot{q}_n$. In other words, T_2 is the part of T that is the sum of terms involving $\dot{q}_1^2$, $\dot{q}_1\dot{q}_2$, $\dot{q}_2^2$, $\dot{q}_1\dot{q}_3$, etc. For any specific problem, such a part is simply recognized by looking at the expression that is obtained for the kinetic energy T.

- A part, denoted by $T_1 \overset{\Delta}{=} \underline{b}^T(\underline{q}, t)\dot{\underline{q}}$, where $\underline{b} = (\partial \underline{r}/\partial \underline{q})^T M (\partial \underline{r}/\partial t)$ is a column matrix, that is a linear function of the generalized velocities $\dot{q}_1, \dot{q}_2, \ldots, \dot{q}_n$. In other words, T_1 is of the form $b_1\dot{q}_1 + b_2\dot{q}_2 + \ldots + b_n\dot{q}_n$. Again, for any specific problem, such a part is simply recognized by looking at the expression that is obtained for the kinetic energy T.

- Finally, a part, denoted by T_0, which does not depend on the generalized velocities (but only on the generalized coordinates, and, maybe, on time explicitly). The term T_0 is also recognizable by inspection of any expression for T.

The classification of the different terms in the kinetic energy as T_2, T_1, and T_0 terms happens to be extremely convenient for obtaining the expression for the *Hamiltonian* of the motion. As will be seen, the Hamiltonian is one of the functions that is tested when one looks for possible integrals of the differential equations of motion of a system. It may also be a good candidate for testing stability of an equilibrium using Lyapunov's method (which was presented in Chapter 1). In dynamics terminology, it is common to refer to systems for which $T = T_2$ as *natural systems*.

7.5   Conservation Principles: Classical Integrals of Motion

7.5.1   What is an Integral of the Motion?

A second-order differential equation in a variable $q(t)$ has two integrals of motion. A *first integral of motion* is a relationship of the form $f_1(q, \dot{q}; t) = \text{constant} \stackrel{\Delta}{=} C_1$; i.e., f_1 is a function of the variable q, of its first time derivative, and maybe is also an explicit function of the independent variable t. Thus, a first integral of the motion reduces the problem of finding $q(t)$ to the integration of a differential equation $f_1(q, \dot{q}, ; t) = C_1$, whose order is one less than the original second-order differential equation. For example, the differential equation $\ddot{q} + q = 0$ has the first integral of motion $\frac{1}{2}(\dot{q}^2 + q^2) = \text{constant}$.

Repeating the process for the equation $f_1(q, \dot{q}; t) = \text{constant}$, a *second integral of motion* is a relationship that involves derivatives up to one order smaller than the highest derivative in this latter differential equation. In this case, the second integral of motion is, then, a relationship of the form $f_2(q; t) = \text{constant} \stackrel{\Delta}{=} C_2$ that only involves q and t. The solution to the original second-order differential equation is then the solution to this latter "algebraic" equation. The constants C_1 and C_2 are simply the functions f_1 and f_2, respectively, evaluated at the value of t when $q(t)$ and $\dot{q}(t)$ are given. The problem is said to be an *initial value problem* when initial values of q and $\dot{q}$, say $q(t_0)$ and $\dot{q}(t_0)$, are given. When the known conditions are split between two different values of t [as, for example, when $q(t_0)$ and $\dot{q}(t_{\text{final}})$ are given], the problem is called a *two-point boundary value problem*. Classical dynamics problems are initial value problems. Problems in structural dynamics, for example, are, in general, two-point boundary value problems.

Lagrange's equation, Newton's equation $\vec{F} = m\vec{a}$, and Euler's equation $\vec{M}_C = d\vec{H}_C/dt$ are equations that involve second derivatives, and no higher. The motion of such dynamical systems is governed by a set of n second-order differential equations, where n is called the *number of degrees of freedom* of the system. Therefore, the differential equations of motion for such systems exhibit $2n$ constants of the motion, n of them being *first integrals*, and the other n being *second integrals* (which are integrals of motion of the first integrals). In some cases, it is possible to find some of the integrals of motion of a dynamical system but, in general, it has been an impossible task to find all of them or even some of them for most cases. The ones that can be found are very useful, though, for investigating the motion.

7.5.2 Integrals of Motion Obtained from Lagrange's Equation

Under certain conditions (listed subsequently in Cases 1 and 2 of this section), several integrals of motion can be obtained from Lagrange's equation, and the search for them turns out to be a relatively simple process in such cases. To begin the search, start by isolating all the terms in δW that can be obtained from a potential energy so that δW becomes

$$\delta W \overset{\Delta}{=} -\delta(U) + (\delta W)_{nc} \tag{7.5.1}$$

When this is done, Eq. (7.4.17) can then be written as:

$$\frac{d}{dt}\frac{\partial \mathcal{L}}{\partial \dot{q}_k} - \frac{\partial \mathcal{L}}{\partial q_k} = (Q_{q_k})_{nc} \qquad \text{for } k = 1, 2, \ldots, n \tag{7.5.2}$$

where

$$\mathcal{L} \overset{\Delta}{=} T - U = T_2 + T_1 + T_0 - U \tag{7.5.3}$$

is called the *Lagrangian* of the motion.[6] There are two classical searches for integrals of motion.

CASE 1 A GENERALIZED MOMENTUM INTEGRAL OF MOTION

This is the simplest of the searches for an integral of motion. Equation (7.5.2) immediately discloses that $\partial \mathcal{L}/\partial \dot{q} = $ constant for any generalized coordinate q for which $\partial \mathcal{L}/\partial q = 0$ and also when the nonconservative generalized force associated with that generalized coordinate, i.e., $(Q_q)_{nc}$, is equal to zero. In summary, $\boxed{\dfrac{\partial \mathcal{L}}{\partial \dot{q}} = \text{constant for any } q \text{ for which } \dfrac{\partial \mathcal{L}}{\partial q} = 0 \text{ and } (Q_q)_{nc} = 0.}$

This integral of motion, if it exists for a particular problem, is called the *generalized momentum integral* because the quantity $p \overset{\Delta}{=} \partial \mathcal{L}/\partial \dot{q}$ is either a linear momentum or an angular momentum. Any generalized coordinate q that does not appear in the Lagrangian (i.e., $\partial \mathcal{L}/\partial q = 0$) is called an *ignorable coordinate*.

---

[6]Since the letter L is already used in this book for denoting a distance, the symbol $\mathcal{L}$ (a calligraphic L) is used for denoting the Lagrangian to try to avoid duplication of symbols (although no confusion should exist even if one wishes to use the same letter for denoting either quantity).

CASE 2: AN ENERGY-LIKE INTEGRAL OF MOTION

To search for another integral of motion, let us see what happens if we take the time derivative of the Lagrangian. Since $\mathcal{L} = \mathcal{L}(\underline{q}, \underline{\dot{q}}; t)$, the total time derivative of $\mathcal{L}$ is

$$\frac{d\mathcal{L}(\underline{q}, \underline{\dot{q}}; t)}{dt} = \frac{\partial \mathcal{L}}{\partial \underline{q}} \underline{\dot{q}} + \frac{\partial \mathcal{L}}{\partial \underline{\dot{q}}} \underline{\ddot{q}} + \frac{\partial \mathcal{L}}{\partial t} \qquad (7.5.4)$$

By replacing the term $\partial \mathcal{L}/\partial \underline{q}$ in this equation for its equivalent expression obtained from Eq. (7.5.2), Eq. (7.5.4) can then be manipulated as follows:

$$\frac{d\mathcal{L}\left(\underline{q}, \underline{\dot{q}}; t\right)}{dt} = \left[\frac{d}{dt}\frac{\partial \mathcal{L}}{\partial \underline{\dot{q}}} - (\underline{Q}^T)_{nc}\right]\underline{\dot{q}} + \frac{\partial \mathcal{L}}{\partial \underline{\dot{q}}}\underline{\ddot{q}} + \frac{\partial \mathcal{L}}{\partial t}$$

$$\underbrace{=}_{\substack{\text{grouping the first} \\ \text{and third terms}}} \frac{d}{dt}\left(\frac{\partial \mathcal{L}}{\partial \underline{\dot{q}}}\underline{\dot{q}}\right) - (\underline{Q}^T)_{nc}\underline{\dot{q}} + \frac{\partial \mathcal{L}}{\partial t} \qquad (7.5.5)$$

By defining the following function $\mathcal{H}$, which is called the *Hamiltonian* of the motion,[7]

$$\boxed{\mathcal{H} \triangleq \frac{\partial \mathcal{L}}{\partial \underline{\dot{q}}}\underline{\dot{q}} - \mathcal{L}} = \frac{\partial \mathcal{L}}{\partial \dot{q}_1}\dot{q}_1 + \frac{\partial \mathcal{L}}{\partial \dot{q}_2}\dot{q}_2 + \ldots + \frac{\partial \mathcal{L}}{\partial \dot{q}_n}\dot{q}_n - \mathcal{L} \qquad (7.5.6)$$

from which it can be verified that $\partial \mathcal{H}/\partial t = -\partial \mathcal{L}/\partial t$, Eq. (7.5.5) yields

$$\boxed{\frac{d\mathcal{H}}{dt} = \underline{Q}_{nc}^T\underline{\dot{q}} + \frac{\partial \mathcal{H}}{\partial t}} = (Q_{q_1})_{nc}\dot{q}_1 + (Q_{q_2})_{nc}\dot{q}_2 + \ldots + (Q_{q_n})_{nc}\dot{q}_n + \frac{\partial \mathcal{H}}{\partial t}$$

$$(7.5.7)$$

Thus, as indicated by this boxed equation, it is an easy task to take the time derivative of a Hamiltonian.

Equation (7.5.7) discloses that the Hamiltonian is constant when the Lagrangian $\mathcal{L}$ is not an *explicit* function of time (i.e., when $\partial \mathcal{L}/\partial t = 0$) and also when $\underline{Q}_{nc}^T\underline{\dot{q}} = 0$. Since the generalized coordinates are independent of each other, this last condition requires each one of the nonconservative generalized

---

[7] Again, to try to avoid, as much as possible, duplication of symbols, the symbol $\mathcal{H}$ is used in the book for denoting the Hamiltonian since the letter H is already used for denoting angular momentum.

forces associated with each of the independent coordinates $q_1, \ldots, q_n$ to be equal to zero. The integral of the motion $\mathcal{H} = $ constant is called the *Jacobi integral*. In summary,

$$\mathcal{H} = \text{constant} \qquad \text{when}$$

$$\frac{\partial \mathcal{L}}{\partial t} = 0 \quad \text{and} \quad (Q_{q_1})_{nc} = (Q_{q_2})_{nc} = \ldots = (Q_{q_n})_{nc} = 0$$

The product

$$\underline{Q}_{nc}^T \underline{\dot{q}} = (Q_{q_1})_{nc}\dot{q}_1 + (Q_{q_2})_{nc}\dot{q}_2 + \ldots + (Q_{q_n})_{nc}\dot{q}_n$$

which is the sum of the product of a nonconservative generalized force (whose unit is either a force or a moment) by the corresponding generalized velocity, is the power into the system due to the generalized nonconservative forces.

The generalized momentum integral and the Jacobi integral, if they exist for a particular problem, are the two *classical* integrals of motion that are known for dynamical systems.

CASE 3: NONCLASSICAL INTEGRALS OF MOTION

An obvious nonclassical integral of motion of Eq. (7.5.2) is immediately obtained when a generalized coordinate, say q, does not appear in the expression for the Lagrangian $\mathcal{L} = T - U$ (i.e., when $\partial \mathcal{L}/\partial q = 0$) and *also* when the generalized nonconservative force $(Q_q)_{nc}$ associated with q is the total time derivative of a function. This will most likely happen in the simpler case when $(Q_q)_{nc}$ is due to viscous damping and is of the form $(Q_q)_{nc} = -c\dot{q}$, where c is a constant viscous friction coefficient. In such a case, one obtains $\partial \mathcal{L}/\partial \dot{q} + cq = $ constant. There is no particular name for this integral of motion.

Other integrals of motion exist in very special cases that have been reported in the technical journals using methods that are beyond the scope of this book (see, for example, Reference [14] listed in Appendix G). Beyond that work and those mentioned in that reference, little progress has been made in attempts to find additional integrals of motion of dynamical systems.

7.5.3   General Form of the Hamiltonian for Any Problem

The classification of the different terms in the kinetic energy as T_2, T_1, and T_0 terms turns out to be extremely convenient for obtaining the expression for the Hamiltonian of the motion for any problem. It is shown here that the Hamiltonian turns out to be simply $\mathcal{H} = T_2 + U - T_0$ and, thus, it can be obtained simply by using the expression for the potential energy U and the appropriate terms collected by inspection of the expression for the kinetic energy T. That is certainly much less "algebra" than computing it using its definition given by Eq. (7.5.6).

To show that the Hamiltonian is simply $\mathcal{H} = T_2 + U - T_0$, let us make use of the general expression for the kinetic energy, and then generate the expression for $\mathcal{H}$. Since, for any specific problem, the expression for the kinetic energy is of the form given by Eq. (7.4.18), i.e.,

$$T = \frac{1}{2}\underline{\dot{q}}^T A(\underline{q}; t)\underline{\dot{q}} + \underline{b}^T(\underline{q}; t)\underline{\dot{q}} + T_0(\underline{q}; t) \overset{\Delta}{=} T_2 + T_1 + T_0$$

where A is always a symmetric matrix, the Hamiltonian $\mathcal{H}$ is then equal to

$$\mathcal{H} \overset{\Delta}{=} \frac{\partial T}{\partial \underline{\dot{q}}}\underline{\dot{q}} - \mathcal{L} = \left[\frac{1}{2}\underline{\dot{q}}^T\left(A + A^T\right) + \underline{b}^T\right]\underline{\dot{q}} - (T - U)$$

$$= \underline{\dot{q}}^T A\underline{\dot{q}} + \underline{b}^T\underline{\dot{q}} - (T_2 + T_1 + T_0 - U) = (2T_2 + T_1) - (T_2 + T_1 + T_0 - U)$$

or, simply

$$\boxed{\mathcal{H} = T_2 + U - T_0} \tag{7.5.8}$$

The result expressed by Eq. (7.5.8) shows that, once the expressions for the kinetic and potential energies are generated, no further mathematical manipulations are needed to obtain the Hamiltonian for any particular problem. To generate the expression for the Hamiltonian $\mathcal{H}$ for a dynamical system, one only needs to recognize the different types of terms T_2, T_1, and T_0 that appear in the kinetic energy and then selectively extract the T_2 and T_0 terms to form the expression for the Hamiltonian $\mathcal{H}$ according to Eq. (7.5.8).

Notice that U and T_0 are of the same form $U(\underline{q}; t)$ and $T_0(\underline{q}; t)$. Thus, it is impossible to separate from $U - T_0$ what is U and what is T_0 when one is given an expression for a Hamiltonian $\mathcal{H}$ without being shown the expression for either T or U. In some cases, however, one might still be able to separate U from T_0 based on physical arguments. However, such a separation is not important at all. The function $U - T_0$ is called the *kinetic potential* of the motion.

As disclosed by Eq. (7.5.8), the Hamiltonian is an energy function and, in general, it is different from the total mechanical energy $E = T + U = T_2 + T_1 + T_0 + U$ of the motion. The Hamiltonian $\mathcal{H}$ is equal to the total mechanical energy $E = T + U$ only when $T_1 = 0$ and $T_0 = 0$, i.e., when $T = T_2$. In such cases, the Jacobi integral, if it exists for a particular problem being analyzed, has the physical interpretation of being the total mechanical energy $T + U$ of the motion.

7.6   A Brief Summary

Here is a brief summary of the material that is sufficient for one to formulate the differential equations of motion using Lagrange's equation in terms of n independent generalized coordinates $q_1, q_2, \ldots, q_n$ and to test for the existence of classical first integrals of the motion.

1. Lagrange's equation is

$$\frac{d}{dt}\frac{\partial T}{\partial \dot{q}_k} - \frac{\partial T}{\partial q_k} = Q_{q_k} \qquad k = 1, \ldots, n \qquad (7.6.1)$$

where Q_{q_k} is the generalized force associated with the generalized coordinate q_k (which is simply the coefficient of δq_k in the expression for δW).

The generalized forces $Q_{q_1}, \ldots, Q_{q_n}$ are determined from the expression for the virtual work δW done by all forces acting on the system. Such an expression is always of the form shown in Eq. (7.2.4) (p. 507), which may also be written as Eq. (7.2.5) (p. 508) by grouping terms that are derivable from a potential energy $U(q_1, q_2, \ldots, q_n; t)$.

Equation (7.6.1) yields n second-order differential equations for the n independent variables $q_1, \ldots, q_n$. The resulting equations are, in general, nonlinear and coupled to each other, but they can be integrated numerically with no major difficulty. The differential equations of motion are linear in rare cases (for some simpler problems involving linear springs, for example).

2. The expression of the kinetic energy for any dynamics problem is always of the form $T = T_2 + T_1 + T_0$ (see Section 7.4.3, p. 525).

3. Equation (7.6.1) admits the following integrals of motion (i.e., functions involving the generalized coordinates $q_1, \ldots, q_n$, and their first time derivatives, that remain constant for all times):

a. For any generalized coordinate that does not appear in the expression of the kinetic and potential energies, say q_ℓ, it follows that $\partial(T-U)/\partial\dot{q}_\ell =$ constant if the generalized nonconservative force associated with q_ℓ is equal to zero. Such an integral of motion has units of either linear or angular momentum and, for this reason, it is called the *generalized momentum integral*. One also obtains $\partial(T-U)/\partial\dot{q}_\ell + cq_\ell =$ constant if the nonconservative force $(Q_{q_\ell})_{nc}$ is of the form $(Q_{q_\ell})_{nc} = -c\dot{q}_\ell$, with c being a constant viscous damping coefficient.

b. The Hamiltonian of the motion is constant if it does not depend explicitly on the time (i.e., when one inspects its expression and does not see time appearing by itself) and if the nonconservative generalized forces associated with the generalized coordinates $q_1, \ldots, q_n$ are zero. The Hamiltonian is defined by Eq. (7.5.6), but it is always equal to $H = T_2 + U - T_0$. Such an integral of the motion is called the *Jacobi integral*.

7.7   Solved Example Problems

The differential equations of motion for the examples in this section were already formulated throughout the book by direct application of Newton's second law. They are considered here using the methods presented in this chapter so that the difference in the two methodologies can be seen.

Example 7.7.1 Differential Equations of Motion for Example 2.5.8

In this example, a body of mass m is thrown from an inertial point A with an initial absolute velocity of its center of mass P as shown in Fig. 7.5. In Example 2.5.8 (p. 122), air resistance is neglected, and the only force acting on the body is the constant vertical gravitational force whose magnitude is mg. Here, in addition to the vertical force $-mg\hat{y}$, include the effect of air resistance modeled as $-c_1\vec{v} - c_2|\vec{v}|\vec{v}$. Obtain the differential equations of motion using Lagrange's equation, and check for the existence of any classical integrals of the motion. Also obtain the maximum height reached by the center of mass of the body.

■ Solution

Using the independent variables x and y as the generalized coordinates, the following is then obtained in terms of the unit vectors $\hat{x}$ and $\hat{y}$ shown in Fig. 7.5:

Absolute position vector for P: $\vec{r} = x\hat{x} + y\hat{y}$

Absolute velocity of P: $\vec{v} = \dfrac{d\vec{r}}{dt} = \dot{x}\hat{x} + \dot{y}\hat{y}$

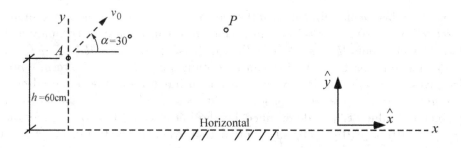

Figure 7.5: Figure for Example 7.7.1.

Resultant force acting on P: $\qquad \vec{F} = -mg\hat{y} - c_1(\dot{x}\hat{x} + \dot{y}\hat{y}) - c_2\sqrt{\dot{x}^2 + \dot{y}^2}(\dot{x}\hat{x} + \dot{y}\hat{y})$

The kinetic energy is then

$$T = \frac{1}{2}m\vec{v} \bullet \vec{v} = \frac{1}{2}m(\dot{x}^2 + \dot{y}^2) \qquad (7.7.1a)$$

which is of the form $T = T_2$, i.e., $T_1 = T_0 = 0$.

Inspection of the expression for $\vec{v}$ discloses that the virtual displacement of P is $\delta\vec{r} = (\delta x)\hat{x} + (\delta y)\hat{y}$. Therefore, the virtual work δW done by $\vec{F}$, and the generalized forces Q_x and Q_y, are obtained as follows:

$$\begin{aligned}
\delta W &= \vec{F} \bullet (\delta\vec{r}) = -\left(c_1\dot{x} + c_2\dot{x}\sqrt{\dot{x}^2 + \dot{y}^2}\right)\delta x - \left(mg + c_1\dot{y} + c_2\dot{y}\sqrt{\dot{x}^2 + \dot{y}^2}\right)\delta y \\
&\triangleq Q_x\delta x + Q_y\delta y \qquad (7.7.1b)
\end{aligned}$$

Thus,

$$\frac{\partial T}{\partial \dot{x}} = m\dot{x} \qquad \frac{\partial T}{\partial \dot{y}} = m\dot{y} \qquad \frac{\partial T}{\partial x} = \frac{\partial T}{\partial y} = 0 \qquad (7.7.1c)$$

and Lagrange's equation $d(\partial T/\partial\dot{q})/dt - \partial T/\partial q = Q_q$ yields, with $q = x$ and $q = y$, the following differential equations of motion:

$$m\ddot{x} = -c_1\dot{x} - c_2\dot{x}\sqrt{\dot{x}^2 + \dot{y}^2} \qquad (7.7.1d)$$

$$m\ddot{y} = -mg - c_1\dot{y} - c_2\dot{y}\sqrt{\dot{x}^2 + \dot{y}^2} \qquad (7.7.1e)$$

Integrals of Motion

There are no classical integrals of the motion, unless $c_1 = c_2 = 0$. In such a case, there are two classical first integrals of the motion, which are the generalized momentum $\partial T/\partial \dot{x} = m\dot{x}$ (thus, $\dot{x}$ = constant, which also can be integrated analytically) and the Hamiltonian $\mathcal{H} = T_2 + U - T_0$, which is the same as $\mathcal{H} = T + U$ since $T = T_2$. The only part of δW that can be obtained as $-\delta(U)$ is the term $-mg\delta y$, which gives, except for an irrelevant constant of integration, the potential energy $U = mgy$. In such a case, $\mathcal{H}$ = constant yields $\mathcal{H} = \frac{1}{2}(\dot{x}^2 + \dot{y}^2) + mgy$ = constant. The constant value of $\mathcal{H}$ is determined by the initial conditions of the motion as $\mathcal{H} = \frac{1}{2}m[\dot{x}^2(0) + \dot{y}^2(0)] + mgy(0)$. Notice that when $c_2 = 0$ and $c_1 \neq 0$, two nonclassical integrals of motion are obtained as $m\dot{x} + c_1 x$ = constant $\overset{\Delta}{=} C_1$, and $m\dot{y} + c_1 y + mgt$ = constant $\overset{\Delta}{=} C_2$.

When $c_2 = 0$, the differential equations of motion can be integrated analytically. In such a case, the maximum value of y can be obtained without the need for numerical integration. Numerical integration may also be used, of course.

When $c_2 \neq 0$, the differential equations have to be solved numerically because an analytical solution can no longer be obtained. The maximum value of y, for example, has to be determined from the numerical integration in such a case.

Example 7.7.2 Differential Equations of Motion for Example 2.5.9
In this example, as in Example 2.5.9 (p. 125), a block P_2 of mass m_2 slides on a face of a sliding block P_1, which is an inclined plane of mass m_1. P_2 is connected to P_1 by a linear spring of stiffness k N/m and unstressed length L_u m. Both blocks move on the same vertical plane. Friction in the system is viscous, with friction coefficients c_2 between P_2 and P_1, and c_1 between P_1 and the horizontal surface. The system, and two sets of unit vectors, are shown in Fig. 7.6, which is the same as Fig. 2.16 (p. 125), and is repeated here for convenience. Obtain the differential

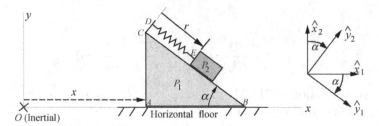

Figure 7.6: A block sliding on a sliding block for Example 7.7.2 (same as for Example 2.5.9).

equations of motion using the Lagrangian formulation and check for the existence of classical integrals of motion.

■ Solution

To obtain the differential equations of motion using the Lagrangian formulation we need to develop expressions for the kinetic energy of the motion of the entire system, and for the generalized forces acting on the system. To generate the expression for the kinetic energy, we need to have the expression of the absolute velocity of each of the two blocks. We then have the following.

Absolute position vector $\vec{r}_1$, and absolute velocity $\vec{v}_1$ for block P_1:

$$\vec{r}_1 = x\hat{x}_1 \qquad \vec{v}_1 = \frac{d\vec{r}_1}{dt} = \dot{x}\hat{x}_1 \qquad (7.7.2a)$$

Absolute position vector $\vec{r}_2$ and absolute velocity $\vec{v}_2$ for block P_2:

$$\vec{r}_2 = x\hat{x}_1 + r\hat{y}_1 \qquad \vec{v}_2 = \frac{d\vec{r}_2}{dt} = \dot{x}\hat{x}_1 + \dot{r}\hat{y}_1 \qquad (7.7.2b)$$

The kinetic energy T is then

$$T = \frac{1}{2}m_1\vec{v}_1 \bullet \vec{v}_1 + \frac{1}{2}m_2\vec{v}_2 \bullet \vec{v}_2 = \frac{1}{2}m_1\dot{x}^2 + \frac{1}{2}m_2(\dot{x}^2 + \dot{r}^2 + 2\dot{x}\dot{r}\underbrace{\hat{x}_1 \bullet \hat{y}_1}_{=\cos\alpha}) \quad (7.7.2c)$$

The generalized forces associated with the chosen generalized coordinates x and r are obtained from the expression for the virtual work δW of the forces acting on the system. The forces for which the virtual work is non-zero are

- The gravitational force $-m_2 g\hat{x}_2$ acting on P_2. The virtual work of the gravitational force $-m_1 g\hat{x}_2$ acting on block P_1 is zero because the virtual displacement of P_1 is $(\delta x)\hat{x}_1$, which is perpendicular to $\hat{x}_2$.
- The viscous friction force $-c_1\dot{x}\hat{x}_1$ acting on block P_1 due to contact with the horizontal surface.
- The viscous friction force that is proportional to $\dot{r}$ and acts parallel to the plane of contact between P_2 and P_1. Such a force does work only when one body moves relative to the other. The virtual displacement of P_2 relative to P_1 is $\delta\vec{r}_2 - \delta\vec{r}_1$, and since such a force is in the direction opposite to that relative displacement, its virtual work is equal to $-c_2\dot{r}\hat{y}_1 \bullet (\delta\vec{r}_2 - \delta\vec{r}_1)$.
- The force due to the spring that connects block P_2 to P_1. Such a force also does work only when one body moves relative to the other. Since the virtual displacement $\delta\vec{r}_2 - \delta\vec{r}_1$ of P_2 relative to P_1 deforms the spring (see Fig. 7.6), the spring force due to such a displacement is equal to $-k(r - L_u)\hat{y}_1$. Therefore, the virtual work of the spring force is equal to $-k(r - L_u)\hat{y}_1 \bullet (\delta\vec{r}_2 - \delta\vec{r}_1)$.

The total virtual work δW is the sum of the virtual work of the forces just listed. Therefore, we have

$$
\begin{aligned}
\delta W &= \underbrace{-m_2 g\,\hat{x}_2 \bullet (\delta \vec{r}_2)}_{= \hat{x}_2 \bullet [(\delta x)\hat{x}_1 + (\delta r)\hat{y}_1]} \underbrace{-c_1 \dot{x}\hat{x}_1 \bullet \delta \vec{r}_1}_{=(\delta x)\hat{x}_1} \underbrace{-c_2 \dot{r}\hat{y}_1 \bullet (\delta \vec{r}_2 - \delta \vec{r}_1)}_{= (\delta r)\hat{y}_1} \\
&= -m_2 g (\delta r)\hat{x}_2 \bullet \hat{y}_1 = (m_2 g \sin \alpha)\delta r \\
&\quad -k(r - L_u)\hat{y}_1 \bullet (\delta \vec{r}_2 - \delta \vec{r}_1)
\end{aligned}
$$

and the virtual work is

$$
\delta W = [m_2 g \sin \alpha - c_2 \dot{r} - k(r - L_u)]\delta r - c_1 \dot{x}\delta x \stackrel{\Delta}{=} Q_r \delta r + Q_x \delta x \qquad (7.7.2\mathrm{d})
$$

The coefficients of δr and δx that appear in this expression are the generalized forces Q_r and Q_x associated with the generalized coordinates r and x, respectively.

Lagrange's equation, $d(\partial T / \partial \dot{q})dt - \partial T / \partial q = Q_q$, yields the following differential equations of motion for the system, with $q = x$ and $q = r$:

For $q = x$:

$$
\frac{d}{dt}[(m_1 + m_2)\dot{x} + m_2 \dot{r} \cos \alpha] = Q_x = -c_1 \dot{x}
$$

or

$$
(m_1 + m_2)\dot{x} + m_2 \dot{r} \cos \alpha + c_1 x = \text{constant} \stackrel{\Delta}{=} C_1 \qquad (7.7.2\mathrm{e})
$$

For $q = r$:

$$
\frac{d}{dt}[m_2(\dot{r} + \dot{x} \cos \alpha)] = Q_r = m_2 g \sin \alpha - c_2 \dot{r} - k(r - L_u)
$$

or

$$
m_2(\ddot{r} + \ddot{x} \cos \alpha) + c_2 \dot{r} + k(r - L_u) - m_2 g \sin \alpha = 0 \qquad (7.7.2\mathrm{f})
$$

Equations (7.7.2e) and (7.7.2f) are the differential equations of motion for the system, with Eq. (7.7.2f) being exactly the same as Eq. (2.5.9g) (p. 126). Equations (7.7.2e) and (7.7.2f) are equivalent to Eqs. (2.5.9g) and (2.5.9i) that were obtained in Chapter 2 via Newton's second law. To verify the equivalence, see Problem 7.3 in Section 7.10.

Integrals of Motion

The potential energy for this system is

$$
U = -mgr \sin \alpha + \frac{k}{2}(r - L_u)^2 \qquad (7.7.2\mathrm{g})
$$

and, because $T = T_2$, as seen from Eq. (7.7.2c), the Hamiltonian is

$$\mathcal{H} = T_2 + U - T_0 = T + U \tag{7.7.2h}$$

The variable x is an ignorable coordinate because it does not appear explicitly in the Lagrangian $\mathcal{L} = T - U$, but the nonconservative generalized force associated with x is $Q_x = -c_1 \dot{x}$. Thus, the generalized momentum $\partial \mathcal{L}/\partial \dot{x}$ is constant only if $c_1 = 0$. However, since Q_x can be integrated with respect to time, the function $\partial \mathcal{L}/\partial \dot{x} + c_1 x = (m_1 + m_2)\dot{x} + m_2 \dot{r} \cos\alpha + c_1 x$ is constant, as already indicated by Eq. (7.7.2e).

The total time derivative of the Hamiltonian $\mathcal{H}$ is

$$\frac{d\mathcal{H}}{dt} = \underbrace{\frac{\partial \mathcal{H}}{\partial t}}_{=0} + (Q_x)_{nc}\dot{x} + (Q_r)_{nc}\dot{r} = -c_1 \dot{x}^2 - c_2 \dot{r}^2 \tag{7.7.2i}$$

and, thus, the Hamiltonian is constant only if $c_1 = c_2 = 0$. The Hamiltonian is a good function for testing the stability of the equilibrium of this system using Lyapunov's. This testing is left as an exercise for the reader (see Problem 7.29 at the end of this chapter).

Example 7.7.3 Differential Equations of Motion for Example 2.7.6
The system of Example 2.7.6 (p. 154) is shown in Fig. 7.7. At point B there is a particle with mass m. The following modifications are made here to make this example more general:

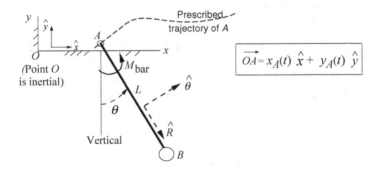

Figure 7.7: A pendulum mounted on a moving base for Example 7.7.3.

■ Point A is given a known motion $\overrightarrow{OA} = x_A(t)\hat{x} + y_A(t)\hat{y}$ in any direction on the vertical plane.

■ Link AB is a uniform bar of mass m_{bar} and length L. It is still subjected to the torque M_{bar} shown in the figure, which is applied by a motor.

Obtain all classical integrals of the motion, if any, using the generalized coordinate θ shown in Fig. 7.7. Also, obtain the differential equation of motion for the system using Lagrange's equation.

■ **Solution**

The center of mass (C) of bar AB is in the middle of the bar. The absolute velocity $\vec{v}_C$ of C and the virtual variation $\delta(\overrightarrow{OC})$ are

$$\vec{v}_C = \frac{d}{dt}\overrightarrow{OC} = \frac{d}{dt}\overrightarrow{OA} + \frac{d}{dt}\left(\frac{L}{2}\hat{R}\right) = \dot{x}_A\hat{x} + \dot{y}_A\hat{y} + \frac{L}{2}\dot{\theta}\hat{\theta}$$

$$\delta\left(\overrightarrow{OC}\right) = \delta\left(\overrightarrow{OA}\right) + \frac{L}{2}(\delta\theta)\hat{\theta} \tag{7.7.3a}$$

where $\delta(\overrightarrow{OA}) = 0$ because the motion of point A is prescribed.

The absolute velocity $\vec{v}_B$ of point B, and the virtual variation $\delta(\overrightarrow{OB})$, are

$$\vec{v}_B = \frac{d}{dt}\overrightarrow{OB} = \frac{d}{dt}\overrightarrow{OA} + \frac{d}{dt}(L\hat{R}) = \dot{x}_A\hat{x} + \dot{y}_A\hat{y} + L\dot{\theta}\hat{\theta}$$

$$\delta(\overrightarrow{OB}) = \delta(\overrightarrow{OA}) + L(\delta\theta)\hat{\theta} = L\,(\delta\theta)\,\hat{\theta} \tag{7.7.3b}$$

Therefore, the kinetic energy T of the motion is:

$$T = \underbrace{\frac{1}{2}m_{\text{bar}}\vec{v}_C \bullet \vec{v}_C + \frac{1}{2}I_C\dot{\theta}^2}_{} + \frac{1}{2}m\vec{v}_B \bullet \vec{v}_B$$

This is the kinetic energy for the bar

$$= \frac{1}{2}m_{\text{bar}}(\dot{x}_A^2 + \dot{y}_A^2 + \frac{L^2}{4}\dot{\theta}^2 + L\dot{\theta}\dot{x}_A \underbrace{\hat{x} \bullet \hat{\theta}}_{= \cos\theta} + L\dot{\theta}\dot{y}_A \underbrace{\hat{y} \bullet \hat{\theta}}_{= \sin\theta}) + \frac{1}{2}I_C\dot{\theta}^2$$

$$+ \frac{1}{2}m(\dot{x}_A^2 + \dot{y}_A^2 + L^2\dot{\theta}^2 + 2L\dot{\theta}\dot{x}_A \underbrace{\hat{x} \bullet \hat{\theta}}_{= \cos\theta} + 2L\dot{\theta}\dot{y}_A \underbrace{\hat{y} \bullet \hat{\theta}}_{= \sin\theta}) = T_2 + T_1 + T_0 \tag{7.7.3c}$$

where I_C is the moment of inertia of bar AB about an axis that passes through its center of mass C and is perpendicular to the plane of the motion, i.e.,

$$I_C = \frac{1}{12}m_{\text{bar}}L^2$$

The generalized force associated with the chosen generalized coordinate θ is obtained from the expression for the virtual work δW of the forces acting on the system. The forces and moments for which the virtual work is nonzero are:

- The gravitational force $-m_{\text{bar}}g\hat{y}$ acting at the center of mass of bar AB, and the constant gravitational force $-mg\hat{y}$ acting at point B. The virtual work of these forces is then $-m_{\text{bar}}g\hat{y} \bullet \delta(\overrightarrow{OC}) - mg\hat{y} \bullet \delta(\overrightarrow{OB}) = -g(m_{\text{bar}}L/2 + mL)(\sin\theta)\,\delta\theta$. The virtual work of these gravitational forces could also be obtained more quickly as $-\delta(U)$, where $U = -m_{\text{bar}}g(L/2)\cos\theta - mgL\cos\theta$ (see Fig. 7.7). The quantities $(L/2)\cos\theta$ and $L\cos\theta$ are the vertical distances that the center of mass C and point B, respectively, are below point A in Fig. 7.7 (thus, the physical interpretation of the minus sign in U).

- The moment applied to bar AB by a motor. With $\hat{z} \overset{\Delta}{=} \hat{R} \times \hat{\theta} = \hat{x} \times \hat{y}$ (see Fig. 7.7) and the absolute angular velocity of bar AB being $\dot{\theta}\hat{z}$, the infinitesimal work due to that moment is $(dW)_{\text{motor}} = (M_{\text{bar}}\hat{z}) \bullet (\dot{\theta}\hat{z}) = M_{\text{bar}}d\theta$. Therefore, the virtual work due to that moment is simply $(\delta W)_{\text{motor}} = M_{\text{bar}}\delta\theta$.

The total virtual work δW is the sum of the virtual work of the forces just listed. Therefore, we have

$$\delta W = \left[M_{\text{bar}} - \left(\frac{1}{2}m_{\text{bar}} + m \right)gL\sin\theta \right]\delta\theta \overset{\Delta}{=} Q_\theta\delta\theta = -\delta(U) + M_{\text{bar}}\delta\theta \quad (7.7.3\text{d})$$

where $U = -m_{\text{bar}}g(L/2)\cos\theta - mgL\cos\theta$. Notice that, since the moment M_{bar} is unspecified, it is not (and could not be) included in the expression for the potential energy U. Thus, it is regarded as a nonconservative moment. If M_{bar} were specified as a function that only depends on θ, $M_{\text{bar}}(\theta)$, for example, one would introduce the potential energy $U = -m_{\text{bar}}g(L/2)\cos\theta - mgL\cos\theta - \int M_{\text{bar}}d\theta$ and that moment would then be a conservative moment.

Tests for the Classical Integrals of the Motion

Generalized momentum integral. Since θ appears explicitly in the Lagrangian $\mathcal{L} = T - U$, it follows that $\partial\mathcal{L}/\partial\theta \neq 0$. Therefore, there is no momentum integral.

Testing the Hamiltonian. The Hamiltonian is $\mathcal{H} = T_2 + U - T_0$, where

$$T_0 = \frac{1}{2}(m_{\text{bar}} + m)(\dot{x}_A^2 + \dot{x}_B^2)$$

and

$$T_2 = \frac{1}{2}\left(I_C + m_{\text{bar}}\frac{L^2}{4} + mL^2 \right)\dot{\theta}^2$$

are obtained by inspection of Eq. (7.7.3c). As per Eq. (7.5.7), the time derivative of $\mathcal{H}$ is

$$\frac{d\mathcal{H}}{dt} = (Q_\theta)_{nc}\dot{\theta} + \frac{\partial \mathcal{H}}{\partial t} = M_{\text{bar}}\dot{\theta} + \frac{\partial \mathcal{H}}{\partial t} = M_{\text{bar}}\dot{\theta} - \frac{1}{2}(m_{\text{bar}} + m)\frac{d}{dt}(\dot{x}_A^2 + \dot{x}_B^2)$$

Since the right-hand side of the expression for $d\mathcal{H}/dt$ is not zero when either (or both) $M_{\text{bar}} \neq 0$ or $\partial \mathcal{H}/\partial t = -\frac{1}{2}(m_{\text{bar}} + m)d(\dot{x}_A^2 + \dot{y}_A^2)/dt \neq 0$, the Hamiltonian is not constant.

When is the Hamiltonian $\mathcal{H}$ constant? That will be the case when $M_{\text{bar}} = 0$ at all times, and $\dot{x}_A^2 + \dot{y}_A^2 = \text{constant}$.

Differential Equation of Motion for the System
The differential equation of motion is readily obtained from Lagrange's equation $d(\partial T/\partial\dot{\theta})/dt - \partial T/\partial\theta = Q_\theta$.
We have:

$$\frac{\partial T}{\partial \dot{\theta}} = \left[I_C + m_{\text{bar}}\frac{L^2}{4} + mL^2\right]\dot{\theta} + \left(\frac{1}{2}m_{\text{bar}} + m\right)L\left(\dot{x}_A\cos\theta + \dot{y}_A\sin\theta\right)$$

$$(7.7.3e)$$

$$\frac{\partial T}{\partial \theta} = \left(\frac{1}{2}m_{\text{bar}} + m\right)L\left(\dot{y}_A\cos\theta - \dot{x}_A\sin\theta\right)\dot{\theta} \qquad (7.7.3f)$$

Therefore, Lagrange's equation yields the following differential equation of motion:

$$\left[m_{\text{bar}}\left(\frac{L^2}{4} + \frac{L^2}{12}\right) + mL^2\right]\ddot{\theta} + \left(\frac{1}{2}m_{\text{bar}} + m\right)gL\sin\theta$$

$$= M_{\text{bar}} - \left(\frac{1}{2}m_{\text{bar}} + m\right)L\left(\ddot{x}_A\cos\theta + \ddot{y}_A\sin\theta\right) \qquad (7.7.3g)$$

When $m_{\text{bar}} = 0$ and $\ddot{y}_A(t) = 0$, which corresponds to the simpler case considered in Chapter 2, this differential equation becomes Eq. (2.7.6d) (p. 156), which was obtained by direct application of Newton's second law. The analysis of the motion is done in the manner presented in Chapter 2.

Example 7.7.4 Differential Equations of Motion for Example 2.7.8
The system of Example 2.7.8 (p. 162) is shown in Fig. 7.8, where motion takes place on a vertical plane. The spring is linear, with stiffness k N/m and unstressed length L_u m, and there is no friction in the system. Obtain all classical integrals of the motion, if any, using the generalized coordinates r and θ shown in Fig. 7.8. Also, obtain the differential equations of motion for the system using Lagrange's equation.

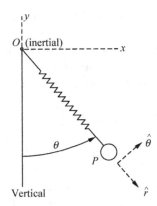

Figure 7.8: The spring-pendulum for Example 7.7.4 (same as for Example 2.7.8).

■ Solution

Since the absolute velocity of point P is

$$\vec{v}_P = \frac{d}{dt}(\overrightarrow{OP}) = \frac{d}{dt}(r\hat{r}) = \dot{r}\hat{r} + r\dot{\theta}\hat{\theta} \tag{7.7.4a}$$

the kinetic energy of the motion is

$$T = \frac{1}{2}m\vec{v}_P \bullet \vec{v}_P = \frac{1}{2}m(\dot{r}^2 + r^2\dot{\theta}^2) = T_2 \tag{7.7.4b}$$

Inspection of Fig. 7.8 discloses that the potential energy U for this system is

$$U = -mgr\cos\theta + \frac{k}{2}(r - L_u)^2 \tag{7.7.4c}$$

and since there are no other forces acting on the system that do virtual work, this then yields the following expression for the virtual work δW:

$$
\begin{aligned}
\delta W &= -\delta(U) = -\left[\frac{\partial U}{\partial r}\delta r + \frac{\partial U}{\partial \theta}\delta\theta\right] \\
&= [mg\cos\theta - k(r - L_u)]\,\delta r - (mgr\sin\theta)\,\delta\theta \overset{\triangle}{=} Q_r\delta r + Q_\theta\delta\theta \quad (7.7.4d)
\end{aligned}
$$

The expressions for the generalized forces Q_r and Q_θ are, respectively, the coefficients of δr and $\delta\theta$ in the expression for δW. Notice that writing the expression for potential energy by inspection (please see the note on page 512), when it is possible to do so, provides the quickest way to obtain the conservative generalized forces acting on the system.[8]

The only classical integral of motion for this system is the Hamiltonian $\mathcal{H} = T_2 + U - T_0$. Since $T = T_2$ by inspection of Eq. (7.7.4b), the Hamiltonian is the same as the total mechanical energy $T + U$. We then have

$$\mathcal{H} = \frac{1}{2}m(\dot{r}^2 + r^2\dot{\theta}^2) - mgr\cos\theta + \frac{k}{2}(r - L_u)^2 = \text{constant} \qquad (7.7.4e)$$

The quantities $\partial\mathcal{L}/\partial\dot{r}$ and $\partial\mathcal{L}/\partial\dot{\theta}$, where $\mathcal{L} = T - U$, are not constants because $\mathcal{L}$ depends explicitly on both r and θ.

The differential equations of motion are readily obtained from Lagrange's equation $d(\partial T/\partial\dot{q})/dt - \partial T/\partial q = Q_q$, with $q = r$ and $q = \theta$. Since

$$\frac{\partial T}{\partial\dot{r}} = m\dot{r} \qquad \frac{\partial T}{\partial\dot{\theta}} = mr^2\dot{\theta} \qquad \frac{\partial T}{\partial r} = mr\dot{\theta}^2 \qquad \frac{\partial T}{\partial\theta} = 0$$

Lagrange's equation yields the following differential equations of motion for this system:

$$m(\ddot{r} - r\dot{\theta}^2) = mg\cos\theta - k(r - L_u) \qquad (7.7.4f)$$
$$m(r\ddot{\theta} + 2\dot{r}\dot{\theta}) = -mg\sin\theta \qquad (7.7.4g)$$

which are exactly of the same form as Eqs. (2.7.8f) and (2.7.8g) (p. 164). The analysis of the motion was presented in Example 2.7.8 in Chapter 2.

Example 7.7.5 Differential Equations of Motion for Example 2.7.9
In this example, a rigid body of mass $m = 0.5$ kg moves inside a horizontal tube of length $R = 2$ m, filled with oil and subjected to a moment M_{motor} applied by a motor as shown in Fig. 7.9. In Example 2.7.9 (p. 171) the body was a mass particle, and the tube was approximated as massless. Here the dimensions of the body and of the tube are taken into account. The vector $\vec{r}$ shown in Fig. 7.9 is the absolute position vector for the center of mass of the body that is moving inside the rotating tube. The tube is always in contact with a fixed horizontal table and is driven by

---

[8]The direct way to obtain the expression for δW is repeated here, in case of need.

$$\delta W = \left[\underbrace{mg(\hat{r}\cos\theta - \hat{\theta}\sin\theta)}_{\text{gravitational force}} \underbrace{-k(r - L_u)\hat{r}}_{\text{spring force}} \right] \bullet \underbrace{\left[(\delta r)\hat{r} + (r\,\delta\theta)\hat{\theta}\right]}_{\text{virtual displacement of P}}$$

$$= [mg\cos\theta - k(r - L_u)]\delta r - (mgr\sin\theta)\,\delta\theta \triangleq Q_r\delta r + Q_\theta\delta\theta$$

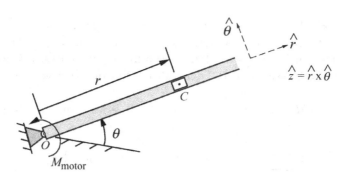

Figure 7.9: A constrained body in a rotating tube for Example 7.7.5 (same as for Example 2.7.9).

a motor with a prescribed absolute angular velocity $\dot{\theta}\hat{z} \overset{\Delta}{=} \Omega(t)\hat{z}$ rad/s. End O of the tube is inertial. Because of the viscosity of the oil, the body is subjected to a linear viscous friction force with friction coefficient $c = 1$ N/(m/s). For this, let $(I_O)_{\text{tube}}$ be the moment of inertia of the tube about an axis that passes through point O and is perpendicular to the plane of the motion (which is the plane of the figure). Also, let $(I_C)_{\text{body}}$ be the moment of inertia of the body about a similar axis, but passing through the center of mass of the body. Obtain all the classical integrals of the motion, if any, and determine the time the center of mass of the body reaches the end of the tube if $\Omega = \text{constant} = 5$ rad/s.

■ **Solution**

For this example, the expression for the kinetic energy is

$$
\begin{aligned}
T &= \frac{m}{2}(\dot{r}^2 + r^2\dot{\theta}^2) + \frac{1}{2}(I_C)_{\text{body}}\dot{\theta}^2 + + \frac{1}{2}(I_O)_{\text{tube}}\dot{\theta}^2 \\
&= \frac{m}{2}[\dot{r}^2 + r^2\Omega^2(t)] + \frac{1}{2}[(I_C)_{\text{body}} + (I_O)_{\text{tube}}]\Omega^2(t) = T_2 + T_0 \quad (7.7.5a)
\end{aligned}
$$

which, as indicated in Eq. (7.7.5a), is of the form $T_2 + T_0$, where $T_2 = m\dot{r}^2/2$ and T_0 is the rest of the expression for T.

The only generalized force for which the virtual work is not equal to zero is the viscous friction force $-c\dot{r}\hat{r}$, and the virtual work of such a force is simply

$$
\delta W = -c\dot{r}\hat{r} \bullet [(\delta r)\hat{r}] = -c\dot{r}\delta r \overset{\Delta}{=} Q_r \delta r \quad (7.7.5b)
$$

If the moment M_{motor} shown in Fig. 7.9 were prescribed, instead of $\dot{\theta}$, its virtual work would be $M_{\text{motor}}\delta\theta$. However, since $\dot{\theta}$ is what is prescribed, $d\theta = \dot{\theta}\,dt$ and, thus, $\delta\theta = \dot{\theta}\,\delta t = 0$ as $\delta t = 0$ by definition of a virtual variation.

As for the classical integrals of motion (i.e., a generalized momentum equal to a constant, and the Hamiltonian also equal to a constant), there are none here. The generalized momentum $\partial T/\partial \dot{r} = m\dot{r}$ (which is a linear momentum) is not constant because $\partial T/\partial r \neq 0$. In addition, the generalized nonconservative force $-c\dot{r}$ is not zero either.

The Hamiltonian $\mathcal{H} = T_2 + U - T_0$, where $U = 0$, is not constant either. Its time derivative $d\mathcal{H}/dt$ is

$$\frac{d\mathcal{H}}{dt} = \frac{\partial \mathcal{H}}{\partial t} + \underbrace{(Q_r)_{nc}}_{= -c\dot{r}} \dot{r} = -[mr^2 + (I_C)_{\text{body}} + (I_O)_{\text{tube}}]\Omega\dot{\Omega} - c\dot{r}^2 \qquad (7.7.5\text{c})$$

The Hamiltonian would be constant only if $c = 0$ and $\Omega = $ constant. Notice that the Hamiltonian $\mathcal{H} = T_2 + U - T_0$ is not the same as the total mechanical energy $E \overset{\Delta}{=} T + U = T_2 + T_1 + T_0 + U$, where $T_1 = 0$ for this example. The two are the same only if $T = T_2$, i.e., if $T_1 = 0$ and $T_0 = 0$. In this example, the external motor must "pump" energy into the system to force the tube to rotate with a prescribed angular velocity $\dot{\theta} = \Omega(t)$, and, thus, the total mechanical energy E cannot be constant. Even if there were no damping in the system, the motor would still have to pump energy into the system to force the tube to rotate in the prescribed manner.

If the Hamiltonian were constant, the equation $\mathcal{H} = $ constant would be a first-order differential equation. Thus, in such a case, there would be no need to work with a second-order differential equation generated either by using Lagrange's equation or Newton's second law. However, even with $\Omega = $ constant, the Hamiltonian is not constant for this problem because the damping coefficient c is not equal to zero. We then need to generate the second-order differential equation of motion to solve the problem at hand.

We have

$$\frac{\partial T}{\partial \dot{r}} = m\dot{r} \qquad \frac{\partial T}{\partial r} = mr\dot{\theta}^2 = mr\Omega^2(t) \qquad (7.7.5\text{d})$$

and the following differential equation of motion is immediately obtained from Lagrange's equation:

$$m(\ddot{r} - \Omega^2 r) = -c\dot{r} \qquad (7.7.5\text{e})$$

As presented in Example 2.7.9, the solution to this linear differential equation is

$$r(t) = A_1 e^{s_1 t} + A_2 e^{s_2 t} \qquad (7.7.5\text{f})$$

For $\Omega = 5$ rad/s, $c = 1$ N/(m/s), and $m = 0.5$ kg, one obtains $A_1 \approx 0.6$ m, $A_2 \approx 0.4$ m, $s_1 \approx 4.1$ s^{-1}, and $s_2 \approx -6.1$ s^{-1}. For $r = 2$ m, Eq. (7.7.5f) yields $t \approx 0.29$ s.

Example 7.7.6 Differential Equations of Motion for Example 2.7.13
In this example, as in Example 2.7.13 (p. 184), a point P of mass m slides on
a circular frictionless track in a homogeneous ring of radius R and mass m_{ring}.
The ring is forced, by a motor, to rotate with a given absolute angular velocity
$\dot{\phi} \overset{\Delta}{=} \Omega(t)$ about a vertical axis AA that passes through its center C as shown in
Fig. 7.10. The center C of the ring is inertial (the unit vector triad $\{\hat{r}, \hat{\theta}, \hat{n} \overset{\Delta}{=} \hat{r} \times \hat{\theta}\}$
shown in the figure will be used in the solution to the problem). With θ as the
generalized coordinate, obtain the differential equation of motion for the system
using Lagrange's equation, and the classical integrals of the motion (if any).

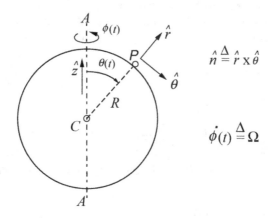

Figure 7.10: A particle moving on a rotating ring for Example 7.7.6.

■ **Solution**
Notice that this is a problem with a constraint, namely, $\dot{\phi} \overset{\Delta}{=} \Omega(t)$, where $\Omega(t)$ is
any given function of time. However, this is a simple constraint, in the sense that
the dynamics of this system is governed by a single independent variable, which is
chosen to be the angle $\theta(t)$.

Since the absolute velocity of point P is

$$\vec{v}_P = \frac{d}{dt}(R\hat{r}) = [\Omega(\hat{r}\cos\theta - \hat{\theta}\sin\theta) + \dot{\theta}\hat{n}] \times (R\hat{r}) = R\dot{\theta}\hat{\theta} + (R\sin\theta)\dot{\phi}\,\hat{n} \quad (7.7.6a)$$

and the ring is rotating with absolute angular velocity $\dot{\phi}\hat{z} = \Omega\hat{z}$, the expression for
the kinetic energy of the motion of this system is

$$T = \frac{mR^2}{2}[\dot{\theta}^2 + \dot{\phi}^2\sin^2\theta] + \frac{1}{2}(I_{AA})_{\text{ring}}\dot{\phi}^2$$

$$= \underbrace{\frac{1}{2}mR^2\dot{\theta}^2}_{=T_2} + \underbrace{\frac{1}{2}[(I_{AA})_{\text{ring}} + mR^2\sin^2\theta]\Omega^2(t)}_{==T_0} \qquad (7.7.6\text{b})$$

where $(I_{AA})_{\text{ring}} = 2\int_0^\pi (R\sin\theta)^2[m_{\text{ring}}/(2\pi R)]R\,d\theta = \frac{1}{2}m_{\text{ring}}R^2$ is the moment of inertia of the ring about axis AA.

As in Example 7.7.5, the virtual work of the moment applied by the motor is equal to zero. Notice that the actual infinitesimal work $dW = M_{\text{motor}}d\phi$ done by the motor is not zero, but that is not relevant for the approach taken in analytical dynamics. The only force for which the virtual work in this system with a frictionless track is not zero is the gravitational force $-mg\hat{z}$, and its virtual work can be obtained either as $\delta W = (-mg\hat{z}) \bullet [\delta(R\hat{r})] = (-mg\hat{z}) \bullet [(R\delta\theta)\hat{\theta}]$ or from the potential energy $U = mg(R\cos\theta)$ as $\delta W = -\delta(U)$. This yields the following expression for the virtual work, from which the expression for the generalized force Q_θ is extracted:

$$\delta W = (mgR\sin\theta)\delta\theta \overset{\Delta}{=} Q_\theta\delta\theta \qquad (7.7.6\text{c})$$

We then have

$$\frac{\partial T}{\partial \dot{\theta}} = mR^2\dot{\theta} \qquad \frac{\partial T}{\partial \theta} = mR^2\Omega^2(\sin\theta)\cos\theta \qquad (7.7.6\text{d})$$

Lagrange's equation $d(\partial T/\partial\dot{\theta})/dt - \partial T/\partial\theta = Q_\theta$ yields the following differential equation of motion:

$$mR^2\ddot{\theta} - mR^2\Omega^2(\sin\theta)\cos\theta = Q_\theta = mgR\sin\theta \qquad (7.7.6\text{e})$$

As for the classical integrals of motion (i.e., a generalized momentum equal to a constant, and the Hamiltonian also equal to a constant), the generalized momentum $\partial T/\partial\dot{\theta}$ (which is an angular momentum) is not constant because $\partial T/\partial\theta \neq 0$.

The Hamiltonian

$$\mathcal{H} = T_2 + U - T_0 = \frac{1}{2}mR^2\dot{\theta}^2 + mgR\cos\theta - \frac{1}{2}[(I_{AA})_{\text{ring}} + mR^2\sin^2\theta]\Omega^2(t) \qquad (7.7.6\text{f})$$

is not constant either. Its time derivative $d\mathcal{H}/dt$ is

$$\frac{d\mathcal{H}}{dt} = \frac{\partial\mathcal{H}}{\partial t} + \underbrace{(Q_\theta)\;nc}_{=0}\dot{\theta} = -[(I_{AA})_{\text{ring}} + mR^2\sin^2\theta]\Omega\dot{\Omega} \qquad (7.7.6\text{g})$$

The Hamiltonian is constant only if $\Omega = \text{constant}$. Notice that, as in Example 7.7.5, the Hamiltonian $\mathcal{H} = T_2 + U - T_0$ is not the same as the total mechanical energy $E \overset{\Delta}{=} T + U = T_2 + T_1 + T_0 + U$, where $T_1 = 0$ for this example. The two are the same only if $T = T_2$, i.e., if $T_1 = 0$ *and* $T_0 = 0$. In this example, the external motor must also pump energy into the system to force the ring to rotate with a prescribed angular velocity $\dot{\phi} = \Omega(t)$ and, thus, the total mechanical energy E cannot be constant.

A very convenient point to notice is that when $\Omega = $ constant, the equation $\mathcal{H} = T_2 + U - T_0 = $ constant is a first-order differential equation. Thus, there is no need to generate the second-order differential equation of motion because the equation $\mathcal{H} = $ constant is an integral of that second-order differential equation. Thus, all the work involving in generating that equation can be bypassed in such a case.

Example 7.7.7 A Modified Form of Example 7.7.6

It is interesting and instructive to look at the same problem of Example 7.7.6 by considering a known moment $M_{\text{motor}}\hat{z}$ applied to the rotating ring, instead of specifying that the ring is to rotate with a prescribed angular velocity. In such a case, the ring will rotate with an unknown angular velocity $\dot{\phi}$, and the system has, then, two degrees of freedom: θ and ϕ. The differential equations of motion for this situation are also easily and readily obtained using the Lagrangian formulation, and the only modification to be made is in the expression for the virtual work δW, which now becomes

$$\delta W = (mgR\sin\theta)\delta\theta + M_{\text{motor}}\delta\phi \qquad (7.7.7a)$$

The expression for the kinetic energy remains the same as Eq. (7.7.6b), but now $T = T_2$ (i.e., a purely quadratic form in the generalized velocities $\dot{\theta}$ and $\dot{\phi}$) since neither θ nor ϕ are known a priori. To avoid confusion, that expression is repeated here.

$$T = \frac{mR^2}{2}(\dot{\theta}^2 + \dot{\phi}^2\sin^2\theta) + \frac{1}{2}(I_{AA})_{\text{ring}}\dot{\phi}^2 = T_2 \qquad (7.7.7b)$$

Since the expression for the generalized force Q_θ also did not change, Eq. (7.7.6e) is one of the differential equations of motion, with Ω^2 in that equation being the unknown $\dot{\phi}^2$. Again to avoid confusion, that equation is repeated here.

$$mR^2\ddot{\theta} - mR^2\dot{\phi}^2(\sin\theta)\cos\theta = Q_\theta = mgR\sin\theta \qquad (7.7.7c)$$

The other differential equation of motion is obtained from Lagrange's equation $d(\partial T/\partial\dot{\phi})/dt - \partial T/\partial\phi = Q_\phi$, where $Q_\phi = M_{\text{motor}}$, as seen by inspection of Eq. (7.7.7a).

Since

$$\frac{d}{dt}\left\{\left[(I_{AA})_{\text{ring}} + mR^2 \sin^2\theta\right]\dot\phi\right\} = M_{\text{motor}} \qquad (7.7.7\text{d})$$

Lagrange's equation then yields the following second differential equation of motion:

$$\frac{d}{dt}\left\{\left[(I_{AA})_{\text{ring}} + mR^2 \sin^2\theta\right]\dot\phi\right\} = M_{\text{motor}} \qquad (7.7.7\text{e})$$

For this form of the problem, M_{motor} is specified; i.e., it is a known function of time, and both Eqs. (7.7.7c) and (7.7.7e) have to be integrated simultaneously to obtain the solutions for $\theta(t)$ and $\phi(t)$. As for the classical integrals of the motion for this modified form of the problem, neither $\partial T/\partial\dot\phi$ is constant, of course (unless $M_{\text{motor}} = 0$, i.e., if the ring, with frictionless bearings, were not driven) nor $\partial T/\partial\dot\theta$ is constant. Since T is now $T = T_2$, the Hamiltonian, which we shall denote by $\mathcal{H}_{\text{new}}$ to not confuse it with the $\mathcal{H}$ in Example 7.7.6, is

$$\mathcal{H}_{\text{new}} = T + U \qquad \text{(i.e., the total mechanical energy of the system)} \qquad (7.7.7\text{f})$$

and its time derivative is

$$\frac{d\mathcal{H}_{\text{new}}}{dt} = \underbrace{\frac{\partial\mathcal{H}_{\text{new}}}{\partial t}}_{=\,0} + \dot\theta\underbrace{(Q_\theta)_{nc}}_{=\,0} + \dot\phi\underbrace{(Q_\phi)_{nc}}_{=\,M_{\text{motor}}} = M_{\text{motor}}\dot\phi \qquad (7.7.7\text{g})$$

which is the power pumped by the motor into the system. Notice that $\partial\mathcal{H}_{\text{new}}/\partial t = 0$ because $\mathcal{H}_{\text{new}}$ is not an *explicit* function of time (i.e., the time t is not seen explicitly in the expression for $\mathcal{H}_{\text{new}}$).

Notice that when $\dot\phi$, instead of M_{motor}, is prescribed, one could modify the problem by ignoring the constraint $\dot\phi = \text{constant}$, for example, applying the moment M_{motor} to the system, and then treating M_{motor} as an unknown. In such a case the virtual work associated with that moment would be $M_{\text{motor}}\delta\phi$, instead of zero as in the case when $\delta\phi = 0$ (i.e., when ϕ is prescribed). This modified approach yields a differential equation associated with $\delta\phi$, which is of the form $I\ddot\phi + \ldots = M_{\text{motor}}$. This may be done for any other problem and, if it is done, the kinetic energy will automatically be of the form $T = T_2$, and the Hamiltonian (i.e., $\mathcal{H}_{\text{new}}$) will automatically be equal to the total mechanical energy $T + U$. At the end of the formulation of the differential equations, the condition $\dot\phi = \text{constant}$ is then imposed in such a modified problem as being a solution to the differential equation $I\ddot\phi + \ldots = M_{\text{motor}}$. Thus, such an equation allows the motor torque M_{motor} to be determined. If the determination of the motor torque is of no interest, then there is no advantage in using such a modified approach.

Example 7.7.8 Example 2.8.8 with Viscous Friction Included
In this example, as in Example 2.8.8 (p. 201), a small body P, modeled as a particle
of mass m, is constrained to slide on a frictionless parabolic vertical track $y = ax^2$ m
(see Fig. 7.11), where a is a given constant. The particle is also connected to a spring,
and the other end of the spring is attached to an inertial point O. The spring is
linear, with stiffness k N/m and unstressed length L_u m. Obtain the differential
equation of motion for the system, including the effect of linear viscous friction with
friction coefficient c N/(m/s), and check for classical integrals of motion.

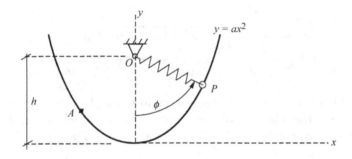

Figure 7.11: A mass particle constrained to move on a parabolic track for Example
7.7.8.

■ Solution
Notice that this is also a problem with a constraint, namely, that particle P is
constrained to move on the parabolic track $y = ax^2$. However, this is also a simple
constraint that makes it possible to describe the motion using a single independent
variable. Such types of constraints, and other types that require the introduction of
additional variables, are discussed in Section 7.8 (p. 553).

In the absence of damping as done in Example 2.8.8 (i.e., for $c = 0$), the infinitesimal work dW for this problem happens to be an exact differential $d(W)$.
This fact was then used in Chapter 2 to determine the maximum height reached by
the particle for the undamped problem in Example 2.8.8 taking advantage of the
basic (but limited in scope) equation $dW = d(T)$. That led to $T + U = \text{constant}$,
which is the integral of the motion that would have been obtained from Lagrange's
equation. As mentioned in Chapter 2, use of the relation $dW = d(T)$ is actually
very restricted for many problems. Such a restriction is overcome by the variational formulation presented in this chapter, which led to Hamilton's principle and
to Lagrange's equation.

With the velocity of point P relative to the parabolic track $y = ax^2$ being equal to

$$\vec{v}_P = \dot{x}\hat{x} + \dot{y}\hat{y} = \dot{x}\hat{x} + 2ax\dot{x}\hat{y} \tag{7.7.8a}$$

the kinetic energy of the motion is

$$T = \frac{1}{2}m(\dot{x}^2 + \dot{y}^2) = \frac{1}{2}m\dot{x}^2(1 + 4a^2x^2) = T_2 \tag{7.7.8b}$$

The expression for the potential energy, which is obtained by inspection of Fig.7.11, is

$$\begin{aligned} U &= mgy + \frac{k}{2}[\sqrt{x^2 + (y-h)^2} - L_u]^2 \\ &= mgax^2 + \frac{k}{2}[\sqrt{x^2 + (ax^2 - h)^2} - L_u]^2 \end{aligned} \tag{7.7.8c}$$

The virtual work done by the gravitational force and by the spring force is equal to $-\delta(U)$. As for the viscous friction force, since the track is not rotating, the velocity of P relative to the track is also equal to $\vec{v}_P$. Thus, the viscous friction force is

$$\vec{F}_{\text{friction}} = -c\vec{v}_P = -c\dot{x}(\hat{x} + 2ax\hat{y}) \tag{7.7.8d}$$

With $\vec{v}_P$ as given previously, the virtual displacement of point P is then $(\delta x)\hat{x} + \delta y\hat{y} = (\hat{x} + 2ax\hat{y})\delta x$. Therefore, the virtual work done by the friction force is

$$(\delta W)_{nc} = -c\dot{x}(1 + 4a^2x^2)\delta x \overset{\Delta}{=} (Q_x)_{nc}\delta x \tag{7.7.8e}$$

As indicated by this equation, the nonconservative generalized force $(Q_x)_{nc}$ is, by definition, the coefficient of δx in the expression for $(\delta W)_{nc}$.

The generalized momentum $\partial T/\partial \dot{x}$ is not constant whether c is zero or not, because x appears explicitly on T. Also, if $c \neq 0$, the Hamiltonian, which is the same as the total mechanical energy $T + U$ for this problem (because T is a purely quadratic form, T_2, in the generalized velocity $\dot{x}$), is not constant. Notice that the total time derivative of the Hamiltonian is

$$\frac{d\mathcal{H}}{dt} = \underbrace{\frac{\partial \mathcal{H}}{\partial t}}_{=\,0} + (Q_x)_{nc}\dot{x} = -c\dot{x}^2(1 + 4a^2x^2) \tag{7.7.8f}$$

where $\partial \mathcal{H}/\partial t = 0$ because t does not appear explicitly in the expression for $\mathcal{H}$.

Lagrange's equation $d(\partial T/\partial \dot{x})/dt - \partial T/\partial x = Q_x$ yields the following differential equation of motion:

$$\frac{d}{dt}\underbrace{[m\dot{x}(1+4a^2x^2)]}_{\text{This is } \frac{\partial T}{\partial \dot{x}}} - \underbrace{4ma^2\dot{x}^2x}_{\text{This is } \frac{\partial T}{\partial x}} = Q_x = -c\dot{x}(1+4a^2x^2) - \frac{\partial U}{\partial x}$$

(7.7.8g)

or

$$\left(1+4a^2x^2\right)\ddot{x} + 4a^2x\dot{x}^2 + \frac{k}{m}\left[1 - \frac{L_u}{\sqrt{x^2+(ax^2-h)^2}}\right]\left[2a\left(ax^2-h\right)+1\right]x$$
$$+\frac{c}{m}\dot{x}\left(1+4a^2x^2\right) + 2agx = 0 \qquad (7.7.8h)$$

Equation (7.7.8h) is the second-order differential equation that governs the motion of the system. Its numerical integration was presented in Example 2.8.9 (p. 204).

Notice that when $c=0$, the integral of the motion $\mathcal{H} = \text{constant}$ is a first-order differential equation. To determine $x(t)$ numerically in this case, one may integrate either Eq. (7.7.8h), which is a second-order differential equation, or simply integrate the first-order differential equation $T+U = \text{constant}$, which is more convenient since it only requires one integration.

Example 7.7.9 Differential Equations of Motion for Example 2.7.10
In this example, as in Example 2.7.10 (p. 174), two particles P_1 and P_2 of masses m_1 and m_2, respectively, are connected by a massless inextensible string of length L that passes through a small hole on a fixed frictionless horizontal table as shown in Fig. 7.12. P_2 moves along the vertical. The system has only one degree of freedom because the string is inextensible. For $m_2 = 2m_1$, $L = 1$ m, $r(0) = 0.5$ m, $\dot{r}(0) = 1$ m/s, and $\dot{\theta}(0) = 2$ rad/s, determine the maximum and the minimum values of $r(t)$ by making use of the constants of the motion of the system obtained with the Lagrangian formulation. Bypass the derivation of a second-order differential equation of motion.

■ **Solution**
Again, this is also a problem with a constraint, namely, that the length of the string connecting the two particles P_1 and P_2 is constant. However, just like the motion of a pendulum of constant length, this is also a simple constraint that makes it possible to describe the motion using a single independent variable, which is chosen to be the distance $r(t) = |\overrightarrow{OP_1}|$ shown in Fig. 7.12.

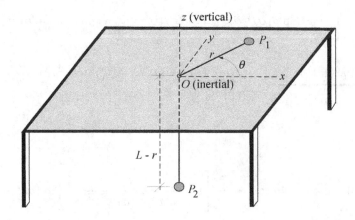

Figure 7.12: Two constrained particles for Example 7.7.9 (same as for Example 2.7.10).

The kinetic and potential energies of the system are readily obtained by inspection of Fig. 7.12 as

$$T = \frac{1}{2}m_1\left[\dot{r}^2 + (r\dot{\theta})^2\right] + \frac{1}{2}m_2\left[\frac{d}{dt}(L-r)\right]^2 = \frac{1}{2}(m_1 + m_2)\dot{r}^2 + \frac{1}{2}m_1 r^2\dot{\theta}^2 = T_2$$

(7.7.9a)

$$U = -m_2 g(L-r)$$

(7.7.9b)

The entire expression for the kinetic energy is of the T_2 form. Since t does not appear explicitly in both T and U, the Hamiltonian $\mathcal{H} = T_2 + U - T_0$, which is $\mathcal{H} = T + U$ for this problem, is constant. Also, since $\partial\mathcal{L}/\partial\theta = 0$, where $\mathcal{L} = T - U$, the generalized momentum $\partial T/\partial\dot{\theta}$ is constant.

We then have (leaving out the irrelevant constant term $-m_2 g L$ in the expression for U):

$$\mathcal{H} = T_2 + U - T_0 = \frac{1}{2}[(m_1 + m_2)\dot{r}^2 + m_1 r^2\dot{\theta}^2] + m_2 g r = \text{constant}$$

$$= \frac{1}{2}[(m_1 + m_2)\dot{r}_0^2 + m_1 r_0^2\dot{\theta}_0^2] + m_2 g r_0$$

(7.7.9c)

$$\frac{\partial T}{\partial\dot{\theta}} = m_1 r^2\dot{\theta} = \text{constant} = m_1 r_0^2\dot{\theta}_0$$

(7.7.9d)

Equation (7.7.9d) can be readily solved for $\dot{\theta}$ in terms of r. When the result is substituted into Eq. (7.7.9c), the following first-order differential equation for r is obtained:

$$\frac{1}{2}(m_1 + m_2)\dot{r}^2 + \frac{1}{2}m_1\frac{(r_0^2\dot{\theta}_0)^2}{r^2} + m_2gr = \frac{1}{2}[(m_1 + m_2)\dot{r}_0^2 + m_1r_0^2\dot{\theta}_0^2] + m_2gr_0$$

$$(7.7.9e)$$

Equation (7.7.9e) is nonlinear, but it can be easily integrated numerically. That is the differential equation of motion for r. There is no need, of course, to generate the second-order differential equations of motion, either using Newton's second law or Lagrange's approach, because Eq. (7.7.9e) is a first integral of those second-order differential equations. For didactic reasons, however, those second-order differential equations are given in a footnote here.[9]

The maximum and minimum values for r, say r_m, can also be obtained from Eq. (7.7.9e), without any numerical integration, just by setting $\dot{r} = 0$ and solving the resulting cubic polynomial equation for r_m. Only real and nonnegative roots should be accepted since they are the only ones that have physical meaning. *Suggestion:* Choose your own values for the initial conditions and determine the resulting maximum and minimum values of r. Also numerically integrate Eq. (7.7.9e) to verify the answers.

7.8 Constraints, and Lagrange's Equation with Constraints

Equation (7.4.17) (p. 524) is valid only if the n variables $q_1, \ldots, q_n$ are independent of each other. If some of the n variables are related to each other, then one has, in general, to work with Eq. (7.4.16) (p. 524), unless the constraint equations are very simple so that one can solve them to eliminate all the "extra" variables. To make this clear, consider the following examples.

■ The motion of a bar of constant length L, such as the pendulum of Example 7.7.3 (p. 537). One certainly does not need to introduce an extra variable $r(t)$ with the constraint $r = \text{constant} = L$ and then formulate the problem using

---

[9]The following differential equations are obtained from Lagrange's equation $d(\partial T/\partial \dot{q})/dt - \partial T/\partial q = Q_q$, with $q = r$ and $q = \theta$, where $Q_r = -\partial U/\partial r = -m_2g$ and $Q_\theta = -\partial U/\partial \theta = 0$:

$$(m_1 + m_2)\ddot{r} - m_1r\dot{\theta}^2 + m_2g = 0$$

$$\frac{d}{dt}(m_1r^2\dot{\theta}) = 0 \qquad (\text{thus, } r^2\dot{\theta} = \text{constant})$$

For this problem, these equations happen to be exactly of the same form as Eqs. (2.7.10m) and (2.7.10k) that were obtained in Example 2.7.10 in Chapter 2, involving additional work that required free-body diagrams and consideration of internal forces.

two variables r and θ to obtain the differential equation for θ. The same was
done for the two particles connected by an inextensible string of length L in
Example 7.7.9 (p. 551).

■ The motion of a particle P constrained to move on a parabolic track $y = ax^2$,
as in Example 7.7.8 (p. 549). For such a constraint, the variable y is easily
eliminated from the problem as done in that example.

7.8.1   Constraints

The simple constraints just exemplified are special cases of constraints of the form
that follows, where some of the variables $q_1, q_2, \ldots, q_n$ may be related by m equations
of constraint (where $m < n$).

$$
\begin{aligned}
a_1(q_1, q_2, \ldots, q_n; t) &= 0 \\
a_2(q_1, q_2, \ldots, q_n; t) &= 0 \\
&\vdots \\
a_m(q_1, q_2, \ldots, q_n; t) &= 0
\end{aligned}
\tag{7.8.1}
$$

Constraint equations of this form are *integral constraints* that, in dynamics, are
called *holonomic constraints* (the word holonomic is of Greek origin). The time t may
or may not appear explicitly in an equation of constraint. Each one of the equations
of (7.8.1) represents a *constraint surface* in the n-dimensional space $q_1, \ldots, q_n$, which
also has a counterpart in a physical space of rectangular x-y-z coordinates. For
example, for a particle moving on a sphere of radius $r = $ constant $= R$, centered at
the origin of such a coordinate system, the constraint surface $x^2 + y^2 + z^2 - R^2 = 0$
in polar coordinates is simply $r - R = 0$. The force that keeps the particle on
the constraint surface is a *constraint force*, and that force is perpendicular to the
constraint surface at the point of application of that constraint force. Therefore, it
can also be said that the constraint force is parallel to the *gradient vector* of the
constraint surface since, as known from vector calculus and analytic geometry, such
a vector is perpendicular to the tangent to that surface (an illustration is given
following the next paragraph).

A constraint surface is fixed in space if time t does not appear in a constraint
equation, as in $a_1(q_1, q_2, \ldots, q_n) = 0$. Otherwise, the surface of constraint is moving
if the constraint is of the form $a_1(q_1, q_2, \ldots, q_n; t) = 0$, i.e., if time appears explicitly
in the equation of constraint. In either case, the virtual work of the constraint force
is equal to zero by definition of a constraint force. Such a force is perpendicular to
the constraint surface, and the virtual displacement of the point of application of
such a force is tangent to that surface.

Figure 7.13 illustrates the notion of the *gradient vector*, using a function of two
variables which, for simplicity, is taken to be a circle of radius L centered at the

origin O' of a nonrotating (but translating) rectangular coordinate system x-y. An example of a system for which this constraint applies is a pendulum of length L, in planar motion, able to rotate about point O' which, in turn, is forced to move in the x-y plane with any desired known motion. The $\hat{x}$-$\hat{y}$ and $\hat{r}$-$\hat{n}$ unit vectors shown in Fig. 7.13 will be used later.

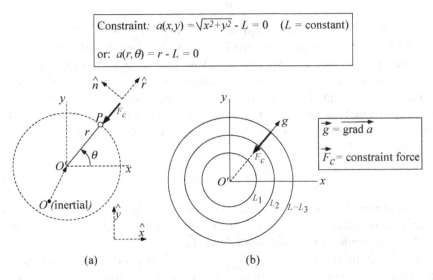

(a) $\qquad\qquad\qquad$ (b)

Figure 7.13: The gradient vector of a constraint and the force of constraint.

Point P of the pendulum shown in Fig. 7.13a is constrained to move, relative to point O', on the dashed circle shown. The constraint is holonomic in this case and is expressed either in terms of the x-y coordinates or the polar r-θ coordinates as follows:

$$a(x, y) = \sqrt{x^2 + y^2} - L = 0 \qquad (7.8.2)$$

or

$$a(r) = r - L = 0 \qquad (7.8.3)$$

In terms of the x-y coordinates, the gradient vector of the function $a(x, y) = \sqrt{x^2 + y^2} - L$ is, from vector calculus and analytic geometry,

$$\overrightarrow{\text{grad } a} = \frac{\partial a}{\partial x}\hat{x} + \frac{\partial a}{\partial y}\hat{y} = \frac{x\hat{x} + y\hat{y}}{\sqrt{x^2 + y^2}} \qquad (7.8.4)$$

while, using polar coordinates, and the simplest representation of the position vector of O relative to O', i.e., $\overrightarrow{OO'} = r\hat{r}$, the gradient vector of the function $a(r) = r - L$ is obtained as

$$\overrightarrow{\text{grad}\, a} = \frac{\partial a}{\partial r}\hat{r} = \hat{r} \qquad\qquad (7.8.5)$$

Equations (7.8.4) and (7.8.5) are equivalent to each other, as can be easily verified with the coordinate transformations (see Fig. 7.13) $x = r\cos\theta$ and $y = r\sin\theta$, and the transformation of triads $\hat{x} = \hat{r}\cos\theta - \hat{n}\sin\theta$ and $\hat{y} = \hat{r}\sin\theta + \hat{n}\cos\theta$. Clearly, the gradient vector is normal to the constraint surface, and so is the force of constraint $\vec{F}_c$, which is shown in Fig. 7.13. Therefore, it is concluded that a constraint force is always proportional to the gradient of the constraint surface, and one can then write, with $\lambda = \lambda(t)$ being a proportionality factor (whose magnitude and sign are unknown),

$$\vec{F}_c = \lambda\,\overrightarrow{\text{grad}\, a} \qquad\qquad (7.8.6)$$

A more convenient form of this expression is $\vec{F}_c = \lambda\,\overrightarrow{\text{grad}\, a}/|\overrightarrow{\text{grad}\, a}|$, where the vector $\overrightarrow{\text{grad}\, a}$ is normalized by its magnitude (to obtain a unit vector) so that $|\lambda|$ becomes the magnitude of the constraint force $\vec{F}_c$.

Returning now to the equations of (7.8.1), each one of those equations of a holonomic constraint represents a constraint surface in the space $q_1, \ldots, q_n$, which, in turn, represents an actual physical surface in three-dimensional space in case $n = 3$. This is illustrated in Fig. 7.14 for the single-mass spring-pendulum system

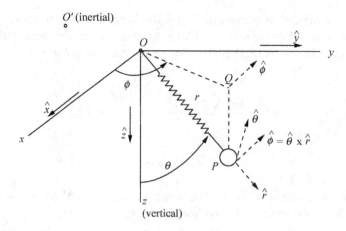

Figure 7.14: The spring-pendulum in three-dimensional space.

where the support point O is given a known motion $\overrightarrow{O'O}(t)$ and, thus, the virtual displacement $\delta(\overrightarrow{O'O})$ is equal to zero. If the pendulum of Fig. 7.14 is constrained so that $r = \text{constant} \overset{\Delta}{=} L$, point P is forced to move on a spherical constraint surface $a_r \overset{\Delta}{=} r - L = 0$ centered at point O. Because of the motion of point O, the spherical constraint surface is moving in space. The gradient of that surface is $\hat{r}$ (as shown in Fig. 7.14), and the constraint force associated with that constraint surface is $\vec{F}_{c_1} = \lambda_1 \hat{r}$. If, in addition, the motion is also restricted so that $\phi = \text{constant} \overset{\Delta}{=} \phi_0$, point P is also constrained to move on the plane that contains the vertical that passes through O and line OP. This second constraint surface is $a_\phi \overset{\Delta}{=} \phi - \phi_0 = 0$ (which is also a moving surface if O is in motion). Its gradient is $\hat{\phi}$ (as shown in Fig. 7.14), and the constraint force associated with this latter constraint surface is $\vec{F}_{c_2} = \lambda_2 \hat{\phi}$. In the presence of the two constraints $a_r = r - L = 0$ and $a_\phi = \phi - \phi_0 = 0$, the motion of P is forced to be in the intersection of the two surfaces of constraint. For simple problems, such as the one involving a single mass particle (as just presented), it is seen that the λ multiplier method for finding the constraint forces can yield those forces without ambiguity. However, this is not the case for more complicated problems.

Analytical dynamics is not the best approach for finding constraint forces. In fact, the idea behind such an approach is to eliminate them right at the beginning of the formulation of any dynamics problem. In general, if one needs to determine constraint forces, it is easier, and more convenient, to go back to the individual free-body diagrams for each mass particle or each rigid body of a multibody system and reformulate the equations using the internal constraint forces to determine them. Such forces will clearly show up in those individual free-body diagrams and there should be no danger of misinterpreting them, which is likely to happen when one tries to find them using the λ multiplier method with Lagrange's equation without a free-body diagram.

Let us now go back to the main problem at hand, which is how to obtain the differential equations of motion of a system in the presence of constraint equations. To do so, one may start by solving the equations of (7.8.1) analytically for the m constrained variables, say, $q_1, q_2, \ldots, q_m$, in terms of the remaining variables $q_{m+1}, q_{m+2}, \ldots, q_n$, *if* that can be done. This, then, eliminates the constrained variables from the formulation, and one is then left with the $n - m$ independent variables $q_{m+1}, q_{m+2}, \ldots, q_n$. But there is a big *if* here; i.e., the equations of constraint may be complicated enough to prevent one from analytically eliminating variables as indicated. For relatively simple illustrations where it is possible to analytically eliminate variables, such as Example 7.7.5 (p. 542), for which the variable θ is prescribed as $\dot{\theta} = \Omega(t)$, and Example 7.7.6 (p. 545), for which the variable ϕ is prescribed as

CHAPTER 7   *Analytical Dynamics*

$\dot{\phi} = \Omega(t)$, where $\Omega(t)$ is a given function of time in both examples. The given constraint equations in both examples could be integrated to yield $\theta(t)$ and $\phi(t)$ for those examples. That is, both constraints were holonomic, despite being given in differential form, and easily solvable for the constrained variables. Also, another problem is that the constraint equations may also involve velocities in a manner that the equations cannot be integrated to yield constraint equations of the form given by the equations of (7.8.1). This is discussed next.

Generally, constraint equations that may appear in a dynamics problem are of the form shown in the m equations of (7.8.7) that involve a total of n variables q_1, $q_2, \ldots, q_m, q_{m+1}, q_{m+2}, \ldots, q_n$, but only the $n - m$ variables $q_{m+1}, q_{m+2}, \ldots, q_n$ are independent of each other.

$$
\begin{aligned}
B_{11}\dot{q}_1 + B_{12}\dot{q}_2 + \ldots + B_{1n}\dot{q}_n &= a_1(q_1, q_2, \ldots, q_n; t) \\
B_{21}\dot{q}_1 + B_{22}\dot{q}_2 + \ldots + B_{2n}\dot{q}_n &= a_2(q_1, q_2, \ldots, q_n; t) \\
&\vdots \\
B_{m1}\dot{q}_1 + B_{m2}\dot{q}_2 + \ldots + B_{mn}\dot{q}_n &= a_m(q_1, q_2, \ldots, q_n; t)
\end{aligned}
\tag{7.8.7}
$$

The equations of (7.8.7) represent m first-order differential equations of constraint. The constraints are called *nonholonomic* if the equations of (7.8.7) cannot be integrated either directly or by introducing an integrating factor.[10] In general, the coefficients $B_{11}, B_{12}, \ldots, B_{mn}$ and the functions $a_1, a_2, \ldots, a_m$ depend on the variables $q_1, q_2, \ldots, q_n$, and also may explicitly depend on the independent variable t.

Holonomic constraints, i.e., integral constraints, are of the general form given by the equations of (7.8.1) and define a surface in space. Nonholonomic constraints involve expressions that cannot be integrated, without the knowledge of the motion, to be put in the form of the equations of (7.8.1). Thus, since the equation of a surface in the space $q_1, \ldots, q_n$ does not involve time derivatives of these variables, one cannot associate surfaces of constraint with nonholonomic constraints. The virtual work of the constraint forces (i.e., the sum of the virtual work of each constraint force) is still equal to zero, though, by definition of constraint forces.

---

[10]For example, the *expression* $2q_2 dq_1 + q_1 dq_2$ is not an exact differential since $\partial(2q_2)/\partial q_2 \neq \partial(q_1)/\partial q_1$. But the *equation* $2q_2 dq_1 + q_1 dq_2 = 0$, when multiplied by q_1, can be integrated because $\partial(2q_1 q_2)/\partial q_2 = \partial q_1^2/\partial q_1 = 2q_1$. The factor q_1 is the *integrating factor*, and the integrated form of constraint equations such as $2q_2 dq_1 + q_1 dq_2 = 0$ or $2q_2 \dot{q}_1 + q_1 \dot{q}_2 = 0$ is $q_1^2 q_2 = \text{constant} \overset{\Delta}{=} C$. Such a constraint, which at first glance seemed to be nonholonomic, is actually holonomic. This means that either variable q_1 or q_2 could be eliminated from a problem where two variables q_1 and q_2 were introduced, but constrained as $2q_2 dq_1 + q_1 dq_2 = 0$. In such a problem, one could simply choose to work with q_1 and then eliminate q_2 by using the relation $q_2 = C/q_1^2$.

7.8.2   Lagrange's Equation in the Presence of Constraints

The equations of constraint must be taken into consideration either in Hamilton's extended principle, i.e., Eq. (7.4.8) (p. 522), or in Eq. (7.4.16) (p. 524) that was obtained directly from that principle. A clever way to do this, which was devised by Lagrange, is to start by obtaining the virtual form of the constraint equations involving the virtual variations $\delta q_1, \ldots, \delta q_n$, as required by Eq. (7.4.16). The virtual form of the nonholonomic constraint equations of (7.8.7) is simply

$$
\begin{aligned}
B_{11}\delta q_1 + B_{12}\delta q_2 + \ldots + B_{1n}\delta q_n &= 0 \\
B_{21}\delta q_1 + B_{22}\delta q_2 + \ldots + B_{2n}\delta q_n &= 0 \\
&\vdots \\
B_{m1}\delta q_1 + B_{m2}\delta q_2 + \ldots + B_{mn}\delta q_n &= 0
\end{aligned}
\tag{7.8.8}
$$

Lagrange's method of dealing with constraints consists in introducing m new variables $\lambda_1, \lambda_2, \ldots, \lambda_m$ (i.e., one new variable for each equation of constraint) and then adding the sum

$$
\lambda_1(B_{11}\delta q_1 + B_{12}\delta q_2 + \ldots + B_{1n}\delta q_n) + \ldots + \lambda_m(B_{m1}\delta q_1 + B_{m2}\delta q_2 + \ldots + B_{mn}\delta q_n)
$$

to Eq. (7.4.16). This certainly does not change Eq. (7.4.16) because only zero is being added to it, but the cleverness of the method is in treating the quantities λ_1, $\lambda_2, \ldots, \lambda_m$ as m new independent variables that are chosen so that

$$
\frac{d}{dt}\frac{\partial T}{\partial \dot{q}_k} - \frac{\partial T}{\partial q_k} = Q_{q_k} + \lambda_1 B_{1k} + \lambda_2 B_{2k} + \ldots + \lambda_m B_{mk}
$$

$$
\text{for } k = 1, 2, \ldots, n
\tag{7.8.9}
$$

The n second-order differential equations represented by Eq. (7.8.9), together with the m first-order differential equations of (7.8.7), form a set of $n+m$ equations in the $n+m$ variables $q_1, q_2, \ldots, q_n, \lambda_1, \lambda_2, \ldots, \lambda_m$. In general, such equations will have to be integrated numerically to obtain the solutions for $q_1, \ldots, q_n$, which are the variables of interest to a specific problem. The m auxiliary variables $\lambda_1, \lambda_2, \ldots, \lambda_m$ that are introduced to handle the constraint equations are called the *Lagrange multipliers*.

This same procedure may also be used when the constraints are of the form given by the equations of (7.8.1). In such a case, Eq. (7.8.9) becomes

$$\frac{d}{dt}\frac{\partial T}{\partial \dot{q}_k} - \frac{\partial T}{\partial q_k} = Q_{q_k} + \lambda_1 \frac{\partial a_1}{\partial q_k} + \lambda_2 \frac{\partial a_2}{\partial q_k} + \ldots + \lambda_m \frac{\partial a_m}{\partial q_k} \qquad \text{for } k = 1, 2, \ldots, n$$

$$(7.8.10)$$

Example 7.9.3 in Section 7.9 illustrates a case where a reasonably complicated holonomic constraint is handled in this manner. The resulting equations consist of n second-order differential equations and m algebraic equations of constraint. In the dynamics and mathematics literature, such a system of equations is called a set of *differential-algebraic equations*.

For holonomic constraints, each equation of constraint defines a constraint surface, and the Lagrange multipliers are proportional to a constraint force, which is normal to that surface. Such correspondence was presented in Section 7.8.1 (p. 554). However, either for multiparticle or multibody systems, the best and safer way to determine constraint forces, if one needs or desires to determine them, is to draw the free-body diagram for each component of the system and then reformulate the equations for those components using Newton's second law and Euler's equation. If one does not need to determine the constraint forces, Lagrange's multiplier method is a clever way to obtain the differential equations of motion of constrained systems.

Constraints can also appear in the form of inequalities. They occur when the motion of a system is limited to some region in space, such as the motion of a particle being limited to the region above the surface of the earth or above the surface defined by an inclined plane.

Generally, the analytical dynamics formulation is the most expeditious way to obtain the differential equations of motion of a dynamical system. Such an approach also has the advantage of providing a methodical search for integrals of the motion that is done in a relatively simple manner when one is able to work with a set of independent variables. The analytical dynamics approach is also the simplest way for obtaining the differential equations of motion of constrained systems when one does not need to determine the constraint forces. However, if one needs to determine the forces of constraint in a constrained system, it would be actually better to use the Newtonian-Eulerian formulation since it unambiguously yields the constraint forces, both in magnitude and in direction, which would clearly appear in the free-body diagrams that one is required to draw in the vectorial formulation. A detailed example of such a situation was presented in Section 5.6 (p. 371), whose specific objective was to determine the internal forces on each element of a four-bar mechanism. In some problems, the best approach might be a combination of the analytical dynamics methods to obtain the differential equations and integrals of motion, and of the Newtonian-Eulerian formulation to obtain the expressions involving the desired constraint forces in specific free-body diagrams that would be drawn showing those forces.

7.9 Solved Example Problems with Constraints

Example 7.9.1 A rigid body of mass m is thrown from a point A so that the initial velocity of its center of mass C has magnitude v_0 and is directed at $\alpha = 30°$ with the horizontal as shown in Fig. 7.15. In addition to the gravitational force with constant magnitude mg, the body is also subjected to a force $\vec{F}$ that constrains the absolute velocity of C so that $\dot{y} = f_1(t)\dot{x} + f_2(t)$. The variables x and y are the coordinates of C in the rectangular nonrotating x-y coordinate system shown in Fig. 7.15, and $f_1(t)$ and $f_2(t)$ are given functions of time, with $v_0 \sin \alpha = f_1(0)v_0 \cos \alpha + f_2(0)$.

a. Obtain the differential equations that govern the motion of C.

b. For $v_0 = 5$ m/s, $x(0) = y(0) = 0$, $f_1(t) = 0.5 \sin t$, and $f_2(t) = v_0 \sin \alpha - 0.85t$, plot $x(t)$ and $y(t)$ versus t and determine, from the plot, the maximum height reached by C.

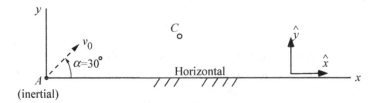

Figure 7.15: Figure for Example 7.9.1.

■ **Solution**

a. Referring to Fig. 7.15, the absolute position vector $\vec{r}_C = \overrightarrow{AC}$ (assuming that point A is inertial) of the center of mass C may be expressed in terms of the nonrotating unit vectors $\hat{x}$ and $\hat{y}$ as

$$\vec{r}_C = x\hat{x} + y\hat{y} \tag{7.9.1a}$$

The kinetic energy for the translational motion of C, and the potential energy U, are then

$$T = \frac{m}{2}(\dot{x}^2 + \dot{y}^2) \tag{7.9.1b}$$

$$U = mgy \qquad (7.9.1c)$$

The constraint force $\vec{F}$ may be expressed as

$$\vec{F} = F_x\hat{x} + F_y\hat{y} \qquad (7.9.1d)$$

and its virtual work $(\delta W)_{\text{due to } \vec{F}} = \vec{F} \bullet \delta\vec{r}_C$ is equal to zero by definition of a constraint force. The virtual work δW done by all forces acting on the system is then

$$\delta W = -\delta(U) + (\delta W)_{\text{due to } \vec{F}} = -mg\delta y \overset{\Delta}{=} Q_x\delta x + Q_y\delta y \qquad (7.9.1e)$$

where $Q_x = 0$.

Unless $f_1(t)$ is constant, the given equation of constraint, $\dot{y} = f_1(t)\dot{x} + f_2(t)$, is nonholonomic because it cannot be converted into an integral form $f(x, y; t) = 0$. Because of the constraint imposed on the motion, the virtual displacements δx and δy are not independent of each other. Their relationship is obtained from the virtual form of the constraint equation. Since the differential form of the constraint equation $\dot{y} - f_1(t)\dot{x} - f_2(t) = 0$ is $dy - f_1(t)dx - f_2(t)dt = 0$, its virtual form is then

$$\delta y - f_1(t)\delta x = 0 \qquad (7.9.1f)$$

We can now obtain the differential equation of motion either from Eq. (7.4.16) (p. 524), or using a Lagrange multiplier as indicated by Eq. (7.8.9) (p. 559) in Section 7.8. For completeness, both approaches, as well as direct application of Newton's second law, are shown here. All approaches yield the same results, which prove their validity.

Using Eq. (7.4.16)

For two variables x and y, Eq. (7.4.16) is:

$$\left(\frac{d}{dt}\frac{\partial T}{\partial \dot{x}} - \frac{\partial T}{\partial x}\right)\delta x + \left(\frac{d}{dt}\frac{\partial T}{\partial \dot{y}} - \frac{\partial T}{\partial y}\right)\delta y - \delta W = 0 \qquad (7.9.1g)$$

With $\delta y = f_1(t)\delta x$, as per Eq. (7.9.1f), this then yields

$$m\ddot{x} + (m\ddot{y} + mg)f_1(t) = 0 \qquad (7.9.1h)$$

Equation (7.9.1h) and the constraint equation

$$\dot{y} = f_1(t)\dot{x} + f_2(t) \qquad (7.9.1i)$$

are the differential equations of motion for the system. They can be combined to yield the following differential equation that only involves the variable x:

$$[1 + f_1^2(t)]\ddot{x} + [\dot{x}\dot{f}_1(t) + \dot{f}_2(t) + g]f_1(t) = 0 \qquad (7.9.1\text{j})$$

Using a Lagrange Multiplier

Only one Lagrange multiplier λ is introduced since there is only one equation of constraint. For the two variables x and y, Eq. (7.8.9) then implies

$$\frac{d}{dt}\frac{\partial T}{\partial \dot{x}} - \frac{\partial T}{\partial x} = Q_x + \lambda B_x \qquad (7.9.1\text{k})$$

$$\frac{d}{dt}\frac{\partial T}{\partial \dot{y}} - \frac{\partial T}{\partial y} = Q_y + \lambda B_y \qquad (7.9.1\text{l})$$

where $B_y = 1$ and $B_x = -f_1(t)$, as disclosed by the virtual form of the constraint given by Eq. (7.9.1f). Also, $Q_x = 0$ and $Q_y = -mg$, as disclosed by the expression for δW given by Eq. (7.9.1e). Equations (7.9.1k) and (7.9.1l) then yield

$$m\ddot{x} = -\lambda f_1(t) \qquad (7.9.1\text{m})$$

$$m\ddot{y} = -mg + \lambda \qquad (7.9.1\text{n})$$

or, by eliminating λ,

$$m\ddot{x} = -(m\ddot{y} + mg)f_1(t)$$

which is the same as Eq. (7.9.1h).

Using Newton's Second Law

Direct application of Newton's second law yields the following vector equation:

$$F_x\hat{x} + (F_y - mg)\hat{y} = m(\ddot{x}\hat{x} + \ddot{y}\hat{y}) \qquad (7.9.1\text{o})$$

This vector equation yields the following two scalar equations:

$$m\ddot{x} = F_x \qquad (7.9.1\text{p})$$

$$m\ddot{y} = F_y - mg \qquad (7.9.1\text{q})$$

Equations (7.9.1p) and (7.9.1q), together with the equation of constraint, Eq. (7.9.1i), form a set of three equations in four unknowns, namely, x, y, F_x, and F_y. Thus, a fourth equation is needed. Either F_x or F_y, or a combination of both, must be specified for the problem to be completely solvable. The problem statement specified that the only forces acting on the body are the gravitational force and the constraint force $\vec{F} = F_x\hat{x} + F_y\hat{y}$. Since the virtual work of the constraint force is zero, that *is* the fourth equation, i.e.,

$$F_x\delta x + F_y\delta y = F_x\delta x + F_y f_1(t)\delta x = 0 \qquad (7.9.1\text{r})$$

from which one obtains

$$F_x = -f_1(t)F_y \qquad (7.9.1\text{s})$$

It can be readily verified that Eqs. (7.9.1p), (7.9.1q), and (7.9.1s) yield Eq. (7.9.1j). Also, comparison of Eqs. (7.9.1m) and (7.9.1p), and Eqs. (7.9.1n) and (7.9.1q), discloses that the Lagrange multiplier λ is equal to the force F_y, and that $F_x = -\lambda f_1(t)$.

A note of caution

One should resist the temptation to make shortcuts when dealing with non-holonomic constraints, such as substituting velocity (or velocities) given by the constraint equation (or equations) into the kinetic energy and then using Eq. (7.4.16) (p. 524), because this will yield wrong equations. This can be verified in this example by substituting $\dot{y} = f_1(t)\dot{x} + f_2(t)$ into Eq. (7.9.1b) and then applying Eq. (7.4.16) to the result. By comparing the resulting differential equation with that obtained here by three methods, including direct application of Newton's second law, it is immediately seen that it is wrong. For nonholonomic constraints, such a shortcut is inconsistent with the formulation that yields Eq. (7.4.16). The formulation is based on the use of all the constrained and unconstrained variables that are introduced in the problem and the variational form of the equations of constraint. The shortcut works for holonomic constraints because, in such a case, constraint equations of the form $(\partial a_i/\partial q_1)\delta q_1 + \ldots + (\partial a_i/\partial q_n)\delta q_n = 0$, for $i = 1, \ldots, m$, imply the integrated equations of (7.8.1), while the equations of (7.8.8) for nonholonomic constraints cannot be expressed in the integrated form of the equations of (7.8.7). In other words, the equations of (7.8.7) imply the equations of (7.8.8), while the converse is not true.

---

b. Figure 7.16 shows the result of the numerical integration of Eqs. (7.9.1i) and (7.9.1j) with $f_1 = 0.5\sin t$ and $f_2 = v_0\sin\alpha - 0.85t$, for $v_0 = 5$ m/s, $\alpha = 30°$, and the initial conditions $x(0) = y(0) = 0$. As disclosed by Fig. 7.15, the initial conditions for $\dot{x}$ and $\dot{y}$ are $\dot{x}(0) = v_0\cos\alpha$ and $\dot{y}(0) = v_0\sin\alpha$. The maximum value for $y(t)$, obtained from Fig. 7.16a, is $y_{max} \approx 4.2$ m. The trajectory $y(t)$ versus $x(t)$ followed by the center of mass of the body is shown in Fig. 7.16b, with the arrow indicating how the curve is traced. The curve resembles the trajectory of a boomerang.

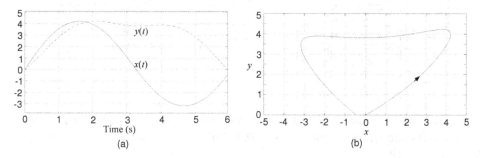

Figure 7.16: (a) Displacements $x(t)$ and $y(t)$ and (b) the trajectory $y(t)$ versus $x(t)$ for Example 7.9.1.

$\implies$ Two similar programs written as functions named *boomerang*, listed in Appendix E, integrate the differential equations of motion, generate the plots shown in Fig. 7.16 (except for the arrow shown in Fig. 7.16b), and animate the motion. One is for Scilab users (which Scilab would save as *boomerang.sci*) and the other is for Matlab users (which Matlab would save as *boomerang.m*). An example of their usage, with the initial conditions $x(0) = 0$, $\dot{x}(0) = 5\cos(\pi/6)$, and $y(0) = 0$, is given in the first few comment lines of both programs. It consists of simply typing either
`my_output = boomerang([0:0.01:6], 9.81, 0, 5*cos(%pi/6), 0);`
in the Scilab console or
`my_output = boomerang([0:0.002:6], 9.81, 0, 5*cos(pi/6), 0);`
in the Matlab console. As mentioned in other parts of the book, different time intervals (0.1 for Scilab and 0.002 for Matlab) are used because of the different execution speeds of the two software. After either program stops running, the variable *my_output* will be a *len* by 4 matrix whose columns are t, x, $\dot{x}$, and y, where *len* is the number of time points in the given array t; the array t is the first argument of the program *boomerang*. It is of interest only if one wishes to generate additional plots, such as `plot(my_output(:, 1), my_output(:, 3))` to plot $\dot{x}$ versus t for example.

Those who have Simulink and prefer to use it may want to construct the block diagram model shown in Fig. 7.17, and use the initial conditions mentioned in the previous paragraph. After the simulation stops, an animation of the

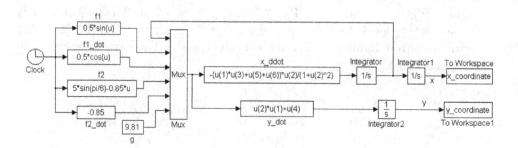

Figure 7.17: A Simulink model for integrating Eqs. (7.9.1i) and (7.9.1j). For the model inputs shown, the time array [0:0.002:6] is suggested for the *Simulation* → *Parameters* settings.

motion is generated by typing
`animate_nbars(x_coordinate, y_coordinate)`
in the Matlab command window. For this, the Matlab program `animate_nbars`, also listed in Appendix E, is needed.

Example 7.9.2 Constraint Equations for a Classical Example of a Nonholonomic Constraint

A classical example of a nonholonomic constraint is a vertical disk of radius R rolling without slipping on a horizontal surface, with the disk being able to turn about a vertical axis as shown in Fig. 7.18. The more complicated situation, and more realistic, of course, is when the disk also tilts so that it does not remain vertical at all times. The case shown in Fig. 7.18 illustrates the simpler nonholonomic constraint when the disk remains vertical. The path shown in Fig. 7.18, followed by the point of contact of the disk with the x-y plane, may or may not be prescribed. The angle ϕ shown in the figure is the angle between the reference x-axis and the tangent to that path at the point of contact. The angle θ is used for tracking the orientation of line CP relative to the reference z-axis, where C is the center of the disk and P is a point fixed to its periphery. What are the constraint equations for this problem?

For rolling without slipping, the virtual displacement of point C must match that of the point of contact of the disk. Therefore, with x_C and y_C denoting the x and y coordinates of point C, the rolling with no slipping condition implies the following virtual equations of constraint (see Fig. 7.18):

$$\delta x_C = (R\,\delta\theta)\cos\phi \tag{7.9.2a}$$

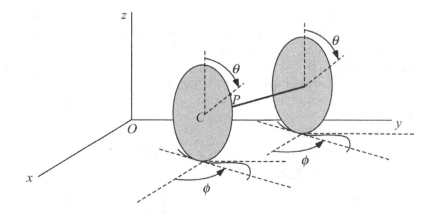

Figure 7.18: A vertical disk rolling, without slipping, on a horizontal surface.

$$\delta y_C = (R\,\delta\theta)\sin\phi \qquad\qquad (7.9.2b)$$

Notice that in terms of velocity, if the velocity of the center C of the disk is to be a vector that is in the plane of the disk and parallel to the tangent line shown in Fig. 7.18 (i.e., no lateral slip either), such a condition implies that

$$\dot{x}_C - R\dot{\theta}\cos\phi = 0 \qquad\qquad (7.9.2c)$$
$$\dot{y}_C - R\dot{\theta}\sin\phi = 0 \qquad\qquad (7.9.2d)$$

and these equations are of the form of the general equations of constraint [i.e., the equations of (7.8.7)] with their right-hand sides equal to zero.

Equations (7.9.2c) and (7.9.2d) cannot be expressed in integral form, i.e., in a form that involves only the variables x_C, y_C, ϕ, and θ, but not their time derivatives. Therefore, these constraints are nonholonomic. One is, then, forced to work with the four constrained variables x_C, y_C, ϕ, and θ. Equations (7.9.2c) and (7.9.2d) can be expressed in integral form only if $\phi = $ constant, i.e., if the disk rolls along a straight line. The case where $\phi = $ constant is a simpler classical case of rolling, and it was analyzed in detail in Section 5.4 (p. 354) of Chapter 5.

Example 7.9.3 Dynamics of a Four-Bar Mechanism

The four-bar mechanism shown in Fig. 7.19 is driven by a motor that applies to bar AB a known moment $M_{\text{motor}}(t)$. The motion takes place on a horizontal plane and is started from rest [i.e., with $\dot{\theta}_2(0) = 0$] with a given initial value of θ_2. The mechanism consists of three moving thin homogeneous bars AB, BC, and CD of lengths L_2, L_3, and L_4, and masses m_2, m_3, and m_4, respectively. Obtain the differential equation of motion for this constrained system in terms of θ_2, and integrate it numerically. For $L_1 = 1$ m, $L_2 = 0.4$ m, $L_3 = 0.8$ m, $L_4 = 0.7$ m, $m_2 = 3.75$ kg, $m_3 = 3$ kg, $m_4 = 3$ kg, and $M_{\mathrm{motor}}(t) = \pi/6 + 1.5\pi \sin(kt)$ N·m (where $k = 2$ rad/s), plot θ_4 and θ_2 versus t for t in a reasonable range, determine the maximum and minimum values of $\theta_4(t)$, and show an animation of the motion. The motion is started with $\theta_2(0) = 30°$.

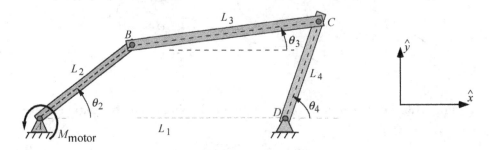

Figure 7.19: A four-bar mechanism subjected to a known moment M_{motor} applied to bar AB.

■ **Solution**

In terms of the unit vectors $\hat{x}$ and $\hat{y}$ shown in Fig. 7.19, the absolute velocity $\vec{v}_{C_3}$ of the center of mass C_3 of bar BC may be written as:[11]

$$
\begin{aligned}
\vec{v}_{C_3} &= \frac{d}{dt}\overrightarrow{DC} + \frac{d}{dt}\overrightarrow{CC_3} = \frac{d}{dt}\left[L_4\left(\hat{x}\cos\theta_4 + \hat{y}\sin\theta_4\right)\right] \\
&\quad + \frac{d}{dt}\left[\frac{L_3}{2}\left(-\hat{x}\cos\theta_3 - \hat{y}\sin\theta_3\right)\right] \\
&= \left(\frac{L_3}{2}\dot{\theta}_3\sin\theta_3 - L_4\dot{\theta}_4\sin\theta_4\right)\hat{x} + \left(L_4\dot{\theta}_4\cos\theta_4 - \frac{L_3}{2}\dot{\theta}_3\cos\theta_3\right)\hat{y} \quad (7.9.3a)
\end{aligned}
$$

The following expression is then obtained for the kinetic energy of the motion:

$$
\begin{aligned}
T &= \frac{1}{2}I_A\dot{\theta}_2^2 + \frac{1}{2}I_D\dot{\theta}_4^2 + \frac{1}{2}I_3\dot{\theta}_3^2 + \frac{1}{2}m_3\vec{v}_{C_3}\bullet\vec{v}_{C_3} = \frac{1}{2}I_A\dot{\theta}_2^2 + \frac{1}{2}I_D\dot{\theta}_4^2 + \frac{1}{2}I_3\dot{\theta}_3^2 \\
&\quad + \frac{1}{2}m_3\left[(L_4\dot{\theta}_4\cos\theta_4 - \frac{L_3}{2}\dot{\theta}_3\cos\theta_3)^2 + \left(L_4\dot{\theta}_4\sin\theta_4 - \frac{L_3}{2}\dot{\theta}_3\sin\theta_3\right)^2\right] \\
&= \frac{1}{2}I_A\dot{\theta}_2^2 + \frac{1}{2}I_D\dot{\theta}_4^2 + \frac{1}{2}I_3\dot{\theta}_3^2 + \frac{1}{2}m_3\left[L_4^2\dot{\theta}_4^2 + \frac{1}{4}L_3^2\dot{\theta}_3^2 - L_3L_4\dot{\theta}_3\dot{\theta}_4\cos(\theta_4 - \theta_3)\right]
\end{aligned}
$$

$$(7.9.3b)$$

---

[11]The expression for $\vec{v}_{C_3}$ could also have been written as $\vec{v}_{C_3} = d\overrightarrow{AB}/dt + d\overrightarrow{BC_3}/dt$. It makes no difference which form is used.

In Eq. (7.9.3b), $I_A = m_2 L_2^2/3$ is the moment of inertia of bar AB about the vertical axis that passes through A, $I_D = m_4 L_4^2/3$ is the moment of inertia of bar CD about the vertical axis that passes through D, and $I_3 = m_3 L_3^2/12$ is the moment of inertia of bar BC about the vertical axis that passes through its center of mass. Using these expressions for the moments of inertia imply that the thin bars are being approximated as lines AB, CD, and BC, respectively.

The angles θ_2, θ_3, and θ_4 are dependent on each other. As first presented in Chapter 3, the following constraint relations between them are obtained by inspection of Fig. 7.19:

$$\underbrace{L_2 \cos\theta_2 + L_3 \cos\theta_3 - L_4 \cos\theta_4 - L_1}_{\triangleq\, a_1(\theta_2, \theta_3, \theta_4)} = 0 \qquad (7.9.3c)$$

$$\underbrace{L_2 \sin\theta_2 + L_3 \sin\theta_3 - L_4 \sin\theta_4}_{\triangleq\, a_2(\theta_2, \theta_3, \theta_4)} = 0 \qquad (7.9.3d)$$

The virtual work δW done by all forces acting on the system is simply

$$\delta W = M_{\text{motor}} \delta\theta_2 \triangleq Q_{\theta_2} \delta\theta_2 \qquad (7.9.3e)$$

Generally, it is more convenient to use Lagrange multipliers to handle constraint equations. Since there are two of such equations, two multipliers $\lambda_1(t)$ and $\lambda_2(t)$ are needed. For the three variables θ_2, θ_3, and θ_4, Eq. (7.8.10) (p. 560) then implies

$$\frac{d}{dt}\frac{\partial T}{\partial \dot\theta_2} - \frac{\partial T}{\partial \theta_2} = Q_{\theta_2} + \lambda_1 \frac{\partial a_1}{\partial \theta_2} + \lambda_2 \frac{\partial a_2}{\partial \theta_2} \qquad (7.9.3f)$$

$$\frac{d}{dt}\frac{\partial T}{\partial \dot\theta_3} - \frac{\partial T}{\partial \theta_3} = Q_{\theta_3} + \lambda_1 \frac{\partial a_1}{\partial \theta_3} + \lambda_2 \frac{\partial a_2}{\partial \theta_3} \qquad (7.9.3g)$$

$$\frac{d}{dt}\frac{\partial T}{\partial \dot\theta_4} - \frac{\partial T}{\partial \theta_4} = Q_{\theta_4} + \lambda_1 \frac{\partial a_1}{\partial \theta_4} + \lambda_2 \frac{\partial a_2}{\partial \theta_4} \qquad (7.9.3h)$$

where $Q_{\theta_2} = M_{\text{motor}}$; $Q_{\theta_3} = Q_{\theta_4} = 0$, and the functions $a_1(\theta_2, \theta_3, \theta_4)$ and $a_2(\theta_2, \theta_3, \theta_4)$ are defined in Eqs. (7.9.3c) and (7.9.3d).

Equations (7.9.3f) to (7.9.3h) yield the following second-order differential equations:

$$\frac{1}{3} m_2 L_2^2 \ddot\theta_2 = M_{\text{motor}} - \lambda_1 L_2 \sin\theta_2 + \lambda_2 L_2 \cos\theta_2 \qquad (7.9.3i)$$

$$\frac{1}{3} m_3 L_3^2 \ddot\theta_3 - \frac{1}{2} m_3 L_3 L_4 \left[\ddot\theta_4 \cos(\theta_4 - \theta_3) - \dot\theta_4^2 \sin(\theta_4 - \theta_3) \right]$$
$$= -\lambda_1 L_3 \sin\theta_3 + \lambda_2 L_3 \cos\theta_3 \qquad (7.9.3j)$$

$$\left(\frac{1}{3}m_4 + m_3\right)L_4^2\ddot{\theta}_4 - \frac{1}{2}m_3L_3L_4\left[\ddot{\theta}_3\cos(\theta_4 - \theta_3) + \dot{\theta}_3^2\sin(\theta_4 - \theta_3)\right]$$
$$= \lambda_1 L_4 \sin\theta_4 - \lambda_2 L_4 \cos\theta_4 \tag{7.9.3k}$$

Equations (7.9.3i) to (7.9.3k) involve five unknowns, which are θ_2, θ_3, θ_4, and the two Lagrange multipliers λ_1 and λ_2. They can be solved by augmenting them with Eqs. (7.9.3l) and (7.9.3m) that follow, which are obtained by taking the second time derivative of the algebraic constraint equations. The five equations are combined into a single matrix equation that can be solved using matrix algebra. In Eq. (7.9.3n), c43 stands for $\cos(\theta_4 - \theta_3)$; this symbol is used here to fit the left side of the equation on the page.

$$L_2\ddot{\theta}_2\sin\theta_2 + L_3\ddot{\theta}_3\sin\theta_3 - L_4\ddot{\theta}_4\sin\theta_4 + L_2\dot{\theta}_2^2\cos\theta_2 + L_3\dot{\theta}_3^2\cos\theta_3 - L_4\dot{\theta}_4^2\cos\theta_4 = 0 \tag{7.9.3l}$$

$$L_2\ddot{\theta}_2\cos\theta_2 + L_3\ddot{\theta}_3\cos\theta_3 - L_4\ddot{\theta}_4\cos\theta_4 - L_2\dot{\theta}_2^2\sin\theta_2 - L_3\dot{\theta}_3^2\sin\theta_3 + L_4\dot{\theta}_4^2\sin\theta_4 = 0 \tag{7.9.3m}$$

$$\begin{bmatrix} \frac{1}{3}m_2L_2^2 & 0 & 0 & L_2\sin\theta_2 & -L_2\cos\theta_2 \\ 0 & \frac{1}{3}m_3L_3^2 & -\frac{1}{2}m_3L_3L_4c43 & L_3\sin\theta_3 & -L_3\cos\theta_3 \\ 0 & -\frac{1}{2}m_3L_3L_4c43 & (\frac{1}{3}m_4 + m_3)L_4^2 & -L_4\sin\theta_4 & L_4\cos\theta_4 \\ L_2\sin\theta_2 & L_3\sin\theta_3 & -L_4\sin\theta_4 & 0 & 0 \\ L_2\cos\theta_2 & L_3\cos\theta_3 & -L_4\cos\theta_4 & 0 & 0 \end{bmatrix}\begin{bmatrix} \ddot{\theta}_2 \\ \ddot{\theta}_3 \\ \ddot{\theta}_4 \\ \lambda_1 \\ \lambda_2 \end{bmatrix}$$

$$= \begin{bmatrix} M_{motor} \\ -\frac{1}{2}m_3L_3L_4\dot{\theta}_4^2\sin(\theta_4 - \theta_3) \\ \frac{1}{2}m_3L_3L_4\dot{\theta}_3^2\sin(\theta_4 - \theta_3) \\ L_4\dot{\theta}_4^2\cos\theta_4 - L_3\dot{\theta}_3^2\cos\theta_3 - L_2\dot{\theta}_2^2\cos\theta_2 \\ -L_4\dot{\theta}_4^2\sin\theta_4 + L_3\dot{\theta}_3^2\sin\theta_3 + L_2\dot{\theta}_2^2\sin\theta_2 \end{bmatrix} \tag{7.9.3n}$$

The implementation of this strategy is summarized in the flowchart shown in Fig. 7.20.

⟹ The strategy indicated in Fig. 7.20 is implemented in two similar programs named *fourbar_mech_dynamics* listed in Appendix E. One is for Scilab users (which Scilab would save as *fourbar_mech_dynamics.sci*) and the other is for Matlab users (which Matlab would save as *fourbar_mech_dynamics.m*. Both animate the motion, and plot θ_4 and θ_2 versus time t. An example of their use is given in the first few comment lines at the beginning of each program. As indicated in the tutorials given in Appendix A, plots can easily be done using the data generated by the programs listed in Appendix E.

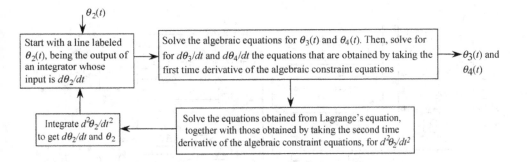

Figure 7.20: Flowchart for solving the differential-algebraic equations for the dynamics of a four-bar mechanism when Lagrange multipliers are used.

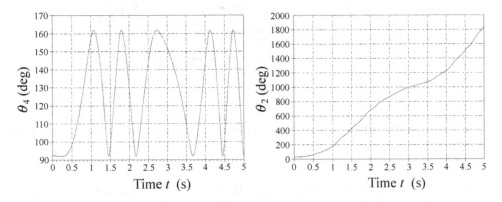

Figure 7.21: Plots of $\theta_4(t)$ and $\theta_2(t)$.

Figure 7.21 shows a solution for $\theta_4(t)$ and $\theta_2(t)$, starting with the configuration defined by "solution number 1" given by Eq. (3.2.14), p. 231, with the data specified at the beginning of this example and with $\dot{\theta}_2(0) = 0$. The maximum and minimum values that are obtained for θ_4 are $\theta_{4_{\max}} \approx 162°$ and $\theta_{4_{\min}} \approx 92°$. The plots and the animation show that θ_4 oscillates between these two values, while θ_2 continually increases. This happens because the average value of the motor torque M_{motor} is positive.

Those who have Simulink and prefer to use it may want to construct the block diagram model shown in Fig. 7.22 instead of using either one of the programs just mentioned. Such a model can be constructed by modifying the model shown in Fig. 3.9 in Chapter 3 (p. 239). By comparing Figs. 3.9 and

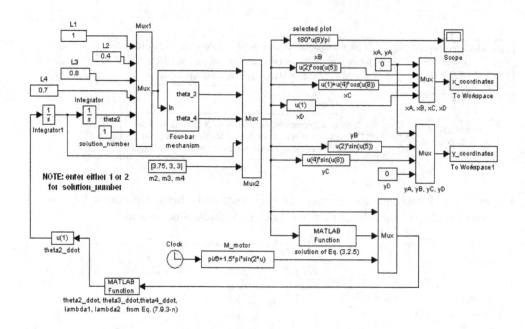

Figure 7.22: A Simulink model implementing the solution strategy shown in Fig. 7.20.

7.22, it is seen that one of the modifications made for constructing the model shown in Fig. 7.22 consists of the inclusion of two integrators for generating $\dot{\theta}_2$ and θ_2 using the expression for $\ddot{\theta}_2$ obtained from Eq. (7.9.3n). Conforming to Eq. (7.9.3n), that expression is the first of the five outputs of the Matlab Function block labeled "theta2_ddot, ..., lambda from Eq. (7.9.3n)". That output is selected in the block named *theta2_ddot* in Fig. 7.22 and then connected to the input of the integrator labeled "Integrator1", thus closing the loop in the model of Fig. 7.22 and in the flowchart of Fig. 7.20.

7.10 Problems

7.1 Obtain the differential equations of motion for the double pendulum in Problem 2.43 (p. 218) and all classical integrals of motion, using Lagrange's formulation.

7.2 Show that the differential equations of motion you obtained for Problem 2.43 in Chapter 2 are equivalent to those obtained in Problem 7.1 using Lagrange's formulation. In other words, if they look different, show that an appropriate combination of the equations obtained in one problem yields the equations obtained in the other problem.

7.3 Show that Eqs. (7.7.2e) (p. 536) and (7.7.2f), which were derived using Lagrange's equation, are equivalent to Eqs. (2.5.9g) (p. 126) and (2.5.9i), derived using Newton's second law. In other words, show that an appropriate combination of the equations in one set yields the equations in the other set.

7.4 Obtain the differential equation of motion for Problem 2.26 (p. 213) using Lagrange's equation, and all the classical integrals of motion.

7.5 Obtain the differential equations of motion for Problem 2.27 (p. 213) using Lagrange's equation, and all the classical integrals of motion.

7.6 A small block of mass m is able to slide inside a frictionless groove that rotates about a vertical axis with a constant angular velocity $\dot{\theta} \triangleq \Omega$ as shown in Fig. 7.23. The block is connected by a linear spring to a fixed point A at a

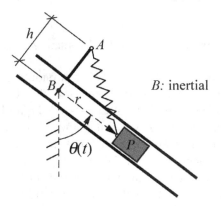

Figure 7.23: Figure for Problem 7.6.

constant distance h above a fixed point B on the axis of rotation, as shown in Fig. 7.23. Motion takes place on a horizontal plane. Disregard the moment of inertia of the small block about its center of mass.

The spring stiffness is k N/m, and its unstressed length is $L_u = ah$ m, where a is given. The motion of the block is started with the initial conditions

$r(0) = 0$ and $\dot{r}(0) = \Omega h$, where $r(t)$ is shown in Fig. 7.23. Making use of classical integrals of motion obtained with Lagrange's formulation, determine, for $m\Omega^2/k = 1/2$, and for $a = 2$ and $a = 1/2$, the maximum and minimum values of $r(t)/h$. Also, analyze the stability of the equilibrium $r_e = 0$ by: (1) linearizing the differential equation of motion for small $r(t)$, and (2) using Lyapunov's method.

7.7   Consider the double pendulum shown in Fig. 7.24 where a linear torsional dashpot (i.e., a viscous torsional damper) whose damping coefficient is c

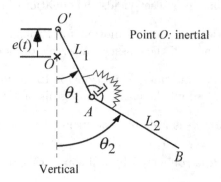

Figure 7.24: Figure for Problem 7.7.

N·m/(rad/s) is added in parallel to a linear torsional spring with stiffness k N·m/rad. The spring is unstressed when $\theta_2 = \theta_1$. Links $O'A$ and AB are modeled as thin homogeneous bars of masses m_1 kg and m_2 kg, respectively. The bars can only move on the vertical plane. Point O' represents a hinge that is forced to move (by an external force not shown in the figure) with a known vertical displacement $e(t)$. Obtain the expressions for the potential energy of the motion, for the virtual work, for the generalized conservative and nonconservative forces associated with θ_1 and θ_2, and the differential equations of motion for the system using Lagrange's equation. Also, for $c \geq 0$, obtain all classical integrals of the motion. For what type of $e(t) \neq 0$ is the Hamiltonian a constant?

7.8   Consider the system of Example 2.8.8 (p. 201), where the vertical parabolic track is spun about the vertical y-axis shown in Fig. 2.50 (p. 201) with a constant angular velocity $\Omega = 50$ rpm relative to inertial space. Neglect friction in the track. Making use of classical integrals of motion obtained with Lagrange's formulation, determine the maximum (y_{max}) and the minimum

$(y_{\min})$ values of the height y reached by the particle. The particle is released with $x(0) = -1.5$ m, and $\sqrt{\dot{x}^2(0) + \dot{y}^2(0)} = 6$ m/s. What are the corresponding values of the distance x? Use the data given in Example 2.8.8 in Chapter 2.

7.9 For the system in Problem 7.8, plot all the equilibrium solutions $x = $ constant $\overset{\Delta}{=} x_e$ versus Ω, for $0 \leq \Omega \leq 50$ rpm. Use the following data: $m = 20$ grams, $a = 1/3$ m^{-1}, $k = 9$ N/m, $L_u = h = 1.2$ m. Also, use the expression for the integral of motion obtained in that problem to draw an accurate phase plane plot $\dot{x}$ versus x, with $\Omega = 20$ and $\Omega = 30$ rpm. Superimpose all the plots drawn for the same value of Ω. Draw the plots using $\dot{x}(0) = 0$ and several values of $x(0)$ near $x = 0$ to obtain different curves. Determine, from any of your plots, the value of Ω when multiple equilibrium solutions start to occur. How does that value compare with the one that is obtained from the linearized differential equation about the equilibrium $x_e = 0$?

7.10 Verify the results of Problems 7.8 and 7.9 by numerical integration of the differential equation of motion of the system.

7.11 Consider the system shown in Fig. 2.40 (p. 174) described in Example 2.7.10. However, instead of particle P_2 being constrained to move along a vertical, let it move in three-dimensional space. Introduce two angles to describe the orientation of line OP_2, and then use Lagrange's formulation to obtain the differential equations of motion for the system and all classical integrals of the motion. Neglect friction in the system.

7.12 Use Lagrange's formulation to obtain the differential equations of motion and all the classical integrals of motion for the two-body problem described in Example 4.3.2 (p. 276). Use the variables r and θ shown in Fig. 4.5 (p. 277).

7.13 The Geneva wheel mechanism shown in Fig. 3.12 (p. 241) is driven by a motor that applies to bar AB a known moment $M_{\text{motor}}(t)$. The motion takes place on a vertical plane. The mechanism consists of a homogeneous bar AB of constant length r_1 and mass m_1, with a pin at B that slides on a slotted bar CP of length L_2 and mass m_2. As shown in Fig. 3.12, the distance L between the fixed pins at A and C is constant. Neglect friction in the system. Using Lagrange's formulation, obtain the equations that govern the motion of that constrained system.

7.14 The driver in the slider crank mechanism shown in Fig. 3.15 (p. 251) is a homogeneous bar of length L_1 and mass m_1. Bar AB, of length L_2, is homogeneous and its mass is m_2. The mass of the block at B is m_B, and $D = 0$.

A motor applies to bar OA a known moment $M_{\text{motor}}(t)$. The motion takes place on a horizontal plane. Neglect friction in the system. Using Lagrange's formulation, obtain the equations that govern the motion of that constrained system.

7.15 The scotch yoke mechanism shown in Fig. 3.21 (p. 260) consists of a thin ho-mogeneous wheel of radius R and mass m_1, with a small pin P at its periphery, and a slotted body of mass m_2 that slides along a straight track. Motion is on a horizontal plane. Using Lagrange's formulation, obtain the differential equation of motion for the system when the wheel is driven by a motor at O that applies a known moment $M_{\text{motor}}(t)$ to the system. Take point O to be in-ertial, and neglect friction in the system. For $M_{\text{motor}} = M \sin\tau$, where $\tau = \Omega t$ and M and Ω are constants, obtain a numerical solution for the angle θ shown in Fig. 3.21 and plot it versus the nondimensional time τ when $m_2 = m_1/5$ and $M/(m_1\Omega^2 R^2) = 2$. Start the motion of the system from rest and with $\theta(0) = 0$.

7.16 Consider the mechanism described in Problem 3.27 (p. 265) when it is driven by a motor, instead of θ_2 being a known function of time. The mechanism is shown in Fig. 3.27 (p. 266), and the motor applies a moment $M_{\text{motor}} = 3\sin(\Omega t)$ N·m to bar AB, where $\Omega = 2$ rad/s. Bars AB, PC, and CD are very thin and homogeneous, and their masses are m_2, m_3, and m_4, respectively, while their lengths are L_2, L_3, and L_4. The quantity L_3 shown in the figure is the variable distance between B and C. Points A and D are inertial. Us-ing Lagrange's formulation, obtain the equations that govern the constrained motion of the mechanism when motion is on a horizontal plane.

7.17 Using Lagrange's formulation, obtain the differential equations of motion and all the classical integrals of motion for the systems shown in Figs. D.2 and D.4 in Appendix D. Both of them are described in that appendix.

7.18 Use Lagrange's formulation to obtain the differential equations of motion and all the classical integrals of motion for the restricted three-body problem pre-sented in Chapter 4 (i.e., for the motion of particle P_3 shown in Fig. 4.12, p. 297). Use the variables r, θ, and z, shown in Fig. 4.12a for describing the motion of P_3.

7.19 Obtain the differential equations of motion and all the classical integrals of motion for Example 5.3.12 (p. 344) using Lagrange's formulation.

7.20 Obtain the differential equations of motion and all the classical integrals of motion for Example 5.3.13 (p. 346) using Lagrange's formulation. Are there any nonclassical integrals of motion that can be easily obtained?

7.21 Obtain the differential equation of motion and all the classical integrals of motion for Problem 5.27 (p. 388) using Lagrange's formulation. For this, neglect friction in the system and consider the case when the moment M_{motor} is a given function of θ (in practice, such a function is generated with a feedback control system).

7.22 Obtain the differential equation of motion and all the classical integrals of motion for Examples 5.4.1 (p. 356) and 5.4.3 (p. 365) using Lagrange's formulation. The cylinder rolls without slipping in both cases.

7.23 Using Lagrange's formulation obtain all the classical integrals of motion for Example 6.7.1 (p. 455) when $\dot{\phi} = \text{constant} \overset{\Delta}{=} \Omega$. The principal moment of inertia of the rotating platform about the vertical axis that passes through point O is I_O. What is the second-order differential equation of motion for the system obtained from Lagrange's equation?

7.24 Obtain the differential equation of motion and all the classical integrals of motion for Example 6.7.2 (p. 462) using Lagrange's formulation.

7.25 By making use of classical integrals of motion obtained with Lagrange's formulation, solve part b of Problem 6.10 (p. 489).

7.26 Obtain all the classical integrals of motion for the system shown in Fig. 6.37 (p. 494), which is described in Problem 6.16. What are the differential equations of motion for the system?

7.27 Using Lagrange's formulation, obtain all the classical integrals of motion for the symmetric spin stabilized satellite in circular orbit that was considered in Example 6.8.4 (p. 482). Use those same orientation angles. What is the expression for the potential energy U associated with the virtual work δW of the gravitational moment $\vec{M}_C$? Is the Hamiltonian the same as the total mechanical energy $E = T + U$ for such a problem? *Hint:* The virtual work associated with $\vec{M}_C$ is obtained by generating the expression for $\vec{M}_C \bullet (\vec{\omega}\, dt)$ and then replacing "d" by "δ" in the resulting expression (which involves terms in $d\theta_1$, $d\theta_2$, $d\theta_3$, and $\Omega\, dt$). The expressions for $\vec{M}_C$ and for the absolute angular velocity $\vec{\omega}$ were formulated in Examples 6.8.3 and 6.8.4.

7.28 Verify that $(\dot{q}^2 + cq\dot{q} + kq^2)e^{ct}$ is an integral of the motion governed by the differential equation $\ddot{q} + c\dot{q} + kq = 0$, where c and k are constants. This is a case found in some of the references mentioned in Reference 14 in Appendix G.

7.29 Test the stability of the equilibrium solution of Example 7.7.2 (p. 534) with Lyapunov's method using the Hamiltonian as a testing function. *Hint:* Start by showing that the stability analysis reduces to the test for positive definiteness of the potential energy U.

CHAPTER 8

Vibrations/Oscillations of Dynamical Systems

The subject matter of vibrations/oscillations of dynamical systems was already introduced in the book in a natural manner, starting in Chapter 1, including basic material needed for analyzing the motion of dynamical systems. Numerical integration of differential equations was also widely used throughout the book. This short chapter complements the rest of the book. Two important problems, namely the response of a vibration absorber device and the analysis of a Foucault pendulum (described on p. 96 in Chapter 2), are analyzed in Sections 8.3 and 8.4. More advanced and specific material in the field of vibrations/oscillations can be found in References 11 and 18 listed in Appendix G, and in many other books that deal specifically with the subject matter.

8.1   What Was Already Presented on Vibrations/Oscillations in Previous Chapters

The equations that govern the motion of dynamical systems are second-order differential equations in terms of the motion coordinates and, with rare exceptions, are nonlinear. For motion governed by a differential equation of the form $\ddot{x} + f(x) = 0$, it was shown in Chapter 1 that a first integral of the motion, namely, $\dot{x}^2/2 + \int f(x)\,dx =$ constant, is readily obtained by use of the chain rule of differentiation. Such an integral of motion yields important information about the motion even when $x(t)$ itself is not known. The search for integrals of motion is better handled by the more advanced and more powerful analytical dynamics methodology presented in Chapter 7.

To be able to investigate the motion using analytical techniques, so that as much as possible can be learned about why the system behaves in the way it does, approximations need to be introduced in the nonlinear differential equations of motion so that techniques that have been already discovered can be applied. Such analytical techniques are applicable to simpler equations that are obtained by restricting oneself to the analysis of "small" motion about a particular solution of the nonlinear equations. A technique that is widely used involves *linearization* of the differential equations for such small motions about each *equilibrium* exhibited by them. Some of the more advanced techniques that deal with nonlinearities and are used for investigating nonlinear oscillations can be found in Reference 18 listed in Appendix G.

Linearization was introduced in Section 1.14 (p. 59), which begins with a presentation about Taylor series of a function of either one or several variables. Several examples were presented in that section. Other examples were presented in Sections 2.5 and 2.7 of Chapter 2, including a reasonably detailed analysis of a two-degree-of-freedom system (the spring-pendulum, in Example 2.7.8, p. 162). A reasonably detailed analysis of a more complicated three-degree-of-freedom system was presented in Section 4.6 of Chapter 4 (the restricted three-body problem, p. 295). Stability of linear systems with many degrees of freedom was presented in Section 1.16 (p. 66), including a presentation of the Routh-Hurwitz criterion (p. 68). Chapter 1 also has a section on the Lyapunov method for analyzing stability (p. 74), including several examples of application of the use of such a method.

Before proceeding with the analysis of a vibration absorber device and the analysis of a Foucault pendulum, let us look at the simpler analysis of the response of a linear system to a sinusoidal forcing function.

8.2   Response of a Linear System to Sinusoidal Forcing Functions

Consider the linear differential equation $m\ddot{x} + c\dot{x} + kx = f(t)$ with $m > 0$, $c \geq 0$, and $k > 0$. As seen in previous chapters, such an equation describes, for example, the translational motion of a mass connected to a linear spring and to a linear dashpot, when the motion takes place along the vertical and a known force $f(t)$ is applied to the mass.[1] However, it is generally obtained after linearization of a nonlinear differential equation about a stable equilibrium of a system, such as the motion of a pendulum. As seen in Chapter 1, the equilibrium is asymptotically stable if $c > 0$. The case $c = 0$ occurs when there is no damping in the system. The function $f(t)$ is also referred to either as the *input* or the *excitation* applied to the system. The analysis of the response $x(t)$ to an input that is either $f(t) = F\sin(\Omega t)$ or $f(t) = F\cos(\Omega t)$ is considered in the following. Without loss of generality, one can consider

---

[1]See Fig. 1.16 (p. 58) for such a system without a dashpot.

either one of such inputs since $\sin(\Omega t) = \cos(\Omega t - \pi/2)$ and $\cos(\Omega t) = \sin(\Omega t + \pi/2)$. Let us use the following equation:

$$m\ddot{x} + c\dot{x} + kx = F\cos(\Omega t) \tag{8.2.1}$$

The solution to Eq. (8.2.1) is the sum of the homogeneous solution $x_h = Be^{st}$, as presented in Chapter 1 (p. 44), and of a particular solution $x_p(t)$. For the homogeneous solution, the exponent s is determined from $ms^2 + cs + k = 0$. Let us write the particular solution as $x_p(t) = A_1\cos(\Omega t) + A_2\sin(\Omega t)$, where A_1 and A_2 are constants. It satisfies the differential equation. It turns out to be more convenient to write the particular solution as either $x_p = A\cos(\Omega t + \alpha)$ or $x_p = A\sin(\Omega t + \beta)$ as $|A|$ is the maximum value of x_p [which one would have to calculate if $x_p(t) = A_1\cos(\Omega t) + A_2\sin(\Omega t)$ were used]. Thus, by expressing the general solution to Eq. (8.2.1) as

$$x = Be^{st} + A\cos(\Omega t + \alpha) \tag{8.2.2}$$

and by substituting such an expression in Eq. (8.2.1), the following equation is obtained:

$$\underbrace{\left(ms^2 + cs + k\right)}_{=0} Be^{st} + A\left[\left(k - m\Omega^2\right)\cos(\Omega t + \alpha) - c\Omega\sin(\Omega t + \alpha)\right] = F\cos\left(\Omega t\right)$$

$$\tag{8.2.3}$$

To collect sine and cosine terms that have the same arguments in Eq. (8.2.3), one may express $\cos(\Omega t)$ as

$$\cos\left(\Omega t\right) = \cos\left(\Omega t + \alpha - \alpha\right) = \left[\cos\left(\Omega t + \alpha\right)\right]\cos\alpha + \left[\sin\left(\Omega t + \alpha\right)\right]\sin\alpha \tag{8.2.4}$$

Then, collecting terms of $\cos\left(\Omega t + \alpha\right)$ and $\sin\left(\Omega t + \alpha\right)$ on both sides of the resulting equation, the following simple equations for the unknowns A and α are obtained:

$$A\left(k - m\Omega^2\right) = F\cos\alpha \tag{8.2.5}$$
$$-c A\Omega = F\sin\alpha \tag{8.2.6}$$

Since $\sin^2\alpha + \cos^2\alpha = 1$, the solution for A is readily obtained by squaring these equations and adding the results. This yields

$$A = \pm\frac{F}{\sqrt{(k - m\Omega^2)^2 + (c\Omega)^2}} \tag{8.2.7}$$

The absolute value of A is called the *amplitude* of the particular solution. Since the motion is oscillatory, there is no loss of generality by using only the $+$ sign to calculate A from Eq. (8.2.7). The angle α is readily obtained from Eqs. (8.2.5) and (8.2.6) as

$$\alpha = \arctan \frac{\Omega c}{m\Omega^2 - k} \tag{8.2.8}$$

Figure 8.1 shows the plot of the nondimensional amplitude $|A|/(F/k)$ versus the nondimensional frequency Ω/ω_n for several values of the nondimensional damping coefficient $c\omega_n/k$, where $\omega_n = \sqrt{k/m}$ is the undamped natural frequency of the system.

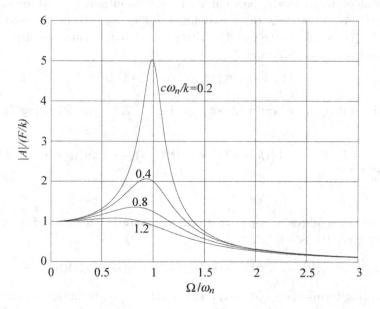

Figure 8.1: Nondimensional amplitude $|A|/(F/k)$ versus Ω/ω_n for several values of the nondimensional damping coefficient $c\omega_n/k$.

The plot in Fig. 8.1 is based on the following equation, which is obtained simply by rearranging Eq. (8.2.7).

$$|A| = \frac{F/k}{\sqrt{(1 - \Omega^2/\omega_n^2)^2 + [(c\omega_n/k)(\Omega/\omega_n)]^2}} \tag{8.2.9}$$

The advantage of presenting the plot in terms of the nondimensional quantities previously indicated is that only one parameter is necessary to generate such plots, namely $c\omega_n/k$, instead of having to specify values for F, k, m, and c as they appear in Eq. (8.2.7).

From these results, we obtain the following conclusions:

Conclusion 1. The particular solution is periodic and has the same frequency Ω of the periodic input to the system. Its amplitude depends on the excitation frequency Ω and the viscous damping coefficient c. For $c > 0$, its maximum amplitude occurs at a frequency smaller than the natural frequency ω_n of the system. The response amplitude tends to infinity as $\Omega \to \omega_n$ and $c \to 0$.

Conclusion 2. The amplitude of the periodic response given by the particular solution is proportional to the amplitude of the input excitation to the system.

Conclusion 3. The particular solution is in phase with the input (i.e., $\alpha = 0$) if $c = 0$ (i.e., no damping) but out of phase if $c \neq 0$. Its phase angle is independent of the excitation amplitude F. Notice that when $c > 0$ the homogeneous solution tends to zero as $t \to \infty$ and the response tends to a steady periodic state that is the same as the particular solution determined earlier.

Figure 8.2 shows typical responses of Eq. (8.2.1) to a sinusoidal excitation with $F/k = 1$ and $\Omega = 1.2\omega_n$, where $\omega_n = \sqrt{k/m}$ as indicated earlier. They were obtained by numerical integration of the normalized form of Eq. (8.2.1), which is

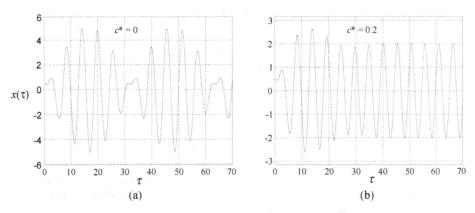

(a)   (b)

Figure 8.2: Solution of Eq. (8.2.10) with $F/k = 1$, $\Omega/\omega_n = 1.2$, and the initial conditions $x(0) = 0.5$ and $\dot{x}(0) = 0$, for (a) $c^* = 0$ and (b) $c^* = 0.2$.

$$\frac{d^2x}{d\tau^2} + c^*\frac{dx}{d\tau} + x = \frac{F}{k}\sin(\Omega^*\tau) \qquad (8.2.10)$$

where $\tau = \omega_n t$, $\Omega^* = \Omega/\omega_n$, and $c^* = c\omega_n/k$.

Notice how the presence of the homogeneous solution causes the response to be quite different from the steady-state solution, which is observed in the presence of damping after the transient homogeneous solution is dissipated. Also notice that the amplitude of the steady-state solution, which is seen in Fig. 8.2b after the homogeneous solution is dissipated, is the same as obtained from Fig. 8.1.

The conclusions presented earlier apply only to linear systems. If Eq. (8.2.1) were obtained by linearization of a nonlinear differential equation about an equilibrium of the system, nonlinear terms that were disregarded in such a process would, of course, affect the motion. Such an effect would become more accentuated as the amplitude of the excitation is increased. However, even for very small excitation amplitudes, the nonlinearities can have a significant effect if the excitation frequency Ω is near values such as ω_n, $2\omega_n$, $3\omega_n$, $\omega_n/3$, to name a few. Such cases are called *nonlinear external resonances*, and their analyses are done with methods that are known as *perturbation methods*. For such analyses, the reader is referred to some of the books that deal with the subject matter, such as Reference 18 listed in Appendix G. It should also be noted that many nonlinear effects, such as the ones previously mentioned, can be essentially eliminated by increasing the damping in the system.

8.3   Analysis of a Vibration Absorber Device

Figure 8.3 is a model of a system that consists of a block of mass m_1 mounted on an

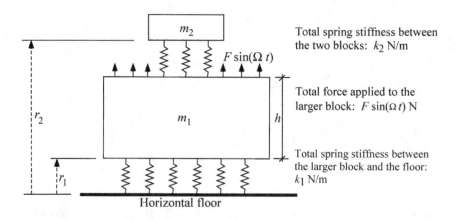

Figure 8.3: A vibration absorber.

elastic foundation that is represented by several linear springs that are equivalent to a single spring of stiffness k_1 N/m and unstressed length L_{u_1} m. A system consisting of a smaller block of mass m_2 connected to a set of identical springs is attached to the larger block. Those springs are also approximated as linear. They are equivalent to a single spring of stiffness k_2 N/m and unstressed length L_{u_2} m. Motion of the system takes place along the vertical and is tracked by the distances r_1 and r_2 shown in the figure. The larger block is subjected to a periodic force $F\sin(\Omega t)$ N, causing the system to vibrate. The objective is to investigate the motion of the system to see if it is possible to reduce, or even eliminate, the vibrations of the larger block by choosing appropriate values of m_2 and k_2 to accomplish such a task.

The differential equations of motion for the system may be obtained by drawing the free-body diagram for each one of the blocks shown in Fig. 8.3 and then using Newton's second law for each block. They may also be readily obtained by using Lagrange's equation, as presented in Chapter 7. The differential equations of motion are presented here; their derivation is left as an exercise for the reader.

$$m_1\ddot{r}_1 + k_1(r_1 - L_{u_1}) + k_2(r_1 - r_2 + h + L_{u_2}) \;=\; F\sin(\Omega t) - m_1 g \qquad (8.3.1)$$
$$m_2\ddot{r}_2 + k_2(r_2 - r_1 - h - L_{u_2}) \;=\; -m_2 g \qquad (8.3.2)$$

To analyze the motion, first determine the equilibrium solution(s) to Eqs. (8.3.1) and (8.3.2), and then perturb the equilibrium. For the equilibrium solution, $r_1 =$ constant $\overset{\Delta}{=} r_{1_e}$ and $r_2 =$ constant $\overset{\Delta}{=} r_{2_e}$, where r_{1_e} and r_{2_e} must satisfy the following equations that are obtained from Eqs. (8.3.1) and (8.3.2):

$$k_1(r_{1_e} - L_{u_1}) + k_2(r_{1_e} - r_{2_e} + h + L_{u_2}) \;=\; -m_1 g \qquad (8.3.3)$$
$$k_2(r_{2_e} - r_{1_e} - h - L_{u_2}) \;=\; -m_2 g \qquad (8.3.4)$$

These equations can be easily solved for r_{1_e} and r_{2_e}. Since the equations are linear, they have only one solution, instead of multiple solutions.

By letting, for convenience, $r_1 \overset{\Delta}{=} r_{1_e} + x_1(t)$ and $r_2 \overset{\Delta}{=} r_{2_e} + x_2(t)$, substitution of these expressions into Eqs. (8.3.1) and (8.3.2) yield the following differential equations for the perturbations $x_1(t)$ and $x_2(t)$:

$$m_1\ddot{x}_1 + k_1 x_1 + k_2(x_1 - x_2) \;=\; F\sin(\Omega t) \qquad (8.3.5)$$
$$m_2\ddot{x}_2 + k_2(x_2 - x_1) \;=\; 0 \qquad (8.3.6)$$

The solution to Eqs. (8.3.5) and (8.3.6) is the sum of the homogeneous solution (i.e., the one obtained with $F = 0$) and the particular solution. The homogeneous solution determines the stability of the equilibrium. Before anything else is done, it is important to look at the question of stability to ascertain that we are dealing with an equilibrium that is not unstable.

8.3.1   Homogeneous Solution and Stability of the Equilibrium

With $F = 0$, the solution to Eqs. (8.3.5) and (8.3.6) is of the form $x_1 = A_1 e^{st}$ and $x_2 = A_2 e^{st}$. As indicated in Section 1.16 of Chapter 1 (p. 66), the exponent s is obtained from

$$\begin{vmatrix} m_1 s^2 + k_1 + k_2 & -k_2 \\ -k_2 & m_2 s^2 + k_2 \end{vmatrix} = m_1 m_2 \left[s^4 + \left(\frac{k_1 + k_2}{m_1} + \frac{k_2}{m_2} \right) s^2 + \frac{k_1}{m_1} \frac{k_2}{m_2} \right] = 0$$

(8.3.7)

The roots of Eq. (8.3.7) are

$$s^2 = -\frac{1}{2} \left(\frac{k_1 + k_2}{m_1} + \frac{k_2}{m_2} \right) \pm \sqrt{\frac{1}{4} \left(\frac{k_1 + k_2}{m_1} + \frac{k_2}{m_2} \right)^2 - \frac{k_1}{m_1} \frac{k_2}{m_2}}$$

(8.3.8)

Since the radicand in Eq. (8.3.8) is a positive quantity[2] and its square root is smaller than $[(k_1 + k_2)/m_1 + k_2/m_2]/2$, both roots s^2 of Eq. (8.3.8) are negative. Therefore, the four quantities $s = \pm\sqrt{s^2}$ are purely imaginary, and the equilibrium is stable.

With $s = i\omega$, where $i = \sqrt{-1}$, the squares of the natural frequencies of the free undamped oscillation of the system are then

$$\left. \begin{matrix} \omega_1^2 \\ \omega_2^2 \end{matrix} \right\} = \frac{1}{2} \left(\frac{k_1 + k_2}{m_1} + \frac{k_2}{m_2} \right) \mp \sqrt{\frac{1}{4} \left(\frac{k_1 + k_2}{m_1} + \frac{k_2}{m_2} \right)^2 - \frac{k_1}{m_1} \frac{k_2}{m_2}}$$

(8.3.9)

The free oscillation is the solution of Eqs. (8.3.5) and (8.3.6) with $F = 0$, and is of the form $a_1 \cos(\omega_1 t + \alpha_1) + a_2 \cos(\omega_2 t + \alpha_2)$, where a_1, a_2, α_1, and α_2 are constants determined by the initial conditions of the motion.

8.3.2   Particular Solution: Response to a Sinusoidal Force

Similar to what was presented in Section 8.2, the particular solutions to Eqs. (8.3.5) and (8.3.6) are also sinusoidal. Since there is no damping in the system, the response is in phase with the excitation $F \sin(\Omega t)$. Thus, by expressing $x_1(t)$ and $x_2(t)$ as

$$x_1(t) = B_1 \sin(\Omega t)$$ (8.3.10)
$$x_2(t) = B_2 \sin(\Omega t)$$ (8.3.11)

Equations (8.3.5) and (8.3.6) yield

$$\begin{bmatrix} k_1 + k_2 - m_1 \Omega^2 & -k_2 \\ -k_2 & k_2 - m_2 \Omega^2 \end{bmatrix} \begin{bmatrix} B_1 \\ B_2 \end{bmatrix} = \begin{bmatrix} F \\ 0 \end{bmatrix}$$

(8.3.12)

---

[2]This can be shown by simply expanding that radicand, which becomes $\frac{1}{4}[(k_1/m_1 - k_2/m_2)^2 + (k_2/m_1)^2 + 2(k_1/m_1 + k_2/m_2)k_2/m_1]$.

The solution to Eq. (8.3.12) is

$$B_1 = \frac{\left(k_2 - m_2\Omega^2\right) F}{D} = \left(\frac{k_2/m_2 - \Omega^2}{\left(\Omega^2 - \omega_1^2\right)\left(\Omega^2 - \omega_2^2\right)}\right)\frac{F}{m_1} \qquad (8.3.13)$$

$$B_2 = \frac{k_2 F}{D} = \left(\frac{k_2/m_2}{\left(\Omega^2 - \omega_1^2\right)\left(\Omega^2 - \omega_2^2\right)}\right)\frac{F}{m_1} \qquad (8.3.14)$$

where

$$D = m_1 m_2 \Omega^4 - \left[m_1 k_2 + m_2\left(k_1 + k_2\right)\right]\Omega^2 + k_1 k_2 = m_1 m_2 \left(\Omega^2 - \omega_1^2\right)\left(\Omega^2 - \omega_2^2\right)$$
$$(8.3.15)$$

is the determinant of the 2×2 matrix that appears in Eq. (8.3.12).

> Inspection of Eq. (8.3.13) immediately discloses that the forced response of the larger block (i.e., x_1) can be suppressed (i.e., $B_1 = 0$) by tuning the m_2-k_2 system so that $k_2/m_2 = \Omega^2$.

Also, notice that both B_1 and B_2 tend to zero as $\Omega \to \infty$ (i.e., very high frequency excitation) and tend to $k_2 F/(m_1 m_2 \omega_1^2 \omega_2^2) = F/k_1$ [since, as disclosed by Eq. (8.3.15), $m_1 m_2 \omega_1^2 \omega_2^2 = k_1 k_2$] as $\Omega \to 0$. Figure 8.4 shows a typical amplitude-frequency response curve for the system. The practical region of operation of the device is in a frequency range where $\sqrt{k_2/m_2}$ is near Ω, for which the response amplitude $|B_1|$ is a small acceptable value dictated by the user of the device.

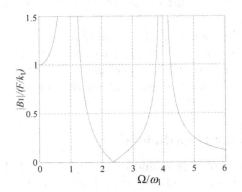 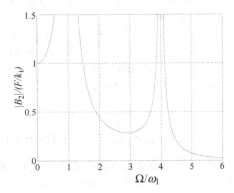

Figure 8.4: A typical amplitude-frequency response curve for the system of Fig. 8.3.

In practice, there is always some amount of damping in the system. It turns out that the presence of damping has the undesirable effect of preventing the amplitude $|B_1|$ of $x_1(t)$ from becoming zero. This can be verified either analytically or by numerical integration of the differential equations after including a damping term on them, such as $c_1\dot{x}_1 + c_2(\dot{x}_1 - \dot{x}_2)$ in the left-hand side of Eq. (8.3.5) and $c_2(\dot{x}_2 - \dot{x}_1)$ in the left-hand side of Eq. (8.3.6). For simplicity, one could take $c_1 = c_2 \overset{\Delta}{=} c$.

8.4   Analysis of the Foucault Pendulum

8.4.1   Problem Description

In 1851, an important experiment was performed by the French physicist Jean-Bernard-Léon Foucault (1819-1868), in which he used a special pendulum constructed for demonstrating the rotation of the earth about its own axis. Foucault was a physicist at the Paris observatory and, for his numerous experimental contributions to mechanics, especially his remarkable experiment with his pendulum, was awarded the title of officer of the French Legion of Honor and also elected a foreign member of the Royal Society of London. Most of his work was compiled by his mother in *Recueil des travaux scientifiques de Léon Foucault* (Gauthier-Villars, Paris, 1878).

When the pendulum was given a small displacement on a vertical plane, it was observed that the oscillating pendulum rotated with respect to the earth with a period that was approximately equal to $24/\sin\lambda$ h, with λ being the latitude of the pendulum's location on the surface of the earth. The pendulum originally used by Foucault consisted of a 28 kg ball suspended by a 67 m wire. He repeated his experiment at the Panthéon in Paris. Today, Foucault's pendulum can be seen in many museums throughout the world.

The differential equations that govern the motion of the pendulum relative to the rotating earth are formulated in this section. The equations are then used for analyzing the effect of the spin of the earth on the rotation of the oscillating pendulum and to compare the results with the assertion made about the period of the rotation.

8.4.2   Formulation Using Rectangular Coordinates

Consider a pendulum consisting of a particle P of mass m and a wire (approximated as massless) of length L, and whose support is at a point Q of latitude λ near the surface of the earth as shown in Fig. 8.5. For simplicity, the vertical distance h between point O', which is on the surface of the earth, and the support at Q is taken to be $h = L$.

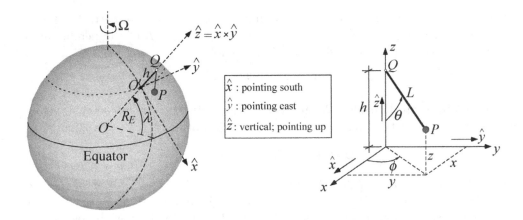

Figure 8.5: Foucault's pendulum (a pendulum for demonstrating the rotation of the earth).

The scale of Fig. 8.5 is grossly exaggerated since the length L of the pendulum is much smaller that the radius R_E of the earth, with the earth being approximated as a homogeneous sphere of mass M_E. It is also assumed that the only forces acting on P are the tension in the wire QP, of magnitude N, and the gravitational force of attraction of the earth. The magnitude of the gravitational force acting on P is approximated as $GM_E m/R_E^2 = mg$, where $g \approx 9.81$ m/s^2 is the acceleration of gravity at the earth's surface. Also, since L is very small compared to R_E, the gravitational force acting on P is practically in the $-\hat{z}$ direction, where $\hat{z}$ is a unit vector directed "upward" along the local vertical line OO'. The rectangular coordinates x, y, and z shown in Fig. 8.5 will be used for analyzing the effect of Ω on the motion of the pendulum relative to the rotating earth.

Note: At this point, a natural question arises, namely: "Why work with rectangular coordinates instead of the more natural independent variables θ and ϕ shown in Fig. 8.5?" The answer has to do with the complexity of the resulting differential equations of motion that are obtained in terms of either (θ, ϕ) or (x, y). To see this, one has to develop both types of equations. When this is done, it can then be seen that the equations developed with rectangular coordinates are simpler to use for determining the effect of the earth spin on the motion of the pendulum.

In terms of the earth-fixed unit vector triad $\{\hat{x}, \hat{y}, \hat{z} = \hat{x} \times \hat{y}\}$ shown in Fig. 8.5, the resultant force $\vec{F}$ acting on P is then approximately equal to

$$\vec{F} \approx -mg\hat{z} - N\frac{\overrightarrow{QP}}{|\overrightarrow{QP}|} = -mg\hat{z} - N\left(\frac{x\hat{x} + y\hat{y} + (z - L)\hat{z}}{L}\right) \qquad (8.4.1)$$

where x, y, and z are the rectangular coordinates of P relative to earth-fixed x-y-z axes centered at point O'. Notice that, since the length L of the pendulum is constant, the variables x, y, and z are related as follows:

$$x^2 + y^2 + (z - L)^2 = L^2 \tag{8.4.2}$$

With λ being the latitude at location O' on the earth, the absolute angular velocity $\vec{\omega}$ of the $\{\hat{x}, \hat{y}, \hat{z}\}$ triad fixed to the earth as shown in Fig. 8.5 is approximated as

$$\vec{\omega} \approx \Omega \left(\hat{z} \sin \lambda - \hat{x} \cos \lambda \right) \tag{8.4.3}$$

where $\Omega = 2\pi$ rad/day $\approx 7.27 \times 10^{-5}$ rad/s. In Eq. (8.4.3), the rotation of the line from the sun to the center O of the earth, with angular velocity of 2π rad/year, is neglected since it is much smaller than the spin Ω of the earth about its own axis.

With the position vector from O' to P expressed as $\overrightarrow{O'P} = x\hat{x} + y\hat{y} + z\hat{z}$ and the center O of the earth now being approximated as inertial, the absolute position vector for P is then

$$\vec{r}_P = R_E \hat{z} + x\hat{x} + y\hat{y} + z\hat{z}$$

Therefore, the absolute velocity $\vec{v}_P$ and acceleration $\vec{a}_P$ of P are

$$\begin{aligned}
\vec{v}_P &= \frac{d\vec{r}_P}{dt} = \dot{x}\hat{x} + \dot{y}\hat{y} + \dot{z}\hat{z} + \vec{\omega} \times \vec{r}_P \\
&= \left[\dot{x} - \Omega y \sin \lambda \right] \hat{x} + \left[\dot{y} + \Omega x \sin \lambda + \Omega \left(R_E + z \right) \cos \lambda \right] \hat{y} + \left[\dot{z} - \Omega y \cos \lambda \right] \hat{z}
\end{aligned} \tag{8.4.4}$$

$$\begin{aligned}
\vec{a}_P &= \frac{d\vec{v}_P}{dt} = \left(\frac{d\vec{v}_P}{dt} \right)_{\{\hat{x}, \hat{y}, \hat{z}\}} + \vec{\omega} \times \vec{v}_P \\
&= \left[\ddot{x} - 2\Omega \dot{y} \sin \lambda - \Omega^2 x \sin^2 \lambda - \Omega^2 \left(R_E + z \right) (\sin \lambda) \cos \lambda \right] \hat{x} \\
&\quad + \left[\ddot{y} + 2\Omega \dot{x} \sin \lambda + 2\Omega \dot{z} \cos \lambda - \Omega^2 y \right] \hat{y} \\
&\quad + \left[\ddot{z} - 2\Omega \dot{y} \cos \lambda - \Omega^2 x (\sin \lambda) \cos \lambda - \Omega^2 \left(R_E + z \right) \cos^2 \lambda \right] \hat{z}
\end{aligned} \tag{8.4.5}$$

With $\vec{F}$ and $\vec{a}_P$ expressed as shown in Eqs. (8.4.1) and (8.4.5), the following scalar equations are obtained directly from $\vec{F} = m\vec{a}_P$:

$$m \left[\ddot{x} - 2\Omega \dot{y} \sin \lambda - \Omega^2 x \sin^2 \lambda - \Omega^2 \left(R_E + z \right) (\sin \lambda) \cos \lambda \right] = -\frac{x}{L} N \tag{8.4.6}$$

$$m \left[\ddot{y} + 2\Omega \dot{x} \sin \lambda + 2\Omega \dot{z} \cos \lambda - \Omega^2 y \right] = -\frac{y}{L} N \tag{8.4.7}$$

$$m \left[\ddot{z} - 2\Omega \dot{y} \cos \lambda - \Omega^2 x (\sin \lambda) \cos \lambda - \Omega^2 \left(R_E + z \right) \cos^2 \lambda \right] = -mg - \frac{z - L}{L} N \tag{8.4.8}$$

Equations (8.4.6) to (8.4.8), together with the constraint Eq. (8.4.2), govern the motion of the pendulum relative to the rotating earth. The effect of Ω on the small motion of the pendulum is now addressed in detail.

Since $\Omega = 2\pi$ rad/day $\approx 7.27 \times 10^{-5}$ rad/s is very small, approximations are now made to reduce the complexity of Eqs. (8.4.6) to (8.4.8), but still retaining the effect of Ω, of course. One simplification consists of neglecting the much smaller Ω^2 terms. When this is done, the equilibrium solutions $x = $ constant $\overset{\Delta}{=} x_e$, $y = $ constant $\overset{\Delta}{=} y_e$, and $z = $ constant $\overset{\Delta}{=} z_e$ to those equations are $x_e = 0$, $y_e = 0$, $z_e = 0$ and $x_e = 0$, $y_e = 0$, $z_e = L$, with $z_e = 0$ and $z_e = L$ obtained from the constraint equation, which is Eq. (8.4.2). The equilibrium for which $z_e = L$ corresponds to the upside down pendulum and is not of interest.

The next approximation consists of limiting the analysis to small motions about the equilibrium $x_e = 0$, $y_e = 0$, $z_e = 0$ and linearizing the equations for small x, y, and z. By noticing that Eq. (8.4.2) is simply $x^2 + y^2 + z^2 - 2zL = 0$, its linearization yields $z \approx 0$, and Eq. (8.4.8) yields $N \approx m(g - 2\Omega \dot{y} \cos \lambda)$. With this expression for N, Eqs. (8.4.6) and (8.4.7) then immediately yield the following linearized differential equations involving the small independent variables $x(t)$ and $y(t)$, with Ω^2 terms neglected:

$$\ddot{x} - (2\Omega \sin \lambda)\,\dot{y} + \frac{g}{L}x \;=\; 0 \tag{8.4.9}$$

$$\ddot{y} + (2\Omega \sin \lambda)\,\dot{x} + \frac{g}{L}y \;=\; 0 \tag{8.4.10}$$

To see the effect Ω has on the motion, it helps to integrate Eqs. (8.4.9) and (8.4.10) numerically. A result of the integration with $\Omega \sin \lambda = 0.0727$ rad/s, which is a value chosen for the sole purpose of speeding up the rotation of the plane of oscillation,[3] $g/L = 9.81/10$ s^{-2}, and the initial conditions $x(0)/L = 0.01$, $y(0) = 0$, and $\dot{x}(0) = \dot{y}(0) = 0$, is shown in Fig. 8.6.

$\Longrightarrow$ For Scilab users, the program *foucault_pendulum.sci* listed in Appendix E integrates Eqs. (8.4.9) and (8.4.10), generates Fig. 8.6, and animates the motion. For Matlab users, the similar program *foucault_pendulum.m* that is also listed in Appendix E will do the same. As indicated in the first few comment lines of those programs, the command to use them, typed either in the Scilab console or in the Matlab command window, is

```
my_output = foucault_pendulum([0 : Δt : 30], 0.0727, 0.981, 0.01, 0, 0, 0);
```

(try $\Delta t = 0.01$ for Scilab, and 0.001 for Matlab).

---

[3]With $\Omega = 7.27 \times 10^{-5}$ rad/s, such a rotation is very slow.

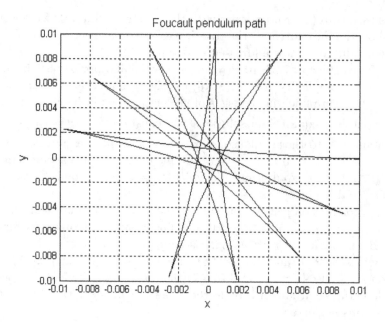

Figure 8.6: Typical result of the numerical integration of Eqs. (8.4.9) and (8.4.10).

Those who have Simulink and prefer to use it may want to construct the block diagram model shown in Fig. 8.7, setting the *Simulation* → *Parameters* menu for this model with time t of the form $[0:\Delta t:30]$, and using the same initial conditions mentioned earlier. To see an animation of the motion after the simulation stops, type `animate_nbars(x_coordinate, y_coordinate)` in the Matlab command window if you have the Matlab program `animate_nbars` stored in your computer.

Figure 8.6 shows that the oscillating pendulum rotates slowly about the vertical line $O'Q$ seen in Fig. 8.5. The rotation is clockwise for a positive value of $\Omega \sin \lambda$ (i.e., in the northern hemisphere) and counterclockwise for negative values (i.e., in the southern hemisphere).

In view of the observed motion shown in Fig. 8.6, let us see what happens if the motion is expressed in terms of coordinates x_1-y_1 that are rotated by a time-varying angle $\alpha(t)$ as shown in Fig. 8.8. According to that figure, x and y are expressed in terms of x_1 and y_1 as follows:

$$x(t) = x_1 \cos \alpha + y_1 \sin \alpha \qquad (8.4.11)$$
$$y(t) = y_1 \cos \alpha - x_1 \sin \alpha \qquad (8.4.12)$$

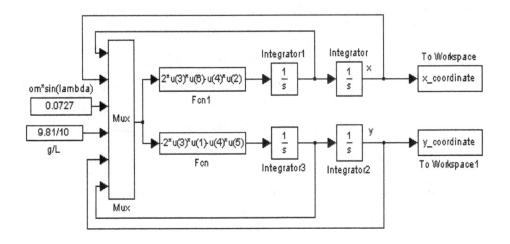

Figure 8.7: A Simulink model for integrating Eqs. (8.4.9) and (8.4.10).

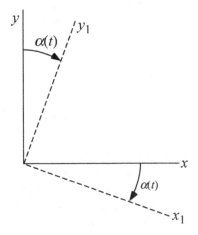

Figure 8.8: The x_1-y_1 axes, rotated by $\alpha(t)$ relative to the x-y earth-fixed axes.

These equations then yield the following expressions for $\dot{x}$ and $\dot{y}$:

$$\dot{x}(t) = (\dot{x}_1 + \dot{\alpha}y_1)\cos\alpha + (\dot{y}_1 - \dot{\alpha}x_1)\sin\alpha \qquad (8.4.13)$$
$$\dot{y}(t) = (\dot{y}_1 - \dot{\alpha}x_1)\cos\alpha - (\dot{x}_1 + \dot{\alpha}y_1)\sin\alpha \qquad (8.4.14)$$

Now, it is not difficult to verify that, in terms of x_1 and y_1, Eqs. (8.4.9) and (8.4.10) become

$$\left[\ddot{x}_1 + \ddot{\alpha} y_1 + 2\left(\dot{\alpha} - \Omega \sin \lambda\right) \dot{y}_1 + \left(\frac{g}{L} - \dot{\alpha}^2 + 2\dot{\alpha}\Omega \sin \lambda\right) x_1\right] \cos \alpha$$
$$+ \left[\ddot{y}_1 - \ddot{\alpha} x_1 + 2\left(\Omega \sin \lambda - \dot{\alpha}\right) \dot{x}_1 + \left(\frac{g}{L} - \dot{\alpha}^2 + 2\dot{\alpha}\Omega \sin \lambda\right) y_1\right] \sin \alpha = 0 \quad (8.4.15)$$

$$\left[\ddot{y}_1 - \ddot{\alpha} x_1 + 2\left(\Omega \sin \lambda - \dot{\alpha}\right) \dot{x}_1 + \left(\frac{g}{L} - \dot{\alpha}^2 + 2\dot{\alpha}\Omega \sin \lambda\right) y_1\right] \cos \alpha$$
$$- \left[\ddot{x}_1 + \ddot{\alpha} y_1 + 2\left(\dot{\alpha} - \Omega \sin \lambda\right) \dot{y}_1 + \left(\frac{g}{L} - \dot{\alpha}^2 + 2\dot{\alpha}\Omega \sin \lambda\right) x_1\right] \sin \alpha = 0 \quad (8.4.16)$$

These two equations are satisfied only if

$$\ddot{x}_1 + \ddot{\alpha} y_1 + 2\left(\dot{\alpha} - \Omega \sin \lambda\right) \dot{y}_1 + \left(\frac{g}{L} - \dot{\alpha}^2 + 2\dot{\alpha}\Omega \sin \lambda\right) x_1 = 0 \quad (8.4.17)$$
$$\ddot{y}_1 - \ddot{\alpha} x_1 + 2\left(\Omega \sin \lambda - \dot{\alpha}\right) \dot{x}_1 + \left(\frac{g}{L} - \dot{\alpha}^2 + 2\dot{\alpha}\Omega \sin \lambda\right) y_1 = 0 \quad (8.4.18)$$

Notice that the angle $\alpha(t)$ is arbitrary. By observing Eqs. (8.4.17) and (8.4.18), it is seen that they can be decoupled by choosing α so that

$$\dot{\alpha} = \Omega \sin \lambda \quad (8.4.19)$$

and this yields

$$\ddot{x}_1 + \left(\frac{g}{L} + \underbrace{\frac{\Omega^2 \sin^2 \lambda}{\text{negligible}}}\right) x_1 = 0 \quad \text{and} \quad \ddot{y}_1 + \left(\frac{g}{L} + \underbrace{\frac{\Omega^2 \sin^2 \lambda}{\text{negligible}}}\right) y_1 = 0 \quad (8.4.20)$$

The equations of (8.4.20) describe two uncoupled oscillators, and they govern the motion of the pendulum as viewed by an observer fixed to the x_1-y_1 axes. As seen from Fig. 8.8, such axes are rotating with respect to the x-y earth-fixed axes. As indicated earlier, in the northern hemisphere, where $\sin \lambda > 0$, the rotation is clockwise. It is counterclockwise in the southern hemisphere, where $\sin \lambda < 0$. The period $T_{2\pi}$ of such a rotation is

$$T_{2\pi} = \frac{2\pi}{\dot{\alpha}} = \frac{2\pi}{\Omega \sin \lambda} = \frac{24}{\sin \lambda} \text{ h}$$

which is in complete agreement with the results originally observed and explained by Foucault. At the poles, the period $T_{2\pi}$ is a minimum and equal to 24 h.

Summary   For an observer fixed to the rotating earth, the oscillating pendulum rotates relative to the earth, about the vertical, with a period $T_{2\pi} = 2\pi/(\Omega \sin \lambda) =$

$24/\sin\lambda$ h. The rotation is clockwise if the pendulum is in the northern hemisphere and counterclockwise in the southern hemisphere. At the equator ($\lambda = 0$), no rotation due to the earth's spin occurs when the very small terms in Ω^2 are neglected, while at the poles the period $T_{2\pi}$ is minimum and equal to 24 hours. At those locations, the oscillating pendulum rotates $360/4 = 90°$ every $24/4 = 6$ h. In Washington, D.C. (United States), Paris (France), and Rio de Janeiro (Brazil), a rotation of $90°$ takes approximately 9 h and 30 min, 8 h, and 15 h and 25 min, respectively.

Scilab, Matlab, and Simulink Tutorials

A.1   Introduction to Scilab and Matlab

Both Scilab and Matlab are used at a number of universities and research institutions, and this section presents a brief tutorial on them. There are many more complete tutorials that are available on the web, and they can be searched by using the words `Scilab tutorial` or `Matlab tutorial`, for example. The following are two examples of such sites:

> `http://www.scilab.org/resources`
> `http://www.mathworks.com`

Scilab can be downloaded for free at `www.scilab.org`. It has versions for Microsoft *Windows*, for *MAC*, and for *GNU/Linux*. By clicking on `download`, a file `scilab-xxx.exe`, where `xxx` is used here to represent the version number, is saved on your computer. To install the software, double-click the computer mouse on that file and follow the instructions that will appear on the computer screen. To launch Scilab, look for its icon and double click on it.

The programs listed in Appendix E were prepared using Scilab version 5.5, and also successfully tested with Scilab version 6. For versions newer than scilab-6, there is a small probability that Scilab Enterprises may either change or eliminate some of the commands used in the programs listed in Appendix E. If that happens, you either need to make small appropriate changes in those programs or try to use one of the previous versions of Scilab found at www.scilab.org/download/Previous-Scilab-Versions. As indicated in the Preface, Scilab is free and open source software for numerical computation and simulation developed and published by Scilab Enterprises. Scilab is a registered trademark.

Matlab is a commercial software and, as such, is not free. To install a personal copy on your computer, follow the instructions that are on the disk on which the software is distributed.

Figure A.1 shows a typical Scilab "console" that appears on the computer screen when Scilab is started, and Fig. A.2 shows a typical Scilab editor, which opens either by typing `editor` in the console or clicking the mouse on the leftmost icon that appears under the word `File` in the console. Do not be concerned if the windows that open in your computer look somewhat different.

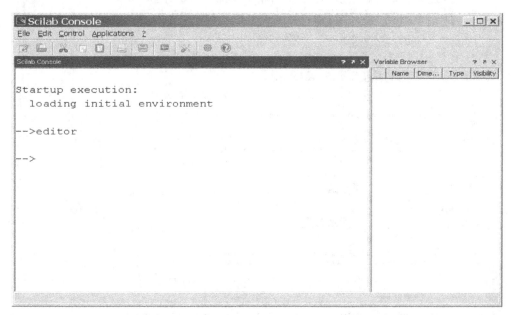

Figure A.1: A typical Scilab console.

Commands to be executed (processed) by Scilab are typed in the Scilab console. The Scilab editor seen in Fig. A.2 shows several tabs with the names of the files that were put there by the author for this demonstration, and the contents of one of those files. Files are displayed in the Scilab editor by clicking on `File` → `Open` in the editor window, selecting (in the window that opens up) the directory where you stored your Scilab files (called the "home directory"), and the Scilab file (or files) you want to display and, then, clicking on the `Open` icon in that window. The Scilab editor will display the file you just opened. All the files you opened in this manner will be in the editor, which will display tabs with their names. To display the contents of any of such files, you only need to click on the tab with the name of the

```
   spring_pendulum.sci                                              _ □ ×

 File  Edit  Format  Options  Window  Execute  ?

 spring_pendulum.sci                                                        ?

 animate_nbars.sci    foucault_pendulum.sci    sim_model_ex5_3_11.sci    spring_pendulum.sci

 1  function output=spring_pendulum(t, mg_over_kre, rs_0, dotrs_0, theta_0, dottheta_0)
 2  // Animation of the spring-pendulum analyzed in Example 2.7.8
 3  // Example of usage [with the initial conditions
 4  //           rs(0)=0.1; rsdot(0)=0; theta(0)=0.01; thetadot(0)=0]:
 5  // my_output = spring_pendulum([0:1:350], 0.5^2,  0.1, 0,  0.01, 0);
 6
 7  disp('NOTE: IN ADDITION TO ANIMATING THE SPRING-PENDULUM ANALYZED IN EXAMPLE 2.7.10,')
 8  disp('AND PLOTTING theta and rs versus time, A CALL TO THE FUNCTION spring_pendulum')
 9  disp(' (SUCH AS my_output = spring_pendulum([0:1:200], 0.5^2, 0.1, 0,  0.01, 0] )')
10  disp('RETURNS 5 COLUMNS CONTAINING  t, rs,  d(rs)/dt, theta, and d(theta)/dt')
11  disp('TO PLOT d(rs)/dt, for example, VERSUS time, TYPE, IN THE SCILAB CONSOLE:')
12  disp('      figure(); plot(my_output(:,1), my_output(:,3)) ')
13  disp('OR:  figure(); plot2d(my_output(:,1), my_output(:,3));')
14  disp('   To put a grid on the plot, type  xgrid ')
15
 1  function dy_dt=rhs_of_diff_eqs(t, y)
 2      // Note:  y(1)=rs, y(2)=d(rs)/dt, y(3)=theta, y(4)=d(theta)/dt
 3    dy_dt(1)=y(2)
 4    dy_dt(2)=(1+y(1))*y(4)^2-mg_over_kre*(1-cos( y(3) ) )-y(1)  // this is d2(rs)/dt2
 5    dy_dt(3)=y(4)
 6    dy_dt(4)=-(2*y(2)*y(4)+mg_over_kre*sin( y(3) ) )/(1+y(1))  // this is d2(theta)/dt2
 7  endfunction
23
24  initial_conditions=[rs_0; dotrs_0; theta_0; dottheta_0]; y=ode(initial_conditions, 0, t, rhs_of_diff_eqs)
25  // pause    // NOTE: this pause command is here only to explain, in Appendix A, what it does
26
27  xdel(winsid())  // deleting all existing figures
28  figure('background', 8)   //  'background' 8 forces the figure to be white
29  plot(t, y(3,:)),   xgrid   // plot of theta versus time t
30  xlabel(['Nondimensional time', '$\tau$'], 'fontsize', 3), ylabel('$\theta$', 'fontsize',4)
31
32  figure('background', 8)
33  plot(t, y(1,:)),   xgrid   // plot of rs versus t
34  xlabel(['Nondimensional time', '$\tau$'], 'fontsize', 3)
35  ylabel('$r_s$', 'fontsize', 5)
36
37  output=[t', y']   // this is a n by 5 matrix, where n is the number of points in t
38  // column 1: time, column 2: rs, column 3: d(rs)/dt, column 4: theta; column 5: d(theta)/dt
39
40  // ANIMATION: using a bar AB from the point (0,0) to (1+rs)*(sin_theta, -cos_theta)
41  len=length(:,4),   // this is how many elements the row matrix t has
42  rs=output(:,2), theta=output(:,4)   // these are len by 1 matrices
43  xA=zeros(len,1), yA=xA,   xB=(1+rs).*sin(theta), yB=-(1+rs).*cos(theta)   // bar AB
44  x_coordinates=[xA, xB], y_coordinates=[yA, yB], animate_nbars(x_coordinates, y_coordinates)
45  endfunction       // end of the file spring_pendulum.sci
```

Figure A.2: The Scilab editor, showing the file spring_pendulum.sci.

file you want to see. To start a new file, choose File → New, and you can then write your new program in the window that opens up. When you finish preparing your program, click either File → Save or File → Save as and Scilab will save it with the extension sci. If your program is a function (i.e., equivalent to a subroutine in Fortran, for example), as the one shown in Fig. A.2, Scilab will automatically save it with the same name of that function (which is spring_pendulum in the illustration seen in Fig. A.2). To use any new function you create, you need to load it into the

Scilab console, and this is done by clicking on Execute → Save and execute. To load all the files that you put in the editor, click on Execute → Save and execute all files (which is more convenient than loading one by one).

After installation, Scilab will be in some default directory when it is started (to see the pathname of that directory, type pwd in the Scilab console). It is convenient to force Scilab to make your home directory be the default directory every time after it starts. For that to happen, prepare, using the Scilab editor, a file such as the one shown in Fig. A.3 and save it (by clicking on File → Save As) with the name scilab.ini **in the directory whose pathname shows up when you type** SCIHOME in the console. As indicated on the paragraphs that start on the seventh line from the bottom of p. 603, it is convenient to also save it in your home directory. Restart Scilab after you do this for the first time. The first line of the scilab.ini file should be the command chdir followed by the pathname of your home directory. The example in Fig. A.3 shows the pathname of a directory *Dynamics_programs* created by a user "john" in the directory (also known as "folder") *Documents*. In some computer installations the pathname for that home directory might be simply C:/Documents/Dynamics_programs, for example. If needed, consult your system installation manual or your system administrator if you need any help on what to use for the correct pathname in your computer. The remaining lines of the scilab.ini file should be of the form exec('file_name.sci', 0), using apostrophes (double quotes may also be used in Scilab), where file_name is the name of each Scilab file you stored in your home directory. The files whose names are shown in Fig. A.3 are the Scilab programs listed in Appendix E (see Appendix E for more details). Type help exec in the Scilab console if you want to see some more details about the exec command.

```
chdir C:/Users/john/Documents/Dynamics_programs
exec('animate_nbars.sci',0);   exec('animate3D_nbars.sci',0);
exec('ballistics_nbars.sci',0);   exec('boomerang_nbars.sci',0);
exec('example6_7_1.sci',0);   exec('foucault_pendulum.sci',0);
exec('fourbar_kinematics.sci',0);   exec('fourbar_mechanism.sci',0);
exec('fourbar_mech_dynamics.sci',0);   exec('fourbarforces.sci',0);
exec('geneva_wheel.sci',0);   exec('inclined_plane.sci',0);
exec('parabola_track.sci',0);   exec('pend_moving_base.sci',0);
exec('restricted_3body.sci',0);   exec('scotch_yoke.sci',0);
exec('sim_model_ex5_3_11.sci',0);   exec('sim_model_ex5_3_13.sci',0);
exec('slider_crank.sci',0);   exec('spring_pendulum.sci',0);
exec('top_dynamics.sci',0);   exec('two_body_simul.sci',0);
exec('two_particles_table.sci',0);
```

Figure A.3: A typical scilab.ini file for this book.

For those who prefer to use Matlab, Fig. A.4 shows a typical window that opens when Matlab is started, displaying the pathname of the directory that is automatically created by Matlab during its installation. Commands to be executed (processed) by Matlab are typed in the Matlab *Command Window*. Do not be concerned if the window that opened in your computer looks somewhat different or if it is not displaying the additional windows shown on the left part of the figure.

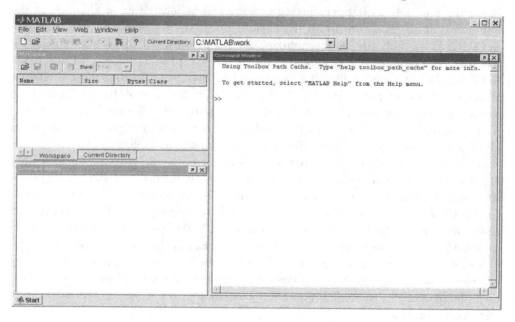

Figure A.4: A typical Matlab window.

It is convenient to force Matlab to be in your Matlab home directory (i.e., the directory you created for storing all your Matlab (.m) files, and your Simulink (.mdl) files if you use Simulink) right after it starts. For that to happen when you start Matlab, prepare, using the Matlab editor (as explained a little later), a file with two command lines. The first line should tell Matlab to change directory to your Matlab home directory, as exemplified in the first line seen in Fig. A.3, **but using cd instead of chdir** (the command chdir is for Scilab, and the equivalent command in Matlab is **cd**). It is also convenient to add a second line with the command **addpath** followed by the same home directory you used in the first line. On computers with Windows, save such a file, naming it **startup** (Matlab will save it as **startup.m**), in the directory **matlabroot/toolbox/local**, with the word "matlabroot" replaced by the pathname of the directory where Matlab is installed in your computer. For Unix computers, consult your system administrator. If you are using Matlab remotely, as in the case of installations on university campuses, for example, you may have to issue the cd command (from the Matlab command window), as indicated above, every time you start Matlab.

To open a Matlab editor, which is where either Matlab programs are written by the user or the contents of existing files are displayed, click the mouse on File → New → M-file. The Matlab editor seen in Fig. A.5 (which is "opened" by

```
1    function output=spring_pendulum(t, mg_over_kre, rs_0, dotrs_0, theta_0, dottheta_0)
2    % Animation of the spring-pendulum analyzed in Example 2.7.8
3    % Example of usage ( using the initial conditions  rs_0=0.1; rsdot_0=0; theta_0=0.01; thetadot_0=0 ):
4    % my_output = spring_pendulum([0:0.02:350], 0.5^2, 0.1, 0, 0.01, 0);
5
6    disp(' ')
7    disp('NOTE: IN ADDITION TO ANIMATING THE SPRING-PENDULUM ANALYZED IN EXAMPLE 2.7.10,')
8    disp('AND PLOTTING theta and rs versus time, A CALL TO THE FUNCTION spring_pendulum ')
9    disp('SUCH AS my_output = spring_pendulum([0:1:200], 0.5^2, [0.1; 0;  0.01; 0]) )')
10   disp('RETURNS 5 COLUMNS CONTAINING  t, rs,  d(rs)/dt, theta, and d(theta)/dt')
11   disp('TO PLOT d(rs)/dt, FOR EXAMPLE, VERSUS time, TYPE, IN THE MATLAB COMMAND WINDOW,');
12   disp('THE COMMANDS:    figure(); plot(my_output(:,1), my_output(:,3)) ')
13   disp('   TO PUT A GRID ON THE PLOT, IF DESIRED, THEN TYPE  grid')
14
15   close all   % closing (i.e., deleting) all existing figures
16   len=length(t);   % this is how many elements the row matrix t has
17
18   initial_conditions=[rs_0; dotrs_0; theta_0; dottheta_0];
19   [t, y]=ode45(@rhs_of_diff_eqs, t, initial_conditions, [], mg_over_kre);
20   figure('color', 'white');  plot(t, y(:,3)); grid       % plot of theta versus time t
21   xlabel(['Nondimensional time  ', '\tau'], 'fontsize', 12);    ylabel('\theta', 'fontsize', 12);
22
23   figure('color', 'white');  plot(t, y(:,1));   grid;    % plot of rs versus t
24   xlabel(['Nondimensional time  ', '\tau'], 'fontsize', 12);    ylabel('r_s', 'fontsize', 12);
25
26   output=[t, y];   % this is a n by 5 matrix, where n is the number of points in the time array t;
27       % column 1: time, column 2 : rs, column 3 : d(rs)/dt; column 4: theta; column 5: d(theta)/dt
28
29   % ANIMATION: using a bar AB from the point (0,0) to (1+rs)*(sin_theta, -cos_theta)
30   rs=output(:,2); theta=output(:,4);   % these are len by 1 matrices
31   x0=zeros(len,1); y0=x0;   xP=(1+rs).*sin(theta);  yP=-(1+rs).*cos(theta);
32   x_coordinates=[x0, xP];   y_coordinates=[y0, yP];  animate_nbars(x_coordinates, y_coordinates);
33
34
35   function dy_dt = rhs_of_diff_eqs(t, y, mg_over_kre)
36   % Note:  y(1)=rs, y(2)=d(rs)/dt, y(3)=theta, y(4)=d(theta)/dt
37   dy_dt=[y(2); (1+y(1))*y(4)^2-mg_over_kre*(1-cos( y(3) )-y(1); y(4); -(2*y(2)*y(4)+mg_over_kre*sin(y(3)) )/(1+y(1))];
38   % end of the file spring_pendulum.m
39
```

Figure A.5: The Matlab editor window showing the program spring_pendulum, which Matlab saves as spring_pendulum.m.

clicking the mouse on File → Open and clicking on the name of the file you want to display in the editor) shows the Matlab file spring_pendulum.m written by the author, which contains two functions called `spring_pendulum` and `rhs_of_diff_eqs`, that will do the same thing as the Scilab function `spring_pendulum` (which is saved by Scilab as `spring_pendulum.sci`) shown in Fig. A.2. By comparing the two figures you will see the programming similarities and differences between Scilab and Matlab.

Scilab and Matlab use the self-explanatory mathematical operators $+$, $-$, $*$, $/$ (for division), $\wedge$ (for raising a variable to a power), and $=$ (for assigning a value to a variable). Enter the commands in the respective software command window (i.e., the Scilab console, for Scilab, or the Matlab command window, for Matlab), and then press the **Return** (or **Enter**) key. For example, typing `a=2` creates the variable a with a value of 2 assigned to it. Both software display $a = 2$ after the **Return** (or **Enter**) key is pressed and released. They are both case sensitive and, therefore, a and A are different variables. Using a semicolon after an assignment suppresses the display. As in other mathematical packages, they also have commands such as `sqrt`, for the square root of a number [as in `sqrt(23.2)`], `exp`, for exponentiation [as in `exp(3.2)`], `log` (for the natural logarithm), `log10` (for the logarithm in base 10), `sin`, `cos`, and `asin` (for `arcsin`), to name a few. There are several predefined variables in Scilab and in Matlab, such as π (`%pi` in Scilab, and `pi` in Matlab), $\sqrt{-1}$ (`%i` in Scilab, and `i` in Matlab), and inf (`%inf` in Scilab, and `inf` in Matlab) for infinity. The format of a numeric display can be chosen by the user, if desired. Type `help format` in either Scilab or Matlab for more information. The commands **det** and **trace** are for obtaining the determinant and the trace of a matrix, respectively. The commands to find the eigenvalues of a matrix are **spec** and **eig**, for Scilab and Matlab, respectively.

For any Scilab or Matlab command, `help command_name` (where `command_name` is the name of the command for which you want to get more information about) will provide information about a command. In Scilab, the command `head_comments function_name` will display all the comment lines from the second line of the function's program (here called `function_name`) to the comment line that precedes either a blank line or the first executable command in that program. The Matlab command that does the same thing is `help function_name`. In Figs. A.2 and A.5, for example, these are lines 2 to 5, and 2 to 4, respectively. Comment lines in Scilab are those that start with a double slash, and in Matlab they are those that start with a percentage sign. Anything written after these symbols, even when a line does not start with them, is interpreted as comments by the software.

The syntax for specifying a list of values starts and ends with a square bracket. The elements of the list are separated either by a comma (a space is also acceptable) or by a semicolon. The list stands for a row when a comma (or a space) is used, and for a column when a semicolon is used. For each of the following command lines, type the commands shown and then press the **Return** (or **Enter**) key at the end of the line for either Scilab or Matlab to process them.

$$a = 2, \; b = 5, \; c = [2, \, 4, \, 8], \; d = [5; \, 10; \, 20]$$

The quantities c and d are a 1×3 row matrix (usually called a "row vector") and a 3×1 column matrix (usually called a "column vector"), respectively. The commas between the commands are for separating them. Semicolons can also be used. With a comma, the result of the operation will be displayed, while a semicolon inhibits the display (which is always stored in memory). A semicolon after the assignments a=2 and d $= [5; \, 10; \, 20]$, for example, would inhibit the display of the obvious result. The commands max and min give the maximum and minimum elements of a matrix, no matter what the dimensions of the matrix are. The commands max(c) and min(c) will display 8 and 2, respectively. To get additional information, type help max and help min.

The elements of c are $c(1) = 2, c(2) = 4$, and $c(3) = 8$. If you type c(2), for example, the software returns the value of that element of c. For m $= [2, \, 4, \, 8, \, 10]$, for example, if you type m$(2 : 4)$, the software returns elements 2 to 4.

The line

$$d = a/b, \; e = 1/a, \; f = c/a, \; g = a * b, \; h = a \wedge 3, \; k = a \wedge (-1)$$

illustrates the division, multiplication, and exponentiation operations. There are three operators that start with a dot, as .*, ./, and .$\wedge$; they are for element by element multiplication, division, and exponentiation, respectively, of arrays that have the same dimension. For example, $[2, 4, 8]. * [1, 3, 5]$ yields $[2, 12, 40]$, while $[2, 4, 8] * [1, 3, 5]$ is an inconsistent operation. Similarly, $[2, 4, 8]./[1, 3, 5]$ yields $[2, 1.3333, 1.6]$, and $[2, 4, 8]. \wedge [1, 3, 5]$ yields $[2, 64, 32768]$. The need for using these operators can be seen, for example, in the Scilab and Matlab similar programs *fourbar_kinematics* listed in Appendix E.

In the line

$$\text{clear}, \; a = [2, \, 4, \, 8], \; b = [1, \, 2; \, 5, \, 8], \; c = a. \wedge (-1), \; d = b. \wedge (-1)$$

the command clear removes all variables from the workspace. In Scilab, that will also remove all the functions you have written and made available to be executed in the Scilab console as indicated earlier. If you use the clear command in Scilab you will have to make your functions recognizable to the Scilab console again. The easiest and quickest way to do this is to have a copy of your scilab.ini file (where you listed your functions using the exec command) in your home directory and simply type exec('scilab.ini', 0) in the Scilab console. In Matlab, the command clear will remove only variables from the workspace (type help clear in the Matlab command window to see the other forms of the Matlab command clear).

The commands c = a. $\wedge$ (−1) and d = b. $\wedge$ (−1) generate arrays whose elements are the inverse of each element of the arrays a and b, respectively. Notice that matrix d is not the inverse of matrix b.

Matrix algebra is illustrated in the following line, where b was defined earlier.

    a=[1;2], c=inv(b), d=b$\wedge$(-1), e=a', f=b*a, g=[5,6,7;8,9,10]

Both Scilab and Matlab use an apostrophe for the transpose operation. To invert a matrix, such as matrix b, one may use either inv(b) (type help inv for additional information) or d=b$\wedge$(-1). Check to see that c = d, and that they are equal to the inverse of b by multiplying them either as c*b or b*c. The operation b*a yields the column [5; 21], and the operation a'*b yields the row [11, 18], as expected. Both Scilab and Matlab will display the names of your current variables if you type who either in the Scilab console or in the Matlab command window, respectively. Type help who for additional information.

The command size gives a 1 by 2 row containing the number of rows and columns in a matrix. For example, size(a) displays 2 and 1, and size(g) displays 2 and 3. Type help size, and also help length for more details. The command [answer_1, answer_2]=size(variable_name) assigns the number of rows and columns of variable_name to the variables answer_1 and answer_2, respectively. For example, [num_of_rows, num_of_columns]=size(g) yields 2 for the variable num_of_rows and 3 for the variable num_of_columns. This is used in the Scilab and Matlab programs animate_nbars (and also in animate3d_nbars) listed in Appendix E to determine the number of rows in a given matrix, and the result is used in a loop that animates the motion of the system described in that program.

The elements of g are g(1, 1) = 5, g(1, 2) = 6, etc. If you type g(2, 1), for example, the value of that element of g, which is 8, is displayed. The command g(:, 2) returns g(1, 2) and g(2, 2), i.e., all the elements in column 2 of g. The command g(1, :) returns all the elements of row 1 of g. The value of any element of a matrix can be changed by assigning a new desired value to that element. Type, for example, the command g(2, 2) = 15. By typing g after that, the new matrix g is displayed.

Use either Scilab or Matlab to obtain the solution to the following equation, with $\theta = \pi/6$ radians.

$$\begin{bmatrix} \cos\theta & \sin\theta \\ -\sin\theta & \cos\theta \end{bmatrix} x = \begin{bmatrix} 3 \\ 4 \end{bmatrix}$$

The approximate answer should be

                    ans   =

                    0.5981

                    4.9641

Write the command so that the software assigns the answer to a variable x.

A row with any number of equally spaced elements may be generated using colons, as in theta $= 0 : 0.1 : 5$ or in theta $= [0 : 0.1 : 5]$. The first and last values are the first and last elements of the row to be created, and the middle one is the step size. The default step size is 1 if it is skipped, as in theta=[0:5] or in theta=0:5. The transpose of the variable theta, just defined, is a column.

A.1.1 Plotting

Two-dimensional plots are generated in Scilab and in Matlab with the same command plot(x, y), to plot a variable y versus a variable x. The command plot(x, y1, x, y2) generates a plot of y1 versus x and y2 versus x in the same graph. The command plot(x, y1, '-', x, y2, '--') generates a plot of y1 versus x with a solid line, and y2 versus x with a dashed line (do not leave any space inside the quote marks that specify the line type). Type help plot for more details. Color lines may also be specified. For example, the command plot(x, y1, 'k-.', x, y2, 'r--') generates a plot of y1 versus x with a dash-dotted black line, and y2 versus x with a dashed red line. The symbols for the colors are k (black), r (red), y (yellow), g (green), etc. Type help LineSpec in Scilab to see the specifications that are available for lines; in Matlab, typing help plot will display a complete information associated with the plot command. Scilab has another command for two-dimensional plots, which is plot2d. Type help plot2d for information. To put a grid on a Scilab plot, type either xgrid or set(gca( ), 'grid', [1, 1]) in the Scilab console. To remove the grid, type set(gca( ), 'grid', [-1, -1]). In Matlab, the command grid acts like an on-off switch; typing grid puts a grid on the plot, and typing grid again, removes the grid.

In Matlab, another way to superimpose plots is to do a single plot first, such as plot(x, y1), "hold" the plot after that (by typing the command hold on), and then issuing the additional plot commands, such as plot(x, y2), plot(x, y3), etc.

Here is an example of a two-dimensional plot. The semicolons after the first three commands are used for suppressing the displays, especially of the array theta that has 1001 elements. The commands xlabel and ylabel are for inserting labels in the x- and y-axes, respectively, of a plot, and the command title inserts a centered title in the upper part of the plot window. The use of the option fontsize illustrates how the size of a font can be changed from its default value. In Scilab, a greek symbol either in a label or in the title of a figure, is generated as '$\symbol\_name$', as in 'α', for example. This is actually from the LaTex language.[1] Do not

---

[1]See, for example, the web site http://en.wikipedia.org/wiki/LaTeX for some information. This book was typeset using LaTex.

use spaces either before the first $ sign or after the second one. In Matlab, omit the $ signs.

In Scilab (with several alternatives for displaying labels and title):
```
theta=[0:0.01:10]; x=40*sin(theta); y=(theta.^2).*sin(theta);
figure('background',8); plot(theta,x,'-', theta,y,'--'); xgrid;
xlabel(['$\theta$','(radians)'],'fontsize', 4);
ylabel(['two functions of', '$\theta$'],'fontsize', 4)
// or: ylabel(['$40sin\theta\; and\;\theta∧2sin\theta$'])
// or: ylabel(['${\rm{40sin\theta\; and\;\theta∧2sin\theta}}$'])
title(['${\rm{solid:40sin\theta\; ; \;dashed:\theta∧2sin\theta}}$'])
// or:
// title(['${\rm{$-$:40sin\theta\; ; \;$--$:\theta∧2sin\theta}}$'])
```

The backslash followed by a semicolon (i.e., \;) seen in the title command is used for generating a space in a mathematical expression (i.e., an expression that is enclosed by $ signs), where needed; use more than one of such combination to increase the spacing between words or symbols in a string, if desired. The command `figure('background', 8)` creates a figure window with a white background.[2] The equivalent command in Matlab is `figure('color', 'white')`.

In Matlab:
```
theta=[0:0.01:10]; x=40*sin(theta); y=(theta.^2).*sin(theta);
figure('color', 'white'); plot(theta,x,'-', theta,y,'--'); grid;
xlabel('\theta (radians)', 'fontsize', 12);
ylabel('two functions of \theta', 'fontsize', 12);
title(['-:40sin\theta ; --:\theta∧2sin\theta'], 'fontsize', 12)
```

Figures A.6a and A.6b show the plots generated with the commands just presented. Figure A.6a was generated with Scilab and Fig. A.6b with Matlab. You may want to click on **Edit** in the figure window and explore the editing possibilities that are offered by both software.

Matlab has a command **gtext** for placing a text on a figure in a location chosen with the mouse pointer. If you are using Matlab, try, for example, the command `gtext('40 sin\theta')`, and then move the mouse pointer to the figure window and click on the location where you want to put the text. To see a more detailed explanation, type **help gtext** in the Matlab command window. By assigning a name to a text command, such as `text1=gtext('40 sin\theta')`, instead of simply typing `gtext('40 sin\theta')`, Matlab creates the "object" called **text1** whose properties (such as **FontSize** and **String**, for example), can be changed by using the

---

[2]Such a command was issued before the command **plot** to create a figure window specifically for the desired plot so that plots you may have created in other figure windows are not affected.

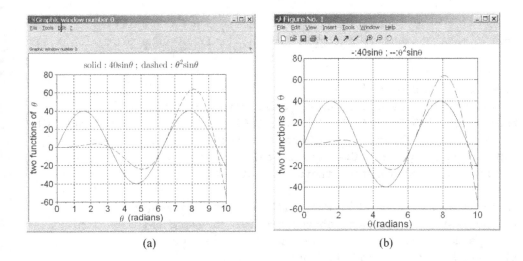

Figure A.6: Figures generated with (a) Scilab and (b) Matlab.

Matlab **set** command. To see the properties of the object **text1** (which are accessed using the **set** command), type **get(text1)** in the Matlab command window. For more details, type **help gtext**, **help get**, and **help set** in the Matlab command window.

As another example of plotting, consider a plot similar to the one shown in Fig. 1.13 (p. 53), where θ and $d\theta/d\tau$ satisfy the following equation.

$$\frac{(d\theta/d\tau)^2}{2} - \cos(\theta) = \text{constant} \overset{\Delta}{=} C$$

In Matlab:
```
fig=figure('color', 'white'); % to create a new figure window
hold on; % to allow multiple plots in the same window
theta=[-2*pi:0.01:2*pi]; C=[-0.8, 0.62, 1, 1.205]; steps=length(C);
for k=1:steps, % starting a loop; for details, type help for
  dtheta_dtau=sqrt(2*( C(k)+cos(theta) ));
  plot(theta, dtheta_dtau, theta, -dtheta_dtau);
end
grid; axis([-2*pi, 2*pi, -2.5, 2.5]); % for details, type help axis
xlabel('\theta (rad)', 'fontsize', 24);
ylabel('d\theta/d\tau','fontsize', 24);
```

gtext($'C = -0.8'$); gtext($'C = 0.62'$); gtext($'C = 1'$); gtext($'C = 1.205'$);
box; % this command is a toggle switch; it frames the plot with a box

In Scilab:
```
fig=figure('background', 8);
// The above command creates a figure with a white background
theta=[-2*%pi:0.01:2*%pi]; C=[-0.8, 0.62, 1, 1.205]; steps=length(C);
for k=1:steps, // starting a loop; for details, type help for
  dtheta_dtau=sqrt(2*( C(k)+cos(theta) ));
  plot(theta, dtheta_dtau, theta, -dtheta_dtau);
end
xgrid; gca_handles=gca( ), gca_handles.tight_limits='on';
// the last two commands, above, are for changing the axes limits
// type help gca to get information about the gca command
xlabel(['$\theta$', '(rad)'], 'fontsize', 6);
ylabel(['$d\theta/d\tau$'], 'fontsize', 6)
```

Figure A.7 shows the plots obtained with Scilab and with Matlab. On the plot generated by Matlab one can click on the arrow that points northwest on the figure window and then double-click on a text in the figure to change the text, itself, and its location. The selected text can be moved by dragging it, with the mouse, to another desired location.

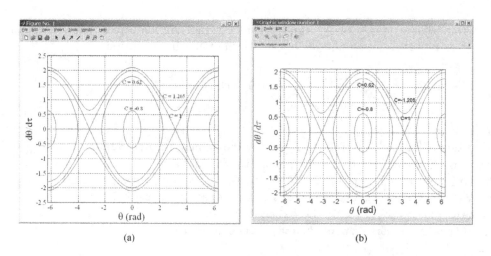

(a)   (b)

Figure A.7: Phase plane plot $d\theta/d\tau$ versus θ created with (a) Matlab and (b) with Scilab.

The strings seen in the plot generated by Scilab were inserted with the following commands
xstring(-0.5,0.7,$'C = -0.8'$), xstring(-0.5,1.5,$'C = 0.62'$),
xstring(2.8, 0.4,$'C = 1'$), xstring(2.3, 1,$'C = 1.205'$)
where the coordinates for the left lower corner of each string were chosen after looking at the plot before inserting the strings. One may change the string in the Scilab plot by first clearing it by clicking on Edit $\rightarrow$ Axes properties in the figure window, clicking on a text labeled Text(i), where i $= 1, 2, \ldots$, and, then, on the tab labeled Data & Mode to change the C=1, for example, to a blank surrounded by double quotes and pressing the Return (or Enter) key on the computer keyboard. After doing that, one would then type the xstring command in the Scilab console, with a new desired string and values for the coordinates, as mentioned earlier.

Three-dimensional plots are created in Scilab with the commands plot3d(x, y, z) and param3d(x, y, z), and in Matlab with the command plot3(x, y, z). The Scilab command plot3d(x,y,z) is for generating three-dimensional surfaces; for plotting lines in three-dimensional space, use param3d. The program animate3d_nbars.sci, listed in Appendix E, uses param3d (see the first *NOTE* in that program). For additional information, type help plot3d and help param3d in the Scilab console, or help plot3 in the Matlab command window.

See the programs listed in Appendix E for examples of usage of some additional Scilab and Matlab commands.

A.1.2 Numerical Solution of Ordinary Differential Equations

Numerical solvers require that an ordinary differential equation of order n be written as a set of $2n$ first-order differential equations. An ordinary differential equation of order n, $d^n x/dt^n = g(x, dx/dt, \ldots, d^{n-1}x/dt^{n-1}; t)$, where g is, in general, a nonlinear function of its arguments, may be written as a set of $2n$ first-order differential equations of the form $dy(i)/dt = f_i(y(1), y(2), \ldots, y(n); t)$ where the $y(i)$'s, for $i = 1, 2, \ldots, n$, are called the *state variables*. For example, by choosing $y(1) = x$, $y(2) = dx/dt, \ldots, y(n) = d^{n-1}x/dt^{n-1}$, one obtains the following set of first-order ordinary differential equations:

$$\frac{dy(1)}{dt} = y(2) \tag{A.1.1}$$

$$\frac{dy(2)}{dt} = y(3) \tag{A.1.2}$$

$$\vdots \tag{A.1.3}$$

$$\frac{dy(n)}{dt} = g(y(1), y(2) \ldots, y(n); t) \tag{A.1.4}$$

The choice of state variables is not unique, and a set of n linearly independent combinations of the state variables shown also qualifies as an alternative set of state variables. The choice shown here is used in most of the literature that deals with differential equations. Our attention here is focused on the use of appropriate subroutines (called *functions* in either Scilab or Matlab) for integrating a set of first-order differential equations (thus, differential equations of any order).

The Scilab command for integrating a set of m first-order differential equations of the form indicated earlier, with specified initial conditions $y_1(0)$, $y_2(0)$, ..., $y_m(0)$ is ode (ode is actually an interface to various solvers). Its simplest syntax is y = ode(y0, t0, t, rhs_of_diff_eqs), where y0 is a column containing the initial conditions $y_1(0)$, $y_2(0)$, ..., $y_m(0)$ (i.e., y0=$[y_1(0); y_2(0); \ldots; y_m(0)]$), t0 (usually, t0=0) is the initial time of the integration, t is a row specifying the times the solution is to be generated, and rhs_of_diff_eqs is a function of the form ydot=rhs_of_diff_eqs(t,y), written by the user, specifying the right-hand side of the differential equations. The function rhs_of_diff_eqs must return a column array[3], which Scilab puts into a matrix where each row is the value of the state variables at the time instants specified in the array t. The following is an example of a Scilab function for such, typed in the Scilab console, using Eqs. (2.7.8f) and (2.7.8g), p. 164, that govern the motion of a spring-pendulum. The last line of any Scilab function must be the command endfunction. Type help ode for additional information on other ways of calling ode, including the use of a Fortran subroutine or a C function for rhs_of_diff_eqs.

Example for Scilab users:
```
function dy_dt=rhs_of_diff_eqs(t, y)
  // Note:  y(1)=rs, y(2)=d(rs)/dt, y(3)=theta, y(4)=d(theta)/dt
  dy_dt(1)=y(2)
  dy_dt(2)=(1+y(1))*y(4)^2-mg_over_kre*( 1-cos(y(3)) )-y(1)
  dy_dt(3)=y(4)
  dy_dt(4)=-( 2*y(2)*y(4)+mg_over_kre*sin(y(3)) )/(1+y(1))
endfunction
```

Suppose we now want to generate a numerical solution of the given differential equations for mg_over_kre = 0.5^2 = 0.25, with the initial conditions y(1)=0.1, y(2)=0, y(3)=0.01, y(4)=0, and obtain the values of the state variables y(1) to y(4) at the instants t=0, 0.1, ..., 350. The following commands, also typed in the Scilab console, will generate the desired data points and store them in the vari-

---

[3]Commands such as b(1)=1; b(2)=2; generate a column b in Scilab, and a row b in Matlab.

able we are naming y, i.e., y=[y(1), y(2), y(3), y(4)]. Notice the use of the semicolon, at the end of a command, to suppress the display (especially the long displays for t and y).

```
t=[0:0.1:350];mg_over_kre=0.5∧2;initial_conditions=[0.1;0;0.01;0];
y=ode(initial_conditions, 0, t, rhs_of_diff_eqs);
```

To plot the normalized length of the pendulum versus time, for example, type the command plot(t, 1+y(1,:)). The example presented here constitutes most of the Scilab program spring_pendulum (which Scilab saves as spring_pendulum.sci) listed in Appendix E (and also shown in Fig. A.2). The last part of that program consists of instructions for animating the motion, whose basic commands are presented in the last part of this Section. There is a comment line in that program, immediately after the differential equations are integrated, with a pause command that was inserted only to illustrate the use of that command. That command is executed only if the user removes the double slash (//) from that line. Scilab interrupts the execution of a program when it encounters a pause command. This is a convenient command to use either for debugging a new program (to try to find in which line an error occurs) or to look at some intermediate results before proceeding with the execution of the program. To proceed with the execution when a pause command is inserted, type, in the Scilab console, either resume or abort to continue or to abort the execution, respectively. Type, in the Scilab console, help pause for a more detailed information.

Matlab users have to call a specific function for integrating a set of m first-order differential equations of the same form indicated earlier. The function ode45, for example, uses a variable step Runge-Kutta method, and is usually a good choice. Type, help ode45 in the Matlab command window for more details, and to see the other choices.

The simplest syntax for calling ode45 is [t, y] = ode45(rhs_of_diff_eqs, t, y0), where rhs_of_diff_eqs and t are the same as described earlier, and y0 is a column containing the values for the initial conditions of the column of state variables y=[y(1); y(2); ...; y(m)]. The syntax [t, y] = ode45(rhs_of_diff_eqs, t, y0, [], P1, P2, ..., Pk) allows the passing of k parameters P1, P2, ..., Pk (such as the parameter mg_over_kre seen earlier) to the function rhs_of_diff_eqs (type help ode45 for more details). In such a case, the syntax for the function rhs_of_diff_eqs must be ydot=rhs_of_diff_eqs(t, y, P1, P2, ..., Pk). The function rhs_of_diff_eqs must return a column array (see the preceding footnote), which is put into a matrix where each row will be the value of the state variables at the time instants specified in the array t.

The following is the same example presented earlier, but typed in the Matlab command window. There is no **end** or **endfunction** statement in a Matlab function. Therefore, in Matlab, there must be a blank line between the last statement of a function and the next executable statement in a Matlab program.

Example for Matlab users:
```
function dy_dt = rhs_of_diff_eqs(t, y, mg_over_kre)
% Note:  y(1)=rs, y(2)=d(rs)/dt, y(3)=theta, y(4)=d(theta)/dt
dy_dt(1)=y(2);
dy_dt(2)=(1+y(1))*y(4)^2-mg_over_kre*( 1-cos(y(3)) )-y(1);
dy_dt(3)=y(4);
dy_dt(4)=-( 2*y(2)*y(4)+mg_over_kre*sin(y(3)) )/(1+y(1));
dy_dt=[dy_dt(1); dy_dt(2); dy_dt(3); dy_dt(4)];
% The previous command transforms dy_dt into a column array
```

Let us generate a numerical solution to the given differential equations for `mg_over_kre` $= 0.5^2$, with the initial conditions `y(1)=0.1`, `y(2)=0`, `y(3)=0.01`, `y(4)=0`, and obtain the values of the state variables `y(1)` to `y(4)` at the instants `t=0, 0.02, ..., 350`. The following commands, also typed in the Matlab command window, will generate the desired data points and store them in the variable we are also naming y, i.e., `y=[y(1); y(2); y(3); y(4)]`. After Matlab finishes integrating the differential equations, t is a column array whose length is the number m of time data points (351 in this example), and y is a matrix with m rows and as many columns as the number of state variables. Notice the need for using the symbol @ in the call to **ode45**, and, again, the use of the semicolon to suppress a long display.

```
t=[0:0.02:350];mg_over_kre=0.5^2;initial_conditions=[0.1;0;0.01;0];
[t, y]=ode45(@rhs_of_diff_eqs, initial_conditions, [ ], mg_over_kre);
```

All the Matlab commands shown for this example constitute most of a program written by the author and saved by Matlab as **spring_pendulum.m**. That program is shown in Fig. A.5, and is also listed in Appendix E. It does the same as its Scilab counterpart **spring_pendulum.sci**, mentioned earlier. The last part of both programs consists of instructions for animating the motion.

A.1.3   Animation

The details involving animation are shown in the similar Scilab and Matlab functions **animate_nbars** (and in the functions **animate3D_nbars**) listed in Appendix E. An extensive set of comment lines with explanations that are meant to be for those that might be interested in such details are in the Scilab function `animate_nbars`.[4] Such functions are for animating two-dimensional (and three-dimensional) motion of a system consisting of an arbitrary number of connected bars. Their usage consists only in the call command `animate_nbars(x_coordinates, y_coordinates)` [or `animate_nbars(x_coordinates, y_coordinates, z_coordinates)`], where the arguments `x_coordinates` and `y_coordinates` (and `z_coordinates`) are the x and y (and z) rectangular coordinates of the endpoints A, B, C, etc., of an arbitrary number of connected bars AB, BC, etc. Such a call is in most of the programs listed in Fig. A.3. There are very few comments in the two similar functions `animate3d_nbars` (one for Scilab and the other for Matlab users) since the steps that lead to the animation in three-dimensional space are the same as explained in the Scilab function `animate_nbars`.

A.2   Introduction to Simulink

Simulink is a graphical extension of MATLAB, where one works with simulation blocks, such as blocks for integration, and for function generation, among many others. This section is, of course, for those users who might prefer to use Simulink. The user formulates the equations that govern the motion of the dynamical system under study and then construct a Simulink block diagram model to do the calculations by connecting the necessary blocks in the appropriate manner. Such diagrams are much like the ones many readers have seen in books on dynamical systems and on control systems. The tutorial presented in this section is relatively brief since there are a number of others available on the web, which can be found by using the words `Simulink tutorial` when searching.

To start Simulink, type `simulink` in the Matlab command window shown in Fig. A.4. This will open a "Simulink Library Browser" window, such as the one illustrated in Fig. A.8. Do not be concerned if the windows on your computer screen do not look exactly like the ones shown here. It makes no difference how they appear on your computer screen. The Library Browser window may be used to search for a specific block by typing its name in the *Find* part of that window and then pressing the `Return` (or `Enter`) key on the computer keyboard.

The icons labeled `Continuous`, `Sources`, `Sinks`, etc., seen in Fig. A.8 represent Simulink library directories that contain a number of different blocks used for constructing Simulink models of dynamical systems. Figure A.8 shows a typical contents of the `Continuous` library. Do not pay any attention to the *controls sym-*

---

[4]For complete information on Graphics and Graphical User Interfaces in general (for Scilab, Matlab, etc.), those who are interested may want to look at (especially Chapters 3 and 5) Marchand, Patrick (1996). Graphics and GUIs with MATLAB. CRC Press.

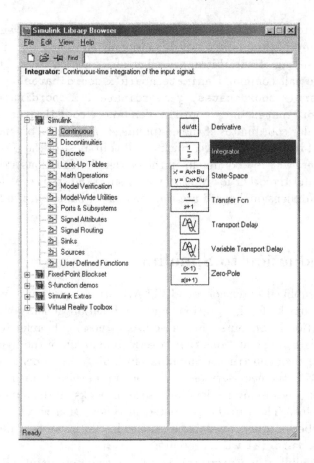

Figure A.8: A typical Simulink Library Browser window.

$bol \frac{1}{s}$ (which is the transfer function of an integrator, for those who are familiar with control theory) that appears on the integrator icon. The contents of the other libraries (`Sources`, `Sinks`, etc.) are displayed by clicking on their names.

A Simulink block diagram model is constructed in a Simulink "model window," which is open by choosing `File` → `New` → `Model` in the Simulink Library Browser. A model window labeled "untitled," such as the one illustrated in Fig. A.9, will then appear on the computer screen. That window, and others, may be resized and moved on the computer screen the same way you resize or move any other window on your computer. The model window also has several menus (`File`, `Edit`, etc.). Any Simulink model (which is saved by Matlab with the extension .mdl) you have may be put in a model window by clicking on `File` $\to$ `Open` and then choosing the directory and selecting the desired model in that directory.

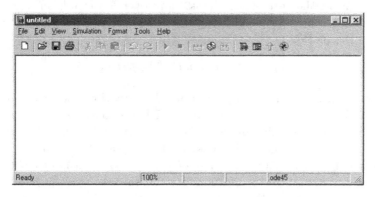

Figure A.9: A typical Simulink model window.

Let us construct a Simulink model to obtain a numerical solution to the differential equation of motion of a dynamical system. For this, let us use the spring-mass systems shown in Fig. A.10.

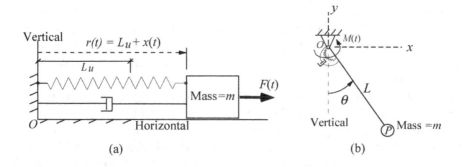

Figure A.10: Two examples of a spring-mass-dashpot system.

The system in Fig. A.10a consists of a spring (which will be modeled as massless) of unstressed length L_u m, a block of mass m kg that is subjected to a known applied force $F(t)$, and a viscous damper (called a *dashpot*). Motion is on a horizontal plane. During the motion, the actual length $r(t)$ of the spring changes by $x(t)$ m from its unstressed length L_u. Let the reaction force applied by the spring be of the form $k_1 x + k_3 x^3$ (with k_1 and k_3 normally determined from laboratory measurements), where k_1 is in N/m and k_3 in N/m^3. Also, let the reaction force applied by the dashpot be proportional to the velocity $\dot{x}$, with the proportionality constant being c N/(m/s). As in the rest of this book, dots over a variable in a mathematical expression denote the time derivative of that variable. The differential equation that governs the motion of this system is

$$m\ddot{x} + c\dot{x} + k_1 x + k_3 x^3 = F(t) \qquad (A.2.1)$$

The system in Fig. A.10b is a pendulum whose motion takes place on a vertical plane. It consists of a bar (which will be modeled as massless) of length L m and subjected to a known applied moment $M(t)$, a torsional spring that is unstressed when $\theta(t) = 0$, a block of mass m kg (which will be modeled as a point mass P), and a viscous torsional damper (called a *torsional dashpot*). The dashpot reaction moment is proportional to $\dot{\theta}$, with the proportionality constant being c N·m/(rad/s). Let the reaction moment applied by the torsional spring also be of the form $k_1\theta + k_3\theta^3$, where k_1 is in N·m/rad and k_3 in N·m/rad^3. With $g = 9.81$ m/s^2 being the acceleration of gravity, the differential equation that governs the motion of this system is

$$mL^2\ddot{\theta} + c\dot{\theta} + k_1\theta + k_3\theta^3 + mgL\sin\theta = M(t) \qquad (A.2.2)$$

Since Eq. (A.2.1) can be viewed as a particular case of Eq. (A.2.2), simply by setting $L = 1$, leaving the g-term out, and replacing $M(t)$ by $F(t)$, we only need to concentrate on Eq. (A.2.2). Notice that the equation is linear if $k_3 = 0$ and g is set to zero (in which case it can be solved analytically), and nonlinear otherwise. It is desired to plot $M(t)/(mL^2)$ and $\theta(t)$ in the same graph, and see how the system responds to a square wave for $M(t)$, which is available from the **Signal Generator** Simulink block.

To construct a Simulink model, one starts by copying some of the blocks from the Simulink libraries to a Simulink model window. A block is copied to a model window by clicking the computer mouse on it in the Library Browser, and dragging it to the model window. Figure A.11 shows a model window with blocks needed for the simulation of the system under consideration. A description of what a block does appears if one right-clicks the computer mouse on the block and selects **Help** in a menu that appears on the computer screen.

Since a second-order differential equation has to be integrated, one needs two copies of the integrator block, one to integrate $\ddot{\theta}(t)$ to generate $\dot{\theta}(t)$, and another to integrate $\dot{\theta}(t)$ to give us $\theta(t)$. Copies of a block may be obtained by selecting the block (with a left-click of the computer mouse) that is already in the model window, clicking on **Edit** $\rightarrow$ **Copy**, and then on **Edit** $\rightarrow$ **Paste**. A **Scope** (i.e., an "oscilloscope") and a **To Workspace** block are also shown in Fig. A.11. The scope, if one desires to include it in a model, is for displaying a plot of a variable, and the **To**

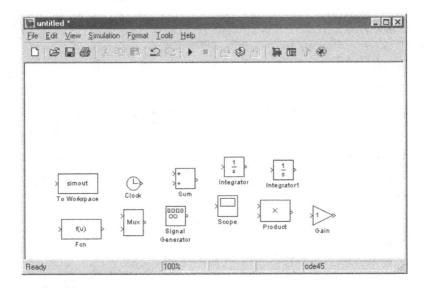

Figure A.11: Blocks that will be used to prepare a Simulink model to integrate Eq. (A.2.2).

`Workspace` block is for sending a variable to the Matlab workspace, where it can then be used (such as plotting it using the `plot` command described in Section A.1.1). Both were included in the figure window shown in Fig. A.11 for demonstration purposes only. One would also send data to the Matlab workspace to animate the motion using the `animate_nbars` listed in Appendix E. The default name of the variable that is sent to the workspace by the `To Workspace` block may be changed in a little window that appears on the computer screen after double-clicking the computer mouse on that block.

One possible Simulink model for solving the problem at hand is obtained by rearranging the needed blocks (by dragging them to desired locations in the model window) and connecting them as shown in Fig. A.12. The integrator whose input is $\dot{\theta}(t)$ generates the output $\theta(t)$, where $\dot{\theta}(t)$ is generated by the second integrator whose input must be the expression for $\ddot{\theta}$, which is $\ddot{\theta} = M(t)/(mL^2) - [c/(mL^2)]\dot{\theta} - [k_1/(mL^2)]\theta - [k_3/(mL^2)]\theta^3 - (g/L)\sin\theta$. The sum of these five terms is the output of the SUM block (double-click the block to change the default list of signs to $+----$).

Notice that the four `Gain` blocks in the Simulink model shown in Fig. A.12 are "flipped" (which is done by clicking on a block and then on `Format` → `Flip block`) so that the model is nicely organized (the name of the block may also be flipped with `Format` → `Flip name`). Also, their gains were changed to the values shown in the

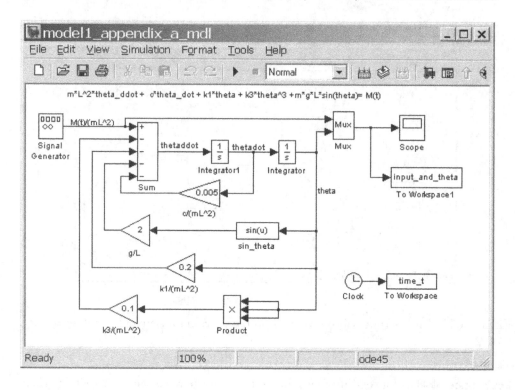

Figure A.12: Simulink model (a fully functional block diagram) for solving Eq. (A.2.2).

figure (which is done after double-clicking a `Gain` block and typing a new value in the window that opens up), and their labels were also changed (after clicking on the labels, one by one, and making changes in the same manner one would do in a word processor) to $c/(mL^2)$, g/L, $k1/(mL^2)$, and $k3/(mL^2)$ for easier identification.

The input to the scope is connected to the output of a `MUX` block [whose inputs are labeled $M(t)/(mL^2)$ and `theta`] so that the plots for $M(t)/(mL^2)$ and θ are superimposed on the same graph. The output of that `MUX` block is also sent to the Matlab workspace by a `To Workspace` block that was dragged to the model window from the Library Browser. Connecting one block to another is done by clicking on the output of one of them and dragging the mouse to the input of the other (or vice versa). To connect the input of a block to a line, drag the mouse from that input to the desired line (orthogonal lines are obtained by changing the direction of the dragging).

To make the block diagram easier to read, the strings `thetadot` (for $\dot{\theta}$) and `thetaddot` (for $\ddot{\theta}$) were inserted near the input lines of the two integrators shown in Fig. A.12. This is done by placing the mouse pointer at the desired location for the text, double-clicking it, and then typing the text. Clicking the mouse elsewhere on the model window terminates this process. An entire block of text can be moved to another desired location in the model window by dragging it with the mouse to that location. For demonstration purposes, a text showing the expression for the second-order differential equation being considered was inserted in the model window. In addition, some of the blocks shown in Fig. A.12 were resized for convenience. Resizing a block is done the same way as resizing a window on a computer screen.

A model is saved by pointing the mouse cursor at `File` and choosing either `File` $\rightarrow$ `Save as` (for the first time a model is saved) or `File` $\rightarrow$ `Save` with a click. It is recommended that you save your models in the same "home directory" mentioned earlier. If you write a Matlab program, and save it as, say, `my_dynamics_file` (which is saved by Matlab with the extension `.m`), and a Simulink model to do the same thing, use a different, but suggestive, name for saving your Simulink model. You might want to save it as `my_dynamics_file_mdl`, say. A Simulink model is automatically saved with the `.mdl` extension. Using this convention, the Simulink model that would do the same thing as the program `ballistics.m` listed in Appendix E, for example, would be saved as `ballistics_mdl.mdl`.

In the model shown in Fig. A.12, the variables `input_and_theta` and `time_t` are sent to the Matlab workspace. Double-click the blocks labeled `To Workspace1` and `To Workspace` and choose `Array` in the `Save format` option that appears in the menus for them. When this is done, the variables `input_and_theta` and `time_t` will be $m \times 2$ (m by 2) and $m \times 1$ arrays, respectively, where m is the number of time samples used by the software during the simulation. Also, type a large value, such as 30000, or even `inf`, for the selection to limit the number of data points.

Double-click the `Signal Generator` to choose one of the available waveforms (sine wave, square wave, etc.), its amplitude, and its frequency. Let us choose a square wave, with the parameter named `amplitude` set to -1 and frequency equal to 0.2 Hz (or cycles per second). This will generate a square wave that starts at $+1$, changes to -1 at the end of half a period,[5] when $t = \frac{1}{2} \times (1/0.2) = 2.5$ time units, and then repeats itself.

Before a simulation is started, the desired initial conditions $\theta(0)$ and $\dot{\theta}(0)$ should be input into the integrators. This is done by double-clicking the integrators in the Simulink model and entering the desired values in the appropriate place of a window

---

[5]If the parameter called "amplitude" were set to $+1$ (which would really be the amplitude of the wave since *amplitude* is, by definition, a positive quantity), the square wave generated by the *Signal Generator* block would start at -1, and then change to $+1$ at the end of half a period.

that opens up. Since the pendulum is driven by an external moment, let us use the default values $\theta(0) = 0$ and $\dot{\theta}(0) = 0$ for both integrators in the model shown in Fig. A.12. Also, several simulation parameters need to be set in the window that appears on the computer screen after clicking on `Simulation` in the model window, and selecting `Simulation Parameters` (or `Configuration Parameters` in some versions of the software). By doing so, one can choose the integration stop time, the numerical algorithm for the integration, the maximum step size, to save data to the workspace, etc. Time is also being sent to the workspace in the manner shown in Fig. A.12 for demonstration purposes only. The default choice for the integration algorithm is probably set as a variable step ode45 algorithm, which is a Runge-Kutta algorithm that is a good choice for most problems. One may also choose either to let the software adjust the intermediate times an output is generated or to force it to generate output (also referred to as "export") at specific times specified in a time array of the form `[t0:`Δ`t:tfinal]`, as seen throughout the book. After the simulation stops, one may want to experiment with different values for the relative and/or absolute error tolerances of the simulation to see if the results are noticeably affected.

One may need to double-click the scope icon to make its display visible during the simulation. To start the simulation, select `Simulation` $\rightarrow$ `Start` in the `Simulation` menu of the model window; the simulation may be stopped manually by selecting `Stop`.

Figure A.13 shows the input $M(t)/(mL^2)$ and the solution $\theta(t)$ in the same plot. Figure A.13a shows the `Scope` display, which becomes visible after double-

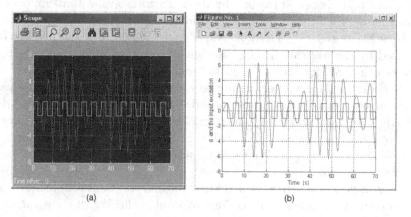

(a)   (b)

Figure A.13: Output of the simulation for $c/(mL^2) = 0.005$, $k_1/(mL^2) = 0.2$, $k_3/(mL^2) = 0.1$, and $g/L = 2$, as shown (a) by the `Scope`, and (b) using the `plot` command.

clicking the Scope block and clicking on the binoculars icon for automatically scaling the plot. Several icons to the left of the binoculars are for zooming in on desired parts of a plot. Figure A.13b is generated by typing the command `plot(time_t, input_and_theta)` in the Matlab command window, with labels inserted as indicated in Section A.1.1.

There are several ways one can construct a block diagram model for solving the same problem. Figure A.14 shows a more compact model for solving Eq. (A.2.2), with a different input $M(t)$, making use of an $f(u)$ block (which is the one labeled `theta_ddot`) and two `Matlab Function` blocks. They are found in

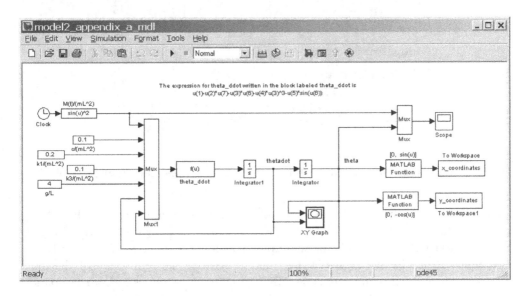

Figure A.14: Another Simulink model for solving Eq. (A.2.2).

the `User-Defined Functions` block library in the Simulink Library Browser (see Fig. A.8). Figure A.14 also shows the expression for $\ddot{\theta}$, given by Eq. (A.2.2), as it should be typed after double-clicking the block labeled `theta_ddot`. The mathematical expressions associated with the two `Matlab Function` blocks are shown in their labels, which were inserted as explained earlier. Figure A.15 shows a typical menu that appears on the computer screen after double-clicking a `Matlab Function` block, and what is typed in that menu for one of those blocks. Under the label `Output dimensions`, one specifies the number of elements of the output signal of the block (which is 2 for the [0, sin(u)] array, for example).

Two `To Workspace` blocks that send to the Matlab workspace the nondimen-

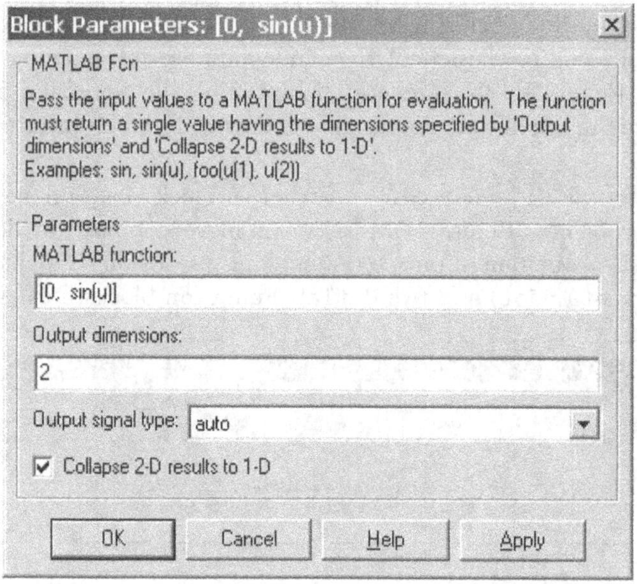

Figure A.15: Menu that appears after double-clicking a `Matlab Function` block.

sional x- and y-coordinates of points O and P of the pendulum shown in Fig. A.10 are also included in the model. These coordinates are the arrays `[0, sin(u)]` and `[0, -cos(u)]` (where u stands for the input, which is the angle θ), respectively, as also indicated in Fig. A.14. After the simulation stops, an animation of the motion will appear on the computer screen by typing `animate_nbars(x_coordinates, y_coordinates)` in the Matlab command window. For this, you need to have in your computer the Matlab program `animate_nbars` listed in Appendix E.

Sometimes, one wants to group a collection of blocks in a Simulink model into a single block (which is called a "subsystem" in Simulink), especially when a model starts to become too big to be displayed in a clear manner on the computer screen. The creation of a subsystem is illustrated with the aid of the model shown in Fig. A.14. To do this, first select an area of the model containing the blocks to be grouped together by dragging the mouse, with its left button pressed, and then releasing the mouse button after the desired area is defined, as illustrated in Fig. A.16.[6] Then, use the mouse to choose `Edit` → `Create Subsystem` in the `Edit` menu of the model window. The result is a model similar to the one shown in Fig. A.17

---

[6]The desired blocks and lines may also be selected one by one by holding down the `Shift` key on the keyboard while clicking the computer mouse on the selected items.

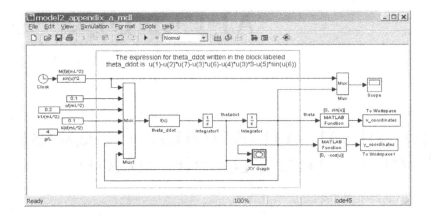

Figure A.16: How to make a subsystem.

(after resizing and rearranging some of the blocks and lines to make the diagram look nicer). The block arrangement of a subsystem can be seen by double-clicking it. Its input labels In1, In2, etc., and its output labels Out1 and Out2 may be changed by the user, if desired.

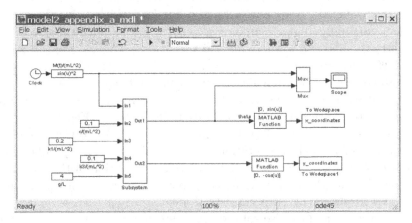

Figure A.17: The model with the subsystem that was made based on Fig. A.16.

Sequential Rotations, Angular Velocity, and Acceleration

B.1 Sequential Rotations and Angular Velocity

The orientation of a line in three-dimensional space can be described by a set of angles, as first illustrated by Fig. 1.1 on p. 3. Unless the rotations are about the same axis (a case which can be replaced by a single rotation about the same line), the final orientation of the line depends on the sequence of the rotations when the rotations are finite instead of infinitesimal. This is illustrated in Fig. B.1 where three orthogonal axes (labeled x_1, x_2, and x_3) and a line (in boldface) are shown.

The line is originally aligned with the x_3 axis. Imagine, now, three axes x_1', x_2', and x_3' fixed to the line and initially aligned with x_1, x_2, and x_3, respectively. Two sequences of 90° successive rotations are then performed, both involving a clockwise rotation θ_1 about one of the primed axes and a counterclockwise[1] rotation θ_2 about the new orientation of a different primed axis. It is suggested that you use either a book or a pencil to reproduce the demonstration illustrated in Fig. B.1.

The top part on the right of Fig. B.1 shows what happens to the line when the sequence consists of a clockwise rotation $\theta_1 = 90°$ about $x_1' = x_1$, followed by a counterclockwise rotation $\theta_2 = 90°$ about the updated position of the x_2' axis. The line that started along the horizontal ends up vertical. However, if the sequence of rotations is now reversed, i.e., if it consists of a counterclockwise rotation $\theta_2 = 90°$ about x_2, followed by a clockwise rotation $\theta_1 = 90°$ about the updated position of the x_1' axis, the line ends up pointing toward you.

This demonstration serves the purpose of showing that finite rotations cannot be represented as vectors because, as seen from Fig. B.1, the final direction of the line that underwent the rotation is not the vector sum of two rotations. If it were,

---

[1] There is no special relevance about choosing the rotations to be clockwise or counterclockwise. These choices were made arbitrarily.

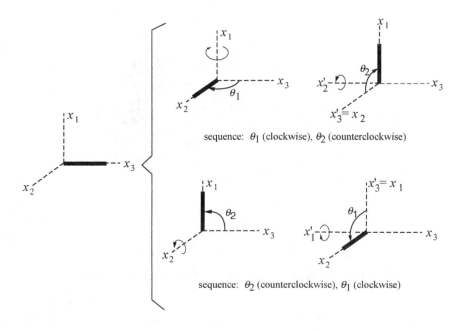

Figure B.1: Spatial orientation of a line using two sequential rotations.

the order of the rotations would be irrelevant as in the sum of two vectors.

Only infinitesimally small rotations (again, *not* finite rotations) can be represented as vectors because, as it turns out, the order of rotations for such a case is irrelevant. As you will see in the rest of this section, such a fact plays a decisive role in the determination of angular velocity when any line rotates in any manner in three-dimensional space.

To show the commutative property of infinitesimal rotations, consider a vector $\vec{r}$, of constant magnitude, undergoing a rotation $\Delta\theta_1$ about a line AA, followed by a rotation $\Delta\theta_2$ about line BB as shown in Fig. B.2. The unit vectors $\hat{a}$ and $\hat{b}$ are parallel to lines AA and BB, respectively. After the first rotation $\vec{r}$ becomes $\vec{r}_1$, and then it becomes $\vec{r}_2$ after the second rotation. The following expressions can be written by inspection of Fig. B.2:

$$\vec{r}_1 \approx \vec{r} + (\Delta\theta_1)\,\hat{a} \times \vec{r} \tag{B.1.1}$$

$$\vec{r}_2 \approx \vec{r}_1 + (\Delta\theta_2)\,\hat{b} \times \vec{r}_1 = \vec{r} + (\Delta\theta_1)\,\hat{a} \times \vec{r} + (\Delta\theta_2)\,\hat{b} \times [\vec{r} + (\Delta\theta_1)\,\hat{a} \times \vec{r}]$$

$$= \vec{r} + \left[(\Delta\theta_1)\,\hat{a} + (\Delta\theta_2)\,\vec{b}\right] \times \vec{r} + (\Delta\theta_1)\,(\Delta\theta_2)\,\vec{b} \times (\hat{a} \times \vec{r}) \tag{B.1.2}$$

When the rotations are infinitesimally small, the second-order infinitesimal quantity

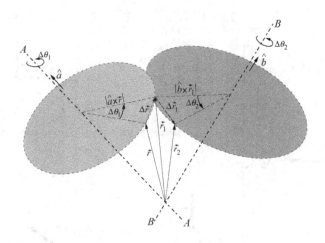

Figure B.2: Commutativity of infinitesimal rotations.

in the last term of Eq. (B.1.2) can be neglected. By denoting the infinitesimal rotations as $d\theta_1$ and $d\theta_2$, respectively, we then have as $\Delta t \to 0$,

$$\vec{r}_2 = \vec{r} + \left[(d\theta_1)\hat{a} + (d\theta_2)\vec{b}\right] \times \vec{r} \tag{B.1.3}$$

If the sequence of the rotations is reversed, so that the rotation $\Delta\theta_2$ about axis BB precedes the rotation $\Delta\theta_1$ about axis AA, the expression for the vector $\vec{r}_2$ as $\Delta t \to 0$ now becomes

$$\vec{r}_2 = \vec{r} + \left[(d\theta_2)\hat{b} + (d\theta_1)\vec{a}] \times \vec{r}\right] \tag{B.1.4}$$

A comparison of Eqs. (B.1.3) and (B.1.4) reveals that the infinitesimal rotations $(d\theta_1)\hat{a}$ and $(d\theta_2)\hat{b}$ satisfy the commutative property $(d\theta_1)\hat{a} + (d\theta_2)\hat{b} = (d\theta_2)\hat{b} + (d\theta_1)\hat{a}$. Since angular velocity is calculated based on infinitesimal changes of angles, angular velocity is indeed a vector, and an angular velocity vector can be formulated even when rotation angles about several axes in space are used to describe the orientation of a line or of a body in space. The angular velocity (relative to the reference frame in which the rotations are referred to) associated with the infinitesimal rotation vector $(d\theta_1)\hat{a} + (d\theta_2)\hat{b}$ is simply $\vec{\omega} = \dot{\theta}_1\hat{a} + \dot{\theta}_2\hat{b}$.

B.2   A Note on the General Form of Acceleration Expressions

The problem of determining the absolute acceleration of a point, expressed in a rotating unit vector triad, was addressed in Sections 2.6 (p. 138) and 2.7 (p. 142) of Chapter 2. To summarize the general procedure, consider a particle P moving in space and a triad $\{x\} = \{\hat{x}_1, \hat{x}_2, \hat{x}_3\}$ that is rotating with angular velocity $\vec{\omega}$ relative to inertial space (i.e., relative to a nonrotating frame of reference). The chosen triad $\{x\}$ may be fixed, for example, to a rotating piece of machinery. In general, both the magnitude and the direction of $\vec{\omega}$ will be changing with time. Let us now go through the entire process of formulating the expression for the absolute velocity and absolute acceleration of P, expressing both vectors in the triad $\{x\}$ that one chooses to work with. The first thing to do is to start with the absolute position vector for P, i.e., with the vector $\vec{r} = \overrightarrow{OP}$ from an inertial point O to P as shown in Fig. B.3.

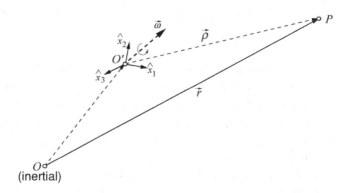

Figure B.3: Using an intermediate point when expressing the position vector of a point P.

In general, for any particular problem under consideration, it is more convenient to express the vector $\vec{r}$ in terms of several "intermediate vectors" such as from O to a point O' whose motion may be known, then from O' to some other point O_2, etc. To account for such situations, let us then express the vector $\vec{r}$ as follows (see Fig. B.3):

$$\vec{r} = \overrightarrow{OO'} + \overrightarrow{O'P} \triangleq \overrightarrow{OO'} + \vec{\rho} \tag{B.2.1}$$

By expressing the position vector $\vec{\rho}$ of P relative to O' in terms of the chosen unit vectors $\{\hat{x}_1, \hat{x}_2, \hat{x}_3\}$, a general expression of the form $\vec{\rho} = \rho_{x_1}\hat{x}_1 + \rho_{x_2}\hat{x}_2 + \rho_{x_3}\hat{x}_3$ is obtained, as seen throughout this book. The absolute velocity of P is then obtained as

$$\vec{v} = \frac{d\vec{r}}{dt} = \frac{d\,\overrightarrow{OO'}}{dt} + \frac{d\vec{\rho}}{dt} = \vec{v}_{O'} + \left(\frac{d\vec{\rho}}{dt}\right)_{\{x\}} + \vec{\omega} \times \vec{\rho} \tag{B.2.2}$$

where $\vec{v}_{O'} \stackrel{\Delta}{=} d\overrightarrow{OO'}/dt$ denotes the absolute velocity of point O' and $[d(\ )/dt]_{\{x\}}$ denotes the time derivative without taking into account the rotation of the unit vectors, such as $(d\vec{\rho}/dt)_{\{x\}} = \dot{\rho}_{x_1}\hat{x}_1 + \dot{\rho}_{x_2}\hat{x}_2 + \dot{\rho}_{x_3}\hat{x}_3$. The absolute velocity $\vec{v}_{O'}$ may be known and, in such a case, that is a reason for the vector $\vec{r}$ to be expressed as the sum of two parts. From Eq. (B.2.2), the expression for the absolute acceleration of P is then obtained as

$$\begin{aligned}
\vec{a} &= \frac{d\vec{v}}{dt} = \frac{d\,\vec{v}_{O'}}{dt} + \left[\left(\frac{d^2\vec{\rho}}{dt^2}\right)_{\{x\}} + \vec{\omega} \times \left(\frac{d\vec{\rho}}{dt}\right)_{\{x\}}\right] + \left[\frac{d}{dt}(\vec{\omega} \times \vec{\rho})_{\{x\}} + \vec{\omega} \times (\vec{\omega} \times \vec{\rho})\right] \\
&= \vec{a}_{O'} + \left(\frac{d^2\vec{\rho}}{dt^2}\right)_{\{x\}} + 2\vec{\omega} \times \left(\frac{d\vec{\rho}}{dt}\right)_{\{x\}} + \left(\frac{d\vec{\omega}}{dt}\right)_{\{x\}} \times \vec{\rho} + \vec{\omega} \times (\vec{\omega} \times \vec{\rho})
\end{aligned}$$

$$\tag{B.2.3}$$

where $\vec{a}_{O'} \stackrel{\Delta}{=} d\vec{v}_{O'}/dt$ denotes the absolute acceleration of point O'.

Equation (B.2.3) discloses that when a position vector $\overrightarrow{OP}$ is "split," for convenience, into two parts as $\overrightarrow{OO'} + \overrightarrow{O'P}$, where the motion of O' may be known, and the vector $\vec{\rho} = \overrightarrow{O'P}$ is expressed in a rotating unit vector triad $\{x\}$ whose absolute angular velocity is $\vec{\omega}$, the resulting expression that is obtained for the absolute acceleration contains five different types of terms.

The term $(d^2\vec{\rho}/dt^2)_{\{x\}}$ in Eq. (B.2.3) is interpreted as the acceleration of point P as seen by an observer fixed to the rotating triad $\{x\}$. The term $2\vec{\omega} \times (d\vec{\rho}/dt)_{\{x\}}$ is known as the *Coriolis acceleration* of P relative to O', while the term $\vec{\omega} \times (\vec{\omega} \times \vec{\rho})$ is known as the *centripetal acceleration* of P relative to O'. The negative of the centripetal acceleration is known as the *centrifugal acceleration*. The Coriolis acceleration is named after the French engineer Gaspard-Gustave Coriolis (1792-1843).

The equations that govern the motion of a particle P of mass m are simply obtained by equating the respective components of each side of the equation $\vec{F} = d(m\vec{v})/dt$, where $\vec{F}$ is the vector sum of all the *actual forces* (i.e., contact forces and field forces) acting on the particle, such as gravitational forces and reaction forces due to contact with another body (which includes friction). Such forces are obtained directly from a free-body diagram for the particle.

One may also write Newton's second law for a particle as $\vec{F} - m d\vec{v}/dt = \vec{F} - m\vec{a} = 0$, refer to the term $-m\vec{a}$ as an *inertial force*, and then proceed to include in the free-body diagram the five types of inertial forces as obtained from Eq. (B.2.3). Two of such forces have received specific names. They are the *Coriolis force*, which is the term $-2m\vec{\omega} \times (d\vec{\rho}/dt)_{\{\hat{x},\hat{y},\hat{z}\}}$, and the *centrifugal force*, which is the term $-m\vec{\omega} \times (\vec{\omega} \times \vec{\rho})$ (the negative of the centrifugal force is known as the centripetal force). Clearly, such terms depend on the particular choice one makes for the $\{\hat{x}_1, \hat{x}_2, \hat{x}_3\}$ triad in which the acceleration is being expressed in, and are nonzero only if the chosen triad is rotating. The use of the "formula" given by Eq. (B.2.3) would require one either to memorize or to consult it. Its use is avoided in this book, as seen by the many examples that were presented. For each problem one has to solve, all one needs to do to obtain the absolute acceleration $\vec{a}$ is to take the time derivative of the absolute velocity vector $\vec{v}$ (which, in turn, is obtained by taking the time derivative of the absolute position vector $\vec{r}$).

Properties of the Inertia Matrix of a Body

As seen in Chapter 6, the determination of the directions of the principal axes of inertia of a body, and of the corresponding principal moments of inertia, involves solving an eigenvector-eigenvalue problem of matrix algebra. The solution to such a problem was presented in Chapter 6. The inertia matrix of a body is always real and symmetric, and this appendix presents important properties associated with such a matrix. The following quantities were used in Chapter 6.

■ $[I_C]$. The inertia matrix of the body about a set of three orthogonal axes x, y, and z centered at the center of mass C of the body

■ $(\hat{x}, \hat{y}, \hat{z} \stackrel{\Delta}{=} \hat{x} \times \hat{y})$. Three orthogonal unit vectors parallel to the x-y-z axes

■ $\vec{\omega} = \omega_x \hat{x} + \omega_y \hat{y} + \omega_z \hat{z}$. Absolute angular velocity of the body

■ λ. Any of the three principal moments of inertia of the body

The determination of the direction of the principal axes and the principal moments of inertia leads to the matrix equation $[I_C]\underline{\omega} = \lambda \underline{\omega}$, i.e.,

$$
\begin{bmatrix}
I_{xx} - \lambda & I_{xy} & I_{xz} \\
I_{yx} & I_{yy} - \lambda & I_{yz} \\
I_{zx} & I_{zy} & I_{zz} - \lambda
\end{bmatrix}
\begin{bmatrix}
\omega_x \\
\omega_y \\
\omega_z
\end{bmatrix}
=
\begin{bmatrix}
0 \\
0 \\
0
\end{bmatrix}
\tag{C.1}
$$

This yields the following polynomial equation for λ:

$$
\begin{vmatrix}
I_{xx} - \lambda & I_{xy} & I_{xz} \\
I_{yx} & I_{yy} - \lambda & I_{yz} \\
I_{zx} & I_{zy} & I_{zz} - \lambda
\end{vmatrix}
\stackrel{\Delta}{=} P(\lambda) = -\left[\lambda^3 - a_2 \lambda^2 + a_1 \lambda - a_0\right] = 0 \tag{C.2}
$$

where

$$
a_2 = I_{xx} + I_{yy} + I_{zz} \tag{C.3}
$$

$$a_1 = \begin{vmatrix} I_{xx} & I_{xy} \\ I_{yx} & I_{yy} \end{vmatrix} + \begin{vmatrix} I_{xx} & I_{xz} \\ I_{zx} & I_{zz} \end{vmatrix} + \begin{vmatrix} I_{yy} & I_{yz} \\ I_{zy} & I_{zz} \end{vmatrix} \tag{C.4}$$

$$a_0 = |[I_C]| \tag{C.5}$$

Equation (C.2) admits three solutions $\lambda = \lambda_1$, $\lambda = \lambda_2$, and $\lambda = \lambda_3$.

The following properties of the inertia matrix $[I_C]$ are obtained from Eqs. (C.1) to (C.5).

■ Since the polynomial $P(\lambda)$ given by Eq. (C.2) can also be written in terms of the principal moments of inertia λ_1, λ_2, and λ_3, which only depend on the mass distribution of the body, as

$$\begin{aligned} P(\lambda) &= -\left(\lambda^3 - a_2\lambda^2 + a_1\lambda - a_0\right) \\ &= -(\lambda - \lambda_1)(\lambda - \lambda_2)(\lambda - \lambda_3) \\ &= -\left[\lambda^3 - (\lambda_1 + \lambda_2 + \lambda_3)\lambda^2 + (\lambda_1\lambda_2 + \lambda_1\lambda_3 + \lambda_2\lambda_3)\lambda - \lambda_1\lambda_2\lambda_3\right] \end{aligned} \tag{C.6}$$

it follows that the three coefficients a_2, a_1, and a_0 do not depend on the orientation of the orthogonal axes $\{x, y, z\}$ that are centered at the center of mass C of the body. Therefore, we conclude that the trace (which is the sum of the diagonal elements of the inertia matrix), the determinant, and the sum of the principal 2×2 minor determinants [which are the determinants shown in Eq. (C.4)] of the inertia matrix of a body referred to any set of orthogonal axes $\{x, y, z\}$ fixed to the body are invariant upon rotation of such axes. Notice that one of such conclusions has already been obtained in Section 6.3.1 [see Eq. (6.3.11), p. 405]. The existence of such invariants is useful, for example, for checking the results of the calculations involved in the determination of the principal moments of inertia.

■ As mentioned in Chapter 6, the principal axes of inertia of a body corresponding to distinct eigenvalues of the inertia matrix are orthogonal to each other. To prove this statement, consider two distinct eigenvalues λ_1 and λ_2 of the inertia matrix $[I_C]$ and the corresponding eigenvectors $\underline{\omega}_1 = [\omega_{1_x}, \omega_{1_y}, \omega_{1_z}]^T$ and $\underline{\omega}_2 = [\omega_{2_x}, \omega_{2_y}, \omega_{2_z}]^T$, respectively. We can then write the following relations:

$$[I_C]\underline{\omega}_1 = \lambda_1\underline{\omega}_1 \tag{C.7}$$

$$[I_C]\underline{\omega}_2 = \lambda_2\underline{\omega}_2 \tag{C.8}$$

By premultiplying Eq. (C.7) by the transpose of $\underline{\omega}_2$, and Eq. (C.8) by the transpose of $\underline{\omega}_1$, the two results may be combined as

$$\underline{\omega}_2^T[I_C]\underline{\omega}_1 - (\underline{\omega}_1^T[I_C]\underline{\omega}_2)^T = \lambda_1\underline{\omega}_2^T\underline{\omega}_1 - (\lambda_2\underline{\omega}_1^T\underline{\omega}_2)^T$$

or

$$\underline{\omega}_2{}^T[I_C]\underline{\omega}_1 - \underline{\omega}_2{}^T[I_C]^T\underline{\omega}_1 \;=\; (\lambda_1 - \lambda_2)\underline{\omega}_2{}^T\underline{\omega}_1$$

Since the inertia matrix $[I_C]$ is symmetric, this equation yields

$$(\lambda_1 - \lambda_2)\underline{\omega}_2{}^T\underline{\omega}_1 = 0 \qquad\qquad\qquad \text{(C.9)}$$

This equation immediately discloses that $\underline{\omega}_2{}^T\underline{\omega}_1 = 0$ if $\lambda_1 \neq \lambda_2$. Since $\underline{\omega}_2{}^T\underline{\omega}_1$ is the same as the dot product of the vectors $\vec{\omega}_1 = \omega_{1x}\hat{x} + \omega_{1y}\hat{y} + \omega_{1z}\hat{z}$ and $\vec{\omega}_2 = \omega_{2x}\hat{x} + \omega_{2y}\hat{y} + \omega_{2z}\hat{z}$, Eq. (C.9) discloses that $\vec{\omega}_1$ and $\vec{\omega}_2$ are perpendicular to each other if $\lambda_1 \neq \lambda_2$. If, on the other hand, $\lambda_1 = \lambda_2$, Eq. (C.9) is identically satisfied regardless of the angle between $\vec{\omega}_1$ and $\vec{\omega}_2$.

■ The eigenvalues of the inertia matrix (or, more generally, of any real symmetric matrix), which are the solutions of the polynomial equation (C.2), are always real. To prove this statement, assume that Eq. (C.1) is satisfied by a complex eigenvalue λ, to which corresponds a complex eigenvector $\vec{\omega}$, which we represented in matrix form as $\underline{\omega}$. Since any complex solution to Eq. (C.1) must occur in complex conjugate pairs, we must have, with an $*$ denoting the complex conjugate,

$$[I_C]\underline{\omega} \;=\; \lambda\underline{\omega} \qquad\qquad\qquad \text{(C.10)}$$
$$[I_C]\underline{\omega}^* \;=\; \lambda^*\underline{\omega}^* \qquad\qquad\qquad \text{(C.11)}$$

By premultiplying Eq. (C.10) by $\underline{\omega}^{*T}$ and Eq. (C.11) by $\underline{\omega}^T$, the result may be combined as

$$\underbrace{\underline{\omega}^{*T}[I_C]\underline{\omega} - (\underline{\omega}^T[I_C]\underline{\omega}^*)^T}_{=0} = \lambda\underline{\omega}^{*T}\underline{\omega} - (\lambda^*\underline{\omega}^T\underline{\omega}^*)^T$$

or

$$0 = (\lambda - \lambda^*)\underline{\omega}^{*T}\underline{\omega} \qquad\qquad\qquad \text{(C.12)}$$

The product $\underline{\omega}^{*T}\underline{\omega}$ in Eq. (C.12) is real and positive since the product of a complex number $a + \sqrt{-1}\,b$ by its complex conjugate $a - \sqrt{-1}\,b$ is equal to $a^2 + b^2$. Therefore, Eq. (C.12) is satisfied only if

$$\lambda^* = \lambda$$

which implies that the eigenvalue λ is real.

■ An interesting visualization associated with the moments of inertia of a body can be generated by manipulating either Eq. (C.7) or Eq. (C.8), which are of the form $[I_C]\underline{\omega} = \lambda\underline{\omega}$, as $\underline{\omega}^T[I_C]\underline{\omega} = \lambda\underline{\omega}^T\underline{\omega}$ to obtain for any eigenvalue λ

$$I_{xx}\omega_x^2 + I_{yy}\omega_y^2 + I_{zz}\omega_z^2 + 2I_{xy}\omega_x\omega_y + 2I_{xz}\omega_x\omega_z + 2I_{yz}\omega_y\omega_z = \lambda\left(\omega_x^2 + \omega_y^2 + \omega_z^2\right)$$
(C.13)

By normalizing the vector $\vec{\omega} = \omega_x\hat{x} + \omega_y\hat{y} + \omega_z\hat{z}$ so that $\omega_x^2 + \omega_y^2 + \omega_z^2 = 1$ (in which case ω_x, ω_y, and ω_z become the direction cosines of the line parallel to $\vec{\omega}$), and by defining the following variables (with $\lambda \neq 0$)

$$x \stackrel{\Delta}{=} \omega_x/\sqrt{\lambda} \qquad y \stackrel{\Delta}{=} \omega_y/\sqrt{\lambda} \qquad z \stackrel{\Delta}{=} \omega_z/\sqrt{\lambda}$$
(C.14)

Eq. (C.13) becomes

$$I_{xx}x^2 + I_{yy}y^2 + I_{zz}z^2 + 2I_{xy}xy + 2I_{xz}xz + 2I_{yz}yz = 1$$
(C.15)

which is the equation of an ellipsoid in the x-y-z space. Such an ellipsoid and the x-y-z axes are shown in Fig. C.1. It is called the *ellipsoid of inertia* of the body. For principal axes, the products of inertia of the body are equal to zero and the equation for the ellipsoid of inertia takes the simpler form $I_{xx}x^2 + I_{yy}y^2 + I_{zz}z^2 = 1$.

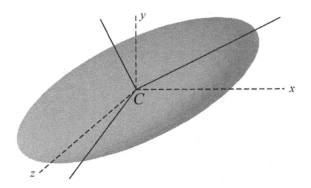

Figure C.1: The inertia ellipsoid of a body.

The ellipsoid of inertia is the geometric representation of the inertia matrix of a body, and its principal axes, shown as three solid lines in Fig. C.1, are the principal axes of the body. The elongated axis shown in that figure corresponds to the principal axis of minimum inertia of the body.

In general, the inertia ellipsoid is not an ellipsoid of revolution. It is an ellipsoid of revolution only if two of the principal moments of inertia of the body are equal to each other. The ellipsoid of inertia of a pencil-shaped body is thin and elongated, while that of a body with three equal principal moments of inertia is a sphere. Such bodies are said to have spherical symmetry.

Two different bodies that have the same inertia ellipsoid (or, equivalently, the same inertia matrix) experience the same rotational motion when they are subjected to the same external moment. They will also experience the same translational motion if they have the same mass and are subjected to the same resultant force.

APPENDIX D

Suggested Computer Lab Assignments

Lab Assignment 1
Part 1

1. (10 points) A small projectile P is launched from the earth's surface with a velocity $\vec{v}_0$ that makes an angle α degrees with the horizontal and whose magnitude is $v_0 = |\vec{v}_0|$ m/s. Because of air resistance, the projectile is subjected to a *drag force*; the drag force is proportional to the square of the projectile's speed [with the proportionality coefficient being equal to c N/(m^2/s^2)] and is always directed opposite to the projectile's velocity vector $\vec{v}$. Using rectangular coordinates, obtain the differential equations that govern the motion of the projectile by treating the earth's gravitational force of attraction as a constant, with the acceleration of gravity being equal to $g = 9.81$ m/s^2. Show the free-body diagram for P, properly labeling all quantities, and show all the steps of the formulation that leads to the differential equations, as done in the book. What are the initial conditions for your differential equations?

2. (40 points) A small block P of mass m is connected to a fixed pin O by a linear spring of stiffness k N/m and unstressed length L_u m. The block is constrained to move along a horizontal guide as shown in Fig. D.1. Friction between the block and the guide is of the linear viscous type, with the friction coefficient equal to c N/(m/s). The perpendicular distance from O to the guide is h m. The plane defined by the guide and the spring is vertical. Obtain the differential equation of motion of block P in terms of the variable $x(t)$ shown in the figure (for this, clearly show the free-body diagram for P, properly labeling all quantities, and show all the steps of the formulation that leads to the differential equation, as done in the book). For $c = 0$, $L_u = h$, $k/m = 0.5$, $x(0) = 5h$, and

$\dot{x}(0) = Ah$ (with $A = 10$ s^{-1}), determine (analytically) the maximum and minimum values of x/h. Also, generate a plot of x_{max}/h and x_{min}/h versus $\dot{x}(0)/h$ for $c = 0$, $L_u = h$, $k/m = 0.5$, $x(0) = 5h$, and $\dot{x}(0) = Ah$, for $-10 \leq A \leq 10$.

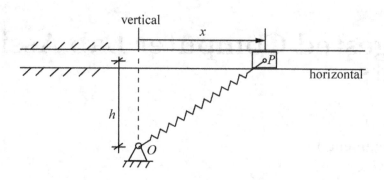

Figure D.1: The system model for Lab Assignment 1.

Part 2 (50 points)

Integrate numerically the differential equation of motion for Problem 2 in part 1 with $c/m = 0$ and $c/m = 0.3$, and plot x/h versus time, and $\dot{x}/h$ versus x/h for both cases. What are the maximum and minimum values of x/h for both values of c/m? Compare the analytical answer obtained in part 1 (for $c = 0$) with that obtained from the numerical integration. Choose your own appropriate value for the final integration time.

Carefully observe your plots, and the motion on the computer screen if you decide to animate it, and write a report about your work. Include your observations in the comments section of your report. Your report should be complete enough to allow another engineer to reproduce your work.

Lab Assignment 2

Essentially, every dynamical system in nature is nonlinear, i.e., its motion is governed by one or more differential equations that are nonlinear. One characteristic of nonlinear systems is that they may exhibit more than one equilibrium.

A system is said to be in *equilibrium* when a variable that is used to describe its motion remains constant for all times. For example, a pendulum (which everyone is familiar with) exhibits two physically distinct equilibria: one corresponding to the pendulum bar hanging-down in the vertical position, and the other corresponding to the upside-down position. If the pendulum is displaced, even slightly, from the equilibrium position, a motion will occur. Consider now a small displacement from the equilibrium. What happens is that the motion will always stay in a small neighborhood of the equilibrium for the case of the hanging-down equilibrium, but that will never happen for the upside-down equilibrium. In technical language, the former equilibrium is said to be *stable*, while the latter is said to be *unstable*. For the stable equilibrium, the perturbed motion is an oscillation about that equilibrium, while for the unstable equilibrium it is not (i.e., the motion that started in a small neighborhood of the unstable equilibrium never stays in that neighborhood). When damping is added to the pendulum, the motion eventually dies out, getting farther and farther from the unstable equilibrium and tending, as time increases, to the stable equilibrium of the pendulum. The equilibrium that was classified as stable is now classified as *asymptotically stable*.

In this lab assignment you will play the role of an engineer investigating the motion of a mechanical device. The system is shown in Fig. D.2. It consists of a bar of length L m, small blocks pinned to the ends of the bar, a spring, a vertical track, and a horizontal track. The connecting bar is very light, and the effect of its mass on the motion of the system can be neglected.

The motion of the system is to be investigated using the model shown in Fig. D.2b. Block A (with mass m_1 kg) slides on the vertical track, and block B (with mass m_2 kg) slides on the horizontal track. Friction in the system is viscous, with friction coefficients c_1 N/(m/s) between A and the vertical track, and c_2 N/(m/s) between B and the horizontal track. Block B is attached to a horizontal linear spring of stiffness k N/m and unstretched length L_u m; the constant distance between the vertical track and end O of the spring is equal to $L \cos \theta_u + L_u$, where θ_u is the value of the angle θ when the spring is unstretched.

1. (50 points) Clearly show all the free-body diagrams, symbols, unit vectors, etc.

   a. Show that the differential equation of motion for the system is

   $$\left(\cos^2 \theta + \frac{m_2}{m_1} \sin^2 \theta \right) \frac{L}{g} \ddot{\theta} + \left(\frac{m_2}{m_1} - 1 \right) (\sin \theta)(\cos \theta) \frac{L}{g} \dot{\theta}^2 + \cos \theta$$

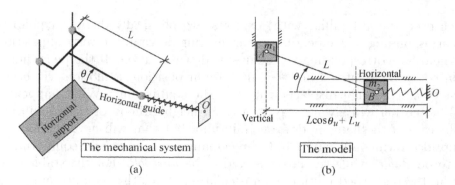

Figure D.2: (a) The system and (b) its model for Lab Assignment 2.

$$+ (c_1^* \cos^2 \theta + c_2^* \sin^2 \theta)\sqrt{\frac{L}{g}}\,\dot{\theta} + \frac{kL}{m_1 g}(\cos\theta_u - \cos\theta)\sin\theta = 0$$

where $c_1^* = (c_1/m_1)\sqrt{L/g}$ and $c_2^* = (c_2/m_1)\sqrt{L/g}$ are nondimensional damping coefficients.

b. Show that by working with a nondimensional time τ, defined as $\tau = t\sqrt{g/L}$, the differential equation in item *a* becomes

$$\left(\cos^2\theta + \frac{m_2}{m_1}\sin^2\theta\right)\frac{d^2\theta}{d\tau^2} + \left(\frac{m_2}{m_1} - 1\right)(\sin\theta)(\cos\theta)\left(\frac{d\theta}{d\tau}\right)^2$$
$$+ \cos\theta + \left(c_1^*\cos^2\theta + c_2^*\sin^2\theta\right)\frac{d\theta}{d\tau} + \frac{kL}{m_1 g}(\cos\theta_u - \cos\theta)\sin\theta = 0$$

c. Using the differential equation of motion for the system, obtain the equation that will allow you to determine the equilibrium values θ_e of the angle θ in terms of the nondimensional parameter $m_1 g/(kL)$, and θ_u.

2. (50 points) This is the numerical investigation part.

a. For $\theta_u = 50°$, $55°$, and $60°$, plot (on the same graph, and properly identifying the curves) θ_e (in degrees) versus the nondimensional parameter $m_1 g/(kL)$ for θ_e in the range $-70° \le \theta_e \le 70°$. *Note:* If you do these plots by hand, they must be presented in a neat, professional manner. For $\theta_u = 55°$, what are the values of θ_e for $m_1 g/(kL) = 0.15$?

b. Integrate numerically the nondimensional differential equation of motion for the system with $m_2/m_1 = 1$, $\theta_u = 55°$, and $m_1 g/(kL) = 0.15$. Do this for $c_1^* = c_2^* = 0$ and for $c_1^* = c_2^* = 0.2$. For each case, show plots of the trajectories $\theta' = d\theta/d\tau$ versus θ. In each plot, mark with an × the location of each equilibrium point found in item a for $\theta_u = 55°$. For the undamped case, plot all curves in the same graph (this will allow for a better understanding of what goes on). Use the following sets of initial conditions to generate the plots:

$$[\theta(0) = 24°, \theta'(0) = 0], \qquad [\theta(0) = 23°, \theta'(0) = 0],$$
$$[\theta(0) = 23°, \theta'(0) = 0.1], \qquad [\theta(0) = 50°, \theta'(0) = 0)],$$
$$[\theta(0) = -50°, \theta'(0) = 0]$$

Animate the motion and carefully observe it on the computer screen, and the plots. Write a report about your work and include, in a comments section of the report, your observations about the motions and about stability (indicating which equilibrium solutions you think are stable and which ones you think are unstable). Put a little arrow on each θ' versus θ trajectory to indicate how it is actually traced (i.e., clockwise or counterclockwise) during the motion.

Lab Assignment 3

The system shown in Fig. D.3 consists of a bar of length $2L$, pinned at a fixed point O, a motor located at O, two equal lumped masses (each of mass m kg), and two linear springs (each of stiffness k N/m and unstressed length L_u m). Each of the lumped masses is attached to the end of the bar, and the bar is approximated as massless. The springs are connected as shown, where points P and Q are fixed. The springs are unstretched when $\theta = 0$. The motor applies a moment M_{motor} to the bar, and there is viscous friction at the pin contact at O, with friction coefficient c N·m/(rad/s). The motion takes place on the horizontal plane.

1. (50 points) Clearly show all the free-body diagrams, symbols, unit vectors, etc.

   a. Obtain the differential equation of motion for the system, and then show that, in terms of a nondimensional time $\tau = t\sqrt{k/m}$ (where t

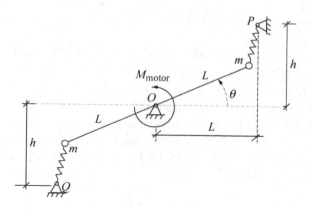

Figure D.3: The system model for Lab Assignment 3.

is dimensional time), the differential equation is

$$\frac{d^2\theta}{d\tau^2} + c^*\frac{d\theta}{d\tau} + \left(\sin\theta - \frac{h}{L}\cos\theta\right)$$

$$\times \left[1 - \frac{1}{\sqrt{1 - 2\left(L/h\right)\sin\theta + 2(L/h)^2(1 - \cos\theta)}}\right] = \frac{M_{\text{motor}}}{2kL^2}$$

In this differential equation, c^* is a nondimensional viscous damping coefficient. What is the relation between c^* and c?

    Note: You can use either one of four ways to formulate the differential equation of motion: (1) directly from Newton's second law, (2) the moment equation (which is a consequence of Newton's second law), (3) the approach based on $dW = d(T)$ (which is also a consequence of Newton's second law), or (4) Lagrange's equation presented in Chapter 7. Each one of these approaches requires a different amount of effort on your part to arrive at the differential equation of motion.

b. With $M_{\text{motor}} = 0$, determine all the equilibrium values θ_e of the angle θ for $h/L = 2$.

c. One of the equilibrium values should be $\theta_e = 0$. Show that for very small angles θ (so that $\sin\theta \approx \theta$ and $\cos\theta \approx 1$, and products of small quantities, such as θ^2, are neglected), a Taylor series expansion about $\theta = 0$ of the nonlinear terms in the differential equation yields the following linearized differential equation of motion:

$$\frac{d^2\theta}{d\tau^2} + c^*\frac{d\theta}{d\tau} + \theta = \frac{M_{motor}}{2kL^2}$$

For $M_{motor} = 0$ and $c^* = 0$, what is the period of these small motions?

2. (50 points) Integrate numerically the nondimensional differential equation of motion for the system with $M_{motor} = 0, c = 0, h/L = 2$, and starting the system from rest. For $\theta(0)$ in the range $(0, 60]$ degrees, generate an *accurate* plot of the period of the oscillation versus the initial condition $\theta(0)$ in degrees. Write a report about your work describing how you gathered the data to generate such a plot, and how you plotted the data. How does your plot compare with your answer for the period for item 1c?

Lab Assignment 4

An earth observation satellite is put in a circular orbit at an altitude of $h = 400$ km above the earth's surface. After a few orbits, one of the attitude control jets of the satellite, and the control system that controls the direction of the thrust produced by that jet, malfunctions. On board instruments determined that the jet is firing continuously, that it is applying to the satellite a constant thrust that is always pointing toward the center of the earth, and that there is no moment about the satellite's center of mass. The magnitude of that applied thrust is equal to mF_0 N, where m is the mass of the satellite, and $F_0/g = 1/128$ (where $g = 9.81$ m/s^2).

Part 1:

1. (25 points) Determine, analytically (and accurately), the maximum and the minimum altitude (i.e., distance from the earth's surface) of the satellite's center of mass after the malfunction. Does the satellite crash on the earth's surface?

2. (25 points) For what range of the nondimensional parameter g/F_0 will the satellite crash on the earth's surface?

Note: Neglect the effect of the earth's atmosphere in your calculations, and treat the earth as a homogeneous sphere of radius $R_E = 6378$ km. Clearly show any free-body diagram you need to have, symbols, unit vectors, etc. Also, make sure that you clearly show all the steps that lead to the differential equations of motion, and all the steps of your analysis.

Part 2 (50 points)

Integrate numerically the differential equations that govern the motion of the center of mass of the satellite. Plot, for $F_0/g = 1/128$ and for two more values of F_0/g of your choice, with one (not both) of the values causing the satellite to crash on the earth's surface:

a. The trajectory of the satellite's center of mass (i.e., a plot of the position of the center of mass, using rectangular coordinates with origin at the center of the earth)

b. The altitude of the center of mass of the satellite (in kilometers) versus time

How much time elapsed (measured from the instant the malfunction occurred) until the satellite crashed? How much (in minutes) is the orbital period when the satellite does not crash? Include in your report your comments and conclusions, and an explanation of how the crash could have been avoided if the satellite designer had limited the value of F_0/g to be outside the range you determined in item 2 in part 1.

Lab Assignment 5

End B of a thin uniform bar AB of length L m and mass m kg is connected to a linear spring of unstressed length L_u m and stiffness k N/m. That end is free to slide along a frictionless horizontal floor, while the other end of the spring is pinned to a fixed point (see Fig. D.4). End A of the bar is free to slide along a frictionless vertical wall. The motion of the system takes place on a vertical plane, and the spring is unstressed when $\theta = \theta_u$ degrees. As seen in Fig. D.4, the floor and the vertical wall limit the angle θ to the range $0 \leq \theta \leq 90°$.

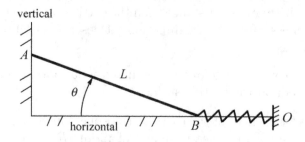

Figure D.4: The system model for Lab Assignment 5.

1. (15 points) Show that the differential equation of motion for the system is (see the note at the end of this assignment)

$$\frac{d^2\theta}{dt^2} + \frac{3g}{L}\left[\frac{1}{2}\cos\theta + \frac{kL}{mg}\left(\cos\theta_u - \cos\theta\right)\sin\theta\right] = 0$$

2. (5 points) For $\theta_u = 50°$ and $kL/(mg) = 5$, determine all the physically possible equilibrium values θ_e for the angle θ (this can be done using a hand-held calculator).

3. (15 points) By letting $\theta(t) = \theta_e + \theta_s(t)$, linearize the differential equation for small values of $\theta_s(t)$. Do not use any symbolic computation software to do this. Do it by hand, clearly showing all your steps, thus demonstrating that you have understood how to obtain a Taylor series and how to linearize a differential equation.

4. (5 points) Which of the equilibrium values determined in item 2 is stable? For $\theta_u = 50°$ and $kL/(mg) = 5$, what is the frequency of the linearized θ_s oscillation about the stable equilibrium? Your answer will be in terms of g/L.

5. (10 points) For the initial conditions $\theta(0) = \theta_0$ and $\dot\theta(0) = 0$, obtain the solution for $\theta(t) = \theta_e + \theta_s(t)$ using the linearized differential equation of motion (you are going to use this in item 6).

6. Rewrite the given nonlinear differential equation in terms of the nondimensional time $\tau = t\sqrt{g/L}$ and, with $\theta_u = 50°$ and $kL/(mg) = 5$, numerically integrate the resulting equations (9 points for the integration) to address the following items.

   a. (9 points) With $\theta(0) = 43°$, plot $\theta(\tau)$ versus τ for $0 \le \tau \le 15$, and then superimpose, with dashed lines, the plot of your solution to the linearized equation for the given value of $\theta(0)$. Carefully observe the motion and/or your plot, and make notes for your report.

   b. (9 points) Repeat the work in item a for $\theta(0) = 46°$. Again, carefully observe the motion and your plot, and make notes for your report.

   c. (9 points) Repeat the work in item a for $\theta(0) = 51.4°$ and $51.5°$. For this item, use $0 \le \tau \le \tau_{max}$, with a value for τ_{max} of your choice. Again, carefully observe the motion and/or your plot, and make notes for your report.

   d. (14 points) Prepare a concise but meaningful report with your observations about what happened to the motion in items a to c. Specifically, you should comment on the following: How the nonlinear and the linearized motions compare in items a and b (and your reasons why they are what you saw); your explanation of what happened when you used $\theta(0) = 51.4°$; and your explanation of what happened when you used $\theta(0) = 51.5°$.

Note: As you have seen in the book, there are several ways to formulate the differential equations of motion of a dynamical system. They include a combination of Newton's second law and the moment/angular-momentum equation (i.e., the Newton-Euler approach), and Lagrange's equation. Each one of these approaches requires a different amount of effort to arrive at the differential equation(s) of motion. You may choose to use any of them. Clearly show all the free-body diagrams you may need, symbols and unit vectors you choose to work with, etc.

Scilab and Matlab Programs

If you are interested in using any of these programs, you can either:

- Type the programs you want in the appropriate editor (see Figs. A.2 and A.5 in Appendix A), and then save them in your "home directory" (i.e., the directory you have created for storing them). Scilab will save the programs in Section E.1 with the extension `.sci`, and Matlab will save the ones in Section E.2 with the extension `.m`. You can exclude the comment lines if you want to. **Please see the** *Note* **on p. 238.**

  Or

- Download them from the Dover website (www.doverpublications.com). Look for them by typing `Crespo da Silva` in the search box that appears on the website page. Disregard/delete either the Matlab programs if you use Scilab or the Scilab programs if you use Matlab (if you use Simulink, you need the Matlab files `animate_nbars.m` and `animate3D_nbars.m`). Save them in your "home directory" (i.e., the directory you created for storing them). If you use Scilab, your `scilab.ini` file should be saved in the directory whose pathname is displayed when you type `SCIHOME` in the Scilab console the first time you use it, and also in your home directory (please see the Scilab tutorial in Appendix A). If you use Matlab, and if it is installed in your computer, prepare the `startup.m` file as indicated in Appendix A.

  NOTE: For the benefit of those who prefer to use Simulink, the models shown in this book are also included in the Matlab program package when you download it. You can delete them from your computer if you do not want them.

> If you want to use Scilab, which is free, see Appendix A (p. 596) for information on how to download it.

Most of the programs listed here consist only of a few lines of commands. Lines that start with a double slash (//) in the Scilab programs, and with a percentage sign (%) in the Matlab programs, are comments. They are directed at those users that may want to have some further explanation about some of the commands. The three-dot-sequence (...) that may appear in any line of a program listed here is interpreted by both Scilab and Matlab as a continuation of the command in that line. The comment lines from the second line of each program to the line before the first blank line in the program serve as a help information for each program, which includes an example of usage. As indicated in Appendix A, those comment lines are displayed on the computer screen when you type `head_comments function_name` in Scilab or `help function_name` in Matlab, with `function_name` replaced by the name of the function of interest (as `animate_nbars` and `four_bar_mechanism`, for example).

See either the **NOTE** on p. 124 or p. 168 for information on how to slow down or speed up an animation in your computer.

If you are interested in the programming details that lead to an animation, they are given as comment lines in the Scilab `animate_nbars` program. The `animate_nbars` program is used by other programs such as `ballistics`, `boomerang`, `four_bar_mechanism`, and many others, for animating the two-dimensional motion of the system. Its usage simply consists in the command `animate_nbars(x_coordinates, y_coordinates)`, where `x_coordinates` and `y_coordinates` are the x and y rectangular coordinates of the endpoints of the bars in a system composed of any number of connected bars. The similar programs `animate3D_nbars`, one for Scilab users and the other for Matlab users, are for animation in three-dimensional space. There are very few comments in them as the technique that leads to the animation is the same as explained in the Scilab program `animate_nbars`.

IMPORTANT: Both open quotes (') and close quotes (') (as in 'on', or in t') that are seen in any Scilab and Matlab command in all the programs listed here are meant to be the prime symbol ' (the apostrophe on the keyboard) seen either in some mathematical expressions or in some Scilab and Matlab commands that appear elsewhere in the book. **Also, again, please see the** *Note* **on p. 238.**

E.1   Scilab Programs

```
function animate_nbars(x_coordinates, y_coordinates)
//        2D animation of any number of connected bars
// When given the lists x_coordinates=[xA, xB, xC, etc] and y_coordinates(yA, yB, yC, etc)
// of  the x and y coordinates of several points A, B, C, etc, a call to the function
// animate_nbars.sci generates a set of straight lines by connecting the points in the
// given sequence A, B, C, etc, with a black dot at each point. Thus a call such as
// animate_nbars([xA, xB, xC], [yA, yB, yC]), typed in the SCILAB console, creates a figure
// window with two lines AB and BC, with black dots at the ends of each line. When xA, xB,
// xC, etc, and yA, yB, yC, etc, are n by m matrices, an animation is created displaying the
// lines whose end point coordinates are specified in each row of the matrix. This function
// is called by several others of the type .sci (such as ballistics.sci, boomerang.sci,
// fourbar_mechanism.sci, etc). For information on the animation of the four bar mechanism,
// for example, type     head_comments fourbar_mechanism

// ********************************************************************************
// NOTE: Comment out this part if you don't want to see this message after you get used to it
disp(' '); disp('  =====>  IMPORTANT MESSAGE:  <=====')
disp(' IF THE SIMULATION RUNS EITHER TOO FAST OR TOO SLOW IN YOUR COMPUTER,')
disp(' change the time_step in the time array t=[t0:time_step:tfinal] of your simulation ')
disp(' ---> EITHER to a smaller value to slow down the animation ')
disp(' ---> OR to a larger value to speed up the animation')
// ********************************************************************************

[number_of_time_points, len]=size(x_coordinates);  // size gives the dimensions of a matrix

// minimum and maximum values of x_coordinates and y_coordinates:
xcoord_min=min(x_coordinates);  xcoord_max=max(x_coordinates);
ycoord_min=min(y_coordinates);  ycoord_max=max(y_coordinates);

// Create a figure window for the animation. Start by determining the computer monitor size
// so that the animation window is properly placed and sized in any computer screen.
screen_width_and_height=get(0,'screensize_px') // width and height of the screen, in pixels
PC_screen_width =screen_width_and_height(3)    // this is the width of the computer monitor
PC_screen_height=screen_width_and_height(4)    // this is the height of the computer monitor

x_topleft= 0.65*PC_screen_width // choice for the top left x-coordinate of the figure window
y_topleft= 0.3*PC_screen_height // choice for the top left y-coordinate of the figure window
top_left_coordinates=[x_topleft, y_topleft]
size_of_fig_window = [0.32*PC_screen_width, 0.47*PC_screen_height]
// the above are the width and height chosen for the animation window
// Now, create the figure window. NOTE: 'background', 8  is for a white background
fig=figure(300,'figure_position',top_left_coordinates,'figure_size', size_of_fig_window, ...
           'figure_name','Animation window', 'background', 8)

// The variable fig is a "handle" (so called in analogy with the handles of a cabinet, say)
// of type "Figure", that has several properties that can be modified by the user (to see
// those properties, type the above command in the Scilab console; you will see that one of
// the properties is named children, which is also a handle with its own properties).
// The command h_gce=gce()   (for "get current entity"), or h_gce=fig.children typed after
// creating a figure, shows the properties of the handle of type "Axes" of the figure window
// that is  displayed on the computer screen. Three of such properties are named isoview,
// mark_mode, and mark_size. The isoview property is set to off by default. If it is left as
// off, a circle, if plotted, will show on the computer screen as an ellipse. To force it to
// appear as a circle, isoview needs to be set to  on  (enclosed in quotes).
```

```
// The commands that will do this are the following:

h_gce=gce()
h_gce.isoview='on'      // we could have used, instead, fig.children.isoview='on'

// Now, initialize (say, with the initial position of all the bars) the 2D plot with a call
// to plot. This will modify the "current entity" to account for the creation of the lines
// by the plot command.
plot(x_coordinates(1,:), y_coordinates(1,:))
    // The first row array is [xA1, xB1, xC1, etc], and the second is [yA1, yB1, yC1, etc]

// If you now type gce(), it will be seen that the new entity (the lines the plot command
// created) is a handle of type "Compound" for which two of its properties are parent: Axes
// and Children: Polyline. We need to "get the current entity" [the command for this is
// gce( )] to be able to modify some of its properties that are needed for an animation.
// Reusing the same variable name h_gce, we can do this with the following command:

h_gce=gce( )      // we could have used, instead, h_gce=fig.children.children

// Now, access the handle h_gce.children, which has a property named data that stores the
// x and y coordinates of the ends of the lines that were created (i.e., such data is stored
// in h_gce.children.data) as a 2-column matrix. For an animation, this, then, will be put
// in a loop so that all the lines AB, BC, etc, are plotted in sequence. Before we start
// that loop, we will put a dot (by choice) at the end of each bar. This is done with the
// following two commands to change the settings of two other properties of h_gce.children
// named mark_mode and mark_size. The first command makes it possible to add the dots,
// and the second command is for specifying a size for the dot [if you do not want such
// markings, comment out the following two command lines by putting a double slash (//)
// at the beginning of their lines].

h_gce.children.mark_mode='on'; // set to on so that we can put a dot at the ends of the bars
h_gce.children.mark_size=10;    // this is the size we chose for the black dot

// The final steps now consist in telling the software to "get the current axes" (which
// are the x and y axes that were created by the plot command) and limiting them to the
// maximum and minimum values xcoord_min, xcoord_max, ycoord_min, and ycoord_max obtained
// earlier. The software stores such values in the property named data_bounds of the
// current axes. Such axes are accessed with the command gca( ).

h_axes = gca();      // gca stands for "get current axes"; this accesses the current axes
h_axes.data_bounds = [xcoord_min, ycoord_min; xcoord_max, ycoord_max]  // the axes limits

// Now, the loop to generate the animation after changing h_gce.children.data at each step.
// NOTE: there is a transpose operation in the loop because h_gce.children.data stores the
//       data specified in the command plot([xA, xB, xC], [yA, yB, yC]), for example,
//       in a 3 by 2 matrix as [xA, yA; xB, yB; xC, yC], which is the same as
//       [ [xA, xB, xC]', [yA, yB, yC]' ]
//       (i.e., h_gce.children.data needs to be a 2-column matrix)

   for itime=1:number_of_time_points
      drawlater();
      h_gce.children.data = [x_coordinates(itime,:)', y_coordinates(itime,:)'];
      drawnow(); // type help drawnow for information on the pair drawlater drawnow
   end
endfunction      // end of the file  animate_nbars.sci
// /////////////////////////////////////////////////////////////////////////////////
```

```
function animate3d_nbars(x_coordinates, y_coordinates, z_coordinates)
//         3D animation of any number of connected bars
// When given the lists x_coordinates=[xA, xB, xC, etc], y_coordinates(yA, yB, yC, etc),
// and z_coordinates(zA, zB, zC, etc)
// of the x-y-z coordinates of several points A, B, C, etc, a call to the function
// animate3d_nbars.sci generates a set of straight lines by connecting the points in the
// given sequence A, B, C, etc, with a black dot at each point. Thus a call such as
// animate3d_nbars([xA, xB, xC], [yA, yB, yC], [zA, zB, zC]), typed in the Scilab console,
// creates a figure window with two lines AB and BC, with black dots at the ends of each
// line. When xA, xB, xC, etc; yA, yB, yC, etc; and zA, zB, zC, etc, are n by m arrays an
//  animation is displayed.
//  This function is called by top_dynamics.sci and two_particles_table.sci
// to animate the 3D motion. See those programs and run them to watch the motion.

// ***********************************************************************************
// NOTE: Comment out this part if you don't want to see this message after you get used to it
disp(' '); disp('  =====>  IMPORTANT MESSAGE:   <=====')
disp(' IF THE SIMULATION RUNS EITHER TOO FAST OR TOO SLOW IN YOUR COMPUTER,')
disp(' change the time_step in the time array t=[t0:time_step:tfinal] of your simulation ')
disp(' ---> EITHER to a smaller value to slow down the animation ')
disp(' ---> OR to a larger value to speed up the animation')
// ***********************************************************************************

[number_of_time_points, len]=size(x_coordinates);
// minimum and maximum value of x_coordinates, y_coordinates, and z_coordinates:
xcoord_min=min(x_coordinates); xcoord_max=max(x_coordinates);
ycoord_min=min(y_coordinates); ycoord_max=max(y_coordinates);
zcoord_min=min(z_coordinates); zcoord_max=max(z_coordinates);

// Create a figure window for the animation. Start by determining the computer monitor size
// so that the animation window is properly placed and sized in any computer screen.
screen_width_and_height=get(0,'screensize_px') // width and height of the screen, in pixels
PC_screen_width =screen_width_and_height(3)    // this is the width of the computer monitor
PC_screen_height=screen_width_and_height(4)    // this is the height of the computer monitor

x_topleft=0.65*PC_screen_width  // choice for the top left x-coordinate of the figure window
y_topleft=0.3*PC_screen_height  // choice for the top left y-coordinate of the figure window
top_left_coordinates=[x_topleft, y_topleft]
size_of_fig_window = [0.32*PC_screen_width, 0.47*PC_screen_height]
// the above are the width and height chosen for the animation window
// Now, create the figure window.  NOTE: 'background', 8  is for a white background
fig=figure(300, 'figure_position',top_left_coordinates,'figure_size', size_of_fig_window, ...
         'figure_name','Animation window', 'background', 8);  xgrid

fig.children.isoview='on';    col_of_zeros=zeros(len,1);
param3d(col_of_zeros, col_of_zeros, col_of_zeros); // this initializes the plot
    // NOTE: Scilab's plot3d draws a filled surface, instead of lines in 3D
    //       To draw lines in 3D, the Scilab command to use is param3d
h_gce = gce();    // h_gce.data  is the data that will be plotted
h_gce.mark_mode='on';    h_gce.mark_size=1;
// NOTE: Put a // at the beginning of the line, above, if no dots are desired

h_axes = gca();    // need this to set axes limits, rotation_angles for viewing the fig, etc
h_axes.data_bounds=[xcoord_min, ycoord_min, zcoord_min; xcoord_max, ycoord_max, zcoord_max];
```

```
h_axes.rotation_angles=[60,-120];
// NOTE: Comment out the above line if the default rotation_angles is desired

 for itime=1:number_of_time_points        // this is the animation loop
    temp=[ ];
    for k=1:len
      temp=[temp; x_coordinates(itime,k), y_coordinates(itime,k), z_coordinates(itime,k)];
    end
    drawlater();
       h_gce.data=temp;
     drawnow();
   end
endfunction
// end of the file animate3d_nbars.sci
// ///////////////////////////////////////////////////////////////////////////


function output = ballistics(t, d2x_dt2, d2y_dt2,  x_0, xdot_0, y_0, ydot_0)
// Example 2.5.8
// Example of usage:
// my_output = ballistics([0:0.01:2.5], 0, -9.81,  0, 10, 0, 10);

function dvar_dt=rhs_of_diff_eqs(t, var)
    // Note:  var(1)=x, var(2)=dx/dt, var(3)=y, var(4)=dy/dt
    dvar_dt(1)= var(2),    // this is dx/dt
    dvar_dt(2)= d2x_dt2    // this is d2(x)/dt2
    dvar_dt(3)= var(4)     // this is dy/dt
    dvar_dt(4)= d2y_dt2 // this is d2(y)/dt2
endfunction

initial_conditions=[x_0; xdot_0; y_0; ydot_0];
state_variables=ode(initial_conditions, 0, t, rhs_of_diff_eqs)

output=[t', state_variables']
   // output is a n by 5 matrix, where n is the number of points in the array t
   // column 1: t, column 2: x, column 3: dx/dt, column 4: y,column 5: dy/dt

disp('NOTE: The output of a call to the function ballistics.sci is a n by 5 matrix, ')
disp('where n is the number of points in the time array t')
disp('1st column: t, 2nd column: x, 3rd column: dx/dt, 4th column: y, 5th column: dy/dt')

xdel(winsid())  // this command deletes all existing figures
x_coordinates=output(:,2);  y_coordinates=output(:,4);
animate_nbars(x_coordinates, y_coordinates)  // Animation (the "bar" will be a point)
endfunction
// end of the file ballistics.sci
// ///////////////////////////////////////////////////////////////////////////


function output = boomerang(t, g,  x_0, xdot_0, y_0)
// Example 7.9.1; using the state variables var(1)=x, var2=dx/dt, var(3)=y
// Example of usage [using the initial conditions x(0)=0, dx_dt(0)=5*cos(%pi/6),  y(0)=0]
// my_output =  boomerang([0:0.01:6], 9.81, 0, 0, 5*cos(%pi/6), 0);

function dvar_dt=rhs_of_diff_eqs(t, var)
```

```
// Note:  var(1)=x, var(2)=d(x)/dt, var(3)=y;  The diff. eqs. are Eqs. (7.9.1j) and (7.9.1i)
        f1=0.5*sin(t);  df1_dt=0.5*cos(t);    f2=5*sin(%pi/6)-0.85*t; df2_dt=-0.85;
    dvar_dt(1)=var(2),     // this is d(theta)/dt
    dvar_dt(2)=-(var(2)*df1_dt+df2_dt+g)*f1/(1+f1^2) // this is Eq. (7.9.1j)
    dvar_dt(3)= f1*var(2)+f2      // this is Eq. (7.9.1i)
endfunction


initial_conditions=[x_0; xdot_0; y_0];
state_variables=ode(initial_conditions, 0, t, rhs_of_diff_eqs),
output=[t' , state_variables']    // ' is the transpose operator
    // output is a n by 4 matrix, where n is the number of points in the array t
    // column 1: t, column 2: x, column 3: dx/dt; column 4: y

x=output(:,2), y=output(:,4)

xdel(winsid())  // deleting all existing figures
figure('background', 8)    // 'background' 8 forces the figure background to be white
plot(t', x,  t', y, '--'),   xgrid        // plot of x and y versus time t
xlabel('t', 'fontsize', 3), ylabel('solid line: x   ;   dashed line: y', 'fontsize', 3),
title('Example 7.9.1','fontsize', 3)

figure('background', 8);   plot(x, y);   xgrid        // plot of y versus x
xlabel('x', 'fontsize', 3), ylabel('y', 'fontsize', 3), title('Example 7.9.1','fontsize',3)

animate_nbars(x, y);    // This call animates the motion
endfunction
// end of the file  boomerang.sci
// /////////////////////////////////////////////////////////////////////////////


function output=example6_7_1(tau, omegasqL_over_g, R_over_L, theta_0, thetadot_0)
// Example 6.7.1 analyzed in Section 6.7
// Example of usage ( using the initial conditions theta(0)=30*%pi/180 rad, thetadot(0)=0 ):
// my_output = example6_7_1([0:0.1:10], 1.5, 1/3, 30*%pi/180, 0);

function dy_dtau=rhs_of_diff_eqs(tau, y)
    // Note:  y(1)=theta, y(2)=d(theta)/d(tau), where tau=t*sqrt(g/L)
    sin_theta=sin(y(1))
    dy_dtau(1)=y(2),      // the next expression is obtained from Eq. (6.7.1q) with b=0
    dy_dtau(2)=(3*R_over_L/2+sin_theta)*omegasqL_over_g*cos(y(1))-3*sin_theta/2
endfunction

initial_conditions=[theta_0; thetadot_0];
y=ode(initial_conditions, 0, tau, rhs_of_diff_eqs)
output=[tau', y']
        // output is a n by 3 matrix, where n is the number of points in the array tau
        // column 1: tau;  column 2: theta;  column 3: d(theta)/d(tau)
theta=output(:,2), dtheta_dtau=output(:,3),  cos_theta=cos(theta),  sin_theta=sin(theta)

// Normalized components A1, A2, and A3 of the reaction force at point A:
  A1=-cos_theta-(dtheta_dtau.^2)/2-omegasqL_over_g*(R_over_L+sin_theta/2).*sin_theta
  A2=sqrt(omegasqL_over_g)*dtheta_dtau.*cos_theta
  A3=(sin_theta-R_over_L*omegasqL_over_g*cos_theta)/4
// Magnitude of the normalized reaction force and moment at point A:
  reaction_force_at_A  = sqrt(A1.^2+A2.^2+A3.^2),   reaction_moment_at_A = 2*A2/3
```

```
// Plots:
 xdel(winsid())  // deleting all existing figures
 figure('background', 8);  plot(tau', theta*180/%pi),   xgrid
 xlabel(['Nondimensional time ', '$\tau$'], 'fontsize', 3)
 ylabel(['$\theta$', ' (degrees)'], 'fontsize', 3)

 figure('background', 8);  plot(tau', reaction_force_at_A),  xgrid
 xlabel(['Nondimensional time ', '$\tau$'], 'fontsize', 3)
 ylabel('$\sqrt{A^2_1+A^2_2+A^2_3}/(mg)$', 'fontsize', 3)

 figure('background', 8);  plot(tau', reaction_moment_at_A),  xgrid
 xlabel(['Nondimensional time ', '$\tau$'], 'fontsize', 3)
 ylabel('$M_3/(mgL)$', 'fontsize', 3)
endfunction
// end of the file example6_7_1.sci
// /////////////////////////////////////////////////////////////////////////


function output=foucault_pendulum(t,omega_times_sin_lambda,g_over_L,x_0, xdot_0,y_0,ydot_0)
// Animation of the Foucault pendulum analyzed in Chapter 8
// Example of usage [using the initial conditions x(0)=0.01;x_dot(0)=0; y(0)=0; y_dot(0)=0]:
// my_output = foucault_pendulum([0:0.01:30], 0.0727, 0.981,   0.01, 0, 0, 0);

disp('NOTE: THE FUNCTION foucault_pendulum RETURNS 5 COLUMNS WITH t, x, dx/dt, y, and dy/dt')

function dy_dt=rhs_of_diff_eqs(t, y)
    // Note:  y(1)=x, y(2)=dx/dt, y(3)=y, y(4)=dy/dt
  dy_dt(1)=y(2)
  dy_dt(2)=2*omega_times_sin_lambda*y(4)-g_over_L*y(1)      // this is d2x/dt2
  dy_dt(3)=y(4)
  dy_dt(4)=-2*omega_times_sin_lambda*y(2)-g_over_L*y(3)     // this is d2y/dt2
endfunction

initial_conditions=[x_0; xdot_0; y_0; ydot_0];
state_variables=ode(initial_conditions, 0, t, rhs_of_diff_eqs)
output=[t', state_variables']    // ' is the transpose operator
// the variable output is a n by 5 matrix, where n is the number of points in the array t
// column 1: t, column 2: x, column 3: dx/dt; column 4: y; column 5: dy/dt

x=output(:,2),  y=output(:,4)

xdel(winsid())  // deleting all existing figures
figure('background',8); plot(x, y), xgrid     // plot of y versus x
xlabel('x', 'fontsize', 3), ylabel('y', 'fontsize', 3)
title('Foucault pendulum path','fontsize', 3)

 animate_nbars(x, y)   // ANIMATION
endfunction
// end of the file foucault_pendulum.sci
// /////////////////////////////////////////////////////////////////////////


function output = fourbar_kinematics(t, L1, L2, L3, L4, theta2, dtheta2_dt, ...
          d2theta2_dt2, solution_number)   // ... is for continuing a command  line
// Kinematics of the four-bar mechanism in Section 3.2.3
// Animation is done in the program fourbar_mechanism.sci
```

```
// Example of usage [copy the next three lines, without the //, to the Scilab console]:
// t=[0:0.1:5]; theta2=%pi/6+1.5*%pi*sin(2*t); dtheta2_dt=3*%pi*cos(2*t);  ...
//                                        d2theta2_dt2=-6*%pi*sin(2*t);
// my_output=fourbar_kinematics(t, 1, 0.45, 0.8, 0.7, theta2, dtheta2_dt, d2theta2_dt2, 1);
//   NOTE: any name can be used for the variable to the left of the equal sign in line 1
//   The variable output is a n by 10 column, where n is the number of points in the array t
//
// The mechanism is defined by four points A, B, C, and D, whose x-y coordinates are:
// point A: (0,0),     point B: ( L2*cos(theta2), L2*sin(theta2) )
// point C( L1+L4*cos(theta4), L4*sin(theta4) ),    point D: (L1, 0 )
// The angles theta3 and theta4 are theta3=output(2,:) and theta4=output(3,:)
// The fourth bar is a fictitious one between the fixed points A and D.


len=length(t);  temp=ones(1,len);  // This assures that theta2, dtheta2_dt, and d2theta2_dt2
//     are rows with the same length as the row t
theta2=theta2.*temp; dtheta2_dt=dtheta2_dt.*temp; d2theta2_dt2=d2theta2_dt2.*temp;

if solution_number ==1 | solution_number == 2 then
    cos_theta2=cos(theta2);  L=sqrt( L1^2+L2^2-2*L1*L2*cos_theta2 ); // L from Eq. (3.2.10)
    sin_theta2=sin(theta2);  sin_alpha=L2*sin_theta2./L;   // need to use ./, instead of /
    if abs(sin_alpha) <=1 then
        cos_alpha = (L1-L2*cos(theta2))./L;        // from the equations of (3.2.9)
        angle_alpha=atan(sin_alpha, cos_alpha);    // angle angle_alpha
        cos_beta=(L3^2+L.^2-L4^2)./(2*L*L3);
        if abs(cos_beta)<=1 then
            sin_beta=(3-2*solution_number)*sqrt(1-cos_beta.^2); // Eqs. (3.2.14) and (3.2.15)
            angle_beta= atan(sin_beta, cos_beta);
            theta3=angle_beta-angle_alpha;         // angle theta3 from Eq. (3.2.16)
            sin_gamma=L3*sin_beta/L4;              // sin(gamma) from Eq. (3.12.11)
            if abs(sin_gamma)<=1 then
                cos_gamma=(L-L3*cos_beta)/L4;      //  cos(gamma) from Eq. (3.12.12)
                angle_gamma= atan(sin_gamma, cos_gamma);
                theta4=%pi-angle_alpha-angle_gamma;
            else disp('THE MECHANISM CANNOT OPERATE because |sin(gamma)|>1'),output=[],return
            end
        else disp('THE MECHANISM CANNOT OPERATE because |cos(beta)|>1'), output=[],return
        end
    else disp('THE MECHANISM CANNOT OPERATE because |sin(alpha))|>1'), output=[], return
    end

// Angular velocities dtheta3_dt=d(theta3)/dt and dtheta4_dt=d(theta4)/dt from Eq. (3.2.5):
den=L3*L4*sin(theta4-theta3); // this is the determinant of the 2 by 2 matrix in Eq. (3.2.5)
temp=-L2*dtheta2_dt./den;
// Solution of Eq. (3.2.5):
dtheta3_dt=L4*temp.*sin(theta4-theta2);  dtheta4_dt=L3*temp.*sin(theta3-theta2);

// Elements a11, a12, a21, and a22 of the inverse of the matrix in Eq. (3.2.5):
cos_theta3=cos(theta3); sin_theta3=sin(theta3);
cos_theta4=cos(theta4); sin_theta4=sin(theta4);
a11=-L4*cos_theta4./den; a12=L4*sin_theta4./den;
a21=-L3*cos_theta3./den;  a22=L3*sin_theta3./den;

// Angular accelerations d2theta3_dt2 and d2theta4_dt2 from Eq. (3.2.8):
v1 = L4*(dtheta4_dt.^2).*cos_theta4 -L3*(dtheta3_dt.^2).*cos_theta3  ...
        -L2*( d2theta2_dt2.*sin_theta2 + (dtheta2_dt.^2).*cos_theta2 );
v2=-L4*(dtheta4_dt.^2).*sin_theta4 +L3*(dtheta3_dt.^2).*sin_theta3 ...
```

```
              - L2*( d2theta2_dt2.*cos_theta2 - (dtheta2_dt.^2).*sin_theta2 );
d2theta3_dt2=a11.*v1+a12.*v2;   d2theta4_dt2 = a21.*v1+a22.*v2;
 // d2theta3_dt2 and d2theta4_dt2 are the solutions of Eq. (3.2.8)

// Store t, all the angles, angular velocities, and angular accelerations in the variable
// output so that they are available to the user:
temp1 = [t; theta2;theta3;theta4; dtheta2_dt;dtheta3_dt;dtheta4_dt]';
temp2 = [d2theta2_dt2;d2theta3_dt2;d2theta4_dt2]';
output = [temp1, temp2];

  else disp(' -----> SOLUTION_NUMBER HAS TO BE  1 or 2 <----- '),    output=[ ], return
end
endfunction
// end of the file fourbar_kinematics.sci
// //////////////////////////////////////////////////////////////////////////////


function output = fourbar_mechanism(t, L1, L2, L3, L4, theta2, dtheta2_dt, ...
            d2theta2_dt2, solution_number)   // ... is for continuing a command line
// Animation of the four-bar mechanism in Section 3.2
// Example of usage [copy the next three lines, without the //, to the Scilab console]:
// t=[0:0.002:5]; theta2=%pi/6+1.5*%pi*sin(2*t); dtheta2_dt=3*%pi*cos(2*t); ...
//                             d2theta2_dt2=-6*%pi*sin(2*t);
// my_output=fourbar_mechanism(t, 1, 0.45, 0.8, 0.7, theta2,  dtheta2_dt, d2theta2_dt2, 1);
//    NOTE: any name can be used for the variable to the left of the equal sign in line 1
//
// To slow down the animation, choose a smaller time step for the input array t
// Conversely, to speed up the animation, choose a larger time step
// For example, the array t=[0:0.002:5] yields a slower animation than the array t=[0:0.02:5]
//
// IN ADDITION TO ANIMATING THE MECHANISM, SUCH A CALL TO THE FUNCTION fourbar_mechanism.sci
// GENERATES A len (WHERE len IS THE NUMBER OF POINTS IN THE ARRAY t) BY 10 MATRIX CALLED
// my_output, WHOSE COLUMNS ARE   t, theta2, theta3, theta4, dtheta2_dt, dtheta3_dt,
// dtheta4_dt, d2theta2_dt2, d2theta3_dt2, d2theta4_dt2
//  my_output(:,3), my_output(:,4)*180/%pi, FOR EXAMPLE, GIVES theta3 AND theta4 (IN DEGREES).
//  TO PLOT theta4 VERSUS TIME t, FOR EXAMPLE, TYPE, IN THE SCILAB CONSOLE, THE COMMANDS
//             figure(); plot(my_output(:,1), my_output(:,4)*180/(2*%pi) )
// To put a grid on the plot, type   xgrid

disp(' MESSAGE FROM THE FUNCTION fourbar_mechanism.sci:')
disp('IN ADDITION TO ANIMATING THE MECHANISM, A CALL TO THE FUNCTION fourbar_mechanism')
disp('RETURNS 10 COLUMNS CONTAINING t, theta2, theta3, theta4, dtheta2_dt, dtheta3_dt, ')
disp(' dtheta4_dt, d2theta2_dt2, d2theta3_dt2, d2theta4_dt2')

output = fourbar_kinematics(t,L1,L2,L3,L4, theta2,dtheta2_dt,d2theta2_dt2,solution_number)
theta2=output(:,2);   theta4=output(:,4);
len=length(t);  temp=ones(len,1);
xA= zeros(len,1), yA=xA, xB=L2*cos(theta2), yB=L2*sin(theta2),  // points A and B of bar AB
xC=L1+L4*cos(theta4),  yC=L4*sin(theta4),      // point C of bar BC
xD=L1*temp,  yD=yA,    // point D

xdel(winsid())  // deleting all existing figures
x_coordinates=[xA, xB, xC, xD], y_coordinates=[yA, yB, yC, yD],
animate_nbars(x_coordinates, y_coordinates) // ANIMATION (done by the program animate_nbars)
endfunction
```

```
// end of the file fourbar_mechanism.sci
// ////////////////////////////////////////////////////////////////////////////////


function  output = fourbarforces(t, L1, L2, L3, L4,  theta2, dtheta2_dt, d2theta2_dt2, ...
                    solution_number, m2, m3, m4)     // ... is for continuing a command line
// Simulation of the problem in Section 5.6 of Chapter 5
// Example of usage  [copy the next three lines to the Scilab console, without the //]:
// t=[0:0.1:5]; theta2=%pi/6+1.5*%pi*sin(2*t);  dtheta2_dt=3*%pi*cos(2*t); ...
//                    d2theta2_dt2=-6*%pi*sin(2*t);
// output=fourbarforces(t,1,0.45,0.8,0.7,theta2,dtheta2_dt,d2theta2_dt2,1,0.15,0.225,0.1875);
//
// THE FOLLOWING CALL GIVES THE SAME RESULTS SHOWN IN SECTION 5.6 OF CHAPTER 5:
// output=fourbarforces(1.e-6, 5.1592,2,3,2.5, %pi/2-0.5902,-3*%pi, 0,1,0.15,0.225,0.1875);
//
// SUCH CALLS TO THE FUNCTION fourbarforces.sci GENERATE A len (WHICH IS THE NUMBER OF
// POINTS IN THE SPECIFIED ARRAY t) BY 10 MATRIX CALLED output, WHOSE COLUMNS ARE
//       t, Mext, Ar2, An2, Dr4, Br2, Bn2, Cr4, Dn4, Cn4
//
//   output(:,3), AND  output(:,4), FOR EXAMPLE, GIVE THE REACTIONS Ar2 AND An2
//   TO PLOT Ar2 VERSUS TIME t, FOR EXAMPLE, TYPE, IN THE SCILAB CONSOLE, THE COMMANDS
//     figure();   plot( output(:,1), output(:,3) )

disp(' ')
disp(' NOTE: THE 10 VARIABLES t, Mext, Ar2, An2, Dr4, Br2, Bn2, Cr4, Dn4, and Cn4, AT THE')
disp('SPECIFIED TIMES, ARE STORED IN A 10-COLUMN MATRIX WHOSE NAME IS THE ONE YOU USED IN')
disp('THE CALL TO THE FUNCTION fourbarforces.sci, SUCH AS   output=fourbarforces(....)' )
disp(' TO PLOT Ar2, FOR EXAMPLE, VERSUS time, USE THE COMMANDS ')
disp('       figure(); plot(output(:,1), output(:,3))')
disp('TO PUT A GRID ON THE PLOT, IF DESIRED, THEN TYPE  xgrid')
disp(' --> To see an animation of the motion, use the function fourbar_mechanism <---')
disp(' ')

// First call four_bar_kinematics to get    theta2, theta3, theta4, dtheta2_dt, dtheta3_dt,
//                            dtheta4_dt,, d2theta2_dt2, d2theta3_dt2, d2theta4_dt2
fourbar=fourbar_kinematics(t,L1,L2,L3,L4,theta2,dtheta2_dt,d2theta2_dt2,solution_number)

if fourbar==[] then output=[], return, end

theta2=fourbar(:,2); theta3=fourbar(:,3); theta4=fourbar(:,4);
dtheta2_dt=fourbar(:,5); dtheta3_dt=fourbar(:,6); dtheta4_dt=fourbar(:,7);
d2theta2_dt2=fourbar(:,8); d2theta3_dt2=fourbar(:,9); d2theta4_dt2=fourbar(:,10);
// The variables theta2, dtheta2_dt, ...,  d2theta4_dt2 are columns

// We can now calculate the support reactions at each time instant specified in the array t
Cn4=m4*L4*d2theta4_dt2/3;  Dn4=Cn4/2   // From Eqs. (5.6.10) and (5.6.11)

// Now, determine the reactions Br2, Bn2, and Cr4 from Eq. (5.6.12). We will use the
// analytical expression for the inverse of the matrix that appears in Eq. (5.6.12),
// which is obtained after some algebra that can be done by hand. Its elements are
// a11, a12, a13, etc, a33, shown below.

cos32=cos(theta3-theta2);      sin32=sin(theta3-theta2);
cos42=cos(theta4-theta2);      sin42=sin(theta4-theta2);
cos43=cos(theta4-theta3);      sin43=sin(theta4-theta3);
cos432=cos(theta4+theta3-2*theta2); sin432=sin(theta4+theta3-2*theta2);
```

```
den=4*sin43;   // this is the determinant of the matrix in Eq. (5.6.12).
// The elements of the inverse of that matrix are:
a11=(sin432+3*sin43)./den;    a12=-(cos43+cos432)./den;    a13=2*cos42./den;
a21=(cos43-cos432)./den;      a22=(3*sin43-sin432)./den;   a23=2*sin42./den;
a31=-2*sin32./den;            a32=2*cos32./den;            a33=-2*ones(length(t),1)./den;

dtheta2_dt_sq=(dtheta2_dt.^2); dtheta3_dt_sq=(dtheta3_dt.^2); dtheta4_dt_sq=(dtheta4_dt.^2);
// Elements of the column array seen in Eq. (5.6.12):
v1= m3*L3*(  dtheta3_dt_sq.*cos32 + d2theta3_dt2.*sin32 )/2 ...
                + m3*L2*dtheta2_dt_sq + Cn4.*sin42;  // ... is for continuing a command line
v2=-m3*L3*( -dtheta3_dt_sq.*sin32 + d2theta3_dt2.*cos32 )/2 ...
                - m3*L2*d2theta2_dt2  - Cn4.*cos42;
v3= m3*L3*d2theta3_dt2/6 + Cn4.*cos43;

// We then have for the reactions Br2, Bn2, and Cr4, from Eq. (5.6.12):
Br2=a11.*v1+a12.*v2+a13.*v3;  Bn2= a21.*v1+a22.*v2+a23.*v3;  Cr4= a31.*v1+a32.*v2+a33.*v3;

// The remaining reactions are:
Mext= m2*L2^2*d2theta2_dt2/3 - L2*Bn2;      // From Eq. (5.6.3)
Ar2= -m2*L2*dtheta2_dt_sq/2-Br2;            // From Eq. (5.6.1)
An2= m2*L2*d2theta2_dt2/2-Bn2;              // From Eq. (5.6.2)
Dr4= -m4*L4*dtheta4_dt_sq/2-Cr4;            // From Eq. (5.6.4)

// disp(' ')
len=length(t);
mprintf('At t= %5.3f, the reactions are: ', t($)) // t($) is the last element of the array t
disp(' ')
mprintf('Mext=%5.3f, Ar2=%f, An2=%f, Dr4=%f', Mext(len), Ar2(len), An2(len), Dr4(len) )
disp(' ')
mprintf('Br2=%f,Bn2=%f,Cr4=%f,Dn4=%f,Cn4=%f',Br2(len),Bn2(len),Cr4(len),Dn4(len),Cn4(len))
disp(' ')

  output=[t', Mext, Ar2, An2, Dr4, Br2, Bn2, Cr4, Dn4, Cn4];
endfunction
// end of the file  fourbarforces.sci
// /////////////////////////////////////////////////////////////////////////


function output=fourbar_mech_dynamics(t, L1, L2, L3, L4, m2, m3, m4, theta2_0, ...
        dtheta2_dt_0, solution_number)  // ... is for continuing a command line
// Simulation for Example 7.9.3 in Section 7.9 of Chapter 7
// Strings used for names of variables are limited in Scilab to 24 characters.
// That is why a name like fourbar_mechanism_dynamics has to be shortened
// Example of usage [copy the line below to the Scilab console, without the //]:
// my_output=fourbar_mech_dynamics([0:0.002:5], 1, 0.4, 0.8, 0.7, 3.75, 3, 3, %pi/6, 0, 1);
//
// NOTE: for these values of L1, L3, and L4, the mechanism locks shortly after t=1.2 if
// L2 > approximately 0.5. If the mechanism locks, a standard error message will appear
// on the computer screen, and the computation ends. If this occurs, ignore the message
// but redesign the mechanism by changing the length of at least one of its bars to
// prevent it from locking.
//
// IN ADDITION TO ANIMATING THE MECHANISM, SUCH A CALL TO THE FUNCTION
// fourbar_mech_dynamics.sci GENERATES A len (WHICH IS THE NUMBER OF POINTS
// IN THE SPECIFIED ARRAY t) BY 10 MATRIX CALLED output (OR ANY OTHER NAME
```

```
// YOU WANT TO USE BEFORE THE EQUAL SIGN SEEN IN LINE 1), WHOSE COLUMNS
// ARE  t, theta2, theta3, theta4, d(theta2)/dt, d(theta3)/dt, d(theta4)/dt, d2(theta2)/dt2,
//          d2(theta3)/dt2, d2(theta4)/dt2
//
//  my_output(:,4), my_output(:,2)*180/%pi, FOR EXAMPLE, GIVE theta4 (IN RAD) AND theta2
//  (IN DEGREES). TO PLOT theta4 IN DEG VERSUS TIME t, FOR EXAMPLE, TYPE,
//  IN THE SCILAB CONSOLE, THE COMMANDS
//    figure(); plot(my_output(:,1), my_output(:,4)*180/(2*%pi) )
//    To put a grid on the plot, if desired, type the command   xgrid
//
// In this problem theta2 and d(theta2)/dt are not given; only their initial conditions
// theta2_0 and dtheta2_dt_0. The system is driven by a motor with a torque M_motor that is
// a specified function of time in the function M_motor(t), which can be changed by the user.
//
// The variables theta2 and d(theta2)/dt are determined by integrating a second-order
// differential equation obtained from the first row of the solution of Eq. (7.9.3n). The
// integration is done by a  call to the Scilab function ode [which calls the function
// rhs_diff_eqs(t, y) that specifies the right hand side of the equations
//    dtheta2/dt= ---, and d(dtheta2/dt)/dt= ---].

if solution_number ~=1 & solution_number ~= 2 then
    disp('--> SOLUTION_NUMBER HAS TO BE 1 or 2 <---'), output=[], return, end

disp(' '),    disp('The function fourbar_mech_dynamics called fourbar_mechanism.')

function M_motor_of_t= M_motor(t)
 M_motor_of_t=%pi/6+1.5*%pi*sin(2*t)    // NOTE: change this expresssion, if desired
endfunction

function dy_dt = rhs_diff_eqs(t, y) // this is called by ode to integrate the diff. eq.
    //state variables: y(1)=theta2,  y(2)=dtheta2_dt
 theta2=y(1),   dtheta2_dt=y(2)
 d2theta2_dt2=ang_acceleration(t, L1, L2, L3, L4, m2, m3, m4, theta2, ...
                dtheta2_dt,solution_number) // ... is for continuing a command line
 dy_dt=[y(2), d2theta2_dt2]
endfunction         // end of the rhs_diff_eqs function

function d2theta2_dt2 = ang_acceleration(t, L1, L2, L3, L4,  m2, m3, m4, theta2,  ...
                  dtheta2_dt, solution_number)
 fourbar = fourbar_kinematics(t, L1, L2, L3, L4, theta2, dtheta2_dt, ...
                  d2theta2_dt2, solution_number)
// Here we have valid values for theta2, theta3, theta4, d(theta2)/dt, d(theta3)/dt,
// and d(theta4)/dt
// Ignore d2theta3_dt2 and d2theta4_dt2 obtained by fourbar_kinematics

 theta2=fourbar(:,2);   theta3=fourbar(:,3);   theta4=fourbar(:,4);
 sin_theta2=sin(theta2);  sin_theta3=sin(theta3);  sin_theta4=sin(theta4);
 cos_theta2=cos(theta2);  cos_theta3=cos(theta3);  cos_theta4=cos(theta4);
 sin43=sin(theta4-theta3);  cos43=cos(theta4-theta3);
 dtheta2_dt=fourbar(:,5); dtheta3_dt=fourbar(:,6); dtheta4_dt=fourbar(:,7);
 dtheta2_dt_sq=dtheta2_dt^2, dtheta3_dt_sq=dtheta3_dt^2, dtheta4_dt_sq=dtheta4_dt^2

// angular acceleration d2theta2_dt2=d2(theta2)/dt2 from Eq. (7.9.3n)
 M=[m2*L2^2/3,        0,                    0,               L2*sin_theta2, -L2*cos_theta2;
    0,            m3*L3^2/3,        -m3*L3*L4*cos43/2,       L3*sin_theta3, -L3*cos_theta3;
    0,           -m3*L3*L4*cos43/2,  (m4/3+m3)*L4^2,        -L4*sin_theta4,  L4*cos_theta4;
```

```
     L2*sin_theta2,   L3*sin_theta3,    -L4*sin_theta4,    0,        0;
     L2*cos_theta2,   L3*cos_theta3,    -L4*cos_theta4,    0,        0];
  Minv=inv(M);      // M is the matrix seen in Eq. (7.9.3n)), and Minv is its inverse

if abs(rcond(M))<= 1.e-7 then
   disp('THE MECHANISM IS ABOUT TO LOCK. SUGGESTION: STOP THE SIMULATION.'),
output=[], return, end

  v1= M_motor(t);    v2=-m3*L3*L4*dtheta4_dt_sq*sin43/2;  v3=m3*L3*L4*dtheta3_dt_sq*sin43/2;
  v4= L4*dtheta4_dt_sq*cos_theta4 -L3*dtheta3_dt_sq*cos_theta3 -L2*dtheta2_dt_sq*cos_theta2;
  v5= -L4*dtheta4_dt_sq*sin_theta4 +L3*dtheta3_dt_sq*sin_theta3 +L2*dtheta2_dt_sq*sin_theta2;
  d2theta2_dt2=Minv(1,1)*v1 +Minv(1,2)*v2 +Minv(1,3)*v3 +Minv(1,4)*v4 +Minv(1,5)*v5;
endfunction        // end of the ang_acceleration function

  // Integration of the second-order diff. eq. for theta2. An arbitrary value is used for
  // d2theta2_dt2 because one is needed in a call to fourbar_mechanism
  d2theta2_dt2=0; // this is an arbitrary value
  state_variables = ode([theta2_0; dtheta2_dt_0], 0, t, rhs_diff_eqs)
  theta2=state_variables(1,:); dtheta2_dt=state_variables(2,:);    // this is the solution

  // We now determine the column of valid values for d2theta2/dt2. After that is done,
  // call fourbar_mechanism to complete the solution and animate the motion
  len=length(theta2); temp=[];
  for k=1:len,
    temp=[temp, ang_acceleration(t(k), L1, L2, L3, L4,  m2, m3, m4, theta2(k), ...
                      dtheta2_dt(k), solution_number)]
  end
  d2theta2_dt2=temp;
  output=fourbar_mechanism(t,L1,L2,L3,L4, theta2,dtheta2_dt,d2theta2_dt2,solution_number)

disp(' ')
disp('MESSAGE FROM THE FUNCTION fourbar_mech_dynamics.sci:')
disp('THE 10 COLUMN MATRIX GENERATED BY THE FUNCTION fourbar_mech_dynamics.sci')
disp('CONTAINS: t, theta2, theta3, theta4, dtheta2_dt, dtheta3_dt, ')
disp('   dtheta4_dt, d2theta2_dt2, d2theta3_dt2, d2theta4_dt2 ')
endfunction     // end of file fourbar_mech_dynamics
// ////////////////////////////////////////////////////////////////////////////


function  output=geneva_wheel(t, L, r1, theta1, dtheta1_dt, d2theta1_dt2)
// Animation of the geneva wheel mechanism (if a bar is always engaged) in Section 3.3
// To slow down the animation, choose a smaller time step for the input array t
// Conversely, to speed up the animation, choose a larger time step
// For example, the array t=[0:0.002:5] yields a slower animation than the array t=[0:0.02:5]
//
// Example of usage [copy the next two lines, without the //, to the Scilab console]:
// t=[0:0.002:5];theta1=%pi/2+3*%pi*sin(t);dtheta1_dt=3*%pi*cos(t);d2theta1_dt2=-3*%pi*sin(t);
// my_output=geneva_wheel(t, 1, 0.3, theta1, dtheta1_dt, d2theta1_dt2);
//
// IN ADDITION TO ANIMATING THE MECHANISM, SUCH A CALL TO THE FUNCTION geneva_wheel.sci
// GENERATES A len (WHERE len IS THE NUMBER OF POINTS IN THE ARRAY t) BY 8 MATRIX CALLED
// my_output, WHOSE COLUMNS ARE
//      t, theta1, theta2, r2, dr2_dt, dtheta2_dt, d2r2_dt2, d2theta2_dt2
// my_output(:,2), my_output(:,3)*180/%pi, FOR EXAMPLE,
// GIVES theta1 (IN RADIANNS) AND theta2 (IN DEGREES). TO PLOT theta2 VERSUS TIME t,
// FOR EXAMPLE, TYPE, IN THE SCILAB CONSOLE, THE COMMANDS
```

```
//   figure(); plot( my_output(:,1), my_output(:,3)*180/(2*%pi) )
// To put a grid on the plot, type   xgrid

disp('NOTE: IN ADDITION TO ANIMATING THE MECHANISM, THE FUNCTION geneva_wheel ')
disp('      RETURNS 8 COLUMNS CONTAINING ')
disp(' t, theta1, theta2, r2, d(r2)/dt, d(theta2)/dt, d2(r2)/dt2, d2(theta2)/dt2')

len_barPC=L+r1;  cos_theta1=cos(theta1); sin_theta1=sin(theta1);

r2=sqrt( L^2+r1^2-2*L*r1*cos_theta1 );      // r2 from Eq. (3.3.3)
sin_theta2=r1*sin_theta1./r2; cos_theta2=(L-r1.*cos_theta1)./r2; // Eqs. (3.3.2) and (3.3.1)
theta2=atan(sin_theta2, cos_theta2);

// Now generate the coordinate matrices needed for the function animate_nbars:
len=length(t),
xA=zeros(len,1), yA=xA,      xB=r1*cos_theta1', yB=r1*sin_theta1'    // bar AB
xP=L-len_barPC*cos_theta2', yP=len_barPC*sin_theta2',   xC=L*ones(len,1), yC=xA   // bar PC
x_coordinates=[xA, xB, xP, xC], y_coordinates=[yA, yB, yP, yC]

// Delete existing figure windows, and call the function animate_nbars:
xdel(winsid());   animate_nbars(x_coordinates, y_coordinates)

// Velocities and Accelerations:
temp=r1*dtheta1_dt; theta2=atan(sin_theta2, cos_theta2);
sin_theta1plustheta2 = sin(theta1+theta2);    cos_theta1plustheta2 = cos(theta1+theta2);
dr2_dt = temp.*sin_theta1plustheta2; dtheta2_dt = temp.*cos_theta1plustheta2./r2;
d2r2_dt2 = r1*d2theta1_dt2.*sin_theta1plustheta2 + r2.*(dtheta2_dt.^2) ...
           + r1*(dtheta1_dt.^2).*cos_theta1plustheta2;  // ... is for continuing a line
d2theta2_dt2 = ( r1*d2theta1_dt2.*cos_theta1plustheta2 - 2*dr2_dt.*dtheta2_dt ...
           - r1*(dtheta1_dt.^2).*sin_theta1plustheta2 )./r2;

output=[t; theta1; theta2; r2; dr2_dt; dtheta2_dt; d2r2_dt2; d2theta2_dt2]'
endfunction
// end of the file  geneva_wheel.sci
// /////////////////////////////////////////////////////////////////////////////


function output=inclined_plane(t, m1_over_m2, k_over_m2, c2_over_m2, c1_over_m2, alpha, ...
         g, Lu, x_0, xdot_0, r_0, rdot_0)    // ... is for continuing a line
// Animation for Example 2.5.9. The animation draws five bars, AB, BC, AC,, CD, and DE,
// each one with a small dot at their ends.
// Example of usage ( using the initial conditions  x(0)=1; xdot(0)=0; r(0)=2; rdot(0)=0 )
// my_output = inclined_plane([0:0.01:20], 0.1,  5, 0, 0, %pi/4, 9.81, 1,  1, 0, 2, 0);

sin_alpha=sin(alpha), cos_alpha=cos(alpha), temp=m1_over_m2+sin_alpha^2

function dy_dt=rhs_of_diff_eqs(t, y)
   // Note:  y(1)=x, y(2)=dx/dt, y(3)=r, y(4)=dr/dt
  dy_dt(1)=y(2)     // the expression on the next line is d2(x)/dt2 from Eq. (2.5.9i)
  dy_dt(2)= ( (c2_over_m2*y(4)+k_over_m2*(y(3)-Lu))*cos_alpha-g*sin(alpha)*cos(alpha) ...
              -c1_over_m2*y(2) )/temp
  dy_dt(3)=y(4)           // the following expression is d2(r)/dt2 from Eq. (2.5.9j)
  dy_dt(4)=( c1_over_m2*cos_alpha*y(2)+(1+m1_over_m2)*( g*sin_alpha-(c2_over_m2*y(4) ...
              +k_over_m2*(y(3)-Lu)) ) )
endfunction
```

```
initial_conditions=[x_0; xdot_0; r_0; rdot_0]
state_variables=ode(initial_conditions, 0, t, rhs_of_diff_eqs)

output= [t', state_variables']     // ' is the transpose operator
   // this is a len by 5 matrix, where len is the number of points in the array t
   // column 1: t, column 2: x, column 3: dx/dt; column 4: r; column 5: dr/dt

// ANIMATION:
len=length(t)    // this is the number of points in the array t
x=output(:,2), r=output(:,4)   // these are len by 1 columns
col_of_ones=ones(len,1)
xA=x, yA=zeros(len,1),   xB=x+3, yB=yA,    xC=x, yC=3*tan(alpha)*col_of_ones
xD=x+0.05*sin(alpha),    yD=(3*tan(alpha)+0.05*cos(alpha))*col_of_ones
xE=xD+r*cos(alpha),      yE=yD-r*sin(alpha)
x_coordinates=[xC, xA, xB, xC, xD, xE]; y_coordinates=[yC, yA, yB, yC, yD, yE]

// Delete all figures, and call the function animate_nbars
xdel(winsid());  animate_nbars(x_coordinates, y_coordinates)


endfunction
// end of the file  inclined_plane.sci
// //////////////////////////////////////////////////////////////////////////////


function output=parabola_track(t, a, k_over_m, c_over_m, Lu, h, g, x_0, xdot_0)
// Example 2.8.9
// Example of usage (with the initial conditions   x(0)=-1,5; xdot(0)=3*sqrt(2))):
// my_output = parabola_track([0:0.005:3], 1/3, 9/0.02, 1.4, 1.2, 1.2, 9.81, -1.5, 3*sqrt(2));
// NOTE: my_output is a matrix. The first column is the transpose of t, the second column is
//       x(t) for the time instants defined in the array t, and the third column is dx/dt

function dy_dt=rhs_of_diff_eqs(t, y)
  // Note:  y(1)=x, y(2)=dx/dt
  x_squared=y(1)^2
  dy_dt(1)=y(2)     // the next expression is d2x_dt2 from Eq. (2.8.9c)
  dy_dt(2)=(k_over_m*(1-Lu/sqrt(y(1)^2+(a*y(1)^2-h)^2))*(2*a*(h-a*y(1)^2)-1)*y(1) ...
          -2*a*g*y(1)-4*a^2*y(1)*y(2)^2)/(1+4*(a^2)*y(1)^2)-c_over_m*y(2)
endfunction

initial_conditions=[x_0; xdot_0]
state_variables=ode(initial_conditions, 0, t, rhs_of_diff_eqs)
output=[t', state_variables']
// The variable output is a n by 3 matrix, where n is the number of points in the array t
// column 1: t;  column 2: x;  column 3: dx/dt

xdel(winsid())  // deleting all existing figures
// Create a figure window and plot x(t) versus time t
figure('background', 8);  plot(t, state_variables(1,:)),   xgrid
xlabel('t  (seconds)', 'fontsize', 3), ylabel('x(t)  (meters)', 'fontsize', 3)
title('Example 2.8.9', 'fontsize', 3)
N of the ball moving on the parabolic track y=a*x^2
   x=output(:,2),  y=a*x.^2,    animate_nbars(x, y)
endfunction
// end of the file  parabola_track.sci
// //////////////////////////////////////////////////////////////////////////////
```

```
function output = pend_moving_base(t, g_over_L, c_over_mLsquare,  theta_0, thetadot_0)
// Example 2.7.6; Use item_number=2 for Part c, and item_number=1 for Part d in that example
//  NOTE:  if desired, change the expression for aA_over_L
// Example of usage ( using the initial conditions  theta_0=0; thetadot_0=0 ):
// my_output = pend_moving_base([0:0.005:15], 9.81, 1,   0, 0);

disp(' Enter the item number for Example 2.7.6 in Chapter 2')
disp('  USE 2 for Part c, 1 for Part d')
item_number=input(' ')
if  item_number==2 then message='Part c', elseif item_number==1 then message='Part d',  end

if item_number ~=1 & item_number ~=2 then         // ~= means    not equal; & means    and
  disp('item_number MUST BE EITHER 1 OR 2')
  disp('use item_number=2 to do Part c of Example 2.7.6, and 1 to do Part d')
  output=[], return
end

function dy_dt=rhs_of_diff_eqs(t, y)
  // Note:  y(1)=theta, y(2)=d(theta)/dt
 if item_number==2  then aA_over_L=2*9.81 // the constant acceleration input for Part c
   elseif item_number==1 then
   aA_over_L=5*t; if t>=5 then aA_over_L=0, end  // the ramp acceleration input for Part d
 end
dy_dt(1)=y(2)     // this is d(theta)/dt and the next line is d2(theta)/dt2
dy_dt(2)=-g_over_L*sin(y(1)) + aA_over_L*cos(y(1)) - c_over_mLsquare*y(2)
endfunction

initial_conditions=[theta_0; thetadot_0]
state_variables=ode(initial_conditions, 0, t, rhs_of_diff_eqs)
output=[t', state_variables']
// The variable output is a n by 3 matrix, where n is the number of points in the array t
// column 1: t, column 2: theta, column3: d(theta)/dt

xdel(winsid())  // deleting all existing figures
fig=figure('background', 8)    // 'background' 8 forces the figure background to be white
plot(t', output(:,2)*180/%pi),   xgrid        // plot of theta versus time t
xlabel('t  (seconds)', 'fontsize', 3), ylabel(['$\theta$', ' (degrees)'], 'fontsize', 3)
title('Example 2.7.6','fontsize', 3)
uicontrol(fig,'style','text', 'position',[10,10,70,30],  'string', message, 'fontsize', 18)
  // type   help uicontrol   in the Scilab console for a description of the command uicontrol

// ANIMATION: using a simple model consisting of a bar AB, with
// A(0,0) and B(sin_theta, -cos_theta)       (lengths are normalized by L)
  len=length(t),    theta=output(:,2)        // theta is a len by 1 column
  xA=zeros(len,1), yA=xA,   xB=sin(theta), yB=-cos(theta)  // these are len by 1 columns
  x_coordinates=[xA, xB];   y_coordinates=[yA, yB]
animate_nbars(x_coordinates, y_coordinates)    // this creates the animation
endfunction
// end of the file pend_moving_base.sci
// ////////////////////////////////////////////////////////////////////////


function output = restricted_3body(t,m1_over_m2,r_0,rdot_0,theta_0,thetadot_0,z_0,zdot_0)
// Integration of the differential equations for the restricted three-body problem
// analyzed in Section 4.6 (Eqs. 4.6.13 to 4.6.15).  All distances are normalized
```

```
//  by L (which is the distance between bodies P1 and P2)
// Example of usage ( using the initial conditions
//  r_0/L=1.001; rdot_0/L=0; theta_0=59*%pi/180; thetadot_0=0; z_0/L=0.01; zdot_0=0 ):
// my_output = restricted_3body([0:0.02:100], 80, 1.001, 0, 59*%pi/180, 0, 0.01, 0);

function dy_dt=rhs_of_diff_eqs(t, y)
// Note:  y(1)=r/L, y(2)=d(r/L)/dt, y(3)=theta, y(4)=d(theta)/dt, y(5)=z/L, y(6)=d(z/L)/dt
    cos_theta=cos(y(3)), sin_theta=sin(y(3))
    r31_cube= (y(1)^2+y(5)^2)^(3/2),  r32_cube= ( 1+y(1)^2+y(5)^2 -2*y(1)*cos_theta )^(3/2)
    dy_dt(1)= y(2)    // this is d(r/L)/dt
    dy_dt(2)= y(1)*(1+y(4))^2+((cos(y(3))-y(1))/r32_cube-cos(y(3)) ...
      -m1_over_m2*y(1)/r31_cube)/(1+m1_over_m2) // this is d2(r/L)/dt2 from Eq. (4.6.13)
    dy_dt(3)= y(4)
    dy_dt(4)= ((1-1/r32_cube)*sin_theta/(1+m1_over_m2)-2*y(2)*(1+y(4)))/y(1)
    dy_dt(5)= y(6)
    dy_dt(6)= -(m1_over_m2/r31_cube+1/r32_cube)*y(5)/(1+m1_over_m2)
endfunction

initial_conditions=[r_0; rdot_0; theta_0; thetadot_0; z_0; zdot_0]
state_variables=ode(initial_conditions, 0, t, rhs_of_diff_eqs)
output=[t', state_variables']
// The variable output is a n by 7 matrix, where n is the number of points in the array t
// column 1: t, column 2: r/L, column 3: d(r/L)/dt; column 4: theta, column 5: d(theta)/dt;
//  column 6: z/L, column 7: d(z/L)/dt
r_over_L=output(:,2);  theta=output(:,4);  z_over_L=output(:,6)

xdel(winsid())  // deleting all existing figures
// Plots:
figure('background', 8); plot(t', r_over_L),  xgrid    // plot of r/L versus time t
xlabel('t    (seconds)', 'fontsize', 3), ylabel('r/L', 'fontsize', 3)
title('Restricted three-body problem','fontsize', 3)

figure('background', 8); plot(t', theta*180/%pi),  xgrid  // plot of theta versus time t
xlabel('t    (seconds)', 'fontsize', 3), ylabel(['$\theta$', ' (degrees)'], 'fontsize', 3)
title('Restricted three-body problem','fontsize', 3)

figure('background', 8); plot(t', z_over_L),  xgrid    // plot of z/L versus t
xlabel('t    (seconds)', 'fontsize', 3), ylabel('z/L', 'fontsize', 3)
title('Restricted three-body problem','fontsize', 3)

// Plot of y/L versus x/L:
figure('background', 8); plot(r_over_L.*cos(theta), r_over_L.*sin(theta)),  xgrid
xlabel('x/L', 'fontsize', 3),   ylabel('y/L', 'fontsize', 3)
title('Restricted three-body problem','fontsize', 3)

  x=r_over_L.*cos(theta), y=r_over_L.*sin(theta), animate_nbars(x, y)  // animation
endfunction
// end of the file  restricted_3body.sci
// ///////////////////////////////////////////////////////////////////////////


function output = scotch_yoke(t, R, theta, dtheta_dt, d2theta_dt2)
// Animation of the Scotch yoke mechanism in Section 3.5
// Example of usage  [copy the next two lines, without the //, to the Scilab console]:
// t=[0:0.01:10]; theta=%pi/2+2*t; dtheta_dt=2; d2theta_dt2=0;
// my_output=scotch_yoke(t, 1, theta, dtheta_dt, d2theta_dt2);
```

```
//
// To slow down the animation, choose a smaller time step for the input array t
// Conversely, to speed up the animation, choose a larger time step
// For example, the array t=[0:0.01:10] yields a slower animation than the array t=[0:0.1:10]
//
// IN ADDITION TO ANIMATING THE MECHANISM, SUCH A CALL TO THE FUNCTION
// scotch_yoke.sci GENERATES A len (WHERE len IS THE NUMBER OF POINTS IN THE
// ARRAY t) BY 4 MATRIX CALLED my_output, WHOSE COLUMNS ARE  time t, theta, dxA_dt, d2xA_dt2
// my_output(:,3), FOR EXAMPLE, GIVES dxA_dt. TO PLOT IT, TYPE, IN THE SCILAB CONSOLE,
//  THE COMMANDS   figure(); plot( my_output(:,1), my_output(:,3) )
// To put a grid on the plot, type   xgrid

disp('NOTE: IN ADDITION TO ANIMATING THE SCOTCH YOKE MECHANISM, THIS FUNCTION')
disp('RETURNS 4 COLUMNS CONTAINING t, theta, d(xA)/dt, d2(xA)/dt2')

// Make sure that theta, dtheta_dt, and d2theta_dt2 have the same length as t
len=length(t);  temp=ones(1,len);
theta=theta.*temp; dtheta_dt=dtheta_dt.*temp; d2theta_dt2=d2theta_dt2.*temp;

sin_theta = sin(theta);   cos_theta = cos(theta);
x0=zeros(len,1), xP=R*cos_theta'; xB=xP; xC=xP
y0=x0, yP=R*sin_theta', yB=1.5*R*ones(len,1), yC=-yB

xdel(winsid());   // deleting all figures
x=[x0, xP, xB, xC],  y=[y0, yP, yB, yC], animate_nbars(x, y) // this animates the motion

// Velocity dxA_dt=d(xA)/dt  and acceleration d2xA_dt2=d2(xA)/dt2, where xA=R*cos(theta):
  dxA_dt = -R*dtheta_dt.*sin_theta;
d2xA_dt2=-R*(d2theta_dt2.*sin_theta+(dtheta_dt.^2).*cos_theta);

  output = [t; theta; dxA_dt; d2xA_dt2]';
endfunction
// end of the file scotch_yoke.sci
// ////////////////////////////////////////////////////////////////////////////


function output = sim_model_ex5_3_11(tau, L2_over_L1, xdoubledot_over_g, Mbar_over_mgL1, ...
                  cstar, theta_0, thetaprime_0)        //  ... is for continuing a line
// Example 5.3.11 in Chapter 5, where  tau=t*sqrt(g/L1), Mbar_over_mgL1=mbar/(m*g*L1),
//   and  cstar=c/(m*L1*sqrt(g*L1))
// Example of usage ( using the initial conditions  theta_0=90*%pi/180; thetaprime_0=0 ):
// tau=[0:0.005:10]; xdoubledot_over_g=-2;  Mbar_over_mgL1=0;
// my_output=sim_model_ex5_3_11(tau,2, xdoubledot_over_g, Mbar_over_mgL1, 0, 90*%pi/180, 0);

disp('NOTE: IN ADDITION TO ANIMATING THE PENDULUM ANALYZED IN EXAMPLE 5.3.11,')
disp('AND PLOTTING theta versus time tau, A CALL TO THE FUNCTION sim_model_ex5_3_11,')
disp(' AS EXEMPLIFIED IN THE FIRST 5 COMMENT LINES OF THIS FUNCTION,')
disp(' RETURNS 3 COLUMNS CONTAINING  tau, theta, and d(theta)/dtau')
disp('TO PLOT d(theta)/dtau VERSUS theta, FOR EXAMPLE, USE THE COMMANDS:')
disp('      figure(); plot(my_output(:,2), my_output(:,3)) ')
disp('   To put a grid on the plot, type  xgrid ')

function dy_dtau=rhs_of_diff_eqs(tau, y)
// Note:  y(1)=theta, y(2)=d(theta)/dtau
  dy_dtau(1)=y(2),    // this is d(theta)/dtau
  dy_dtau(2)=(3*(1+L2_over_L1)*(Mbar_over_mgL1 -cstar*y(2)) -3*( 1+2*L2_over_L1 ...
```

```
       - xdoubledot_over_g*L2_over_L1^2 )*sin(y(1))/2 -3*( L2_over_L1^2 ...
       +xdoubledot_over_g*(1+2*L2_over_L1) )*cos(y(1))/2 )/(1+3*L2_over_L1+L2_over_L1^3)
// dy_dtau(2) is d2(theta)/dtau2 from Eq. (5.3.11c)
endfunction

initial_conditions=[theta_0; thetaprime_0]
theta_and_dthetadtau=ode(initial_conditions, 0, tau, rhs_of_diff_eqs),

output=[tau', theta_and_dthetadtau']    // ' is the transpose operator
// The variable output is a n by 3 matrix, where n is the number of points in the array tau
// column 1: tau, column 2: theta, column 3: d(theta)/dtau

xdel(winsid( ))  // deleting all existing figures before plotting theta versus tau
figure('background', 8);   plot(tau, theta_and_dthetadtau(1,:)*180/%pi),   xgrid
xlabel(['Normalized time ', '$\tau$'], 'fontsize', 3)
ylabel(['$\theta$', ' (degrees)'], 'fontsize', 3),   title('Example 5.3.11', 'fontsize', 3)

// ANIMATION (distances are normalized by L1):
  len=length(tau)
  theta=output(:,2),   sin_theta=sin(theta), cos_theta=cos(theta)
  xA=zeros(len,1) , yA=xA,   xB=sin_theta,  yB=-cos_theta
// The seven variables defined above are len by 1 columns
  xD=xB+L2_over_L1*cos_theta,  yD= L2_over_L1*sin_theta-cos_theta
  x_coordinates=[xA, xB, xD]; y_coordinates=[yA, yB, yD];
animate_nbars(x_coordinates, y_coordinates) // this call animates the motion
endfunction
// end of the file sim_model_ex5_3_11.sci
// //////////////////////////////////////////////////////////////////////////////


function output = sim_model_ex5_3_13(t, k_factor, c1_star, c2_star, phi_0, phidot_0, ...
                    theta_0, thetadot_0)    // ... is for continuing command a line
// Example 5.3.13 (t here is actually the normalized time tau)
// Example of usage [using the initial conditions phi_0=20*%pi/180;
//                         phidot_0=0; theta_0=32*%pi/180; thetadot_0=50]:
// my_output = sim_model_ex5_3_13([0:0.005:50], 12, 0.1, 0.1, 20*%pi/180, 0, 32*%pi/180, 50);

disp('The k_factor is 12*(1+m1/(3*m2)/[(b/L)^2+(h/L)^2)]')
disp('Change it, if desired, to any other  appropriate value')
disp('The k_factor is 12*(1+m1/(3*m2)/[(b/L)^2+(h/L)^2)]')
disp('NOTE: IN ADDITION TO ANIMATING THE SYSTEM ANALYZED IN EXAMPLE 5.3.13,')
disp('AND PLOTTING phi AND theta versus time, A CALL TO THE FUNCTION sim_model_ex5_3_13,')
disp(' AS EXEMPLIFIED IN THE FIRST 4 COMMENT LINES OF THIS FUNCTION,')
disp(' RETURNS 5 COLUMNS CONTAINING  t, phi, d(phi)/dt, theta, and d(theta)/dt')
disp(' PLOTS CAN BE DONE USING THE plot COMMAND')

function dy_dt=rhs_of_diff_eqs(t, y)
// Note:  y(1)=phi, y(2)=d(phi)/dt, y(3)=theta, y(4)=d(theta)/dt
  dy_dt(1)=y(2), // this is d(phi)/dt, and the next one is is d2(phi)/dt2 from Eq. (5.3.13j)
  dy_dt(2)= -sin(y(1))-c1_star*y(2)-c2_star*(y(2)+y(4))
  dy_dt(3)=y(4)
  dy_dt(4)= -k_factor*c2_star*(y(2)+y(4))  // this is from Eq. (5.3.13k)
endfunction

initial_conditions=[phi_0; phidot_0; theta_0; thetadot_0]
state_variables=ode(initial_conditions, 0, t, rhs_of_diff_eqs)
```

```
output=[t', state_variables']
// The variable output is a n by 5 matrix, where n is the number of points in the array t
// column 1: t, column 2: phi, column 3: d(phi)/dt;  column 4: theta,
// column 5: d(theta)/dt
phi=output(:,2) , theta=output(:,4)

xdel(winsid( )) // deleting all existing figures
figure('background', 8); plot(t', phi*180/%pi), xgrid  // plot of phi (in degrees) versus tau
xlabel(['Nondimensional time', '$\tau$'], 'fontsize', 3)
ylabel(['$\phi$', ' (degrees)'], 'fontsize',3), title('Example 5.3.13','fontsize', 3)

figure('background', 8); plot(t', theta*180/%pi), xgrid // plot of theta (in deg) versus tau
xlabel(['Nondimensional time', '$\tau$'], 'fontsize', 3),
ylabel(['$\theta$', ' (degrees)'], 'fontsize',3), title('Example 5.3.13','fontsize', 3)

// ANIMATION (distances are normalized by L):
  len=length(t), // this is the number of time occurrences in the given array t
  col_of_zeros=zeros(len,1),   // this is a len by 1 column of zeros
  sin_phi=sin(phi), cos_phi=cos(phi),  sin_theta=sin(theta), cos_theta=cos(theta)
  x0=col_of_zeros, y0=x0,      xC2=sin_phi, yC2=-cos_phi
  xN=xC2+0.25*sin_theta, yN= yC2+0.25*cos_theta
  xM=xC2-0.25*sin_theta, yM= yC2-0.25*cos_theta
  x_coordinates=[x0, xC2, xN, xM]; y_coordinates=[y0, yC2, yN, yM]
  animate_nbars(x_coordinates, y_coordinates)
endfunction
// end of the file sim_model_ex5_3_13.sci
// ////////////////////////////////////////////////////////////////////////////


function  output=slider_crank(t,D,L1,L2,theta1,dtheta1_dt,d2theta1_dt2,solution_number)
// Animation of the slider_crank mechanism in Section 3.4 using the function animate_nbars
// Examples of usage:
// t=[0:0.002:5]; theta1=2*%pi*t-%pi/6; dtheta1_dt=2*%pi; d2theta1_dt2=0;
// my_output = slider_crank(t, -0.5, 1, 2.5, theta1, dtheta1_dt, d2theta1_dt2,  1);
// Example 2:
// t=1/12; theta1=2*%pi*t-%pi/6; dtheta1_dt=2*%pi; d2theta1_dt2=0;
// my_output = slider_crank(t, -0.5, 1, 2.5, theta1, dtheta1_dt, d2theta1_dt2,  1);
// Example 3:
// t=[0:0.002:0.5]; theta1=2*%pi*t-%pi/6; dtheta1_dt=2*%pi; d2theta1_dt2=0;
// my_output = slider_crank(t, -0.5, 1, 2.5, theta1, dtheta1_dt, d2theta1_dt2,  1);
//
// To slow down the animation, choose a smaller time step for the input array t
// Conversely, to speed up the animation, choose a larger time step
//
// IN ADDITION TO ANIMATING THE MECHANISM, SUCH A CALL TO THE FUNCTION
// slider_crank.sci GENERATES A len (WHERE len IS THE NUMBER OF
// POINTS IN THE ARRAY t) BY 8 MATRIX CALLED my_output, WHOSE COLUMNS
// ARE time, theta1, theta2, xB, dtheta2_dt, dxB_dt, d2theta2_dt2, d2xB_dt2
// my_output(:,3)*180/%pi, FOR EXAMPLE, GIVES theta2 (IN DEGREES).
// TO PLOT IT VERSUS TIME t, TYPE, IN THE SCILAB CONSOLE, THE COMMANDS
//     figure(); plot( my_output(:,1), my_output(:,3)*180/(2*%pi) )
// To put a grid on the plot, type   xgrid

disp('NOTE: IN ADDITION TO ANIMATING THE MECHANISM, A CALL SUCH AS')
disp('t=[0:0.02:5]; theta1=2*%pi*t-%pi/6; dtheta1_dt=2*%pi; d2theta1_dt2=0;')
disp('output=slider_crank(t, -0.5, 1, 2.5, theta1, dtheta1_dt, d2theta1_dt2,1,1);')
```

```
disp('RETURNS, IN THE VARIABLE output, 8 COLUMNS CONTAINING')
disp(' t, theta1, theta2, xB, dtheta2_dt, dxB_dt, d2theta2_dt2, d2xB_dt2')
disp(' (the time array t only has those times for which the mechanism does not lock)')

if solution_number ==1 | solution_number == 2 then
    // First, make sure that theta1, dtheta1_dt, and d2theta1_dt2 have the same length as t
    len=length(t); temp=ones(1,len);  newlen=len
theta1=theta1.*temp; dtheta1_dt=dtheta1_dt.*temp; d2theta1_dt2=d2theta1_dt2.*temp

// Make the simulation proceed only while |sin_theta2| <=1
sin_theta2=(L1*sin(theta1)-D)/L2                // sin(theta2) from Eq. (3.4.6)
// Let's see if this is between -1 and 1:
temp=find( abs(sin_theta2)>1 )
if temp ~= [ ] then
    temp1=temp(1)-1
    if temp1 == 1 then
        output = []
        disp('The mechanism, as specified, does not work because the condition')
        disp('|L1*sin(theta1)-D|>=L2 is violated at the initial time')
        return
    else
        newlen=temp1,   tnew=t(1:newlen)   // redefining the time array
        mprintf('The mechanism, as specified, stops after t= %5.3f', tnew(newlen) )
        mprintf(' because |L1*sin(theta1)-D|>L2 sometime thereafter.')
    end
end

// Resize all the proper lists and do the computations.
tnew=t(1:newlen)
theta1=theta1(1:newlen);dtheta1_dt=dtheta1_dt(1:newlen);d2theta1_dt2=d2theta1_dt2(1:newlen)
sin_theta1 = sin(theta1);    cos_theta1 = cos(theta1)
sin_theta2=(L1*sin(theta1)-D)/L2 // sin(theta2) from Eq. (3.4.6), with the resized theta1
cos_theta2=(3-2*solution_number).*sqrt(1-sin_theta2.^2)
// The above expression for cos(theta2) is from Eqs. (3.4.8) and (3.4.9)
xB=L1*cos_theta1+L2*cos_theta2    // xB from Eq. (3.4.7)
theta2= atan(sin_theta2, cos_theta2)

xdel(winsid()) // deleting all figure windows
// Now, generate the coordinate matrices needed for the function animate_nbars:
x0=zeros(newlen,1), y0=x0,    xA=L1*cos_theta1', yA=L1*sin_theta1'   // bar OA
x_coordinates=[x0, xA,  xB'], y_coordinates=[y0, yA, D*ones(newlen,1)]
animate_nbars(x_coordinates, y_coordinates)   // animation

// Angular velocity dtheta2_dt=d(theta2)/dt and velocity dxB_dt=d(xB)/dt:
dtheta2_dt=L1*dtheta1_dt.*cos_theta1./(L2*cos_theta2)    // d(theta2)/dt from Eq. (3.4.10)
dxB_dt=-L1*dtheta1_dt.*sin_theta1-L2*dtheta2_dt.*sin_theta2 // d(xB)/dt from Eq. (3.4.11)

// d2(theta2)/dt2 from Eq. (3.4.12), and d2(xB)/dt2 from Eq. (3.4.13):
d2theta2_dt2 = ( L1*(d2theta1_dt2.*cos_theta1-(dtheta1_dt.^2).*sin_theta1)/L2 + ...
                (dtheta2_dt.^2).*sin_theta2 )./cos_theta2 // ... is for continuing a line
d2xB_dt2=-L1*(d2theta1_dt2.*sin_theta1+(dtheta1_dt.^2).*cos_theta1) - ...
            L2*(d2theta2_dt2.*sin_theta2+(dtheta2_dt.^2).*cos_theta2)

output=[tnew; theta1; theta2; xB; dtheta2_dt; dxB_dt; d2theta2_dt2; d2xB_dt2]'

else disp(' ----> SOLUTION_NUMBER HAS TO BE  1 or 2 <----- '),  output=[ ], return
```

```
end

endfunction
// end of the file  slider_crank.sci
// /////////////////////////////////////////////////////////////////////////


function output=spring_pendulum(t, mg_over_kre, rs_0, dotrs_0, theta_0, dottheta_0)
// Animation of the spring-pendulum analyzed in Example 2.7.8
// Example of usage [with the initial conditions
//                          rs(0)=0.1;rsdot(0)=0; theta(0)=0.01;thetadot(0)=0]:
// my_output = spring_pendulum([0:0.1:350], 0.5^2,  0.1, 0,  0.01, 0);

disp('NOTE: IN ADDITION TO ANIMATING THE SPRING-PENDULUM ANALYZED IN EXAMPLE 2.7.8,')
disp('AND PLOTTING theta and rs versus time, A CALL TO THE FUNCTION spring_pendulum')
disp(' (SUCH AS my_output = spring_pendulum([0:1:350], 0.5^2, 0.1, 0,  0.01, 0) )')
disp('RETURNS 5 COLUMNS CONTAINING  t, rs,  d(rs)/dt, theta, and d(theta)/dt')
disp('TO PLOT d(rs)/dt, FOR EXAMPLE, VERSUS time, TYPE, IN THE SCILAB CONSOLE:');
disp('          figure(); plot(my_output(:,1), my_output(:,3)) ')
disp('OR:  figure( ); plot2d(my_output(:,1), my_output(:,3));')
disp('   To put a grid on the plot, type  xgrid ')

function dy_dt=rhs_of_diff_eqs(t, y)
   // Note: y(1)=rs, y(2)=d(rs)/dt, y(3)=theta, y(4)=d(theta)/dt
  dy_dt(1)=y(2)
  dy_dt(2)=(1+y(1))*y(4)^2-mg_over_kre*(1-cos( y(3) ) )-y(1)  // this is d2(rs)/dt2
  dy_dt(3)=y(4)
  dy_dt(4)=-(2*y(2)*y(4)+mg_over_kre*sin( y(3) ) )/(1+y(1))  // this is d2(theta)/dt2
endfunction

initial_conditions=[rs_0; dotrs_0; theta_0; dottheta_0];
y=ode(initial_conditions, 0, t, rhs_of_diff_eqs)
// pause  // NOTE: this pause command is here only to explain, in Appendix A, what it does

xdel(winsid())  // deleting all existing figures
figure('background', 8)    // 'background' 8 forces the figure background to be white
plot(t, y(3,:)),   xgrid   // plot of theta versus time t
xlabel(['Nondimensional time', '$\tau$'], 'fontsize', 3), ylabel('$\theta$', 'fontsize',4)

figure('background', 8), plot(t, y(1,:)),   xgrid      // plot of rs versus t
xlabel(['Nondimensional time', '$\tau$'], 'fontsize', 3),  ylabel('$r_s$', 'fontsize', 5)

output=[t', y]
// The variable output is a n by 5 matrix, where n is the number of points in the array t
// column 1: t, column 2: rs, column 3: d(rs)/dt; column 4: theta; column 5: d(theta)/dt

// ANIMATION: using a bar AB from the point (0,0) to (1+rs)*(sin_theta, -cos_theta)
 len=length(t),   // this is the number of points in the array t
 rs=output(:,2), theta=output(:,4)      // these are len by 1 columns
 xA=zeros(len,1), yA=xA,   xB=(1+rs).*sin(theta), yB=-(1+rs).*cos(theta)    // bar AB
 x_coordinates=[xA, xB], y_coordinates=[yA, yB], animate_nbars(x_coordinates, y_coordinates)
endfunction
// end of the file spring_pendulum.sci
// /////////////////////////////////////////////////////////////////////////
```

```
function output=top_dynamics(t,Iz_over_In,mgL_over_In,phidot_0,psidot_0,theta_0,thetadot_0)
// Top dynamics analyzed in Section 6.6.3
// Example of usage ( using the initial conditions
//          phidot(0)=5.03; psidot(0)=45.65; theta(0)=%pi/6; thetadot(0)=0 ):
// my_output = top_dynamics([0:0.002:10], 27/68, 27*98.1/34, 5.03, 45.65, 30*%pi/180, 0);

disp('NOTE: IN ADDITION TO ANIMATING THE MOTION OF THE TOP, A CALL SUCH AS ')
disp(' output = top_dynamics([0:0.025:10], 27/68, 27*98.1/34, 5.03, 45.65, 30*%pi/180, 0);')
disp('RETURNS, IN THE VARIABLE output, 4 COLUMNS CONTAINING  t, phi, theta, dtheta_dt')
disp(' ')

function dy_dt=rhs_of_diff_eqs(t, y)
// y(1)=phi, y(2)=theta, y(3)=d(theta)/dt
  cos_theta0=cos(theta_0), sin_theta0=sin(theta_0), sin_theta=sin(y(2)), cos_theta=cos(y(2))
  omegaz = psidot_0+phidot_0*cos_theta0
  Hz0_over_In = phidot_0*(sin_theta0^2)+Iz_over_In*omegaz*cos_theta0
  dy_dt(1)= (Hz0_over_In - Iz_over_In*omegaz*cos_theta)/(sin_theta^2) // this is d(phi)/dt
  dy_dt(2)=y(3),     // this is d(theta)/dt and the one that follows is d2(theta)/dt2
  dy_dt(3)=(mgL_over_In + (dy_dt(1)*cos_theta - Iz_over_In*omegaz)*dy_dt(1) )*sin_theta
endfunction

phi_0=0;  phi_theta_thetadot=ode([phi_0; theta_0; thetadot_0], 0, t, rhs_of_diff_eqs)
output=[t', phi_theta_thetadot']
phi=output(:,2), theta=output(:,3),   sin_theta=sin(theta)

xdel(winsid())  // deleting all existing figures
// Plot of phi (in degrees) versus time
  figure('background', 8)     // 'background' 8 forces the figure background to be white
  plot(t', phi*180/%pi),   xgrid // plot of phi (in degrees) versus time t
  xlabel('t (seconds)', 'fontsize', 3), ylabel(['$\phi$', '(degrees)'], 'fontsize', 3)

// Plot of theta (in degrees) versus time
  figure('background', 8), plot(t', theta*180/%pi),   xgrid
  xlabel('t (seconds)', 'fontsize', 3), ylabel(['$\theta$', '(degrees)'], 'fontsize', 3)

// ANIMATION:
  len=length(t)   // number of points in the given array t
  x0=zeros(len,1), y0=x0, z0=x0  // these are columns of zeros
  xC=sin_theta.*cos(phi), yC=sin_theta.*sin(phi),  zC=cos(theta)
  x_coordinates=[x0, xC], y_coordinates=[y0, yC], z_coordinates=[z0, zC]
  animate3d_nbars(x_coordinates, y_coordinates, z_coordinates)
endfunction
// end of the file top_dynamics.sci
// ////////////////////////////////////////////////////////////////////////////////


function output = two_body_simul(t, Re, g, m2_over_m1_plus_1, r_0, v_0, flight_path_angle_0)
// Animation of the two-body problem analyzed in Example 4.3.2 and in Section 4.4
// (Re: in km, g: in m/s^2, r_0: in km, v_0: in km/hour, flight_path_angle_0: in degrees)
// Example of usage (with the initial conditions r_0=6578; v_0=28040; flight_path_angle_0=0):
// my_output = two_body_simul([0:0.005:30], 6378, 9.81, 1,   6578, 28040, 0);
// also try:  my_output = two_body_simul([0:0.005:30], 6378, 9.81, 1,   6578, 38040, 0);
//       and: my_output = two_body_simul([0:0.005:30], 6378, 9.81, 1,   6578, 27040, 0);

disp('NOTE: IN ADDITION TO ANIMATING THE TWO-BODY PROBLEM, A CALL SUCH AS ')
disp(' output=two_body_simul(t, Re, g, m2_over_m1_plus_1, r_0, v_0, flight_path_angle_0)')
```

```
disp('RETURNS, IN THE VARIABLE output, 4 COLUMNS CONTAINING  t, r, dr/dt, and theta')
// ***************************************************************************
disp(' '), disp('  =====>  IMPORTANT MESSAGE:  <=====')
disp(' IF THE SIMULATION RUNS EITHER TOO FAST OR TOO SLOW IN YOUR COMPUTER,')
disp(' change the time_step in the time array t=[t0:time_step:tfinal] of your simulation ')
disp(' ---> EITHER to a smaller value to slow down the animation ')
disp(' ---> OR to a larger value to speed up the animation')
// ***************************************************************************
disp(' ')

function dy_dt=rhs_diff_eqs(t, y)
// y(1)=r, y(2)=dr/dt, y(3)=theta; unit of distance: km, unit of time: hour; g: in m/s^2
      r_sq=y(1)^2
      dtheta_dt = r_0*v_0*cos(flight_path_angle_0*%pi/180)/r_sq
      d2r_dt2 = y(1)*dtheta_dt^2 - m2_over_m1_plus_1*g*3600*3.6*(Re^2)/r_sq
   dy_dt(1) = y(2),          // this is dr/dt
   dy_dt(2) = d2r_dt2        // this is d2(r)/dt2 from Eq. (4.3.2h)
   dy_dt(3) = dtheta_dt      // this is d(theta)/dt from Eq. (4.3.2i)
endfunction

dr_dt_0=v_0*sin(flight_path_angle_0*%pi/180), theta0=0
r_drdt_and_theta=ode([r_0; dr_dt_0; theta0], 0, t, rhs_diff_eqs)
output=[t', r_drdt_and_theta'],    r=output(:,2), theta=output(:,4)

// ************* Let's see if the satellite crashed ***************************
number_of_time_points=length(t),   newlen=number_of_time_points,  temp=find(r <= Re)

tnew=t;
if temp ~= [ ] then
    temp1=temp(1)-1, newlen=temp(1)-1
    tnew=t(1:newlen), number_of_time_points=newlen
    disp(' '), mprintf('THE SATELLITE CRASHES AFTER  t= %5.3f  hours', tnew(newlen))
end
// Resize all the proper lists and then proceed with the new reduced time list tnew
t=tnew, r=r(1:newlen,1), theta=theta(1:newlen,1), output=[tnew', r, theta]
// ***************************************************************************

temp1=r.*cos(theta), temp2=r.*sin(theta)
x_min=min(temp1), x_max=max(temp1),  y_min=min(temp2), y_max=max(temp2)
if x_min==x_max,
        disp('THE STEP SIZE IS TOO BIG; DECREASE IT TO RUN THE SIMULATION'), return,
end

xdel(winsid())  // deleting all existing figures
// Create a figure window for the animation. Start by determining the computer monitor size
// so that the animation window is properly placed and sized in any computer screen.
screen_width_and_height=get(0,'screensize_px')  // width and height of the screen, in pixels
PC_screen_width =screen_width_and_height(3)      // this is the width of the computer monitor
PC_screen_height=screen_width_and_height(4)      // this is the height of the computer monitor

x_topleft= 0.65*PC_screen_width  // choice for the top left x-coordinate of the figure window
y_topleft= 0.3*PC_screen_height  // choice for the top left y-coordinate of the figure window
top_left_coordinates=[x_topleft, y_topleft]
size_of_fig_window = [0.32*PC_screen_width, 0.47*PC_screen_height]
// the above are the width and height chosen for the animation window
// Now, create the figure window. NOTE: 'background', 8  is for a white background)
```

```
animation_fig=figure('figure_position',top_left_coordinates,'figure_size', ...
         size_of_fig_window, 'figure_name','Animation window', 'background', 8)

animation_fig.children.isoview='on'
xlabel('x (in km) from the planet center', 'fontsize', 3)
ylabel('y (in km) from the planet center', 'fontsize', 3)
title('The two-body problem', 'fontsize', 3)

h_axes = gca(),  h_axes.data_bounds =[x_min, y_min; x_max, y_max]

th=0:0.02:2*%pi, plot(Re*cos(th), Re*sin(th)) // a circle representing the attracting planet
plot(1.05*Re, 0, 'o'), satellite=gce()  // this will be the satellite
plot([0,0], [0, 0]),  rot_planet=gce()
         // rot_planet will be a line representing the spinning attracting planet

// record the Earth-day number in the figure window
   om=2*%pi/24,   tau=om*t'
   day_number=ceil(tau/(2*%pi)),  dia=mtlb_num2str(day_number)
   uicontrol(animation_fig,'style','text', 'position',[10,10,40,15], 'string', 'day')
   day_info=uicontrol(animation_fig,'style','text', 'position',[40,10,40,15], 'string', ' ')
   // type  help uicontrol  in the Scilab console for information about the uicontrol command

tT=t';
   for itime=1:number_of_time_points
    drawlater();
    if r(itime)<= Re then disp('THE SATELLITE CRASHES ON THE PLANET'), output=[], return, end
    satellite.children.data=[r(itime,1)*cos(theta(itime,1)), r(itime,1)*sin(theta(itime,1))]
    rot_planet.children.data= [ 0,0; Re*cos(om*tT(itime,1)),Re*sin(om*tT(itime,1))]
    drawnow();
    day_info.string= dia(itime)
    end
endfunction
// end of the file two_body_simul.sci
// ///////////////////////////////////////////////////////////////////////////


function output=two_particles_table(t, m2_over_m2plusm1, g_over_L, r_0, rdot_0, theta_0, ...
                          thetadot_0)    // ... is for continuing a line
// Animation of Example 2.7.10. Here, the distance denoted by r is actually r/L
// State variables: y(1)=r_over_L, y(2)=d(r_over_L)/dt, y(3)=theta, y(4)=d(theta)/dt
// Example of usage (with the initial conditions for the state variables: [0.5; 0;  0; 2] ):
// my_output = two_particles_table([0:0.005:10], 0.5, 9.81, 0.5, 0,  0, 2);

disp('NOTE: IN ADDITION TO ANIMATING THE SYSTEM IN EXAMPLE 2.7.10, A CALL TO THE ')
disp('FUNCTION two_particles_table (SUCH AS')
disp('   my_output = two_particles_table([0:0.1:10], 0.5, 9.81, 0.5, 0, 0, 2) ')
disp('RETURNS 5 COLUMNS CONTAINING t, r_over_L,  d(r_over_L)/dt, theta, and d(theta)/dt')
disp(' TO PLOT r_over_L, FOR EXAMPLE, VERSUS time, USE THE COMMANDS ');
disp('        figure(); plot(my_output(:,1), my_output(:,2)) ')
disp('TO PUT A GRID ON THE PLOT, IF DESIRED, THEN TYPE  xgrid ')

function dy_dt=rhs_of_diff_eqs(t, y)
    // Note:  y(1)=r_over_L, y(2)=d(r_over_L)/dt, y(3)=theta, y(4)=d(theta)/dt
   dy_dt(1)=y(2)
   dy_dt(2)=(1-m2_over_m2plusm1)*y(1)*y(4)^2 -m2_over_m2plusm1*g_over_L
   dy_dt(3)=y(4)                  // dy_dt(2), above, is d2(r_over_L)/dt2; Eq. (2.7.101)
```

```
  dy_dt(4)=-2*y(2)*y(4)/y(1)    // this is d2(theta)/dt2 obtained from Eq. (2.7.10k)
endfunction

initial_conditions=[r_0; rdot_0; theta_0; thetadot_0]
state_variables=ode(initial_conditions, 0, t, rhs_of_diff_eqs), output=[t', state_variables']
// output is a len by 5 matrix, where len is the number of points in the time array t
// column 1: t, column 2: r/L; column 3: d(r/L)/dt; column 4: theta; column 5: d(theta)/dt

xdel(winsid())  // deleting all existing figures
// ANIMATION: We have two bars, O_P1 and O_P2
//     the total normalized length of the system of bars is r/L+(L-r)/L=1
  r=output(:,2), theta=output(:,4)  // these are len by 1 columns
  x_O=zeros(length(t),1), y_O=x_O, z_O=x_O, x_P1=r.*cos(theta), y_P1=r.*sin(theta), z_P1=x_O
      // the above are the ends of bar O_P1
  x_P2=x_O,   y_P2=x_O, z_P2=r-1      // end P2 of bar O_P2
  x_coordinates=[x_P2, x_O, x_P1]; y_coordinates=[y_P2, y_O, y_P1]
  z_coordinates=[z_P2, z_O, z_P1]
  animate3d_nbars(x_coordinates, y_coordinates, z_coordinates)
endfunction
// end of the file two_particles_table.sci
// /////////////////////////////////////////////////////////////////////////////
```

E.2 Matlab Programs

```
function animate_nbars(x_coordinates, y_coordinates)
%        2D animation of any number of connected bars
% Given the lists x_coordinates=[xA,xB,xC, etc] and y_coordinates(yA, yB, yC, etc) of the x
% and y coordinates of several points A, B, C, etc, a call to the function animate_nbars.m
% generates a set of straight lines by connecting the points in the given sequence A, B, C,
% etc, with a black dot at each point. Thus a call such as animate_nbars([xA, xB, xC],
% [yA, yB, yC]), typed in the Matlab command window, creates a figure window with two lines
% AB and BC, with black dots at the ends of each line. When xA, xB,xC, etc, and yA, yB, yC,
% etc, are n by m arrays, an animation is displayed. This  function is called by several
% routines of the type .m (such as ballistics.m, boomerang.m, fourbar_mechanism.m, etc).

% *****************************************************************************************
% NOTE: Comment out this part if you don't want to see this message after you get used to it
disp(' '); disp(' =====>  IMPORTANT MESSAGE:  <=====')
disp(' IF THE SIMULATION RUNS EITHER TOO FAST OR TOO SLOW IN YOUR COMPUTER,')
disp(' change the time_step in the time array t=[t0:time_step:tfinal] of your simulation ')
disp(' ---> EITHER to a smaller value to slow down the animation ')
disp(' ---> OR to a larger value to speed up the animation')
% *****************************************************************************************

[number_of_time_points, len]=size(x_coordinates);

% minimum and maximum value of x_coordinates and y_coordinates:
xcoord_min=min(min(x_coordinates)); xcoord_max=max(max(x_coordinates));
ycoord_min=min(min(y_coordinates)); ycoord_max=max(max(y_coordinates));

% Create a figure for the animation, first deleting it, if it exists
% (to prevent new plots to be superimposed in the same window)
delete( findobj('type','figure','name', 'Animation window') )
% Start by determining the computer monitor size so that the animation window
```

```
% is properly placed and sized in any computer screen.
screen_width_and_height=get(0,'screensize');    % width and height of the screen, in pixels
PC_screen_width =screen_width_and_height(3);     % this is the width of the computer monitor
PC_screen_height=screen_width_and_height(4);     % this is the height of the computer monitor

x_topleft=0.65*PC_screen_width;  % choice for the top left x-coordinate of the figure window
y_topleft=0.17*PC_screen_height; % choice for the top left y-coordinate of the figure window
top_left_coordinates=[x_topleft, y_topleft];
size_of_fig_window = [0.32*PC_screen_width, 0.47*PC_screen_height];
% the above are the width and height chosen for the animation window
% Now, create the figure window.
fig=figure('position',[x_topleft, y_topleft,size_of_fig_window(1),size_of_fig_window(2)], ...
        'name', 'Animation window', 'color', 'white');

fig_axis = axes('Xlim', [xcoord_min, xcoord_max], 'Ylim', [ycoord_min, ycoord_max], ...
        'dataaspectratio', [1,1,1], 'NextPlot', 'add');    % ... is for continuing a line
  % Note: remove the    'NextPlot', 'add'   (and the comma before it) to see what it does

anim=plot([0 0],[0 0], 'linewidth', 2,  'marker', '.', 'markersize', 40, ...
        'MarkerFaceColor', 'black', 'EraseMode','xor');
  % Note: With EraseMode set to xor, Matlab erases the object that was plotted before
  %   redrawing it. A setting to   none   would not do the erasing.

% The following loop changes the xdata and ydata handles of the object anim
for itime=1:number_of_time_points
   set(anim,'xdata', x_coordinates(itime,:),  'ydata', y_coordinates(itime,:));
   drawnow    % type   help drawnow   to see what it does
end
% end of the file animate_nbars.m
% %%%%%%%%%%%%%%%%%%%%%%%%%%%%%%%%%%%%%%%%%%%%


function animate3d_nbars(x_coordinates, y_coordinates, z_coordinates)
%          3D animation of any number of connected bars
% When given the lists x_coordinates=[xA, xB, xC, etc], y_coordinates(yA, yB, yC, etc),
% and z_coordinates(zA, zB, zC, etc)
% of the x-y-z coordinates of several points A, B, C, etc, a call to the function
% animate3d_nbars.m generates a set of straight lines by connecting the points in the
% given sequence A, B, C, etc, with a black dot at each point. Thus a call such as
% animate3d_nbars([xA, xB, xC], [yA, yB, yC], [zA, zB, zC]), typed in the Matlab command
% window, creates a figure window with two lines AB and BC, with black dots at the ends
% of each line. When xA, xB, xC, etc; yA, yB, yC, etc; and zA, zB, zC, etc, are n by m
% arrays an animation is displayed.
%  This function is called by top_dynamics.m and two_particles_table.m
% to animate the 3D motion. See those programs and run them to watch the motion.

% ****************************************************************************
% NOTE: Comment out this part if you don't want to see this message after you get used to it
disp(' '); disp('  =====>  IMPORTANT MESSAGE:  <=====')
disp(' IF THE SIMULATION RUNS EITHER TOO FAST OR TOO SLOW IN YOUR COMPUTER,')
disp(' change the time_step in the time array t=[t0:time_step:tfinal] of your simulation ')
disp(' ---> EITHER to a smaller value to slow down the animation ')
disp(' ---> OR to a larger value to speed up the animation')
% ****************************************************************************
```

```
[number_of_time_points, len]=size(x_coordinates):
% minimum and maximum value of x_coordinates, y_coordinates, and z_coordinates:
xcoord_min=min(min(x_coordinates));  xcoord_max=max(max(x_coordinates));
ycoord_min=min(min(y_coordinates));  ycoord_max=max(max(y_coordinates));
zcoord_min=min(min(z_coordinates));  zcoord_max=max(max(z_coordinates));

% Create a figure for the animation, first deleting it, if it exists
% (to prevent new plots to be superimposed in the same window)
delete( findobj('type','figure','name', 'Animation window') )
% Start by determining the computer monitor size so that the animation window
% is properly placed and sized in any computer screen.
screen_width_and_height=get(0,'screensize');  % width and height of the screen, in pixels
PC_screen_width =screen_width_and_height(3);  % this is the width of the computer monitor
PC_screen_height=screen_width_and_height(4);  % this is the height of the computer monitor

x_topleft=0.65*PC_screen_width; % choice for the top left x-coordinate of the figure window
y_topleft=0.17*PC_screen_height; % choice for the top left y-coordinate of the figure window
top_left_coordinates=[x_topleft, y_topleft];
size_of_fig_window = [0.32*PC_screen_width, 0.47*PC_screen_height];
% the above are the width and height chosen for the animation window
% Now, create the figure window.
fig=figure('position', [x_topleft,y_topleft,size_of_fig_window(1),size_of_fig_window(2)], ...
           'name','Animation window', 'color', 'white');

azimuth=-240; elevation=30;  % these are the azimuth (from the -y axis) and elevation
% (from the xy plane) viewing angles in degrees
fig_axis = axes('Xlim', [xcoord_min, xcoord_max], 'Ylim', [ycoord_min, ycoord_max], ...
    'Zlim', [zcoord_min, zcoord_max], 'dataaspectratio', [1, 1, 1], 'NextPlot','add', ...
    'View', [azimuth, elevation], 'box', 'on');  % ... is for continuing a line
anim=plot3([0 0 0], [0 0 0], [0 0 0], 'linewidth', 2, 'marker', '.', 'markersize', 40, ...
       'MarkerFaceColor', 'black', 'EraseMode','xor');
xlabel('x'); ylabel('y')

for itime=1:number_of_time_points    % this is the animation loop
   set(anim,'xdata', x_coordinates(itime,:),  'ydata', y_coordinates(itime,:),  ...
            'zdata', z_coordinates(itime,:) );
   drawnow;
end
% end of the file animate3d_nbars.m
% %%%%%%%%%%%%%%%%%%%%%%%%%%%%%%%%%%%%%%%%%%%%%%


function output = ballistics(t, d2x_dt2, d2y_dt2,  x_0, xdot_0, y_0, ydot_0)
% Example 2.5.8
% Example of usage:
% my_output = ballistics([0:0.002:2.5], 0, -9.81,  0, 10, 0, 10);

disp('NOTE: The output of a call to the function ballistics.m is a')
disp('n by 5 matrix, where n is the number of points in the time array t')
disp('1st column: t, 2nd column: x, 3rd column: dx/dt, 4th column: y, 5th column: dy/dt')
disp(' ')

initial_conditions=[x_0; xdot_0; y_0; ydot_0];
[t, var]=ode45(@rhs_of_diff_eqs, t, initial_conditions, [], d2x_dt2, d2y_dt2);
```

```
output=[t, var];
% output is a n by 5 matrix, where n is the number of points in the time array t
% column 1: t, column 2: x, column 3: dx/dt, column 4: y, column 5: dy/dt

% ANIMATION (the "bar" is a point):
x_coordinates=output(:,2); y_coordinates=output(:,4);
animate_nbars(x_coordinates, y_coordinates);

function dvar_dt=rhs_of_diff_eqs(t, var, d2x_dt2, d2y_dt2)
    % Note:  var(1)=x, var(2)=dx/dt, var(3)=y, var(4)=dy/dt
  dvar_dt=[var(2); d2x_dt2; var(4); d2y_dt2];
% end of the file ballistics.m
% %%%%%%%%%%%%%%%%%%%%%%%%%%%%%%%%%%%%%%%%%%%%


function output = boomerang(t, g,  x_0, xdot_0, y_0)
% Example 7.9.1; using the state variables var(1)=x, var2=dx/dt, var(3)=y
% Example of usage [ using the initial conditions x(0)=0, dx_dt(0)=5*cos(pi/6),  y(0)=0 ]
% my_output = boomerang([0:0.002:6], 9.81,  0, 5*cos(pi/6), 0);

initial_conditions=[x_0; xdot_0; y_0];
[t, y]=ode45(@rhs_of_diff_eqs, t, initial_conditions, [], g);
output=[t, y];
% output is a n by 4 matrix, where n=length(t) is the number of points in the array t
% column 1: t, column 2: x, column 3: dx/dt; column 4: y

x=output(:,2); y=output(:,4);

close all; % close (i.e., delete) all figure windows
figure('color', 'white');  plot(t, x, t, y, '--'); grid;  % plot of x and y versus t
xlabel('t', 'fontsize', 12); ylabel('solid line: x   ;   dashed line: y', 'fontsize', 12);
title('Example 7.9.1','fontsize', 12);

figure('color', 'white');  plot(x, y); grid          % plot of y versus x
xlabel('x', 'fontsize', 12); ylabel('y', 'fontsize', 12);
title('Example 7.9.1','fontsize', 12);

animate_nbars(x, y);  % this call animates the motion

function dvar_dt=rhs_of_diff_eqs(t, var, g)
% Note:  var(1)=x, var(2)=d(x)/dt, var(3)=y;  The diff. eqs. are Eqs. (7.9.1j) and (7.9.1i)
    f1=0.5*sin(t);  df1_dt=0.5*cos(t);     f2=5*sin(pi/6)-0.85*t;  df2_dt=-0.85;
    dvar_dt=[var(2); -(var(2)*df1_dt+df2_dt+g)*f1/(1+f1^2); f1*var(2)+f2];
% end of the file boomerang.m
% %%%%%%%%%%%%%%%%%%%%%%%%%%%%%%%%%%%%%%%%%%%%


function output=example6_7_1(tau, omegasqL_over_g, R_over_L, theta_0, thetadot_0)
% Example 6.7.1 analyzed in Section 6.7
% Example of usage ( using the initial conditions theta(0)=30*pi/180 rad, thetadot(0)=0 );
% my_output = example6_7_1([0:0.1:10], 1.5, 1/3, 30*pi/180, 0);

initial_conditions=[theta_0; thetadot_0];
[tau, y]=ode45(@rhs_of_diff_eqs, tau, initial_conditions, [ ], omegasqL_over_g, R_over_L);
output=[tau, y];
% output is a n by 3 matrix, where n is the number of points in the array tau
```

```
% column 1: tau;   column 2: theta;   column 3: d(theta)/d(tau)
theta=output(:,2); dtheta_dtau=output(:,3); cos_theta=cos(theta); sin_theta=sin(theta);

% Normalized omponents A1, A2, and A3 of the reaction force at point A:
 A1=-cos_theta-(dtheta_dtau.^2)/2-omegasqL_over_g*(R_over_L+sin_theta/2).*sin_theta;
 A2=sqrt(omegasqL_over_g)*dtheta_dtau.*cos_theta;
 A3=(sin_theta-R_over_L*omegasqL_over_g*cos_theta)/4;
% Magnitude of the normalized reaction force and moment at point A:
 reaction_force_at_A  = sqrt(A1.^2+A2.^2+A3.^2);   reaction_moment_at_A = 2*A2/3;

% Plots:
 close all  % closing (i.e., deleting) all existing figures
 figure('color', 'white'); plot(tau, theta*180/pi);   grid
 xlabel(['Nondimensional time ', '\tau'], 'fontsize', 12),
 ylabel(['\theta', ' (degrees)'], 'fontsize', 12)

 figure('color', 'white'); plot(tau, reaction_force_at_A); grid
 xlabel(['Nondimensional time ', '\tau'], 'fontsize', 12)
 ylabel('(A^2_1+A^2_2+A^2_3)^{1/2}/(mg)', 'fontsize', 12)

 figure('color', 'white'); plot(tau, reaction_moment_at_A), grid
 xlabel(['Nondimensional time ', '\tau'], 'fontsize', 12)
 ylabel('M_3/(mgL)', 'fontsize', 12)

 function dy_dtau=rhs_of_diff_eqs(tau, y, omegasqL_over_g, R_over_L)
% Note:  y(1)=theta, y(2)=d(theta)/d(tau), where tau=t*sqrt(g/L)
 sin_theta=sin(y(1));
 dy_dtau(1)=y(2);      % the next expression is d2x_dt2 from Eq. (6.7.1q) with b=0
 dy_dtau(2)=(3*R_over_L/2+sin_theta)*omegasqL_over_g*cos(y(1))-3*sin_theta/2;
 dy_dtau=[dy_dtau(1); dy_dtau(2)];
% end of the file example6_7_1.m
% %%%%%%%%%%%%%%%%%%%%%%%%%%%%%%%%%%%%%%%%%%%%%%


 function output=foucault_pendulum(t,omega_times_sin_lambda,g_over_L,x_0, xdot_0,y_0,ydot_0)
% Animation of the Foucault pendulum analyzed in Chapter 8
% Example of usage [with the initial conditions x(0)=0.01, x_dot(0) =0, y(0)=0, y_dot(0)=0]:
% my_output = foucault_pendulum([0:0.001:30], 0.0727, 0.981, 0.01, 0,  0, 0);

disp(' ')
disp(' NOTE: THE FUNCTION foucault_pendulum RETURNS 5 COLUMNS CONTAINING')
disp(' t, x, dx/dt, y, and dy/dt:')
disp(' ')

initial_conditions=[x_0; xdot_0; y_0; ydot_0];
[t, y]=ode45(@rhs_of_diff_eqs, t, initial_conditions, [], omega_times_sin_lambda, g_over_L);

close all
figure('color', 'white'); plot(y(:,1), y(:,3));   grid         % plot of y versus x
xlabel('x', 'fontsize', 12), ylabel('y', 'fontsize', 12)
title('Foucault pendulum path','fontsize', 12)

output=[t, y];
% output is a n by 5 matrix, where n is the number of points in the time array t;
% column 1: t, column 2: x, column 3: dx/dt; column 4: y; column 5: dy/dt
```

```
x_coordinates=y(:,1);  y_coordinates=y(:,3);
animate_nbars(x_coordinates, y_coordinates);     % ANIMATION

function dy_dt=rhs_of_diff_eqs(t, y,  omega_times_sin_lambda, g_over_L)
    % Note:  y(1)=x, y(2)=d(x)/dt, y(3)=y, y(4)=d(y)/dt
  dy_dt(1)=y(2);
  dy_dt(2)=2*omega_times_sin_lambda*y(4)-g_over_L*y(1);     % this is d2(x)/dt2
  dy_dt(3)=y(4);
  dy_dt(4)=-2*omega_times_sin_lambda*y(2)-g_over_L*y(3);    % this is d2(y)/dt2
  dy_dt=[dy_dt(1); dy_dt(2); dy_dt(3); dy_dt(4)];
% end of the file foucault_pendulum.m
% %%%%%%%%%%%%%%%%%%%%%%%%%%%%%%%%%%%%%%%%%%%%


function output = fourbar_kinematics(t, L1, L2, L3, L4, theta2, dtheta2_dt, ...
            d2theta2_dt2, solution_number)   % ... is for continuing a command line
% Kinematics of the four-bar mechanism in Section 3.2.3
% Animation is done in the program  fourbar_mechanism.m
% Example of usage [copy the next three lines, without the %, to the Matlab command window]:
% t=[0:0.1:5]; theta2=pi/6+1.5*pi*sin(2*t);  dtheta2_dt=3*pi*cos(2*t);  ...
%                             d2theta2_dt2=-6*pi*sin(2*t);
% my_output=fourbar_kinematics(t, 1, 0.45, 0.8, 0.7, theta2,  dtheta2_dt, d2theta2_dt2, 1);
%  NOTE: any name can be used for the variable to the left of the equal sign in line 1
%  The variable output is a n by 10 column, where n is the number of points in the array t
%
% The mechanism is defined by four points A, B, C, and D, whose x-y coordinates are:
% point A: (0,0),   point B: ( L2*cos(theta2), L2*sin(theta2) )
% point C( L1+L4*cos(theta4), L4*sin(theta4) ),   point D: (L1, 0 )
% The angles theta3 and theta4 are theta3=output(:,2) and theta4=output(:,3)
% The fourth bar is a fictitious one between the fixed points A and D.

len=length(t);  temp=ones(1,len); % This assures that theta2, dtheta2_dt, and d2theta2_dt2
        % are rows with the same length as the row t
theta2=theta2.*temp; dtheta2_dt=dtheta2_dt.*temp; d2theta2_dt2=d2theta2_dt2.*temp;

if solution_number ==1 | solution_number == 2
    cos_theta2=cos(theta2);  L=sqrt( L1^2+L2^2-2*L1*L2*cos_theta2 ); % L from Eq. (3.2.10)
    sin_theta2=sin(theta2);  sin_alpha=L2*sin_theta2./L;  % need to use ./, instead of /
    if abs(sin_alpha) <=1
        cos_alpha = (L1-L2*cos(theta2))./L;        % from the equations of (3.2.9)
        angle_alpha=atan2(sin_alpha, cos_alpha);    % angle angle_alpha
        cos_beta=(L3^2+L.^2-L4^2)./(2*L*L3);
        if abs(cos_beta)<=1
            sin_beta=(3-2*solution_number)*sqrt(1-cos_beta.^2); % Eqs. (3.2.14) and (3.2.15)
            angle_beta= atan2(sin_beta, cos_beta);
            theta3=angle_beta-angle_alpha;          % angle theta3 from Eq. (3.2.16)
            sin_gamma=L3*sin_beta/L4;               % sin(gamma) from Eq. (3.12.11)
            if abs(sin_gamma)<=1
                cos_gamma=(L-L3*cos_beta)/L4;        %  cos(gamma) from Eq. (3.12.12)
                angle_gamma= atan2(sin_gamma, cos_gamma);
                theta4=pi-angle_alpha-angle_gamma;
            else disp('THE MECHANISM CANNOT OPERATE because |sin(gamma)|>1'),output=[],return
                end
            else disp('THE MECHANISM CANNOT OPERATE because |cos(beta)|>1'), output=[],return
                end
        else disp('THE MECHANISM CANNOT OPERATE because |sin(alpha)|>1'), output=[], return
```

```
    end

% Angular velocities dtheta3_dt=d(theta3)/dt and dtheta4_dt=d(theta4)/dt from Eq. (3.2.5):
den=L3*L4*sin(theta4-theta3); % this is the determinant of the 2 by 2 matrix in Eq. (3.2.5)
temp=-L2*dtheta2_dt./den;
% Solution of Eq. (3.2.5):
dtheta3_dt=L4*temp.*sin(theta4-theta2);   dtheta4_dt=L3*temp.*sin(theta3-theta2);

% Elements a11, a12, a21, and a22 of the inverse of the matrix in Eq. (3.2.5):
cos_theta3=cos(theta3); sin_theta3=sin(theta3);
cos_theta4=cos(theta4); sin_theta4=sin(theta4);
a11=-L4*cos_theta4./den; a12=L4*sin_theta4./den;
a21=-L3*cos_theta3./den;  a22=L3*sin_theta3./den;

% Angular accelerations d2theta3_dt2 and d2theta4_dt2 from Eq. (3.2.8):
v1 =  L4*(dtheta4_dt.^2).*cos_theta4 -L3*(dtheta3_dt.^2).*cos_theta3 ...
            - L2*( d2theta2_dt2.*sin_theta2 + (dtheta2_dt.^2).*cos_theta2 );
v2 = -L4*(dtheta4_dt.^2).*sin_theta4 +L3*(dtheta3_dt.^2).*sin_theta3 ...
            - L2*( d2theta2_dt2.*cos_theta2 - (dtheta2_dt.^2).*sin_theta2 );
d2theta3_dt2=a11.*v1+a12.*v2; d2theta4_dt2=a21.*v1+a22.*v2;
% d2theta3_dt2 and d2theta4_dt2 are the solutions of Eq. (3.2.8)

% Store t, all the angles, angular velocities, and angular accelerations in the variable
% output so that they are available to the user:
output=[t; theta2; theta3; theta4; dtheta2_dt; dtheta3_dt; dtheta4_dt; ...
                    d2theta2_dt2; d2theta3_dt2; d2theta4_dt2]';

    else disp(' -----> SOLUTION_NUMBER HAS TO BE  1 or 2 <----- '),    output=[], return
end
% end of the file fourbar_kinematics.m
% %%%%%%%%%%%%%%%%%%%%%%%%%%%%%%%%%%%%%%%%%%%%%%%%%%


function output = fourbar_mechanism(t, L1, L2, L3, L4, theta2, dtheta2_dt, ...
            d2theta2_dt2, solution_number)   % ... is for continuing a command line
% Animation of the four-bar mechanism in Section 3.2
% Example of usage
% [copy the next three lines, without the %, to the Matlab command window]:
% t=[0:0.001:5]; theta2=pi/6+1.5*pi*sin(2*t); dtheta2_dt=3*pi*cos(2*t);  ...
%                                             d2theta2_dt2=-6*pi*sin(2*t);
% my_output=fourbar_mechanism(t, 1, 0.45, 0.8, 0.7, theta2,  dtheta2_dt, d2theta2_dt2, 1);
%    NOTE: any name can be used for the variable to the left of the equal sign in line 1
%
% To slow down the animation, choose a smaller time step for the input array t
% Conversely, to speed up the animation, choose a larger time step
% For example, the array t=[0:0.001:5] yields a slower animation than the array t=[0:0.01:5]
%
% IN ADDITION TO ANIMATING THE MECHANISM, SUCH A CALL TO THE FUNCTION fourbar_mechanism.m
% GENERATES A len (WHERE len IS THE NUMBER OF POINTS IN THE ARRAY t) BY 10 MATRIX CALLED
% my_output, WHOSE COLUMNS ARE    t, theta2, theta3, theta4, dtheta2_dt, dtheta3_dt,
% dtheta4_dt, d2theta2_dt2, d2theta3_dt2, d2theta4_dt2
% my_output(:,3), my_output(:,4)*180/pi, FOR EXAMPLE, GIVE theta3 AND theta4 (IN DEGREES).
%  TO PLOT theta4 VERSUS TIME t, FOR EXAMPLE, TYPE, IN THE MATLAB COMMAND WINDOW,
% THE COMMANDS       figure(); plot(my_output(:,1), my_output(:,4)*180/(2*pi) )
% To put a grid on the plot, type   grid
```

```
disp(' MESSAGE FROM THE FUNCTION fourbar_mechanism.m:')
disp('IN ADDITION TO ANIMATING THE MECHANISM, A CALL TO THE FUNCTION fourbar_mechanism')
disp('RETURNS 10 COLUMNS CONTAINING t, theta2, theta3, theta4, dtheta2_dt, dtheta3_dt, ')
disp(' dtheta4_dt, d2theta2_dt2, d2theta3_dt2, d2theta4_dt2')

 output=fourbar_kinematics(t,L1,L2,L3,L4,theta2,dtheta2_dt,d2theta2_dt2,solution_number);
 theta2=output(:,2);  theta4=output(:,4);
 len=length(t);  temp=ones(len,1);
 xA= zeros(len,1); yA=xA; xB=L2*cos(theta2); yB=L2*sin(theta2);  % points A and B of bar AB
 xC=L1+L4*cos(theta4); yC=L4*sin(theta4);      % point C of bar BC
 xD=L1*temp;  yD=yA;    % point D

 close all  % deleting all existing figures
 x_coordinates=[xA, xB, xC, xD]; y_coordinates=[yA, yB, yC, yD];
 animate_nbars(x_coordinates, y_coordinates)  % ANIMATION (done by the program animate_nbars)
% end of the file fourbar_mechanism.m
% %%%%%%%%%%%%%%%%%%%%%%%%%%%%%%%%%%%%%%%%%%%%%%%%%


function  output = fourbarforces(t, L1, L2, L3, L4, theta2, dtheta2_dt, d2theta2_dt2,  ...
                    solution_number,  m2, m3, m4)
% Simulation of the problem in Section 5.6 of Chapter 5
%  Examples of usage [copy the next 3 lines to the Matlab command window, without the %]:
% t=[0:0.1:5];theta2=pi/6+1.5*pi*sin(2*t);dtheta2_dt=3*pi*cos(2*t);  ...
%            d2theta2_dt2=-6*pi*sin(2*t);
% output=fourbarforces(t,1,0.45,0.8,0.7,theta2,dtheta2_dt,d2theta2_dt2,1,0.15,0.225,0.1875);
%
% THE FOLLOWING CALL GIVES THE SAME RESULTS SHOWN IN SECTION 5.6 OF CHAPTER 5:
% output=fourbarforces(1.e-6,5.1592, 2,3,2.5, pi/2-0.5902,-3*pi,0,1,0.15,0.225,0.1875);
%
% SUCH CALLS TO THE FUNCTION fourbarforces.m GENERATE A len (WHICH IS THE NUMBER OF
% POINTS IN THE SPECIFIED ARRAY t) BY 10 MATRIX CALLED output, WHOSE COLUMNS ARE
%      t, Mext, Ar2, An2, Dr4, Br2, Bn2, Cr4, Dn4, Cn4
%
%  output(:,3), AND  output(:,4), FOR EXAMPLE, GIVE THE REACTIONS Ar2 AND An2
%  TO PLOT Ar2 VERSUS TIME t, FOR EXAMPLE, TYPE, IN THE MATLAB COMMAND WINDOW,
%  THE COMMANDS          figure();  plot( output(:,1), output(:,3) )

disp(' ')
disp(' NOTE: THE 10 VARIABLES t, Mext, Ar2, An2, Dr4, Br2, Bn2, Cr4, Dn4, and Cn4, AT THE')
disp('SPECIFIED TIMES, ARE STORED IN A 10-COLUMN MATRIX WHOSE NAME IS THE ONE YOU USED IN')
disp('THE CALL TO THE FUNCTION fourbarforces.m (SUCH AS output=fourbarforces(....) )')
disp(' TO PLOT Ar2, FOR EXAMPLE, VERSUS time, USE THE COMMANDS ');
disp('       figure( ); plot(output(:,1), output(:,3))')
disp('TO PUT A GRID ON THE PLOT, IF DESIRED, THEN TYPE grid')
disp(' --> To see an animation of the motion, use the function fourbar_mechanism <---')
disp(' ')

% First call four_bar_kinematics to get theta2, theta3, theta4, dtheta2_dt, dtheta3_dt,
%                  dtheta4_dt, d2theta2_dt2, d2theta3_dt2, and d2theta4_dt2
fourbar=fourbar_kinematics(t,L1,L2,L3,L4,theta2,dtheta2_dt,d2theta2_dt2,solution_number);

if isempty(fourbar), output=[]; return, end

theta2=fourbar(:,2); theta3=fourbar(:,3); theta4=fourbar(:,4);
dtheta2_dt=fourbar(:,5); dtheta3_dt=fourbar(:,6); dtheta4_dt=fourbar(:,7);
```

```
d2theta2_dt2=fourbar(:,8); d2theta3_dt2=fourbar(:,9); d2theta4_dt2=fourbar(:,10);
% The variables theta2, dtheta2_dt, ..., d2theta4_dt2 are columns

% We can now calculate the support reactions at each time instant specified in the array t
Cn4=m4*L4*d2theta4_dt2/3;  Dn4=Cn4/2;  % From Eqs. (5.6.10) and (5.6.11)

% Now, determine the reactions Br2, Bn2, and Cr4 from Eq. (5.6.12). We will use the
% analytical expression for the inverse of the matrix that appears in Eq. (5.6.12),
% which is obtained after some algebra that can be done by hand. Its elements are
% a11, a12, a13, etc, a33, shown below.

cos32=cos(theta3-theta2);     sin32=sin(theta3-theta2);
cos42=cos(theta4-theta2);     sin42=sin(theta4-theta2);
cos43=cos(theta4-theta3);     sin43=sin(theta4-theta3);
cos432=cos(theta4+theta3-2*theta2); sin432=sin(theta4+theta3-2*theta2);
den=4*sin43;  % this is the determinant of the matrix in Eq. (5.6.12).
% The elements of the inverse of that matrix are:
a11=(sin432+3*sin43)./den;  a12=-(cos43+cos432)./den;  a13=2*cos42./den;
a21=(cos43-cos432)./den;    a22=(3*sin43-sin432)./den;  a23=2*sin42./den;
a31=-2*sin32./den;          a32=2*cos32./den;           a33=-2*ones(length(t),1)./den;

dtheta2_dt_sq=(dtheta2_dt.^2); dtheta3_dt_sq=(dtheta3_dt.^2); dtheta4_dt_sq=(dtheta4_dt.^2);
% Elements of the column array seen in Eq. (5.6.12):
v1= m3*L3*(  dtheta3_dt_sq.*cos32 + d2theta3_dt2.*sin32 )/2 ...
                + m3*L2*dtheta2_dt_sq + Cn4.*sin42;  % ... is for continuing a command line
v2=-m3*L3*( -dtheta3_dt_sq.*sin32 + d2theta3_dt2.*cos32 )/2 ...
                - m3*L2*d2theta2_dt2  - Cn4.*cos42;
v3= m3*L3*d2theta3_dt2/6 + Cn4.*cos43;

% We then have for the reactions Br2, Bn2, and Cr4, from Eq. (5.6.12):
Br2=a11.*v1+a12.*v2+a13.*v3;  Bn2= a21.*v1+a22.*v2+a23.*v3;  Cr4= a31.*v1+a32.*v2+a33.*v3;

% The remaining reactions are:
Mext= m2*L2^2*d2theta2_dt2/3 - L2*Bn2;     % From Eq. (5.6.3)
Ar2= -m2*L2*dtheta2_dt_sq/2-Br2;           % From Eq. (5.6.1)
An2= m2*L2*d2theta2_dt2/2-Bn2;             % From Eq. (5.6.2)
Dr4= -m4*L4*dtheta4_dt_sq/2-Cr4;           % From Eq. (5.6.4)

len=length(t);
disp(['The reactions at t= ', num2str(t(len)), ' are: '])
disp(['Mext= ',num2str(Mext(len)),', Ar2= ',num2str(Ar2(len)), ',An2= ', num2str(An2(len))] )
disp(['Dr4= ',num2str(Dr4(len)),  ', Br2= ',num2str(Br2(len)), ',Bn2= ', num2str(Bn2(len))] )
disp(['Cr4= ',num2str(Cr4(len)),  ', Dn4= ',num2str(Dn4(len)), ',Cn4= ', num2str(Cn4(len))] )

   output=[t', Mext, Ar2, An2, Dr4, Br2, Bn2, Cr4, Dn4, Cn4];
%  end of the file fourbarforces.m
% %%%%%%%%%%%%%%%%%%%%%%%%%%%%%%%%%%%%%%%%%%%%%%%%%


function output=fourbar_mech_dynamics(t, L1, L2, L3, L4, m2, m3, m4, theta2_0, ...
        dtheta2_dt_0, solution_number)  % ... is for continuing a line
% Simulation for Example 7.9.3 in Section 7.9 of Chapter 7
% Example of usage [copy the line below, without the %, to the Matlab command window]:
% my_output=fourbar_mech_dynamics([0:0.001:5], 1, 0.4, 0.8, 0.7, 3.75, 3, 3, pi/6, 0, 1);
%
% NOTE: for these values of L1, L3, and L4, the mechanism locks shortly after t=1.2 if
```

```
% L2 > approximately 0.5.  If the mechanism locks, a standard error message will appear
% on the computer screen, and the computation ends. If this occurs, ignore the message
% but redesign the mechanism by changing the length of at least one of its bars to
% prevent it from locking.
%
% IN ADDITION TO ANIMATING THE MECHANISM, SUCH A CALL TO THE FUNCTION
% fourbar_mech_dynamics.m GENERATES A len (WHICH IS THE NUMBER OF POINTS
% IN THE SPECIFIED ARRAY t) BY 10 MATRIX CALLED output (OR ANY OTHER NAME
% YOU WANT TO USE BEFORE THE EQUAL SIGN SEEN IN LINE 1), WHOSE COLUMNS
% ARE  t, theta2, theta3, theta4, d(theta2)/dt, d(theta3)/dt, d(theta4)/dt, d2(theta2)/dt2,
%           d2(theta3)/dt2, d2(theta4)/dt2
%
%  my_output(:,4), my_output(:,2)*180/pi, FOR EXAMPLE, GIVE theta4 (IN RAD) AND theta2
% (IN DEGREES). TO PLOT theta4 IN DEGREES VERSUS TIME t, FOR EXAMPLE, TYPE,
%  IN THE MATLAB COMMAND WINDOW, THE COMMANDS
%           figure();  plot(my_output(:,1), my_output(:,4)*180/(2*pi) )
%   To put a grid on the plot, if desired, type the command      grid
%
% In this problem theta2 and d(theta2)/dt are not given; only their initial conditions
% theta2_0 and dtheta2_dt_0. The system is driven by a motor with a torque M_motor that is
% a specified function of time in the function M_motor(t), which can be changed by the user.
%
% The variables theta2 and d(theta2)/dt are determined by integrating a second-order
% differential equation obtained from the first row of the solution of Eq. (7.9.3n). The
% integration is done by a  call to the Matlab function ode45 [which calls the function
% rhs_diff_eqs(t, y) that specifies the right hand side of the equations
%   dtheta2/dt= ---, and d(dtheta2/dt)/dt= ---].

if solution_number ~=1 & solution_number ~= 2,
        disp(' -----> SOLUTION_NUMBER HAS TO BE  1 or 2 <----- '), output=[]; return,
end

disp(' '),    disp('The function fourbar_mech_dynamics called fourbar_mechanism.')

global d2theta2_dt2
   % Since this is needed in the function rhs_diff_eqs, Matlab requires it to be
   % declared global here and in that function. Otherwise the variable d2theta2_dt2
   % would not be known by the function rhs_diff_eqs. This is not the case with Scilab.

   % Integration of the second-order diff. eq. for theta2. An arbitrary value is used for
% d2theta2_dt2 because one is needed in a call to fourbar_mechanism
d2theta2_dt2=0;    % this is an arbitrary value
options=odeset('RelTol', 1.e-6, 'AbsTol', 1.e-6);
% NOTE: with options=[ ] the solution is not accurate
[t, y] = ode45(@rhs_diff_eqs, t, [theta2_0; dtheta2_dt_0], options, L1, L2, L3, L4, ...
                                  m2, m3, m4, solution_number);
theta2=y(:,1)';   dtheta2_dt=y(:,2)';  t=t';
   % NOTE: transposed the variables in [t, y] to obtain rows because rows, instead of columns,
   % are needed for these quantities in the call to the function fourbar_mechanism a few lines
   % later. Now, determine the row of valid values for d2theta2/dt2. After that is done,
   % call fourbar_mechanism to complete the solution and animate the motion
   len=length(theta2); temp=[];
   for k=1:len,
       temp_prev= ang_acceleration(t(k), L1, L2, L3, L4,  m2, m3, m4, theta2(k), ...
                                       dtheta2_dt(k), solution_number);
       temp=[temp, temp_prev];   % this is a 1 by len row
```

```
    end
    d2theta2_dt2=temp;
    output=fourbar_mechanism(t,L1,L2,L3,L4,theta2,dtheta2_dt,d2theta2_dt2,solution_number);

disp(' ')
disp('MESSAGE FROM THE FUNCTION fourbar_mech_dynamics.m:')
disp('THE 10 COLUMN MATRIX GENERATED BY THE FUNCTION fourbar_mech_dynamics.m')
disp(' CONTAINS:  t, theta2, theta3, theta4, dtheta2_dt, dtheta3_dt, ')
disp('   dtheta4_dt, d2theta2_dt2, d2theta3_dt2, d2theta4_dt2 ')

% The following are the three functions needed for the calculations involved
% with this problem:

function M_motor_of_t= M_motor(t),  M_motor_of_t=pi/6+1.5*pi*sin(2*t);
% NOTE: change, if desired, the pi/6+1.5*pi*sin(2*t) to any other expression

function dy_dt = rhs_diff_eqs(t, y, L1, L2, L3, L4, m2, m3, m4, solution_number)
    % state variables: y(1)=theta2,  y(2)=dtheta2_dt
  theta2=y(1);  dtheta2_dt=y(2);
  d2theta2_dt2=ang_acceleration(t,L1,L2,L3,L4,m2,m3,m4,theta2,dtheta2_dt,solution_number);
  dy_dt=[y(2); d2theta2_dt2];


function d2theta2_dt2 = ang_acceleration(t, L1, L2, L3, L4,  m2, m3, m4, theta2, ...
                                         dtheta2_dt, solution_number)
global d2theta2_dt2 % In Matlab, this declaration is needed for passing d2theta2_dt2
                    % from the main program to here
  fourbar=fourbar_kinematics(t,L1,L2,L3,L4,theta2,dtheta2_dt,d2theta2_dt2,solution_number);
    % here we have valid values for theta2, theta3, theta4, d(theta2)/dt, d(theta3)/dt, and
    % d(theta4)/dt; ignore d2theta3_dt2 and d2theta4_dt2 obtained by fourbar_kinematics
%
  theta2=fourbar(:,2);  theta3=fourbar(:,3);  theta4=fourbar(:,4);
  sin_theta2=sin(theta2);  sin_theta3=sin(theta3);  sin_theta4=sin(theta4);
  cos_theta2=cos(theta2);  cos_theta3=cos(theta3);  cos_theta4=cos(theta4);
  sin43=sin(theta4-theta3);  cos43=cos(theta4-theta3);
  dtheta2_dt=fourbar(:,5);  dtheta3_dt=fourbar(:,6);  dtheta4_dt=fourbar(:,7);
  dtheta2_dt_sq=dtheta2_dt^2; dtheta3_dt_sq=dtheta3_dt^2; dtheta4_dt_sq=dtheta4_dt^2;
%
% angular acceleration d2theta2_dt2=d2(theta2)/dt2 from Eq. (7.9.3n)
  M=[m2*L2^2/3,          0,                   0,            L2*sin_theta2,  -L2*cos_theta2;
     0,                  m3*L3^2/3,      -m3*L3*L4*cos43/2, L3*sin_theta3,  -L3*cos_theta3;
     0,             -m3*L3*L4*cos43/2,   (m4/3+m3)*L4^2,   -L4*sin_theta4,   L4*cos_theta4;
     L2*sin_theta2,   L3*sin_theta3,     -L4*sin_theta4,    0,               0;
     L2*cos_theta2,   L3*cos_theta3,     -L4*cos_theta4,    0,               0];
  Minv=inv(M);   % M is the matrix seen in Eq. (7.9.3n)), and Minv is its inverse
%
  v1= M_motor(t);   v2=-m3*L3*L4*dtheta4_dt_sq*sin43/2;  v3=m3*L3*L4*dtheta3_dt_sq*sin43/2;
  v4= L4*dtheta4_dt_sq*cos_theta4 -L3*dtheta3_dt_sq*cos_theta3 -L2*dtheta2_dt_sq*cos_theta2;
  v5= -L4*dtheta4_dt_sq*sin_theta4 +L3*dtheta3_dt_sq*sin_theta3 +L2*dtheta2_dt_sq*sin_theta2;
  d2theta2_dt2=Minv(1,1)*v1 +Minv(1,2)*v2 +Minv(1,3)*v3 +Minv(1,4)*v4 +Minv(1,5)*v5;
% end of the file fourbar_mech_dynamics.m
% %%%%%%%%%%%%%%%%%%%%%%%%%%%%%%%%%%%%%%%%%%%%%%%%%


function  output=geneva_wheel(t, L, r1, theta1, dtheta1_dt, d2theta1_dt2)
% Animation of the geneva wheel mechanism (if a bar is always engaged) in Section 3.3
```

```
% To slow down the animation, choose a smaller time step for the input array t
% Conversely, to speed up the animation, choose a larger time step
% For example, the array t=[0:0.001:5] yields a slower animation than the array t=[0:0.01:5]
%
% Example of usage [copy the next two lines, without the %, to the Matlab command window]:
% t=[0:0.001:5]; theta1=pi/2+3*pi*sin(t); dtheta1_dt=3*pi*cos(t); d2theta1_dt2=-3*pi*sin(t);
% my_output=geneva_wheel(t, 1, 0.3, theta1, dtheta1_dt, d2theta1_dt2);
%
% IN ADDITION TO ANIMATING THE MECHANISM, SUCH A CALL TO THE FUNCTION geneva_wheel.m
% GENERATES A len (WHERE len IS THE NUMBER OF POINTS IN THE ARRAY t) BY 8 MATRIX
% CALLED my_output, WHOSE COLUMNS ARE
%           t, theta1, theta2, r2, dr2_dt, dtheta2_dt, d2r2_dt2, d2theta2_dt2
% my_output(:,2), my_output(:,3)*180/pi, FOR EXAMPLE, GIVE theta1 (IN RADIANS) AND
% theta2 (IN DEGREES). TO PLOT theta2 VERSUS TIME t, FOR EXAMPLE, TYPE, IN THE MATLAB
% COMMAND WINDOW, THE COMMANDS
%             figure();   plot( my_output(:,1), my_output(:,3)*180/(2*pi) )
% To put a grid on the plot, type    grid

disp(' ')
disp('NOTE: IN ADDITION TO ANIMATING THE MECHANISM, THE FUNCTION ')
disp('geneva_wheel  RETURNS 8 COLUMNS CONTAINING ')
disp(' t, theta1, theta2, r2, d(r2)/dt, d(theta2)/dt, d2(r2)/dt2, d2(theta2)/dt2')

len_barPC=L+r1;  cos_theta1=cos(theta1); sin_theta1=sin(theta1);
r2=sqrt( L^2+r1^2-2*L*r1*cos_theta1 );      % r2 from Eq. (3.3.3)
sin_theta2=r1*sin_theta1./r2;  cos_theta2=(L-r1.*cos_theta1)./r2; % Eqs. (3.3.2) and (3.3.1)
theta2=atan2(sin_theta2, cos_theta2);

% Now generate the coordinate matrices needed for the function animate_nbars:
len=length(t);
xA=zeros(len,1); yA=xA;      xB=r1*cos_theta1'; yB=r1*sin_theta1';   % bar AB
xP=L-len_barPC*cos_theta2'; yP=len_barPC*sin_theta2'; xC=L*ones(len,1);  yC=xA; % bar PC
x_coordinates=[xA, xB, xP, xC];  y_coordinates=[yA, yB, yP, yC];

% Delete existing figure windows, and call the function animate_nbars:
close all;  animate_nbars(x_coordinates, y_coordinates)

% Velocities and accelerations:
temp=r1*dtheta1_dt;  theta2=atan2(sin_theta2, cos_theta2);
sin_theta1plustheta2 = sin(theta1+theta2);      cos_theta1plustheta2 = cos(theta1+theta2);
dr2_dt = temp.*sin_theta1plustheta2;  dtheta2_dt = temp.*cos_theta1plustheta2./r2;
d2r2_dt2 = r1*d2theta1_dt2.*sin_theta1plustheta2 + r2.*(dtheta2_dt.^2) ...
      + r1*(dtheta1_dt.^2).*cos_theta1plustheta2;  % ... is for continuing a line
d2theta2_dt2 = ( r1*d2theta1_dt2.*cos_theta1plustheta2 - 2*dr2_dt.*dtheta2_dt ...
      - r1*(dtheta1_dt.^2).*sin_theta1plustheta2 )./r2;

   output=[t; theta1; theta2; r2; dr2_dt; dtheta2_dt; d2r2_dt2; d2theta2_dt2]';
% end of the file geneva_wheel.m
% %%%%%%%%%%%%%%%%%%%%%%%%%%%%%%%%%%%%%%%%%%%%%


function output=inclined_plane(t, m1_over_m2, k_over_m2, c2_over_m2, c1_over_m2, alpha, ...
      g, Lu, x_0, xdot_0, r_0, rdot_0)   % ... is for continuing a line
% Animation for Example 2.5.9. The animation draws five bars, AB, BC, AC, CD, and DE,
%  each one with a small dot at their ends.
% Example of usage ( using the initial conditions  x(0)=1; xdot(0)=0; r(0)=2; rdot(0)=0 )
```

```
% my_output = inclined_plane([0:0.005:20], 0.1,  5, 0, 0, pi/4, 9.81, 1, 1, 0,  2, 0);

initial_conditions=[x_0; xdot_0; r_0; rdot_0];
[t, y]=ode45(@rhs_of_diff_eqs, t, initial_conditions,[], m1_over_m2, k_over_m2,  ...
             c2_over_m2, c1_over_m2, alpha, g, Lu);
% This is a 5 by len matrix, where len is the number of points in the time array t
% 1st column: t,  2nd column: x, 3rd column: dx/dt;  4th column: r; 5th column: dr/dt
output= [t, y];

% ANIMATION:
   len=length(t);    col_of_ones=ones(len,1);
   x=output(:,2); r=output(:,4);
   xA=x;    xB=x+3;   xC=x;   xD=x+0.05*sin(alpha);   xE=xD+r*cos(alpha);
   yA=zeros(len,1);   yB=yA;   yC=3*tan(alpha)*col_of_ones;
 yD=(3*tan(alpha)+0.05*cos(alpha))*col_of_ones;  yE=yD-r*sin(alpha);
   x_coordinates=[xC, xA, xB, xC, xD, xE]; y_coordinates=[yC, yA, yB, yC, yD, yE];
   animate_nbars(x_coordinates, y_coordinates);

function dy_dt=rhs_of_diff_eqs(t,y,m1_over_m2,k_over_m2,c2_over_m2,c1_over_m2,alpha,g,Lu)
% Note:  y(1)=x, y(2)=dx/dt, y(3)=r, y(4)=dr/dt
   sin_alpha=sin(alpha); cos_alpha=cos(alpha); temp=m1_over_m2+sin_alpha^2;
   dy_dt(1)=y(2);
   dy_dt(2) = ( (c2_over_m2*y(4)+k_over_m2*(y(3)-Lu))*cos_alpha-g*sin(alpha)*cos(alpha) ...
               -c1_over_m2*y(2) )/temp;   % this is d2(x)/dt2 from Eq. (2.5.9i)
   dy_dt(3)=y(4);
   dy_dt(4)=( c1_over_m2*cos_alpha*y(2)+(1+m1_over_m2)*( g*sin_alpha-(c2_over_m2*y(4) ...
            +k_over_m2*(y(3)-Lu)) ) );   % and this is d2(r)/dt2 from Eq. (2.5.9j)
   dy_dt=[dy_dt(1); dy_dt(2); dy_dt(3); dy_dt(4)];
% end of the file inclined_plane.m
% %%%%%%%%%%%%%%%%%%%%%%%%%%%%%%%%%%%%%%%%%%%%%%%


function output=parabola_track(t, a, k_over_m, c_over_m, Lu, h, g,  x_0, xdot_0)
% Example 2.8.9
% Example of usage (with the initial conditions   x_0=-1,5; xdot_0=3*sqrt(2))):
% my_output=parabola_track([0:0.001:3], 1/3, 9/0.02, 1.4, 1.2, 1.2, 9.81, -1.5, 3*sqrt(2));
% NOTE:
% my_output is a n by 3 matrix, where n is the number of points in the time array t
%  1st column: t; 2nd column: x;  3rd column: dx/dt;

initial_conditions=[x_0; xdot_0];
[t, y]=ode45(@rhs_of_diff_eqs, t, initial_conditions, [],  a, k_over_m, c_over_m, Lu, h, g);
output=[t, y];        x=y(:,1);   y=a*x.^2;

close all
% Create a figure window and plot x(t) versus time t
figure('color', 'white');  plot(t, x),   grid
xlabel('t (seconds)', 'fontsize', 12), ylabel('x(t) (meters)', 'fontsize', 12)
title('Example 2.8.9','fontsize', 12)

animate_nbars(x, y)   % ANIMATION of the ball moving on the parabolic track y=a*x^2

function dy_dt=rhs_of_diff_eqs(t, y,  a, k_over_m, c_over_m, Lu, h, g)
   % Note:  y(1)=x, y(2)=dx/dt
   x_squared=y(1)^2;
   dy_dt(1)=y(2);      % the next expression is d2x_dt2 from Eq. (2.8.9c)
```

```
   dy_dt(2)=(k_over_m*(1-Lu/sqrt(y(1)^2+(a*y(1)^2-h)^2))*(2*a*(h-a*y(1)^2)-1)*y(1)...
             -2*a*g*y(1)-4*a^2*y(1)*y(2)^2)/(1+4*(a^2)*y(1)^2)-c_over_m*y(2);
   dy_dt=[dy_dt(1); dy_dt(2)];
% end of the file parabola_track.m
% %%%%%%%%%%%%%%%%%%%%%%%%%%%%%%%%%%%%%%%%%%%%%%%%


function output = pend_moving_base(t, g_over_L, c_over_mLsquare,  theta_0, thetadot_0)
% Example 2.7.6, and animation;  Use item_number=2 for Part c,
%                                and item_number=1 for Part d in that example
%    NOTE:  if desired, change the expressions for aA_over_L
% Example of usage ( using the initial conditions  theta_0=0, thetadot_0=0 ):
% my_output = pend_moving_base([0:0.002:15], 9.81, 1,   0, 0);

disp(' Enter the item number for Example 2.7.6 in Chapter 2 ');
disp(' USE 2 for Part c, 1 for Part d ');
item_number=input('  ');
if  item_number==2, message='Part c'; elseif item_number==1,  message='Part d';  end

if item_number ~=1 & item_number ~=2,   % ~= means  not equal; & means  and
   disp(' '),  disp('item_number MUST BE EITHER 1 OR 2;')
   disp(' use item_number=2 to do Part c of Example 2.7.6, and 1 to do Part d')
   output=[]; return
end


initial_conditions=[theta_0; thetadot_0];
[t, y]=ode45(@rhs_of_diff_eqs, t,initial_conditions,[],g_over_L,c_over_mLsquare,item_number);

close all
fig1=figure('color', 'white');   % 'background' 8 forces the figure background to be white
plot(t, y(:,1)*180/pi),  grid       % plot of theta versus time t
xlabel('t  (seconds)', 'fontsize', 12), ylabel(['\theta', ' (degrees)'], 'fontsize', 12),
title('Example 2.7.6','fontsize', 12)
uicontrol(fig1,'style','text', 'position',[10,10,50,20], 'string', message, 'fontsize',12)
% type   help uicontrol   in the Matlab command window for a description
% of the command uicontrol

% ANIMATION: using a simple model consisting of a bar AB, with
% A(0,0) and B(sin_theta, -cos_theta)        (lengths are normalized by L)
   len=length(t);    theta=y(:,1);      % theta is a len by 1 column
   xA=zeros(len,1); yA=xA;   xB=sin(theta); yB=-cos(theta);  % these are len by 1 columns
   x_coordinates=[xA, xB];  y_coordinates=[yA, yB];
   animate_nbars(x_coordinates, y_coordinates)

output=[t, y];
% The variable output is a n by 3 matrix, where n is the number of points in the array t
% 1st column: t, 2nd column: theta; 3rd column: d(theta)/dt

function dy_dt=rhs_of_diff_eqs(t, y,  g_over_L, c_over_mLsquare, item_number)
% Note:  y(1)=theta, y(2)=d(theta)/dt
   if item_number==2,  aA_over_L=2*9.81; % this is the constant aceleration input for Part c
      elseif item_number==1
      aA_over_L=5*t; if t>=5, aA_over_L=0; end % this is the ramp acceleration input for Part d
      end
   end
 dy_dt(1)=y(2);     % this is d(theta)/dt and the next one is d2(theta)/dt2
```

```
    dy_dt(2)=-g_over_L*sin(y(1)) + aA_over_L.*cos(y(1)) - c_over_mLsquare*y(2);
    dy_dt=[dy_dt(1); dy_dt(2)];
% end of the file pend_moving_base.m
% %%%%%%%%%%%%%%%%%%%%%%%%%%%%%%%%%%%%%%%%%%%%%%%%%%


function output=restricted_3body(t, m1_over_m2, r_0, rdot_0, theta_0,thetadot_0, z_0,zdot_0)
% Integration of the differential equations for the restricted three-body problem
% analyzed in Section 4.6 (Eqs. 4.6.13 to 4.6.15). All distances are normalized
% by L (which is the distance between bodies P1 and P2)
% Example of usage ( using the initial conditions
% r_0/L=1.001; rdot_0/L=0; theta_0=59*pi/180; thetadot_0=0; z_0/L=0.01; zdot_0=0 ):
% my_output = restricted_3body([0:0.01:100], 80,  1.001, 0, 59*pi/180, 0, 0.01, 0);

initial_conditions=[r_0; rdot_0; theta_0; thetadot_0; z_0; zdot_0];
[t, y]=ode45(@rhs_of_diff_eqs, t, initial_conditions, [],  m1_over_m2);
output=[t, y];
% The variable output is a n by 7 matrix, where n is the number of points in the array t
% 1st column: t, 2nd column: r/L, 3rd column: d(r/L)/dt; 4th column: theta,
% 5th column: d(theta)/dt; 6th column: z/L, 7th column: d(z/L)/dt
r_over_L=y(:,1); theta=y(:,3); z_over_L=y(:,5);

close all  % closing (i.e., deleting) all figure windows
% Plots:
figure('color', 'white');   plot(t, r_over_L); grid     % plot of r/L versus time t
xlabel('t   (seconds)', 'fontsize', 12), ylabel('r/L', 'fontsize', 12)
title('Restricted three-body problem','fontsize', 12)

figure('color', 'white');   plot(t, theta*180/pi); grid   % plot of theta versus time t
xlabel('t   (seconds)', 'fontsize', 12), ylabel(['\theta', ' (degrees)'], 'fontsize', 12)
title('Restricted three-body problem','fontsize', 12)

figure('color', 'white');   plot(t, z_over_L);   grid % plot of z/L versus t
xlabel('t   (seconds)', 'fontsize', 12), ylabel('z/L', 'fontsize', 12)
title('Restricted three-body problem','fontsize', 12)

%  Plot of y/L versus x/L:
figure('color', 'white'); plot(r_over_L.*cos(theta), r_over_L.*sin(theta)),  grid
xlabel('x/L', 'fontsize', 12),    ylabel('y/L', 'fontsize', 12),
title('Restricted three-body problem','fontsize', 12)

x=r_over_L.*cos(theta); y=r_over_L.*sin(theta);  animate_nbars(x, y)  % animation

function dy_dt=rhs_of_diff_eqs(t, y,  m1_over_m2)
% Note:  y(1)=r/L, y(2)=d(r/L)/dt, y(3)=theta, y(4)=d(theta)/dt, y(5)=z/L, y(6)=d(z/L)/dt
    cos_theta=cos(y(3)); sin_theta=sin(y(3));
    r31_cube= (y(1)^2+y(5)^2)^(3/2);  r32_cube= ( 1+y(1)^2+y(5)^2 -2*y(1)*cos_theta )^(3/2);
    dy_dt(1)= y(2);    % this is d(r/L)/dt
    dy_dt(2)= y(1)*(1+y(4))^2+((cos(y(3))-y(1))/r32_cube-cos(y(3)) ...
      -m1_over_m2*y(1)/r31_cube)/(1+m1_over_m2); % this is d2(r/L)/dt2 from Eq. (4.6.13)
    dy_dt(3)= y(4);
    dy_dt(4)= ((1-1/r32_cube)*sin_theta/(1+m1_over_m2)-2*y(2)*(1+y(4)))/y(1);
    dy_dt(5)= y(6);
    dy_dt(6)= -(m1_over_m2/r31_cube+1/r32_cube)*y(5)/(1+m1_over_m2);
    dy_dt=[dy_dt(1); dy_dt(2); dy_dt(3); dy_dt(4); dy_dt(5); dy_dt(6)];
% end of the file restricted_3body.m
```

```
% %%%%%%%%%%%%%%%%%%%%%%%%%%%%%%%%%%%%%%%%%%

function output = scotch_yoke(t, R, theta, dtheta_dt, d2theta_dt2)
% Animation of the Scotch yoke mechanism in Section 3.5
% Example of usage:
% (copy the next two lines, without the %, to the Matlab command window)
% t=[0:0.002:10]; theta=pi/2+2*t; dtheta_dt=2; d2theta_dt2=0;
% my_output=scotch_yoke(t, 1, theta, dtheta_dt, d2theta_dt2);
%
% To slow down the animation, choose a smaller time step for the input array t
% Conversely, to speed up the animation, choose a larger time step
% Example: the array t=[0:0.002:10] yields a slower animation than the array t=[0:0.02:10]
%
% IN ADDITION TO ANIMATING THE MECHANISM, SUCH A CALL TO THE FUNCTION
% scotch_yoke.m GENERATES A len (WHERE len IS THE NUMBER OF POINTS IN THE
% ARRAY t) BY 4 MATRIX CALLED my_output, WHOSE COLUMNS ARE  t, theta, dxA_dt, d2xA_dt2
% my_output(:,3), FOR EXAMPLE, GIVES dxA_dt. TO PLOT IT, TYPE, IN THE MATLAB
% COMMAND WINDOW, THE COMMANDS      figure();  plot( my_output(:,1), my_output(:,3) )
% To put a grid on the plot, type   grid

disp('NOTE: IN ADDITION TO ANIMATING THE SCOTCH YOKE MECHANISM, THIS FUNCTION')
disp('RETURNS 4 COLUMNS CONTAINING t, theta, d(xA)/dt, d2(xA)/dt2')

% Make sure that theta, dtheta_dt, and d2theta_dt2 have the same length as t
len=length(t);  temp=ones(1,len);
theta=theta.*temp; dtheta_dt=dtheta_dt.*temp; d2theta_dt2=d2theta_dt2.*temp;

sin_theta = sin(theta);    cos_theta = cos(theta);
x0=zeros(len,1);  xP=R*cos_theta';  xB=xP;  xC=xP;
y0=x0;  yP=R*sin_theta';  yB=1.5*R*ones(len,1);  yC=-yB;

close all;     % deleting all existing figures
x=[x0, xP, xB, xC];   y=[y0, yP, yB, yC]; animate_nbars(x, y); % this animates the motion

% Velocity dxA_dt=d(xA)/dt and acceleration d2xA_dt2=d2(xA)/dt2, where xA=R*cos(theta):
dxA_dt=-R*dtheta_dt.*sin_theta;
d2xA_dt2=-R*(d2theta_dt2.*sin_theta+(dtheta_dt.^2).*cos_theta);

  output = [t; theta; dxA_dt; d2xA_dt2]';
% end of the file scotch_yoke.m
% %%%%%%%%%%%%%%%%%%%%%%%%%%%%%%%%%%%%%%%%%%%%


function output = sim_model_ex5_3_11(tau, L2_over_L1, xdoubledot_over_g, Mbar_over_mgL1, ...
                    cstar, theta_0, thetaprime_0)      % ... is for continuing a line
% Example 5.3.11 in Chapter 5, where  tau=t*sqrt(g/L1), Mbar_over_mgL1=mbar/(m*g*L1),
%    and  cstar=c/(m*L1*sqrt(g*L1))
% Example of usage ( using the initial conditions  theta_0=90*pi/180; thetaprime_0=0 ):
% tau=[0:0.002:10]; xdoubledot_over_g=-2;  Mbar_over_mgL1=0;
% my_output = sim_model_ex5_3_11(tau,2, xdoubledot_over_g, Mbar_over_mgL1, 0, 90*pi/180, 0);

disp(' ')
disp('NOTE: IN ADDITION TO ANIMATING THE PENDULUM ANALYZED IN EXAMPLE 5.3.11,')
disp('AND PLOTTING theta versus time tau, A CALL TO THE FUNCTION sim_model_ex5_3_11,')
disp(' AS EXEMPLIFIED IN THE FIRST 5 COMMENT LINES OF THIS FUNCTION,')
```

```
disp(' RETURNS 3 COLUMNS CONTAINING  tau, theta, and d(theta)/dtau')
disp('TO PLOT d(theta)/dtau VERSUS theta, FOR EXAMPLE, USE THE COMMANDS:')
disp('        figure();  plot(my_output(:,2), my_output(:,3)) ')
disp('   To put a grid on the plot, type   grid ')

   initial_conditions=[theta_0; thetaprime_0];
   [tau, y]=ode45(@rhs_of_diff_eqs, tau, initial_conditions, [], L2_over_L1, ...
    xdoubledot_over_g,Mbar_over_mgL1, cstar); % ... is for continuing a command line

   output=[tau, y];
% The variable output is a n by 3 matrix, where n is the number of points in the array tau
% 1st column: tau, 2nd column: theta, 3rd column: d(theta)/dtau

   close all  % closing (i.e., deleting) all existing figures
   figure('color', 'white'); plot(tau, y(:,1)*180/pi); grid % plot of theta versus tau
   xlabel('Normalized time \tau', 'fontsize', 12)
ylabel(['\theta', ' (degrees)'], 'fontsize', 12);   title('Example 5.3.11','fontsize', 12)

   % ANIMATION (distances are normalized by L1):
   len=length(tau);
   theta=output(:,2); sin_theta=sin(theta); cos_theta=cos(theta);
   xA=zeros(len,1);  yA=xA;      xB=sin_theta;  yB=-cos_theta;
% The seven variables defined above are len by 1 columns
   xD=xB+L2_over_L1*cos_theta; yD= L2_over_L1*sin_theta-cos_theta;
   x_coordinates=[xA, xB,  xD]; y_coordinates=[yA, yB, yD];
animate_nbars(x_coordinates, y_coordinates)

function dy_dtau=rhs_of_diff_eqs(tau, y,L2_over_L1, xdoubledot_over_g,Mbar_over_mgL1,cstar)
% Note:  y(1)=theta, y(2)=d(theta)/dtau
   dy_dtau(1)=y(2);   % this is d(theta)/dtau
   dy_dtau(2)=(3*(1+L2_over_L1)*(Mbar_over_mgL1 -cstar*y(2)) -3*( 1+2*L2_over_L1 ...
        -xdoubledot_over_g*L2_over_L1^2 )*sin(y(1))/2 -3*( L2_over_L1^2 ...
        +xdoubledot_over_g*(1+2*L2_over_L1) )*cos(y(1))/2 )/(1+3*L2_over_L1+L2_over_L1^3);
% dy_dtau(2) is d2(theta)/dtau2 from Eq. (5.3.11c)
   dy_dtau=[dy_dtau(1); dy_dtau(2)];
% end of the file sim_model_ex5_3_11.m
% %%%%%%%%%%%%%%%%%%%%%%%%%%%%%%%%%%%%%%%%%%%%%%%%%%%%


function output = sim_model_ex5_3_13(t, k_factor, c1_star, c2_star, phi_0, phidot_0, ...
                    theta_0, thetadot_0)   % ... is for continuing a line
% Example 5.3.13 (t here is actually the normalized time tau)
% Example of usage [using the initial conditions phi_0=20*pi/180;
%                    phidot_0=0; theta_0=32*pi/180; thetadot_0=50]:
% my_output = sim_model_ex5_3_13([0:0.002:50], 12, 0.1, 0.1, 20*pi/180, 0, 32*pi/180, 50);

disp(' ')
disp('The k_factor is 12*(1+m1/(3*m2)/[(b/L)^2+(h/L)^2)])')
disp('Change it, if desired, to any other appropriate value')
disp('NOTE: IN ADDITION TO ANIMATING THE SYSTEM ANALYZED IN EXAMPLE 5.3.13,')
disp('AND PLOTTING phi AND theta versus time, A CALL TO THE FUNCTION sim_model_ex5_3_13,')
disp(' AS EXEMPLIFIED IN THE FIRST 4 COMMENT LINES OF THIS FUNCTION,')
disp(' RETURNS 5 COLUMNS CONTAINING  t, phi, d(phi)/dt, theta, and d(theta)/dt')
disp(' PLOTS CAN BE DONE USING THE plot COMMAND')

initial_conditions=[phi_0; phidot_0; theta_0; thetadot_0];
```

```
[t, y]=ode45(@rhs_of_diff_eqs, t, initial_conditions,  [ ],  k_factor, c1_star, c2_star);
output=[t, y];
 % The variable output is a n by 5 matrix, where n is the number of points in the array t
 % column 1: t, column 2: phi, column 3: d(phi)/dt; column 4: theta, column 5: d(theta)/dt
phi=output(:,2);   theta=output(:,4);

close all  % closing (i.e., deleting) all existing figures
figure('color', 'white'); plot(t, phi*180/pi); grid  % plot of phi versus  tau
xlabel(['Nondimensional time', ' \tau'], 'fontsize', 12),
ylabel(['\phi', ' (degrees)'], 'fontsize', 12),  title('Example 5.3.13','fontsize', 12)

figure('color', 'white'); plot(t, theta*180/pi); grid  % plot of theta versus tau
xlabel(['Nondimensional time', ' \tau'], 'fontsize', 12)
ylabel(['\theta', ' (degrees)'], 'fontsize', 12),  title('Example 5.3.13','fontsize', 12)

% ANIMATION (distances are normalized by L):
len=length(t);  % this is the number of time occurrences in the given array t
col_of_zeros=zeros(len,1);   % this is a len by 1 column of zeros
sin_phi=sin(phi); cos_phi=cos(phi);  sin_theta=sin(theta); cos_theta=cos(theta);
x0=col_of_zeros; y0=x0;       xC2=sin_phi;  yC2=-cos_phi;
xN=xC2+0.25*sin_theta;   yN= yC2+0.25*cos_theta;
xM=xC2-0.25*sin_theta;   yM= yC2-0.25*cos_theta;
x_coordinates=[x0, xC2, xN, xM];  y_coordinates=[y0, yC2, yN, yM];
animate_nbars(x_coordinates, y_coordinates)

function dy_dt=rhs_of_diff_eqs(t, y,  k_factor, c1_star, c2_star)
% Note:  y(1)=phi, y(2)=d(phi)/dt, y(3)=theta, y(4)=d(theta)/dt
  dy_dt(1)=y(2);  % this is d(phi)/dt, and the next one is is d2(phi)/dt2 from Eq. (5.3.13j)
  dy_dt(2)= -sin(y(1))-c1_star*y(2)-c2_star*(y(2)+y(4));
  dy_dt(3)=y(4);
  dy_dt(4)= -k_factor*c2_star*(y(2)+y(4));   % this mis from Eq. (5.3.13k)
  dy_dt=[dy_dt(1); dy_dt(2); dy_dt(3); dy_dt(4)];
% end of the file sim_model_ex5_3_13.m
% %%%%%%%%%%%%%%%%%%%%%%%%%%%%%%%%%%%%%%%%%%%%


function output =slider_crank(t,D,L1,L2,theta1,dtheta1_dt,d2theta1_dt2,solution_number)
% Animation of the slider_crank mechanism in Section 3.4
% Examples of usage:
% t=[0:0.001:5]; theta1=2*pi*t-pi/6; dtheta1_dt=2*pi; d2theta1_dt2=0;
% my_output = slider_crank(t, -0.5, 1, 2.5, theta1, dtheta1_dt, d2theta1_dt2,   1);
% Example 2:
% t=1/12; theta1=2*pi*t-pi/6; dtheta1_dt=2*pi; d2theta1_dt2=0;
% my_output = slider_crank(t, -0.5, 1, 2.5, theta1, dtheta1_dt, d2theta1_dt2, 1);
% Example 3:
% t=[0:0.001:0.5]; theta1=2*pi*t-pi/6; dtheta1_dt=2*pi; d2theta1_dt2=0;
% my_output = slider_crank(t, -0.5, 1, 2.5, theta1, dtheta1_dt, d2theta1_dt2,   1);
%
% To slow down the animation, choose a smaller time step for the input array t
% Conversely, to speed up the animation, choose a larger time step
%
% IN ADDITION TO ANIMATING THE MECHANISM, SUCH A CALL TO THE FUNCTION
% slider_crank.m GENERATES A len (WHERE len IS THE NUMBER OF
% POINTS IN THE ARRAY t) BY 8 MATRIX CALLED my_output, WHOSE COLUMNS
% ARE time, theta1, theta2, xB, dtheta2_dt, dxB_dt, d2theta2_dt2, d2xB_dt2
% my_output(:,3)*180/pi, FOR EXAMPLE, GIVES theta2 (IN DEGREES).
```

```
% TO PLOT IT VERSUS TIME t, TYPE, IN THE MATLAB COMMAND WINDOW, THE
% COMMANDS      figure();  plot( my_output(:,1), my_output(:,3)*180/(2*pi) )
% To put a grid on the plot, type    grid

disp(' ')
disp('NOTE: IN ADDITION TO ANIMATING THE MECHANISM, A CALL SUCH AS')
disp('t=[0:0.001:5]; theta1=2*pi*t-pi/6; dtheta1_dt=2*pi; d2theta1_dt2=0;')
disp('output=slider_crank(t, -0.5, 1, 2.5, theta1, dtheta1_dt, d2theta1_dt2,1,1);')
disp('RETURNS, IN THE VARIABLE output, 8 COLUMNS CONTAINING')
disp(' time, theta1, theta2, xB, dtheta2_dt, dxB_dt, d2theta2_dt2, d2xB_dt2')
disp(' (the time array t only has those times for which the mechanism does not lock)')
disp(' ')

if solution_number ~=1 & solution_number ~= 2,
    disp(' -----> SOLUTION_NUMBER HAS TO BE  1 or 2 <----- '),    output=[], return, end

% First, make sure that theta1, dtheta1_dt, and d2theta1_dt2 have the same length as t
len=length(t); temp=ones(1,len);  newlen=len;
theta1=theta1.*temp; dtheta1_dt=dtheta1_dt.*temp; d2theta1_dt2=d2theta1_dt2.*temp;

% Make the simulation proceed only while |sin_theta2|<=1.
% For this, redefine the array t, if necessary.
sin_theta2=(L1*sin(theta1)-D)/L2;          % sin(theta2) from Eq. (3.4.6)
% Let's see if this is between -1 and 1:
temp=find( abs(sin_theta2)>1 );
if ~isempty(temp)
    temp1=temp(1)-1;
    if temp1 == 1
        output = [ ];
        disp('The mechanism, as specified, does not work because the condition')
disp('          |L1*sin(theta1)-D|>=L2 is violated at the initial time')
        return
    else
        newlen=temp1;   tnew=t(1:newlen);
        mprintf('The mechanism, as specified, stops after t= %5.3f', tnew(newlen) )
        mprintf(' because |L1*sin(theta1)-D|>L2 sometime thereafter.')
    end
end

% Resize all the proper lists and do the computations.
tnew=t(1:newlen);
theta1=theta1(1:newlen);dtheta1_dt=dtheta1_dt(1:newlen);d2theta1_dt2=d2theta1_dt2(1:newlen);
sin_theta1 = sin(theta1);    cos_theta1 = cos(theta1);
sin_theta2=(L1*sin(theta1)-D)/L2;
% sin(theta2) is from Eq. (3.4.6) using the resized theta1
cos_theta2=(3-2*solution_number).*sqrt(1-sin_theta2.^2);
% cos(theta2) ) is from Eqs. (3.4.8) and (3.4.9)
xB=L1*cos_theta1+L2*cos_theta2;          % xB from Eq. (3.4.7)
theta2= atan2(sin_theta2, cos_theta2);

% Now, generate the coordinate matrices needed for the function animate_nbars:
x0=zeros(newlen,1);  y0=x0;  xA=L1*cos_theta1';  yA=L1*sin_theta1';  % bar OA
x_coordinates=[x0, xA, xB'];  y_coordinates=[y0, yA, D*ones(newlen,1)];

close all   % closing (i.e., deleting) all existing figures
% Call the function animate_nbars to animate the motion
```

```
animate_nbars(x_coordinates, y_coordinates);

% Angular velocity dtheta2_dt=d(theta2)/dt and velocity dxB_dt=d(xB)/dt:
dtheta2_dt=L1*dtheta1_dt.*cos_theta1./(L2*cos_theta2);   % d(theta2)/dt from Eq. (3.4.10)
dxB_dt=-L1*dtheta1_dt.*sin_theta1-L2*dtheta2_dt.*sin_theta2; % d(xB)/dt from Eq. (3.4.11)

% d2(theta2)/dt2 from Eq. (3.4.12), and d2(xB)/dt2 from Eq. (3.4.13):
d2theta2_dt2 = ( L1*(d2theta1_dt2.*cos_theta1-(dtheta1_dt.^2).*sin_theta1)/L2  ...
            + (dtheta2_dt.^2).*sin_theta2 )./cos_theta2;
d2xB_dt2=-L1*(d2theta1_dt2.*sin_theta1+(dtheta1_dt.^2).*cos_theta1)  ...
            - L2*(d2theta2_dt2.*sin_theta2+(dtheta2_dt.^2).*cos_theta2);

  output = [tnew; theta1; theta2; xB; dtheta2_dt; dxB_dt; d2theta2_dt2; d2xB_dt2]';
% end of the file  slider_crank.m
% %%%%%%%%%%%%%%%%%%%%%%%%%%%%%%%%%%%%%%%%%%%%%


function output=spring_pendulum(t, mg_over_kre, rs_0, dotrs_0, theta_0, dottheta_0)
% Animation of the spring-pendulum analyzed in Example 2.7.8
% Example of usage [with the initial conditions
                            rs_0=0.1;rsdot_0=0;theta_0=0.01;thetadot_0=0]:
% my_output = spring_pendulum([0:0.02:350], 0.5^2,  0.1, 0, 0.01, 0);

disp(' ')
disp('NOTE: IN ADDITION TO ANIMATING THE SPRING-PENDULUM ANALYZED IN EXAMPLE 2.7.8,')
disp('AND PLOTTING theta and rs versus time, A CALL TO THE FUNCTION spring_pendulum ')
disp('SUCH AS my_output = spring_pendulum([0:0.02:350], 0.5^2,  0.1, 0,  0.01, 0) )')
disp('RETURNS 5 COLUMNS CONTAINING  t, rs,  d(rs)/dt, theta, and d(theta)/dt')
disp('TO PLOT d(rs)/dt, FOR EXAMPLE, VERSUS time, TYPE, IN THE MATLAB COMMAND WINDOW,');
disp('THE COMMANDS:    figure();  plot(my_output(:,1), my_output(:,3)) ')
disp('   TO PUT A GRID ON THE PLOT, IF DESIRED, THEN TYPE  grid')

close all   % closing (i.e., deleting) all existing figures
len=length(t);   % this is the number of points in the array t

initial_conditions=[rs_0; dotrs_0; theta_0; dottheta_0];
[t, y]=ode45(@rhs_of_diff_eqs, t, initial_conditions, [ ], mg_over_kre);
figure('color', 'white');   plot(t, y(:,3)); grid        % plot of theta versus time t
xlabel(['Nondimensional time  ', '\tau'], 'fontsize', 12);   ylabel('\theta', 'fontsize', 12);

figure('color', 'white');   plot(t, y(:,1));   grid;      % plot of rs versus t
xlabel(['Nondimensional time  ', '\tau'], 'fontsize', 12); ylabel('r_s', 'fontsize', 12);

output=[t, y];
% The variable output is a n by 5 matrix, where n is the number of points in the array t
% column 1: time, column 2: rs, column 3: d(rs)/dt; column 4: theta; column 5: d(theta)/dt

% ANIMATION: using a bar AB from the point (0,0) to (1+rs)*(sin_theta, -cos_theta)
rs=output(:,2); theta=output(:,4);   % these are len by 1 columns
x0=zeros(len,1); y0=x0;    xP=(1+rs).*sin(theta);   yP=-(1+rs).*cos(theta);
x_coordinates=[x0, xP];  y_coordinates=[y0, yP]; animate_nbars(x_coordinates, y_coordinates);

function dy_dt = rhs_of_diff_eqs(t, y, mg_over_kre)
% Note:  y(1)=rs, y(2)=d(rs)/dt, y(3)=theta, y(4)=d(theta)/dt
  dy_dt=[y(2); (1+y(1))*y(4)^2-mg_over_kre*(1-cos( y(3)) )-y(1); y(4);  ...
        -(2*y(2)*y(4)+mg_over_kre*sin(y(3)) )/(1+y(1))]; % ... is for continuing a line
```

```
% end of the file spring_pendulum.m
% %%%%%%%%%%%%%%%%%%%%%%%%%%%%%%%%%%%%%%%%%%%%


function output = top_dynamics(t, Iz_over_In, mgL_over_In,  ...
                 phidot_0, psidot_0, theta_0, thetadot_0)  % ... is for continuing a command line
% Top dynamics analyzed in Section 6.6.3
% Example of usage ( using the initial conditions
%      phidot_0=5.03; psidot_0=45.65; theta_0=pi/6; thetadot_0=0 ):
% my_output = top_dynamics([0:0.001:10], 27/68, 27*98.1/34, 5.03, 45.65, 30*pi/180, 0);

disp(' ')
disp('NOTE: IN ADDITION TO ANIMATING THE MOTION OF THE TOP, A CALL SUCH AS ')
disp(' output = top_dynamics([0:0.001:10], 27/68, 27*98.1/34, 5.03, 45.65, 30*pi/180, 0);')
disp('RETURNS, IN THE VARIABLE output, 4 COLUMNS CONTAINING  t, phi, theta, dtheta_dt')
disp(' ')

phi_0=0;
[t, y]=ode45(@rhs_of_diff_eqs, t, [phi_0; theta_0; thetadot_0], [], Iz_over_In, ...
                      mgL_over_In, phidot_0, psidot_0, theta_0, thetadot_0);
output=[t, y];
% The variable output is a n by 4 matrix, where n is the number of points in the array t
% 1st column: t,  2nd column: phi, 3rd column: theta; 4th column: d(theta)/dt

phi=output(:,2); theta=output(:,3); sin_theta=sin(theta); cos_theta=cos(theta);
sin_phi=sin(phi);  cos_phi=cos(phi);

close all  % deleting all existing figures
% Plot of phi versus time
figure('color', 'white'); plot(t, phi*180/pi), grid   % plot of phi versus time t
xlabel('t  (seconds)', 'fontsize', 12), ylabel(['\phi  ', '(degrees)'], 'fontsize', 12)
title('Top dynamics','fontsize', 12)

% Plot of theta versus time
figure('color', 'white'); plot(t, theta*180/pi), grid  % plot of theta versus t
xlabel('t  (seconds)', 'fontsize', 12), ylabel(['\theta  ', '(degrees)'], 'fontsize', 12)

% ANIMATION:
len=length(t);   % number of points in the given array t
x0=zeros(len,1); y0=x0; z0=x0;   % these are columns of zeros
xC=sin_theta.*cos_phi; yC=sin_theta.*sin_phi; zC=cos_theta;
x_coordinates=[x0, xC]; y_coordinates=[y0, yC]; z_coordinates=[z0, zC];
animate3d_nbars(x_coordinates, y_coordinates, z_coordinates)

function dy_dt=rhs_of_diff_eqs(t, y,  Iz_over_In, mgL_over_In, phidot_0,  ...
                         psidot_0, theta_0, thetadot_0)
% y(1)=phi, y(2)=theta, y(3)=d(theta)/dt,
    cos_theta0=cos(theta_0); sin_theta0=sin(theta_0); sin_theta=sin(y(2));
    cos_theta=cos(y(2)); omegaz = psidot_0+phidot_0*cos_theta0 ;
    Hz0_over_In = phidot_0*(sin_theta0^2)+Iz_over_In*omegaz*cos_theta0;
    dy_dt(1)= (Hz0_over_In - Iz_over_In*omegaz*cos_theta)/(sin_theta^2); % this is d(phi)/dt
    dy_dt(2)=y(3);   % this is d(theta)/dt
    dy_dt(3)= ( mgL_over_In + (dy_dt(1)*cos_theta - Iz_over_In*omegaz)*dy_dt(1) )*sin_theta;
  % the above is d2(theta)/dt2
    dy_dt=[dy_dt(1); dy_dt(2); dy_dt(3)];
% end of the file top_dynamics.m
```

```
% %%%%%%%%%%%%%%%%%%%%%%%%%%%%%%%%%%%%%%%%%%%%%%%


function output = two_body_simul(t, Re, g, m2_over_m1_plus_1, r_0, v_0, flight_path_angle_0)
% Animation of the two-body problem analyzed in Sections 4.4 and 4.5 of Chapter 4
% NOTE: Re: in km, g: in m/s^2, r_0: in km, v_0: in km/hour, flight_path_angle_0: in degrees
% Example of usage (with the initial conditions r_0=6578; v_0=28040; flight_path_angle_0=0):
% my_output = two_body_simul([0:0.001:30], 6378, 9.81, 1,   6578, 28040, 0);
% also try:  my_output = two_body_simul([0:0.001:30], 6378, 9.81, 1,   6578, 38040, 0);
%     and:   my_output = two_body_simul([0:0.001:30], 6378, 9.81, 1,   6578, 27040, 0);

disp(' ')
disp('NOTE: IN ADDITION TO ANIMATING THE TWO-BODY PROBLEM, A CALL SUCH AS ')
disp(' output=two_body_simul(t, Re, g, m2_over_m1_plus_1, r_0, v_0, flight_path_angle_0)')
disp('RETURNS, IN THE VARIABLE output, 4 COLUMNS CONTAINING  time, r, dr/dt, and theta')
% ************************************************************************
disp(' '), disp('  =====>  IMPORTANT MESSAGE:  <=====')
disp(' IF THE SIMULATION RUNS EITHER TOO FAST OR TOO SLOW IN YOUR COMPUTER,')
disp(' change the time_step in the time array t=[t0:time_step:tfinal] of your simulation ')
disp(' ---> EITHER to a smaller value to slow down the animation ')
disp(' ---> OR to a larger value to speed up the animation')
% ************************************************************************
disp(' ')

dr_dt_0=v_0*sin(flight_path_angle_0*pi/180); theta0=0;
[t, y]=ode45(@rhs_diff_eqs, t, [r_0; dr_dt_0; theta0], [ ], Re, ...
       g, m2_over_m1_plus_1, r_0, v_0, flight_path_angle_0); % ... is for continuing a line
output=[t, y];    r=output(:,2); theta=output(:,4);

% ************* Let's see if the satellite crashed *******************************
number_of_time_points=length(t);   newlen=number_of_time_points;   temp=find(r <= Re);
if ~isempty(temp)
    temp1=temp(1)-1; newlen=temp(1)-1;
    tnew=t(1:newlen); number_of_time_points=newlen;
    disp(' '), disp(['THE SATELLITE CRASHES AFTER  t= ', num2str(tnew(newlen)), ' hours'])
end
% Resize all the proper lists and then proceed with the new reduced time list tnew
tnew=t(1:newlen); t=tnew; r=r(1:newlen,1); theta=theta(1:newlen,1); output=[tnew, r, theta];
% ************************************************************************

temp1=r.*cos(theta); temp2=r.*sin(theta);
x_min=min(temp1); x_max=max(temp1); y_min=min(temp2); y_max=max(temp2);
if x_min==x_max,
    disp('THE STEP SIZE IS TOO BIG; DECREASE IT TO RUN THE SIMULATION'), return,
end

close all  % deleting all existing figures
% Create a figure window for the animation. Start by determining the computer monitor size
% so that the animation window is properly placed and sized in any computer screen.
screen_width_and_height=get(0,'screensize');   % width and height of the screen, in pixels
PC_screen_width =screen_width_and_height(3);   % this is the width of the computer monitor
PC_screen_height=screen_width_and_height(4);   % this is the height of the computer monitor

x_topleft=0.65*PC_screen_width; % choice for the top left x-coordinate of the figure window
y_topleft=0.17*PC_screen_height; % choice for the top left y-coordinate of the figure window
top_left_coordinates=[x_topleft, y_topleft];
```

```
size_of_fig_window = [0.32*PC_screen_width, 0.47*PC_screen_height];
% the above are the width and height chosen for the animation window
% Now, create the figure window.
animation_fig=figure('position', [x_topleft, y_topleft,size_of_fig_window(1), ...
  size_of_fig_window(2)],'name','Animation window','color','white','Doublebuffer','on');
% NOTE: if 'DoubleBuffer' is 'off' (the default), this animation will flicker,
%              which is not desirable

xlabel('x (in km) from the planet center', 'fontsize', 12)
ylabel('y (in km) from the planet center', 'fontsize', 12)
title('The two-body problem', 'fontsize', 12)

h_axes=gca; set(h_axes, 'xlim', [-abs(x_min), x_max], 'ylim', [-abs(y_min), y_max], ...
        'dataaspectratio', [1, 1, 0.1], 'NextPlot','add')

th=0:0.02:2*pi; plot(Re*cos(th), Re*sin(th)); % a circle representing the attracting planet
satellite= plot(1.05*Re, 0, 'o');
rot_planet=plot([0,0], [0, 0]);
    % rot_planet will be a line that will represent the spinning central attracting planet

% record the Earth-day number in the figure window
om=2*pi/24; tau=om*t; day_number=ceil(tau/(2*pi)); dia=num2str(day_number);
uicontrol(animation_fig,'style','text', 'position',[10,10,40,15], 'string', 'day');
daynumber=uicontrol(animation_fig,'style','text', 'position',[40,10,40,15], 'string', ' ');

for itime=1:number_of_time_points
   set(satellite, 'xdata', r(itime,1)*cos(theta(itime,1)), ...
                'ydata', r(itime,1)*sin(theta(itime,1)) );
   set(rot_planet, 'xdata', [0, Re*cos(om*t(itime,1))], 'ydata', [0, Re*sin(om*t(itime,1))]);
   drawnow;
   set(daynumber, 'string', dia(itime) )
end

function dy_dt=rhs_diff_eqs(t, y, Re, g, m2_over_m1_plus_1, r_0, v_0, flight_path_angle_0)
% State variabes: y(1)=r, y(2)=dr/dt, y(3)=theta
% unit of distance: km, unit of time: hour; g: m/s^2
      r_sq=y(1)^2;
      dtheta_dt = r_0*v_0*cos(flight_path_angle_0*pi/180)/r_sq;
      d2r_dt2 = y(1)*dtheta_dt^2 - m2_over_m1_plus_1*g*3600*3.6*(Re^2)/r_sq;
   dy_dt(1) = y(2);        % this is dr/dt
   dy_dt(2) = d2r_dt2;     % this is d2(r)/dt2
   dy_dt(3) = dtheta_dt;   % this is d(theta)/dt
   dy_dt=[dy_dt(1); dy_dt(2); dy_dt(3)];
% end of the file two_body_simul.m
% %%%%%%%%%%%%%%%%%%%%%%%%%%%%%%%%%%%%%%%%%%%%%%


function output=two_particles_table(t, m2_over_m2plusm1, g_over_L, r_0, rdot_0, ...
                     theta_0, thetadot_0)     % ... is for continuing a command line
% Animation of Example 2.7.10. Here, the distance denoted by r is actually r/L
% State variables: y(1)=r_over_L, y(2)=d(r_over_L)/dt, y(3)=theta, y(4)=d(theta)/dt
% Example of usage  (using the initial conditions for the state variables: [0.5; 0;  0; 2] ):
% my_output=two_particles_table([0:0.001:10], 0.5, 9.81, 0.5, 0,  0, 2);

disp(' ')
disp('NOTE: IN ADDITION TO ANIMATING THE SYSTEM IN EXAMPLE 2.7.10, A CALL TO THE ')
```

```
disp('FUNCTION  two_particles_table (SUCH AS')
disp('  my_output = two_particles_table([0:0.001:10],0.5,9.81,0.5,0,0;2)  )')
disp('RETURNS 5 COLUMNS CONTAINING t, r_over_L, d(r_over_L)/dt, theta, and d(theta)/dt')
disp('TO PLOT r_over_L, FOR EXAMPLE, VERSUS time, USE THE COMMANDS ');
disp('        figure( ); plot(my_output(:,1), my_output(:,2)) ')
disp('TO PUT A GRID ON THE PLOT, IF DESIRED, THEN TYPE  grid')
disp(' ')


initial_conditions=[r_0; rdot_0; theta_0; thetadot_0];
[t, y]=ode45(@rhs_of_diff_eqs, t, initial_conditions, [],  m2_over_m2plusm1, g_over_L);
output=[t, y];
% the variable output is a len by 5 matrix, where len is the number of points in the array t
% column 1: t, column 2: r/L, column 3: d(r/L)/dt; column 4: theta; column 5: d(theta)/dt


% ANIMATION: We have two bars, O_P1 and O_P2, and the total normalized length
% of the system of bars is r/L+(L-r)/L=1
len=length(t);
r=output(:,2); theta=output(:,4); sin_theta=sin(theta); cos_theta=cos(theta);
x_O=zeros(len,1);y_O=x_O;z_O=x_O; x_P1=r.*cos_theta;y_P1=r.*sin_theta;z_P1=x_O; % bar O_P1
x_P2=x_O;  y_P2=x_O;  z_P2=r-1;  % end P2 of bar O_P2
x_coordinates=[x_P2, x_O, x_P1];  y_coordinates=[y_P2, y_O, y_P1];
z_coordinates=[z_P2, z_O, z_P1];
animate3d_nbars(x_coordinates, y_coordinates, z_coordinates);


function dy_dt=rhs_of_diff_eqs(t, y,  m2_over_m2plusm1, g_over_L)
    % Note:  y(1)=r_over_L, y(2)=d(r_over_L)/dt, y(3)=theta, y(4)=d(theta)/dt
    dy_dt(1)=y(2);      % the next expression is d2(r_over_L)/dt2 from Eq. (2.7.10l)
    dy_dt(2)=(1-m2_over_m2plusm1)*y(1)*y(4)^2 -m2_over_m2plusm1*g_over_L;
    dy_dt(3)=y(4);
    dy_dt(4)=-2*y(2)*y(4)/y(1);   % this is d2(theta)/dt2 obtained from Eq. (2.7.10k)
    dy_dt=[dy_dt(1); dy_dt(2); dy_dt(3); dy_dt(4)];
% end of the file two_particles_table.m
% %%%%%%%%%%%%%%%%%%%%%%%%%%%%%%%%%%%%%%%%%%%%%%
```

Answer to Selected Problems

Chapter 1

1.3 $\pm 14.8°$

1.4 0.093

1.7 $\vec{c} \approx 2.49\hat{x}_1 + 2.07\hat{x}_2 + 3.49\hat{x}_3$

1.11 $3.294\hat{x} + 0.824\hat{y}$

1.17 $\int P(x)\,dx + k\cos\theta +$ a constant

1.18 for $k = 2\Omega^2$, $\theta_{max} \approx 39.6°$ and $\theta_{min} = -\theta_{max}$

1.20 (a) Either $\ddot{x}_s + k\dot{x}_s/(1 - x_e) + (\cos x_e)x_s + kx_s\dot{x}_s/(1 - x_e)^2 + \ldots = 0$
or $(1 - x_e - x_s)[\ddot{x}_s + x_s\cos\theta_e] + k\dot{x}_s + \ldots = 0$

1.23 $x_{min} \approx -0.48$, $x_{max} \approx 0.36$

1.26c 1.4 and $\sqrt{2}$

1.28 $x_{max} \approx 0.174$

1.32 Unstable if $k < 0$; asymptotically stable if $k > 0$

1.40 $\ddot{x} = -10x + 3 - 2\dot{x}$

Chapter 2

2.3 (a) $\vec{v} \approx 2.297\hat{x} - 4.795\hat{y}$ m/s, and $\vec{a} \approx 27.769\hat{x} + 0.919\hat{y}$ m/s^2
(b) $\vec{v} \approx 2.297\hat{x}$ m/s and $\vec{a} \approx 26.81\hat{x}$ m/s^2

2.5 (b) 0.866 km^{-1}

2.8 $\Delta t \approx 2.63$ s

2.11 Constant gravitational force approximation: $y_{max} \approx 157.6$ km after 189 s;
Using the inverse square law: $y_{max} \approx 162$ km after 200 s

2.15 $y_{max} \approx 40.7$ km after $t \approx 68$ s

2.17 $\alpha \approx 17.7°$ and also $52.3°$

2.19 Range ≈ 934.5 m when $c/m = 10^{-4}$ m^{-1}, and 600.6 m when $c/m = 10^{-3}$ m^{-1}

2.21 0.33 s

2.24 (a) 10 m; (b) $t \approx 4.2$ s; $r(3) \approx 4.9$ m

2.27 $r_e = 0.2$ m, $\omega \approx 3.5$ rad/s

2.33 $\theta \approx 48.2°$; P strikes the ground at $x \approx 1.125R$ to the right of point O.

2.34 $\theta \approx 43.3°$

2.37 There is only one equilibrium when $L_u \leq h$, and three when $L_u > h$. Linearization is not valid when the coefficient of $x_s(s) \triangleq x(t) - x_e$ in the linearized differential equation for x_s is equal to zero.

2.39 For $c/\sqrt{mk} = 0.2$, $x_{max} \approx 10.96h$ and $x_{min} \approx -7.92h$

2.45 $x_{max} \approx 1.37$ m and $x_{min} \approx -1.5$ m

2.49 $r_{max} \approx 60.3$ cm, $r_{min} \approx 11.2$ cm

Chapter 3

3.4 $\omega_{BC} \approx 0.14$ rad/s (clockwise), $\omega_{CD} \approx 6.94$ rad/s (counterclockwise)

3.8 $\omega_{BC} \approx 3.03$ rad/s (counterclockwise); $\omega_{CD} \approx 4.05$ rad/s (clockwise)

3.18 $\dot{\omega}_{BC} \approx 15.18$ rad/s^2 (counterclockwise); $\dot{\omega}_{CD} \approx 6.72$ rad/s^2 (counterclockwise)

3.22 $\dot{\omega}_{BC} \approx 39.6$ rad/s^2 (counterclockwise); $\dot{\omega}_{CD} \approx 48.05$ rad/s^2 (counterclockwise)

3.25 When $\theta_1 = 45°$: $\dot{\theta}_2 \approx 0.108$ rad/s, $\ddot{\theta}_2 \approx -0.35$ rad/s^2, $\dot{r}_2 \approx -0.88$ m/s, $\ddot{r}_2 \approx -0.728$ m/s^2

3.28 $\dot{\theta}_3 = 100$ rpm, $\dot{\theta}_4 \approx 12.57$ rad/s

3.34 The range for the mechanism to operate without locking is $-55.16° < \phi_2 < 122.8°$. Locking may be avoided with $L_4 = 4.2$ m, for example.

Chapter 4

4.2 $m \approx 4.362M_E$, where M_E is the mass of the earth.

4.4 $\Delta t \approx 53.5$ s. The initial speed should be increased.

4.6 $r_{max} \approx 6697$ km, $r_{min} \approx 6678$ km; $e \approx 0.0014$

4.11 (b) For $c_1/m = 0.0001$ h^{-1} and $c_2/m = 5 \times 10^{-9}$ km^{-1}, $t_{final} \approx 36$ h. The satellite crashes on the earth.

4.12 for $C_D A/m = 5 \times 10^{-3}$ m^2/kg, $t_{\text{crash}} \approx 8.17$ h

Chapter 5

5.6 $I_C \approx 49.2$ kg·m^2; period ≈ 4.3 s when the plate is hung from B.

5.8 44.9 cm $\leq x(0) \leq 73.2$ cm

5.11 The system does not oscillate when $c/(mL_1^2) = 4.5$ s^{-1}. Its motion is overdamped.

5.16 With $c_1^* = c_2^* = 0.1$, $[F_1/(m_2 g)]_{\text{max}} \approx 11.35$, and $[F_2/(m_2 g)]_{\text{max}} \approx 0.27$

5.17 $|M_{\text{ext}}| \approx 1.886$ N·m (clockwise),

$A_{r_2} \approx -10.57$ N, $A_{n_2} \approx -4.715$ N; $\sqrt{A_{r_2}^2 + A_{n_2}^2} \approx 11.57$ N,

$B_{r_2} \approx 9.916$ N, $B_{n_2} \approx 4.715$ N; $\sqrt{B_{r_2}^2 + B_{n_2}^2} \approx 10.98$ N,

$C_{r_4} \approx -6.463$ N, $C_{n_4} \approx 1.653$ N; $\sqrt{C_{r_2}^2 + C_{n_2}^2} \approx 6.67$ N,

$D_{r_4} \approx 5.805$ N, $D_{n_4} \approx 0.827$ N; $\sqrt{D_{r_2}^2 + D_{n_2}^2} \approx 5.86$ N

5.21 $|M_{ext}| \approx 15.7$ N·m (clockwise), $A_{r1} \approx -46$ N, $A_{n1} \approx -39.3$ N, $B_{r2} = 0$, $C_{r2} \approx -8.3$ N, $C_{n2} \approx 17.7$ N

5.29 $\omega \approx 18.4$ rad/s

5.34 $\theta_e \approx -3.9°$ and $\theta_e \approx 176.1°$; $\omega \approx 2.3$ rad/s

5.35 $\theta_e \approx -23.4°$ and $\theta_e \approx 156.6°$, $|\vec{a}_C| = 2g/5$, $\omega \approx 6.9$ rad/s when $b = 35$ cm, and $\omega \approx 8.2$ rad/s when $b = 25$ cm

5.38 $\omega \approx 2.5$ rad/s

Chapter 6

6.3 With x', y', and z' being axes parallel to x, y, and z, respectively:

$I_{x'} \approx 282.9$ kg · m^2, $I_{y'} \approx 61.5$ kg · m^2, $I_{z'} \approx 294.4$ kg · m^2.

The center of mass C is at $\overrightarrow{OC} = -0.25\hat{x} + 0.608\hat{y} + 0.2\hat{z}$ m

6.4 Only the last two.

6.7 $T = k\sin(\omega t)+$ a constant

6.10b $\theta_{\text{max}} \approx 8.1°$ when $\Omega^2 L/g = 1$, and $\theta_{\text{max}} \approx 60.2°$ when $\Omega^2 L/g = 2$. For both cases, $\theta_{\text{min}} = -\theta_{\text{max}}$

6.12 56.5°

6.14b (b) $\Omega_{\text{max}} \approx 28.6$ rad/s ≈ 273 rpm

6.17 $\pm 107.2°$, $\pm 26.4°$, and $\pm 70.6°$, respectively.

Chapter 7

7.6 The Hamiltonian $\mathcal{H} = T_2 + U - T_0 = \frac{m}{2}(\dot{r}^2 - \Omega^2 r^2) + \frac{k}{2}(\sqrt{r^2 + h^2} - L_u)^2$ is constant. For $a = 2$, $r_{\max}/h \approx 7.09$; $r_{\min} = -r_{\max}$

7.8 The Hamiltonian $\mathcal{H} = T_2 + U - T_0 = \frac{m}{2}[1 + \frac{1}{4ay}]\dot{y}^2 + \frac{k}{2}(\sqrt{\frac{y}{a} + (h - y)^2} - L_u)^2 - \frac{m\Omega^2 y}{2a} + mgy$ is constant. For the data given in Example 2.8.8, $y_{\max} \approx 92.5$ cm, $y_{\min} = 0$, $x_{\max} \approx 1.67$ m, $x_{\min} = -x_{\max}$

7.9 $\Omega_{\text{critical}} = \sqrt{2ga} \approx 2.56$ rad/s ≈ 24.4 rpm, obtained from the plots and from the linearized differential equation.

7.18 The Hamiltonian $\mathcal{H} = T_2 + U - T_0$ is the only classical integral of motion for this case, where

$$T_2 = \frac{1}{2}m_3\left[\dot{r}^2 + r^2\dot{\theta}^2 + \dot{z}^2\right]$$

$$T_0 = \frac{1}{2}m_3\left[r^2 + \left(\frac{m_2 L}{m_1 + m_2}\right)^2 - \frac{2m_2 Lr\cos\theta}{m_1 + m_2}\right]\Omega^2$$

$$U = -\frac{Gm_1 m_3}{\sqrt{r^2 + z^2}} - \frac{Gm_2 m_3}{\sqrt{L^2 + r^2 + z^2 - 2rL\cos\theta}}$$

The equation $\mathcal{H} = T_2 + U - T_0$ can be used as one of the differential equations of motion. Lagrange's equation yields Eqs. (4.6.13) to (4.6.15) shown in Section 4.6 in Chapter 4.

7.25 The Hamiltonian $\mathcal{H} = T_2 + U - T_0$ is constant. This gives:

$$\left(\frac{mL^2}{8} + \frac{1}{2}I_x\right)\dot{\theta}^2 - mg\frac{L}{2}\cos\theta - \frac{\Omega^2}{2}\left(I_y\sin^2\theta + I_z\cos^2\theta\right)$$

$$-\frac{mL^2}{8}\Omega^2\sin^2\theta = \text{constant}$$

where $I_x = mL^2/12$, $I_z = mb^2/12$, and $I_y = I_x + I_z$.

For values of $\theta_{\max}$ and $\theta_{\min}$, see the answer to Problem 6.10.

7.26 There are three classical integrals of motion: the Hamiltonian $\mathcal{H} = I_x(\dot{\theta}^2 + \dot{\phi}^2\sin^2\theta)/2 + I_z(\dot{\psi} + \dot{\phi}\cos\theta)^2/2 + mgL\cos\theta$, and the generalized momenta $\partial T/\partial\dot{\psi} = I_z(\dot{\psi} + \dot{\phi}\cos\theta)$ and $\partial T/\partial\dot{\phi} = I_x\dot{\phi}\sin^2\theta + I_z(\dot{\psi} + \dot{\phi}\cos\theta)\cos\theta$. Since there are three variables, these serve as the three differential equations of motion, and are closer to the solutions θ, ϕ, and ψ than the corresponding second-order differential equations.

APPENDIX G

Some References for Advanced Studies

The following books and journal papers are some of the suggestions for more advanced studies. Some of these books are *classics*, including those of Pars, Lanczos, and La Salle and Lefschetz.

1. Pars, L. A. (1979). *A Treatise on Analytical Dynamics*. Ox Bow Press, Connecticut (first published in 1965 by Wiley).

2. Rosenberg, R. M. (1977). *Analytical Dynamics of Discrete Systems*. Plenum Press, New York.

3. Lanczos, C. (1966). *The Variational Principles of Mechanics*. University of Toronto Press, Toronto, Canada.

4. Goldstein, H. (1965). *Classical Mechanics*. Addison-Wesley, Reading, Massachusetts.

5. Arnold, R. N. and Maunder, L. (1961). *Gyrodynamics and its Engineering Applications*. Academic Press, New York.

6. Gray, A. (1959). *A Treatise on Gyrostatics and Rotational Motion: Theory and Applications*. Dover, New York.

7. Deimel, R. F. (1950). *Mechanics of the Gyroscope: the Dynamics of Rotation*. Dover, New York.

8. Perry, J. (1957). *Spinning Tops and Gyroscopic Motion: a Popular Exposition of Dynamics of Rotation*. Dover, New York.

9. Meirovitch, L. (1970). *Methods of Analytical Dynamics*. McGraw-Hill, New York.

10. Appell, P. (1924). *Traité de Mécanique Rationnelle*, Vol. 2. Gauthier-Villars, Paris, France.

11. Meirovitch, L. (2001). *Fundamentals of Vibrations*. McGraw-Hill, New York.

12. La Salle, J. and Lefschetz, S. (1961). *Stability by Liapunov's Direct Method with Applications*. Academic Press, New York.

13. Hughes, P. C. (1986). *Spacecraft Attitude Dynamics*. Wiley, New York.

14. Crespo da Silva, M. R. M. (1974). *A transformation approach for finding first integrals of motion of dynamical systems*. Int. J. Non-linear Mechanics, Vol. 9, pp. 241-250.

15. Crespo da Silva, M. R. M. (1970). *Attitude stability of a gravity-stabilized gyrostat satellite*. Celestial Mechanics, Vol. 2, No. 2, pp. 147-165.

16. Crespo da Silva, M. R. M. (1972). *Non-linear resonant attitude motions in gravity-stabilized gyrostat satellites*. Int. J. Non-linear Mechanics, Vol. 7, pp. 621-641.

17. Crespo da Silva, M. R. M. (1978). *Attitude dynamics, stabilization and control of spacecrafts*. Marks' Standard Handbook for Mechanical Engineers, Eight Edition, Astronautics Section (Organizer: Neil A. Armstrong), pp. 11-117 to 11-119, McGraw-Hill.[1]

18. Nayfeh, A. H. and Mook, D.T. (1979). *Nonlinear Oscillations*. Wiley-Interscience, New York.

---

[1]This publication also appears in newer editions of the handbook.

Index